Lecture Notes in Computer Science 8660

Commenced Publication in 1973
Founding and Former Series Editors:
Gerhard Goos, Juris Hartmanis, and Jan van Leeuwen

T0212626

Vladimir P. Gerdt   Wolfram Koepf
Werner M. Seiler   Evgenii V. Vorozhtsov (Eds.)

Computer Algebra in Scientific Computing

16th International Workshop, CASC 2014
Warsaw, Poland, September 8-12, 2014
Proceedings

 Springer

Volume Editors

Vladimir P. Gerdt
Joint Institute of Nuclear Research
Laboratory of Information Technologies (LIT)
Dubna, Russia
E-mail: gerdt@jinr.ru

Wolfram Koepf
Universität Kassel, Institut für Mathematik
Kassel, Germany
E-mail: koepf@mathematik.uni-kassel.de

Werner M. Seiler
Universität Kassel, Institut für Mathematik
Kassel, Germany
E-mail: seiler@mathematik.uni-kassel.de

Evgenii V. Vorozhtsov
Russian Academy of Sciences
Institute of Theoretical and Applied Mechanics
Novosibirsk, Russia
E-mail: vorozh@itam.nsc.ru

ISSN 0302-9743   e-ISSN 1611-3349
ISBN 978-3-319-10514-7   e-ISBN 978-3-319-10515-4
DOI 10.1007/978-3-319-10515-4
Springer Cham Heidelberg New York Dordrecht London

Library of Congress Control Number: 2014946203

LNCS Sublibrary: SL 1 – Theoretical Computer Science and General Issues

Typesetting: Camera-ready by author, data conversion by Scientific Publishing Services, Chennai, India

Printed on acid-free paper

Springer is part of Springer Science+Business Media (www.springer.com)

Preface

Since their start in 1998, the International Workshops on Computer Algebra in Scientific Computing have provided excellent international forums for sharing knowledge and results in computer algebra (CA) methods, systems, and CA applications in scientific computing. The aim of the 16th CASC Workshop was to provide a platform to the researchers and practitioners from both academia as well as industry to meet and share cutting-edge development in the field.

Last autumn, Ernst Mayr, a co-founder of CASC and co-chair of the CASC conference series since its foundation in 1997 and holding the first conference in April 1998 retired from his positions as General Chair and Proceedings Editor of CASC. He decided to step down to spend more time on his other professional activities and with his family. The importance of his contributions to establishing CASC as a renowned conference series and to its evident progression from 1998 until today cannot be overestimated. When Ernst informed his co-chairs about his retirement plans, he suggested Werner M. Seiler as his successor and expressed his readiness to assist in future CASC activities. All chairs of CASC wish Ernst good health and continuing success in his professional work and hope to see him again at many further CASC conferences. Werner gladly accepted the honorable invitation to take over as General Chair and Co-editor of the proceedings. He will be supported in his work by Andreas Weber, who has agreed to assume the new position of Publicity Chair of CASC. They will try to keep CASC in the great spirit created by Ernst.

This year the CASC conference was held in Poland, where research in the field of CA becomes more and more popular. At present, research on the development and application of methods, algorithms, and programs of CA is performed at universities of Białystok, Kraków, Lublin, Siedlce, Toruń, Warsaw, Zielona Góra and others. In many universities, for example, AGH University of Science and Technology in Kraków, Nicolaus Copernicus University in Toruń, Warsaw University and Warsaw University of Life Sciences, regular courses have been introduced for students on symbolic computation and application of computer algebra to the theory of differential equations, dynamical systems, coding theory, and cryptography. In connection with the above, it was decided to hold the 16th CASC Workshop in Warsaw to draw the attention of young researchers to this interesting and important field of applied mathematics and computer science.

This volume contains 33 full papers submitted to the workshop by the participants and accepted by the Program Committee after a thorough reviewing process. Additionally, the volume includes two invited talks.

Studies in polynomial algebra are represented by contributions devoted to factoring sparse bivariate polynomials using the priority queue, the construction of irreducible polynomials by using the Newton index, real polynomial root finding by means of matrix and polynomial iterations, application of the eigenvalue method with symmetry for solving polynomial systems arising in the vibration analysis of mechanical structures with symmetry properties, application of Gröbner systems for computing the (absolute) reduction number of polynomial ideals, the application of cylindrical algebraic decomposition for solving the quantifier elimination problems, certification of approximate roots of overdetermined and singular polynomial systems via the recovery of an exact rational univariate representation from approximate numerical data, and new parallel algorithms for operations on univariate polynomials (multi-point evaluation, interpolation) based on subproduct tree techniques.

Several papers are devoted to using CA for the investigation of various mathematical and applied topics related to ordinary differential equations (ODEs): application of CAS *Mathematica* for the investigation of movable singularities of the complex valued solutions of ODEs, the decidability problem for linear ODE systems with variable coefficients, conversion of nonlinear ODEs to integral equations for the purpose of parameter estimation.

The invited talk by L. Plaskota deals with the information-based complexity (IBC) as a branch of computational complexity that studies continuous problems, for which available information is partial, noisy, and priced. The basic ideas of IBC are presented, and some sample results on optimal algorithms, complexity, and tractability of such problems are given. The focus is on numerical integration of univariate and multivariate functions.

A number of papers deal with applications of symbolic and symbolic-numeric computations for investigating and solving partial differential equations (PDEs) in mathematical physics: symbolic solution of algebraic PDEs, testing uniqueness of analytic solutions of PDE with boundary conditions, construction of high-order difference schemes for solving the Poisson equation, symbolic-numeric solution of the Burgers and Korteweg–de Vries–Burgers equations at very high Reynolds numbers, and derivation of new analytic solutions of the PDEs governing the motion of a special Cosserat rod-like fiber.

Application of symbolic and symbolic-numeric algorithms in mechanics and physics is represented by the following themes: analytic calculations in Maple for modeling smoothly irregular integrated optical waveguide structures, the integrability of evolutionary equations in the restricted three-body problem with variable masses, quantum tunneling problem of diatomic molecule through repulsive barriers, symbolic-numeric solution of the parametric self-adjoint Sturm–Liouville problem, obtaining new invariant manifolds in the classic and generalized Goryachev–Chaplygin problem of rigid body dynamics.

The invited talk by G. Regensburger, article written jointly with S. Müller, addresses a recent extension of chemical reaction network theory (CRNT), called generalized mass-action systems, where reaction rates are allowed to be power-laws in the concentrations. In particular, the kinetic orders (the real exponents) can differ from the corresponding stoichiometric coefficients. As with mass action kinetics, complex balancing equilibria (determined by the graph Laplacian of the underlying network) can be characterized by binomial equations and parameterized by monomials. However, uniqueness and existence for all rate constants and initial conditions additionally depend on sign vectors of the stoichiometric and kinetic-order subspaces. This leads to a generalization of Birch's theorem, which is robust with respect to certain perturbations in the exponents. Finally, the occurrence of multiple complex balancing equilibria is discussed. The presentation is focused on a constructive characterization of positive real solutions to generalized polynomial equations with real and symbolic exponents.

The other topics include the application of the CAS GAP (Groups, Algorithms, Programming) for construction of directed strongly regular graphs, the application of CAS *Mathematica* for derivation of a four-points piecewise-quadratic interpolant, the efficient calculation of the determinant of a generalized Vandermonde matrix with CAS *Mathematica*, the solution of systems of linear inequalities by combining virtual substitution with learning strategies with the use of the REDUCE package REDLOG, finding a generic position for an algebraic space curve, the application of CAS RELVIEW for solving problems of voting systems, computation of the truncated annihilating ideals for algebraic local cohomology class attached to isolated hypersurface singularities, optimal estimations of Seiffert-type means by some special Gini means, solving parametric sparse linear systems by local block triangularization, computation of the topology of an arrangement of algebraic plane curve defined by implicit and parametric equations, the use of CASs Maxima and *Mathematica* for studying the coherence and large-scale pattern formation in coupled logistic-map lattices, and enumeration of all Schur rings over the groups of orders up to 62.

Our particular thanks are due to the dean of the Faculty of Applied Informatics and Mathematics, Arkadiusz Orłowski, and the members of the CASC 2014 local Organizing Committee in Warsaw (Warsaw University of Life Sciences), i.e., Alexander Prokopenya (chair of the local Organizing Committee), and Ryszard Kozera, Luiza Ochnio, and Artur Wilinski, who have ably handled all the local arrangements in Warsaw. Furthermore, we want to thank all the members of the Program Committee for their thorough work. Finally we are grateful to W. Meixner for his technical help in the preparation of the camera-ready manuscript for this volume and the design of the conference poster.

July 2014

V.P. Gerdt
W. Koepf
W.M. Seiler
E.V. Vorozhtsov

Organization

CASC 2014 was organized jointly by the Institute of Mathematics at the University of Kassel, Kassel, Germany, and the Faculty of Applied Informatics and Mathematics, Warsaw University of Life Sciences (SGGW), Warsaw, Poland.

Workshop General Chairs

Vladimir P. Gerdt (Dubna)
Werner M. Seiler (Kassel)

Program Committee Chairs

Wolfram Koepf (Kassel)
Evgenii V. Vorozhtsov (Novosibirsk)

Program Committee

François Boulier (Lille)
Hans-Joachim Bungartz (Munich)
Jin-San Cheng (Beijing)
Victor F. Edneral (Moscow)
Dima Grigoriev (Lille)
Jaime Gutierrez (Santander)
Sergey A. Gutnik (Moscow)
Jeremy Johnson (Philadelphia)
Victor Levandovskyy (Aachen)
Marc Moreno Maza (London, Canada)
Alexander Prokopenya (Warsaw)

Georg Regensburger (Linz)
Eugenio Roanes-Lozano (Madrid)
Valery Romanovski (Maribor)
Markus Rosenkranz (Canterbury)
Doru Stefanescu (Bucharest)
Thomas Sturm (Saarbrücken)
Jan Verschelde (Chicago)
Stephen M. Watt
    (W. Ontario, Canada)
Andreas Weber (Bonn)
Kazuhiro Yokoyama (Tokyo)

Additional Reviewers

Parisa Alvandi
Hirokazu Anai
Alexander Batkhin
Johannes Blömer
Charles Bouillaguet
Jürgen Bräckle

Jorge Caravantes
Francisco-Jesus Castro-Jimenez
Changbo Chen
Colin Denniston
Gema M. Diaz-Toca
Matthew England

Ruyong Feng
Mario Fioravanti
Mark Giesbrecht
Joris van der Hoeven
Martin Horvat
Silvana Ilie
Fredrik Johansson
Wolfram Kahl
Manuel Kauers
Irina Kogan
Marek Kosta
Ryszard Kozera
Dmitry Kulyabov
François Lemaire
Wei Li
Gennadi Malaschonok
Thanos Manos
Michael Monagan

Bernard Mourrain
Hirokazu Murao
Ioana Necula
Masayuki Noro
Alina Ostafe
Alfredo Parra
Roman Pearce
Marko Petkovšek
Daniel Robertz
Vikram Sharma
Takeshi Shimoyama
Kristof Unterweger
Luis Verde-Star
Konrad Waldherr
Uwe Waldmann
Uli Walther
Hitoshi Yanami
Hangzhou Zhejiang

Local Organization

Alexander Prokopenya (Chair)
Ryszard Kozera

Luiza Ochnio
Artur Wilinski

Publicity Chair

Andreas Weber (Bonn)

Website

http://wwwmayr.in.tum.de/CASC2014/

Table of Contents

Computable Infinite Power Series in the Role of Coefficients of Linear Differential Systems

Sergei A. Abramov[1,*] and Moulay A. Barkatou[2]

[1] Computing Centre of the Russian Academy of Sciences, Vavilova, 40,
Moscow 119333, Russia
sergeyabramov@mail.ru
[2] Institut XLIM, Département Mathématiques et Informatique,
Université de Limoges; CNRS, 123, Av. A. Thomas,
87060 Limoges cedex, France
moulay.barkatou@unilim.fr

Abstract. We consider linear ordinary differential systems over a differential field of characteristic 0. We prove that testing unimodularity and computing the dimension of the solution space of an arbitrary system can be done algorithmically if and only if the zero testing problem in the ground differential field is algorithmically decidable. Moreover, we consider full-rank systems whose coefficients are computable power series and we show that, despite the fact that such a system has a basis of formal exponential-logarithmic solutions involving only computable series, there is no algorithm to construct such a basis.

1 Introduction

Linear ordinary differential systems with variable coefficients appear in various areas of mathematics. Power series are very important objects in the representation of the solutions of such systems as well as of the systems themselves. The representation of infinite series lies at the core of computer algebra. A general formula that expresses the coefficients of a series is not always available and may even not exist. One natural way to represent the series is the algorithmic one, i.e., providing an algorithm which computes its coefficients. Such algorithmic representation of a concrete series is not, of course, unique. This non-uniqueness is one of the reasons for undecidability of the zero testing problem for such computable series.

At first glance, it may seem that if we cannot decide algorithmically whether a concrete coefficient of a system is zero or not, then we will not be able to solve any more or less interesting problem related to the search of solutions. However, this is not completely right: at least, if we know in advance that the system is of full rank then some of the problems can still be solved. For example, we can find

---

* Supported in part by the Russian Foundation for Basic Research, project no. 13-01-00182-a. The first author thanks also Department of Mathematics and Informatics of XLIM Institute of Limoges University for the hospitality during his visits.

V.P. Gerdt et al. (Eds.): CASC Workshop 2014, LNCS 8660, pp. 1–12, 2014.

Laurent series [3] and regular [6] solutions. Some non-trivial characteristics can be computed as well, e.g., the so called "width" of the system [3]. Nevertheless, many of the problems are undecidable. For example, we cannot answer algorithmically the following question: does a given full-rank system with power series coefficients have a formal exponential-logarithmic solution which is not regular? We prove this undecidability in the present paper. It is also shown that if exponential-logarithmic solutions of a given full-rank system exist then there exists a basis of the space of those solutions such that all the series which appear in the elements of the basis are computable; the exact formulation is given in Proposition 7 of this paper.

So, we know that there exists a basis of the solution space which consists of computable objects, but we are not able to find this basis algorithmically. This is analogous to some facts of constructive mathematical analysis. In fact, the notion of a constructive real number (computable point) is fundamental in that discipline: "... an algorithm which finds the zeros of any alternating, continuous, computable function is impossible. At the same time, there cannot be a computable function that assumes values of different signs at the ends of a given interval and does not vanish at any computable point of this interval (a priori, it is impossible to rule out the existence of computable alternating functions whose zeros are all 'noncomputable'). These results are due to Tseitin [21] ..." ([14, p. 5], see also [16, §24]).

We prove in the same direction that testing unimodularity, i.e., the invertibility of the corresponding operator and computing the dimension of the solution space of an arbitrary system can be done algorithmically if and only if the zero testing problem in the ground differential field is algorithmically decidable. As a consequence, these problems are undecidable when the coefficients are power series or Laurent series which are represented by arbitrary algorithms.

If the algorithmic way of series representation is used then some of the problems related to linear ordinary systems are decidable while others are not. Note that the above mentioned algorithms for finding Laurent series solutions and regular solutions are implemented in Maple [23]. The implementation is described in [3,6] and, is available at http://www.ccas.ru/ca/doku.php/eg.

The rest of the paper is organized as follows: After stating some preliminaries in Section 2, we give in Section 3 a review of some results related to systems whose coefficients belong to a field K of characteristic zero. The field K is supposed to be a constructive differential field, i.e., there exist algorithms for the field operations, differentiation, and for zero testing. The problems that are listed in Section 3 can be solved algorithmically. On the other hand, we show in Section 4 that the same problems are algorithmically undecidable, if the field K is semi-constructive, i.e., there exist algorithms for the field operations and differentiation but there is no algorithm for zero testing. Finally, we consider in Section 5 semi-constructive fields of computable formal Laurent series in the role of coefficient field of systems of linear ordinary differential systems.

The results of this paper supplement known results on the zero testing problem and some algorithmically undecidable problems related to differential equations (see, e.g., [10], [13]).

2   Preliminaries

The ring of $m \times m$ matrices with entries belonging to a ring R is denoted by $\mathrm{Mat}_m(R)$. We use the notation $[M]_{i,*}$, $1 \leqslant i \leqslant m$, for the $1 \times m$-matrix which is the ith row of an $m \times m$-matrix M. The notation M^T is used for the transpose of a matrix (vector) M.

If F is a differential field with derivation ∂ then $\mathrm{Const}\,(F) = \{c \in F \mid \partial c = 0\}$ is the *constant field* of F.

2.1   Differential Universal and Adequate Field Extensions

Let K be a differential field of characteristic 0 with derivation $\partial = '$.

Definition 1. *An* adequate differential extension Λ *of* K *is a differential field extension* Λ *of* K *such that any differential system*

$$\partial y = Ay, \tag{1}$$

with $A \in \mathrm{Mat}_m(K)$ *has a solution space of dimension* m *in* Λ^m *over* $\mathrm{Const}\,(\Lambda)$.

If $\mathrm{Const}\,(K)$ is algebraically closed then there exists a unique (up to a differential isomorphism) adequate differential extension Λ such that $\mathrm{Const}\,(\Lambda) = \mathrm{Const}\,(K)$ which is called the *universal differential field extension* of K [18, Sect. 3.2]. For any differential field K of characteristic 0 there exists a differential extension whose constant field is algebraically closed. Indeed, this is the algebraic closure $\bar{K}$ with the derivation obtained by extending the derivation of K in the natural way. In this case, $\mathrm{Const}\,(\bar{K}) = \overline{\mathrm{Const}\,(K)}$ (see [18, Exercises 1.5, 2:(c),(d)]). Existence of the universal differential extension for $\bar{K}$ implies that there exists an adequate differential extension for K, i.e., for an arbitrary differential field of characteristic zero.

In the sequel, we denote by Λ a fixed adequate differential extension of K, and we suppose that the vector solutions of systems in the form (2) lie in Λ^m.

In addition to the first-order systems of the form (1), we also consider the differential systems of arbitrary order $r \geqslant 1$. Each of these systems can be represented, e.g., in the form

$$A_r y^{(r)} + A_{r-1} y^{(r-1)} + \cdots + A_0 y = 0, \tag{2}$$

where the matrices $A_0, A_1, \ldots, A_r$ belong to $\mathrm{Mat}_m(K)$, $m \geqslant 1$, and A_r (the *leading matrix* of the system) is non-zero. The system (2) can be written as $L(y) = 0$ where

$$L = A_r \partial^r + A_{r-1} \partial^{r-1} + \cdots + A_0. \tag{3}$$

The number r is the *order* of L (we write $r = \mathrm{ord}\, L$). The operator (3) can be alternatively represented as a matrix in $\mathrm{Mat}_m(K[\partial])$:

$$\begin{pmatrix} L_{11} & \cdots & L_{1m} \\ \cdots & \cdots \cdots & \cdots \\ L_{m1} & \cdots & L_{mm} \end{pmatrix}, \tag{4}$$

$L_{ij} \in K[\partial]$, $i,j = 1,\ldots,m$, with $\max_{i,j} \operatorname{ord} L_{ij} = r$. We say that the operator $L \in \operatorname{Mat}_m(K[\partial])$ (as well as the system $L(y) = 0$) is of *full rank*, if the rows $(L_{i1},\ldots,L_{im})$, $i = 1,\ldots,m$, of matrix (4) are linearly independent over $K[\partial]$. The matrix A_r is the leading matrix of both the system $L(y) = 0$ and operator L, regardless of representation form.

2.2   Universal Differential Extension of Formal Laurent Series Field

Let K_0 be a subfield of the complex number field $\mathbb{C}$ and K be the field $K_0((x))$ of formal Laurent series with coefficients in K_0, equipped with the derivation $\partial = \frac{d}{dx}$. As it is well known [20, Sect. 110], if K_0 is algebraically closed then the universal differential field extension Λ is the quotient field of the ring generated by expressions of the form

$$e^{P(x)} x^{\gamma} (\psi_0 + \psi_1 \ln x + \cdots + \psi_s (\ln x)^s), \tag{5}$$

where in any such expression

- $P(x) \in K_0[x^{-1/q}]$, q is a positive integer,
- $\gamma \in K_0$,
- s is a non-negative integer and

$$\psi_j \in K_0[[x^{1/q}]], \tag{6}$$

$j = 0, 1, \ldots, s$.

In fact, system (1) has m linearly independent solutions $b_1(x), \ldots, b_m(x)$ such that

$$b_i(x) = e^{P_i(x)} x^{\gamma_i} \Psi_i(x), \tag{7}$$

where the factor $e^{P_i(x)} x^{\gamma_i}$ is common for all components of b_i, and

$\gamma_i \in K_0$, q_i is a positive integer, $P_i(x) \in K_0[x^{-1/q_i}]$, $\Psi_i(x) \in K_0^m[[x^{1/q_i}]][\ln x]$,

$i = 1, \ldots, m$.

Definition 2. *Solutions of the form (7) will be called* (formal) exponential-logarithmic *solutions. If $q = 1$ and $P(x) = 0$ then the solutions (7) are called* regular.

Remark 1. *If K_0 is not algebraically closed then there exists a simple algebraic extension K_1 of K_0 (specific for each system) such that system (1) has m linearly independent solutions of the form (7) with $\gamma_i \in K_1$, $P_i(x) \in K_1[x^{-1/q_i}]$, $\Psi_i(x) \in K_1^m[[x^{1/q_i}]][\ln x]$, $i = 1, \ldots, m$.*

2.3   Row Frontal Matrix and Row Order

Let a full-rank operator $L \in \mathrm{Mat}_m(K[\partial])$ be of the form (3). If $1 \leqslant i \leqslant m$ then define $\alpha_i(L)$ as the biggest integer k, $0 \leqslant k \leqslant r$, such that $[A_k]_{i,*}$ is a nonzero row. The matrix $M \in \mathrm{Mat}_m(K)$ such that $[M]_{i,*} = [A_{\alpha_i(L)}]_{i,*}$, $i = 1, \ldots, m$, is the *row frontal matrix* of L. The vector $(\alpha_1(L), \ldots, \alpha_m(L))$ is the *row order* of L. We will write simply $(\alpha_1, \ldots, \alpha_m)$, when it is clear which operator is considered.

Definition 3. *An operator* $U \in \mathrm{Mat}_m(K[\partial])$ *is* unimodular *(or* invertible*) if there exists* $\bar{U} \in \mathrm{Mat}_m(K[\partial])$ *such that* $\bar{U}U = U\bar{U} = I_m$. *An operator in* $\mathrm{Mat}_m(K[\partial])$ *is* row reduced *if its row frontal matrix is invertible.*

The following proposition is a consequence of [9, Thm. 2.2]:

Proposition 1. *Let* $L \in \mathrm{Mat}_m(K[\partial])$ *then there exist* $U, \check{L} \in \mathrm{Mat}_m(K[\partial])$ *such that* U *is unimodular and* $\check{L}$ *defined by*

$$\check{L} = UL \tag{8}$$

and represented in the form (4), has k zero rows, where $0 \leqslant k \leqslant m$, and the row frontal matrix of $\check{L}$ is of rank $m - k$ over K. The operator L is of full rank if and only if $k = 0$, and in this case the operator $\check{L}$ in (8) is row reduced.

We will say that the system (2) is unimodular whenever the corresponding matrix (4) is.

3   When K Is a Constructive Field

Definition 4. *A* ring (field) K *is said to be* constructive *if there exist algorithms for performing the ring (field) operations and an algorithm for zero testing in K*

This definition is close to the definition of an *explicit* field given in [11].

Suppose that K is a constructive field. Then the proof of the already mentioned theorem [9, Thm. 2.2] gives an algorithm for constructing $U, \check{L}$. We will refer to this algorithm as RR (*Row-Reduction*).

3.1   The Dimension of the Solution Space of a Given Full Rank System

Proposition 2. *([1]) Let* $L \in \mathrm{Mat}_m(K[\partial])$ *be row reduced, and denote by* $\alpha = (\alpha_1, \ldots, \alpha_m)$ *its row order. Then the dimension of its solution space V_L is given by:* $\dim V_L = \sum_{i=1}^m \alpha_i$.

Hence, when the field K is constructive we can apply algorithm RR, and compute, by Proposition 2, the dimension of the solution space of a given full-rank system.

Note that in the case when K is the field of rational functions of x over a field of characteristic zero with $\partial = \frac{d}{dx}$, some inequalities close to the formula given in Proposition 2 can be derived from the results of [12].

3.2 Recognizing the Unimodularity of an Operator and Computing the Inverse Operator

The following property of unimodular operators is a direct result of Proposition 2.

Proposition 3. *[2] Let $L \in \mathrm{Mat}_m(K[\partial])$ be of full rank. Then L is unimodular if and only if $\dim V_L = 0$. Moreover, in the case when the row frontal matrix of L is invertible, L is unimodular if and only if $\mathrm{ord}\, L = 0$.*

Algorithm RR allows one to compute a unimodular $U \in \mathrm{Mat}_m(K[\partial])$ such that the operator $\check{L} = UL$ has an invertible row frontal matrix. Proposition 3 implies that L is unimodular if and only if $\check{L}$ is an invertible matrix in $\mathrm{Mat}_m(K)$. In this case $(\check{L})^{-1}UL = I_m$, i.e., $(\check{L})^{-1}U$ is the inverse of L. Hence the following proposition holds (taking into account Proposition 1, we need not assume that L is of full rank):

Proposition 4. *Let K be constructive and $L \in \mathrm{Mat}_m(K[\partial])$. One can recognize algorithmically whether L is unimodular or not, and compute the inverse operator if it is.*

4   When the Zero Testing Problem in K Is Undecidable

It is easy to see that if the zero testing problem in K is undecidable then the problem of recognizing whether a given $L \in \mathrm{Mat}_m(K[\partial])$ is of full rank is undecidable. Indeed, let $u \in K$, then the operator

$$L = \begin{pmatrix} u\partial & \partial \\ 0 & 1 \end{pmatrix} = \begin{pmatrix} u & 1 \\ 0 & 0 \end{pmatrix}\partial + \begin{pmatrix} 0 & 0 \\ 0 & 1 \end{pmatrix}$$

is of full rank if and only if $u \neq 0$, and any algorithm to recognize whether a given $L \in \mathrm{Mat}_m(K[\partial])$ is of full rank can be used for zero testing in K.

Furthermore, it turns out that if the zero testing problem in K is undecidable then even with a prior knowledge that operators under consideration are of full rank, many questions about those operators remain undecidable.

Proposition 5. *Let the zero testing problem in K be undecidable. Then for $m \geqslant 2$ the following problems about a full-rank operator $L \in \mathrm{Mat}_m(K[\partial])$ are undecidable:*
  (a) computing $\dim V_L$,
  (b) testing unimodularity of L.

Proof. (a) Let $u \in K$ and

$$L = \begin{pmatrix} u\partial + 1 & \partial \\ 0 & 1 \end{pmatrix} = \begin{pmatrix} u & 1 \\ 0 & 0 \end{pmatrix}\partial + \begin{pmatrix} 1 & 0 \\ 0 & 1 \end{pmatrix}. \tag{9}$$

If $u = 0$ then L is unimodular:

$$\begin{pmatrix} 1 & \partial \\ 0 & 1 \end{pmatrix}^{-1} = \begin{pmatrix} 1 & -\partial \\ 0 & 1 \end{pmatrix}$$

and, therefore, $\dim V_L = 0$. If $u \neq 0$ then $\dim V_L = 1$ by Proposition 2. We have

$$\dim V_L = \begin{cases} 0 & \text{if } u = 0, \\ 1 & \text{if } u \neq 0. \end{cases}$$

This implies that if we have an algorithm for computing the dimension then we have an algorithm for the zero testing problem.

(b) As we have seen the operator L of the form (9) is unimodular if and only if $u = 0$.

As a consequence of Propositions 4, 5 we have the following:

Testing unimodularity and determining the dimension of the solution space of an arbitrary full-rank system can be done algorithmically if and only if the zero testing problem in K can be solved algorithmically.

One of the general causes of difficulties in the zero testing problem in K may be associated with non-uniqueness of representation of the elements of K [11, Sect. 2]. This is illustrated in Section 5.1.

5 Computable Power Series

5.1 Semi-constructive Fields

Let K be the field $K_0((x))$ where K_0 is a constructive field of characteristic 0. The field K contains the set $K|_C$ of *computable* series, whose sequences of coefficients can be represented algorithmically. That is to say that for each series $a(x) \in K|_C$ there exists an algorithm Ξ_a to compute the coefficient $a_i \in K_0$ for a given i; arbitrary algorithms which are applicable to integer numbers and return elements of K_0 are allowed. For this set to be considered as a constructive differential subfield of K, it would be necessary to define algorithmically on $K|_C$ the field operations of the field K, the unary operation $\frac{d}{dx}$, and a zero testing algorithm as well. However, in accordance with the classical results of Turing [22], we are not able to solve algorithmically the zero testing problem in $K|_C$. As mentioned in Section 4, the undecidability of the zero testing problem is quite often associated with the fact that the elements of the field (or ring) under consideration can be represented in various ways, and for some of which the test is evident while for the others is not. This holds for $K|_C$ as well.

Remark 2. *The field $K|_C$ is smaller than the field K because not every sequence of coefficients can be represented algorithmically. Indeed, the set of elements of $K|_C$ is countable (each of the algorithms is a finite word in some fixed alphabet) while the cardinality of the set of elements of K is uncountable.*

If the only information we possess about the elements of $K|_C$ is an algorithm to compute their coefficients then the problem of finding the valuation of a given $a(x) \in K|_C$, $\operatorname{val} a(x)$, is undecidable even in the case when it is known in advance that $a(x)$ is not the zero series. This implies that when we work with elements of $K|_\mathrm{C}$, i.e., with computable Laurent series, we cannot compute $a^{-1}(x)$ for a given non-zero $a(x) \in K|_\mathrm{C}$, since the coefficient of x^{-1} of the series $a'(x)a^{-1}(x) \in K|_\mathrm{C}$ is equal to $\mathrm{val}\,a(x)$, i.e., is equal to the value that we are not able to find algorithmically knowing only $\varXi_a$. This means that a suitable representation has to contain some additional information besides a corresponding algorithm. The value $\mathrm{val}\,a(x)$ cannot close the gap, since we have no algorithm to compute the valuation of the sum of two series. However, we can use a lower bound of the valuation instead: observe that if we know that a series $a(x)$ is non-zero then using a valuation lower bound we can compute the exact value of $\mathrm{val}\,a(x)$. Thus, we can use as the representation of $a(x) \in K|_\mathrm{C}$ a pair of the form

$$(\varXi_a, \mu_a), \tag{10}$$

where $\varXi_a$ is an algorithm for computing the coefficient $a_i \in K_0$ for a given i, and the integer μ_a is a lower bound for the valuation of $a(x)$. A computable Laurent series $a(x)$, represented by a pair of the form (10) is equal to $\sum_{i=\mu_a}^{\infty} \varXi_a(i)x^i$.

Of course, there exist other ways to represent computable Laurent series. For example, one can use a pair $(\varXi_a, p_a(x))$, where the algorithm $\varXi_a$ represents a power series that is the regular part of $a(x)$ while $p_a(x) \in K_0[x^{-1}]$ represents explicitly its singular part. We can also represent each Laurent series as a fraction of two power series (the latter are represented algorithmically, this is possible as the field of Laurent series is the quotient field of the ring of power series). So a Laurent series can be represented as a couple $(a(x), b(x))$ of power series with $b(x)$ nonzero.

We can define the field structure on $K|_\mathrm{C}$: all field operations can be performed algorithmically. Since we do not have an algorithm for solving the zero testing problem in $K|_\mathrm{C}$, we use for $K|_\mathrm{C}$ the term "semi-constructive field" instead.

Definition 5. *A ring (field) is* semi-constructive *if there are algorithms to perform the ring (field) operations, but there exists no algorithm to solve the zero testing problem.*

Observe that if the standard representation form is used for rational functions, i.e., for elements in $K_0(x)$, then the field $K_0(x)$ is constructive.

Remark 3. *Consider for the ring $R = K_0[[x]]$ its semi-constructive sub-ring $R|_\mathrm{C}$ of computable power series. In this case we do not need to include a lower bound for the valuation into a representation of a series $a(x) \in R|_\mathrm{C}$, since 0 is such a bound.*

5.2   Systems with Computable Power Series Coefficients

Below we suppose that K_0 is a constructive field of characteristic 0, $K = K_0((x))$, $R = K_0[[x]]$, and

$$K|_\mathrm{C}, \quad R|_\mathrm{C}$$

are a semi-constructive field and, resp., a semi-constructive ring as in Section 5.1. We will consider systems of the form

$$L(y) = 0, \quad L \in \mathrm{Mat}_m \left(R|_{\mathbb{C}} \left[\frac{d}{dx} \right] \right). \tag{11}$$

It follows from Proposition 5 that the problems (a) and (b) listed in that proposition are undecidable if L is as in (11). At first glance, it seems that such undecidability is mostly caused by the inability to distinguish zero and nonzero coefficients of operators and systems. However, even if we know in advance which of the coefficients of an operator L are null, we, nevertheless, cannot solve problems (a) and (b) of Proposition 5 algorithmically. Let $u(x) \in R|_{\mathbb{C}}$ and

$$L = \begin{pmatrix} (u(x)x + 1)\frac{d}{dx} + 1 & \frac{d}{dx} \\ 1 & 1 \end{pmatrix} = \begin{pmatrix} u(x)x + 1 & 1 \\ 0 & 0 \end{pmatrix} \frac{d}{dx} + \begin{pmatrix} 1 & 0 \\ 1 & 1 \end{pmatrix}.$$

For such an operator, we know in advance which of its coefficients are equal to zero, but we do not know whether the power series $u(x)$ is equal to zero. It is easy to see that

$$\dim V_L = \begin{cases} 0 & \text{if } u(x) = 0, \\ 1 & \text{if } u(x) \neq 0. \end{cases}$$

5.3   On Formal Exponential-Logarithmic Solutions

In [3,6], it was proven that the problems of existence of Laurent series solutions and regular solutions (see Definition 2) for a given system (11) are decidable. A regular solution has the form $x^\gamma w(x)$, where $\gamma \in \bar{K}_0$, and $w(x) \in \bar{K}_0((x))^m [\ln x]$; in the context of [6], $w(x) \in \left(\bar{K}_0((x))|_{\mathbb{C}} \right)^m [\ln x]$. In those papers, it was proven also that if non-zero Laurent series or regular solutions exist then we can construct them, i.e., find a lower bound for valuations of all involved Laurent series as well as any number of terms of the series; for regular solutions we also find the corresponding values of γ, the degrees of polynomials in $\ln x$ etc. It was shown also that instead of $\bar{K}_0$ which is the algebraic closure of K_0 some simple algebraic extension K_1 of K_0 may be used.

Remark 4. *The power series which appear in [3,6] as coefficients of a given system can be represented not only by algorithms as described above but also as "black boxes", i.e., by procedures of unknown internal form.*

Proposition 6. *Let m be an integer, $m \geqslant 2$, and K_0 be a constructive subfield of $\mathbb{C}$. Then for a given full-rank system of the form (11),*

(i) the question whether nonzero Laurent series solutions exist as well as the question whether nonzero regular solutions exist are algorithmically decidable;

(ii) the question whether nonzero formal exponential-logarithmic solutions exist is algorithmically undecidable;

(iii) the question whether nonzero formal exponential-logarithmic solutions which are not regular solutions exist is algorithmically undecidable.

Proof. (i) This follows from [3,6], as it was explained in the beginning of this section.

(ii) A given L is unimodular if and only if the system (11) has no non-zero formal exponential-logarithmic solution, and the claim follows from Proposition 5 (problem (b)).

(iii) A full-rank operator L is evidently unimodular if and only if it has no regular solution and no exponential-logarithmic solution which is not regular. By (i), we can test whether the system $L(y) = 0$ has no regular solution. Thus, if we are able to test whether this system has no exponential-logarithmic solution which is not regular then we can test whether L is unimodular or not. However, this is an undecidable problem by Proposition 5 (problem (b)).

Proposition 7. *Let m be an integer number, $m \geqslant 2$, K_0 be a constructive subset of $\mathbb{C}$. Let $L(y) = 0$ be a full-rank system of the form (11), and $d = \dim V_L$. Then V_L has a basis $b_1(x), \ldots, b_d(x)$ consisting of exponential-logarithmic solutions such that any $\Psi_i(x)$ from (7) is of the form $\Psi_i(x) = \Phi_i(x^{1/q_i})$ where q_i is a non-negative integer,*

$$\Phi_i(x) \in ((K_1[[x]]) \mid_c)^m [\ln x], \tag{12}$$

and K_1 is a simple algebraic extension of K_0. In addition to (12), $\gamma_i \in K_1$, $P_i(x) \in K_1[x^{-1/q_i}]$, $i = 1, \ldots, d$.

Proof. It follows from, e.g., [4,5,8], that for any operator L of full rank there exists an operator F such that the leading matrix of FL is invertible. The system $FL(y) = 0$ is equivalent to a first order system of the form $y' = Ay$, $A \in \mathrm{Mat}_{ms}(K((x)))$, $s = \mathrm{ord}\, FL$. It is known ([7]) that for a first-order system there exists a simple algebraic extension K_1 of K_0 such that those γ_i and the coefficients of $P_i(x)$ which appear in its solutions of the form (7), belong to K_1. The field K_1 is constructive since K_0 is. Obviously, $q_i \in \mathbb{N}$.

The substitution

$$x = t^{q_i}, \quad y(t^{q_i}) = z(t)e^{P_i(t^{q_i})},$$

$P_i(t^{q_i}) \in K_1[1/t]$, into the original system $L(y) = 0$ transforms it into a full-rank system which can be represented as

$$\tilde{L}(z) = 0, \quad \tilde{L} \in \mathrm{Mat}_m \left((K_1[[t]]) \mid_c \left[\frac{d}{dt} \right] \right).$$

The Laurent series that appear in the regular solutions of this new system can be taken to be computable, as it follows from [3,6] (see the beginning of this section).

Thus, the series that appear in the representation of solutions are computable (Proposition 7), but we cannot find them algorithmically (Proposition 6). In fact, Proposition 7 guarantees existence. However, the operator F mentioned therein cannot be constructed algorithmically.

Remark 5. *In the case of first-order systems of the form (1), the questions formulated in Proposition 6(ii, iii) are decidable. This follows from the fact that* *for constructing exponential-logarithmic solutions of a system of this form one needs only a finite number of terms of the entries (which are Laurent series) of A, and the number of those terms can be computed in advance ([15,7,17]). This holds also for higher-order systems whose leading matrices are invertible.*

It is proven ([19]) that if the dimension d of the space of exponential-logarithmic solutions is known in advance then the basis $b_1, \ldots, b_d$ which is mentioned in Proposition 7 can be constructed algorithmically. The corresponding algorithm is implemented in Maple.

As we see, if the algorithmic representation of series is used and if arbitrary algorithms representing series are admitted then some of the problems related to linear ordinary differential systems are decidable, while others are not. There is a subtle border between them, and a careful formulation of each of the problems under consideration is absolutely necessary. A small change in the formulation of a decidable problem can transform it into an undecidable one, and vice versa.

Acknowledgments. The authors are thankful to S. Maddah, M. Petkovšek, A. Ryabenko, M. Rybowicz, S. Watt for interesting discussions, and to anonymous referees for their useful comments.

References

1. Abramov, S.A., Barkatou, M.A.: On the dimension of solution spaces of full rank linear differential systems. In: Gerdt, V.P., Koepf, W., Mayr, E.W., Vorozhtsov, E.V. (eds.) CASC 2013. LNCS, vol. 8136, pp. 1–9. Springer, Heidelberg (2013)
2. Abramov, S.A., Barkatou, M.A.: On solution spaces of products of linear differential or difference operators. ACM Communications in Computer Algebra (accepted)
3. Abramov, S.A., Barkatou, M.A., Khmelnov, D.E.: On full-rank differential systems with power series coefficients. J. Symbolic Comput. (accepted)
4. Abramov, S.A., Khmelnov, D.E.: Desingularization of leading matrices of systems of linear ordinary differential equations with polynomial coefficients. In: International Conference "Differential Equations and Related Topics" Dedicated to I.G.Petrovskii, Moscow, MSU, May 30-June 4, p. 5. Book of Abstracts (2011)
5. Abramov, S.A., Khmelnov, D.E.: On singular points of solutions of linear differential systems with polynomial coefficients. J. Math. Sciences 185(3), 347–359 (2012)
6. Abramov, S.A., Khmelnov, D.E.: Regular solutions of linear differential systems with power series coefficients. Programming and Computer Software 40(2), 98–106 (2014)
7. Barkatou, M.A.: An algorithm to compute the exponential part of a formal fundamental matrix solution of a linear differential system. Applicable Algebra in Engineering, Communication and Computing 8, 1–23 (1997)
8. Barkatou, M.A., El Bacha, C., Labahn, G., Pflügel, E.: On simultaneous row and column reduction of higher-order linear differential systems. J. Symbolic Comput. 49(1), 45–64 (2013)
9. Beckermann, B., Cheng, H., Labahn, G.: Fraction-free row reduction of matrices of Ore polynomials. J. Symbolic Comput. 41(5), 513–543 (2006)
10. Denef, J., Lipshitz, L.: Power series solutions of algebraic differential equations. Math. Ann. 267, 213–238 (1984)

11. Frölich, A., Shepherdson, J.C.: Effective procedures in field theory. Phil. Trans. R. Soc. Lond. 248(950), 407–432 (1956)
12. Grigoriev, D.: NC solving of a system of linear differential equations in several unknowns. Theor. Comput. Sci. 157(1), 79–90 (1996)
13. van der Hoeven, J., Shackell, J.R.: Complexity bounds for zero-test algorithms. J. Symbolic Comput. 41(4), 1004–1020 (2006)
14. Kushner, B.A.: Lectures on Constructive Mathematical Analysis (Translations of Mathematical Monographs) Amer. Math. Soc. (1984)
15. Lutz, D.A., Schäfke, R.: On the identification and stability of formal invariants for singular differential equations. Linear Algebra and Its Applications 72, 1–46 (1985)
16. Martin-Löf, P.: Notes on Constructive Mathematics. Almquist & Wiskell, Stokholm (1970)
17. Pflügel, E.: Effective formal reduction of linear differential systems. Applicable Algebra in Engineering, Communication and Computation 10(2), 153–187 (2000)
18. van der Put, M., Singer, M.F.: Galois Theory of Linear Differential Equations. Grundlehren der mathematischen Wissenschaften, vol. 328. Springer, Heidelberg (2003)
19. Ryabenko, A.: On exponential-logarithmic solutions of linear differential systems with power series coefficients (In preparation)
20. Schlesinger, L.: Handbuch der Theorie der linearen Differentialgleichungen, vol. 1. Teubner, Leipzig (1895)
21. Tseitin, G.S.: Mean-value Theorems in Constructive Analysis. Problems of the Constructive Direction in Mathematics. Part 2. Constructive Mathematical Analysis. Collection of Articles: Trudy Mat. Inst. Steklov, Acad. Sci. USSR 67, 362–384 (1962)
22. Turing, A.: On computable numbers, with an application to the Entscheidungs-problem. Proc. London Math. Soc., Series 2 42, 230–265 (1936)
23. Maple online help, http://www.maplesoft.com/support/help/

Relation Algebra, RelView, and Plurality Voting

Rudolf Berghammer

Institut für Informatik, Universität Kiel, Olshausenstraße 40, 24098 Kiel, Germany
rub@informatik.uni-kiel.de

Abstract. We demonstrate how relation algebra and a supporting tool can be combined to solve problems of voting systems. We model plurality voting within relation algebra and present relation-algebraic specifications for some computational tasks. They can be transformed immediately into the programming language of the BDD-based Computer Algebra system RelView, such that this tool can be used to solve the problems in question and to visualize the computed results. The approach is extremely formal, very flexible and especially appropriate for prototyping, experimentation, scientific research, and education.

1 Introduction

Since centuries systematic experiments are an accepted means for doing science. In the meantime they have also become important in formal fields, such as mathematics, scientific computing, and theoretical computer science. Here computer support proved to be very useful, e.g., for computing results and for discovering mathematical relationships by means of visualization and animation. Used are general Computer Algebra systems, like Maple and Mathematica, but frequently also tools that focus on specific mathematical objects and (in many cases: algebraic) structures and particular applications. RelView (cf. [2,12,15]) is a tool of the latter kind. It can be regarded as *specific purpose Computer Algebra system* and its main purpose is the mechanization of heterogeneous relation algebra in the sense of [13,14] and the visualization of relations. Computational tasks can be expressed by relational functions and programs. These frequently consist of only a few lines that present the relation-algebraic specification of the notions in question. In view of efficiency the implementation of relations via binary decision diagrams (BDDs) proved to be superior to many other well-known implementations, like Boolean matrices or successor lists. Their use in RelView led to an amazing computational power, in particular if hard problems are to solve and this is done by the search of a huge set of objects.

In [4,6] it is demonstrated how relation algebra and RelView can be combined to solve computational problems of voting systems. The papers consider two specific systems, known as *approval voting* and *Condorcet voting*. Our paper constitutes a continuation of this work. We concentrate on *plurality voting*, a further popular voting system that simply selects the alternatives with the highest number of first place votes and is widely used through the world. After the presentation of the relation-algebraic preliminaries we model two versions of this

V.P. Gerdt et al. (Eds.): CASC Workshop 2014, LNCS 8660, pp. 13–27, 2014.
© Springer International Publishing Switzerland 2014

kind of voting within relation algebra. Based on these models, we then present relation-algebraic specifications for computing dominance relations and (sets of) winners. Next, for the second model we show how relation-algebraically to solve hard so-called *control problems*. For a control of an election the assumption is that the authority conducting the election – usually called the *chair* – knows all individual preferences of the voters and tries to achieve (in case of *constructive control*) or to avoid (in case of *destructive control*) win for a specific alternative by a strategic manipulation of the set of voters and alternatives, respectively. The chair's knowledge of all individual preferences and the ability to manipulate by 'dirty tricks' (mistimed meetings, excuses like 'to expensive' or 'legally not allowed', etc.) are worst-case assumptions that are not entirely unreasonable in some settings, for instance, in case of commissions of political institutions. All relation-algebraic specifications we will present can be transformed immediately into the programming language of the Computer Algebra system RELVIEW such that this tool can be used to compute results and to visualize them. Finally, we evaluate our approach in view of its advantages and its drawbacks.

2   Relation-Algebraic Preliminaries

In this section we recall some preliminaries of (typed, heterogeneous) relation algebra. For more details, see [13,14], for example.

Given two sets X and Y we write $[X \leftrightarrow Y]$ for the set of all (binary) relations with source X and target Y, i.e., for the powerset $2^{X \times Y}$. Furthermore, we write $R : X \leftrightarrow Y$ instead of $R \in [X \leftrightarrow Y]$ and call then $X \leftrightarrow Y$ the *type* of R. If the two sets X and Y of R's type are finite, then we may consider R as a Boolean matrix. Since a Boolean matrix interpretation is well suited for many purposes and also used by the RELVIEW tool as the main possibility to visualize relations, in the present paper we frequently use matrix terminology and notation. In particular, we speak about the entries, rows and columns of a relation/matrix and write $R_{x,y}$ instead of $(x, y) \in R$ or $x \, R \, y$. We assume the reader to be familiar with the basic operations on relations, viz. R^T (*transposition*), $\overline{R}$ (*complement*), $R \cup S$ (*union*), $R \cap S$ (*intersection*), and $R; S$ (*composition*), the predicates $R \subseteq S$ (*inclusion*) and $R = S$ (*equality*), and the special relations O (*empty relation*), L (*universal relation*), and I (*identity relation*). In case of O, L, and I we overload the symbols, i.e., avoid the binding of types to them.

By $syq(R, S) := \overline{R^\mathsf{T}; \overline{S}} \cap \overline{\overline{R}^\mathsf{T}; S}$ we define the *symmetric quotient* of two relations $R : X \leftrightarrow Y$ and $S : X \leftrightarrow Z$. From this definition we get the typing $syq(R, S) : Y \leftrightarrow Z$ and that for all $y \in Y$ and $z \in Z$ it holds

$$syq(R, S)_{y,z} \;\Leftrightarrow\; \forall \, x \in X : R_{x,y} \leftrightarrow S_{x,z}. \tag{1}$$

Descriptions of the form (1) with relationships expressed by logical formulae and indices are called *point-wise*. Such descriptions of symmetric quotients and the constructs we introduce now for modeling sets and direct products will play a fundamental role in the remainder of the paper. We will use them to get relation-algebraic (or *point-free*) specifications, which then can be treated by RELVIEW.

Vectors are a first well-known relational means to model sets. For the purpose of this paper it suffices to define them as relations $r : X \leftrightarrow \mathbf{1}$ (we prefer lower case letters in this context) with a specific singleton set $\mathbf{1} = \{\bot\}$ as target. Then relational vectors can be considered as Boolean column vectors. To be consonant with the usual notation, we always omit the second subscript, i.e., write r_x instead of $r_{x,\bot}$. Given $r : X \leftrightarrow \mathbf{1}$ and $Y \in 2^X$ we then define that

$$r \text{ models the subset } Y \text{ of } X :\Leftrightarrow \forall x \in X : x \in Y \leftrightarrow r_x. \tag{2}$$

A *point* is a specific vector with precisely one 1-entry if considered as a Boolean column vector. Consequently, it models a singleton subset $\{x\}$ of X and we then say that it *models the element* x of X. In conjunction with a powerset 2^X we will also use the *membership relation* $\mathsf{M} : X \leftrightarrow 2^X$, point-wisely defined by

$$\mathsf{M}_{x,Y} :\Leftrightarrow x \in Y, \tag{3}$$

for all $x \in X$ and $Y \in 2^X$. The following lemma shows how symmetric quotients and membership relations allow to describe a specific relationship between sets relation-algebraically. The construction $set(R)$ of the lemma is an instance of the *power transpose* construction of [14] and will play a prominent role in the remainder of the paper, too.

Lemma 2.1. *For a given relation $R : X \leftrightarrow Y$ and $\mathsf{M} : X \leftrightarrow 2^X$ as a membership relation we define*

$$set(R) := syq(R, \mathsf{M}) : Y \leftrightarrow 2^X.$$

Then we have $set(R)_{y,z}$ iff $Z = \{x \in X \mid R_{x,y}\}$, for all $y \in Y$ and $Z \in 2^X$.

Proof. Given arbitrary $y \in Y$ and $Z \in 2^X$, we get the result as follows:

$$
\begin{aligned}
set(R)_{y,Z} &\Leftrightarrow syq(R, \mathsf{M})_{y,Z} \\
&\Leftrightarrow \forall x \in X : R_{x,y} \leftrightarrow \mathsf{M}_{x,Z} \qquad \text{by (1)} \\
&\Leftrightarrow \forall x \in X : R_{x,y} \leftrightarrow x \in Z \qquad \text{by (3)} \\
&\Leftrightarrow Z = \{x \in X \mid R_{x,y}\} \qquad\qquad\qquad \square
\end{aligned}
$$

In each of our later applications we will assume the sets in question to be finite. On powersets 2^X of finite sets X we then will use the *size comparison relation* $\mathsf{S} : 2^X \leftrightarrow 2^X$, which is point-wisely defined by

$$\mathsf{S}_{Y,Z} :\Leftrightarrow |Y| \leq |Z|, \tag{4}$$

for all $Y, Z \in 2^X$. In the next result it is shown how to compare for a finite set X the specific sets $\{x \in X \mid R_{x,y}\}$ of Lemma 2.1 w.r.t. their sizes within the language of relation algebra.

Lemma 2.2. *Let relations $Q : X \leftrightarrow Y$ and $R : X \leftrightarrow Z$ be given, where X is finite, and $\mathsf{S} : 2^X \leftrightarrow 2^X$ be a size-comparison relation. Furthermore, let*

$$scomp(Q, R) := set(Q); \mathsf{S}^{\mathsf{T}}; set(R)^{\mathsf{T}} : Y \leftrightarrow Z.$$

Then we have $scomp(Q, R)_{y,z}$ iff $|\{x \in X \mid Q_{x,y}\}| \geq |\{x \in X \mid R_{x,z}\}|$, for all $y \in Y$ and $z \in Z$.

Proof. We get for arbitrary $y \in Y$ and $z \in Z$ the claim as follows:

$$scomp(Q, R)_{y,z}$$
$$\Leftrightarrow (set(Q); \mathsf{S}^\mathsf{T}; set(R)^\mathsf{T})_{y,z}$$
$$\Leftrightarrow \exists U \in 2^X : set(Q)_{y,U} \land \exists V \in 2^X : \mathsf{S}^\mathsf{T}_{U,V} \land set(R)^\mathsf{T}_{V,z}$$
$$\Leftrightarrow \exists U \in 2^X : U = \{x \in X \mid Q_{x,y}\} \land$$
$$\qquad \exists V \in 2^X : |U| \geq |V| \land V = \{x \in X \mid R_{x,z}\} \qquad \text{Lemma 2.1, (4)}$$
$$\Leftrightarrow |\{x \in X \mid Q_{x,y}\}| \geq |\{x \in X \mid R_{x,z}\}| \qquad\qquad\qquad \square$$

As a general assumption, in the remainder of this paper we always assume pairs $u \in X \times Y$ to be of the form (u_1, u_2). To model direct products $X \times Y$ relation-algebraically, the *projection relations* $\pi : X \times Y \leftrightarrow X$ and $\rho : X \times Y \leftrightarrow Y$ have been proven as the convenient means. They are the relational variants of the well-known projection functions and, in point-wise definitions, given by

$$\pi_{u,x} :\Leftrightarrow u_1 = x \qquad\qquad \rho_{u,y} :\Leftrightarrow u_2 = y, \qquad\qquad (5)$$

for all $u \in X \times Y$, $x \in X$ and $y \in Y$. Projection relations enable us to specify the pairing operation of functional programming with relation-algebraic means. Assuming π and ρ as above, the *(right) pairing* (also called *fork*) of $R : Z \leftrightarrow X$ and $S : Z \leftrightarrow Y$ is defined as $[R, S] := R; \pi^\mathsf{T} \cap S; \rho^\mathsf{T}$. This leads to $Z \leftrightarrow X \times Y$ as type of $[R, S]$ and to a point-wise description saying that

$$[R, S]_{z,u} \Leftrightarrow R_{z,u_1} \land S_{z,u_2}, \qquad\qquad (6)$$

for all $z \in Z$ and $u \in X \times Y$. Using the projection relations $\pi : X \times Y \leftrightarrow X$ and $\rho : X \times Y \leftrightarrow Y$, we also are able to define functions which establish an isomorphism between the Boolean lattices $[X \leftrightarrow Y]$ and $[X \times Y \leftrightarrow \mathbf{1}]$. The direction from $[X \leftrightarrow Y]$ to $[X \times Y \leftrightarrow \mathbf{1}]$ is given by the function *vec*, with $vec(R) := (\pi; R \cap \rho); \mathsf{L}$, and that from $[X \times Y \leftrightarrow \mathbf{1}]$ back to $[X \leftrightarrow Y]$ by the inverse function *rel*, with $rel(v) := \pi^\mathsf{T}; (\rho \cap v; \mathsf{L}^\mathsf{T})$. In point-wise descriptions the definitions say that R_{u_1,u_2} iff $vec(R)_u$ and v_u iff $rel(v)_{u_1,u_2}$, for all $u \in X \times Y$.

As from now we assume all constructions of Section 2 to be at hand. Except the functions *set*, *scomp*, *vec* and *rel* they are available in the programming language of RELVIEW as or via pre-defined operations and the implementing BDDs of relations are comparatively small. For example, a size-comparison relation $\mathsf{S} : 2^X \leftrightarrow 2^X$ can be implemented by a BDD with $O(|X|^2)$ nodes and for a membership relation $\mathsf{M} : X \leftrightarrow 2^X$ even $O(|X|)$ nodes suffice (for details, see [11,12]). An implementation of *set*, *scomp*, *vec*, and *rel* in the tool is nothing else than the translation of their relation-algebraic definitions into RELVIEW-code; the RELVIEW-programs for *set* and *scomp* can be found in Section 5.

3   Relation-Algebraic Models of Plurality Voting

In Social Choice Theory (see e.g., [7] for an overview) an election consists of a non-empty and finite set N of *voters* (agents, individuals), normally defined as $N := \{1, \ldots, n\}$, a non-empty and finite set A of *alternatives* (candidates, proposala), the *individual preferences* (choices, wishes) of the voters, and a *voting rule* that aggregates the winners from the individual preferences. A well-known voting rule is the *plurality voting rule*. In its classical form each voter votes for exactly one alternative and the alternatives with the most votes win.

Usually, the individual preferences of the voters are expressed via a function $P : N \to A$ such that $P(i) = a$ iff voter i votes for alternative a, for all $i \in N$ and $a \in A$. But functions are nothing else than unique and total relations and for the *individual preferences relation* $P : N \leftrightarrow A$ then $P(i) = a$ is nothing else than an alternative notation for the relationship $P_{i,a}$. Under this view, a *(classical) plurality election* is a triplet (N, A, P), with P as unique and total individual preferences relation. Its *dominance relation* $D : A \leftrightarrow A$ is point-wisely defined by

$$D_{a,b} :\Leftrightarrow |\{i \in N \mid P_{i,a}\}| \geq |\{i \in N \mid P_{i,b}\}|, \tag{7}$$

for all $a, b \in A$, and its *set of winners* is $W := \{a \in A \mid \forall b \in B : D_{a,b}\}$. At this place it should be mentioned that (7) defines *weak dominance* and the definition of W allows *multiple winners*. There is also a version with *strict dominance* and *single winners*. All results of this section can immediately be transferred to this version (see Section 6). Its disadvantage is that W may become empty.

The following first result shows how classical plurality voting can be modeled with relation-algebraic means, i.e., how the (weak) dominance relation and the set of winners can be specified within the language of relation algebra.

Theorem 3.1. *If (N, A, P) is a plurality election and $D : A \leftrightarrow A$ its dominance relation, then we have $D = \text{scomp}(P, P)$. Furthermore, the vector $\overline{D}; \mathsf{L} : A \leftrightarrow \mathbf{1}$ models the set of winners W as a subset of A.*

Proof. Using (7) and then Lemma 2.2, we show for all $a, b \in A$ that

$$D_{a,b} \Leftrightarrow |\{i \in N \mid P_{i,a}\}| \geq |\{i \in N \mid P_{i.b}\}| \Leftrightarrow \text{scomp}(P, P)_{a,b}.$$

From this property we get the first claim by means of the point-wise description of the equality of relations. The second claim is shown by

$$a \in W \Leftrightarrow \forall b \in A : D_{a,b} \Leftrightarrow \neg \exists b \in A : \overline{D}_{a,b} \wedge \mathsf{L}_b \Leftrightarrow \overline{\overline{D}; \mathsf{L}}_b$$

for all $a \in A$, in combination with definition (2). □

In its classical form it is easy to control a plurality election via the removal of voters – constructively as well as destructively. But a control via the removal of alternatives does not make sense since in such a case the individual preferences relation $P : N \leftrightarrow A$ may become non-total. To be able to investigate the constructive control complexity of plurality elections via the removal of alternatives, in [1] a refined model of plurality voting is considered. In [1] it is assumed that each voter ranks the alternatives from top to bottom, that is, the individual preferences of the voters $i \in N$ are expressed via *linear strict order relations* $>_i : A \leftrightarrow A$, that collectively constitute a so-called *preference profile*. From the latter then a (strict) dominance relation $D : A \leftrightarrow A$ is computed, where $D_{a,b}$ iff a has more first place preferences than b, for all $a, b \in A$, and an alternative is defined as a winner if it (strictly) dominates all other alternatives.

In the following we concentrate again on weak dominance and allow multiple winners. For reasons of simplification we furthermore consider *linear order relations* $\geq_i : A \leftrightarrow A$ as individual preferences instead of linear strict order relations. Hence, we call the triplet $(N, A, (\geq_i)_{i\in N})$ a *ranked plurality election* to distinguish it from the classical notion (N, A, P). In case of a ranked plurality election the (weak) *dominance relation* $D : A \leftrightarrow A$ is point-wisely defined by

$$D_{a,b} \Leftrightarrow |\{i \in N \mid a = \max_i A\}| \geq |\{i \in N \mid b = \max_i A\}|, \tag{8}$$

for all $a, b \in A$, where $\max_i B$ denotes the *greatest element* of the non-empty set B of alternatives w.r.t. the linear order relation $\geq_i$, for all $B \in 2^A$. The *set of winners* is again defined as $\mathcal{W} := \{a \in A \mid \forall b \in B : D_{a,b}\}$.

For a relation-algebraic treatment of ranked plurality voting we first model the individual preferences of the voters as a single relation as follows.

Definition 3.1. *A relation $P : N \leftrightarrow A^2$ is a* model of the ranked plurality election $(N, A, (\geq_i)_{i\in N})$ *iff $P_{i,u}$ is equivalent to $u_1 \geq_i u_2$, for all $i \in N$ and $u \in A^2$.*

In the following RELVIEW picture a model $P : N \leftrightarrow A^2$ of a ranked plurality election is depicted as a Boolean matrix. The labels of the rows and columns indicate that the voters are the natural numbers from 1 to 17 and the alternatives are the letters from a to h. A black square of the matrix means a 1-entry and a white square means a 0-entry so that, for example, the completely black first colum indicates that $a \geq_i a$ is true for all voters $i \in N$ and the black parts of the second colum indicates that $a \geq_i b$ is true for all voters $i \in \{7, 8, 9, 10, 15, 16, 17\}$.

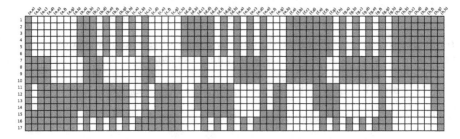

It is rather troublesome to identify from this 17×64 Boolean matrix the voter's individual preferences. But if we select the single rows of the matrix, transpose each of them to obtain 17 vectors of type $A^2 \leftrightarrow \mathbf{1}$ and apply, finally, the function *rel* of Section 2 to the latter, then RELVIEW depicts the individual preferences as 8×8 Boolean matrices, i.e., as relations of type $A \leftrightarrow A$. For the rows 1, 7, 11, 15, and 16 we show in the next pictures, in the same order, the Boolean matrices for the linear order relations $\geq_1, \geq_7, \geq_{11}, \geq_{15}$, and $\geq_{16}$. Note, that the relations $\geq_2$ to $\geq_6$ are equal to $\geq_1$, the relations $\geq_8$ to $\geq_{10}$ are equal to $\geq_7$, and so forth.

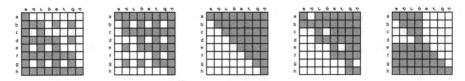

Now, the preferences of the single voters are easy to see: Voters 1 to 6 rank their alternatives from top to bottom as h, f, d, b, g, e, c, a, voters 7 to 10 as a, c, e, g, b, d, f, h, voters 11 to 14 as a, b, c, d, e, f, g, h, voter 15 as b, a, d, c, f, e, h, g, and the remaining voters 16 and 17 as h, g, f, e, a, b, c, d.

The above procedure also indicates how to construct, in general, the model $P : N \leftrightarrow A^2$ from the linear order relations $\geq_i \, : \, A \leftrightarrow A$ of a ranked plurality election $(N, A, (\geq_i)_{i \in N})$ by inverting it. We have to transform each relation $\geq_i$ into the transpose $vec(\geq_i)^\mathsf{T} : \mathbf{1} \leftrightarrow A^2$ of its corresponding vector and to combine these transposed vectors row by row into a Boolean matrix. The latter means that we have to form the *relation-algebraic sum* $vec(\geq_1)^\mathsf{T} + \cdots + vec(\geq_n)^\mathsf{T}$. We won't to go into details with regard to relation-algebraic sums and refer to [14], where a specification of $R + S$ via injection relations is given. Instead, we demonstrate in the next theorem how also ranked plurality voting can be modeled with relation-algebraic means, i.e., how again the dominance relation and the set of winners can be specified within the language of relation algebra.

Theorem 3.2. *If $(N, A, (\geq_i)_{i \in N})$ is a ranked plurality election, $P : N \leftrightarrow A^2$ its model, and $D : A \leftrightarrow A$ its dominance relation, then we get $D = scomp(\overline{P}; \pi, \overline{P}; \pi)$, where $\pi : A^2 \leftrightarrow A$ is the first projection relation of the direct product A^2. Again the vector $\overline{D}; \mathsf{L} : A \leftrightarrow \mathbf{1}$ models the set of winners.*

Proof. First, we prove for all $i \in N$ and $a \in A$ the following property:

$$
\begin{aligned}
\overline{\overline{P}; \pi}_{i,a} &\Leftrightarrow \neg \exists\, u \in A^2 : \overline{P}_{i,u} \wedge \pi_{u,a} \\
&\Leftrightarrow \forall u \in A^2 : P_{i,u} \vee \neg \pi_{u,a} \\
&\Leftrightarrow \forall u \in A^2 : u_1 = a \to u_1 \geq_i u_2 \qquad P \text{ is model}, (5) \\
&\Leftrightarrow \forall c \in A : a \geq_i c \\
&\Leftrightarrow a = \max_i A \qquad\qquad\qquad\qquad \text{definition } \max_i
\end{aligned}
$$

As a consequence, now for all $a, b \in A$ we can calculate as follows:

$$
\begin{aligned}
D_{a,b} &\Leftrightarrow |\{i \in N \mid a = \max_i A\}| \geq |\{i \in N \mid b = \max_i A\}| & \text{by (8)} \\
&\Leftrightarrow |\{i \in N \mid \overline{\overline{P}; \pi}_{i,a}\}| \geq |\{i \in N \mid \overline{\overline{P}; \pi}_{i,b}\}| & \text{see above} \\
&\Leftrightarrow scomp(\overline{P}; \pi, \overline{P}; \pi)_{a,b} & \text{Lemma 2.2}
\end{aligned}
$$

From this we get the first claim again by the point-wise description of the equality of relations. For the second claim, see the proof of Theorem 3.1.   □

The two relation-algebraic specifications of this theorem can be translated immediately into RELVIEW-code (cf. again Section 5). If we apply the corresponding

RELVIEW-programs to the above model of a ranked plurality election, i.e., the above 17×64 Boolean matrix, then RELVIEW yields the following two results for the dominance relation $D : A \leftrightarrow A$ and the vector $\overline{\overline{D}; \mathsf{L}} : A \leftrightarrow \mathbf{1}$.

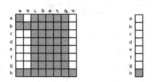

The Boolean vector shows that the alternative h is the only winner. This is in accordance with the matrix since its last black row indicates that h dominates each alternative. To demonstrate another visualization feature of RELVIEW, we show in the next picture how the tool depicts the strict part $D \cap \overline{\mathsf{I}}$ of D as a directed graph, using the built-in hierarchical graph drawing algorithm of [9].

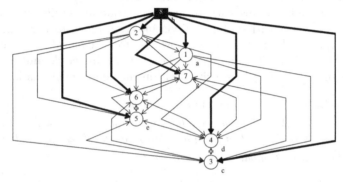

In this drawing the node corresponding to the winner h is highlighted as black square and the relationships which show that h dominates all other alternatives are highlighted as boldface arrows.

4   Relation-Algebraic Solution of Hard Control Problems

Having modeled two kinds of plurality elections relation-algebraically, we now show how to solve hard control problems for the ranked variant with the same means. For the ranked variant the following facts are known (see [1,10]): Constructive and destructive control by a removal or addition of alternatives is computationally hard, but for the constructive and destructive control by the removal or addition of voters there exist efficient algorithms. One says that plurality voting is *resistant* to control by manipulating the alternatives but *vulnerable* to control by manipulating the voters. For reasons of space we only consider control by a removal of alternatives. We start with constructive control.

Usually, the constructive control of plurality voting via the removal of alternatives is formulated as the following minimization-problem: Given a ranked plurality election $(N, A, (\geq_i)_{i \in N})$ and a specific alternative $a^* \in A$, compute a minimum set of alternatives $C \in 2^A$ such that the removal of C from A makes

a^* to a winner. To allow for an easier relation-algebraic solution, we consider the dual maximization-problem. We ask for a maximum set $B \in 2^A$ such that a^* is in B and wins in the ranked plurality election $(N, B, (\geq_i)_{i \in N})$. It is obvious that from B then a desired C is obtained via $C := A \setminus B$.

For the remainder of this section we assume a fixed ranked plurality election $(N, A, (\geq_i)_{i \in N})$ and a fixed alternative $a^* \in A$ that shall made to a winner by means of control. Furthermore, we assume that $P : N \leftrightarrow A^2$ is a model of the ranked plurality election and a^* is modeled by the point $p : A \leftrightarrow \mathbf{1}$ (i.e., we have p_b iff $b = a^*$, for all $b \in A$). As a first step towards a solution of our control problem we prove the following auxiliary result.

Lemma 4.1. *Using the membership relation* $\mathsf{M} : A \leftrightarrow 2^A$ *and* $\pi, \rho : A^2 \leftrightarrow A$ *as the first and second projection relation of the direct product* A^2, *respectively, we define the following relation:*

$$R := \mathrm{scomp}(\overline{\overline{P}; (\pi; p; \mathsf{L} \cap \rho; \mathsf{M})}, \overline{\overline{P}; [\pi, \rho; \mathsf{M}]}) : 2^A \leftrightarrow A \times 2^A$$

Then we have $R_{B,(b,C)}$ *iff* $|\{i \in N \mid a^* \geq_i \max_i B\}| \geq |\{i \in N \mid b \geq_i \max_i C\}|$, *for all* $b \in A$ *and* $B, C \in 2^A$.

Proof. First, we calculate for all $i \in N$ and $B \in 2^A$ as follows:

$$
\begin{aligned}
\overline{\overline{P}; (\pi; p; \mathsf{L} \cap \rho; \mathsf{M})}_{i,B} &\Leftrightarrow \neg \exists\, u \in A^2 : \overline{P}_{i,u} \wedge (\pi; p; \mathsf{L} \cap \rho; \mathsf{M})_{u,B} \\
&\Leftrightarrow \forall\, u \in A^2 : P_{i,u} \vee \neg((\pi; p)_u \wedge (\rho; \mathsf{M})_{u,B}) \\
&\Leftrightarrow \forall\, u \in A^2 : (\pi; p)_u \wedge (\rho; \mathsf{M})_{u,B} \to P_{i,u} \\
&\Leftrightarrow \forall\, u \in A^2 : p_{u_1} \wedge u_2 \in B \to P_{i,u} && \text{by } (3), (5) \\
&\Leftrightarrow \forall\, u \in A^2 : p_{u_1} \wedge u_2 \in B \to u_1 \geq_i u_2 && P \text{ is model} \\
&\Leftrightarrow \forall\, u \in A^2 : \\
&\quad\quad u_1 = a^* \wedge u_2 \in B \to u_1 \geq_i u_2 && p \text{ models } a^* \\
&\Leftrightarrow \forall\, c \in A : c \in B \to a^* \geq_i c \\
&\Leftrightarrow a^* \geq_i \max_i B && \text{def. } \max_i
\end{aligned}
$$

In a rather similar manner we get for all $i \in N$, $b \in A$, and $C \in 2^A$ the following equivalence:

$$
\begin{aligned}
\overline{\overline{P}; [\pi, \rho; \mathsf{M}]}_{i,(b,C)} &\Leftrightarrow \neg \exists\, u \in A^2 : \overline{P}_{i,u} \wedge [\pi, \rho; \mathsf{M}]_{u,(b,C)} \\
&\Leftrightarrow \neg \exists\, u \in A^2 : \overline{P}_{i,u} \wedge \pi_{u,b} \wedge (\rho; \mathsf{M})_{u,C} && \text{by } (6) \\
&\Leftrightarrow \neg \exists\, u \in A^2 : \overline{P}_{i,u} \wedge u_1 = b \wedge u_2 \in C && \text{by } (5), (3) \\
&\Leftrightarrow \forall\, u \in A^2 : P_{i,u} \vee \neg(u_1 = b \wedge u_2 \in C) \\
&\Leftrightarrow \forall\, u \in A^2 : u_1 = b \wedge u_2 \in C \to P_{i,u} \\
&\Leftrightarrow \forall\, u \in A^2 : u_1 = b \wedge u_2 \in C \to u_1 \geq_i u_2 && P \text{ is model} \\
&\Leftrightarrow \forall\, c \in A : c \in C \to b \geq_i c \\
&\Leftrightarrow b \geq_i \max_i C && \text{def. } \max_i
\end{aligned}
$$

Now, we are able to prove for all $b \in A$ and $B, C \in 2^A$ the desired result by

$$R_{B,(b,C)} \Leftrightarrow scomp(\overline{\overline{P}; (\pi; p; \mathsf{L} \cap \rho; \mathsf{M})}, \overline{\overline{P}; [\pi, \rho; \mathsf{M}]})_{B,(b,C)}$$
$$\Leftrightarrow |\{i \in N \mid \overline{\overline{P}; (\pi; p; \mathsf{L} \cap \rho; \mathsf{M})}_{i,B}\}| \geq |\{i \in N \mid \overline{\overline{P}; [\pi, \rho; \mathsf{M}]}_{i,(b,C)}\}|$$
$$\Leftrightarrow |\{i \in N \mid a^* \geq_i \max_i B\}| \geq |\{i \in N \mid b \geq_i \max_i B\}|,$$

where we use Lemma 2.2 in combination with the above results.   □

Using this lemma we now prove a theorem that shows how to specify relation-algebraically a vector that models the set of all solutions of the constructive control problem by a removal of alternatives for the given ranked plurality election and the given specific alternative (modeled by the point p).

Theorem 4.1. *Using the membership relation* $\mathsf{M} : A \leftrightarrow 2^A$, *the first and second projection relation* $\pi : A \times 2^A \leftrightarrow A$ *and* $\rho : A \times 2^A \leftrightarrow 2^A$, *respectively, of the direct product* $A \times 2^A$, *and the relation* $R : 2^A \leftrightarrow A \times 2^A$ *of Lemma 4.1, we define the following vector:*

$$cand := \mathsf{M}^\mathsf{T}; p \cap \overline{(\mathsf{M}^\mathsf{T} \cap \overline{(R \cap \rho^\mathsf{T})}; \pi)}; \mathsf{L} : 2^A \leftrightarrow \mathbf{1}$$

Then cand models the subset $\mathfrak{W}$ *of* 2^A *that consists of the sets* B *for which* a^* *wins the ranked plurality election* $(N, B, (\geq_i)_{i \in N})$. *Furthermore, the vector*

$$csol := cand \cap \overline{\overline{S^\mathsf{T}}; cand} : 2^A \leftrightarrow \mathbf{1}$$

models the subset $\mathfrak{W}^*$ *of* 2^A *that consists of the maximum sets of* $\mathfrak{W}$.

Proof. To prove the first claim, we take an arbitrary set $B \in 2^A$ and calculate as given below:

$$cand_B \Leftrightarrow (\mathsf{M}^\mathsf{T}; p \cap \overline{(\mathsf{M}^\mathsf{T} \cap \overline{(R \cap \rho^\mathsf{T})}; \pi)}; \mathsf{L})_B$$
$$\Leftrightarrow (\mathsf{M}^\mathsf{T}; p)_B \wedge \overline{(\mathsf{M}^\mathsf{T} \cap \overline{(R \cap \rho^\mathsf{T})}; \pi)}; \mathsf{L}_B$$
$$\Leftrightarrow (\exists b \in A : \mathsf{M}_{b,B} \wedge p_b) \wedge$$
$$\qquad \neg \exists b \in A : (\mathsf{M}^\mathsf{T} \cap \overline{(R \cap \rho^\mathsf{T})}; \pi)_{B,b} \wedge \mathsf{L}_b$$
$$\Leftrightarrow \mathsf{M}_{a^*,B} \wedge \neg \exists b \in A : \mathsf{M}_{b,B} \wedge \neg((R \cap \rho^\mathsf{T}); \pi)_{B,b} \qquad p \text{ models } a^*$$
$$\Leftrightarrow \mathsf{M}_{a^*,B} \wedge \forall b \in A : \mathsf{M}_{b,B} \to ((R \cap \rho^\mathsf{T}); \pi)_{B,b}$$
$$\Leftrightarrow a^* \in B \wedge$$
$$\qquad \forall b \in A : b \in B \to \exists u \in A \times 2^A : R_{B,u} \wedge \rho^\mathsf{T}_{B,u} \wedge \pi_{u,b} \qquad \text{by (3)}$$
$$\Leftrightarrow a^* \in B \wedge$$
$$\qquad \forall b \in B : \exists u \in A \times 2^A : R_{B,u} \wedge u_2 = B \wedge u_1 = b \qquad \text{by (5)}$$
$$\Leftrightarrow a^* \in B \wedge \forall b \in B : R_{B,(b,B)}$$

Because $a^* \in B$ implies $\{i \in N \mid a^* \geq_i \max_i B\} = \{i \in N \mid a^* = \max_i B\}$ and $b \in B$ implies $\{i \in N \mid b \geq_i \max_i B\} = \{i \in N \mid b = \max_i B\}$, from the above calculation and Lemma 4.1 we get the following result:

$$cand_B \Leftrightarrow a^* \in B \wedge \forall b \in B : |\{i \in N \mid a^* = \max_i B\}| \geq |\{i \in N \mid b = \max_i B\}|$$

But the right-hand side of this equivalence is precisely the formalization of the fact that a^* is a winner of the ranked plurality election $(N, B, (\geq_i)_{i\in N})$, such that definition (2) shows the claim. The second claim follows from the relation-algebraic specification of greatest elements w.r.t. pre-orders that, for instance, can be found in [13] or [14].   □

When RELVIEW is used to compute the vector $csol : 2^A \leftrightarrow \mathbf{1}$ of Theorem 4.1, then at a first glance it seems to be rather difficult to identify from it the sets of $\mathfrak{W}^*$, i.e., the solutions of our problem. But the relation-algebraic construction of an *embedding function generated by a vector* (that is, e.g., studied in detail in [3] and also available in the tool as a pre-defined operation) offers a simple and elegant solution. In case of the vector $csol$ the embedding function $inj(csol)$ has type $\mathfrak{W}^* \leftrightarrow 2^A$ and it holds $inj(csol)_{B,C}$ iff $B = C$, for all $B \in \mathfrak{W}^*$ and $C \in 2^A$. Hence, in combination with $\mathsf{M} : A \leftrightarrow 2^A$ the relation $\mathsf{M}; inj(csol)^\mathsf{T} : A \leftrightarrow \mathfrak{W}^*$ fulfills $(\mathsf{M}; inj(csol)^\mathsf{T})_{a,B}$ iff $a \in B$, for all $a \in A$ and $B \in \mathfrak{W}^*$. In Boolean matrix terminology the latter means that its columns, when regarded as single vectors, precisely model the single sets of $\mathfrak{W}^*$.

The RELVIEW-programs resulting from Lemma 4.1 and Theorem 4.1 and that for computing the relation $\mathsf{M}; inj(csol)^\mathsf{T} : A \leftrightarrow \mathfrak{W}^*$ are again given in Section 5. Using them, we have solved the constructive control problem via the removal of alternatives for the model P that is presented in Section 3 as 17×64 Boolean matrix. Doing so, we considered all eight alternatives a to h. The following series of eight RELVIEW pictures shows, in the same order, the corresponding column-wise representations of the sets of solutions.

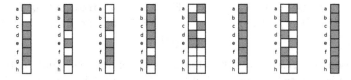

To give three examples, the Boolean vector at position 1 shows that a wins if at least two alternatives are removed and the removal of b, h is the only possibility, the 8×2 Boolean matrix at position 7 shows that g wins if at least four alternatives are removed and the removals of a, c, e, h and of b, d, f, h, respectively, are the only possibilities, and the Boolean universal vector at position 8 shows that the removal of no alternative makes h to the winner.

So far, we only have considered constructive control via the removal of alternatives. But if we negate the second part of the conjunction of the last equivalence of the proof of Theorem 4.1, then the resulting right-hand side

$$a^* \in B \wedge \neg \forall b \in B : |\{i \in N \mid a^* = \max_i B\}| \geq |\{i \in N \mid b = \max_i B\}|$$

or, more clearly, the equivalent formula

$$a^* \in B \wedge \exists b \in B : |\{i \in N \mid a^* = \max_i B\}| < |\{i \in N \mid b = \max_i B\}|$$

specifies that the alternative a^* is not a winner in the ranked plurality election $(N, B, (\geq_i)_{i\in N})$. Hence, to get from the relation $R : 2^A \leftrightarrow A \times 2^A$ of Lemma 4.1 a vector $dsol : 2^A \leftrightarrow \mathbf{1}$ that models the subset of 2^A which consists of all solutions of the *destructive control problem via a removal of alternatives*, only in the definition of the vector *cand* in Theorem 4.1 the expression $(\mathsf{M}^\mathsf{T} \cap \overline{(R \cap \rho^\mathsf{T})}; \pi); \mathsf{L}$ must be altered to $(\mathsf{M}^\mathsf{T} \cap \overline{(R \cap \rho^\mathsf{T})}; \pi); \mathsf{L}$ by deleting the outermost complement operation. We have RELVIEW also used to compute for all alternatives of our running example the solutions of the destructive control problem. The tool reported that win for the seven alternatives $a, b, \ldots, g$ can be avoided by removing no alternative. But in case of h we obtained for *dsol* the empty vector. This means that it is not possible to avoid win for the alternative h by a removal of any subset of $\{a, b, \ldots, g\}$, i.e., there is no solution for the destructive control problem via the removal of alternatives for a^* being h.

5   Implementation in RELVIEW

In the following we present the RELVIEW-code for the relation-algebraic specifications of Section 2 to 4. We start with the code for the functions *set* and *scomp*. In **set** a pre-defined operation **epsi** is used that computes for a vector $r : X \leftrightarrow \mathbf{1}$ the membership relation $\mathsf{M} : X \leftrightarrow 2^X$. To obtain such a vector, the program uses a further pre-defined operation **On1** that computes for a relation $Q : X \leftrightarrow Y$ the empty vector $\mathsf{O} : X \leftrightarrow \mathbf{1}$. In **scomp** a pre-defined operation **cardrel** is applied to get from a vector $r : X \leftrightarrow \mathbf{1}$ the size comparison relation $\mathsf{S} : 2^X \leftrightarrow 2^X$.

```
set(R)                       scomp(Q,R)
  DECL M                       DECL S
  BEG  M = epsi(On1(R))        BEG  S = cardrel(On1(Q))
       RETURN syq(R,M)              RETURN set(Q)*S^*set(R)^
  END.                         END.
```

A comparison of **scomp** and the definition of *scomp* shows that in RELVIEW composition is denoted by the symbol '*' and transposition by the symbol '^'.

Based on **scomp**, the following two programs compute the dominance relation for classical (left) and ranked (right) plurality voting. In **domrel_ranked** the pre-defined operations **PROD** and **p-1** are used to define **AA** as name for the direct product A^2 and to compute the first projection relation of a direct product, respectively. For the definition of a direct product $X \times Y$ via **PROD** the operation needs arbitrary relations of type $X \leftrightarrow X$ and $Y \leftrightarrow Y$ as input. In **domrel_ranked** such relations are provided by the additional parameter $T : A \leftrightarrow A$.

```
domrel(P)                    domrel_ranked(P,T)
  DECL D                       DECL AA = PROD(T,T);
  BEG  D = scomp(P,P)               pi
       RETURN D                BEG  pi = p-1(AA)
  END.                              RETURN scomp(-(-P*pi),-(-P*pi))
                               END.
```

Again a comparison of **domrel_ranked** and Theorem 3.2 shows that in RELVIEW complementation is denoted by the symbol '-'.

The next program compR computes the relation $R : 2^A \leftrightarrow A \times 2^A$. of Lemma 4.1. Here the relation $p; p^\top : A \leftrightarrow A$ is used in PROD for defining the direct product A^2. The program uses four pre-defined operations which we have not mentioned so far, viz. '&' as symbol for intersection, p-2 for computing the second projection relation of a direct product, L1n for computing for a relation $Q : X \leftrightarrow Y$ the transposed universal vector $L : \mathbf{1} \leftrightarrow Y$, and $[\cdot, \cdot |]$ for computing pairings.

```
compR(P,p)
  DECL AA = PROD(p*p^,p*p^);
       pi, rho, M
  BEG  pi = p-1(AA);
       rho = p-2(AA);
       M = epsi(p)
       RETURN scomp(-(-P*(pi*p*L1n(M) & rho*M)),-(-P*[pi,rho*M|]))
  END.
```

In the following program cand for the vector *cand* the input of PROD consists of $p; p^\top : A \leftrightarrow A$ and the size comparison relation $S : 2^A \leftrightarrow 2^A$, such that AxPA denotes the direct product $A \times 2^A$. The pre-defined operation dom computes for a relation $Q : X \leftrightarrow Y$ its so-called *domain* $Q; L : X \leftrightarrow \mathbf{1}$.

```
        cand(P,p)
          DECL AxPA = PROD(p*p^,cardrel(p));
               pi, rho, M, R
          BEG  pi = p-1(AxPA);
               rho = p-2(AxPA);
               M = epsi(p);
               R = compR(P,p)
               RETURN M^*p & -dom(M^ & -((R & rho^)*pi))
          END.
```

And here are, finally, the programs csol and csolList for computing the vector *csol* and the relation $M; inj(csol)^\top$, respectively.

```
csol(P,p)                        csolList(P,p)
  DECL S, ca                       DECL M, so
  BEG  S = cardrel(On1(p));        BEG  M = epsi(p);
       ca = cand(P,p)                   so = csol(P,p)
       RETURN ca & -(-S^*ca)            RETURN M*inj(so)^
  END.                             END.
```

6 Assessment of the Approach and Concluding Remarks

Despite of the use of BDDs our experiments have shown that the RELVIEW-programs csol and csolList of Section 5 are only feasible for elections with at most 25 alternatives, whereas larger sets with several hundreds voters constituted no problem. (In case of the programs domrel and domrel_ranked for computing dominance relations also larger sets of alternatives did not lead to problems.) We also have transferred the approach of [6], that solves control problems for Condorcet voting via so-called *relativized dominance relations*, to plurality voting. But this did not lead to faster programs either. Concerning efficiency, therefore, our general and model-oriented approach can not compete with algorithms that are specifically tailored for such hard problems and implemented in a conventional programming language like C or Java. But we believe that it has two decisive pros, which we will describe now.

For the presentation of the results of this paper we have used the prevalent mathematical theorem-proof-style to enhance readability. But actually we have obtained all relation-algebraic specifications by developing them from the original logical specifications of the problems in question. We regard this formal and goal-oriented development of algorithms from logical specifications, that are correct by construction, as the first main advantage. In respect thereof it is also beneficial that the calculations are formalized in such a way that the danger of making errors is minimized and the use of theorem-provers seems to be possible. Presently we investigate the use of the tool Prover9 (see [16]) for automated verification; see [5]. As the second main advantage we regard the support by means of an appropriate Computer Algebra system. All results we have developed are expressed by very short and concise RELVIEW programs, which are nothing else than their formulation in a specific syntax. The programs are easy to alter in case of slightly changed specifications. Combining this with RELVIEW's possibilities for visualization, animation, and the random generation of relations allows prototyping of specifications, testing of hypotheses, experimentation with concepts, exploration of ideas, generation of examples and counterexamples etc., while avoiding unnecessary overhead. This makes RELVIEW very useful for scientific research and also for teaching. In this regard the BDD-implementation of relations is of immense help since it allows to treat also non-trivial examples.

Concerning slightly changed specifications, we have already mentioned that the step from constructive to destructive control requires only to remove a complement operation. To give a further example, if plurality voting with strict dominance and single winners is to treat, then we only have to replace in the definition of $scomp(Q, R)$ the relation S^T by $\overline{\mathsf{S}}$ to get the relations D of Theorem 3.1 and Theorem 3.2 as strict dominance relations, to replace in the definition of the vector that models the set of winners D by $D \cap \overline{\mathsf{I}}$ to prevent multiple winners, and to replace in the definition of the vector $cand$ in Theorem 4.1 the universal vector L by $\overline{p}$ (i.e., the complement of the point that models a^*) to compute the solutions of the control problems. The correctness proofs are obtained by slight modifications of those we have presented in Section 3 and Section 4, where most of the steps of the calculations can be re-used.

In real live elections it frequently happens that a voter considers certain alternatives as incomparable or to be of the same value. In such a case its preferences are not specified by a linear order relation but by a partial order relation or even a pre-order relation. The first case is studied in [8]. Presently we investigate ranked plurality elections $(N, A, (\succeq_i)_{i \in N})$ with pre-order relations $\succeq_i$. Here it seems to be reasonable to define the dominance relation $D : A \leftrightarrow A$ for all $a, b \in A$ pointwisely by $D_{a,b}$ iff $|\{i \in N \mid a \in \max_i A\}| \geq |\{i \in N \mid b \in \max_i A\}|$, where now $\max_i A := \{a \in A \mid \neg \exists\, b \in A : b \neq a \wedge b \succeq_i a\}$ defines the set of maximal elements of A w.r.t. the pre-order relation $\succeq_i$. This leads to the relation-algebraic specification $D = scomp(\overline{P; (\overline{\pi} \cap \rho)}, \overline{P; (\overline{\pi} \cap \rho)})$, with $\pi, \rho : A^2 \leftrightarrow A$ as the first and second projection relation of the direct product A^2, respectively. For this generalization of plurality voting, among other things, by systematic RELVIEW-experiments we want to find out which of the desirable properties of voting systems (see e.g., [7]) are definitely not satisfied and which of them possibly may be (or even most likely are) satisfied, such that in the latter cases the chance of success of proofs is given.

Acknowledgment. I thank Henning Schnoor for valuable discussions and his comments and suggestions.

References

1. Bartholdi III, J.J., Tovey, C.A., Trick, M.A.: How hard is it to control an election? Mathematical and Computer Modeling 16, 27–40 (1992)
2. Berghammer, R., Neumann, F.: RELVIEW – An OBDD-based Computer Algebra system for relations. In: Ganzha, V.G., Mayr, E.W., Vorozhtsov, E.V. (eds.) CASC 2005. LNCS, vol. 3718, pp. 40–51. Springer, Heidelberg (2005)
3. Berghammer, R., Winter, M.: Embedding mappings and splittings with applications. Acta Informatica 47, 77–110 (2010)
4. Berghammer, R., Danilenko, N., Schnoor, H.: Relation algebra and RELVIEW applied to approval voting. In: Höfner, P., Jipsen, P., Kahl, W., Müller, M.E. (eds.) RAMiCS 2014. LNCS, vol. 8428, pp. 309–326. Springer, Heidelberg (2014)
5. Berghammer, R., Höfner, P., Stucke, I.: Automated verification of relational while-programs. In: Höfner, P., Jipsen, P., Kahl, W., Müller, M.E. (eds.) RAMiCS 2014. LNCS, vol. 8428, pp. 173–190. Springer, Heidelberg (2014)
6. Berghammer, R., Schnoor, H.: Control of Condorcet voting: Complexity and a relation-algebraic approach (Extended abstract). In: Proc. AAMAS 2014 (to appear, 2014)
7. Brandt, F., Conitzer, V., Endriss, U.: Computational social choice. In: Weiss, G. (ed.) Multiagent Systems, 2nd edn., pp. 213–283. MIT Press (2013)
8. Faliszewski, P., Hemaspaandra, E., Hemaspaandra, L., Rothe, J.: Llull and Copeland voting computationally resist bribery and constructive control. Journal of Artificial Intelligence Research 35, 275–341 (2009)
9. Gansner, E.R., Koutsofios, E., North, S.C., Vo, K.P.: A technique for drawing directed graphs. IEEE Transactions on Software Engineering 19, 214–230 (1993)
10. Hemaspaandra, E., Hemaspaandra, L., Rothe, J.: Anyone but him: The complexity of precluding an alternative. Artificial Intelligence 171, 255–285 (2007)
11. Leoniuk, B.: ROBDD-based implementation of relational algebra with applications. Dissertation, Universität Kiel (2001) (in German)
12. Milanese, U.: On the implementation of a ROBDD-based tool for the manipulation and visualization of relations. Dissertation, Universität Kiel (2003) (in German)
13. Schmidt, G., Ströhlein, T.: Relations and graphs. Springer (1993)
14. Schmidt, G.: Relational mathematics. Cambridge University Press (2010)
15. RELVIEW-homepage, http://www.informatik.uni-kiel.de/~progsys/relview/
16. Prover9-homepage, http://www.prover9.org

An Algorithm for Converting Nonlinear Differential Equations to Integral Equations with an Application to Parameter Estimation from Noisy Data

François Boulier[1], Anja Korporal[2], François Lemaire[1], Wilfrid Perruquetti[2,3], Adrien Poteaux[1], and Rosane Ushirobira[2]

[1] Université Lille 1, LIFL, UMR CNRS 8022, Computer Algebra Group,
`FirstName.LastName@univ-lille1.fr`
[2] Inria, Non-A team
`FirstName.LastName@inria.fr`
[3] École Centrale de Lille, LAGIS, UMR CNRS 8219

Abstract. This paper provides a contribution to the parameter estimation methods for nonlinear dynamical systems. In such problems, a major issue is the presence of noise in measurements. In particular, most methods based on numerical estimates of derivations are very noise sensitive. An improvement consists in using integral equations, acting as noise filtering, rather than differential equations. Our contribution is a pair of algorithms for converting fractions of differential polynomials to integral equations. These algorithms rely on an improved version of a recent differential algebra algorithm. Their usefulness is illustrated by an application to the problem of estimating the parameters of a nonlinear dynamical system, from noisy data.

In Engineering, a wide variety of information is not directly obtained through measurement. Various parameters or internal variables are unknown or not measured. In addition, sensor signals are very often distorted and tainted by measurement noises. To simulate, control or supervise such processes, and to extract information conveyed by the signals, a system has to be identified and parameters and variables must be estimated. Most of traditional estimation methods are related to asymptotic statistics. However, there exist some difficulties that have been long known as inherent to these existing methods. Among them, two important limitations can be pointed out: these methods apply essentially to linear systems and they are noise sensitive due to the use of numerical derivation. The parameter estimation problem has been tackled by many different approaches in control theory. Algebraic techniques to this end were notably introduced in the works by M. Fliess et al. [8, 15, 7, 9, 6] and inspired for instance, algebraic methods for the parameter estimation of a multi-sinusoidal waveform signal from noisy data [22].

This paper[1] provides a contribution to these issues. Two algorithms are provided which convert differential equations to integral equations. They rely on a

---

[1] This work was partially supported by the French ANR-10-BLAN-0109 LEDA project.

V.P. Gerdt et al. (Eds.): CASC Workshop 2014, LNCS 8660, pp. 28–43, 2014.

differential algebra [18, 14] algorithm for integrating differential fractions, which was presented in [3, Algorithm 4]. They are applied on the so-called *differential input-output equation* of a given nonlinear dynamical system in order to obtain *integral input-output equations*. For some systems, such as the one considered in this paper, the integral equation does not involve any derivative of the time varying variables. This property implies that its numerical evaluation does not require any numerical derivation. Indeed, numerical experiments confirm that, on white noisy data, integral forms of an input-output equation yield much better estimates of parameters than differential forms. It is well-known that numerical integration process has a filtering property on noisy data.

The paper is organized as follows. First, parameter estimation methods are presented in Section 1, notably by applying the notion of modulating functions. This approach is classical in automatic control theory and perhaps it is not so well-known in other fields and might be of interest for experts in integro-differential algebras, for instance. Section 1 features our new algorithms as well. Basic notions of differential algebra, required to understand the new algorithms, are presented in Section 2. An improved version of [3, Algorithm 4] is presented as Algorithm 3 in Section 3, together with additional properties (Propositions 1 and 2). Finally, two algorithms for computing integral equations are presented in Section 4 as Algorithms 4 and 5.

1   A Parameter Estimation Method

1.1   Problem Formulation

We consider the academic two-compartment model depicted in Figure 1. Compartment 1 represents the blood system and compartment 2 represents some organ. A medical drug is injected in compartment 1 at $t = 0$. It is diffused between the two compartments, following linear laws: the proportionality constants are named k_{12} and k_{21}. The drug exits compartment 1, following a law of Michaelis-Menten type. Such a law indicates an implicit enzymatic reaction and in general, it depends on two constants V_e and k_e. For the sake of simplicity, it is assumed that $k_e = 1$.

The state variables in this system are $x_1(t)$ and $x_2(t)$. They represent the concentrations of drugs in each compartment. The system has no input. Its output, denoted $y(t)$, is equal to $x_1(t)$, meaning that some numerical data are

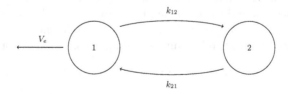

Fig. 1. A two-compartment model featuring three parameters

available for $x_1(t)$. No data is available for $x_2(t)$. To simplify the problem formulation, assume that both compartments have unit volumes. We obtain the
following nonlinear dynamical system, which features three parameters to be
estimated : k_{12}, k_{21} and V_e.

$$\dot{x}_1(t) = -k_{12}\, x_1(t) + k_{21}\, x_2(t) - \frac{V_e\, x_1(t)}{1 + x_1(t)},$$
$$\dot{x}_2(t) = k_{12}\, x_1(t) - k_{21}\, x_2(t),$$
$$y(t) = x_1(t).$$

(1)

Estimation methods are applied on data obtained as follows: some made-
up numerical values are assigned to the three parameters and the two initial
values $x_1(0)$ and $x_2(0)$. A numerical curve is obtained by numerically integrating (1). Some white Gaussian noise, depending on a given standard deviation σ,
is added to the curve, for $\sigma \in [0, 0.2]$. These curves are displayed in Figure 2.

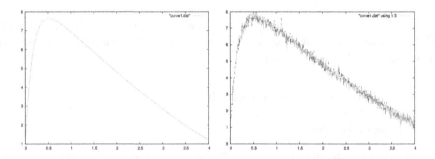

Fig. 2. The leftmost curve is obtained by numerically integrating (1) for $t = [0, 4]$, with
$(x_1(0), x_2(0), k_{12}, k_{21}, V_e) = (1, 10, 1, 5, 3)$. The rightmost one is obtained by adding to
it a white Gaussian noise with standard deviation $\sigma = 0.2$.

1.2   The Input-Output Equations

In general, an input-output equation of a dynamical system is a differential equation, which belongs to the differential ideal defined by the model equations. It
depends only on the input, the output, their derivatives and the model parameters (observe that our example has no input). Such input-output relations are
used for parameter identification since the only measurable variables are the
input and the output. See [5, 8, 9, 6, 7, 17] and references therein. Using a
differential elimination method [2] and a ranking that eliminates state variables:

$$\cdots > \ddot{x}_2 > \ddot{x}_1 > \dot{x}_2 > \dot{x}_1 > x_2 > x_1 > \cdots > \ddot{y} > \dot{y} > y > (k_{12}, k_{21}, V_e),$$

it is possible to automatically compute the differential input-output equation
of (1). Since our new algorithms are all about rewriting a single equation into a

more convenient form, it is important to give the result almost as in the same form as it is returned by the elimination procedure:

$$\ddot{y}(t)\, y(t)^2 + 2\, \ddot{y}(t)\, y(t) + \ddot{y}(t)$$
$$+\, \dot{y}(t)\, y(t)^2\, \theta_2 + 2\, \dot{y}(t)\, y(t)\, \theta_2 + \dot{y}(t)\, \theta_3 + y(t)^2\, \theta_1 + y(t)\, \theta_1 = 0 \tag{2}$$

where the θ_i stand for the blocks of parameters:

$$\theta_1 = k_{21}\, V_e\,, \quad \theta_2 = k_{12} + k_{21}\,, \quad \theta_3 = k_{12} + k_{21} + V_e\,.$$

Dividing (2) by the coefficient of $\ddot{y}(t)$ (its *initial*, in the terminology of differential algebra) and observing it can be factored, one obtains a normalized differential input-output equation:

$$\theta_1\, \frac{y(t)}{y(t)+1} + \theta_2\, \frac{y(t)\, \dot{y}(t)\, (y(t)+2)}{(y(t)+1)^2} + \theta_3\, \frac{\dot{y}(t)}{(y(t)+1)^2} = -\ddot{y}(t) \tag{3}$$

Equation (3) depends on the first and the second derivative of $y(t)$. Before applying our algorithms, let us take a few lines to see what can be done easily or not from this equation, *i.e.* where the issue lies. It is actually easy to decrease by one the order of (3) since $\ddot{y}(t)$ occurs with degree one. Indeed

$$\int_a^t \ddot{y}(t)\, \mathrm{d}t = \int_a^t \frac{\mathrm{d}}{\mathrm{d}\tau}\, \dot{y}(\tau)\, \mathrm{d}\tau = [\dot{y}(\tau)]_a^t = \dot{y}(t) - \dot{y}(a)\,. \tag{4}$$

It is easy to obtain the following equivalent equation by integrating (3):

$$\theta_1 \int_a^t \frac{y(\tau)}{y(\tau)+1}\, \mathrm{d}\tau + \theta_2 \int_a^t \frac{y(\tau)\, \dot{y}(\tau)\, (y(\tau)+2)}{(y(\tau)+1)^2}\, \mathrm{d}\tau$$
$$+\, \theta_3 \int_a^t \frac{\dot{y}(\tau)}{(y(\tau)+1)^2}\, \mathrm{d}\tau = -\dot{y}(t) + \dot{y}(a)\,. \tag{5}$$

Let us stress that in (4), the integral operator and the derivation operator are simplified thanks to the fact that the derivation operator is factored out under the integral sign, *i.e.* the expression has the following form. However, the derivation operators still occuring under the integral signs of (5) do not satisfy this pattern and the simplification cannot be performed easily. There lies the issue.

$$\int_a^t \frac{\mathrm{d}}{\mathrm{d}\tau}\ \text{something (possibly nonlinear)}\, \mathrm{d}\tau\,. \tag{6}$$

A recent algorithm [3, Algorithm 4] applied to equation (3) solves it and returns the following expression. This new equation (7) is a sum of expressions that are prepared to be in the form of (6). Moreover, all differential fractions in the above equation have order zero.

$$\theta_1 \frac{y(t)}{y(t)+1} + \theta_2\, \frac{\mathrm{d}}{\mathrm{d}t}\, \frac{y(t)^2}{y(t)+1} - \theta_3\, \frac{\mathrm{d}}{\mathrm{d}t}\, \frac{1}{y(t)+1} = -\frac{\mathrm{d}^2}{\mathrm{d}t^2}\, y(t) \tag{7}$$

We now describe the two possibilities for computing an integral equation from (7).

First Approach. Apply twice the integration operator on (7). It results an integral input-output equation, that can be used for estimating parameters. This formula still involves a derivative: $\dot{y}(a)$. We consider it as a new parameter to be estimated.

$$
\theta_1 \int_a^t \int_a^{\tau_1} \frac{y(\tau_2)}{y(\tau_2) + 1} \, d\tau_2 \, d\tau_1
$$

$$
+ \theta_2 \left(\int_a^t \frac{y(\tau)^2}{y(\tau) + 1} \, d\tau - \frac{y(a)^2}{y(a) + 1} (t - a) \right) \tag{8}
$$

$$
- \theta_3 \left(\int_a^t \frac{1}{y(\tau) + 1} \, d\tau - \frac{1}{y(a) + 1} (t - a) \right) - \dot{y}(a) (t - a) = -y(t) + y(a)
$$

Second Approach. A second possibility is to use some particular filter functions called *modulating functions*. They were introduced by M. Shinbrot in the 50's for system identification problems [20]. Shinbrot suggested the use of integral transformations on these problems to facilitate the identification for higher-order nonlinear dynamical systems. In addition, the effects of the initial conditions are annihilated by the modulating functions making this method more propitious to applications on noisy signals. Other authors have used modulating functions for estimating parameters for different two-compartment models, for instance A. Pearson applies Fourier modulating functions in [17] and K. Godfrey applies a successive derivatives method for the particular model considered here [11].

A modulating function of order n is a real-valued function $\phi_n(\tau)$ defined on a time interval $[a, t]$ that satisfies the $2n$ end-point conditions:

$$
\frac{d^\ell \phi_n}{d\tau^\ell}(a) = \frac{d^\ell \phi_n}{d\tau^\ell}(t) = 0, \qquad 0 \le \ell < n.
$$

Thus integration by parts yields for any function $f(\tau)$:

$$
\int_a^t \phi_n(\tau) \frac{d}{d\tau} f(\tau) \, d\tau = [\phi_n(\tau) f(\tau)]_a^t - \int_a^t \left(\frac{d}{d\tau} \phi_n(\tau) \right) f(\tau) \, d\tau
$$

$$
= - \int_a^t \left(\frac{d}{d\tau} \phi_n(\tau) \right) f(\tau) \, d\tau .
$$

Hence, multiplying (7) by a modulating function $\phi_2(\tau)$, integrating once and applying integration by parts as many times as needed gives a second integral input-output equation, that can be used for estimating parameters:

$$
\theta_1 \int_a^t \phi_2(\tau) \frac{y(\tau)}{y(\tau) + 1} \, d\tau - \theta_2 \int_a^t \dot{\phi}_2(\tau) \frac{y(\tau)^2}{y(\tau) + 1} \, d\tau
$$

$$
+ \theta_3 \int_a^t \dot{\phi}_2(\tau) \frac{1}{y(\tau) + 1} \, d\tau = - \int_a^t \ddot{\phi}_2(\tau) \, y(\tau) \, d\tau . \tag{9}
$$

For our experiments, we tested three types of modulating functions:

- *Hartley modulating functions* $\phi_n(\tau)$, where $\mu = 0$, $n = 2$ and $\text{cas}(\tau) = \cos(\tau) + \sin(\tau)$, for all τ.

$$\phi_n(\tau) = \sum_{\ell=0}^{n} (-1)^\ell \binom{n}{\ell} \text{cas}\left((n + \mu - \ell)\frac{2\pi}{t}\tau\right), \tag{10}$$

- *Modulating functions based on Hermite polynomials*: they are based on the weight functions for Hermite polynomials by a change of variables $t \mapsto \frac{t-2\tau}{t}$:

$$\phi(t) = e^{-\alpha\left(\frac{t-2\tau}{t}\right)^2}. \tag{11}$$

Remark that the graph of $\phi(t)$ is symmetric with respect to $\tau = \frac{t}{2}$. Strictly speaking, $\phi(t)$ is not a modulating function since it does not vanish at $\tau = 0$ and $\tau = t$. However, its values are arbitrarily close to zero at $\tau = 0$ and $\tau = t$, depending on the factor α. Experiments were performed using $\alpha = \frac{25}{2}$.

- *Modulating polynomial functions defined by interpolation*: we set

$$\phi(t) = a_4\tau^4 + \cdots + a_1\tau + a_0, \tag{12}$$

such that $\phi(0) = \phi(t) = \frac{d\phi}{d\tau}(0) = \frac{d\phi}{d\tau}(t) = 0$ and $\phi\left(\frac{t}{2}\right) = 1$. A polynomial of degree four is sufficient to obtain a modulating function of order $n = 2$.

1.3   Method, Implementation and Results

The method (we borrowed it from [16, 5]) consists in evaluating a chosen input-output equation for many different values of t, over the available data. Thereby, one gets an overdetermined linear system that can be solved by linear least squares. If one chooses a differential equation (3), (7) or the integral equation (9) which relies on modulating functions, the unknowns are the blocks of parameters θ_1, θ_2 and θ_3. If one chooses the integral equation (8), the unknowns are the blocks of parameters plus the extra parameter $\dot{y}(a)$. In our experiments, we picked many different values (about 20) for a. For each new value of a, we had to introduce a new indeterminate.

The different methods were implemented in FORTRAN 77. The code is available at [21]. Linear algebra (least squares) was performed using the LAPACK library. The numerical integration of the dynamical system (needed to produce the experimental data) was performed using the RADAU integrator, borrowed from [13]. The code for evaluating the modulating functions and their derivatives was produced in FORTRAN, from MAPLE, using the `CodeGeneration` package of MAPLE. The input-output equation (2) was produced by the BLAD libraries [1], through the `DifferentialAlgebra` package of MAPLE. The numerical derivations needed when using (3), (7) were computed over degree 3 polynomials best fitting about 20 consecutive data points. The numerical integrations were performed using the composite Simpson's formula.

For each input-output equation (3), (7), (8), (9), the accuracy of the estimation of the blocks of parameters θ_1, θ_2 and θ_3 was tested over noisy data, varying the noise from $\sigma = 0$ to $\sigma = 0.8$. For each value of σ, about 50 different simulations were performed.

Experiments show that the methods based on the two integral equations (8), and (9) give the best results. Modulating functions give simpler formulas and algorithms but need to be tuned carefully. See Figure 3 for details. Tuning was possible here, because we knew in advance the exact values of the parameter blocks. In a real life situation, it would be more complicated.

The methods based on the two different formulations of the differential input-output equation give similar results. We were expecting the one based on (7) to be more accurate, but we observed the opposite phenomenon.

2   Basic Notions of Differential Algebra

The reference books for this Section are [18, 14]. In this paper, we restrict ourselves to ordinary differential rings, which are rings endowed with a single derivation, that we assimilate do $\delta = \mathrm{d}/\mathrm{d}t$. To this derivation, one associates an *independent variable* t, defined by $\delta t = 1$. One denotes $\mathscr{X} = \{t\}$. In order to form differential polynomials, one introduces a set $\mathscr{U} = \{u_1, \ldots, u_n\}$ of n *differential indeterminates*. In Engineering, differential indeterminates would be called *dependent variables*. The derivatives of various orders of the differential indeterminates are simply called *derivatives*. We sometimes write $\dot{u}$ instead of δu and $\ddot{u}$ instead of $\delta^2 u$. The order of a derivative is the number of times it is differentiated: u, $\dot{u}$ and $\ddot{u}$ have respectively orders 0, 1 and 2.

For our concerns, it is crucial that one can also handle *parametric* differential equations. Parameters are nothing but symbolic *constants*, i.e., symbols whose derivatives are zero. Let $\mathscr{C}$ denote the set of constants.

The differential fractions considered in this paper are ratios of differential polynomials taken from the differential ring $\mathscr{R} = \mathbb{Z}[\mathscr{X} \cup \mathscr{C}]\{\mathscr{U}\}$, using the notations of [14]. A differential fraction is said to be *reduced* if its numerator and denominator do not have any common factor. A differential fraction is said to be a *coefficient* if it belongs to the field $\mathscr{K} = \mathbb{Q}(\mathscr{X} \cup \mathscr{C})$. The field $\mathscr{K}$ contains a field of constants, $\mathscr{K}_c = \mathbb{Q}(\mathscr{C})$. A fraction which is not in $\mathscr{K}$ thus depends on, at least, a derivative.

A *ranking* is a total ordering over the set of all derivatives that satisfies the two following axioms:

1. $u \leq \delta^n u$ for any $u \in \mathscr{U}$ and nonnegative integer n,
2. $u < v \Rightarrow \delta^n u < \delta^n v$ for all derivatives u, v and nonnegative integers n.

Rankings are well-orderings, i.e., every strictly decreasing sequence of derivatives is finite [14, §I.8]. Rankings such that $n < m \Rightarrow \delta^n u < \delta^m v$ for all nonnegative integers n, m and differential indeterminates $u, v \in U$ are called *orderly*.

Fix a ranking and consider some differential polynomial $P \notin \mathscr{K}$. The highest derivative v w.r.t. the ranking such that $\deg(P, v) > 0$ is called the *leading*

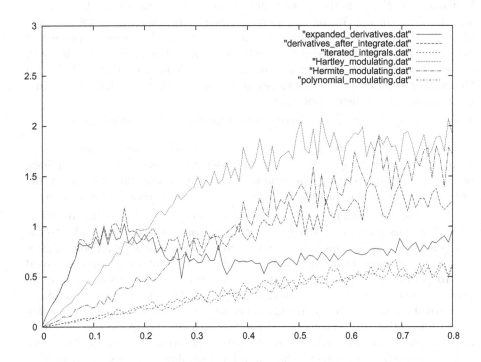

Fig. 3. Estimation of the blocks of parameters θ_i using different forms of the input-output equation. In abscissa, the standard deviation σ used to produce the white gaussian noise. In ordinate, the average value, for 50 different simulations, of the relative error, i.e. the 2-norm of the difference between the vector of estimated values and the vector of known values, divided by the 2-norm of the vector of known values. Therefore, all relative errors above 1 are equivalent, since they all correspond to computed values void of informations. For all formulas, two phases seem to occur: first, the relative error grows linearly with the standard deviation σ ; second, the error reaches some plateau. The best results are obtained using the integral equation (8) and the integral equation (9), with the polynomial modulating function (12). The two intermediate curves (for $\sigma \in [0, 0.15]$) are obtained using the integral equation (9), with modulating functions of Hermite type (11), below, and Hartley type (10), above. The worst results (for $\sigma \in [0, 0.15]$) are obtained by estimating derivatives on the differential equations (3) and (7). The curve corresponding to Equation (3) is strange since the error seems to decrease, temporarily, while σ increases. We suspect an artefact due to the example.

derivative of P. The monomial $v^{\deg(P,v)}$ is called the *rank* of P. The leading coefficient of P w.r.t. v is called the *initial* of P.

Extensions of these definitions to differential fractions were introduced in [3]. The *order* of a reduced fraction $F \notin \mathscr{K}$ is the maximum of the orders of the derivatives it depends on. It is denoted $\mathrm{ord}(F)$. The order of fractions which are not in $\mathcal{K}$ is defined to be zero. The leading derivative of a fraction $F \notin \mathcal{K}$ is denoted $\mathrm{ld}(F)$. It is the highest derivative v, with respect to the ranking, such that $\partial F/\partial v \neq 0$. We do not need to define leading derivatives for elements of $\mathcal{K}$. Let $F = P/Q$ be a fraction which is not in $\mathcal{K}$. Its *degree* is defined as $\deg(F) = \deg(P, \mathrm{ld}(F)) - \deg(Q, \mathrm{ld}(F))$. Its rank is the pair $(\mathrm{ld}(F), \deg(F))$. Ranks of fractions are ordered lexicographically. Though we have not defined the rank of elements of $\mathcal{K}$, it is sufficient to consider that they have lower rank than any fraction which is not in $\mathcal{K}$.

Lemma 1. *If $F \notin \mathcal{K}$ is a reduced fraction then* $\mathrm{ord}(\delta F) = \mathrm{ord}(F) + 1$.

Proof. By [3, Proposition 3], we have $\mathrm{ld}(\delta F) = \delta \, \mathrm{ld}(F)$, whatever the ranking. In particular, this Proposition holds for orderly rankings. For such rankings, $\mathrm{ord}(F) = \mathrm{ord}(\mathrm{ld}(F))$. Thus the lemma is true for orderly rankings. Since the order of a fraction is ranking independent, the lemma is true for any ranking.

The expressions computed by Algorithms 4 and 5 are finite sums of differential fractions, integrated finitely many times from some lower bound a to the independent variable t, multiplied afterwards by elements of $\mathcal{K}$. Though, after a single call to one of our algorithms, a single lower bound a occurs, we would like to permit an expression to feature many different such bounds. Such expressions look like elements of integro-differential algebras. The introduction of [10] provides a nice survey on integro-differential algebras and their relationship with (differential) Rota-Baxter algebras. See also [19, 12]. A precise definition of an algebraic structure containing the output of our algorithms, and its algorithmic properties, could be much helpful for designing sound computer algebra packages. We do not provide it here.

In the sequel, we will use the classical notation for integral operators, with an explicit upper bound t and a trailing symbol $\mathrm{d}t$ which plays the role of a closing parenthesis. To lighten notations, we will however avoid to introduce new variables τ and write :

$$\int_a^t F(t)\,\mathrm{d}t \quad \text{instead of} \quad \int_a^t F(\tau)\,\mathrm{d}\tau \,.$$

3   Improvements of the **integrate** Algorithm

In this section, we recall the specifications of [3, Algorithms 3 and 4] and adapt them to the context of this paper. Moreover, we fix a bug, due to a possibly non-terminating auxiliary function. Two versions of the integrate algorithm were given: [3, Algorithm 3] is the core algorithm while [3, Algorithm 4] is its "iterated" version.

[3, Algorithm 3] gathers as input a differential fraction F_0 and an independent variable, which is t, here. It returns two differential fractions R and W such that

1. $F_0 = \delta \, R + W$,
2. W is zero iff there exists R such that $F_0 = \delta \, R$
3. The fractions $\delta \, R$ and W have ranks lower than or equal to that of F_0.

[3, Algorithm 4] gathers as input a differential fraction F_0 and an independent variable, which is necessarily t, here. It returns a possibly empty list (if is empty if, and only if $F_0 = 0$) of differential fractions $[W_0, W_1, \ldots, W_s]$ such that

1. W_s is nonzero
2. $F_0 = W_0 + \delta W_1 + \cdots + \delta^s W_s$
3. $W_0, W_1, \ldots, W_i$ are zero if, and only if there exists a differential fraction R and an index $i < s$ such that $F_0 = \delta^{i+1} R$
4. The differential fractions $W_0, \delta W_1, \ldots, \delta^s W_s$ have ranks lower than or equal to that of F_0.

Unfortunately, both these algorithms, as they are stated in [3], are flawed, for they rely over an auxiliary algorithm, [3, Algorithm 2, integrateWithRemainder], which may not terminate over some inputs. We take the opportunity of this paper to fix this mistake, by replacing [3, Algorithm 2, integrateWithRemainder] by [4, Mack's linear version of Hermite reduction, page 44]. The new version is provided in Algorithm 1.

Algorithm 1. The integrateWithRemainder (fixed version) Algorithm is nothing but a slight variant of [4, Mack's linear version of Hermite reduction, page 44]

Require: F_0 a reduced fraction and v a variable
Ensure: Two fractions R and W such that (1) $F_0 = \partial R/\partial v + W$; (2) the denominator of W is squarefree; (3) $\deg(W) < 0$.
1: $cont :=$ the content of $\mathrm{denom}(F_0)$ w.r.t. v
2: $A := \mathrm{numer}(F_0)$
3: $D := \mathrm{denom}(F_0)/cont$
4: $G := 0$
5: $D^- :=$ the multivariate gcd of D and $\partial D/\partial v$
6: $D^* := D/D^-$
7: **while** $\deg(D^-, v) > 0$ **do**
8:    $D^{-2} :=$ the multivariate gcd of D^- and $\partial D^-/\partial v$
9:    $D^{-*} := D^-/D^{-2}$
10:   $(B, C) := \mathsf{extendedEuclidean}(-D^* (\partial D^-/\partial v)/D^-, D^{-*}, A, v)$
11:   $A := C - (\partial B/\partial v) D^*/D^{-*}$
12:   $G := G + B/D^-$
13:   $D^- := D^{-2}$
14: **end while**
15: $R := G/cont$
16: $W := A/(D^- cont)$
17: **return** R, W

Since $\alpha/(\alpha + u) + u/(\alpha + u) = 1$ one sees that a fraction whose denominator is free of α may very well be decomposed as a sum of fractions whose denominators depend on α. Therefore, though the next Proposition is not surprising, it cannot be considered as obvious.

Algorithm 2. The extendedEuclidean Algorithm is a restatement of the diophantine version of Euclide's extended algorithm, given in [4, Sect. 1.3, page
13]

Require: P_1, P_2 and A are multivariate polynomials and v is a variable. Viewed as
   univariate polynomials in v, the polynomials P_i are coprime.
Ensure: Two fractions B and C such that $B P_1 + C P_2 = A$
 1: $G :=$ the gcd (a fraction) of P_1 and P_2 viewed as univariate polynomials in v
 2: $S, T :=$ fractions such that $S P_1 + T P_2 = G$
 3: $Q :=$ the quotient of A by G w.r.t. v
 4: $S, T := Q S, Q T$
 5: $Q, R :=$ the quotient and remainder of S by P_2 w.r.t. v
 6: $B := R$
 7: $C := T + Q P_1$
 8: **return** B, C

Proposition 1. *Let α denote either a constant of $\mathscr{C}$ or a derivative, and $\mathscr{R}_\alpha$
the set of all fractions whose denominators are free of α. If $F_0 \in \mathscr{R}_\alpha$ then the
two fractions R, W returned by [3, Algorithm 3] belong to $\mathscr{R}_\alpha$.*

Proof. The set $\mathscr{R}_\alpha$ is stable under addition, multiplication and derivations (as
well δ as $\partial/\partial v$, for any v): it is a differential subring of the field of fractions of $\mathscr{R}$.
Moreover, consider two differential polynomials A and B s.t. $\deg(B, \alpha) = 0$.
Then, given any variable v (v may either be a constant or a derivative), the
quotient and remainder of A by B w.r.t. v are elements of $\mathscr{R}_\alpha$.

We claim that, if $F_0 \in \mathscr{R}_\alpha$ then the two fractions returned by Algorithm 1
belong to $\mathscr{R}_\alpha$. Consider Algorithm 1 and assume $F_0 \in \mathscr{R}_\alpha$. Then, the polynomials $D, D^-, D^*, D^{-2}, D^{-*}$ are always free of α. Thus, Algorithm 2 is always
called with its two first parameters, P_1 and P_2, free of α. Now, if $A \in \mathscr{R}_\alpha$ then,
according to the remarks stated in the previous paragraph, the returned frac-
tions $B, C \in \mathscr{R}_\alpha$. Initially, A is a polynomial, thus an element of $\mathscr{R}_\alpha$. Therefore,
at line 11, we have $B, C \in \mathscr{R}_\alpha$ whence $A \in R_\alpha$ again. Turning this argument into
an inductive proof, we see A is always an element of $\mathscr{R}_\alpha$. A similar argument
proves that $G \in \mathscr{R}_\alpha$. The proof of the claim is then easily completed.

To complete the proof of the Proposition, it is necessary to study the code of
[3, Algorithm 3]. We will not give details. As for Algorithm 1, the argument is a
straightforward application of the remarks stated at the beginning of this proof.

When dealing with elements of $\mathscr{K}$, [3, Algorithm 4] may have a counter-intuitive
behaviour. For instance, applied on a constant a, this algorithm returns the list
$[a]$. However, applied on a differential polynomial $a + u(t)$, it returns $[u(t), a\,t]$.
The reason is informally this one: the algorithm tries to integrate fractions as much
as it can but stops whenever the current fraction is an element of $\mathscr{K}$, since such
an element could be integrated indefinitely. Thus, elements of $\mathscr{K}$ are not handled
the same way when they occur alone or mixed with fractions which are not in $\mathscr{K}$.
In our context, this behaviour is a problem. To overcome it, we introduce a slight
modification of [3, Algorithm 4], in Algorithm 3.

Algorithm 3. The integrate (coefficient free version) Algorithm is a slight variant of [3, Algorithm 4], which forbids the W_i ($i \geq 1$) to be nonzero elements of $\mathscr{K}$

Require: F_0 a reduced differential fraction and an independent variable t
Ensure: A list $[W_0, W_1, \ldots, W_s]$ satisfying the same properties as [3, Algorithm 4] plus the following condition:

$$W_i \notin \mathscr{K}, \text{ unless it is 0, for each } 1 \leq i \leq s \tag{13}$$

1: $[W_0, W_1, \ldots, W_s] :=$ the list of fractions returned by [3, Algorithm 4] over F_0 and t
2: **while** $s \geq 1$ and $W_s \in \mathscr{K}$ **do**
3:    $W_{s-1} := W_{s-1} + \delta W_s$
4:    $s := s - 1$
5: **end while**
6: **for** i from $s - 1$ to 1 by -1 **do**
7:    **if** $W_i \in \mathscr{K}$ **then**
8:       $W_{i-1} := W_{i-1} + \delta W_i$
9:       $W_i := 0$
10:   **end if**
11: **end for**
12: **return** $[W_0, W_1, \ldots, W_s]$

The following Proposition is new.

Proposition 2. *Assume the ranking is orderly and consider some differential fraction F_0. Let $[W_0, W_1, \ldots, W_s]$ be the list returned by the application of Algorithm 3 to F_0 and t. Then $i + \mathrm{ord}(W_i) \leq \mathrm{ord}(F_0)$ for each $0 \leq i \leq s$. In particular, $s \leq \mathrm{ord}(F_0)$.*

Proof. If $F_0 \in \mathscr{K}$ then $s = 0$ and the proposition is satisfied. Assume from now on that $F_0 \notin \mathscr{K}$. If $W_0 \in \mathscr{K}$ then its order is 0 and the condition $0 + \mathrm{ord}(W_0) \leq \mathrm{ord}(F_0)$ holds. If it is not, then none of the nonzero W_i belong to $\mathscr{K}$, by condition (13), stated in the specifications of Algorithm 3. Then, by Lemma 1, we have $\mathrm{ord}(\delta^i W_i) = i + \mathrm{ord}(W_i)$. Since $\delta^i W_i$ has rank lower than, or equal to F_0 (this specification of [3, Algorithm 4] holds for Algorithm 3 also) and the ranking is orderly, one concludes that $i + \mathrm{ord}(W_i) \leq \mathrm{ord}(F_0)$ for each $0 \leq i \leq s$. The last sentence is a consequence of this result, and the fact that Algorithm 3 forbids W_s to be an element of $\mathscr{K}$. $\square$

4   Two Algorithms for Computing Integral Equations

Observe that, in the specifications of Algorithm 4, the differential ring which contains (14) may be a proper differential ring extension of the differential ring $\mathscr{R}$ which contains the input fraction F_0. Indeed, in order to represent the values of the derivatives at $t = a$, one may need to add arbitrary constants [18, chapter III, page 57] to $\mathscr{R}$ and the integration process may generate polynomials in t. Note

---

Algorithm 4. The iteratedIntegral algorithm

Require: F_0, Θ, a where Θ is a set of parameters ; F_0 is a reduced ordinary differential fraction (dependent variable t) which can be written as the product of two differential fractions $F_0 = P/(Q_1 Q_2)$ such that $Q_1 \in K[\Theta]$ and Q_2 does not depend on Θ ; and a is an evaluation point (numeric or symbolic)
Ensure: an expression of the form

$$F = \sum_{i=1}^{N} C_i \underbrace{\int_a^t \cdots \int_a^t}_{\ell_i \text{ times}} G_i \underbrace{dt \cdots dt}_{\ell_i \text{ times}} \tag{14}$$

such that the $C_i \in K(\Theta)$, the differential fractions G_i do not depend on Θ and the nonnegative integers ℓ_i satisfy the following property (in the case of an orderly ranking, we have $\ell_i + \mathrm{ord}(G_i) \leq \mathrm{ord}(F_0)$ for each i):

$$F_0 = \left(\frac{d}{dt}\right)^{\max(\ell_i)} F. \tag{15}$$

1: $[W_0, W_1, \ldots, W_s] :=$ the list of fractions returned by Algorithm 3 over F_0 and t
2: $F := 0$
3: **for** i from 0 to s **do**
4:     decompose $W_i = \sum_{k=1}^{n} C_k T_k$ where the $C_k \in K(\Theta)$ and T_k do not depend on Θ
5:     **for** k from 1 to n **do**
6:         $Q := 0$
7:         **for** j from 0 to s **do**
8:             **if** $j \leq i$ **then**
9:                 $B := \left(\frac{d}{dt}\right)^{i-j} T_k$
10:                $B_a := B(a)$
11:            **else**
12:                $B := \underbrace{\int_a^t \cdots \int_a^t}_{j-i \text{ times}} T_k \underbrace{dt \cdots dt}_{j-i \text{ times}}$
13:                $B_a := 0$
14:            **end if**
15:            $P := Q - B_a$
16:            $Q := \int_a^t P \, dt$
17:        **end for**
18:        $F := F + C_k (B + P)$
19:    **end for**
20: **end for**
21: **return** F

Algorithm 5. The modulatedIntegral algorithm

Require: F_0, Θ, a, ϕ where Θ is a set of parameters ; F_0 is a reduced ordinary differential fraction (dependent variable t) which can be written as the product of two differential fractions $F_0 = P/(Q_1 Q_2)$ such that $Q_1 \in K[\Theta]$ and Q_2 does not depend on Θ ; a is an evaluation point ; and ϕ is a function of t

Ensure: an expression of the form

$$F = \sum_{i=1}^{N} C_i \int_a^t \frac{d^{\ell_i} \phi}{dt^{\ell_i}} G_i \, dt \tag{16}$$

such that the $C_i \in K(\Theta)$, the differential fractions G_i do not depend on Θ and, provided that ϕ is a modulating function of sufficient order, we have

$$F = \int_a^t \phi \, F_0 \, dt . \tag{17}$$

In the case of an orderly ranking, we have $\mathrm{ord}(G_i) \leq \mathrm{ord}(F_0)$ for each i, and the order of the modulating function can be chosen less than or equal to $\mathrm{ord}(F_0)$.

1: $[W_0, W_1, \ldots, W_s] :=$ the list of fractions returned by Algorithm 3 over F_0 and t
2: $F := 0$
3: **for** i from 0 to s **do**
4:     decompose $W_i = \sum_{k=1}^{n} C_k T_k$ where the $C_k \in K(\Theta)$ and T_k do not depend on Θ
5:     **for** k from 1 to n **do**
6:         $F := F + C_k (-1)^i \int_a^t \frac{d^i \phi}{dt^i} T_k \, dt$
7:     **end for**
8: **end for**
9: **return** F

also that one may want to put some of these constants in the set of parameters Θ, as mentioned in the remark following (8).

Proposition 3. *Algorithm 4 terminates and is correct.*

Proof. Termination is clear. Let us address the correctness. According to the specifications of Algorithm 3, recalled in Section 3, we have the following formula. Algorithm 4 applies s times the integration from a to t operator over it:

$$F_0 = W_0 + \frac{d}{dt} W_1 + \cdots + \left(\frac{d}{dt}\right)^s W_s .$$

A difficulty needs to be adressed at line 4 since, without any information on the shape of W_i, the decomposition needs not be possible. First notice that, since Q_1 is a constant, the list of the W_i can be obtained at line 1 by applying first Algorithm 3 over P/Q_2, then dividing all the obtained fractions by Q_1. Then, notice that, with the notations of Proposition 1, the fraction $P/Q_2 \in \mathscr{R}_\alpha$ for any $\alpha \in \Theta$. According to this Proposition, all the W_i have the same shape as F_0 (i.e, if one excepts a possible factor $Q_1 \in K[\Theta]$ at the denominator, all parameters occur at the numerator of the fraction). In such a case, the decomposition performed at line 4 is possible.

Thanks to this special shape of the W_i, the blocks of parameters are moved outside the integral operators.

The exponent $\max(\ell_i)$ occuring in (15) is actually equal to the index s determined at line 1. The statement about orderly rankings follows Proposition 2.

Proposition 4. *Algorithm 5 terminates and is correct.*

Proof. Termination is clear. Let us address the correctness. The proof starts as that of Proposition 3. According to the specifications of Algorithm 3 we have

$$F_0 = W_0 + \frac{\mathrm{d}}{\mathrm{d}t} W_1 + \cdots + \left(\frac{\mathrm{d}}{\mathrm{d}t}\right)^s W_s .$$

The issue at line 4 is solved as in Proposition 3. Algorithm 5 then computes

$$
\begin{aligned}
F &= \int_a^t \phi\, F_0 \, \mathrm{d}t \\
&= \int_a^t \phi\, W_0 \, \mathrm{d}t + \int_a^t \phi\, \frac{\mathrm{d}}{\mathrm{d}t} W_1 \, \mathrm{d}t + \cdots + \int_a^t \phi \left(\frac{\mathrm{d}}{\mathrm{d}t}\right)^s W_s \, \mathrm{d}t \\
&= \int_a^t \phi\, W_0 \, \mathrm{d}t + (-1) \int_a^t \frac{\mathrm{d}\phi}{\mathrm{d}t} W_1 \, \mathrm{d}t + \cdots + (-1)^s \int_a^t \frac{\mathrm{d}^s \phi}{\mathrm{d}t^s} W_s \, \mathrm{d}t .
\end{aligned}
$$

The third formula is obtained by applying the integration by part axiom plus the hypothesis that ϕ is a modulating function of order at least s. Thanks to the shape of the W_i, blocks of parameters are moved outside the integral operators.

The statement concerning orderly rankings follows from Proposition 2 and the fact that the order of the modulating function should be s.

5   Conclusion

We have presented two algorithms for converting nonlinear differential equations to integral equations, taking this opportunity to fix a flaw[2] in a recent algorithm. Such conversions permit to decrease the orders of differential equations, a feature of dramatic importance whenever equations need to be numerically evaluated, since numerical derivation methods can be avoided, at least partially.

An interesting theoretical question, which was asked[3] to us, is left open in this paper: could we characterize the class of dynamical systems for which our algorithms provide order zero integral equations, i.e. equations whose numerical evaluations do not require any numerical derivation?

---

[2] The authors would like to thank Joseph Lallemand, who contributed to fix the flaw.
[3] The authors would like to thank Cédric Join and Mamadou Mboup.

References

[1] Boulier, F.: http://www.lifl.fr/~boulier/BLAD (2004)

[2] Boulier, F., Lazard, D., Ollivier, F., Petitot, M.: Computing representations for radicals of finitely generated differential ideals. AAECC 20(1), 73–121 (2009)

[3] Boulier, F., Lemaire, F., Regensburger, G., Rosenkranz, M.: On the Integration of Differential Fractions. In: ISSAC 2013, pp. 101–108. ACM, New York (2013)

[4] Bronstein, M.: Symbolic Integration I. Springer (1997)

[5] Denis-Vidal, L., Joly-Blanchard, G., Noiret, C.: System identifiability (symbolic computation) and parameter estimation (numerical computation). Numerical Algorithms 34, 282–292 (2003)

[6] Fliess, M., Join, C., Sira-Ramírez, H.: Non-linear estimation is easy. Int. J. Modelling Identification and Control 4(1), 12–27 (2008)

[7] Fliess, M., Mboup, M., Mounier, H., Sira-Ramírez, H.: Questioning some paradigms of signal processing via concrete examples. In: Silva-Navarro, G., Sira-Ramírez, H. (eds.) Algebraic Methods in Flatness, Signal Processing and State Estimation, pp. 1–21. Editorial Lagares (2003)

[8] Fliess, M., Sira-Ramírez, H.: An algebraic framework for linear identification. ESAIM Control Optim. Calc. Variat. 9, 151–168 (2003)

[9] Fliess, M., Sira-Ramírez, H.: Closed-loop parametric identification for continuous-time linear systems via new algebraic techniques. In: Identification of Continuous-Time Models from Sampled Data. Advances in Industrial Control, pp. 362–391 (2008)

[10] Gao, X., Guo, L.: Constructions of Free Commutative Integro-Differential Algebras. In: Barkatou, M., Cluzeau, T., Regensburger, G., Rosenkranz, M. (eds.) AADIOS 2012. LNCS, vol. 8372, pp. 1–22. Springer, Heidelberg (2014)

[11] Godfrey, K.R.: The identifiability of parameters of models used in biomedicine. Mathematical Modelling 7(9-12), 1195–1214 (1986)

[12] Guo, L., Regensburger, G., Rosenkranz, M.: On integro-differential algebras. JPAA 218(3), 456–473 (2014)

[13] Hairer, E.: Homepage, http://www.unige.ch/~hairer (2000)

[14] Kolchin, E.R.: Differential Algebra and Algebraic Groups. Academic Press, New York (1973)

[15] Mboup, M.: Parameter estimation for signals described by differential equations. Applicable Analysis 88, 29–52 (2009)

[16] Noiret, C.: Utilisation du calcul formel pour l'identifiabilité de modèles paramétriques et nouveaux algorithmes en estimation de paramètres. PhD thesis, Université de Technologie de Compiègne (2000)

[17] Pearson, A.E.: Explicit parameter identification for a class of nonlinear input/output differential operator models. In: Proceedings of the 31st IEEE Conference on Decision and Control, vol. 4, pp. 3656–3660 (1992)

[18] Ritt, J.F.: *Differential Algebra*. American Mathematical Society Colloquium Publications, vol. 33. AMS, New York (1950)

[19] Rosenkranz, M., Regensburger, G.: Integro-differential polynomials and operators. In: ISSAC 2008, pp. 261–268. ACM, New York (2008)

[20] Shinbrot, M.: On the analysis of linear and nonlinear dynamical systems from transient-response data. NACA, Washington, D.C (1954)

[21] The Ametista Group (2013), http://www.lifl.fr/Ametista

[22] Ushirobira, R., Perruquetti, W., Mboup, M., Fliess, M.: Algebraic parameter estimation of a multi-sinusoidal waveform signal from noisy data. In: European Control Conference, Zurich (April 2013)

Truth Table Invariant Cylindrical Algebraic Decomposition by Regular Chains

Russell Bradford[1], Changbo Chen[2], James H. Davenport[1], Matthew England[1],
Marc Moreno Maza[3] and David Wilson[1]

[1] University of Bath, Bath, BA2 7AY, UK
[2] CIGIT, Chinese Academy of Sciences, Chongqing, 400714, China
[3] University of Western Ontario, London, Ontario, N6A 5B7, Canada
{R.Bradford,J.H.Davenport,M.England,D.J.Wilson}@bath.ac.uk,
moreno@csd.uwo.ca, changbo.chen@hotmail.com

Abstract. A new algorithm to compute cylindrical algebraic decompositions (CADs) is presented, building on two recent advances. Firstly, the output is truth table invariant (a TTICAD) meaning given formulae have constant truth value on each cell of the decomposition. Secondly, the computation uses regular chains theory to first build a cylindrical decomposition of complex space (CCD) incrementally by polynomial. Significant modification of the regular chains technology was used to achieve the more sophisticated invariance criteria. Experimental results on an implementation in the **RegularChains** Library for MAPLE verify that combining these advances gives an algorithm superior to its individual components and competitive with the state of the art.

Keywords: cylindrical algebraic decomposition, equational constraint, regular chains, triangular decomposition.

1 Introduction

A *cylindrical algebraic decomposition* (CAD) is a collection of cells such that: they do not intersect and their union describes all of $\mathbb{R}^n$; they are arranged *cylindrically*, meaning the projections of any pair of cells are either equal or disjoint; and, each can be described using a finite sequence of polynomial relations.

CAD was introduced by Collins in [16] to solve quantifier elimination problems, and this remains an important application (see [14] for details on how our work can be used there). Other applications include epidemic modelling [8], parametric optimisation [24], theorem proving [27], robot motion planning [29] and reasoning with multi-valued functions and their branch cuts [18]. CAD has complexity doubly exponential in the number of variables. While for some applications there now exist algorithms with better complexity (see for example [5]), CAD implementations remain the best general purpose approach for many.

We present a new CAD algorithm combining two recent advances: the technique of producing CADs via regular chains in complex space [15], and the idea of producing CADs closely aligned to the structure of logical formulae [2]. We continue by reminding the reader of CAD theory and these advances.

V.P. Gerdt et al. (Eds.): CASC Workshop 2014, LNCS 8660, pp. 44–58, 2014.
© Springer International Publishing Switzerland 2014

1.1   Background on CAD

We work with polynomials in ordered variables $\boldsymbol{x} = x_1 \prec \ldots \prec x_n$. The *main variable* of a polynomial (mvar) is the greatest variable present with respect to the ordering. Denote by QFF a *quantifier free Tarski formula*: a Boolean combination $(\wedge, \vee, \neg)$ of statements $f_i \, \sigma \, 0$ where $\sigma \in \{=, >, <\}$ and the f_i are polynomials. CAD was developed as a tool for the problem of quantifier elimination over the reals: given a quantified Tarski formula

$$\Psi(x_1, \ldots, x_k) := Q_{k+1} x_{k+1} \ldots Q_n x_n F(x_1, \ldots, x_n)$$

(where $Q_i \in \{\forall, \exists\}$ and F is a QFF), produce an equivalent QFF $\psi(x_1, \ldots, x_k)$. Collins proposed to build a CAD of $\mathbb{R}^n$ which is *sign-invariant*, so each $f_i \in F$ is either positive, negative or zero on each cell. Then ψ is the disjunction of the defining formulae of those cells $c \in \mathbb{R}^k$ where Ψ is true, which given sign-invariance, requires us to only test one *sample point* per cell.

Collins' algorithm works by first *projecting* the problem into decreasing real dimensions and then *lifting* to build CADs of increasing dimension. Important developments range from improved projection operators [26] to the use of certified numerics when lifting [30] [25]. See for example [2] for a fuller discussion.

1.2   Truth Table Invariant CAD

One important development is the use of *equational constraints* (ECs), which are equations logically implied by a formula. These may be given explicitly as in $(f = 0) \wedge \varphi$, or implicitly as $f_1 f_2 = 0$ is by $(f_1 = 0 \wedge \varphi_1) \vee (f_2 = 0 \wedge \varphi_2)$.

In [26] McCallum developed the theory of a reduced operator for the first projection, so that the CAD produced was sign-invariant for the polynomial defining a given EC, and then sign-invariant for other polynomials only when the EC is satisfied. Extensions of this to make use of more than one EC have been investigated (see for example [9]) while in [3] it was shown how McCallum's theory could allow for further savings in the lifting phase.

The CADs produced are no longer sign-invariant for polynomials but instead *truth-invariant* for a formula. Truth-invariance was defined in [6] where sign-invariant CADs were refined to maintain it. We consider a related definition.

Definition 1 ([2]). *Let* $\Phi = \{\phi_i\}_{i=1}^t$ *be a list of QFFs. A CAD is* Truth Table Invariant *for* Φ *(a TTICAD) if on each cell every* ϕ_i *has constant Boolean value.*

In [2] an algorithm to build TTICADs when each ϕ_i has an EC was derived by extending [26] (which could itself apply in this case but would be less efficient). Implementations in MAPLE showed this offered great savings in both CAD size and computation time when compared to the sign-invariant theory. In [3] this theory has been extended to work on arbitrary ϕ_i, with savings if at least one has an EC. Note that there are two distinct reasons to build a TTICAD:

1. *As a tool to build a truth-invariant CAD:* If a parent formula ϕ^* is built from $\{\phi_i\}$ then any TTICAD for $\{\phi_i\}$ is also truth-invariant for ϕ^*.

    A TTICAD may be the best truth-invariant CAD, or at least the best we can compute. Note that the TTICAD theory allows for more savings than the use of [26] with an implicit EC built as the product of ECs from ϕ_i [2].

2. *When truth table invariance is required:* There are applications which provide a list of formulae but no parent formula. For example, decomposing complex space according to a set of branch cuts for the purpose of algebraic simplification [1] [28] [21]. When the branch cuts can be expressed as semi-algebraic systems a TTICAD provides exactly the required decomposition.

1.3   CAD by Regular Chains

Recently, a radically different method to build CADs has been investigated. Instead of projecting and lifting, the problem is moved to complex space where the theory of triangular decomposition by regular chains is used to build a *complex cylindrical decomposition* (CCD): a decomposition of $\mathbb{C}^n$ such that each cell is cylindrical. This is encoded as a tree data structure, with each path through the tree describing the end leaf as a solution of a regular system [32].

This was first proposed in [15] to build a sign-invariant CAD. Techniques developed for comprehensive triangular decomposition [11] were used to build a sign-invariant decomposition of $\mathbb{C}^n$ which was then refined to a CCD. Finally, real root isolation is applied to refine further to a CAD of $\mathbb{R}^n$. The computation of the CCD may be viewed as an enhanced projection phase since gcds of pairs of polynomials are calculated as well as resultants. The extra work used here makes the second phase, which may be compared to lifting, less expensive. The main advantage is the use of case distinction in the second phase, so that the zeros of polynomials not relevant in a particular branch are not isolated there.

The construction of the CCD was improved in [13]. The former approach built a decomposition for the input in one step using existing algorithms. The latter approach proceeds incrementally by polynomial, each time using purpose-built algorithms to refine an existing tree whilst maintaining cylindricity. Experimental results showed that the latter approach is much quicker, with its implementation in MAPLE's `RegularChains` library now competing with existing state of the art CAD algorithms: QEPCAD [7] and MATHEMATICA [31]. One reason for this improvement is the ability of the new algorithm to recycle subresultant calculations, an idea introduced and detailed in [12] for the purpose of decomposing polynomial systems into regular chains incrementally.

Another benefit of the incremental approach is that it allows for simplification when constructing a CAD in the presence of ECs. Instead of working with polynomials, the algorithm can be modified to work with relations. Then branches in which an EC is not satisfied may be truncated, offering the possibility of a reduction in both computation time and output size. In [13] it was shown that using this optimization allowed the algorithm to process examples which MATHEMATICA and QEPCAD could not.

1.4   Contribution and Outline

In Section 2 we present our new algorithms. Our aim is to combine the savings from an invariance criteria closer to the underlying application, with the savings offered by the case distinction of the regular chains approach. It requires adapting the existing algorithms for the regular chains approach so that they refine branches of the tree data structure only when necessary for truth-table invariance, and so that branches are truncated only when appropriate to do so.

We implemented our work in the **RegularChains** library for MAPLE. In Section 3 we qualitatively compare our new algorithm to our previous work and in Section 4 we present experimental results comparing it to the state of the art. Finally, in Section 5 we give our conclusions and ideas for future work.

2 Algorithm

2.1 Constructing a Complex Cylindrical Tree

Let $\boldsymbol{x} = x_1 \prec \cdots \prec x_n$ be a sequence of ordered variables. We will construct TTICADs of $\mathbb{R}^n$ for a semi-algebraic system sas (Definition 5). However, to achieve this we first build CCDs of $\mathbb{C}^n$ with respect to a complex system.

Definition 2. *Let* $F = \{p_1, \ldots, p_s\}$ *be a finite set from* $\mathbb{Q}[\boldsymbol{x}]$, $G \subseteq F$ *and* $\sigma_i \in \{=, \neq\}$. *Then we define a* complex system *(denoted by cs) as a set*

$$\{p_i \ \sigma_i \ 0 \mid p_i \in G\} \cup \{p_i \mid p_i \in F \setminus G\}.$$

The complex systems we work with will be defined in accordance with a semi-algebraic counterpart (see Definition 5). For $p \in \mathbb{Q}[\boldsymbol{x}]$ we denote the zero set of p in $\mathbb{C}^n$ by $Z_{\mathbb{C}}(p)$, or $Z_{\mathbb{C}}(p = 0)$, and its complement by $Z_{\mathbb{C}}(p \neq 0)$.

We compute CCDs as trees, following [15,13]. Throughout let T be a rooted tree with each node of depth i a polynomial constraint of type either, "any x_i" (with zero set defined as $\mathbb{C}^n$), or $p = 0$, or $p \neq 0$ (where $p \in \mathbb{Q}[x_1, \ldots, x_i]$). For any i denote the induced subtree of T with depth i by T_i. Let Γ be a path of T and define its zero set $Z_{\mathbb{C}}(\Gamma)$ as the intersection of zero sets of its nodes. The zero set of T, denoted $Z_{\mathbb{C}}(T)$, is defined as the union of zero sets of its paths.

Definition 3. T *is a* complete complex cylindrical tree *(complete CCT) of* $\mathbb{Q}[\boldsymbol{x}]$ *if it satisfies recursively:*

1. *If* $n = 1$: *either* T *has one leaf "any* x_1", *or it has* $s + 1$ $(s \geq 1)$ *leaves* $p_1 = 0, \ldots, p_s = 0, \prod_{i=1}^{s} p_i \neq 0$, *where* $p_i \in \mathbb{Q}[x_1]$ *are squarefree and coprime.*
2. *The induced subtree* T_{n-1} *is a complete CCT.*
3. *For any given path* Γ *of* T_{n-1}, *either its leaf* V *has only one child "any* x_n", *or* V *has* $s + 1$ $(s \geq 1)$ *children* $p_1 = 0, \ldots, p_s = 0, \prod_{i=1}^{s} p_i \neq 0$, *where* $p_1, \ldots, p_s \in \mathbb{Q}[\mathbf{x}]$ *are squarefree and coprime satisfying:*

3a. *for any* $\alpha \in Z_{\mathbb{C}}(\Gamma)$, *none of* $\mathrm{lc}(p_j, x_n)$, $j = 1, \ldots, s$, *vanishes at* α, *and*
3b. $p_1(\alpha, x_n), \ldots, p_s(\alpha, x_n)$ *are squarefree and coprime.*

The set $\{Z_{\mathbb{C}}(\Gamma) \mid \Gamma$ is a path of $T\}$ is called the *complex cylindrical decomposition* (CCD) of $\mathbb{C}^n$ associated with T: condition (3b) assures that it is a decomposition. Note that for a complete CCT we have $Z_{\mathbb{C}}(T) = \mathbb{C}^n$. A proper subtree rooted at the root node of T of depth n is called a *partial CCT* of $\mathbb{Q}[\boldsymbol{x}]$. We use CCT to refer to either a complete or partial CCT. We call a complex cylindrical tree T an *initial tree* if T has only one path and T is complete.

Definition 4. *Let T be a CCT of $\mathbb{C}^n$ and Γ a path of T. A polynomial $p \in \mathbb{Q}[x]$ is sign invariant on Γ if either $Z_{\mathbb{C}}(\Gamma) \cap Z_{\mathbb{C}}(p) = \emptyset$ or $Z_{\mathbb{C}}(p) \supseteq Z_{\mathbb{C}}(\Gamma)$. A constraint $p = 0$ or $p \neq 0$ is truth-invariant on Γ if p is sign-invariant on Γ. A complex system cs is truth-invariant on Γ if the conjunction of the constraints in cs is truth-invariant on Γ, and each polynomial in cs is sign-invariant on Γ.*

Example 1. Let $q := (x_2^2 + x_2 + x_1)$ and $p := x_1 q$. The following tree is a CCT such that p is sign-invariant (and $p = 0$ is truth invariant) on each path.

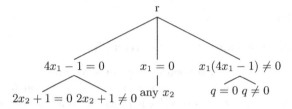

We introduce Algorithm 1 to produce truth-table invariant CCTs, and new sub-algorithms 2 and 3. It also uses `IntersectPath` and `NextPathToDo` from [13]. `IntersectPath` takes: a CCT T; a path Γ; and p, either a polynomial or constraint. When a polynomial it refines T so p is sign-invariant above each path from Γ (still satisfying Definition 3). When a constraint it refines so the constraint is true, possibly truncating branches if there can be no solution. This necessitates the housekeeping algorithm `MakeComplete` which restores to a complete CCT by simply adding missing siblings (if any) to every node. `NextPathToDo` simply returns the next incomplete path Γ of T.

Proposition 1. *Algorithm 1 satisfies its specification.*

Proof. It suffices to show that Algorithm 2 is as specified. First observe that Algorithm 3 just recursively calls `IntersectPath` on constraints and so its correctness follows from that of `IntersectPath`. When called on ECs `IntersectPath` may return a partial tree and so `MakeComplete` must be used in line 6.

Algorithm 2 is clearly correct is its base cases, namely line 2, line 5 and line 9. It also clearly terminates since the input of each recursive call has less constraints. For each path C of the refined Γ, by induction, it is sufficient to show that cs is truth-invariant on C. If $p \neq 0$ on C, then cs is false on C. If $p = 0$ on C, then the truth of cs is invariant since it is completely determined by the truth of $cs' := cs \setminus \{p = 0\}$, invariant on C by induction.

Algorithm 1. TTICCD(L)

Input: A list L of complex systems of $\mathbb{Q}[x]$.
Output: A complete CCT T with each $cs \in L$ truth-invariant on each path.
1 Create the initial CCT T and let Γ be its path;
2 IntersectLCS(L, Γ, T);

Algorithm 2. IntersectLCS(L, Γ, T)

Input: A CCT T of $\mathbb{Q}[\boldsymbol{x}]$. A path Γ of T. A list of complex systems L.
Output: Refinements of Γ and T such that T is complete, and $cs \in L$ is truth-invariant above each path of Γ.

1 if $L = \emptyset$ then
2 │ return;
3 else if $|L| = 1$ then
4 │ Let cs be the only complex system;
5 │ IntersectPolySet(cs, Γ, T);
6 │ MakeComplete(T);
7 else if *no $cs \in L$ has an equational constraint* then
8 │ Let F be the set of polynomials appearing in L;
9 │ IntersectPolySet(F, Γ, T);
10 else
11 │ Let cs be a complex system of L with an EC denoted $p = 0$;
12 │ IntersectPath(p, Γ, T); // Γ may become a tree
13 │ while $C := $ NextPathToDo(Γ) $\neq \emptyset$ do
14 │ │ if $p = 0$ *on C* then
15 │ │ │ $cs' := cs \setminus \{p = 0\}$;
16 │ │ │ IntersectLCS($L \setminus \{cs\} \cup \{cs'\}, C, T$);
17 │ │ else
18 │ │ │ IntersectLCS($L \setminus \{cs\}, C, T$);

Algorithm 3. IntersectPolySet(F, Γ, T)

Input: A CCT T, a path Γ and a set F of polynomials (constraints).
Output: T is refined and Γ becomes a subtree. Each polynomial (constraint) of F is sign (truth)-invariant above each path of Γ.

1 if $F = \emptyset$ then return;
2 Let $p \in F$; $F' := F \setminus \{p\}$;
3 IntersectPath(p, Γ, T); // Γ may become a tree
4 if $F' \neq \emptyset$ then
5 │ while $C := $ NextPathToDo(Γ) $\neq \emptyset$ do
6 │ │ IntersectPolySet(F', C, T);

2.2 Illustrating the Computational Flow

Consider using Algorithm 1 on input of the form
$$L = [cs_1, cs_2] := [\{f_1 = 0, g_1 \neq 0\}, \{f_2 = 0, g_2 \neq 0\}].$$
Algorithm 1 constructs the initial tree and passes to Algorithm 2. We enter the fourth branch of the conditional, let $p = f_1$, and refine to a sign invariant CCT for f_1. This makes a case distinction between $f_1 = 0$ and $f_1 \neq 0$. On the branch $f_1 \neq 0$, we recursively call IntersectLCS on $[cs_2]$ which then passes directly to IntersectPolySet. On the branch $f_1 = 0$, we recursively call IntersectLCS on $[\{g_1 \neq$

$0\}, \{f_2 = 0, g_2 \neq 0\}]$. This time $p = f_2$ and a case discussion is made between $f_2 = 0$ and $f_2 \neq 0$. On the branch $f_2 \neq 0$, we end up calling IntersectPolySet($g_1 \neq 0$) while on the branch $f_2 = 0$ we call IntersectLCS on $[\{g_1 \neq 0\}, \{g_2 \neq 0\}]$, which reduces to IntersectPolySet(g_1, g_2). The case discussion is summarised by:

$$\begin{cases} f_1 = 0 : \begin{cases} f_2 = 0 : g_1, g_2 \\ f_2 \neq 0 : g_1 \neq 0 \end{cases} \\ f_1 \neq 0 : f_2 = 0, g_2 \neq 0 \end{cases} .$$

2.3   Refining to a TTICAD

We now discuss how Section 2.1 can be extended from CCDs to CADs.

Definition 5. *A* semi-algebraic system *of* $\mathbb{Q}[\boldsymbol{x}]$ *(sas) is a set of constraints* $\{p_i \ \sigma_i \ 0\}$ *where each* $\sigma_i \in \{=, >, \geq, \neq\}$ *and each* $p_i \in \mathbb{Q}[\boldsymbol{x}]$. *A corresponding* complex system *is formed by replacing all* $p_i > 0$ *by* $p_i \neq 0$ *and all* $p_i \geq 0$ *by* p_i.
   A sas is truth-invariant *on a cell if the conjunction of its constraints is.*

Note that the ECs of an sas are still identified as ECs of the corresponding cs. Algorithm 4 produces a TTICAD of $\mathbb{R}^n$ for a sequence of semi-algebraic systems.

Algorithm 4. RC – TTICAD(L)

   Input: A list L of semi-algebraic systems of $\mathbb{Q}[\boldsymbol{x}]$.
   Output: A CAD such that each *sas* $\in L$ is truth-invariant on each cell.
 1 Set L' to be the list of corresponding complex systems;
 2 $\mathcal{D} := $ TTICCD(L');
 3 MakeSemiAlgebraic($\mathcal{D}, n$);

Proposition 2. *Algorithm 4 satisfies its specification.*

Proof. Algorithm 4 starts by building the corresponding *cs* for each *sas* in the input. It uses Algorithm 1 to form a CCD truth-invariant for each of these and then the algorithm MakeSemiAlgebraic introduced in [15] to move to a CAD. MakeSemiAlgebraic takes a CCD $\mathcal{D}$ and outputs a CAD $\mathcal{E}$ such that for each element $d \in \mathcal{D}$ the set $d \cap \mathbb{R}^n$ is a union of cells in $\mathcal{E}$. Hence $\mathcal{E}$ is still truth-invariant for each *cs* $\in L'$. It is also a TTICAD for L, (as to change sign from positive to negative would mean going through zero and thus changing cell). The correctness of Algorithm 4 hence follows from the correctness of its sub-algorithms.

   The output of Algorithm 4 is a TTICAD for the formula defined by each semi-algebraic system (the conjunction of the individual constraints of that system). To consider formulae with disjunctions we must first use disjunctive normal form and then construct semi-algebraic systems for each conjunctive clause.

3   Comparison with Prior Work

We now compare qualitatively to our previous work. Quantitative experiments and a comparison with competing CAD implementations follows in Section 4.

3.1   Comparing with Sign-invariant CAD by Regular Chains

Algorithm 4 uses work from [13] but obtains savings when building the complex tree by ensuring only truth-table invariance. To demonstrate this we compare diagrams representing the number of times a constraint is considered when building a CCD for a complex system.

Definition 6. *Let cs be a complex system. We define the* complete *(resp. partial)* combination diagram *for cs, denoted by $\Delta_0(cs)$ (resp. $\Delta_1(cs)$), recursively:*
- *If $cs = \emptyset$, then $\Delta_i(cs)$ ($i = 0, 1$) is defined to be null.*
- *If cs has any ECs then select one, ψ (defined by a polynomial f), and define*

$$\Delta_0(cs) := \begin{cases} f = 0 \; \Delta_0(cs \setminus \{\psi\}) \\ f \neq 0 \; \Delta_0(cs \setminus \{\psi\}) \end{cases}, \qquad \Delta_1(cs) := \begin{cases} f = 0 \; \Delta_1(cs \setminus \{\psi\}) \\ f \neq 0 \end{cases}.$$

- *Otherwise select a constraint ψ (which is either of the form $f \neq 0$, or f) and for $i = 0, 1$ define*

$$\Delta_i(cs) := \begin{cases} f = 0 \; \Delta_i(cs \setminus \{\psi\}) \\ f \neq 0 \; \Delta_i(cs \setminus \{\psi\}) \end{cases}.$$

The combination diagrams illustrate the combinations of relations that must be analysed by our algorithms, with the partial diagram relating to Algorithm 1 and the complete diagram the sign-invariant algorithm from [13].

Lemma 3. *Assume that the complex system cs has s ECs and t constraints of other types. Then the number of constraints appearing in $\Delta_0(cs)$ is $2^{s+t+1} - 2$, and the number appearing in $\Delta_1(cs)$ is $2(2^t + s) - 2$.*

Proof. The diagram $\Delta_0(cs)$ is a full binary tree with depth $s + t$. Hence the number of constraints appearing is the geometric series $\sum_{i=1}^{s+t} 2^i = 2^{s+t+1} - 2$.

$\Delta_1(cs)$ will start with a binary tree for the ECs, with only one branch continuing at each depth, and thus involves $2s$ constraints. The full binary tree for the other constraints is added to the final branch, giving a total of $2^{t+1} + 2s - 2$.

Definition 7. *Let L be a list of complex systems. We define the* complete *(resp. partial)* combination diagram *of L, denoted by $\Delta_0(L)$ (resp. $\Delta_1(L)$) recursively: If $L = \emptyset$, then $\Delta_i(L)$, $i = 0, 1$, is null. Otherwise let cs be the first element of L. Then $\Delta_i(L)$ is obtained by appending $\Delta_i(L \setminus \{cs\})$ to each leaf node of $\Delta_i(cs)$.*

Theorem 4. *Let L be a list of r complex systems. Assume each $cs \in L$ has s ECs and t constraints of other types. Then the number of constraints appearing in $\Delta_0(L)$ is $2^{r(s+t)+1} - 2$ and the number of constraints appearing in $\Delta_1(L)$ is $N(r) = 2(s + 2^t)^r - 2$.*

Proof. The number of constraints in $\Delta_0(L)$ again follows from the geometric series. For $\Delta_1(L)$ we proceed with induction on r. The case $r = 1$ is given by Lemma 3, so now assume $N(r - 1) = 2(s + 2^t)^{r-1} - 2$.

The result for r follows from $C(r) = C(1) + (s + 2^t)C(r - 1)$. To conclude this identity consider the diagram for the first $cs \in L$. To extend to $\Delta_1(L)$ we append $\Delta_1(L \setminus cs)$ to each end node. There are s for cases where an EC was not satisfied and 2^t from cases where all ECs were (and non-ECs were included).

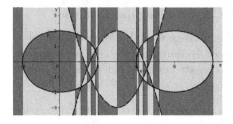 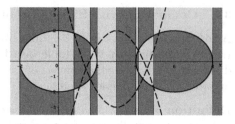

Fig. 1. The left is a sign-invariant CAD, and the right a TTICAD, for (1)

Example 2. We demonstrate these savings by considering

$$f_1 := x^2 + y^2 - 4, \qquad g_1 := (x-3)^2 - (y+3), \quad \phi_1 := f_1 = 0 \wedge g_1 < 0,$$
$$f_2 := (x-6)^2 + y^2 - 4, \quad g_2 := (x-3)^2 + (y-2), \quad \phi_2 := f_2 = 0 \wedge g_2 < 0, \quad (1)$$

and ordering $x \prec y$. The polynomials are graphed in Figure 1 where the solid circles are the f_i and the dashed parabola the g_i. To study the truth of the formulae $\{\phi_1, \phi_2\}$ we could create a sign-invariant CAD. Both the incremental regular chains technology of [13] and QEPCAD [7] do this with 231 cells. The 72 full dimensional cells are visualised on the left of Figure 1, (with the cylinders on each end actually split into three full dimensional cells out of the view).

Alternatively we may build a TTICAD using Algorithm 4 to obtain only 63 cells, 22 of which have full dimension as visualised on the right of Figure 1. By comparing the figures we see that the differences begin in the CAD of the real line, with the sign-invariant case splitting into 31 cells compared to 19. The points identified on the real line each align with a feature of the polynomials. Note that the TTICAD identifies the intersections of f_i and g_j only when $i = j$, and that no features of the inequalities are identified away from the ECs.

3.2 Comparing with TTICAD by Projection and Lifting

We now compare Algorithm 4 with the TTICADs obtained by projection and lifting in [2]. We identify three main benefits which we demonstrate by example.

(I) Algorithm 4 Can Achieve Cell Savings from Case Distinction

Example 3. Algorithm 4 produces a TTICAD for (1) with 63 cells compared to a TTICAD of 67 cells from the projection and lifting algorithm in [2]. The full-dimensional cells are identical and so the image on the right of Figure 1 is an accurate visualisation of both. To see the difference we must compare lower dimensional cells. Figure 2 compares the lifting to $\mathbb{R}^2$ over a cell on the real line aligned with an intersection of f_1 and g_1. The left concerns the algorithm in [2] and the right Algorithm 4. The former isolates both the y-coordinates where $f_1 = 0$ while the latter only one (the single point over the cell where ϕ_1 is true).

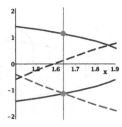
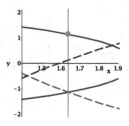

Fig. 2. Comparing TTICADs for (1). The left uses [2] and the right Algorithm 4.

If we modified the problem so the inequalities in (1) were not strict then ϕ_1 becomes true at both points and Algorithm 4 outputs the same TTICAD as [2]. Unlike [2], the type of the non-ECs affects the behaviour of Algorithm 4.

(II) Algorithm 4 Can Take Advantage of More than One EC Per Clause

Example 4. We assume $x \prec y$ and consider

$$f_1 := x^2 + y^2 - 1, \qquad h := y^2 - \tfrac{x}{2}, \qquad g_1 := xy - \tfrac{1}{4}$$
$$f_2 := (x-4)^2 + (y-1)^2 - 1 \quad g_2 := (x-4)(y-1) - \tfrac{1}{4},$$
$$\phi_1 := h = 0 \wedge f_1 = 0 \wedge g_1 < 0, \ \ \phi_2 := f_2 = 0 \wedge g_2 < 0. \tag{2}$$

The polynomials are graphed in Figure 3 where the dashed curves are f_1 and h, the solid curve is f_2 and the dotted curves are g_1 and g_2. A TTICAD produced by Algorithm 4 has 69 cells and is visualised on the right of Figure 3 while a TTICAD produced by projection and lifting has 117 cells and is visualised on the left. This time the differences are manifested in the full-dimensional cells.

The algorithm from [2] works with a single designated EC in each QFF (in this case we chose f_1) and so treats h in the same way as g_1. This means for example that all the intersections of h or g_1 with f_1 are identified. By comparison, Algorithm 4 would only identify the intersection of g_1 with an EC if this occurred at a point where both f_1 and h were satisfied (does not occur here). For comparison, a sign-invariant CAD using QEPCAD or [13] has 611 cells.

To use [2] we had to designate either f_1 or h as the EC. Choosing f_1 gave 117 cells and h 163. Our new algorithm has similar choices: what order should the systems be considered and what order the ECs within (step 11 of Algorithm 2)? Processing f_1 first gives 69 cells but other choice can decrease this to 65 or increase it to 145. See [20] for advice on making such choices intelligently.

(III) Algorithm 4 Will Succeed Given Sufficient Time and Memory
This contrasts with the theory of reduced projection operators used in [2], where input must be *well-oriented* (meaning that certain projection polynomials cannot be nullified when lifting over a cell with respect to them).

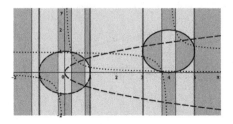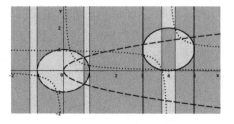

Fig. 3. TTICAD for (2). The left uses [2] and the right Algorithm 4.

Example 5. Consider the identity $\sqrt{z}\sqrt{w} = \sqrt{zw}$ over $\mathbb{C}^2$. We analyse its truth by decomposing according to the branch cuts and testing each cell at its sample point. Letting $z = x + \mathrm{i}y, w = u + \mathrm{i}v$ we see that branch cuts occur when
$$(y = 0 \wedge x < 0) \vee (v = 0 \wedge u < 0) \vee (yu + xv = 0 \wedge xu - yv < 0).$$
We desire a TTICAD for the three clauses joined by disjunction. Assuming $v \prec u \prec x \prec y$ Algorithm 4 does this using 97 cells, while the projection and lifting approach identifies the input as not well-oriented. The failure is triggered by $yu + xv$ being nullified over a cell where $u = x = 0$ and $v < 0$.

4   Experimental Results

We present experimental results obtained on a Linux desktop (3.1GHz Intel processor, 8.0Gb total memory). We tested on 52 different examples, with a representative subset of these detailed in Table 1. The examples and other supplementary information are available from `http://opus.bath.ac.uk/38344/`. One set of problems was taken from CAD papers [10] [2] and a second from system solving papers [11] [13]. The polynomials from the problems were placed into different logical formulations: disjunctions in which every clause had an EC (indicated by †) and disjunctions in which only some do (indicated by ††). A third set was generated by branch cuts of algebraic relations: addition formulae for elementary functions and examples previously studied in the literature.

Each problem had a declared variable ordering (with n the number of variables). For each experiment a CAD was produced with the time taken (in seconds) and number of cells (cell count) given. The first is an obvious metric and the second crucial for applications acting on each cell. T/O indicates a time out (set at 30 minutes), FAIL a failure due to theoretical reasons such as input not being well-oriented (see [26] [2]) and Err an unexpected error.

We start by comparing with our previous work (all tested in MAPLE 18) by considering the first five algorithms in Table 1. RC-TTICAD is Algorithm 4, PL-TTICAD the algorithm from [2], PL-CAD CAD with McCallum projection, RC-Inc-CAD the algorithm from [13] and RC-Rec-CAD the algorithm from [15]. Those starting RC are part of the `RegularChains` library and those starting PL the `ProjectionCAD` package [23]. RC-Rec-CAD is a modification of the algorithm currently distributed with MAPLE; the construction of the CCD is the same but

Table 1. Comparing our new algorithm to our previous work and competing CAD implementations

Problem	n	RC-TTICAD Cells	Time	RC-Inc-CAD Cells	Time	RC-Rec-CAD Cells	Time	PL-TTICAD Cells	Time	PL-CAD Cells	Time	MATHEMATICA Time	QEPCAD Cells	Time	SYNRAC Cells	Time	REDLOG Cells	Time
Intersection†	3	541	1.0	3723	12.0	3723	19.0	579	3.5	3723	29.5	0.1	3723	4.9	3723	12.8	Err	—
Ellipse†	5	71231	317.1	81183	544.9	81193	786.8	FAIL	—	FAIL	—	11.2	500609	275.3	Err	—	Err	—
Solotareff†	4	2849	8.8	54037	209.1	54037	539.0	FAIL	—	54037	407.6	0.1	16603	5.2	Err	—	3353	8.6
Solotareff†‡	4	8329	21.4	54037	226.9	54037	573.4	FAIL	—	54037	414.3	0.1	16603	5.3	Err	—	8367	13.6
2D Ext	2	97	0.2	317	1.0	317	2.6	105	0.6	317	1.8	0.0	249	4.8	317	1.1	305	0.9
2D Ext†‡	2	183	0.4	317	1.1	317	2.6	183	1.1	317	1.8	0.0	317	4.6	317	1.2	293	0.9
3D Ext	3	109	3.5	3497	63.1	3525	1165.7	109	2.9	5493	142.8	0.1	739	5.4	—	T/O	Err	—
MontesS10	7	3643	19.1	3643	28.3	3643	26.6	—	T/O	—	T/O	T/O	—	T/O	—	T/O	Err	—
Wang 93	5	507	44.4	507	49.1	507	46.9	—	T/O	T/O	—	897.1	FAIL	—	Err	—	Err	—
Rose†	3	3069	200.9	7075	498.8	7075	477.1	—	T/O	—	T/O	T/O	FAIL	—	Err	—	Err	—
genLinSyst-3-2†	11	222821	3087.5	—	T/O	—	T/O	FAIL	—	FAIL	—	T/O	FAIL	—	Err	—	Err	—
BC-Kahan	2	55	0.2	409	2.4	409	4.9	55	0.2	409	2.4	0.1	261	4.8	409	1.5	Err	—
BC-Arcsin	2	57	0.1	225	0.9	225	1.9	57	0.2	225	0.9	0.0	225	4.8	225	0.7	161	2.4
BC-Sqrt	4	97	0.2	113	0.5	113	1.3	FAIL	—	113	0.6	0.0	105	4.7	105	0.4	73	0.0
BC-Arctan	4	211	3.5	—	T/O	—	T/O	FAIL	—	—	T/O	T/O	—	T/O	Err	—	—	T/O
BC-Arctanh	4	211	3.5	—	T/O	—	T/O	FAIL	—	—	T/O	T/O	—	T/O	Err	—	—	T/O
BC-Phisanbut-1	4	325	0.8	389	1.8	389	5.8	FAIL	—	389	3.6	0.1	377	4.8	389	2.0	217	0.2
BC-Phisanbut-4	4	543	1.6	2007	13.6	2065	21.5	FAIL	—	51763	932.5	11.9	51763	8.6	Err	—	Err	—

the conversion to a CAD has been improved. Algorithms RC-TTICAD and RC-Rec-CAD are currently being integrated into the `RegularChains` library, which can be downloaded from `www.regularchains.org`.

We see that RC-TTICAD never gives higher cell counts than any of our previous work and that in general the TTICAD theories allow for cell counts an order of magnitude lower. RC-TTICAD is usually the quickest in some cases offering vast speed-ups. It is also important to note that there are many examples where PL-TTICAD has a theoretical failure but for which RC-TTICAD will complete (see point (III) in Section 3.2). Further, these failures largely occurred in the examples from branch cut analysis, a key application of TTICAD.

We can conclude that our new algorithm combines the good features of our previous approaches, giving an approach superior to either. We now compare with competing CAD implementations, detailed in the last four columns of Table 1: MATHEMATICA [31] (V9 graphical interface); QEPCAD-B [7] (v1.69 with options +N500000000 +L200000, initialization included in the timings and implicit EC declared when present); the REDUCE package REDLOG [19] (2010 Free CSL Version); and the MAPLE package SYNRAC (2011 version) [25].

As reported in [2], the TTICAD theory allows for lower cell counts than QEPCAD even when manually declaring an EC. We found that both SYNRAC and REDLOG failed for many examples, (with SYNRAC returning unexpected errors and REDLOG hanging with no output or messages). There were examples for which REDLOG had a lower cell count than RC-TTICAD due to the use of partial lifting techniques, but this was not the case in general. We note that we were using the most current public version of SYNRAC which has since been replaced by a superior development version, (to which we do not have access) and that REDLOG is mostly focused on the virtual substitution approach to quantifier elimination but that we only tested the CAD command.

MATHEMATICA is the quickest in general, often impressively so. However, the output there is not a CAD but a formula with a cylindrical structure [31] (hence cell counts are not available). Such a formula is sufficient for many applications (such as quantifier elimination) but not for others (such as algebraic simplification by branch cut decomposition). Further, there are examples for which RC-TTICAD completes but MATHEMATICA times out. MATHEMATICA's output only references the CAD cells for which the input formula is true. Our implementation can be modified to do this and in some cases this can lead to significant time savings; we will investigate this further in a later publication.

Finally, note that the TTICAD theory allows algorithms to change with the logical structure of a problem. For example, Solotareff† is simpler than Solotareff†† (it has an inequality instead of an equation). A smaller TTICAD can hence be produced, while sign-invariant algorithms give the same output.

5   Conclusions and Further Work

We presented a new CAD algorithm which uses truth-table invariance, to give output aligned to underlying problem, and regular chains technology, bringing the benefits of case distinction and no possibility of theoretical failure from well-orientedness conditions. However, there are still many questions to be considered:

- Can we make educated choices for the order systems and constraints are analysed by the algorithm? Example 4 and [20] shows this could be beneficial.
- Can we use heuristics to make choices such as what variable ordering (see current work in [22] and previous work in [19] [4]).
- Can we modify the algorithm for the case of providing truth invariant CADs for a formula in disjunctive normal form? In this case we could cease refinement in the complex tree once a branch is known to be true.
- Can we combine with other theory such as partial CAD [17] or cylindrical algebraic sub-decompositions [33]?

Acknowledgements. Supported by the CSTC (grant cstc2013jjys0002), the EPSRC (grant EP/J003247/1) and the NSFC (grant 11301524).

References

1. Bradford, R., Davenport, J.H.: Towards better simplification of elementary functions. In: Proc. ISSAC 2002, pp. 16–22. ACM (2002)
2. Bradford, R., Davenport, J.H., England, M., McCallum, S., Wilson, D.: Cylindrical algebraic decompositions for boolean combinations. In: Proc. ISSAC 2013, pp. 125–132. ACM (2013)
3. Bradford, R., Davenport, J.H., England, M., McCallum, S., Wilson, D.: Truth table invariant cylindrical algebraic decomposition. Preprint: arXiv:1401.0645
4. Bradford, R., Davenport, J.H., England, M., Wilson, D.: Optimising problem formulations for cylindrical algebraic decomposition. In: Carette, J., Aspinall, D., Lange, C., Sojka, P., Windsteiger, W. (eds.) CICM 2013. LNCS, vol. 7961, pp. 19–34. Springer, Heidelberg (2013)
5. Basu, S., Pollack, R., Roy, M.F.: Algorithms in Real Algebraic Geometry. Algorithms and Computations in Mathematics, vol. 10. Springer (2006)
6. Brown, C.W.: Simplification of truth-invariant cylindrical algebraic decompositions. In: Proc. ISSAC 1998, pp. 295–301. ACM (1998)
7. Brown, C.W.: An overview of QEPCAD B: A program for computing with semi-algebraic sets using CADs. SIGSAM Bulletin 37(4), 97–108 (2003)
8. Brown, C.W., El Kahoui, M., Novotni, D., Weber, A.: Algorithmic methods for investigating equilibria in epidemic modelling. J. Symb. Comp. 41, 1157–1173 (2006)
9. Brown, C.W., McCallum, S.: On using bi-equational constraints in CAD construction. In: Proc. ISSAC 2005, pp. 76–83. ACM (2005)
10. Buchberger, B., Hong, H.: Speeding up quantifier elimination by Gröbner bases. Technical report, 91-06. RISC, Johannes Kepler University (1991)
11. Chen, C., Golubitsky, O., Lemaire, F., Maza, M.M., Pan, W.: Comprehensive triangular decomposition. In: Ganzha, V.G., Mayr, E.W., Vorozhtsov, E.V. (eds.) CASC 2007. LNCS, vol. 4770, pp. 73–101. Springer, Heidelberg (2007)
12. Chen, C., Moreno Maza, M.: Algorithms for computing triangular decomposition of polynomial systems. J. Symb. Comp. 47(6), 610–642 (2012)
13. Chen, C., Moreno Maza, M.: An incremental algorithm for computing cylindrical algebraic decompositions. In: Proc. ASCM 2012. Springer (2012) (to appear) Preprint: arXiv:1210.5543
14. Chen, C., Moreno Maza, M.: Quantifier elimination by cylindrical algebraic decomposition based on regular chains. In: Proc. ISSAC 2014 (to appear, 2014)

15. Chen, C., Moreno Maza, M., Xia, B., Yang, L.: Computing cylindrical algebraic decomposition via triangular decomposition. In: Proc. ISSAC 2009, pp. 95–102. ACM (2009)
16. Collins, G.E.: Quantifier elimination for real closed fields by cylindrical algebraic decomposition. In: Proc. 2nd GI Conference on Automata Theory and Formal Languages, pp. 134–183. Springer (1975)
17. Collins, G.E., Hong, H.: Partial cylindrical algebraic decomposition for quantifier elimination. J. Symb. Comp. 12, 299–328 (1991)
18. Davenport, J.H., Bradford, R., England, M., Wilson, D.: Program verification in the presence of complex numbers, functions with branch cuts etc. In: Proc. SYNASC 2012, pp. 83–88. IEEE (2012)
19. Dolzmann, A., Seidl, A., Sturm, T.: Efficient projection orders for CAD. In: Proc. ISSAC 2004, pp. 111–118. ACM (2004)
20. England, M., Bradford, R., Chen, C., Davenport, J.H., Maza, M.M., Wilson, D.: Problem formulation for truth-table invariant cylindrical algebraic decomposition by incremental triangular decomposition. In: Watt, S.M., Davenport, J.H., Sexton, A.P., Sojka, P., Urban, J. (eds.) CICM 2014. LNCS, vol. 8543, pp. 45–60. Springer, Heidelberg (2014)
21. England, M., Bradford, R., Davenport, J.H., Wilson, D.: Understanding branch cuts of expressions. In: Carette, J., Aspinall, D., Lange, C., Sojka, P., Windsteiger, W. (eds.) CICM 2013. LNCS, vol. 7961, pp. 136–151. Springer, Heidelberg (2013)
22. England, M., Bradford, R., Davenport, J.H., Wilson, D.: Choosing a variable ordering for truth-table invariant cylindrical algebraic decomposition by incremental triangular decomposition. In: Hong, H., Yap, C. (eds.) ICMS 2014. LNCS, vol. 8592, pp. 450–457. Springer, Heidelberg (2014)
23. England, M., Wilson, D., Bradford, R., Davenport, J.H.: Using the Regular Chains Library to build cylindrical algebraic decompositions by projection and lifting. In: Hong, H., Yap, C. (eds.) ICMS 2014. LNCS, vol. 8592, pp. 458–465. Springer, Heidelberg (2014)
24. Fotiou, I.A., Parrilo, P.A., Morari, M.: Nonlinear parametric optimization using cylindrical algebraic decomposition. In: Proc. Decision and Control, European Control Conference 2005, pp. 3735–3740 (2005)
25. Iwane, H., Yanami, H., Anai, H., Yokoyama, K.: An effective implementation of a symbolic-numeric cylindrical algebraic decomposition for quantifier elimination. In: Proc. SNC 2009, pp. 55–64 (2009)
26. McCallum, S.: On projection in CAD-based quantifier elimination with equational constraint. In: Proc. ISSAC 1999, pp. 145–149. ACM (1999)
27. Paulson, L.C.: Metitarski: Past and future. In: Beringer, L., Felty, A. (eds.) ITP 2012. LNCS, vol. 7406, pp. 1–10. Springer, Heidelberg (2012)
28. Phisanbut, N., Bradford, R.J., Davenport, J.H.: Geometry of branch cuts. ACM Communications in Computer Algebra 44(3), 132–135 (2010)
29. Schwartz, J.T., Sharir, M.: On the "Piano-Movers" Problem: II. General techniques for computing topological properties of real algebraic manifolds. Adv. Appl. Math. 4, 298–351 (1983)
30. Strzeboński, A.: Cylindrical algebraic decomposition using validated numerics. J. Symb. Comp. 41(9), 1021–1038 (2006)
31. Strzeboński, A.: Computation with semialgebraic sets represented by cylindrical algebraic formulas. In: Proc. ISSAC 2010, pp. 61–68. ACM (2010)
32. Wang, D.: Computing triangular systems and regular systems. J. Symb. Comp. 30(2), 221–236 (2000)
33. Wilson, D., Bradford, R., Davenport, J.H., England, M.: Cylindrical algebraic sub-decompositions. Mathematics in Computer Science 8(2), 263–288 (2014)

Computing the Topology of an Arrangement of Implicit and Parametric Curves Given by Values

Jorge Caravantes[1], Mario Fioravanti[2],
Laureano Gonzalez–Vega[2], and Ioana Necula[3],[*]

[1] Universidad Complutense de Madrid
[2] Universidad de Cantabria
[3] Universidad de Sevilla

Abstract. Curve arrangement studying is a subject of great interest in Computational Geometry and CAGD. In our paper, a new method for computing the topology of an arrangement of algebraic plane curves, defined by implicit and parametric equations, is presented. The polynomials appearing in the equations are given in the Lagrange basis, with respect to a suitable set of nodes. Our method is of sweep-line class, and its novelty consists in applying algebra by values for solving systems of two bivariate polynomial equations. Moreover, at our best knowledge, previous works on arrangements of curves consider only implicitly defined curves.

1 Introduction

Given a finite collection $\mathcal{S}$ of geometric objects in $\mathbb{R}^n$, the arrangement associated to $\mathcal{S}$ is the decomposition of $\mathbb{R}^n$ into the connected open cells of dimensions $0, 1, ..., n$ induced by $\mathcal{S}$. The study of arrangements is a relevant problem in Computational Geometry and CAGD, with applications in Geometric Modeling, Computer Vision, Robot Motion Planning, etc. Edelsbrunner studied arrangements of lines and hyperplanes in the 80's [1, 2]. Arrangements of arcs in the plane and of surface patches in higher dimensions, and several applications were presented in [3]. Different algorithms have been proposed for the computation of arrangements of certain kind of plane curves: for conics [4, 5], for cubics [6, 7], for quartics [8], and for non-singular algebraic curves [9]. Recent algorithms for computing arrangements of algebraic curves in the plane use one of two main approaches: subdivision methods (e.g. [10–13]), and sweep-line methods (e.g. [14, 15]). In all these papers, the curves are defined only by their implicit equations.

In this paper, a new method for computing the topology of the arrangement of a set of plane curves, which may be given by implicit or parametric equations, is

---

[*] The authors are partially supported by the Spanish "Ministerio de Economía y Competitividad" and by the European Regional Development Fund (ERDF), under the Project MTM2011-25816-C02-02.

V.P. Gerdt et al. (Eds.): CASC Workshop 2014, LNCS 8660, pp. 59–73, 2014.

presented. The polynomials appearing in the equations are given in the Lagrange basis, with respect to a suitable set of nodes. Our method is of sweep-line class, and its novelty consists in applying algebra by values for solving systems of two bivariate polynomial equations. Using algebra by values means doing all computations (polynomial derivatives, Bezoutians, etc.) in the Lagrange basis (see [16]). There are three main reasons for using algebra by values:

1. Avoiding the manipulation of high degree polynomials.
2. If the initial data is given in Lagrange form, it is well known that the conversion between distinct polynomial bases may be numerically unstable, and this instability increases with the degree (see [17–19]).
3. There is a straightforward method for building a companion matrix pencil associated to a Bezoutian, such that the generalized eigenvalues of the companion matrix pencil give the roots of the determinant of the Bezoutian.

The following example illustrates the reasons why sometimes one may prefer slower computation through values, instead of manipulating expressions in the monomial basis.

Example 1. Consider the polynomial $f(x,y) = (x - 10)^{10} + (y - 10)^{10} - 10$. By working with the expanded version of f, and using 8-digit precision float numbers, numerical inconsistencies are obtained when trying to determine the topology of the curve (in the monomial basis).It is obtained the same result as for $g(x,y) = (x - 10)^{10} + (y - 10)^{10}$, while the topologies are different (f is a curve quite close to a 3×3 square centered at the point $(10, 10)$, while g reduces to the point $(10, 10)$). However, evaluating such polynomials at nodes not too far from $(10, 10)$, the topologies are well differentiated.

In this paper, it is not our intention to analyze in detail numerical aspects of our method. However, we would like to remark some facts: for computing nullspaces it is advisable to use Singular Value Decomposition, regarding efficiency and accuracy [20] ; on the other hand, the issue of solving generalized eigenvalue problems is supported by robust software As part of the input data, the users are allowed to choose two constant bounds, in order to specify which numerical quantities are considered too high (resp. too low) for the given curves, due to the fact that float imprecisions can make intersections in the infinite line to become finite (resp. can make coincident important points to separate); finally, using too low precision can lead to contradictions in the data the algorithm will use to build the graph (depending on the implementation, this can break the algorithm or return impossible phenomena such as branches that disappear or bifurcate). For more details, see [21] and the references therein.

The paper is organized in the following way. In Section 2 some relevant properties of Bezoutians, its expression in the Lagrange basis and generalized eigenvalues are recalled. In Section 3.1 and Section 3.2 the methods for the case of implicit and respectively parametric curves are presented. In the last section, the main conclusions and comments on future work are drawn.

2   Preliminaries

In this section, the definition of the Bezoutian of two polynomials, the structure of its nullspace and the generalized eigenvalue problem associated to a matrix polynomial are reviewed. These are the main algebraic tools to be used in our algorithms. The vector space of polynomials of degree at most d will be denoted by $\mathbb{P}_d$.

2.1   The Bezoutian

Definition 1. *Let $p(t), q(t)$ be two polynomials, $n = \max\{\deg(p(t)), \deg(q(t))\}$. The Cayley quotient of $p(t)$ and $q(t)$ is the polynomial $C_{p,q}$ of degree at most $n - 1$ defined by*

$$C_{p,q}(t, x) = \frac{p(t)q(x) - p(x)q(t)}{t - x} \tag{1}$$

Thus, if $\boldsymbol{\Phi}(t) = \{\phi_1(t), \ldots, \phi_n(t)\}$ is a basis for $\mathbb{P}_{n-1}$ then $C_{p,q}$ can be uniquely written

$$C_{p,q}(t, x) = \sum_{i,j=1}^{n} b_{ij}\phi_i(t)\phi_j(x) = (\phi_1(t), \ldots, \phi_n(t))\, (b_{ij}) \begin{pmatrix} \phi_1(x) \\ \vdots \\ \phi_n(x) \end{pmatrix}.$$

The symmetric matrix $\mathbf{Bez}_{p,q} = (b_{ij})$ is called the Bezoutian in the given polynomial basis $\boldsymbol{\Phi}(t)$.

The following properties of the Bezoutian are well known (see, for example, [22, 23]):

- $\deg(\gcd(p(t), q(t))) = n - \mathrm{rk}(\mathbf{Bez}_{p,q})$.
- The rank of $\mathbf{Bez}_{p,q}$, when computed in monomial basis in increasing degree order, is equal to the order of the largest nonsingular principal minor, when starting from the lower right hand corner.
- Its determinant is proportional to the resultant of $p(t)$ and $q(t)$.

The nullspace of the Bezoutian has an elegant structure presented in the next proposition that can be used to determine the common roots of the given polynomials (for a proof see [24], page 42).

Proposition 1. *The nullspace of $\mathbf{Bez}_{p,q}$, in the monomial basis, is spanned by*

$$\mathrm{nullspace}(\mathbf{Bez}_{p,q}) = [X_1, X_2, \ldots, X_k] \tag{2}$$

where each block X_j corresponds to a different common root of p and q. The dimension of each block is the geometric multiplicity k_j of the common root x_j

(i.e., its multiplicity as a root of the greatest common divisor of p and q). More-over each block can be parameterized by the common root x_j in the form

$$X_j = \begin{pmatrix} 1 & 0 & 0 & \cdots & 0 \\ x_j & 1 & 0 & & 0 \\ x_j^2 & 2x_j & 2 & & \vdots \\ x_j^3 & 3x_j^2 & 6x_j & \ddots & \vdots \\ \vdots & \vdots & \vdots & \ddots & (k_j-1)! \\ & & & & k_j! \, x_j \\ \vdots & \vdots & \vdots & & \vdots \\ x_j^{n-1} & (n-1)x_j^{n-2} & (n-1)(n-2)x_j^{n-3} & \cdots & (n-1)\frac{k_j-1}{} x_j^{n-k_j} \end{pmatrix} \tag{3}$$

where $n\frac{k_j}{} = n(n-1)\cdots(n-k_j+1)$.

Given a list $\boldsymbol{\tau} = (\tau_1, \ldots, \tau_{d+1})$ of distinct numerical values,

$$\boldsymbol{L}(t; \boldsymbol{\tau}) = \{L_1(t; \boldsymbol{\tau}), \ldots, L_{d+1}(t; \boldsymbol{\tau})\}$$

will denote the Lagrange basis of $\mathbb{P}_d$ associated to $\boldsymbol{\tau}$. A polynomial $p(t) \in \mathbb{P}_d$ is *given by values* if a pair of lists $\boldsymbol{\tau}$ and $\mathbf{p} = (p(\tau_1), \ldots, p(\tau_{d+1}))$ are available.

Lemma 1. *[16] Let $p(t), q(t) \in \mathbb{P}_d$ with $d = \max(\deg(p(t)), \deg(q(t)))$. We assume that $p(t)$ and $q(t)$ are given by values, by data $\boldsymbol{\tau}$, $\mathbf{p}$, and $\mathbf{q}$. Suppose that $t^* \in \mathbb{C}$ is a simple common zero of $p(t)$ and $q(t)$. If B denotes $\mathbf{Bez}_{p,q}$ in the Lagrange basis $\boldsymbol{L}(t; \tilde{\boldsymbol{\tau}}) \in [\mathbb{P}_{d-1}]^d$, then $\boldsymbol{L}(t^*; \tilde{\boldsymbol{\tau}})$ is a null vector of B.*

Remark 1. Note that, if $\boldsymbol{BM}$ and $\boldsymbol{BL}$ represent the Bezout matrix computed in monomial and respectively Lagrange basis, the following relation is fulfilled: $\boldsymbol{BL} = \boldsymbol{V}(\boldsymbol{\tau}) \cdot \boldsymbol{BM} \cdot \boldsymbol{V}(\boldsymbol{\tau})^T$, where $\boldsymbol{V}(\boldsymbol{\tau})$ is the Vandermomde matrix,

$$\boldsymbol{V}(\boldsymbol{\tau}) = \left(\tau_i^{j-1}\right), \ 1 \le i, j \le d . \tag{4}$$

As a consequence, when choosing pairwise different nodes, the rank of $\boldsymbol{BL}$ equals the rank of $\boldsymbol{BM}$.

2.2 Generalized Eigenvalue Problem Associated to a Matrix Polynomial

Given a $r \times r$ polynomial matrix $\mathbf{A}(t)$ of degree d, there are two matrices $\mathbf{C}_0, \mathbf{C}_1$, with constant entries, such that $\det \mathbf{A}(t) = \det(t\mathbf{C}_1 - \mathbf{C}_0)$. The pair $(\mathbf{C}_0, \mathbf{C}_1)$ is called a companion matrix pencil, and the solutions of $\det(t\mathbf{C}_1 - \mathbf{C}_0) = 0$ are called generalized eigenvalues of $\mathbf{A}(t)$. If $\mathbf{A}(t)$ is given in the Lagrange basis, the following definition gives an expression for the companion matrix pencil.

Definition 2. *[25] Consider the matrix polynomial*

$$\mathbf{A}(t) = \mathbf{A}_1 L_1(t; \boldsymbol{\tau}) + \mathbf{A}_2 L_2(t; \boldsymbol{\tau}) + \ldots + \mathbf{A}_{d+1} L_{d+1}(t; \boldsymbol{\tau}). \tag{5}$$

where, for $1 \leqslant i \leqslant d+1$, $\mathbf{A}_i$ are $r \times r$ matrices such that $\mathbf{A}(\tau_i) = \mathbf{A}_i$. The corresponding companion matrix pencil is

$$\mathbf{C}_0 = \begin{pmatrix} \tau_1 \mathbf{I} & & & & \mathbf{A}_1 \\ & \tau_2 \mathbf{I} & & & \mathbf{A}_2 \\ & & \ddots & & \vdots \\ & & & \tau_{d+1}\mathbf{I} & \mathbf{A}_{d+1} \\ -\omega_1 \mathbf{I} & -\omega_2 \mathbf{I} & \cdots & -\omega_{d+1}\mathbf{I} & 0 \end{pmatrix}, \quad \mathbf{C}_1 = \begin{pmatrix} \mathbf{I} & & & \\ & \mathbf{I} & & \\ & & \ddots & \\ & & & \mathbf{I} \\ & & & & 0 \end{pmatrix}, \tag{6}$$

where $\mathbf{I}$ and $\mathbf{0}$ are the identity matrix and the zero matrix, respectively, conformal with the $r \times r$ matrices $\mathbf{A}_i$, for $1 \leqslant i \leqslant d+1$. The matrices $\mathbf{C}_0$ and $\mathbf{C}_1$ are of dimension $r(d+2) \times r(d+2)$.

When computing eigenvalues, it is common to obtain some too big (according to the initially chosen bound) eigenvalues, which are considered to be infinite eigenvalues. In addition, clustering methods to combine a set of very close numerical solutions into a single eigenvalue are applied. For more details referring this numerical aspects, see [26].

3 Arrangements of Real Algebraic Plane Curves

In this section the problem of computing the topology of an arrangement of real algebraic plane curves is tackled, considering the following cases: a) the curves are implicitly defined, as presented in Section 3.1, and b) the curves are parametrically defined, as presented in Section 3.2. In both cases, the curves are supposed to be defined by values. The following result can be found in many papers dealing about topology of arrangements of curves (e.g. [27]):

Lemma 2. *Let us assume that $\mathcal{C}_1, \ldots, \mathcal{C}_r$ are real algebraic plane curves, defined either implicitly or parametrically. The topology of the arrangements of these curves is completely determined by the following data:*

- *The critical points (α, β), which are either:*
  - *critical points of a curve $\mathcal{C}_i$ (i.e. singular points or points whose first coordinate is either locally maximum or locally minimum), or*
  - *intersections of a pair of curves $\mathcal{C}_i$ and $\mathcal{C}_j$*
- *The points $(\alpha, y_{\alpha,l})$ which are the intersections of the **critical** lines $x = \alpha$ with the curves (excluding the critical points).*
- *The points $(\gamma, y_{\gamma,l})$ which are intersections of the curves with **noncritical** lines $x = \gamma$, chosen such that two adjacent critical lines are separated by a noncritical one, such that all γ are rational numbers.*

Therefore, a topology computation algorithm can be divided in the following steps: 1) Computation of the critical points of each curve. 2) Computation of the intersection points of each pair of curves. 3) Computation of noncritical points on critical (vertical) lines. 4) Computation of noncritical points on noncritical (vertical) lines. 5) Generating the graph that represents the topology of the arrangement.

In the first case (implicitly defined curves), *generic position* conditions are supposed to be verified, as it will be explained below.

The topology computation of an arrangement of curves defined by values is approached by a sweeping algorithm. For a formal exposition of the theoretical aspects of the algorithm, see [21].

3.1   Arrangements of Implicit Curves

Let the curves C_i be defined by $f_i(x, y) = 0$, for $i = 1, \ldots, r$. The main problem in the case of a single curve is computing intersection points between the curve defined by a polynomial f and one of its partial derivatives, f_y. If we consider the multiple curves case, the unique added problem consists in computing the intersection points of pairs of curves.

Extending the *generic position* conditions presented in [21],the following conditions are supposed to hold:

- $\{f_1, \ldots, f_r\}$ are square-free and pairwise coprime.
- No f_i has vertical asymptotes.
- Any two critical points of the same curve do not share first coordinate.
- Any two intersection points of a given pair of curves do not share first coordinate.

Therefore, for the rest of this section, each $f_i(x, y)$ is a bivariate square-free polynomial that can be, at least, evaluated. Although the expressions $f_i(x, y)$ may be not available, or costly to be computed and/or manipulated, the values $f_i(\tau_k, \sigma_l)$ are supposed to be known.

Let d_i (resp. e_i) be the degree of f_i on the variable x (resp. y). Let $m = \max\{d_i\}$, $n = \max\{e_i\}$, and let $\boldsymbol{\tau} = (\tau_0 < \ldots < \tau_{2m})$, $\boldsymbol{\sigma} = (\sigma_0 < \ldots < \sigma_n)$ be real numbers.

First of all, for all $k = 0, \ldots, 2m$, $l = 0, 1, \ldots, n$, the values of the partial derivatives, w.r.t. y at the given node (τ_k, σ_l) is computed, as in [21, Lemma 3]:

$$\frac{\partial f_i}{\partial y}(\tau_k, \sigma_l) = \sum_{t \neq l} \frac{(f_i(\tau_k, \sigma_t) - f_i(\tau_k, \sigma_l)) \prod_{s \neq t}(\sigma_t - \sigma_s)}{(\sigma_l - \sigma_t)\prod_{s \neq l}(\sigma_l - \sigma_s)} . \tag{7}$$

Once $f_{i,y} = \frac{\partial f_i}{\partial y}$ are known by values, $f_{i,yy} = \frac{\partial f_{i,y}}{\partial y}$ are computed using the same method.

Computing the Critical Points of Each Curve. The Bezoutian of $f_i(x, y)$ and $f_{i,y}(x, y)$, with respect to the variable y is computed by values. It has polynomial entries of degree up to $2d_i$ (this is the reason of the list τ's length) on the variable x, and is called henceforth $B_i(x)$. The Lagrange basis $L(x; \tau) = \{L_0(x; \tau), \dots, L_{2d}(t; \tau)\}$ is considered. Hence, $B_i(x) = \sum_{k=0}^{2d} B_i^k L_{k+1}(x, \tau)$, with $B_i^k = B_i(\tau_k) = (b_i^k(s, t))_{s,t} \in \mathcal{M}_{e_i}(\mathbb{R})$ where, as in [21, Proposition 8]:

$$b_i^k(s, t) = \frac{f_i(\tau_k, \sigma_s) f_{i,y}(\tau_k, \sigma_t) - f_i(\tau_k, \sigma_t) f_{i,y}(\tau_k, \sigma_s)}{\sigma_s - \sigma_t}, \quad \text{if } s \neq t,$$

$$b_i^k(s, s) = f_{i,y}(\tau_k, \sigma_s)^2 - f_i(\tau_k, \sigma_s) f_{i,yy}(\tau_k, \sigma_s), \quad \text{otherwise.} \qquad (8)$$

By solving the generalized eigenvalue problem for $B_i(x)$ as in Definition 2, the real eigenvalues $\alpha_{i,1} < \dots < \alpha_{i,r_i}$ are obtained. According to Proposition 1 and Remark 1, these are the x-coordinates of the intersection points of f_i and $f_{i,y}$. In order to obtain the y-coordinates $\beta_{i,j}$ of the critical points, which is the common root of $f_i(\alpha_{i,j}, y)$ and $f_{i,y}(\alpha_{i,j}, y)$, it is necessary to compute $B_i(\alpha_{i,j})$. Given the fact that f can be evaluated at any point, $f_i(\alpha_{i,j}, \sigma_l)$, $f_{i,y}(\alpha_{i,j}, \sigma_l)$ and $f_{i,yy}(\alpha_{i,j}, \sigma_l)$ are computed (the last two using (7)) for all $l = 0, \dots, n$. Afterwards, $B_i(\alpha_{i,j})$ can be computed by values as in (8) (replacing τ_k by $\alpha_{i,j}$).

According to Proposition 1 and Remark (1), $B_i(\alpha_{i,j})$ is a singular matrix. The following two cases are considered:

1. If $\mathrm{rk} B_i(\alpha_{i,j}) = e_i - 1$, any nonzero vector $[a_1, \dots, a_{e_i}]$ in $\ker B_i(\alpha_{i,j})$ is chosen and the y-coordinate of the critical point is:

$$\beta_{i,j} = \frac{\sigma_1 a_1 + \dots + \sigma_{e_i} a_{e_i}}{a_1 + \dots + a_{e_i}}.$$

2. If $\mathrm{rk} B_i(\alpha_{i,j}) = e_i - k$ with $k > 1$, a basis of $\ker B_i(\alpha_{i,j})$ is chosen and its elements are organized in a matrix $N \in \mathcal{M}_{e_i \times k}(\mathbb{R})$. Let $V = (V_{h,j}) = (\sigma_j^h) \in \mathcal{M}_{(k+1) \times e_i}(\mathbb{R})$, which is a submatrix of the transpose of the Vandermonde matrix. We now compute $Z = VN \in \mathcal{M}_{(k+1) \times e_i}(\mathbb{R})$ and its lower triangulation $\hat{Z} = (\hat{Z}_{s,t})$ (i.e. $\hat{Z}_{i,j} = 0$ when $i < j$) by Gauss method. Then, provided generic position, the y-coordinate of the critical point is $\beta_{i,j} = \hat{Z}_{(k+1),k}/(k\hat{Z}_{k,k})$

Computing Intersection Points of a Pair of Curves. The procedure is similar to the previous subsection, but $f_{i,y}$ is replaced f_j. In this way, the Bezoutian $B_{i,j}(x)$ of f_i and f_j with respect to y is computed by values. Now, $B_{i,j}^k = B_{i,j}(\tau_k) = (b_{i,j}^k(s, t))_{s,t} \in \mathcal{M}_{e_{i,j}}(\mathbb{R})$, such that $e_{i,j} = \max\{e_i, e_j\}$ and:

$$b_{i,j}^k(s, t) = \frac{f_i(\tau_k, \sigma_s) f_j(\tau_k, \sigma_t) - f_i(\tau_k, \sigma_t) f_j(\tau_k, \sigma_s)}{\sigma_s - \sigma_t}, \quad \text{if } s \neq t,$$

$$b_{i,j}^k(s, s) = f_{i,y}(\tau_k, \sigma_s) f_j(\tau_k, \sigma_s) - f_i(\tau_k, \sigma_s) f_{j,y}(\tau_k, \sigma_s), \quad \text{otherwise.} \qquad (9)$$

By solving the generalized eigenvalue problem for $B_{i,j}(x)$, the real eigenvalues $\alpha_{i,j,1} < \dots < \alpha_{i,j,r_i}$ are obtained. According to Proposition 1 and Remark 1, these are the x-coordinates of the intersection points of f_i and f_j. In order to obtain the y-coordinate, which is the common root of $f_i(\alpha_{i,j,k}, y)$ and $f_j(\alpha_{i,j}, y)$, $B_{i,j}(\alpha_{i,j,k})$ is computed. The procedure is similar to the one described in the previous subsection: $f_i(\alpha_{i,j}, \sigma_l)$, $f_j(\alpha_{i,j}, \sigma_l)$, $f_{i,y}(\alpha_{i,j}, \sigma_l)$ and $f_{j,y}(\alpha_{i,j}, \sigma_l)$ are computed (the last two, using (7)), for all $l = 0, \ldots, n$. Afterwards $B_{i,j}(\alpha_{i,j,k})$ is computed by values as in (9) (replacing τ_k by $\alpha_{i,j}$).

According to Proposition 1 and Remark 1, $B_{i,j}(\alpha_{i,j,k})$ is a singular matrix. There are two possibilities:

1. If $\mathrm{rk}B_i(\alpha_{i,j,k}) = e_{i,j} - 1$, any nonzero vector $[a_1, \ldots, a_{e_{i,j}}]$ in $\ker B_i(\alpha_{i,j})$ that, due to Lemma 1, is $(a_1 + \cdots + a_{e_{i,j}})[L_1(\beta_\alpha; \sigma), \ldots, L_n(\beta_\alpha; \sigma)]$, is chosen. Then the y-coordinate of the critical point is:

$$\beta_{i,j,k} = \frac{\sigma_1 a_1 + \cdots + \sigma_{e_{i,j}} a_{e_{i,j}}}{a_1 + \cdots + a_{e_{i,j}}}.$$

2. If $\mathrm{rk}B_{i,j}(\alpha_{i,j,k}) = e_{i,j} - l$ with $l > 1$, proceeding as in the previous subsection and assuming generic position, the y-coordinate of the critical point is $\beta_{i,j,k} = \hat{Z}_{(l+1),l}/(k\hat{Z}_{l,l})$, where $\hat{Z}$ is the lower triangulation of $Z = VN \in \mathcal{M}_{(l+1) \times l}(\mathbb{R})$, and the columns of N form a basis of $\ker B_{i,j}(\alpha_{i,j,k})$.

Checking Conditions. The generic position conditions are checked by using the following remarks:

Remark 2. If f_i had a real vertical asymptote, then the leading coefficient of f_i with respect to y, a polynomial in $\mathbb{R}[x]$, would have a root in the x-coordinate α of such asymptote, and this α would be an eigenvalue of $B_i(x)$. In order to check this situation, the Lagrange basis $\{L_0(x, \tau), \ldots, L_{2m}(x, \tau)\}$ associated to $\tau_0, \ldots, \tau_{2m}$ is computed. The leading coefficient of $f(\alpha, y)$ is, according to [21, Proposition 10], $LC_\alpha = \sum_{l=0}^{e_i} \prod_{s \neq l}(\sigma_l - \sigma_s) \sum_{k=0}^{2m} f(\tau_k, \sigma_l) L_j(\alpha, \tau)$. If it does not vanish, then $x = \alpha$ is not a vertical asymptote.

Remark 3. Given an eigenvalue α of $B_i(x)$, the condition for β_α, as computed above, to be the only common root of $f_i(\alpha, y)$ and $f_{i,y}(\alpha, y)$ is that its multiplicity as a root of the second one is $k = e_i - \mathrm{rk}B_i(\alpha) + 1$. Then β_α is also a root of $f_{i,y^t}(\alpha, y)$ for all $t = 2, \ldots, k$. Through Proposition 1, this happens if and only if, when considering the subspaces $N_t = \bigcap_{s=1}^{t} \mathrm{nullspace}\left(\mathrm{Bez}_y\left(f_i(\alpha, y), \frac{\partial^s f_i}{\partial y^s}(\alpha, y)\right)\right)$ for $t = 1, \ldots, k$, the conditions $\dim(N_t) - \dim(N_{t-1}) = 1$ hold for all $t = 2, \ldots, k$.

Remark 4. Given an eigenvalue α of $B_{i,j}(x)$, the condition for β_α, as computed above, to be the only common root of $f_i(\alpha, y)$ and $f_j(\alpha, y)$ is that the multiplicity of β_α as a root of both polynomials is at least $k = e_{i,j} - \mathrm{rk}B_{i,j}(\alpha)$ for any of the polynomials. Then β_α is root also of $f_{i,y^t}(\alpha, y)$ and $f_{j,y^t}(\alpha, y)$ for all $t = 2, \ldots, k$. Through Proposition 1, this happens if and only if, when considering the subspaces $N_t = \bigcap_{s=1}^{t} \mathrm{nullspace}\left(\mathrm{Bez}_y\left(\frac{\partial^s f_i}{\partial y^s}(\alpha, y), \frac{\partial^s f_j}{\partial y^s}(\alpha_j, y)\right)\right)$ for $t = 1, \ldots, k$, the conditions $\dim(N_t) - \dim(N_{t-1}) = 1$ hold for all $t = 2, \ldots, k$.

Computing Noncritical/Intersection Points on a Critical Line. Let (α, β) be either a critical point of a curve $f_i(x, y)$ or an intersection point with

$f_j(x,y)$. Let also suppose that β is the only root of $f_i(\alpha,y)$ (at most) already known. From previous subsections, we also know the multiplicity m_i of β as a root of $f_i(\alpha,y)$ (we can admit $m_i = 0$, if we do not know any root of $f_i(\alpha,y)$). Our goal now is obtaining all other roots of $f_i(\alpha,y)$.

In order to do this, $\hat{f}(y) = \frac{f_i(\alpha,y)}{(y-\beta)^{m_i}}$ is computed by values. It is a polynomial on y of degree $e_i - k_i$, so $\hat{f}(\sigma_1), ..., \hat{f}(\sigma_{e_i - m_i})$ need to be computed (obviously, if β happens to be one of the nodes, a different value should be chosen). By generating the companion matrices $(e_i - m_i + 1) \times (e_i - m_i + 1)$ with these values, solving the eigenvalue problem as in Definition 2 (where $\mathbf{A}$ is the 1×1 polynomial matrix $\hat{f}$) the remaining roots of $f_i(\alpha,y)$ are obtained.

Remark 5. It may happen that several roots of $f_i(\alpha,y)$ (e.g., a multiple root, a shared (different) root with $f_j(\alpha,y)$, $j \neq i$ and a shared (also different) root with $f_k(\alpha,y)$, $k \neq i,j$ are known. In any case, the multiplicities of all of them is known (in generic position, only one can be multiple) and so nodes can be reduced accordingly.

Computing Noncritical Arcs between Critical Lines. After finding the x-coordinates of all critical points (both of just one curve and intersections), the values are ordered: $\alpha_1 < ... < \alpha_z$. Afterwards, rational numbers γ_k such that $\gamma_0 < \alpha_1 < \gamma_1 < ... < \alpha_z < \gamma_z$ are chosen. Nextly, all the roots of $f_i(\gamma_k,y)$ for all suitable i and k are computed. The method is exactly as in the previous section in the case $m_i = 0$.

Generating the Graph. Let $k \in \{1,...,z\}$. Let $\delta_{k-1,1} < ... < \delta_{k-1,s_{k-1}}$ be the roots of all $f_i(\gamma_{k-1},y)$ (all are simple, since $x = \gamma_{k-1}$ is not a critical line, and each of them annihilates only one polynomial, since $x = \gamma_{k-1}$ is not an intersection line). Let $\delta_{k,1} < ... < \delta_{k,s_k}$ be the roots of all $f_i(\gamma_k,y)$ (same properties as with γ_{k-1}). Let $\beta_{k,1} < ... < \beta_{k,t_k}$ be the roots of all $f_i(\alpha_k,y)$. Let $m_{k,i,j}$ be the multiplicity of $\beta_{k,j}$ as a root of $f_i(\alpha_k,y)$, $n_{k,i,j}$ the multiplicity of $\delta_{k,j}$ as a root of $f_i(\gamma_k,y)$ and $n_{k-1,i,j}$ the multiplicity of $\delta_{k-1,j}$ as a root of $f_i(\gamma_{k-1},y)$. At this moment, the nodes of the graph which shape the topology of the arrangement are available: the dots of types (α,β) and (γ,δ). The next step consists in generating the edges, which will be assigned a color in order to allow the user to distinguish the different curves. The procedure is the usual one in sweeping algorithms. For all $i = 1,...,r$:

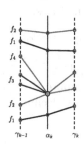

Fig. 1. Edges of four curves:
$m_{k,1} = [1,0,1,0]$,
$m_{k,2} = [0,1,0,1]$,
$m_{k,3} = [0,2,0,0]$,
$m_{k,4} = [0,1,0,0]$

- Due to the generic position, the vector $\mathbf{m}_{k,i} = [m_{k,i,1},...,m_{k,i,t_k}]$ has all coordinates equal to 1 or 0 except for, at most, one entry. On the other hand, $\mathbf{n}_{k,i} = [n_{k,i,1},...,n_{k,i,s_k}]$ and $\mathbf{n}_{k-1,i} = [n_{k-1,i,1},...,n_{k-1,i,s_{k-1}}]$ are binary vectors (see Figure 1).

– If all $m_{k,i,j} \in \{0,1\}$, for $j = 1,,t_k$, then the amount of "ones" in $m_{k,i}$ is the same as in $n_{k,i}$. Then, the following edges are included with the color of f_i: $\overline{(\alpha_k,\beta_{k,j})(\gamma_k,\delta_{k,l})}$ such that $m_{i,j}$ has the same position among the "ones" of $m_{k,i}$ as $n_{k,i,l}$ among the ones of $n_{k,i}$. On the left side, the situation is treated symmetrically, so the following edges are also added: $\overline{(\gamma_{k-1},\delta_{k-1,l})(\alpha_k,\beta_{k,j})}$ such that $m_{k,i,j}$ has the same position among the "ones" of $m_{k,i}$ as $n_{k-1,i,l}$ among the "ones" of $n_{k-1,i}$.

– If there is j_0 such that $m_{k,i,j_0} > 1$, then the amount of "ones" in $m_{k,i}$ is lesser or equal than the amount of "ones" in $n_{k,i}$ (which can also be zero). Then the following nodes are joined with an f_i-colored edge:

$(\alpha_k,\beta_{k,j})$, $(\gamma_k,\delta_{k,l})$ with $j < j_0$, $m_{i,j}$ and $n_{k,i,l}$ of same position among their respective "ones"

$(\alpha_k,\beta_{k,j})$, $(\gamma_k,\delta_{k,l})$ with $j > j_0$, $m_{i,j}$ and $n_{k,i,l}$ of same position among their respective "ones" in reverse order

After this, if there exist l such that $n_{k,i,l} = 1$ and $(\gamma_k,\delta_{k,l})$ is not yet joined, it will be joined to (α_k,β_{k,j_0}).

Some Examples. The method has been tested in some cases getting manageable times with Maple 17 running on an AMD Turion64 × 2 processor at 1.9 GHz with 2GB RAM:

Example 2. The above mentioned computer took 69.521 seconds to provide Figure 2, working with 10 (decimal) digits, with an input of 20 randomly chosen circles.

Example 3. Consider the curves $(y^4 - x^2 + y)$ $((-y^2 + x)^2 - x^4 - \frac{1}{100}) + y^4 x^4 = 0$ and $(x+y)^4 + (x-y)^4 - 2 = 0$. The computer took 222.577 seconds to compute the graph shown in Figure 3, working with 20 (decimal) digits.

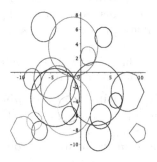

Fig. 2. Graph of 20 random circles

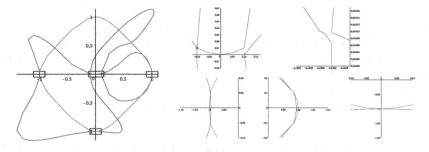

Fig. 3. Topological graph (and some details) for the two curves of Example 3

3.2 Arrangements of Parametric Curves

In this section, the same problem as in Section 3.1 is considered. Nevertheless, the below considered curves are parametrically defined by values. Let $\mathcal{C}_1, \ldots, \mathcal{C}_r$ be a family of real curves parametrized by $\phi_i(t) = (x_i(t), y_i(t))$, $t \in \mathbb{R}$, with $i = 1, \ldots, r$, where $x_i(t)$ and $y_i(t)$ are polynomials described by values. The curves are supposed to be proper (recall that a parametrization is said to be proper if it is injective for almost all the curve points).

The degrees of $x_i(t)$ and $y_i(t)$, with respect to t, are m_i and n_i, respectively, and $d_i = \max(m_i, n_i)$, for $i = 1, \ldots, r$. Let $\boldsymbol{\tau_i} = (\tau_{i,1} < \ldots < \tau_{i,d_i+1})$ and $\boldsymbol{\sigma_i} = (\sigma_{i,1} < \ldots < \sigma_{i,d_i-1})$ be sets of given nodes. Although the parametrizations are not available, or costly to be computed and/or manipulated, the point coordinates $\{(x_i(\tau_{i,j}), y_i(\tau_{i,j})) : 1 \le j \le d_i+1\}$ and $\{(x_i(\sigma_{i,j}), y_i(\sigma_{i,j})) : 1 \le j \le d_i-1\}$ of each curve $\mathcal{C}_i$ are supposed to be known.

Remark 6. Note that in this case, unlike the implicit curves, the possibility of different nodes for each curve is considered. However, the existence of vertical asymptotes is also avoided in the parametric case.

The computation of the topology of a single parametric curve has been studied, in the monomial basis, in [28].

Computing the Critical Points of Each Curve. Our goal in this step is to compute the singular and ramification points (those whose first coordinate meets either a local maximum or a local minimum as a function of the parameter) of the curve $\mathcal{C}_i$. Firstly, the Bezoutian of

$$p(t, s) = \frac{x_i(t) - x_i(s)}{t - s}, \qquad q(t, s) = \frac{y_i(t) - y_i(s)}{t - s}$$

with respect to t is computed by values, as follows: for all $k = 0, \ldots, d_i - 1$, we compute the matrix $B_i^k = B_i(\tau_k) = (b_i^k(u, v))_{u,v} \in \mathcal{M}_{d_i-1}(\mathbb{R})$ such that

$$b_i^k(u, v) =$$
$$\frac{[x_i(\tau_{i,u}) - x_i(\sigma_{i,k})][y_i(\tau_{i,v}) - y_i(\sigma_{i,k})] - [x_i(\tau_{i,v}) - x_i(\sigma_{i,k})][y_i(\tau_{i,u}) - y_i(\sigma_{i,k})]}{(\tau_{i,u} - \sigma_{i,k})(\tau_{i,v} - \sigma_{i,k})(\tau_{i,u} - \tau_{i,v})},$$
$$\text{if } u \ne v,$$

$$b_i^k(u, u) = \frac{x_i'(\tau_{i,u})[y_i(\tau_{i,u}) - y_i(\sigma_{i,k})] - y_i'(\tau_{i,u})[x_i(\tau_{i,u}) - x_i(\sigma_{i,k})]}{(\tau_{i,u} - \sigma_{i,k})^2}, \quad \text{otherwise.}$$

$$(10)$$

The derivatives are computed according to (7), obtaining:

$$x_i'(\tau_{i,u}) = \sum_{v \ne u} \frac{[x_i(\tau_{i,v}) - x_i(\tau_{i,u})]\prod_{l \ne u}(\tau_{i,u} - \tau_{i,l})}{(\tau_{i,u} - \tau_{i,v})\prod_{l \ne v}(\tau_{i,v} - \tau_{i,l})}, \qquad (11)$$

with the obvious modifications for y_i'.

By solving the generalized eigenvalue problem for $B_i(s)$, the real eigenvalues $t^s_{i,1}, \ldots, t^s_{i,s_i}$ are obtained, which are the parameter values of the singular points.

Afterwards, in order to compute the ramification points, the generalized eigenvalue problem associated with $(\tau_{i,u} : 1 \le u \le m_i)$ and $(x'_i(\tau_{i,u}) : 1 \le u \le m_i)$ (as the aim consists in generating the solutions of the equation $x'_i(t) = 0$) is solved, obtaining the real eigenvalues $t^r_{i,1}, \ldots, t^r_{i,r_i}$, which are the parameter values of the ramification points.

Finally, the eigenvalue clustering problem needs to be solved, distinguishing those eigenvalues coming from the same roots, by following the approach presented in [26]. In this way, the parameter values $t_{i,1} < \ldots < t_{i,v_i}$ are finally obtained. In order to compute the corresponding cartesian coordinates, the evaluation procedure for $x_i(t)$ and $y_i(t)$ is used:

$$x_i(t) = \sum_{k=1}^{d_i+1} \frac{x_i(\tau_{i,k}) \prod_{j=1}^{d_i+1} (t - \tau_{i,j})}{(t - \tau_{i,k}) \prod_{j \ne k} (\tau_{i,k} - \tau_{i,j})} \ , y_i(t) = \sum_{k=1}^{d_i+1} \frac{y_i(\tau_{i,k}) \prod_{j=1}^{d_i+1} (t - \tau_{i,j})}{(t - \tau_{i,k}) \prod_{j \ne k} (\tau_{i,k} - \tau_{i,j})}.$$
(12)

In order to compute the correct number of critical points, the x-coordinates of the critical points must be obtained and also there must be identified those points corresponding to self-intersection of the curve, by following the approach presented in [26].

Computing the Intersection Points of a Pair of Curves. Our goal in this step is to compute the intersection points of the curves $\mathcal{C}_i$ and $\mathcal{C}_j$. Each point of intersection $(\tilde{x}, \tilde{y})$ is generated by a pair of real values $\tilde{t}$ and $\tilde{s}$ such that $(\tilde{x}, \tilde{y}) = (x_i(\tilde{t}), y_i(\tilde{t})) = (x_j(\tilde{s}), y_j(\tilde{s}))$. Let us define $X(s,t) = x_j(s) - x_i(t)$ and $Y(s,t) = y_j(s) - y_i(t)$.

By solving the generalized eigenvalue problem for $B(t)$, the Bezoutians of $X(s,t)$ and $Y(s,t)$ seen as polynomials in s whose coefficients are polynomials in t, the real eigenvalues $t^q_{i,1}, \ldots, t^q_{i,q_{i,j}}$ are obtained, which are the parameter values of the intersection points, seen as points of the curve $\mathcal{C}_i$. In order to obtain the parameter values for the other curve, the process is repeated inverting the roles of s and t.

Finally, the corresponding cartesian coordinates are computed by using the evaluation procedure for $x_i(t)$, $y_i(t)$, $x_j(s)$ and $y_j(s)$ (see (12)) .

Computing Noncritical Points on Critical Lines. Firstly, let $\mathcal{X} = \{x_1 < \ldots < x_z\}$ be the x-coordinates of all critical points. For each $x_l \in \mathcal{X}$, there may be other noncritical points on some curve $\mathcal{C}_i$ sharing the same x-coordinates. The parameter coordinates of these points are the real generalized eigenvalues associated with $(\tau_{i,k} : 1 \le k \le m_i + 1)$ and $(x_i(\tau_{i,k}) - x_l : 1 \le k \le m_i + 1)$ (as we are searching the solutions of the equations $x_i(t) = x_l$). Finally, those generalized eigenvalues already computed in the previous steps must be discarded, by following the approach presented in [26].

Computing Noncritical Points on Noncritical Lines. Rational numbers γ_l such that $\gamma_0 < x_1 < \gamma_1 < \ldots x_z < \gamma_z$ are chosen and, in a similar way as in the previous step, for each γ_l the points on the curves C_i whose x-coordinates equals γ_l are computed.

Generating the Graph. All the cartesian points computed in the previous steps are stored together with the index of the curve they lay on and the corresponding parametric coordinate. The segment connection of these points is realized by applying a procedure based on the signs of $x_i'(t_j)$, as described and proven in detail in [28].

This step is extremely important, as if the set of rules describing the segment connection may not be successfully applied, the computing precision must be increased and the computations repeated.

3.3 Example

In this section we present the main graphical results for the arrangement of 3 parametric curves C_1, C_2, C_3, defined by:

$$
\begin{aligned}
x_1(t) &= -t^3 + t^2 + t - 1, \quad y_1(t) = t^2 - 2t - 1, \\
x_2(t) &= 2t^2 - t, \qquad\qquad y_2(t) = -t^3 - t^2 + 3t, \quad (13) \\
x_3(t) &= -t^3 - t^2 + 3t, \quad y_3(t) = \frac{1}{2}t^2 + t - 1.
\end{aligned}
$$

After applying the algorithm presented in the previous section, the topological graph in Figures 4 and 5 are obtained.

The diamond points in the graph correspond to curve points placed on critical lines and the cross points correspond to curve non critical points on non critical lines. The algorithm has been implemented in the Symbolic Computation System `Maple` Version 17. The computing times obtained on a 2.27 GHz Intel Core i5 machine with 4GB of RAM have been:

- singular points: 0.078 s.
- ramification points: 0.078 s.
- intersection points: 1.576 s.
- non critical points on critical lines: 1.248 s.
- non critical points on non critical lines: 1.919 s.

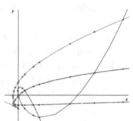

Fig. 4. Graph of C_1, C_2, C_3

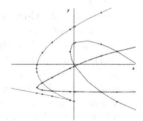

Fig. 5. Zoom near the origin

4 Conclusions and Ongoing Work

A new method for computing the topology of an arrangement of algebraic plane curves, defined by implicit and parametric equations, has been presented. In our approach, the polynomials appearing in the equations are given in the Lagrange basis, with respect to a suitable set of nodes. Our method is of sweep-line class, and its novelty consists in applying algebra by values for the solution of systems of two bivariate polynomial equations. Moreover, at our best knowledge, previous related works considered only implicitly defined curves. The following issues represent main topics of our further work:

- Generalizing the results presented in Section 3.2 to the rational case.
- Solving the arrangement problem for offset curves.
- Solving the arrengement problem for bisector curves.
- Merging all the mentioned cases (implicit, rational parametric, offset and bisector curves) in a unique arrengement problem.

References

1. Edelsbrunner, H.: Algorithms in Combinatorial Geometry. EATCS Monographs on Theoretical Computer Science, vol. 10. Springer (1987)
2. Edelsbrunner, H., O'Rourke, J., Seidel, R.: Constructing arrangements of lines and hyperplanes with applications. SIAM J. Comput. 15, 341–363 (1986)
3. Agarwal, P.K., Sharir, M.: Arrangements and their applications. In: Sack, J.R., Urrutia, J. (eds.) Handbook of Computational Geometry, pp. 49–119. Elsevier (2000)
4. Berberich, E., Eigenwillig, A., Hemmer, M., Hert, S., Mehlhorn, K., Schömer, E.: A computational basis for conic arcs and boolean operations on conic polygons. In: Möhring, R.H., Raman, R. (eds.) ESA 2002. LNCS, vol. 2461, pp. 174–186. Springer, Heidelberg (2002)
5. Wein, R.: High-level filtering for arrangements of conic arcs. In: Möhring, R.H., Raman, R. (eds.) ESA 2002. LNCS, vol. 2461, pp. 884–895. Springer, Heidelberg (2002)
6. Eigenwillig, A., Kettner, L., Schömer, E., Wolpert, N.: Exact, efficient and complete arrangement computation for cubic curves. Computational Geometry 35, 36–73 (2006)
7. Caravantes, J., Gonzalez-Vega, L.: Improving the topology computation of an arrangement of cubics. Computational Geometry 41, 206–218 (2008)
8. Caravantes, J., Gonzalez-Vega, L.: Computing the topology of an arrangement of quartics. In: Martin, R., Sabin, M.A., Winkler, J.R. (eds.) Mathematics of Surfaces 2007. LNCS, vol. 4647, pp. 104–120. Springer, Heidelberg (2007)
9. Wolpert, N.: Jacobi curves: Computing the exact topology of arrangements of non-singular algebraic curves. In: Di Battista, G., Zwick, U. (eds.) ESA 2003. LNCS, vol. 2832, pp. 532–543. Springer, Heidelberg (2003)
10. Plantinga, S., Vegter, G.: Isotopic approximation of implicit curves and surfaces. In: Boissonnat, J.D., Alliez, P. (eds.) Symposium on Geometry Processing. ACM International Conference Proceeding Series, vol. 71, pp. 245–254. Eurographics Association (2004)
11. Hijazi, Y., Breuel, T.: Computing arrangements using subdivision and interval arithmetic. In: Chenin, P., Lyche, T., Schumaker, L. (eds.) Curve and Surface Design: Avignon 2006, pp. 173–182. Nashboro Press (2007)
12. Alberti, L., Mourrain, B., Wintz, J.: Topology and arrangement computation of semi-algebraic planar curves. Computer Aided Geometric Design 25(8), 631–651 (2008)

13. Mourrain, B., Wintz, J.: A subdivision method for arrangement computation of semi-algebraic curves. In: Emiris, I.Z., Sottile, F., Theobald, T. (eds.) Nonlinear Computational Geometry. The IMA Volumes in Mathematics and its Applications, vol. 151, pp. 165–188. Springer (2010)
14. Eigenwillig, A., Kerber, M.: Exact and efficient 2d-arrangements of arbitrary algebraic curves. In: Proceedings of the 9th Annual ACM-SIAM Symposium on Discrete Algorithms, SODA 2008, pp. 122–131. SIAM (2008)
15. Berberich, E., Emeliyanenko, P., Kobel, A., Sagraloff, M.: Exact symbolic-numeric computation of planar algebraic curves. Theoretical Computer Science 491, 1–32 (2013)
16. Shakoori, A.: Bivariate Polynomial Solver by Values. PhD thesis, The University of Western Ontario (2007)
17. Hermann, T.: On the stability of polynomial transformations between Taylor, Bézier, and Hermite forms. Numerical Algorithms 13, 307–320 (1996)
18. Berrut, J., Trefethen, L.: Barycentric Lagrange interpolation. SIAM Review 46(3), 501–517 (2004)
19. Higham, N.J.: The numerical stability of barycentric Lagrange interpolation. IMA Journal of Numerical Analysis 24, 547–556 (2004)
20. Demmel, J.W.: Applied Numerical Linear Algebra. Society for Industrial and Applied Mathematics, Philadelphia (1997)
21. Corless, R., Diaz-Toca, G., Fioravanti, M., Gonzalez-Vega, L., Rua, I., Shakoori, A.: Computing the topology of a real algebraic plane curve whose defining equations are available only "by values". Comput. Aided Geom. Des. 30(7), 675–706 (2013)
22. Helmke, U., Fuhrmann, P.A.: Bezoutians. Linear Algebra and Its Applications 122/123/124, 1039–1097 (1989)
23. Bini, D., Pan, V.: Polynomial and Matrix Computations. Birkhäuser (1994)
24. Heinig, G., Rost, K.: Algebraic methods for Toeplitz-like matrices and operators. Operator Theory: Advances and Applications 13 (1984)
25. Corless, R.M.: On a Generalized Companion Matrix Pencil for Matrix Polynomials Expressed in the Lagrange basis. In: Symbolic-Numeric Computation, pp. 1–18. Birkhäuser (2006)
26. Corless, R., Gonzalez-Vega, L., Necula, I., Shakoori, A.: Topology determination of implicitly defined real algebraic plane curves. In: Proceedings of the 5th International Workshop on Symbolic and Numeric Algorithms for Scientific Computing SYNASC 2003, Universitatea din Timisoara. Analele Universitatii din Timisoara, Matematica - Informatica, vol. XLI, pp. 78–90 (2003)
27. Eigenwillig, A., Kerber, M., Wolpert, N.: Fast and exact geometric analysis of real algebraic plane curves. In: Proceedings ISSAC 2007 (July 2007)
28. Alcazar, J., Diaz-Toca, G.: Topology of 2d and 3d rational curves. Comput. Aided Geom. Des. 27, 483–502 (2010)

Finding a Deterministic Generic Position for an Algebraic Space Curve

Jin-San Cheng and Kai Jin

Key Lab of Mathematics Mechanization
Institute of Systems Science, AMSS, Chinese Academy of Sciences
{jcheng,jinkai}@amss.ac.cn

Abstract. Checking whether an algebraic space curve is in a generic position or not is an important step for computing the topology of real algebraic space curve. In this paper, we present an algorithm to find a deterministic generic position for an algebraic space curve.

Keywords: Algebraic space curve, generic position, weak generic position.

1 Introduction

Algebraic space curves are used in computer aided (geometric) design, and geometric modeling. Computing the topology of algebraic space curves is also a basic step to compute the topology of algebraic surfaces [4,9]. Most of the existing work ([1,5,6,8,12]) of computing the topology of algebraic space curves require the space curve to be in a generic position, though the definitions of the generic position is different. The generic position property ensures the correctness and completeness of the algorithms for computing the topology of algebraic space curves(with less calculations), see [1,5,6,8,12]. But checking whether an algebraic space curve is in a generic position or not is not a trivial task, see [1,6,9,12]. None of the papers give an algorithm to find a deterministic generic position for an algebraic space curve. In this paper, we will give a deterministic algorithm to find a generic position for an algebraic space curve, which is the main contribution of the paper.

Related Works. Let $\mathbb{Q}, \mathbb{R}, \mathbb{C}$ be the fields of rational numbers, real numbers and complex numbers respectively. In the following, we always use $\mathcal{C} = f \wedge g$ to represent the algebraic space curve defined by $f, g \in \mathbb{Q}[x, y, z]$ with $\gcd(f, g) = 1$. Let $\pi_z : (x, y, z) \to (x, y)$, similarly for π_y.

In [1], the generic position of an algebraic space curve $\mathcal{C}$ is defined as below.

Definition A. We say that $\mathcal{C}$ is in a space generic position if $\pi_z(\mathcal{C})$ and $\pi_y(\mathcal{C})$ are in plane generic position w.r.t. the ox axis, and it satisfies the following three properties.

1. The leading coefficient of either f or g w.r.t. z, and the leading coefficient of either f or g w.r.t. y are non-zero constants.

V.P. Gerdt et al. (Eds.): CASC Workshop 2014, LNCS 8660, pp. 74–84, 2014.

2. The projections π_z and π_y of $\mathcal{C}$ are injective up to a finite number of exceptions.

3. $\mathcal{C}$ has no point whose projections onto the xy-plane and the xz-plane are both multiple.

Condition 3 of Definition A implies that the algebraic space curve here is without multiple points. Note that even for algebraic space curves without multiple points, Condition 3 may not be satisfied. The authors present methods to check the conditions in the definition. Checking Condition 1 is trivial. Condition 2 is checked by Corollary 9. Let h denote the square-free part of $\mathrm{Res}_z(f, g)$, and let m denote the square-free part of $\mathrm{Res}_y(f, g)$, then, Condition 3 is satisfied if

$$\gcd(\gcd(\mathrm{Res}_y(h, \frac{\partial h}{\partial x}), \mathrm{Res}_y(h, \frac{\partial h}{\partial y})), \gcd(\mathrm{Res}_z(m, \frac{\partial m}{\partial x}), \mathrm{Res}_z(h, \frac{\partial m}{\partial z}))) = 1.$$

The generic position of plane projection curves is checked by Algorithm 6 in Section 2.

In [12], they study the general case of algebraic space curves whose defining polynomials may be more than two. They give the following definition.

Definition B. Let $\tilde{\mathcal{C}} \subset \mathbb{C}^3$ be an algebraic space curve and $\mathbb{I}(\tilde{\mathcal{C}}) \subset \mathbb{C}[x, y, z]$ be its ideal. We will say that $\tilde{\mathcal{C}}$ is in a generic position w. r. t. the projection on the xy-plane if the following conditions hold.

1. $\mathbb{I}(\tilde{\mathcal{C}}) \cap \mathbb{C}[x, y] = \mathbb{I}(\mathcal{C}_h)$, the projection curve $\mathcal{C}_h$ is in a generic position and the projection $\pi_z : (\alpha, \beta, \gamma) \in \tilde{\mathcal{C}} \to (\alpha, \beta) \in \mathcal{C}_h$ is birational,
2. $\mathbb{C}[\tilde{\mathcal{C}}]$ is integral over $\mathbb{C}[\mathcal{C}_h]$,
3. if (α, β, γ) is a critical point of $\tilde{\mathcal{C}}$ then this point is the only one intersection of $\tilde{\mathcal{C}}$ with the line $x = \alpha, y = \beta$,
4. if (α, β, γ) is a critical nonsingular point of $\tilde{\mathcal{C}}$ then the line $x = \alpha, y = \beta$ is not tangent to $\tilde{\mathcal{C}}$ at this point,
5. if (α, β, γ) is a plane singular point of $\tilde{\mathcal{C}}$ then its tangent plane does not contain the line $x = \alpha, y = \beta$,
6. if (α, β, γ) is nonsingular in $\tilde{\mathcal{C}}$ but (α, β) is singular in $\mathcal{C}_h$, then (α, β) is a node.

They also give a symbolic based algorithm to check the generic position, which requires the computation of generators of the radical of the ideal, that involves Gröbner basis computation. For the case that there are only two defining polynomials, they use subresultant method to check birational property of the curve for the checking. For more details, please see [12].

In the papers [8,9], the authors give another definition (Definition C) of the generic position for an algebraic space curve. In fact, the definition here is similar to that of the Definition B for the case that there are only two defining polynomials. Let $\mathcal{C}^{\mathbb{C}}, \mathcal{C}^{\mathbb{R}}$ present that we consider the space curve in complex space, real space.

The curve $\mathcal{C}^{\mathbb{R}}$ is in a **pseudo-generic position** w.r.t. the xy-plane if and only if almost every point of $\pi_z(\mathcal{C}^{\mathbb{C}})$ has only one geometric inverse-image, i.e. generically, if $(\alpha, \beta) \in \pi_z(\mathcal{C}^{\mathbb{C}})$, then $\pi_z^{-1}(\alpha, \beta)$ consists one point possibly multiple.

Definition C. The curve $\mathcal{C}^{\mathbb{R}}$ is in a generic position w.r.t. the xy-plane if and only if

1. $\mathcal{C}^{\mathbb{R}}$ is in a pseudo-generic position w.r.t. the xy-plane,
2. $D = \pi_z(\mathcal{C}^{\mathbb{R}})$ is in a generic position (as a plane algebraic curve) w.r.t. the x-direction,
3. any apparent singularity (a plane singularity which does not related to a space singularity of $\mathcal{C}^{\mathbb{R}}$.) of $D = \pi_z(\mathcal{C}^{\mathbb{R}})$ is a node.

In their papers, Conditions 1, 2 in Defintion C both are checked by computing subresultant of certain polynomials. The checking of Condition 3 involves testing whether the Hessian matrix of the projection curve of the algebraic space curve at a plane algebraic point is regular or not.

In [6,5], we give a definition of the weak generic position of an algebraic space curve, see Definition 7, which is weaker than the pseudo-generic position of an algebraic space curve given before. In [6], we use the technique of checking pseudo-generic position to certify the weak generic position. But a weak generic position is enough to compute the topology of a given algebraic space curve [5].

None of the methods above give a deterministic way to find a generic position for an algebraic space curve. Of course, since there are only finite non-generic positions, when trying bad positions at most on a fixed times, they can certainly get a generic position for an algebraic space curve. But it means an increasing complexity with a factor of a polynomial in the degree of the defining polynomials comparing to the original ones.

We give a deterministic algorithm to find a weak generic position for an algebraic space curve. In [5], we use the technique in this paper to get a certified weak generic position with which we compute the topology of general algebraic space curves defined by two polynomials. We implement our algorithm to compute the topology of a given algebraic space curve in Maple, which contains the algorithm to compute the weak generic position of an algebraic space curve. The experiments show the efficiency of the whole algorithm. We compute two examples to show the computing times of our methods comparing to one existing method. It is efficient especially for space curves with large defining polynomials.

We also give a definition of the generic position (see Definition 12), which is weaker than Definition C. Condition 3 in Definition C is removed in Definition 12. A deterministic algorithm to find a generic position for an algebraic space curve is given.

The paper is organized as follows. In the next section, we give some definitions and results. In Section 3, we present two algorithms to find a (weak) generic position for an algebraic space curve. We draw a conclusion in the last section.

2   Definitions and Theoretical Results

Let $h(x, y) \in \mathbb{Q}[x, y]$. We denote the plane algebraic curve defined by $h = 0$ as $\mathcal{C}_h$. Let $\mathbf{p} = (x_0, y_0)$ be a point on $\mathcal{C}_h$. We call $\mathbf{p}$ as an x-**critical point** if $h(\mathbf{p}) = \frac{\partial h}{\partial y}(\mathbf{p}) = 0$, a **singular point** if $h(\mathbf{p}) = \frac{\partial h}{\partial y}(\mathbf{p}) = \frac{\partial h}{\partial x}(\mathbf{p}) = 0$.

Let $\mathcal{C}_{f,g}$ denote the algebraic space curve defined by the polynomials $f(x,y,z)$, $g(x,y,z) \in \mathbb{R}[x,y,z]$. We always use $\mathcal{C}$ to replace $\mathcal{C}_{f,g}$ when no ambiguity exists.

The following definitions can be found in [9].

Let $\mathbb{I}(g_1, \ldots, g_s)$ be the radical ideal of the ideal $\mathbb{I}(f,g)$. Let $M(X,Y,Z)$ be the $s \times 3$ Jacobian matrix with $(\frac{\partial g_i}{\partial x}, \frac{\partial g_i}{\partial y}, \frac{\partial g_i}{\partial z})$ as its i-th row. A point $\mathbf{p} \in C$ is **regular (or smooth)** if the rank of $M(\mathbf{p})$ is 2. A point $\mathbf{p} \in C$ which is not regular is called **singular**. A point $\mathbf{p} = (\alpha, \beta, \gamma) \in C$ is x-**critical** (or critical for the projection on the x-axis) if the curve $\mathcal{C}$ is tangent at this point to a plane parallel to the yz-plane.

The following two definitions (Definitions 1, 4) can be found in many references, for example, in [5,6,13].

Definition 1. *Let $u, v \in \mathbb{Q}[x,y]$, $\gcd(u,v) = 1$, $w = \mathrm{Res}_y(u,v)$ is the resultant of the polynomials u, v w.r.t. y. We say u, v are in a* **generic position** *w.r.t. y if the following two conditions are satisfied:*

1) $\gcd(lc(u,y), lc(v,y)) = 1$.

2) $\forall \alpha \in \mathbb{V}_{\mathbb{C}}(w)$, $u(\alpha, y), v(\alpha, y)$ have only one common zero in $\mathbb{C}$,

where $lc(q,y)$ is the leading coefficient of $q \in \mathbb{Q}[x,y]$ w.r.t. y, $\mathbb{V}_{\mathbb{C}}(p)$ is the zeros of p in $\mathbb{C}$ (Similarly for $\mathbb{V}_{\mathbb{R}}(p)$ in the rest of the paper).

We can find out the related result of the following theorem in [11,14].

Theorem 2. *Let two coprime polynomials $u, v \in \mathbb{Q}[x,y]$ such that*

$$\gcd(lc(u,y), lc(v,y)) = 1.$$

Let

$$W(s,x) = \mathrm{Res}_y(u(x + sy, y), v(x + sy, y)),$$

$w(s,x)$ be the square free part of $W(s,x)$, and $D(s)$ be the discriminant of $w(s,x)$ w.r.t. x. If $\exists s_0 \in \mathbb{Q}$, such that

$$D(s_0) \neq 0, \deg_y(q(x + s_0 y, y)) = \deg(q(x,y)), q = u, v,$$

where $\deg$ (or $\deg_y$) means the total degree of the polynomial (or the degree w.r.t. y of the polynomial). Then $u(x + s_0 y, y)$ and $v(x + s_0 y, y)$ are in a generic position w.r.t. y.

The theorem gives a deterministic method to find a generic position for a plane algebraic curve. This technique is already used in [10,7] for bivariate polynomial system solving. We will give an algorithm to find a generic position for a plane algebraic curve below (see Algorithm 6). In fact, there are many other methods to check the generic position for two plane algebraic curves (or a plane algebraic curve), see [2,3,13] for example.

The following lemma gives an easy way to check the generic position for two plane algebraic curves in a special case.

Lemma 3 ([7]). *Let $f, g \in \mathbb{Q}[x,y]$ be square free polynomials, $\gcd(lc(f,y), lc(g,y))$ is a constant, then f, g are in a generic position w.r.t. y if $\mathrm{Res}_y(f,g)$ is square free.*

Definition 4. *Let h be a square free polynomial in $\mathbb{Q}[x, y]$. The real algebraic plane curve defined by h is in a generic position w.r.t. y if the following two conditions are satisfied:*
  1) The leading coefficient of h w.r.t. y is a nonzero constant.
  2) $\forall \alpha \in \mathbb{R}$, $h(\alpha, y)$, $\frac{\partial h}{\partial y}(\alpha, y)$ have at most one common distinct zero in $\mathbb{C}$.

The following corollary is deduced from Lemma 3 and Definition 4.

Corollary 5. *Let $h \in \mathbb{Q}[x, y]$ be a square free polynomial, $lc(h, y)$ is a constant, then the plane curve $\mathcal{C}_h$ is in a generic position w.r.t. y if h, $\frac{\partial h}{\partial y}$ are in a generic position w.r.t. y. Thus if the discriminant of h w.r.t. y is square free, $\mathcal{C}_h$ is in a generic position w.r.t. y.*

The following algorithm to find a generic position for a plane algebraic curve is based on Proposition 3.2 in [11].

Algorithm 6. *Find a generic position of an algebraic plane curve.*
INPUT: A square free polynomial $h \in \mathbb{Q}[x, y]$.
OUTPUT: A univariate polynomial $Q(s)$ whose zeros contain all the possible values s such that $h(x + s\, y, y)$ is not in a generic position w.r.t. y.

1. Let $f := h(x + s\, y, y)$.
2. $q(x, s) = sqrfree(\text{Res}_y(f, \frac{\partial f}{\partial y}))$, where $sqrfree(p)$ is the square free part of the polynomial p.
3. $Q(s) = sqrfree(\text{Res}_x(q, \frac{\partial q}{\partial x}))$.

Choose any $s_0 \in \mathbb{Q} \setminus \mathbb{V}_\mathbb{R}(Q)$, the curve defined by $h(x + s_0\, y, y) = 0$ is in a generic position, where $\mathbb{V}_\mathbb{C}(h)$ ($\mathbb{V}_\mathbb{R}(h)$) is all the complex (real) roots of the polynomial equation $h = 0$.

We can simply denote Algorithm 6 as

$$Q(s) = \mathbf{PCGP}(h, x, y, s). \tag{1}$$

Similarly, the algorithm to compute the univariate polynomial related to no-generic position of two plane curves based on Theorem 2 can be denoted as

$$Q(s) = \mathbf{P2CGP}(f, g, x, y, s). \tag{2}$$

Definition 7 ([5,6]). *Let $f, g \in \mathbb{Q}[x, y, z]$ be coprime, $h = sqrfree(\text{Res}_z(f, g))$. We say f, g are in a **weak generic position** w.r.t. z if*
  1) $\gcd(lc(f, z), lc(g, z)) = 1$.
  2) There are only a finite number of $(\alpha,\ \beta) \in \mathbb{V}_\mathbb{C}(h) \subset \mathbb{C}^2$ such that $f(\alpha, \beta, z)$, $g(\alpha, \beta, z)$ have more than one distinct common zeros in $\mathbb{C}$.

It is close to but weaker than the definition *Pseudo-generic position* in [8]. In the following, we will provide a simple method to find a weak generic position for an algebraic space curve.
  We introduce the map at first. Let

$$
\begin{array}{rccc}
\phi_s : & \mathbb{R}^3 & \longrightarrow & \mathbb{R}^3 \\
& (x, y, z) & \longrightarrow & (x, y + sz, z)
\end{array}
$$

Assume that $\mathcal{C} = f \wedge g$ for $f, g \in \mathbb{Q}[x, y, z]$ is not in a weak generic position and two branches of $\mathcal{C}$, denoted as C_1, C_2, satisfy $\pi_z(C_1) = \pi_z(C_2)$. If C_1 intersects C_2 at a point $\mathbf{p}$ of $\mathcal{C}$, we call $\mathbf{p}$ a **cylindrical singularity** of $\mathcal{C}$.

Theorem 8. *Let $f, g \in \mathbb{Q}[x, y, z]$ be coprime and $\gcd(lc(f, z), lc(g, z)) = 1$. Then $f(x, y, z)$, $g(x, y, z)$ are in a weak generic position w.r.t. z if and only if $\exists \alpha \in \mathbb{C}$, such that*

1. α *is not the first coordinate of any cylindrical singularity of $\mathcal{C}$ if it exists.*
2. $f(\alpha, y, z)$, $g(\alpha, y, z)$ *are in a generic position w.r.t. z.*

Proof. Let $h(x, y) = \operatorname{Res}_z(f(x, y, z), g(x, y, z))$ be the resultant of $f(x, y, z)$ and $g(x, y, z)$ w.r.t. z.

$\Rightarrow$ If $f(x, y, z)$, $g(x, y, z)$ is in a weak generic position w.r.t. z, then there are only a finite number of $(x_i, y_i) \in \mathbb{C}^2$ such that $h(x_i, y_i) = 0$ and $f(x_i, y_i, z)$, $g(x_i, y_i, z)$ have more than one distinct common zeros in $\mathbb{C}$. We choose a complex number α which is distinct from all of those x_i. Let $H(y) = \operatorname{Res}_z(f(\alpha, y, z), g(\alpha, y, z))$ and $\beta \in \mathbb{C}$ such that $H(\beta) = 0$. Actually $h(\alpha, \beta) = 0$ since $u(\alpha, y) \neq 0$. Hence $f(\alpha, \beta, z)$, $g(\alpha, \beta, z)$ have at most one common zero in $\mathbb{C}$, which turns out that $f(\alpha, y, z)$, $g(\alpha, y, z)$ are in a generic position w.r.t. z. Note that there is no cylindrical singularity in this case.

$\Leftarrow$ If $f \wedge g$ is not in a weak generic position, then there exists a polynomial $h_0(x, y) | h(x, y)$ such that almost for any $\mathbf{p} = (x_0, y_0) \in \mathbb{V}_\mathbb{C}(h_0)$, the polynomials $f(x_0, y_0, z), g(x_0, y_0, z)$ have more than one distinct common zeros in $\mathbb{C}$ with a finite number of exceptions by Definition 7. For all these exceptions, the related points correspond to cylindrical singularities by the definition. Thus $\forall p = (\alpha, \beta) \in \mathbb{V}_\mathbb{C}(h_0)$, the system $\{f(\alpha, \beta, z) = g(\alpha, \beta, z) = 0\}$ has at least two distinct solutions in $\mathbb{C}$. This is against to that $f(\alpha, y, z)$, $g(\alpha, y, z)$ are in a generic position w.r.t. z. Thus, the theorem is proved. ∎

Remark. It is not difficult to find out that the image of the cylindrical singularities of $\mathcal{C}$ will be singular points of $\pi_z \circ \phi_s(\mathcal{C})$ under the composite mapping $\pi_z \circ \phi_s$ when $s \neq 0$. Thus when we choose two different s, say s_1, s_2, we can get the x-coordinates of the possible cylindrical singularities of $\mathcal{C}$ from the singularities of $\pi_z \circ \phi_{s_1}(\mathcal{C})$ and $\pi_z \circ \phi_{s_2}(\mathcal{C})$. So we can use this fact to determine whether the condition 1 in Theorem 8 is satisfied or not.

According to the above theorem, we can transform the problem of weak generic position checking of an algebraic space curve into the generic position checking of the intersection of two plane algebraic curves. Thus the original problem is greatly simplified.

The following corollary is an extension of Lemma 3. It is useful for the reduced algebraic space curves. It appears in [1] to check Condition 2 in Definition A.

Corollary 9. *Let $f, g \in \mathbb{Q}[x, y, z]$ be square free polynomials and $\gcd(lc(f, z), lc(g, z)) = 1$. Then f, g are in a weak generic position w.r.t. z if $\operatorname{Res}_z(f, g)$ is square free.*

The following corollary is a much simpler way to check the weak generic position property of an algebraic space curve.

Corollary 10. *Let $f, g \in \mathbb{Q}[x, y, z]$ be square free polynomials and $\gcd(lc(f, z),$ $lc(g, z)) = 1$. For $\alpha \in \mathbb{R}$, if $f(\alpha, y, z), g(\alpha, y, z)$ are in a generic position w.r.t. z and they have only simple roots, then f, g are in a weak generic position w.r.t. z. For $\alpha \in \mathbb{R}$, if $\mathrm{Res}_z(f(\alpha, y, z), g(\alpha, y, z))$ is square free, then f, g are in a weak generic position w.r.t. z.*

Proof. The first part of the corollary can be directly derived from Theorem 8. Note that the two conditions in Theorem 8 hold directly. The second part can de deduced from the first part and Lemma 3.   ∎

3   The Algorithms for Finding a (Weak) Generic Position of an Algebraic Space Curve

In this section, we will present deterministic algorithms to find a (weak) generic position for an algebraic space curve.

The following algorithm is to find a weak generic position for an algebraic space curve.

Algorithm 11. *Finding a weak generic position for an algebraic space curve.*
Input: $f(x, y, z), g(x, y, z) \in \mathbb{Q}[x, y, z]$.
Output: $F \wedge G$ (isotopic to $f \wedge g$) is in a weak generic position w.r.t. z.

1. If $\deg(f) \neq \deg_z(f)$ and $\deg(g) \neq \deg_z(g)$, choose proper $c_i, i = 1, 2$ such that $f := f(x + c_1 z, y + c_2 z, z)$, $g := g(x + c_1 z, y + c_2 z, z)$ satisfy $\deg_z(f) = \deg(f)$ or $\deg_z(g) = \deg(g)$.
2. Let $H := \mathrm{Res}_z(f(x, y, z), g(x, y, z))$, $h := sqrfree(H)$.
   If $H = h$, return $f \wedge g$.
3. Else, do
   (a) Let $q(x) = \mathrm{Res}_y(h, \frac{\partial h}{\partial y})$, choose a rational number $\alpha \in \mathbb{Q} \setminus \mathbb{V}_\mathbb{R}(q)$.
   (b) Denote $Q(s) = \mathbf{P2CGP}(f(\alpha, y, z), g(\alpha, y, z), y, z, s)$, choose $s_0 \in \mathbb{Q} \setminus \mathbb{V}_\mathbb{R}(Q) \cup \{0\}$.
   (c) Let $h_0(x, y) = sqrfree(\mathrm{Res}_z(f(x, y + s_0 z, z), g(x, y + s_0 z, z)))$ and $\bar{q}(x) = discrim(h_0, y)$.
   (d) If $\alpha \notin \mathbb{V}_\mathbb{R}(\bar{q})$, return $f(x, y + s_0 z, z) \wedge g(x, y + s_0 z, z)$;
   (e) Else, do
      i. Choose $\beta \in \mathbb{Q} \setminus \mathbb{V}_\mathbb{R}(q * \bar{q})$.
      ii. $Q_1(s) = \mathbf{P2CGP}(f(\beta, y, z), g(\beta, y, z), y, z, s)$.
      iii. If $s_0 \notin \mathbb{V}_\mathbb{R}(Q_1)$, return $f(x, y + s_0 z, z) \wedge g(x, y + s_0 z, z)$;
         Else, choose $s_1 \in \mathbb{Q} \setminus \mathbb{V}_\mathbb{R}(Q * Q_1)$, return $f(x, y + s_1 z, z) \wedge g(x, y + s_1 z, z)$.

Remark: "Choose" in the algorithm above is to select any number satisfying the condition. Usually we prefer the one with a small bitsize. It is similar for the word in Algorithm 13.

Proof for the Correctness and Termination. The termination of the algorithm is clear. We will prove the correctness of the algorithm. Step 1 ensures that the gcd of the leading coefficients of f, g w.r.t. z is a constant. In Step 2, the algebraic space curve is already in a weak generic position if $H = h$ by Corollary 9. Steps 3.a and 3.c compute the x-coordinates of all possible cylindrical singularities if they exist. Steps 3.a, 3.b, 3.d and 3.e.i ensure that we choose right x-value which is not equal to the x-coordinate of any cylindrical singularity of $\mathcal{C}$ if it exists. Note that we choose two different projections, thus all cylindrical singularities appears as singularities of the projection curves. Steps 3.b, 3.e.ii and 3.e.iii ensure that the two chosen plane curves are in a generic position. So by the Remark of Theorem 8, we prove the correctness of the algorithm.   ∎

We give the definition of generic position for an algebraic space curve below.

Definition 12. *Let $f, g \in \mathbb{Q}[x, y, z]$ be squarefree polynomials. The algebraic space curve defined by f, g, denoted as $\mathcal{C}$, is called in a **generic position** w.r.t. z if*

*1) f, g are in a **weak generic position** w.r.t. z*

2) The projected plane curve $\mathcal{C}_h$ is in a generic position w.r.t. y, where $h = $ sqrfree($\mathrm{Res}_z(f, g)$).

3) For each x-critical point $\mathbf{p}$ of $\mathcal{C}_h$, there is at most one x-critical points of $\mathcal{C}$ whose projection is $\mathbf{p}$.

We give the following deterministic algorithm which finds a generic position for an algebraic space curve.

Algorithm 13. *Finding a deterministic generic position for an algebraic space curve.*

Input: $f(x, y, z), g(x, y, z) \in \mathbb{Q}[x, y, z]$.

Output: $F \wedge G$ (isotopic to $f \wedge g$) is in a generic position w.r.t. z.

1. *If $\deg(f) \neq \deg_z(f)$ and $\deg(g) \neq \deg_z(g)$, choose proper $c_i, i = 1, 2$ such that $f := f(x + c_1 z, y + c_2 z, z), g := g(x + c_1 z, y + c_2 z, z)$ satisfy $\deg_z(f) = \deg(f)$ or $\deg_z(g) = \deg(g)$.*

2. *Let $h_0(y, z, s) = $ sqrfree($\mathrm{Res}_x(f(x, y + s z, z), g(x, y + s z, z))$),*

$$Q(s) = \mathbf{PCGP}(h_0, y, z, s).$$

3. *Let $h := $ sqrfree($\mathrm{Res}_z(f(x, y, z), g(x, y, z))$), $q(x) = \mathrm{Res}_y(h, \frac{\partial h}{\partial y})$, choose a rational number $\alpha \in \mathbb{Q} \setminus \mathbb{V}_{\mathbb{R}}(q)$.*

4. *Denote $Q_1(s) = \mathbf{P2CGP}(f(\alpha, y, z), g(\alpha, y, z), y, z, s)$, choose $s_1 \in \mathbb{Q} \setminus \mathbb{V}_{\mathbb{R}}(Q_1 * Q) \cup \{0\}$.*

5. *Let $h_1(x, y) = $ sqrfree($\mathrm{Res}_z(f(x, y + s_1 z, z), g(x, y + s_1 z, z))$) and $\bar{q}(x) = $ discrim(h_1, y).*

6. *If $\alpha \in \mathbb{V}_{\mathbb{R}}(\bar{q})$, do*

   (a) Choose $\beta \in \mathbb{Q} \setminus \mathbb{V}_{\mathbb{R}}(q * \bar{q})$.

   (b) $Q_2(s) = \mathbf{P2CGP}(f(\beta, y, z), g(\beta, y, z), y, z, s)$.

   (c) If $s_1 \in \mathbb{V}_{\mathbb{R}}(Q_2)$, choose another $s_1 \in \mathbb{Q} \setminus \mathbb{V}_{\mathbb{R}}(Q * Q_2) \cup \{0\}$.

7. *$f := f(x, y + s_1 z, z), g := g(x, y + s_1 z, z)$.*

8. Let $h_2(x, y, s) := \mathrm{Res}_z(f(x + s\,y, y, z), g(x + s\,y, y, z))$.

$$Q_3(s) = \mathbf{PCGP}(h_2, x, y, s),$$

choose $s_2 \in \mathbb{Q} \setminus \mathbb{V}_{\mathbb{R}}(Q_3)$.

9. $f := f(x + s_2\,y, y, z)$, $g := g(x + s_2\,y, y, z)$.

Proof for the Correctness and Termination. The termination of the Algorithm 13 is clear. We need only to prove the correctness of the algorithm. The proof of the correctness of the weak generic position of the algebraic space curve in this algorithm is similar to the proof of Algorithm 11. We will not present the similar proof here again, but we confine it to its main ideas. In Steps 2, 4, 6.b, 6.c also ensure that the x-critical points of the new space curve are with different (x, y)-coordinates since Step 2 ensure that the projection of $f(x, y + s_1\,z, z), g(x, y + s_1\,z, z)$ obtained in Step 7 to the yz-plane is in a generic position w.r.t. z. Step 8 ensures that the projection of the space curve to the xy-plane is in a generic position w.r.t. y. Note that the coordinate transformation in Step 8 does not change the weak generic position property of the space curve and any two x-critical points of the new space curve will not overlap after the coordinate transformation. Note that $h_2(x, y, s)$ can be represented by $k(x + s\,y, y)$, so we can use $\mathbf{PCGP}$ to compute a generic position of the curve $h_2(x, y, s) = 0$ w.r.t. y. Thus the output space curve is in a generic position w.r.t. z. ∎

In the following, we test two examples taken from [5]. We will find a weak generic position for the curves with our new method introduced in this paper. We implemented Algorithm 11 in Maple 15 on a PC with Inter(R)Core(TM)i3-2100 CPU @3.10GHz 3.10GHz, 2G memory and Windows 7 operating system. We also implement the main steps of [9] with Maple by ourselves. It is denoted as SubResultant, the most time-consuming part of it is to compute the subresultant of the two defining polynomials. The core function of SubResultant is SubresultantChain, which is a function in the package ChainTools of Maple.

The first example is a small one, and the algebraic space curve is not in a generic position. The input polynomials are:
$f = x^2 + y^2 + z^2 - 4$, $g = (z - 1)\left(x^2 + y^2 - 3\,z^2\right)$.

It is not in a generic position. Algorithm 11 spends 0.0301 seconds to find a weak generic position w.r.t. z, while SubResultant spends 0.124 seconds to check and search a pseudo-generic position.

The second example is a complicated one, actually the algebraic space curve is in a generic position. The input polynomials are:
$f = 2 + 23\,y + z + 2\,x - y^2 z^3 + y^3 z^3 + y^4 z - x^2 y^3 - x^3 z + x^2 y^2 - x^2 y - x^2 z + yz^2 - xy^3 + xz^3 - y^2 z^2 - yz^3 + x^2 z^4 + yz^4 - xz^5 - xz^4 + y^5 z + y^2 z^4 + x^5 z - xy^5 + x^3 y^2 - yz^5 + 2\,x^2 y^4 + 2\,xy^4 + 2\,x^3 y - 20\,yz + 2\,x^2 z^2 - 72\,xy^2 + 2\,xy + 56\,xz^2 - 2\,x^4 z + 2\,x^3 z^2 - 2\,x^2 z^3 - 2\,y^3 z^2 + 2\,x^3 y^3 - 2\,x^3 z^3 + 2\,x^4 z^2 + 2\,x^5 y - 2\,x^4 y^2 - 2\,x^4 y + 2\,x^3 yz^2 + 2\,xyz^3 + 2\,xy^3 z - 2\,xyz^4 - 2\,xyz - 2\,x^3 y^2 z - 2\,x^2 y^2 z + 2\,xy^2 z - x^2 yz - 2\,xy^3 z^2 + 2\,x^2 y^3 z + 2\,xy^4 z + x^2 yz^3 - xy^2 z^3 - x^2 yz^2 - xy^2 z^2 + x^3 yz + x^4 yz + 2\,z^6 + 2\,x^2 - 11\,y^2 - 2\,z^3 + z^4 + x^6 - 18\,y^3 + y^6 + 2\,x^3 + 2\,z^2 + 2\,z^5 - 2\,x^4 + 2\,y^5 - x^5$,

$$g = -2 + y - 84\,z - x - 2\,xy^3z^3 - 2\,xy^2z^4 - 2\,xy^5z + 2\,x^2yz^4 - 2\,y^3z^3 - y^4z +$$
$$2\,x^3z - 74\,x^2y + 2\,x^2z - yz^2 + xy^3 - y^2z^2 + yz^3 - x^2z^4 - 2\,yz^4 + xz^5 + xz^4 + 2\,y^5z +$$
$$y^2z^4 - x^5z + xy^5 - x^3y^2 - yz^5 - x^2y^4 - xy^4 - 95\,yz - x^2z^2 + xy^2 - 17\,xz^2 - 2\,x^4z +$$
$$2\,x^3z^2 + 2\,x^2z^3 + y^3z^2 + x^3y^3 - 2\,x^3z^3 - x^4z^2 - 2\,x^5y + x^4y^2 - 2\,x^4y - xyz^2 -$$
$$x^2y^2z^2 + 2\,xyz^5 + 2\,x^2y^2z^3 - 2\,x^2y^4z - 2\,x^2y^3z^2 + 2\,x^3yz^2 - 2\,xy^3z - 2\,xyz^4 -$$
$$2\,xyz + 2\,x^3y^2z + x^2y^2z - 2\,xy^2z + 2\,x^2yz - xy^3z^2 - 2\,x^2y^3z - xy^4z - 2\,x^2yz^3 +$$
$$xy^2z^3 + x^2yz^2 - 2\,xy^2z^2 + 2\,x^3yz - 2\,x^4yz - x^3y^2z^2 + x^5yz - x^4y^2z - 2\,y^2 - 2\,z^3 +$$
$$z^4 - 2\,x^6 - y^6 + x^3 + 2\,z^2 - 2\,z^5 + x^4 - 2\,y^5 - x^5 - x^4yz^2 - 2\,x^3y^3z - xy^4z^2 +$$
$$y^4z^2 + x^5z^2 + x^3z^4 - x^3y^4 + x^6y + xz^6 - y^6z - 12\,y^2z - 81\,xz - 2\,y^3z - 2\,x^6z +$$
$$2\,x^4z^3 + 2\,x^2y^5 - 2\,xy^6 + 2\,x^5y^2 - 2\,y^4z^3 - 2\,y^3z^4 - 2\,y^7 + z^7.$$

Based on Corollary 10, we can find that the algebraic space curve is in a generic position w.r.t. z by computing a resultant of two bivariate polynomials. It takes only 0.405 seconds, while SubResultant needs more than 3000 seconds (We stop after 3000 seconds). This is due to the time-consuming computation of subresultant sequence for the input polynomials.

4   Conclusion

We discuss how to compute a generic position for an algebraic space curve in this paper. We present a theorem to certainly find a weak generic position for an algebraic space curve, which transforms the 3D problem into a 2D problem. The finding is deterministic. The other two conditions for generic position of a space curve is certified by deterministically finding a generic position of one (or two) plane algebraic curve(s). We can find that the algorithms given in this paper can be optimized. In our full version of the paper, we will optimize the algorithm and analyze its complexity.

Acknowledgement. The authors would like to thank the anonymous referees for their helpful comments and suggestions, which led to this revised version. The work is partially supported by NKBRPC (2011CB302400), NSFC Grants (11001258, 60821002), SRF for ROCS, SEM.

References

1. Alcázar, J.G., Sendra, J.R.: Computation of the topology of real algebraic space curves. Journal of symbolic Computation 39, 719–744 (2005)
2. Bouzidi, Y., Lazard, S., Pouget, M., Rouillier, F.: Rational Univariate Representations of bivariate systems and applications. In: Preceeding of ISSAC. ACM (2013)
3. Cheng, J.S., Gao, X.S., Guo, L.: Root Isolation of Zero-dimensional Polynomial Systems with Linear Univariate Representation. Journal of Symbolic Computation 47(7), 843–858 (2012)
4. Cheng, J.-S., Gao, X.-S., Li, M.: Determining the topology of real algebraic surfaces. In: Martin, R., Bez, H.E., Sabin, M.A. (eds.) IMA 2005. LNCS, vol. 3604, pp. 121–146. Springer, Heidelberg (2005)
5. Jin, K., Cheng, J.S.: Isotopic ϵ-Meshing of Real Algebraic Space Curves. In: SNC 2014 (to appear, 2014) (preprint)

6. Cheng, J.S., Jin, K., Lazard, D.: Certified Rational Parametric Approximation of Real Algebraic Space Curves with Local Generic Position Method. Journal of Symbolic Computation 58, 18–40 (2013)
7. Cheng, J.S., Jin, K.: A Generic Position Method For Real Roots Isolation Of Zero-Dimensional Polynomial System, arXiv:1312.0462
8. Diatta, D.N., Mourrain, B., Ruatta, O.: On the computation of the topology of a non-reduced implicit space curve. In: ISSAC 2008, pp. 47–54 (2008)
9. Diatta, D.N., Mourrain, B., Ruatta, O.: On the isotopic meshing of an algebraic implicit surface. Journal Symbolic Computation 47, 903–925 (2012)
10. Diochnos, D.I., Emiris, I.Z., Tsigaridas, E.P.: On the asymptotic and practical complexity of solving bivariate systems over the reals. Journal Symbolic Computation, Special issue for ISSAC 2007 44(7), 818–835 (2009)
11. González-Vega, L., El Kahoui, M.: An improved upper complexity bound for the topology computation of a real algebraic plane curve. Journal of Complexity 12(4), 527–544 (1996)
12. El Kahoui, M.: Topology of real algebraic space curves. Journal of Symbolic Computation 43, 235–258 (2008)
13. González-Vega, L., Necula, I.: Efficient topology determination of implicitly defined algebraic plane curves. Computer Aided Geometric Design 19, 719–743 (2002)
14. Sakkalis, T., Farouki, R.: Singular points of algebraic curves. Journal of Symbolic Computation 9(4), 405–421 (1990)

Optimal Estimations of Seiffert-Type Means By Some Special Gini Means

Iulia Costin[1] and Gheorghe Toader[2]

[1] Department of Computer Science, Technical University of Cluj-Napoca, Romania
Iulia.Costin@cs.utcluj.ro
[2] Department of Mathematics, Technical University of Cluj-Napoca, Romania
Gheorghe.Toader@math.utcluj.ro

Abstract. Let us consider the logarithmic mean $\mathcal{L}$, the identric mean $\mathcal{I}$, the trigonometric means $\mathcal{P}$ and $\mathcal{T}$ defined by H. J. Seiffert, the hyperbolic mean $\mathcal{N}$ defined by E. Neuman and J. Sándor, and the Gini mean $\mathcal{J}$. The optimal estimations of these means by power means $\mathcal{A}_p$ and also some of the optimal estimations by Lehmer means $\mathcal{L}_p$ are known. We prove the rest of optimal estimations by Lehmer means and the optimal estimations by some other special Gini means $\mathcal{S}_p$. In proving some of the results we used the computer algebra system *Mathematica*. We believe that some parts of the proofs couldn't be done without the help of such a computer algebra system (at least by following our way of proving those results).

Keywords: logarithmic mean, identric mean, Seiffert type means, power means, Lehmer means, special Gini means, inequalities of means.

1 Introduction

A **mean** is a function $M : \mathbb{R}_+^2 \to \mathbb{R}_+$, with the property

$$\min(a, b) \leq M(a, b) \leq \max(a, b), \ \forall a, b > 0 \ .$$

Each mean is **reflexive**, that is

$$M(a, a) = a, \ \forall a > 0 \ .$$

This is also used as the definition of $M(a, a)$.

A mean is **symmetric** if

$$M(b, a) = M(a, b), \ \forall a, b > 0 \ ,$$

and it is **homogeneous** (of degree 1) if

$$M(ta, tb) = t \cdot M(a, b), \ \forall a, b, t > 0 \ .$$

We shall refer here to the following symmetric and homogeneous means:

V.P. Gerdt et al. (Eds.): CASC Workshop 2014, LNCS 8660, pp. 85–98, 2014.
© Springer International Publishing Switzerland 2014

- the Gini (or sum) means $\mathcal{S}_{p,q}$, defined by

$$\mathcal{S}_{p,q}(a,b) = \begin{cases} \left(\dfrac{a^p+b^p}{a^q+b^q}\right)^{\frac{1}{p-q}} & \text{if } p \neq q \ , \\[4mm] \left(a^{a^p} \cdot b^{b^p}\right)^{\frac{1}{a^p+b^p}} & \text{if } p = q \ ; \end{cases}$$

- the power means $\mathcal{A}_p = \mathcal{S}_{p,0}$;
- the arithmetic mean $\mathcal{A} = \mathcal{A}_1$;
- the geometric mean $\mathcal{G} = \mathcal{A}_0 = \mathcal{S}_{0,0}$;
- the Lehmer means $\mathcal{L}_p = \mathcal{S}_{p+1,p}$;
- the special Gini means $\mathcal{S}_p = \mathcal{S}_{p-1,1}$;
- the logarithmic mean $\mathcal{L}$ defined by

$$\mathcal{L}(a,b) = \frac{a-b}{\ln a - \ln b}, \quad a \neq b \ ;$$

- the identric mean $\mathcal{I}$ defined by

$$\mathcal{I}(a,b) = \frac{1}{e}\left(\frac{a^a}{b^b}\right)^{\frac{1}{a-b}}, \quad a \neq b \ ;$$

- the similar Gini mean $\mathcal{J} = \mathcal{S}_{1,1}$ given by

$$\mathcal{J}(a,b) = \left(a^a b^b\right)^{\frac{1}{a+b}} \ ;$$

- the first Seiffert mean $\mathcal{P}$, defined in [16] by

$$\mathcal{P}(a,b) = \frac{a-b}{2\sin^{-1}\frac{a-b}{a+b}}, \quad a \neq b \ ;$$

- the second Seiffert mean $\mathcal{T}$, defined in [17] by

$$\mathcal{T}(a,b) = \frac{a-b}{2\tan^{-1}\frac{a-b}{a+b}}, \quad a \neq b \ ;$$

- the Neuman-Sándor mean $\mathcal{N}$, defined in [11] by

$$\mathcal{N}(a,b) = \frac{a-b}{2\sinh^{-1}\frac{a-b}{a+b}}, \quad a \neq b \ .$$

As $\mathcal{L}$ can be represented by

$$\mathcal{L}(a,b) = \frac{a-b}{2\tanh^{-1}\frac{a-b}{a+b}}, \quad a \neq b \ ,$$

the four means $\mathcal{P}$, $\mathcal{T}$, $\mathcal{N}$, and $\mathcal{L}$ are very similar and are called Seiffert-type means.

Definition 1. *For two means M and N, we say that M is **less than** N, and we write $M < N$, if $M(a,b) < N(a,b)$ for $\forall a, b > 0$, $a \neq b$.*

For instance, it is known that

$$\mathcal{L} < \mathcal{P} < \mathcal{I} < \mathcal{A} < \mathcal{N} < \mathcal{T} < \mathcal{J}, \tag{1}$$

as it was shown in [16] and [11].

Some complicated means were estimated by families of simpler means. Let us consider a family of means F_p, $p \in \mathbb{R}$. It is an **increasing** family if

$$F_p < F_q \text{ for } p < q .$$

A lower (upper) **estimation** of a given mean M by this family of means assumes the determination of some real index p (respectively q) such that $F_p < M$ (respectively $M < F_q$). A lower estimation is **optimal** if p is the greatest index r such that $F_r < M$. Similarly an upper estimation is **optimal** if q is the smallest index r with the property that $M < F_r$.

Optimal estimations by power means were given for the logarithmic mean in [10]:

$$\mathcal{A}_0 < \mathcal{L} < \mathcal{A}_{1/3} , \tag{2}$$

for the identric mean in [14]:

$$\mathcal{A}_{2/3} < \mathcal{I} < \mathcal{A}_{\ln 2} , \tag{3}$$

for the Gini mean in [12]:

$$\mathcal{A}_2 < \mathcal{J} , \tag{4}$$

and for the first Seiffert mean in [8]:

$$\mathcal{A}_{\ln 2/ \ln \pi} < \mathcal{P} < \mathcal{A}_{2/3} . \tag{5}$$

We consider also the following

Definition 2. *Two means $M < N$ can be **separated** by a family of means F_p, $p \in \mathbb{R}$ if there is an index p such that $M < F_p < N$.*

For example, the first separation of the means $\mathcal{N} < \mathcal{T}$ by power means was given in [4]:

$$\mathcal{N} < \mathcal{A}_{3/2} < \mathcal{T} .$$

Using the above inequalities, the chain of means (1) was separated in [5] by power means with equidistant indices:

$$\mathcal{A}_0 < \mathcal{L} < \mathcal{A}_{1/3} < \mathcal{P} < \mathcal{A}_{2/3} < \mathcal{I} < \mathcal{A}_1 < \mathcal{N} < \mathcal{A}_{4/3} < \mathcal{T} < \mathcal{A}_{5/3} < \mathcal{A}_{6/3} < \mathcal{J} .$$

Optimal estimations by power means were obtained independently in [6] and [18] for the second Seiffert mean:

$$\mathcal{A}_{\ln 2/ \ln(\pi/2)} < \mathcal{T} < \mathcal{A}_{5/3} \tag{6}$$

and in [6], [19] and [2] for the Neuman-Sándor mean:

$$\mathcal{A}_{\ln 2/ \ln(\ln(3+2\sqrt{2}))} < \mathcal{N} < \mathcal{A}_{4/3} . \tag{7}$$

It is easy to see that the optimal estimations by power means can be ordered as:

$$\mathcal{A}_0 < \mathcal{L} < \mathcal{A}_{1/3} < \mathcal{A}_{\ln 2/ \ln \pi} < \mathcal{P} < \mathcal{A}_{2/3} < \mathcal{I} < \mathcal{A}_{\ln 2} < \mathcal{A}_{\ln 2/ \ln(\ln(3+2\sqrt{2}))}$$

$$< \mathcal{N} < \mathcal{A}_{4/3} < \mathcal{A}_{\ln 2/ \ln(\pi/2)} < \mathcal{T} < \mathcal{A}_{5/3} < \mathcal{A}_2 < \mathcal{J} < \mathcal{A}_\infty . \tag{8}$$

In the next sections, we present the known estimations of Seiffert-type means by Lehmer means, we determine the optimal estimations by the other special Gini means $\mathcal{S}_p$, and we also compare the estimations by $\mathcal{A}_p$, $\mathcal{L}_p$ and $\mathcal{S}_p$. Some applications are also given.

2   Estimations by Lehmer Means

The first (lower) estimation of one of the above means by a family of means was published in [9] for the logarithmic mean

$$\mathcal{L}_{-1/3} < \mathcal{L} . \tag{9}$$

In fact, in [9] it is proved that $\mathcal{R}_{1/3} < \mathcal{L}$, where

$$\mathcal{R}_p(a,b) = \frac{ab^p + ba^p}{a^p + b^p},$$

but it is easy to see that

$$\mathcal{R}_p = \mathcal{S}_{1-p,-p} = \mathcal{L}_{-p} .$$

The inequality (9) was shown in [1] to be optimal. A second estimation was given in [1] for the the identric mean:

$$\mathcal{L}_{-1/6} < \mathcal{I} . \tag{10}$$

Recently, optimal estimations for the Seiffert means

$$\mathcal{L}_{-1/6} < \mathcal{P} < \mathcal{L}_0 \tag{11}$$

and

$$\mathcal{L}_0 < \mathcal{T} < \mathcal{L}_{1/3} \tag{12}$$

were given in [21], while in [20] it is proved the optimal inequality

$$\mathcal{N} < \mathcal{L}_{1/6} . \tag{13}$$

As both the power means and the Lehmer means are Gini means, we want to compare the above estimations using the following result proved in [13]. Defining the function

$$k(u,v) = \begin{cases} \frac{|u|-|v|}{u-v}, & \text{if } u \neq v, \\ sgn(u), & \text{if } u = v, \end{cases}$$

Theorem 1. *The inequality $\mathcal{S}_{p,q} \leq \mathcal{S}_{r,s}$ holds if and only if $p + q \leq r + s$ and:*

1. $\min\{p, q\} \leq \min\{r, s\}$ *if* $0 \leq \min\{p, q, r, s\}$;
2. $k(p, q) \leq k(r, s)$ *if* $\min\{p, q, r, s\} < 0 < \max\{p, q, r, s\}$;
3. $\max\{p, q\} \leq \max\{r, s\}$ *if* $\max\{p, q, r, s\} \leq 0$.

Applying this result to $\mathcal{A}_p = \mathcal{S}_{p,0}$ and $\mathcal{L}_p = \mathcal{S}_{p+1,p}$ we get the following

Corollary 1. *For $p \geq 0$ we have:*

1. $\mathcal{A}_p \leq \mathcal{L}_q$ *if and only if $p \geq 1$ and $q \geq (p-1)/2$; and*
2. $\mathcal{L}_q \leq \mathcal{A}_p$ *if and only if $0 \leq p \leq 1$ and $q \leq (p-1)/2$.*

We deduce that:

1. $\mathcal{A}_{5/3} \leq \mathcal{L}_{1/3}$ thus the second part of (6) implies the second part of (12);
2. $\mathcal{L}_{-1/6} \leq \mathcal{A}_{2/3}$ thus the first part of (3) implies (10) and the second part of (5) suggests the first part of (11);
3. $\mathcal{L}_{-1/3} \leq \mathcal{A}_{1/3}$ thus the second part of (2) suggests (9).
4. $\mathcal{A}_{4/3} \leq \mathcal{L}_{1/6}$ thus the second part of (7) implies (13).

In all these cases it must be proven that the estimations are optimal. Other implications among the previous inequalities cannot be deduced this way. The involved means are not comparable. For instance, the first inequality of (6) is $\mathcal{A}_p < \mathcal{T}$ where $p = \ln 2 / \ln(\pi/2) = 1.5349...$ We have $\mathcal{A}_p \leq \mathcal{L}_{(p-1)/2}$ but $\mathcal{L}_{(p-1)/2}$ and $\mathcal{T}$ are not comparable, as it follows from (12) as $(p-1)/2 < 1/3$.

Thus the lower optimal estimations of the mean $\mathcal{T}$ by power means and by Lehmer means are not comparable. The lower optimal estimations of the mean $\mathcal{L}$ by Lehmer means is better than that by power means as $\mathcal{A}_0 = \mathcal{G} = \mathcal{L}_{-1/2} < \mathcal{L}_{-1/3}$. In all the other cases the optimal estimations by power means are better than that by Lehmer means.

We can supplement the inequality (13) in the following:

Theorem 2. *The optimal estimation of the Neuman-Sándor mean $\mathcal{N}$ by Lehmer means is given by*

$$\mathcal{L}_0 < \mathcal{N} < \mathcal{L}_{1/6} \ . \tag{14}$$

Proof. The first part of (14) is also known from (1) as $\mathcal{L}_0 = \mathcal{A}$. We want now to prove that the estimation is optimal. We have to prove that for any $\varepsilon > 0$, the mean $\mathcal{N}$ is not greater than $\mathcal{L}_\varepsilon$. As

$$\lim_{t \to \infty} \frac{\mathcal{L}_\varepsilon(t, 1)}{\mathcal{N}(t, 1)} = \lim_{t \to \infty} \frac{t^{\varepsilon+1} + 1}{t^\varepsilon + 1} \cdot \frac{2 \sinh^{-1} \frac{t-1}{t+1}}{t - 1} = 2 \sinh^{-1} 1 = 2 \ln\left(1 + \sqrt{2}\right) > 1,$$

for t sufficiently large we have $\mathcal{L}_\varepsilon(t, 1) > \mathcal{N}(t, 1)$. □

We underline also similar results for the mean $\mathcal{J}$.

Theorem 3. *The optimal estimations of the Gini mean $\mathcal{J}$ by Lehmer means are given by*

$$\mathcal{L}_{1/2} < \mathcal{J} < \mathcal{L}_1 \ . \tag{15}$$

Proof. The results follow easily from Theorem 1 as $\mathcal{L}_p = \mathcal{S}_{p+1,p}$ and $\mathcal{J} = \mathcal{S}_{1,1}$.

$\square$

Remark 1. It follows that the estimations of the Gini mean $\mathcal{J}$ by Lehmer means are better than by power means. Indeed $\mathcal{A}_2 < \mathcal{L}_{1/2}$ and there is no finite index p such that $\mathcal{J} < \mathcal{A}_p$.

Remark 2. The optimal estimations by Lehmer means can be ordered only in two chains:

$$\mathcal{L}_{-1/3} < \mathcal{L} < \mathcal{L}_0 < \mathcal{N} < \mathcal{L}_{1/6} < \mathcal{L}_{1/2} < \mathcal{J} < \mathcal{L}_1 \ ,$$

and

$$\mathcal{L}_{-1/6} < \mathcal{P} < \mathcal{I} < \mathcal{L}_0 < \mathcal{T} < \mathcal{L}_{1/3} \ .$$

Of course we can write also a single chain

$$\mathcal{L}_{-1/3} < \mathcal{L} < \mathcal{P} < \mathcal{I} < \mathcal{L}_0 < \mathcal{N} < \mathcal{T} < \mathcal{L}_{1/3} < \mathcal{L}_{1/2} < \mathcal{J} < \mathcal{L}_1 \ ,$$

but losing the terms $\mathcal{L}_{-1/6}$ and $\mathcal{L}_{1/6}$ which are not comparable with $\mathcal{L}$ and $\mathcal{T}$ respectively. The means $\mathcal{L} < \mathcal{P} < \mathcal{I}$ and $\mathcal{N} < \mathcal{T}$ cannot be separated by Lehmer means.

3 Estimations by Other Special Gini Means

In [3] we have determined the optimal estimation of the logarithmic mean by the special Gini means

$$\mathcal{S}_{1/3} < \mathcal{L} < \mathcal{S}_1 \ . \tag{16}$$

In what follows we find optimal estimations for the means $\mathcal{P}, \mathcal{I}, \mathcal{N}$, and $\mathcal{T}$, by special Gini means $\mathcal{S}$. As $\mathcal{A}_p = \mathcal{S}_{p,0}$, $\mathcal{L}_q = \mathcal{S}_{p+1,p}$ and $\mathcal{S}_q = \mathcal{S}_{q-1,1}$ we can use Theorem 1 to obtain

Corollary 2. *If $p \geq 0$ we have:*

1. $\mathcal{A}_p \leq \mathcal{S}_q$ *if and only if $q \geq 1$ and $q \geq p$; and*
2. $\mathcal{S}_q \leq \mathcal{A}_p$ *if and only if $q \leq 1$ and $q \leq p$.*

Corollary 3. *We have:*

1. $\mathcal{S}_q \leq \mathcal{L}_p$ *if and only if $q \leq 2p+1$ and one of the following situations holds:*
   (i) $1 \leq q \leq 2, p \geq q-1$; or
   (ii) $q > 2, p \geq 1$; or
   (iii) $-1 < p < 0, q \leq (1+2p)/(1+p)$; and
2. $\mathcal{L}_p \leq \mathcal{S}_q$ *if and only if $q \geq 2p+1$ and one of the following situations holds:*
   (i) $1 \leq q \leq 2, 0 \leq p$; or
   (ii) $q > 2, 0 \leq p \leq 1$; or
   (iii) $-1 < p < 0, q \geq (1+2p)/(1+p)$; or
   (iv) $p \leq -1$.

These results will be used to deduce some estimations by special Gini means knowing the optimal estimations with power means and with Lehmer means.

Theorem 4. *The optimal estimation of the mean $\mathcal{P}$ by Gini means $\mathcal{S}_q$ is given by*

$$\mathcal{S}_{2/3} < \mathcal{P} < \mathcal{S}_1 . \tag{17}$$

Proof. The first inequality of (17) follows from (11) and the inequality $\mathcal{S}_{2/3} < \mathcal{L}_{-1/6}$ which can be proved using Corollary 3. We want to prove that the above estimation is optimal, thus for every $\varepsilon > 0$, $\mathcal{S}_{(2+\varepsilon)/3}$ is not less than $\mathcal{P}$. This happens if there is a $t > 1$ such that $\mathcal{S}_{(2+\varepsilon)/3}(t, 1) > \mathcal{P}(t, 1)$, that is

$$\left(\frac{t+1}{t^{(\varepsilon-1)/3} + 1} \right)^{\frac{3}{4-\varepsilon}} > \frac{t-1}{2 \sin^{-1} \frac{t-1}{t+1}},$$

or, equivalently

$$f(t, \varepsilon) = \sin^{-1} \frac{t-1}{t+1} - \frac{t-1}{2} \left(\frac{t^{(\varepsilon-1)/3} + 1}{t+1} \right)^{\frac{3}{4-\varepsilon}} > 0 .$$

This follows from the Taylor series expansion

$$f(t, \varepsilon) = \frac{\varepsilon}{48} (t-1)^3 + O\left((t-1)^4 \right).$$

Plotting the function $f(t, \varepsilon)$ for $t \in (1.001, 1.01)$ and $\varepsilon \in (10^{-3}, 10^{-5})$, we obtain:

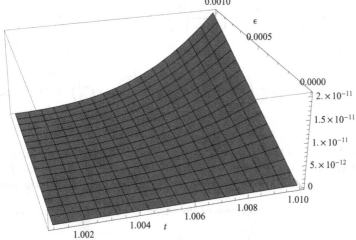

If we plot the function f for $\varepsilon = 10^{-4}$, we obtain:

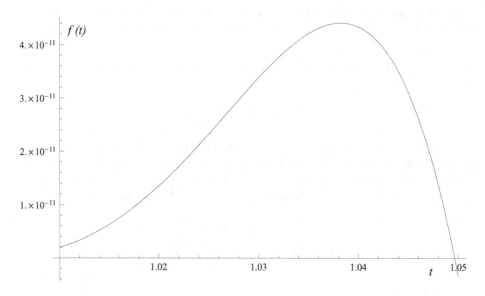

so

$$f(t, 10^{-4}) > 0 \text{ for } t \in (1.01, 1.048) .$$

The second estimation is known from (1) as $\mathcal{S}_1 = \mathcal{A}$. To prove its optimality we have to show that for every $\varepsilon > 0$, $\mathcal{S}_{1-\varepsilon}$ is not greater than $\mathcal{P}$, thus there is a $t > 1$ such that $\mathcal{S}_{1-\varepsilon}(t, 1) < \mathcal{P}(t, 1)$, or

$$\left(\frac{t+1}{t^{-\varepsilon}+1}\right)^{\frac{1}{1+\varepsilon}} < \frac{t-1}{2\sin^{-1}\frac{t-1}{t+1}} .$$

This is equivalent with the condition that the function

$$g(t, \varepsilon) = \sin^{-1}\frac{t-1}{t+1} - \frac{t-1}{2}\left(\frac{t^{-\varepsilon}+1}{t+1}\right)^{\frac{1}{1+\varepsilon}} < 0, \text{ for some } t > 1 .$$

This follows from

$$\lim_{t\to\infty} g(t, \varepsilon) = \sin^{-1}1 - \frac{1}{2}\lim_{t\to\infty}\left(\frac{t}{t+1}\cdot\left(t^{-\varepsilon}+1\right)\right)^{\frac{1}{1+\varepsilon}}\left(1-t^{-1}\right)\cdot t^{\frac{\varepsilon}{1+\varepsilon}} = -\infty .$$

$\square$

Remark 3. We have used the computer algebra system Mathematica to get the Taylor series expansion of g, the plots and the interval of positivity for f. We also used Mathematica in the same way for proving other results from the paper.

Theorem 5. *The optimal estimation of the mean $\mathcal{I}$ by Gini means $\mathcal{S}_q$ is given by*

$$\mathcal{S}_{2/3} < \mathcal{I} < \mathcal{S}_1 . \tag{18}$$

Proof. The first inequality follows from (17) and (1). The second inequality from (18) is known from (1) as $\mathcal{S}_1 = \mathcal{A}$. Their optimality follows from [7], as $\mathcal{I}$ is a Stolarsky mean.   □

Theorem 6. *Optimal estimation of the mean $\mathcal{N}$ by Gini means $\mathcal{S}_q$ is given by*

$$\mathcal{S}_1 < \mathcal{N} < \mathcal{S}_{4/3} \ . \tag{19}$$

Proof. The first inequality follows again from (1) as $\mathcal{S}_1 = \mathcal{A}$. To prove that it is optimal, we have to show that for every $\varepsilon > 0$, $\mathcal{S}_{1+\varepsilon}$ is not less than $\mathcal{N}$, thus there is a $t > 1$ such that $\mathcal{S}_{1+\varepsilon}(t,1) > \mathcal{N}(t,1)$, or

$$\left(\frac{t+1}{t^\varepsilon + 1} \right)^{\frac{1}{1-\varepsilon}} > \frac{t-1}{2 \sinh^{-1} \frac{t-1}{t+1}} \ .$$

This is equivalent with the condition that the function

$$f(t,\varepsilon) = \sinh^{-1} \frac{t-1}{t+1} - \frac{t-1}{2} \left(\frac{t^\varepsilon + 1}{t+1} \right)^{\frac{1}{1-\varepsilon}} > 0, \text{ for some } t > 1 \ .$$

Or

$$\lim_{t \to \infty} f(t,\varepsilon) = \sinh^{-1} 1 - \frac{1}{2} \lim_{t \to \infty} \frac{t-1}{t+1} \left(\frac{t^\varepsilon + 1}{(t+1)^\varepsilon} \right)^{\frac{1}{1-\varepsilon}} = \ln \left(1 + \sqrt{2} \right) - 0.5 = 0.38...$$

The second estimation from (19) follows from (7) and Corollary 2. To prove its optimality we have to show that for every $\varepsilon > 0$, $\mathcal{S}_{(4-\varepsilon)/3}$ is not greater than $\mathcal{N}$, thus there is a $t > 1$ such that $\mathcal{S}_{(4-\varepsilon)/3}(t,1) < \mathcal{N}(t,1)$, or

$$\left(\frac{t+1}{t^{\frac{1-\varepsilon}{3}} + 1} \right)^{\frac{3}{2+\varepsilon}} < \frac{t-1}{2 \sinh^{-1} \frac{t-1}{t+1}} \ .$$

This is equivalent with the condition that the function

$$g(t,\varepsilon) = \frac{t-1}{2} \left(\frac{t^{\frac{1-\varepsilon}{3}} + 1}{t+1} \right)^{\frac{3}{2+\varepsilon}} - \sinh^{-1} \frac{t-1}{t+1} > 0, \text{ for some } t > 1 \ .$$

This follows from the Taylor formula

$$g(t,\varepsilon) = \frac{\varepsilon}{48} (t-1)^3 + O\left((t-1)^4 \right) \ .$$

Plotting the function $g(t,\varepsilon)$ for $t \in (1.001, 1.01)$ and $\varepsilon \in (10^{-3}, 10^{-5})$, we obtain:

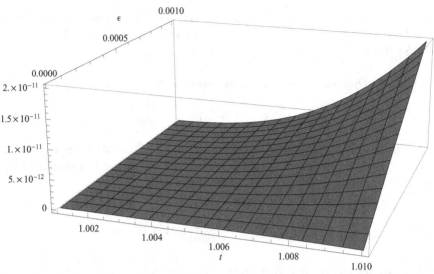

If we plot the function g for $\varepsilon = 10^{-5}$, we obtain:

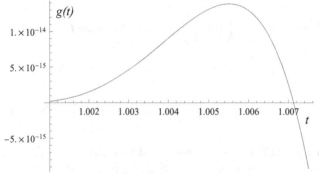

so, for instance

$$g(t, 10^{-5}) > 0 \text{ for } t \in (1.001, 1.007) .$$

$\square$

Theorem 7. *The optimal estimation of the mean $\mathcal{T}$ by Gini means $\mathcal{S}_q$ is given by*

$$\mathcal{S}_1 < \mathcal{T} < \mathcal{S}_{5/3} . \tag{20}$$

Proof. The first inequality follows again from (1) as $\mathcal{S}_1 = \mathcal{A}$. To prove that it is optimal, we have to show that for every $\varepsilon > 0$, $\mathcal{S}_{1+\varepsilon}$ is not less than $\mathcal{T}$, thus there is a $t > 1$ such that $\mathcal{S}_{1+\varepsilon}(t, 1) > \mathcal{T}(t, 1)$, or

$$\left(\frac{t+1}{t^\varepsilon + 1} \right)^{\frac{1}{1-\varepsilon}} > \frac{t-1}{2 \tan^{-1} \frac{t-1}{t+1}} .$$

This is equivalent with the condition that the function

$$f(t, \varepsilon) = \tan^{-1} \frac{t-1}{t+1} - \frac{t-1}{2} \left(\frac{t^\varepsilon + 1}{t+1} \right)^{\frac{1}{1-\varepsilon}} > 0, \text{ for some } t > 1 .$$

Or

$$\lim_{t\to\infty} f(t,\varepsilon) = \tan^{-1} 1 - \frac{1}{2} \lim_{t\to\infty} \frac{t-1}{t+1} \left(\frac{t^{\varepsilon}+1}{(t+1)^{\varepsilon}} \right)^{\frac{1}{1-\varepsilon}} = \frac{\pi}{4} - \frac{1}{2} > 0 \; .$$

The second inequality in (20) follows from (6) and Corollary 2. To prove its optimality we have to show that for every $\varepsilon > 0$, $\mathcal{S}_{(5-\varepsilon)/3}$ is not greater than $\mathcal{T}$, thus there is a $t > 1$ such that $\mathcal{S}_{(5-\varepsilon)/3}(t,1) < \mathcal{T}(t,1)$, or

$$\left(\frac{t+1}{t^{\frac{2-\varepsilon}{3}}+1} \right)^{\frac{3}{1+\varepsilon}} < \frac{t-1}{2 \tan^{-1} \frac{t-1}{t+1}} \; .$$

This is equivalent with the condition that the function

$$g(t,\varepsilon) = \frac{t-1}{2} \left(\frac{t^{\frac{2-\varepsilon}{3}}+1}{t+1} \right)^{\frac{3}{1+\varepsilon}} - \tan^{-1} \frac{t-1}{t+1} > 0, \; \text{for some } t > 1 \; .$$

This follows from the Taylor formula

$$g(t,\varepsilon) = \frac{\varepsilon}{48} (t-1)^3 + O\left((t-1)^4 \right) \; .$$

Plotting the function $g(t,\varepsilon)$ for $t \in (1.001, 1.007)$ and $\varepsilon \in (10^{-3}, 10^{-5})$, we obtain:

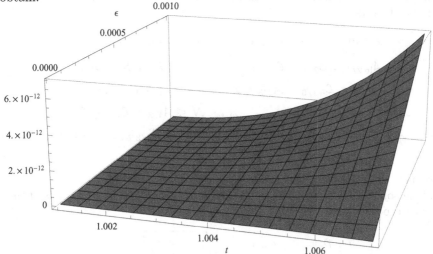

If we plot the function g for $\varepsilon = 10^{-5}$, we obtain:

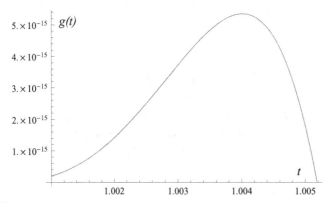

For instance

$$g(t, 10^{-5}) > 0 \text{ for } t \in (1.001, 1.00518...) \ .$$

□

Remark 4. As $\mathcal{J} = \mathcal{S}_2$, we need no estimations for it. We can summarize the results of this chapter by the following two chains of optimal inequalities

$$\mathcal{S}_{1/3} < \mathcal{L} < \mathcal{S}_1 < \mathcal{N} < \mathcal{S}_{4/3} \ ,$$

and

$$\mathcal{S}_{2/3} < \mathcal{P} < \mathcal{I} < \mathcal{S}_1 < \mathcal{N} < \mathcal{T} < \mathcal{S}_{5/3} < \mathcal{J} = \mathcal{S}_2 \ .$$

Let us remark that the pairs of means $\mathcal{L} < \mathcal{P}$, $\mathcal{P} < \mathcal{I}$ and $\mathcal{N} < \mathcal{T}$ cannot be separated by special Gini means.

4   Applications

Using again the corollaries 1, 2 and 3, the above optimal estimations can be ordered as follows:

$$\mathcal{G} = \mathcal{A}_0 < \mathcal{S}_{1/3} < \mathcal{L}_{-1/3} < \mathcal{L} < \mathcal{A}_{1/3} < \mathcal{L}_0 = \mathcal{S}_1 = \mathcal{A} \ ,$$

$$\mathcal{A}_{\ln 2/\ln \pi}, \mathcal{S}_{2/3} < \mathcal{L}_{-1/6} < \mathcal{P} < \mathcal{A}_{2/3} < \mathcal{L}_0 = \mathcal{S}_1 = \mathcal{A} \ ,$$

$$\mathcal{S}_{2/3} < \mathcal{L}_{-1/6} < \mathcal{A}_{2/3} < \mathcal{I} < \mathcal{A}_{\ln 2} < \mathcal{L}_0 = \mathcal{S}_1 = \mathcal{A} \ ,$$

$$\mathcal{A} = \mathcal{L}_0 = \mathcal{S}_1 < \mathcal{A}_{\ln 2/\ln(\ln(3+2\sqrt{2}))} < \mathcal{N} < \mathcal{A}_{4/3} < \mathcal{L}_{1/6} < \mathcal{S}_{4/3} \ ,$$

$$\mathcal{A} = \mathcal{L}_0 = \mathcal{S}_1 < \mathcal{A}_{\ln 2/\ln(\pi/2)} < \mathcal{T} < \mathcal{A}_{5/3} < \mathcal{L}_{1/3} < \mathcal{S}_{5/3} \ , \text{ and}$$

$$\mathcal{A}_2 < \mathcal{L}_{1/2} < \mathcal{S}_2 = \mathcal{J} = \mathcal{S}_2 < \mathcal{L}_1 < \mathcal{A}_\infty \ .$$

We remark that $\mathcal{A}_{\ln 2/\ln \pi}$ is not comparable with $\mathcal{S}_{2/3}$ and $\mathcal{L}_{-1/6}$.

Taking some of these inequalities in the point $(t, 1)$, we get estimations of the following type,

$$\frac{t + t^{1/6}}{1 + t^{1/6}} < \frac{t - 1}{\sin^{-1}\frac{t-1}{t+1}} < \sqrt{\frac{\left(t^{2/3} + 1\right)^3}{2}}, t \neq 1 \tag{21}$$

or

$$t - \sqrt{t} + 1 < t^{\frac{t}{t+1}} < \frac{t^2 + 1}{t + 1}, t \neq 1 \ . \tag{22}$$

But we also have:

$$\mathcal{P}(1 + \sin t, 1 - \sin t) = \frac{\sin t}{t} \ ,$$

$$\mathcal{T}(1 + \tan t, 1 - \tan t) = \frac{\tan t}{t} \ ,$$

$$\mathcal{N}(1 + \sinh t, 1 - \sinh t) = \frac{\sinh t}{t} \ ,$$

$$\mathcal{L}(e^t, e^{-t}) = \frac{\sinh t}{t} \ , \quad \text{and}$$

$$\mathcal{I}(e^t, e^{-t}) = e^{\frac{t}{\tan t} - 1} \ ,$$

so that we can obtain some other trigonometric and hyperbolic inequalities (see also [15] for many other inequalities of these types).

References

1. Alzer, H.: Best possible estimates for special mean values. Prirod.-Mat. Fak. Ser. Mat. 23, 331–346 (1993) (German)
2. Chu, Y.-M., Long, B.-Y.: Bounds of the Neuman-Sándor mean using power and identric mean. Abstract Appl. Anal. Article ID 832591, 6 pages (2013)
3. Costin, I., Toader, G.: Some optimal evaluations of the logarithmic means. Aut. Comput. Appl. Math. 22, 103–112 (2013)
4. Costin, I., Toader, G.: A nice separation of some Seiffert-type means by power means. Int. J. Math. Math. Sc., Article ID 430692, 6 pages (2012)
5. Costin, I., Toader, G.: A separation of some Seiffert-type means by power means. Rev Anal. Num. Th. Approx. 41, 125–129 (2012)
6. Costin, I., Toader, G.: Optimal evaluations of some Seiffert-type means by power means. Appl. Math. Comput. 219, 4745–4754 (2013)
7. Czinder, P., Páles, Z.: Some comparison inequalities for Gini and Stolarsky means. Math. Inequal. Appl. 9, 607–616 (2006)
8. Hästö, P.A.: Optimal inequalities between Seiffert's means and power means. Math. Inequal. Appl. 7, 47–53 (2004)
9. Karamata, J.: Sur quelques problèmes posés par Ramanujan. J. Indian Math. Soc. (N. S.) 24, 343–365 (1960)
10. Lin, T.P.: The power and the logarithmic mean. Amer. Math. Monthly 81, 879–883 (1974)
11. Neuman, E., Sándor, J.: On the Schwab-Borchardt mean. Math. Panon. 14, 253–266 (2003)
12. Neuman, E., Sándor, J.: Comparison inequalities for certain bivariate means. Appl. Anal. Discrete Math. 3, 46–51 (2009)
13. Páles, Z.: Inequalities for sums of powers. J. Math. Anal. Appl. 131, 271–281 (1988)
14. Pittenger, A.O.: Inequalities between arithmetic and logarithmic means. Univ. Beograd. Publ. Elektrotehn. Fak. Ser. Mat. Fiz. 678–715, 15–18 (1980)

15. Sándor, J.: Trigonometric and hyperbolic inequalities arXiv:1105.0859v1 [math.CA], http://arxiv.org/abs/1105.0859
16. Seiffert, H.-J.: Problem 887. Niew Arch. Wisk (Ser. 4) 11, 176–176 (1993)
17. Seiffert, H.: Aufgabe β16. Die Wurzel 29, 221–222 (1995)
18. Yang, Z.: Sharp bounds for the second Seiffert mean in terms of power means, http://arxiv.org/pdf/1206.5494v1.pdf
19. Yang, Z.: Sharp power means bounds for Neuman-Sándor mean, http://arxiv.org/pdf/1208.0895v1.pdf
20. Yang, Z.-H.: Estimates for Neuman-Sándor mean by power means and their relative errors. J. Math. Inequal. 7, 711–726 (2013)
21. Wang, M.-K., Qiu, Y.-F., Chu, Y.-M.: Sharp bounds for Seiffert means in terms of Lehmer means. J. Math. Inequal. 4, 581–586 (2010)

CAS Application to the Construction of High-Order Difference Schemes for Solving Poisson Equation⋆

Grigoriy M. Drozdov[1] and Vasily P. Shapeev[2]

[1] Novosibirsk State University, Novosibirsk, Russia
drozdovgrigoriy@gmail.com
[2] Novosibirsk State University, Novosibirsk, Russia
Khristianovich Institute of Theoretical and Applied Mechanics,
Russian Academy of Sciences, Novosibirsk, Russia
shapeev@itam.nsc.ru

Abstract. In the present work, a computer algebra system (CAS) is applied for constructing a new difference scheme of high-accuracy order for solving boundary-value problem for Poisson equation. The formulae of the difference scheme are constructed in the symbol form in CAS. CAS are used for translation of complex formulae to C++ language operators, calculation of arithmetic values of the constructed scheme coefficients and matrix elements of a system of linear algebraic equations for the discrete problem approximating the initial difference problem. Efficiency of the CAS application and of the schemes constructed with its help is shown.

1 Introduction

Search for the new high-accuracy methods for solving boundary-value problems for partial differential equations is one of the research directions of computational mathematics [1–3, 6, 7, 9–12]. It cannot be a substitute for the direction of multisequencing numerical solutions on supercomputers, but it can essentially complement it because the high-accuracy methods, being realized on computers, possess often certain advantages over the low-accuracy ones. E.g., when a certain high accuracy is to be achieved in a problem solution with limited computer resources, a high-accuracy method is often more efficient than a low-accuracy one. An obvious example is the advantage of Runge–Kutta methods over the Euler one in solving ordinary differential equations. There are many other examples. Besides, the multisequencing does not eliminate the rounding errors, while achieving an acceptable result by a low-accuracy method may require a large number of arithmetic operations, which leads to accumulation of a large rounding error. The latter makes often the numerical result obtained by this method unacceptable. One of the problem requiring application of high-accuracy methods, the direct numerical simulation of instability of gas jets acoustics, is studied in [10]. Another problem is the known test problem on movement

---

⋆ The work was supported by the Russian Foundation for Basic Research (grant No. 13-01-00277).

V.P. Gerdt et al. (Eds.): CASC Workshop 2014, LNCS 8660, pp. 99–110, 2014.

of viscous fluid in 2D and 3D lid-driven cavities. The most accurate solutions of this problem are obtained by the methods of high accuracy. In a number of works, the methods of high accuracy are used to obtain solutions coinciding in the values of the solution characteristic parameters with accuracy up to 10^{-7} – 10^{-8} [1, 2, 7, 11]. The constructed high-accuracy solutions of this problems reproduce the vortices thin structure. Many researchers note that the problems of direct numerical simulation of turbulence (DNS) [8], methods of large-eddy simulation (LES), problems of computational aeroacoustics require high accuracy and resolution of fine scales. Otherwise, the characteristics of the objects under investigation reproduced by the solution will not agree with the reality. Science history confirms the thesis that its progress, beside different circumstances, is defined by accuracy of the measurements and computations. For example, it is the high accuracy of calculations, that primarily allowed Leverrier to discover planet Neptune in 1846. The list of problems studied in the past, being solved at present and requiring application of high-accuracy algorithms, no doubt, can be continued.

The construction of new high-accuracy algorithms faces obvious difficulties. The equations and their derivation are more complicated than in the case of low-accuracy methods. Derivation of equations and study of properties of the high-accuracy numerical methods can be effectively assisted by the systems of computer algebra (CAS) [4,5,13–16]. We consider here the application of CAS for derivation of difference schemes with high order of accuracy of solving boundary-value problems for Poisson equation, which is important in the theory of electric and magnetic fields, theory of minimal surfaces. The latter is widely used in construction of individual critical parts of technical devices, optimal in their volumes.

The method of undetermined coefficients used here is the most general method for construction of difference schemes. Its application shows if there is the scheme of a given order for a given equation and mesh stencil. Moreover, if the scheme does exist, it provides the arbitrary parameters in the scheme's formulae, i.e., it shows the arbitrariness with which one can define the scheme formulae on a given stencil. It neither causes any additional troubles at construction of schemes for equations with variable coefficients.

2   Problem Statement and Description of the Method

The boundary-value problem with Dirichlet conditions for Poisson equation in region $\Omega = [0,1] \times [0,1]$ is considered:

$$\begin{cases} \dfrac{\partial^2 u}{\partial x^2} + \dfrac{\partial^2 u}{\partial y^2} = f, & (x,y) \in \Omega, \\ u|_{\partial \Omega} = g, \end{cases} \tag{1}$$

where $f(x,y)$ and $g(x,y)$ are given functions with required smoothness, $u(x,y)$ is the solution sought. Poisson equation describes electrostatic field, stationary temperature field, pressure field, velocity potential field in hydrodynamics. Efficient methods for solving boundary-value problems for Poisson equation are important for physics. Computational region Ω is covered by uniform grid with nodes (x_i, y_j), where $x_i = ih, i = 0, 1, \ldots, N, y_j = jh, j = 0, 1, \ldots, N, h = \frac{1}{N}$.

We will describe the construction of the mth order difference scheme on n-point stencil for numerical solution of problem (1) by the method of undetermined coefficients. We assume that all necessary derivatives of the solution with respect to variables x and y up to the mth order exist and are continuous. Difference equation approximating differential equation (1) is sought in the form:

$$\sum_{i=0}^{n-1} c_i u_i - \phi(x, y) = 0, \tag{2}$$

where u_i are values of sought difference solution of problem (1) in stencil nodes $(i = 0, \ldots, n-1)$, c_i are sought undetermined coefficients, i is the number of points in the stencil, $\phi(x, y)$ is the nonhomogeneous term of the difference scheme. Stencils of schemes considered in this work are demonstrated in Fig. 1. At first we expand all u_i from the left-hand side of (2) into the truncated Taylor series in the stencil central node (denote this point (x_0, y_0)) taking into account specific increments of arguments x, y with respect to u_0:

$$u(x, y) = \sum_{l+m \leq N} \frac{1}{l! m!} \frac{\partial^{l+m} u(x_0, y_0)}{\partial x^l \partial y^m} (x - x_0)^l (y - y_0)^m + O(h^N). \tag{3}$$

We denote the obtained sum by Σ_1. Then, we differentiate with respect to variables x and y equation (1) resolved with respect to derivative u_{xx}. Then, we eliminate step by step derivatives $\partial^{l+m} u(x_0, y_0)/\partial x^l \partial y^m$ ($l = 2, 3, \ldots, m = 0, 1, \ldots, (l, m) \neq (2, 0)$) using relations obtained. We denote the new sum obtained by Σ_2. In order to find the nonhomogeneous term $\phi(x, y)$, we collect in Σ_2 free (from values of solution $u(x, y)$ and its derivatives) term, which is a linear combination of $f(x_0, y_0)$ and different derivatives of $f(x, y)$, and equate the obtained expression to $\phi(x, y)$. Next, we construct a system to determine c_i on the strength of requirement of difference equation (2) to approximate differential equation (1) on the chosen stencil with the highest possible order. For this purpose, we collect coefficients of u_{xx} and u_{yy} in Σ_2 and equate them to 1 in order to approximate initial equation (1). Then we equate step by step the coefficients of higher-order derivatives of u in Σ_2 to zero, beginning from the least order and until number of linearly independent equations become equal to n, which is the number of unknown c_i. The equalities obtained form the system of linear algebraic equations (SLAE) for coefficients c_i.

A posteriori we establish that in our case coefficients c_i have the order $O(h^{-2})$. Therefore, if we managed to nullify all derivatives of order less than N in the residual

$$(u_{xx} + u_{yy} - f(x, y)) - (\sum_{i=0}^{n-1} c_i u_i - \phi(x, y)) = O(h^m), \tag{4}$$

then m is equal to $N - 2$. Obviously, the more n, one can obtain the more additional equations for determining c_i in such a way, and the more m. In this work, the question was studied, what is the maximal m, which we can achieve at given n. In particular, on a stencil with $n = 13$ points, we managed to construct a scheme of the 10th approximation order ($m = 10$) using the described method, on the stencil with $n = 25$ points – scheme with $m = 18$ order, with $n = 37$ points – 26th order scheme, with $n = 49$ points – 34th order.

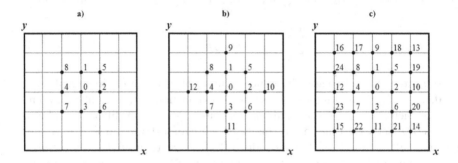

Fig. 1. Multipoint stencils for schemes with: a) 6th order, b) 10th order and c) 18th order

A special attention should be paid to question of approximation for nonho-mogeneous term $\phi(x, y)$, which is a linear combination of different derivatives of $f(x, y)$ and function $f(x, y)$ itself. As a rule, on n-point stencil in case of high approximation order schemes, there is no linear combination of values of function f, approximating $\phi(x, y)$ with order m. Therefore, we have to add k new points to n-point stencil that do not coincide with grid nodes. In this work, additional points for every specific stencil were taken inside the stencil between the grid nodes. Number k is chosen as the least from those for which there exists the set of coefficients $a_i, (i = 0, ..., n + k - 1)$ such that

$$\phi(x, y) = \sum_{i=0}^{n+k-1} a_i f_i + O(h^m). \tag{5}$$

We find coefficients a_i using the above-described method of undetermined coeffi-cients on supplemented stencil. In general, this algorithm allows to approximate any linear combination of derivatives of unknown function. The algorithm of writ-ing the SLAE with respect to coefficients a_i is nearly the same as we described above in the case of coefficients c_i in approximating Poisson equation. Approximation of $f(x, y)$ is sought in form (5). We expand all f_i in the right-hand side of (5) into the truncated Taylor series in the central node. Then, we equate coefficient affect-ing $f(x, y)$ with 1 and nullify all other coefficients. The obtained equations form a

SLAE with respect to sought coefficients a_i. For example, for the 6th order scheme we have obtained the following nonhomogeneous term:

$$\phi(x,y) = f(x,y) + \frac{h^2}{12}\left(f^{(2,0)}(x,y) + f^{(0,2)}(x,y)\right) +$$

$$+\frac{h^4}{360}\left(f^{(4,0)}(x,y) + 4f^{(2,2)}(x,y) + f^{(4,0)}(x,y)\right).$$

It contains the derivatives of $f(x,y)$ up to the 4th order. If we seek approximation for $\phi(x,y)$ in form (5) with $k = 0$, then SLAE is underdetermined. And if we substitute the found coefficients in (5), the power of h in residual term is less than m. But if we supplement the stencil by $k = 4$ additional points, placing it symmetrically in regard to the stencil central node, the SLAE is fully determined, and we obtain an approximation for $\phi(x,y)$ with the necessary order. Note that in our case function $f(x,y)$ does not depend on u and specified in all region Ω, and, therefore, stencil supplementation by additional nodes for approximation $\phi(x,y)$ does not introduce significant changes into the algorithm for solving the problem. Clearly, with increase of the approximation order, which is to be achieved, the number of additional points increase, since in $\phi(x,y)$ more different derivatives of function f appear. For example, for the 6th order scheme, $k = 4$ additional points were introduced, for 10th order scheme it was $k = 16$, for 18th order – $k = 188$.

3 CAS Application

Method of undetermined coefficients of constructing difference schemes applied in this work requires a lot of symbolic operations. In simple problems, they often can be performed without computers. However, in this work they represent a serious difficulty for researcher. CAS allow us to handle such sort of difficulties to save time of researcher and to avoid many mistakes usually made at derivation of formulae, what was proved in works [4,5,13–16].

The algorithm for constructing difference scheme using the method of undetermined coefficients was realized in CAS *Mathematica*. Operator Series [...] in this system allows to expand the function of several variables in the Taylor series up to the required order. This operator was applied here for expanding the left-hand side of (2), as it was described in the previous paragraph. In order to derive differential consequences of equation (1), we differentiate it with respect to variables x and y using the operator Derivative[...] and write obtained equations to system by operator Table[...]. Via operator Solve[...], which is used for solving algebraic equations, we express lower derivatives through higher ones, and substitute them into expansion of the left-hand side of equation (2) in Taylor series. In obtained expression, we collect coefficients of solution derivatives using operator Coefficient[...]. Equating coefficients of u_{xx} and u_{yy} to 1, and coefficients of higher derivatives to 0, we obtain SLAE for sought coefficients of difference scheme. In the code, the SLAE constructed is solved symbolically using again the operator Solve[...]. It is useful to simplify the formulae of scheme coefficients obtained from the SLAE solution to group them and decrease the number of arithmetic operations for their computation. For this purpose, operators `Simplify[...]` and `FullSimplify[...]` are used. Then, using operator `CForm[...]`, we translate the expressions obtained to form the code required for substituting into the program in C++ language for numerical solving of the problem using the difference scheme obtained.

In this work, facilities of CAS *Mathematica* were used for graphical representation of results of numerical solving with the help of operators `Plot[...]` and `Plot3D[...]`, as well as constructing images of difference schemes stencils (Figure 1) and exact solutions graphs (Figure 2,3). Also, it is important, that at deriving new formulae, CAS allows us to facilitate verification of the formulae obtained. For example, in this work, the numerical solution convergence obtained using the constructed difference scheme was preliminarily verified as follows. Solution value in any grid node, in our case, at the center point of computational region, was taken as unknown, and in surrounding grid nodes, the exact solution of equation (2) was specified, and then the obtained difference scheme was applied. Then, the grid step was halved, and the same procedure was applied. Based on the sequence of obtained solution errors in a chosen node, we can preliminarily judge about convergence of solution obtained by the new difference scheme. Moreover, in case of coincidence of convergence order of numerical solution with approximation order of the difference scheme this is a reliable criterion of faultlessness of the constructed formulae.

In the modern conditions of fast development, power and availability of CAS, they show themselves as a universal and effective instrument for creating new algorithms, new numerical methods, in particular, algorithms for the numerical solution of differential equations.

4   Schemes Formulae

Sixth-order approximation for Laplacian, which is the left-hand side of Poisson equation, is obtained relatively easily and looks simple. It is more difficult to obtain the sixth-order approximation for nonhomogeneous term of Poisson equation. We present here all formulae for schemes with 6th, 10th, and 18th approximation order of Poisson equation on grid with square cells. The 6th order scheme on 9-point stencil has the form:

$$c_0 u_{ij} + c_1(u_{i,j+1} + u_{i+1,j} + u_{i,j-1} + u_{i-1,j})+$$

$$+c_2(u_{i+1,j+1} + u_{i+1,j-1} + u_{i-1,j-1} + u_{i-1,j+1}) = \phi(x,y),$$

$$c_0 = -\frac{10}{3h^2}, \quad c_1 = \frac{2}{3h^2}, \quad c_2 = \frac{1}{6h^2},$$

$$\phi(x,y) = f(x,y) + \frac{h^2}{12}\left(f^{(2,0)}(x,y) + f^{(0,2)}(x,y)\right) +$$

$$+\frac{h^4}{360}\left(f^{(4,0)}(x,y) + 4f^{(2,2)}(x,y) + f^{(4,0)}(x,y)\right).$$

Approximation for nonhomogeneous term $\phi(x,y)$ of the 6th order scheme looks as follows:

$$\phi(x,y) = a_0 f_{ij} + a_1\left(f_{i,j+1} + f_{i+1,j} + f_{i,j-1} + f_{i-1,j}\right) + a_2\left(f_{i+1,j+1} + f_{i+1,j-1} + \right.$$

$$\left. + f_{i-1,j-1} + f_{i-1,j+1}\right) + a_3\left(f_{i+\frac{1}{2},j+\frac{1}{2}} + f_{i+\frac{1}{2},j-\frac{1}{2}} + f_{i-\frac{1}{2},j-\frac{1}{2}} + f_{i-\frac{1}{2},j+\frac{1}{2}}\right),$$

$$a_0 = \frac{37}{90}, \quad a_1 = \frac{1}{90}, \quad a_2 = \frac{1}{360}, \quad a_3 = \frac{2}{15}.$$

Residual term of the 6th order scheme is written in form:

$$R(h) = -\frac{h^6}{60480}\left(3u^{(8,0)}(x,y)+\right.$$

$$\left. +28u^{(6,2)}(x,y) + 70u^{(4,4)}(x,y) + 28u^{(2,6)}(x,y) + 3u^{(0,8)}(x,y)\right).$$

Here are 10th order scheme formulae on 13-point stencil:

$$c_0 u_{ij} + c_1\left(u_{i,j+1} + u_{i+1,j} + u_{i,j-1} + u_{i-1,j}\right) + c_2\left(u_{i+1,j+1} + u_{i+1,j-1}+\right.$$

$$\left. + u_{i-1,j-1} + u_{i-1,j+1}\right) + c_3\left(u_{i,j+2} + u_{i+2,j} + u_{i,j-2} + u_{i-2,j}\right) = \phi(x,y),$$

$$c_0 = -\frac{25}{7h^2}, \quad c_1 = \frac{16}{21h^2}, \quad c_2 = \frac{1}{7h^2}, \quad c_3 = -\frac{1}{84h^2},$$

$$\phi(x,y) = f(x,y) + \frac{h^2}{14}\left(f^{(2,0)}(x,y) + f^{(0,2)}(x,y)\right) +$$

$$+\frac{h^4}{1260}\left(f^{(4,0)}(x,y) + 14f^{(2,2)}(x,y) + f^{(4,0)}(x,y)\right) -$$

$$-\frac{h^6}{10080}\left(f^{(6,0)}(x,y) - 5\left(f^{(2,4)}(x,y) + f^{(4,2)}(x,y)\right) + f^{(0,6)}(x,y)\right) -$$

$$-\frac{h^8}{2116800}\left(13f^{(8,0)}(x,y) - 28\left(f^{(2,6)}(x,y) + f^{(6,2)}(x,y)\right) + 13f^{(0,8)}(x,y)\right).$$

Approximation for the nonhomogeneous term $\phi(x,y)$ of the 10th order scheme can be written in form:

$$\phi(x,y) = a_0 f_{ij} + a_1\left(f_{i,j+1} + f_{i+1,j} + f_{i,j-1} + f_{i-1,j}\right) + a_2\left(f_{i+1,j+1}+\right.$$

$$f_{i+1,j-1} + f_{i-1,j-1} + f_{i-1,j+1}\right) + a_3\left(f_{i,j+2} + f_{i+2,j} + f_{i,j-2} + f_{i-2,j}\right) +$$

$$+ a_4\left(f_{i+\frac{1}{2},j+\frac{1}{2}} + f_{i+\frac{1}{2},j-\frac{1}{2}} + f_{i-\frac{1}{2},j-\frac{1}{2}} + f_{i-\frac{1}{2},j+\frac{1}{2}}\right) + a_5\left(f_{i,j+\frac{1}{2}} + f_{i+\frac{1}{2},j}+\right.$$

$$+ f_{i,j-\frac{1}{2}} + f_{i-\frac{1}{2},j}\right) + a_6\left(f_{i,j+\frac{3}{2}} + f_{i+\frac{3}{2},j} + f_{i,j-\frac{3}{2}} + f_{i-\frac{3}{2},j}\right) + a_7\left(f_{i+\frac{1}{4},j+\frac{1}{4}}+\right.$$

$$+ f_{i+\frac{1}{4},j-\frac{1}{4}} + f_{i-\frac{1}{4},j-\frac{1}{4}} + f_{i-\frac{1}{4},j+\frac{1}{4}}\right) + a_8\left(f_{i-\frac{1}{2},j+1} + f_{i+\frac{1}{2},j+1} + f_{i+1,j+\frac{1}{2}}+\right.$$

$$+ f_{i+1,j-\frac{1}{2}} + f_{i+\frac{1}{2},j-1} + f_{i-\frac{1}{2},j-1} + f_{i-1,j-\frac{1}{2}} + f_{i-1,j+\frac{1}{2}}\right),$$

$$a_0 = \frac{83}{9450}, \quad a_1 = -\frac{22}{14175}, \quad a_2 = \frac{2}{2025}, \quad a_3 = -\frac{29}{793800}, \quad a_4 = \frac{632}{14175},$$

$$a_5 = \frac{536}{14175}, \quad a_6 = -\frac{184}{33075}, \quad a_7 = \frac{2048}{14175}, \quad a_8 = \frac{64}{4725}.$$

The residual term of the 10th order scheme looks like the following:

$$R(h) = \frac{h^{10}}{838252800}\left(167u^{(0,12)}(x,y) - 66u^{(2,10)}(x,y) - 495u^{(4,8)}(x,y)-\right.$$

$$\left. -924u^{(6,6)}(x,y) - 495u^{(8,4)}(x,y) - 66u^{(10,2)}(x,y) + 167u^{(12,0)}(x,y)\right).$$

Here are the 18th order scheme formulae on the 25-point stencil:

$$c_0 u_{ij} + c_1(u_{i,j+1} + u_{i+1,j} + u_{i,j-1} + u_{i-1,j}) + c_2(u_{i+1,j+1} + u_{i+1,j-1}+$$

$$+u_{i-1,j-1} + u_{i-1,j+1}) + c_3(u_{i,j+2} + u_{i+2,j} + u_{i,j-2} + u_{i-2,j}) + c_4(u_{i+2,j+2}+$$

$$+u_{i+2,j-2} + u_{i-2,j-2} + u_{i-2,j+2}) + c_5(u_{i-1,j+2} + u_{i+1,j+2} + u_{i+2,j-1}+$$

$$+u_{i+2,j+1} + u_{i+1,j-2} + u_{i-1,j-2} + u_{i-2,j-1} + u_{i-2,j+1}) = \phi(x,y),$$

$$c_0 = -\frac{17459}{5082h^2}, \quad c_1 = \frac{5440}{7623h^2}, \quad c_2 = \frac{3400}{22869h^2}, \quad c_3 = -\frac{95}{15246h^2},$$

$$c_4 = -\frac{1}{182952h^2}, \quad c_5 = \frac{32}{22869h^2}.$$

We do not give here formula for nonhomogeneous term $\phi(x,y)$ of this scheme and a fortiori its approximation because of their bulkiness.

5   Numerical Experiments

For conducting numerical experiments, the exact solution was taken as the function $g(x,y)$ in boundary condition of problem (1). Difference equation was written in every interior grid node. As a result, on the $N \times N$ grid, we have obtained a SLAE with $(N-1)^2$ equations for $(N-1)^2$ unknown values of solution in interior grid nodes. The obtained SLAE was solved by orthogonal method: using decomposition of the matrix into product of orthogonal and upper triangular matrices. It is known that this method reduces the solution of the original system to solving a system with triangular matrix, the condition number of which is equal to the condition number of the original SLAE matrix. Using the 10th and 18th order schemes constructed above for determining solution in near-boundary nodes (adjacent with nodes on region boundary) it is impossible to apply the scheme formulae because in this case, the stencil contains points not lying in region Ω. For solving problem on schemes with symmetric difference equations written on symmetric 13- and 25-point stencils, different approximation of equation (1) in near-boundary nodes is required. In this work, at solving problem by the 10th and 18th approximation order schemes, the 6th order scheme on 9-point stencil was applied for writing difference equation in near-boundary nodes. It does naturally not exclude a possibility of applying nonsymmetric schemes for writing different equations in near-boundary grid nodes. Numerical experiments on solution convergence on a sequence of refining grids with grid step tending to zero were conducted. In applied sequence step h of every grid, beginning from second, was the half of step of previous grid. Numerical solution error was estimated in the norm $\| E(h) \|_C = \max\limits_{(x,y)\in\Omega} |u(x,y) - u_h(x,y)|$. Convergence order was computed using formula $\mu = \log_2 \frac{\|E(h)\|_C}{\|E(h/2)\|_C}$.

The first test solution in this work was $u(x,y) = e^{-\beta(x^2+y^2)}$. We can see its graph in Fig. 2.

It represents an isolated peak with steep slopes. The larger parameter β, the larger gradients this solution has. Here, we compare results at $\beta = 100$ and $\beta = 1000$. The corresponding right-hand side is $f(x,y) = 4(\beta^2(x^2+y^2)-\beta)e^{-\beta(x^2+y^2)}$. Computational region $\Omega = [-0.5, 0.5] \times [-0.5, 0.5]$ was taken in order to place the peak into the region center. Numerical experiments results are shown in Table 1. For comparison, results of solving the problem using the well-known "cross" scheme of the 2nd approximation order are also shown in Table 1.

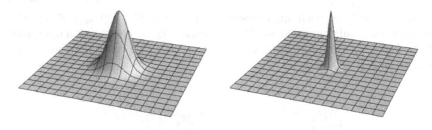

Fig. 2. Exact solution $u(x,y) = e^{-\beta(x^2+y^2)}$ with $\beta = 100$ and $\beta = 1000$

Table 1. Difference solution error $E(h)$ on grid sequence

Scheme order	h	$\beta = 100$		$\beta = 1000$	
		$\|E(h)\|_C$	μ	$\|E(h)\|_C$	μ
2	1/10	$3.843 \cdot 10^{-1}$	—	19.978	—
2	1/20	$6.771 \cdot 10^{-2}$	2.50	3.11	2.68
2	1/40	$1.591 \cdot 10^{-2}$	2.09	$1.959 \cdot 10^{-1}$	3.99
2	1/80	$3.921 \cdot 10^{-3}$	2.02	$4.104 \cdot 10^{-2}$	2.25
6	1/10	$4.942 \cdot 10^{-3}$	—	7.985	—
6	1/20	$3.594 \cdot 10^{-5}$	7.10	$1.13 \cdot 10^{-1}$	6.14
6	1/40	$4.635 \cdot 10^{-7}$	6.28	$7.624 \cdot 10^{-4}$	7.21
6	1/80	$7.110 \cdot 10^{-9}$	6.03	$8.193 \cdot 10^{-6}$	6.54
10	1/10	$3.036 \cdot 10^{-3}$	—	2.936	—
10	1/20	$9.894 \cdot 10^{-7}$	11.58	$3.086 \cdot 10^{-1}$	3.25
10	1/40	$8.473 \cdot 10^{-10}$	10.19	$1.559 \cdot 10^{-4}$	10.95
10	1/80	$8.407 \cdot 10^{-13}$	9.98	$8.78 \cdot 10^{-8}$	10.79
18	1/10	$8.388 \cdot 10^{-2}$	—	154283	—
18	1/20	$2.844 \cdot 10^{-7}$	18.17	307.82	8.97
18	1/40	$1.826 \cdot 10^{-12}$	17.25	$1.19 \cdot 10^{-3}$	17.98
18	1/80	$2.992 \cdot 10^{-12}$	−0.71	$3.92 \cdot 10^{-9}$	18.21

We can see from the table that numerical solution converges with the corresponding order of the schemes approximation. At $\beta = 1000$, on grid with step $1/80$ using the 18th order scheme, error $3.92 \cdot 10^{-9}$ is achieved. And the CPU time was 64.8 seconds, i.e., nearly one minute. While for achieving nearly the same accuracy using the 6th order scheme it was needed to take step $1/320$, and herewith the error was $1.724 \cdot 10^{-9}$ at the CPU time of 12306 seconds (on the same computer), which is 3.4 hours. Applying the 2nd order scheme, it is rather more difficult to achieve high accuracy. For example, on grid with step $1/640$ with the CPU time of nearly 29 hours, only the error $6.11 \cdot 10^{-4}$ was achieved. Therefore, the high approximation order schemes have significant advantages over lower order schemes at solving the problem with a given high accuracy.

The second test solution in this work was $u(x, y) = G((x - a) \cos \varphi + (y - b) \sin \varphi)$, where $G(\xi) = \frac{1}{2} + \frac{1}{\sqrt{\pi}} \int\limits_{0}^{\xi/\sqrt{2\beta}} e^{-x^2} dx$. The corresponding right-hand side is $f(x, y) = -\frac{1}{\sqrt{2\pi}\beta^{3/2}} \xi e^{-\frac{\xi^2}{2\beta}}$, where $\xi = (x - a) \cos \varphi + (y - b) \sin \varphi$. This solution has an interior layer with high gradients in region $\Omega = [0, 1] \times [0, 1]$, when parameter β is small, as we can see from graph in Fig. 3. The solution parameters had the following values: $a = 0.6, b = 0.3, \varphi = \pi/6$. Two cases were considered: $\beta = 0.001$ and $\beta = 0.0001$.

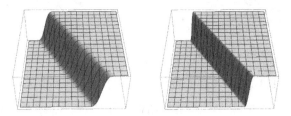

Fig. 3. Exact solution with interior layer with $\beta = 0.001$ and $\beta = 0.0001$

The results of numerical experiments are presented in Table 2. We see that the convergence order in case of this solution is lower than the corresponding approximation order. It happens due to the high condition number of the SLAE corresponding to a given problem. Nevertheless, good results are obtained even on coarse grids. It appeared that the solution obtained by the 10th order scheme has no significant advantage in accuracy and convergence over the solution obtained by the 6th order scheme. The reason is that in the case of the 10th order scheme, we used the 6th order scheme in near-boundary nodes. But this solution has high gradients near boundary as well as inside the computational domain. In order to increase the convergence order of solution using the 10th order scheme, it is required to construct additional difference formulae with higher order (more than 6th) of equation approximation in near-boundary nodes. One of possible

Table 2. Difference solution error $E(h)$ on grid sequence

Scheme order	h	$\beta = 0.001$		$\beta = 0.0001$	
		$\|E(h)\|_C$	μ	$\|E(h)\|_C$	μ
2	1/10	$2.749 \cdot 10^{-1}$	—	9.31	—
2	1/20	$3.96 \cdot 10^{-2}$	2.79	1.178	2.98
2	1/40	$8.575 \cdot 10^{-3}$	2.20	$1.55 \cdot 10^{-1}$	2.93
6	1/10	$9.147 \cdot 10^{-2}$	—	1.685	—
6	1/20	$9.117 \cdot 10^{-4}$	6.65	$4.666 \cdot 10^{-1}$	1.85
6	1/40	$5.531 \cdot 10^{-5}$	4.04	$2.383 \cdot 10^{-2}$	4.29
10	1/10	$1.280 \cdot 10^{-1}$	—	1.98	—
10	1/20	$9.804 \cdot 10^{-4}$	7.03	$6.049 \cdot 10^{-1}$	1.71
10	1/40	$5.514 \cdot 10^{-5}$	4.15	$3.74 \cdot 10^{-2}$	4.01

ways of solving this question is the construction of nonsymmetric formulae on multi-point stencil. Such approach to solving boundary-value problem using high order schemes for elliptic equation with Dirichlet and Neumann conditions in region with curvilinear boundary was applied in the work [12].

6 Conclusions

For constructing new high approximation order schemes, the method of undetermined coefficients was applied. Using this method on multi-point stencils, difference schemes with different approximation orders for the Poisson equation were constructed. Here, for the first time, the 10th and 18th approximation order schemes formulae are given, including formulae for nonhomogeneous term of the 10th order scheme. CAS application allowed to simplify significantly the derivation of schemes formulae and their verification. Numerical experiments on a sequence of refining grids with grid step descended to zero were conducted. It was established that on smooth solutions of the problem, numerical solution constructed using high approximation order schemes has high convergence order and good accuracy even on coarse grids. It was shown that in the issue of constructing solution of problem with a given high accuracy when computational resources are limited, the schemes with high accuracy order have significant advantages over lower order schemes.

At the current stage of development of science and technology, it is difficult to say, where and when it will essentially require the use of good facilities of numerical methods with high accuracy. However, one can assume that in the future, at developing technologies, in which economical high-precision calculation will be required or we will need to control physical (or other) high-speed processes by the way of numerical solution of difficult equations in real time, the numerical methods with high accuracy order will appear irreplaceable. The construction of such methods is the objective of computational mathematics.

References

1. Albensoeder, S., Kuhlmann, H.C.: Accurate three-dimensional lid-driven cavity flow. J. Comput. Phys. 206(2), 536–558 (2005)
2. Botella, O., Peyret, R.: Benchmark spectral results on the lid-driven cavity flow. Comput. Fluids 27(4), 421–433 (1998)
3. Canuto, C., Hussaini, M.Y., Quarteroni, A., Zang, T.A.: Spectral Methods in Fluid Dynamics. Springer (1988)
4. Ganzha, V.G., Mazurik, S.I., Shapeev, V.P.: Symbolic manipulations on a computer and their application to generation and investigation of difference schemes. In: Caviness, B.F. (ed.) ISSAC 1985 and EUROCAL 1985. LNCS, vol. 204, pp. 335–347. Springer, Heidelberg (1985)
5. Gerdt, V.P., Blinkov, Y.A., Mozzhilkin, V.V.: Grobner Bases and Generation of Difference Schemes for Partial Differential Equations. Symmetry, Integrability and Geometry: Methods and Applications (SIGMA) 2, 051 (2006) arXiv:math.RA/0605334
6. Harten, A.: High resolution schemes for hyperbolic conservation laws. J. Comput. Phys. 49, 357–393 (1983)
7. Isaev, V.I., Shapeev, V.P.: High-Accuracy Versions of the Collocations and Least Squares Method for the Numerical Solution of the Navier-Stokes Equations. Comput. Math. and Math. Phys. 50(10), 1670–1681 (2010)
8. Lee, M.K., Malaya, N., Moser, R.D.: Petascale direct numerical simulation of turbulent channel flow on up to 786K cores. In: SC 2013 Proceedings of SC13: International Conference for High Performance Computing, Networking, Storage and Analysis, Denver, CO (2013)
9. Lele, S.K.: Compact finite difference schemes with spectral-like resolution. J. Comput. Phys. 102(1), 16–42 (1992)
10. Lipavskii, M.V., Tolstykh, A.I.: Tenth-order accurate multioperator scheme and its application in direct numerical simulation. Comput. Math. and Math. Phys. 53(4), 455–468 (2013)
11. Shapeev, A.V., Lin, P.: An asymptotic fitting finite element method with exponential mesh refinement for accurate computation of corner eddies in viscous flows. SIAM J. Sci. Comput. 31(3), 1874–1900 (2009)
12. Shapeev, A.V., Shapeev, V.P.: Difference schemes of increased order of accuracy for solving elliptical equations in domain with curvilinear boundary. Comput. Math. and Math. Phys. 40(2), 213–221 (2000)
13. Shapeev, V.P., Vorozhtsov, E.V.: CAS application to the construction of the collocations and least residuals method for the solution of 3D Navier-Stokes equations. In: Gerdt, V.P., Koepf, W., Mayr, E.W., Vorozhtsov, E.V. (eds.) CASC 2013. LNCS, vol. 8136, pp. 381–392. Springer, Heidelberg (2013)
14. Steinberg, S.: A problem solving environment for numerical partial differential equations. In: 6th IMACS Int. Conf. on Applications of Computer Algebra. Abstracts, St.Petersburg, Russia, June 25-28, pp. 98–99 (2000)
15. Stetter, H.J.: Condition analysis of overdetermined algebraic problems. Computer Algebra in Scientific Computing, pp. 345–365. Springer, Berlin (2000)
16. Valiullin, A.N., Shapeev, V.P., et al.: Symbolic manipulations in the methods of mathematical physics. In: Symposium Mathematics for Computer Science, Paris, March 16-18, pp. 431–438 (1982)

On Symbolic Solutions of Algebraic Partial Differential Equations

Georg Grasegger[1,2,*], Alberto Lastra[3,**], J. Rafael Sendra[3,***],
and Franz Winkler[2,†]

[1] Doctoral Program Computational Mathematics and
[2] Research Institute for Symbolic Computation, Johannes Kepler University Linz,
4040 Linz, Austria
[3] Dpto. de Física y Matemáticas, Universidad de Alcalá, 28871 Alcalá de Henares,
Madrid, Spain

Abstract. In this paper we present a general procedure for solving first-order autonomous algebraic partial differential equations in two independent variables. The method uses proper rational parametrizations of algebraic surfaces and generalizes a similar procedure for first-order autonomous ordinary differential equations. We will demonstrate in examples that, depending on certain steps in the procedure, rational, radical or even non-algebraic solutions can be found. Solutions computed by the procedure will depend on two arbitrary independent constants.

Keywords: Partial differential equations, algebraic surfaces, rational parametrizations, radical parametrizations.

1 Introduction

Recently algebraic-geometric solution methods for first-order algebraic ordinary differential equations (AODEs) were investigated. A first result on computing solutions of AODEs using Gröbner bases was presented in [10]. Later in [3] a degree bound of rational solutions of a given AODE is computed. From this one might find a solution by solving algebraic equations. The starting point for algebraic-geometric methods was an algorithm by Feng and Gao [4, 5] which decides whether or not an autonomous AODE, $F(y, y') = 0$ has a rational solution and in the affirmative case computes it. In the algorithm a proper rational

* Supported by the Austrian Science Fund (FWF): W1214-N15, project DK11.
** Supported by the Spanish Ministerio de Economía y Competitividad under the project MTM2012-31439. Member of the group ASYNACS (Ref. CCEE2011/R34).
*** Supported by the Spanish Ministerio de Economía y Competitividad under the project MTM2011-25816-C02-01. Member of the group ASYNACS (Ref. CCEE2011/R34).
† Supported by the Austrian Science Fund (FWF): W1214-N15, project DK11 and by the Spanish Ministerio de Economía y Competitividad under the project MTM2011-25816-C02-01.

V.P. Gerdt et al. (Eds.): CASC Workshop 2014, LNCS 8660, pp. 111–120, 2014.

parametrization of an algebraic curve is used. By means of a special property of this parametrization the existence of a rational solution can be decided. From a rational solution a rational general solution can be deduced.

This result was then generalized by Ngô and Winkler [14–16] to the non-autonomous case $F(x, y, y') = 0$. Here, parametrizations of surfaces play an important role. On the basis of a proper parametrization, the algorithm builds a so called associated system of first-order linear ODEs for which solution methods exist. With the solution of the associated system, a rational general solution of the differential equation is computed.

First results on higher order AODEs can be found in [7–9]. Ngô, Sendra and Winkler [13] also classified AODEs in terms of rational solvability by considering affine linear transformations. Classes of AODEs are investigated which contain an autonomous equation. A generalization to birational transformations can be found in [12].

In [6] a solution method for autonomous AODEs is presented which generalizes the method of Feng and Gao to the computation of radical and also non-radical solutions. Again a crucial tool is the parametrization involved in the process. To the contrary of the previous algorithms also radical parametrizations can be used in this method. However, this method is not complete, for if it does not yield a solution, no conclusion on the solvability of the initial AODE can be drawn.

In this paper we present a generalization of the procedure in [6] to algebraic partial differential equations (APDEs). We restrict to first-order autonomous APDEs in two variables. Solutions computed by the procedure will depend on two arbitrary independent constants. However, the class of functions which may appear in the solution of the procedure is only defined implicitly since the procedure depends on the solution of certain ODEs.

In Sect. 2 we will recall and introduce the necessary definitions and concepts. Then we will present the general procedure for solving APDEs in Sect. 3 and show some examples.

2   Preliminaries

We consider the field of rational functions $\mathbb{K}(x, y)$ for some field $\mathbb{K}$ of characteristic zero. By $\frac{\partial}{\partial x}$ and $\frac{\partial}{\partial y}$ we denote the usual derivative by x and y respectively. Sometimes we might use the abbreviations $u_x = \frac{\partial u}{\partial x}$ and $u_y = \frac{\partial u}{\partial y}$. The ring of differential polynomials is denoted as $\mathbb{K}(x, y)\{u\}$. It consists of all polynomials in u and its derivatives, i. e.

$$\mathbb{K}(x, y)\{u\} = \mathbb{K}(x, y)[u, u_x, u_y, u_{xx}, u_{xy}, u_{yy}, \ldots] \ .$$

Let $\mathbb{K}[x, y]\{u\} \subseteq \mathbb{K}(x, y)\{u\}$ be the elements which are polynomial in the variables x and y. An algebraic partial differential equation (APDE) is given by

$$F(x, y, u, u_x, u_y, u_{xx}, u_{xy}, u_{yy}, \ldots) = 0$$

where $F \in \mathbb{K}[x, y]\{u\}$. In this paper we restrict to the first-order autonomous case, i.e. $F(u, u_x, u_y) = 0$.

Let $\overline{\mathbb{K}}$ be the algebraic closure of $\mathbb{K}$, and $A(\overline{\mathbb{K}})^3$ be the 3-dimensional affine space. An *algebraic surface* $\mathcal{S}$ in $A(\overline{\mathbb{K}})^3$ is a two-dimensional algebraic variety, i.e. $\mathcal{S}$ is the zero set of a squarefree non-constant polynomial $f \in \mathbb{K}[x, y, z]$, $\mathcal{S} = \{(a, b, c) \in A(\overline{\mathbb{K}})^3 \mid f(a, b, c) = 0\}$. We call the polynomial f the *defining polynomial*. An important aspect of algebraic surfaces is their rational parametrizability. We consider an algebraic surface defined by an irreducible polynomial f. A triple of rational functions $\mathcal{P}(s, t) = (p_1(s, t), p_2(s, t), p_3(s, t))$ is called a *rational parametrization* of the surface if $f(p_1(s, t), p_2(s, t), p_3(s, t)) = 0$ for all s and t and the jacobian of $\mathcal{P}$ has generic rank 2. We observe that this condition is fundamental since, otherwise, we are parametrizing a point (if the rank is 0) or a curve on the surface (if the rank is 1). A parametrization can be considered as a dominant map $\mathcal{P}(s, t) : \mathbb{K}^2 \to \mathcal{S}$. By abuse of notation we also call this map a parametrization. We call a parametrization $\mathcal{P}(s, t)$ *proper* if it is a birational map or in other words if for almost every point (a, b, c) on the surface we find exactly one pair (s, t) such that $\mathcal{P}(s, t) = (a, b, c)$ or equivalently if $\mathbb{K}(\mathcal{P}(s, t)) = \mathbb{K}(s, t)$.

Above we have considered rational parametrizations of a surface. However, we might want to deal with more general parametrizations. If so, we will say that a triple of differentiable functions $\mathcal{Q}(s, t) = (q_1(s, t), q_2(s, t), q_3(s, t))$ is a parametrization of the surface if $f(\mathcal{Q}(s, t))$ is identically zero and the jacobian of $\mathcal{Q}(s, t)$ has generic rank 2.

Let $F(u, u_x, u_y) = 0$ be an autonomous APDE. We consider the corresponding algebraic surface by replacing the derivatives by independent transcendental variables, $F(z, p, q) = 0$. Given any differentiable function $u(x, y)$ with $F(u, u_x, u_y) = 0$, then $(u(s, t), u_x(s, t), u_y(s, t))$ is a parametrization. We call this parametrization the *corresponding parametrization of the solution*. We observe that the corresponding parametrization of a solution is not necessarily a parametrization of the associated surface. For instance, let us consider the APDE $u_x = 0$. A solution would be of the form $u(x, y) = g(y)$, with g differentiable. However, this solution generates $(g(t), 0, g'(t))$ that is a curve in the surface; namely the plane $p = 0$. Now, consider the APDE $u_x = \lambda$, with λ a nonzero constant. Hence, the solutions are of the form $u(x, y) = \lambda x + g(y)$. Then, $u(x, y) = \lambda x + y$ generates the line $(\lambda s + t, \lambda, 1)$ while $u(x, y) = \lambda x + y^2$ generates the parametrization $(\lambda s + t^2, \lambda, 2t)$ of the associated plane $p = \lambda$. Clearly a solution of an APDE is a function $u(x, y)$ such that $F(u, u_x, u_y) = 0$. The examples above motivate the following definition.

Definition 1. *We say that a solution $u(x, y)$ of an APDE is* rational *if $u(x, y)$ is a rational function over an algebraic extension of $\mathbb{K}$.*
We say that a rational solution of an APDE is proper *if the corresponding parametrization is proper.*

In the case of autonomous ordinary differential equations, every non-constant solution induces a proper parametrization of the associated curve (see [4]). However, this is not true in general for autonomous APDEs. For instance, the solution

$x + y^3$ of $u_x = 1$, induces the parametrization $(s + t^3, 1, 3t^2)$ which is not proper, although its jacobian has rank 2.

Remark 2. By the definition of a surface parametrization we know that the jacobian of a proper parametrization has generic rank 2.

3   A Method for Solving First-Order Autonomous APDEs

Let $F(u, u_x, u_y) = 0$ be an algebraic partial differential equation. We consider the surface $F(z, p, q) = 0$ and assume it admits a proper (rational) surface parametrization

$$\mathcal{Q}(s, t) = (q_1(s, t), q_2(s, t), q_3(s, t)) \ .$$

An algorithm for computing a proper rational parametrization of a surface can be found for instance in [17]. Here, we will stick to rational parametrizations, but the procedure which we present will work as well with other kinds of parametrizations, for instance radical ones. First results on radical parametrizations of surfaces can be found in [18]. Assume that $\mathcal{L}(s, t) = (p_1(s, t), p_2(s, t), p_3(s, t))$ is the corresponding parametrization of a solution of the APDE. Furthermore we assume that the parametrization $\mathcal{Q}$ can be expressed as

$$\mathcal{Q}(s, t) = \mathcal{L}(g(s, t))$$

for some invertible function $g(s, t) = (g_1(s, t), g_2(s, t))$. This assumption is motivated by the fact that in case of rational algebraic curves every non-constant rational solution of an AODE yields a proper rational parametrization of the associated algebraic curve and each proper rational parametrization can be obtained from any other proper one by a rational transformation. However, in the case of APDEs, not all rational solutions provide a proper parametrization, as mentioned in the remark after Definition 1. Now, using the assumption, if we can compute g^{-1} we have a solution $\mathcal{Q}(g^{-1}(s, t))$.

Let $\mathcal{J}$ be the jacobian matrix. Then we have

$$\mathcal{J}_{\mathcal{Q}}(s, t) = \mathcal{J}_{\mathcal{L}}(g(s, t)) \cdot \mathcal{J}_g(s, t) \ .$$

Taking a look at the first row we get that

$$
\begin{aligned}
\frac{\partial q_1}{\partial s} &= \frac{\partial p_1}{\partial s}(g)\frac{\partial g_1}{\partial s} + \frac{\partial p_1}{\partial t}(g)\frac{\partial g_2}{\partial s} = q_2(s, t)\frac{\partial g_1}{\partial s} + q_3(s, t)\frac{\partial g_2}{\partial s} \ , \\
\frac{\partial q_1}{\partial t} &= \frac{\partial p_1}{\partial s}(g)\frac{\partial g_1}{\partial t} + \frac{\partial p_1}{\partial t}(g)\frac{\partial g_2}{\partial t} = q_2(s, t)\frac{\partial g_1}{\partial t} + q_3(s, t)\frac{\partial g_2}{\partial t} \ .
\end{aligned}
\tag{1}
$$

This is a system of quasilinear equations in the unknown functions g_1 and g_2. In case q_2 or q_3 is zero the problem reduces to ordinary differential equations. Hence, from now on we assume that $q_2 \neq 0$ and $q_3 \neq 0$. First we divide by q_2:

$$
\begin{aligned}
a_1 &= \frac{\partial g_1}{\partial s} + b\frac{\partial g_2}{\partial s} \ , \\
a_2 &= \frac{\partial g_1}{\partial t} + b\frac{\partial g_2}{\partial t}
\end{aligned}
\tag{2}
$$

with

$$a_1 = \frac{\frac{\partial g_1}{\partial s}}{q_2} , \qquad a_2 = \frac{\frac{\partial g_1}{\partial t}}{q_2} , \qquad b = \frac{q_3}{q_2} . \qquad (3)$$

By taking derivatives we get

$$\frac{\partial a_1}{\partial t} = \frac{\partial^2 g_1}{\partial s \partial t} + \frac{\partial b}{\partial t}\frac{\partial g_2}{\partial s} + b\frac{\partial^2 g_2}{\partial s \partial t} ,$$
$$\frac{\partial a_2}{\partial s} = \frac{\partial^2 g_1}{\partial t \partial s} + \frac{\partial b}{\partial s}\frac{\partial g_2}{\partial t} + b\frac{\partial^2 g_2}{\partial t \partial s} . \qquad (4)$$

Subtraction of the two equations yields

$$\frac{\partial b}{\partial t}\frac{\partial g_2}{\partial s} - \frac{\partial b}{\partial s}\frac{\partial g_2}{\partial t} = \frac{\partial a_1}{\partial t} - \frac{\partial a_2}{\partial s} . \qquad (5)$$

This is a single quasilinear differential equation which can be solved by the method of characteristics (see for instance [19]). In case $\frac{\partial b}{\partial t} = 0$ or $\frac{\partial b}{\partial s} = 0$ equation (5) reduces to a simple ordinary differential equation.

Remark 3. If both derivatives of b are zero then b is a constant. Hence, the left hand side of (5) is zero. In case the right hand side is non-zero we get a contradiction, and hence there is no solution. In case the right hand side is zero as well we get from (5) that

$$0 = \frac{\partial a_1}{\partial t} - \frac{\partial a_2}{\partial s} = \frac{\partial}{\partial t}\left(\frac{\frac{\partial g_1}{\partial s}}{q_2}\right) - \frac{\partial}{\partial s}\left(\frac{\frac{\partial g_1}{\partial t}}{q_2}\right)$$
$$= \frac{\frac{\partial^2 g_1}{\partial t \partial s}q_2 - \frac{\partial g_1}{\partial s}\frac{\partial g_2}{\partial t}}{q_2^2} - \frac{\frac{\partial^2 g_1}{\partial s \partial t}q_2 - \frac{\partial g_1}{\partial t}\frac{\partial g_2}{\partial s}}{q_2^2}$$
$$= -\frac{\frac{\partial g_1}{\partial s}\frac{\partial g_2}{\partial t} - \frac{\partial g_1}{\partial t}\frac{\partial g_2}{\partial s}}{q_2^2} ,$$

hence,

$$0 = \frac{\partial q_1}{\partial s}\frac{\partial q_2}{\partial t} - \frac{\partial q_1}{\partial t}\frac{\partial q_2}{\partial s} .$$

Moreover, since b is constant, $q_2 = kq_3$ for some constant k. But this means that the rank of the jacobian of $\mathcal{Q}$ is 1, a contradiction to $\mathcal{Q}$ being proper.

Therefore we assume from now on, that the derivatives of b are non-zero. According to the method of characteristics, we need to solve the following system of first-order ordinary differential equations

$$\frac{ds(t)}{dt} = -\frac{\frac{\partial b}{\partial t}(s(t),t)}{\frac{\partial b}{\partial s}(s(t),t)} ,$$
$$\frac{dv(t)}{dt} = \frac{\frac{\partial a_1}{\partial t}(s(t),t) - \frac{\partial a_2}{\partial s}(s(t),t)}{-\frac{\partial b}{\partial s}(s(t),t)} .$$

The second equation is linear and separable but depends on the solution of the first. The first ODE can be solved independently. Its solution $s(t) = \eta(t, k)$ will depend on an arbitrary constant k. Hence, also the solutions of the second ODE depends on k. Finally, the function g_2 we are looking for is $g_2(s, t) = v(t, \mu(s, t)) + \nu(\mu(s, t))$ where μ is computed such that $s = \eta(t, \mu(s, t))$ and ν is an arbitrary function. In case we are only looking for rational solutions we can use the algorithm of Ngô and Winkler [14–16] for solving these ODEs.

Knowing g_2 we can compute g_1 by using (1) which now reduces to a separable ODE in g_1. The remaining task is to compute h_1 and h_2 such that $g(h_1(s, t), h_2(s, t)) = (s, t)$. Then $q_1(h_1, h_2)$ is a solution of the original PDE.

Finally the method reads as

Procedure 4. Given an autonomous APDE, $F(u, u_x, u_y) = 0$, where F is irreducible and $F(z, p, q) = 0$ is a rational surface with a proper rational parametrization $\mathcal{Q} = (q_1, q_2, q_3)$.

1. Compute the coefficients b and a_i as in (3).
2. If $\frac{\partial b}{\partial s} = 0$ and $\frac{\partial b}{\partial t} \neq 0$ compute $g_2 = \int \frac{\frac{\partial a_1}{\partial t} - \frac{\partial a_2}{\partial s}}{\frac{\partial b}{\partial t}} ds + \kappa(t)$ and go to step 6
   otherwise continue.
   If $\frac{\partial b}{\partial s} = \frac{\partial b}{\partial t} = 0$ return "No proper solution".
3. Solve the ODE $\frac{ds(t)}{dt} = -\frac{\frac{\partial b}{\partial t}(s(t), t)}{\frac{\partial b}{\partial s}(s(t), t)}$ for $s(t) = \eta(t, k)$ with arbitrary constant k.
4. Solve the ODE $\frac{dv(t)}{dt} = \frac{\frac{\partial a_1}{\partial t}(\eta(t,k), t) - \frac{\partial a_2}{\partial s}(\eta(t,k), t)}{-\frac{\partial b}{\partial s}(\eta(t,k), t)}$
   by $v(t) = v(t, k) = \int \frac{\frac{\partial a_1}{\partial t}(\eta(t,k), t) - \frac{\partial a_2}{\partial s}(\eta(t,k), t)}{-\frac{\partial b}{\partial s}(\eta(t,k), t)} dt + \nu(k)$.
5. Compute μ such that $s = \eta(t, \mu(s, t))$ and then $g_2(s, t) = v(t, \mu(s, t))$.
6. Use the second equation of (2) to compute $g_1(s, t) = m(s) + \int a_2 - b\frac{\partial g_2}{\partial t} dt$.
7. Determine $m(s)$ by using the first equation of (2).
8. Compute h_1, h_2 such that $g(h_1(s, t), h_2(s, t)) = (s, t)$.
9. Return the solution $q_1(h_1, h_2)$.

Observe that the proper rational parametrization $\mathcal{Q}$ can be computed applying Schicho's algorithm (see [17]). In addition, we also observe that the procedure can be extended to the non-rational algebraic case, if one has an injective parametrization, in that case non-rational, of the surface defined by $F(z, p, q) = 0$.

In general ν will depend on a constant c_2 and m on a constant c_1. As a special case of the procedure we will fix $\nu = c_2$. This choice is done for simplicity reasons but we may sometimes refer to cases with other choices which are a subject of further research.

Furthermore, the procedure can be considered symmetrically in step 2 for the case that $\frac{\partial b}{\partial t} = 0$ and $\frac{\partial b}{\partial s} \neq 0$. In such a case the rest of the procedure has to be changed symmetrically as well. We will not go into further details.

Theorem 5. *Let $F(u, u_x, u_y) = 0$ be an autonomous APDE. If Procedure 4 returns a function $v(x, y)$ for input F, then v is a solution of F.*

Proof. By the procedure we know that $v(x, y) = q_1(h_1(x, y), h_2(x, y))$ with h_i such that $g(h_1(s, t), h_2(s, t)) = (s, t)$. Since g is a solution of system (1) it fulfills the assumption that $u(g_1, g_2) = q_1$ for a solution u . Hence, v is a solution. We have seen a more detailed description at the beginning of this section.   □

Remark 6. In step 3 and 4 ODEs have to be solved. Depending on the class of functions to which the requested solution should belong, these ODEs do not necessarily have a solution. Furthermore, an explicit inverse (step 8) does not necessarily exist.

It will be a subject of further research, to investigate conditions on cases were the procedure does definitely not fail.

Now, we will show that the result of Procedure 4 does not change if we postpone the introduction of c_1 and c_2 to the end of the procedure. It is easy to show that if $u(x, y)$ is a solution of an autonomous APDE then so is $u(x + c, y + d)$ for any constants c and d. From the procedure we see that in the computation of g_1 we use the derivative of g_2 only (and hence c_2 disappears). We can write

$$g_2 = \bar{g}_2 + c_2 , \qquad\qquad g_1 = \bar{g}_1 + c_1$$

for some functions $\bar{g}_1, \bar{g}_2$ which do not depend on c_1 and c_2. Let $g = (g_1, g_2)$ and $\bar{g} = (\bar{g}_1, \bar{g}_2)$. In the step 8 we are looking for a function h such that $g \circ h = $ id. Now $g \circ h = \bar{g} \circ h + (c_1, c_2)$. Take $\bar{h}$ such that $\bar{g} \circ \bar{h} = $ id. Then $g \circ \bar{h}(s - c_1, t - c_2)) = $ id. Hence, we can introduce the constants at the end.

In case the original APDE is in fact an AODE, the ODE in step 4 turns out to be trivial and the integral in step 7 is exactly the one which appears in the procedure for AODEs [6]. Of course then g is univariate and so is its inverse. In this sense, this new procedure generalizes the procedure in [6]. We do not specify Procedure 4 to handle this case.

In the following we will show some examples which can be solved by Procedure 4. Note, that the examples have more solutions than the one computed below. In Example 7 for instance, other solutions can be found by choosing different ν, e. g. $\nu(x) = c_2 + x^2$. However, the results might not be rational solutions then. In general the procedure, as stated in this paper, will yield only one solution containing two arbitrary independent constants. Hence, it will not be a general solution in the sense of depending on an arbitrary function (compare [11]).

We start with a simple well known APDE which has a rational solution.

Example 7 (Inviscid Burgers' Equation [1, p. 7]). We consider the autonomous APDE

$$F(u, u_x, u_y) = uu_x + u_y = 0 .$$

Since F is of degree one in each of the derivatives, it is easy to compute a parametrization $\mathcal{Q} = \left(-\frac{t}{s}, s, t\right)$. We compute the coefficients

$$a_1 = \frac{t}{s^3} , \qquad\qquad a_2 = -\frac{1}{s^2} , \qquad\qquad b = \frac{t}{s} .$$

In step 3 we find $s(t) = kt$ and in step 4 we compute $v(t) = \frac{1}{kt} + \nu(k)$. Then $\mu(s,t) = \frac{s}{t}$ and hence (with $\nu = c_2$),

$$g_2 = \frac{1}{s} + c_2 \ ,$$

$$g_1 = -\frac{t}{s^2} + m(s) \ .$$

Using step 7 we find out that $m(s) = c_1$. Computing the inverse of g we find

$$h_1 = \frac{1}{t - c_2} \ ,$$

$$h_2 = \frac{-s + c_1}{(t - c_2)^2} \ .$$

Finally, we get the solution $\frac{x - c_1}{y - c_2}$.

Procedure 4 can also handle more complicated APDEs.

Example 8. We consider the APDE

$$0 = F(u, u_x, u_y)$$
$$= u u_x^4 + u_x^3 u_y - u u_x^3 u_y - u_x^2 u_y^2 + u u_x^2 u_y^2 + u_x u_y^3 - u u_x u_y^3 + u u_y^4 \ .$$

Then

$$\mathcal{Q} = \left(-\frac{t\left(1 - t + t^2\right)}{1 - t + t^2 - t^3 + t^4}, t\gamma(s,t), \gamma(s,t) \right) \ ,$$

with $\gamma(s,t) = \frac{t(-10+7t)\left(-9+t^2\right)\left(-1+2t-3t^2+3t^4-2t^5+t^6\right)}{2s(45-63t+5t^2)(1-t+t^2-t^3+t^4)^2}$, is a proper parametrization of the corresponding algebraic surface. This parametrization is not easy to find. It is computed by first using parametrization by lines and then applying a linear transformation in s. Alternatively one could use this parametrization by lines directly. Procedure 4 will find the same solution, but the intermediate steps need more writing space. Using the procedure with the parametrization $\mathcal{Q}$ we get

$$g_1 = s\left(\frac{7}{10 - 7t} - \frac{1}{t} + \frac{2t}{-9 + t^2} \right) \ , \qquad g_2 = \frac{2s\left(45 - 63t + 5t^2\right)}{(-10 + 7t)\left(-9 + t^2\right)} \ ,$$

$$h_1 = -\frac{t\left(-90s^3 - 63s^2 t + 10st^2 + 7t^3\right)}{2s\left(45s^2 + 63st + 5t^2\right)} \ , \qquad h_2 = \frac{-t}{s} \ ,$$

and finally the solution $u(x,y) = \frac{xy\left(x^2 + xy + y^2\right)}{x^4 + x^3 y + x^2 y^2 + xy^3 + y^4}$. As mentioned before, $u(x + c_1, y + c_2)$ with constants c_1 and c_2 is also a solution.

The procedure presented in this paper is, however, not restricted to rational solutions nor to rational parametrizations as we will see in the following examples. We start with an example which has a radical solution.

Example 9 (Eikonal Equation [2, p. 2]). We consider the APDE

$$F(u, u_x, u_y) = u_x^2 + u_y^2 - 1 = 0 .$$

From the rational parametrization of a circle it is easy to see that

$$\mathcal{Q} = \left(s, \frac{1 - t^2}{1 + t^2}, \frac{2t}{1 + t^2} \right)$$

is a parametrization of the corresponding surface. Using the procedure we get some rational g_1 and g_2 which yield

$$h_2 = \frac{-s + c_1 \pm \sqrt{s^2 + t^2 - 2sc_1 + c_1^2 - 2tc_2 + c_2^2}}{t - c_2} ,$$

$$h_1 = \pm \sqrt{s^2 + t^2 - 2sc_1 + c_1^2 - 2tc_2 + c_2^2} .$$

Finally, we get the radical solution

$$u(x, y) = \pm \sqrt{x^2 + y^2 - 2xc_1 + c_1^2 - 2yc_2 + c_2^2} .$$

In a further example we compute an exponential solution of an APDE.

Example 10 (Convection-Reaction Equation [1, p. 7]). We consider the APDE

$$F(u, u_x, u_y) = u_x + cu_y - du = 0 ,$$

where $d \neq 0$. We compute a parametrization $\mathcal{Q} = \left(\frac{s + ct}{d}, s, t \right)$ and the coefficients

$$a_1 = \frac{1}{ds} , \qquad a_2 = \frac{c}{ds} , \qquad b = \frac{t}{s} .$$

Solving the ODEs of steps 3–6 we get

$$g_2 = \frac{c \log(t)}{d} + c_2 , \qquad g_1 = c_1 + \frac{\log(s)}{d} .$$

Computing the inverse of g we find

$$h_1 = e^{ds - dc_1} , \qquad h_2 = e^{\frac{dt}{c} - \frac{dc_2}{c}} .$$

Finally, we get the solution $\frac{e^{d(x - c_1)} + ce^{\frac{d(y - c_2)}{c}}}{d}$.

4   Conclusion

We have introduced a procedure which, in case all steps are computable, yields a solution of the input APDE. In case one step of the procedure is not computable (in a certain class of functions) we cannot give any answer to the question of solvability of the APDE. We have shown examples of APDEs solvable by the procedure. These include rational, radical and exponential solutions. The investigation of rational solutions as well as a possible extension to an arbitrary number of variables is currently subject to further research.

References

1. Arendt, W., Urban, K.: Partielle Differenzialgleichungen. Eine Einführung in analytische und numerische Methoden. Spektrum Akademischer Verlag, Heidelberg (2010)
2. Arnold, V.I.: Lectures on Partial Differential Equations. Springer, Heidelberg (2004)
3. Eremenko, A.: Rational solutions of first-order differential equations. Annales Academiae Scientiarum Fennicae. Mathematica 23(1), 181–190 (1998)
4. Feng, R., Gao, X.S.: Rational General Solutions of Algebraic Ordinary Differential Equations. In: Gutierrez, J. (ed.) Proceedings of the 2004 International Symposium on Symbolic and Algebraic Computation (ISSAC), pp. 155–162. ACM Press, New York (2004)
5. Feng, R., Gao, X.S.: A polynomial time algorithm for finding rational general solutions of first order autonomous ODEs. Journal of Symbolic Computation 41(7), 739–762 (2006)
6. Grasegger, G.: Radical Solutions of First Order Autonomous Algebraic Ordinary Differential Equations. In: Nabeshima, K. (ed.) ISSAC 2014: Proceedings of the 39th International Symposium on International Symposium on Symbolic and Algebraic Computation, pp. 217–223. ACM, New York (2014)
7. Huang, Y., Ngô, L.X.C., Winkler, F.: Rational General Solutions of Trivariate Rational Systems of Autonomous ODEs. In: Proceedings of the Fourth International Conference on Mathematical Aspects of Computer and Information Sciences (MACIS 2011), pp. 93–100 (2011)
8. Huang, Y., Ngô, L.X.C., Winkler, F.: Rational General Solutions of Trivariate Rational Differential Systems. Mathematics in Computer Science 6(4), 361–374 (2012)
9. Huang, Y., Ngô, L.X.C., Winkler, F.: Rational General Solutions of Higher Order Algebraic ODEs. Journal of Systems Science and Complexity 26(2), 261–280 (2013)
10. Hubert, E.: The General Solution of an Ordinary Differential Equation. In: Lakshman, Y.N. (ed.) Proceedings of the 1996 International Symposium on Symbolic and Algebraic Computation (ISSAC), pp. 189–195. ACM Press, New York (1996)
11. Kamke, E.: Differentialgleichungen: Lösungsmethoden und Lösungen II, Leipzig. Akademische Verlagsgesellschaft Geest & Portig K.-G. (1965)
12. Ngô, L.X.C., Sendra, J.R., Winkler, F.: Birational Transformations on Algebraic Ordinary Differential Equations. Tech. Rep. 12–18, RISC Report Series, Johannes Kepler University Linz, Austria (2012)
13. Ngô, L.X.C., Sendra, J.R., Winkler, F.: Classification of algebraic ODEs with respect to rational solvability. In: Computational Algebraic and Analytic Geometry, Contemporary Mathematics, vol. 572, pp. 193–210. American Mathematical Society, Providence (2012)
14. Ngô, L.X.C., Winkler, F.: Rational general solutions of first order non-autonomous parametrizable ODEs. Journal of Symbolic Computation 45(12), 1426–1441 (2010)
15. Ngô, L.X.C., Winkler, F.: Rational general solutions of parametrizable AODEs. Publicationes Mathematicae Debrecen 79(3-4), 573–587 (2011)
16. Ngô, L.X.C., Winkler, F.: Rational general solutions of planar rational systems of autonomous ODEs. Journal of Symbolic Computation 46(10), 1173–1186 (2011)
17. Schicho, J.: Rational Parametrization of Surfaces. Journal of Symbolic Computation 26(1), 1–29 (1998)
18. Sendra, J.R., Sevilla, D.: First steps towards radical parametrization of algebraic surfaces. Computer Aided Geometric Design 30(4), 374–388 (2013)
19. Zwillinger, D.: Handbook of Differential Equations, 3rd edn. Academic Press, San Diego (1998)

Eigenvalue Method with Symmetry and Vibration Analysis of Cyclic Structures

Aurelien Grolet[1], Philippe Malbos[2], and Fabrice Thouverez[1]

[1] LTDS, École centrale de Lyon,
36 avenue Guy de Collongue, 69134 ECULLY cedex, France
[2] Université de de Lyon, ICJ CNRS UMR 5208, Université Claude Bernard Lyon 1,
43 boulevard du 11 novembre 1918, 69622 VILLEURBANNE cedex, France

Abstract. We present an application of the eigenvalue method with symmetry for solving polynomial systems arising in the vibration analysis of mechanical structures with symmetry properties. The search for solutions is conducted by the so called multiplication matrix method in which the symmetry of the system is taken into account by introducing a symmetry group and by working with the set of invariant polynomials under the action of this group. Using this method, we compute the periodic solutions of a simple dynamic system modeling a cyclic mechanical structure subjected to nonlinearities.

1 Introduction

Many engineering problems can be modeled or approximated such that the determination of a solution goes through the resolution of a polynomial system. In this paper, we are interested in computing periodic solutions of nonlinear dynamic equations. It can be shown that the Fourier coefficients of the (approximated) periodic solutions can be obtained by solving multivariate polynomial equations resulting from the application of the harmonic balance method [1, 2]. Moreover, in our applications, the dynamical system is often invariant under some transformations (cyclic permutation, change of sign, ...) due to the presence of symmetry in the mechanical structure. This implies that the polynomial system to be solved is also invariant under some transformations, and so does its solutions.

Most of the time, in mechanical engineering, polynomial systems are solved by numeric methods such as a Newton-like algorithms, which output only one solution of the system depending on the starting point provided. Although the Newton method is an efficient algorithm (quadratic convergence), the search for all solutions of a polynomial system cannot be conducted in a reasonable time using only this method. In the continuation methods framework [3], the study of bifurcations allows to follow new branches of solution, but does not warranty that all solutions are computed (e.g., disconnected solutions).

Homotopy methods [2, 4] are an alternative to the Newton algorithm when searching for all solutions of a multivariate polynomial system. Basically, homotopy methods rely on the continuation of the (known) roots of a starting

V.P. Gerdt et al. (Eds.): CASC Workshop 2014, LNCS 8660, pp. 121–137, 2014.

polynomial Q (easy to solve) to the (unknown) roots of a target polynomial P. The choice of the starting polynomial is a key point on witch depends the efficiency of the method. Indeed, if the starting polynomial has to many roots compared to P, most of the continuations will lead to divergent solutions, thus wasting time and resources. Improvements such as the polyhedral homotopy aims at reducing the number of divergent path by considering a starting polynomial structurally close to the target polynomial [5]. However, the presence of high combinatoric or probabilistic considerations makes the application of the method rather difficult. Moreover, it is not clear how to take into account symmetry properties in the polyhedral homotopy.

In this context, where numerical methods are not entirely satisfactory, computer algebra appears as an attractive alternative, since there exists an efficient method especially developed for solving symmetric polynomial systems. This method, proposed by Gatermann [6], is called eigenvalue method with symmetry. It is based upon the multiplication matrix method [7–9], where solutions of the polynomial system are obtained by solving an eigenvalue problem. Moreover, it takes into account the symmetry of the system. The method is efficient since taking into account symmetry allow for reducing the size of the multiplication matrix such that only one representative of each orbit of solution is computed.

In this paper, we propose a new application of the eigenvalue method with symmetry for computing periodic solutions of nonlinear dynamic systems solved by the harmonic balance method. It constitutes an attempt to evaluates the capabilities of computer algebra methods in the field of mechanical engineering, in which numerical methods are often the norm.

The paper is organized as follow. Section 2 describes systems studied in this work. The motion's equations are presented along with a brief recall of the Harmonic Balance Method, and we derive the polynomial equations solved in this study. Section 3 concentrates on polynomial systems solving. We recall the multiplication matrix method and we describe how to take into account the symmetry of the system. We present our resolution algorithm in this section. Section 4 is dedicated to numerical examples and the paper ends with some concluding remarks.

2 Dynamic System and Periodic Solutions

2.1 System of Interest

We aim at finding periodic solutions of (polynomial) nonlinear mechanical structures with special symmetry. For example, bladed disks subjected to geometric nonlinearities represent such a structure [1]. Here, only a simple cyclic system (which can be seen as a reduced order model of a bladed disk, where all blades have been reduced on their first mode of vibration) will be considered. The model consists in N Duffing oscillators linearly coupled, governed by the following motion equation:

$$m\ddot{u}_i + c\dot{u}_i + (k + 2k_c)u_i - k_c u_{i-1} - k_c u_{i+1} + k_{nl}u_i^3 = f_i, \quad i = 1, \dots, N, \quad (1)$$

where $u_i(t)$ represents the temporal evolution of degree of freedom (dof) number i, and $f_i(t)$ represents the temporal evolution of the excitation force acting on dof number i. If there is no force, note that this dynamic system is invariant under the action of the dihedral group $\mathcal{D}_N$ (group of symmetries of a regular polygon with N sides).

Equation (1) can be written in the following matrix form:

$$\mathbf{M}\ddot{u} + \mathbf{C}\dot{u} + \mathbf{K}u + \mathbf{F}_{nl}(u) = \mathbf{F}_{ex}, \qquad (2)$$

were $u(t)$ is the vector of dof of size N, $\mathbf{M} = m\mathbf{I}$ is the mass matrix, $\mathbf{C} = c\mathbf{I}$ is the damping matrix, $\mathbf{K} = (k + 2k_c)\mathbf{I} - k_c\mathbf{I}^L - k_c\mathbf{I}^U$ is the stiffness matrix, and $\mathbf{F}_{nl}(u) = k_{nl}u^3$, $\mathbf{F}_{ex}(t) = f(t)$ correspond to the nonlinear and excitation forces respectively. The excitation forces will be assumed to be periodic, with period $T = \frac{2\pi}{\omega}$, and we will search for periodic solutions $u(t)$, using the harmonic balance method described hereafter.

2.2 Harmonic Balance Method

The harmonic balance method (HBM) is a widely used method in finding approximation to periodic solutions of nonlinear differential equations such as (2) [1,10]. The solutions $u(t)$ is approximated under the form of a truncated Fourier series, up to harmonic H, and a system of algebraic equations is derived by applying Galerkin projections. Let us recall the main steps of the method. At first, each component $u_i(t)$ of the periodic solution $u(t)$ is approximated by $\widehat{u}_i(t)$ under the following form (H harmonics):

$$\widehat{u}_i(t) = x^{(0)} + \sum_{k=1}^{H} x_i^{(k)} \cos(k\omega t) + y_i^{(k)} \sin(k\omega t), \quad i = 1, \ldots, N. \qquad (3)$$

We substitute (3) in (2) and we project the resulting equations on the truncated Fourier basis:

$$\frac{1}{T}\int_0^T \mathbf{R}(\widehat{u}) \times 1 \, dt = \mathbf{0}, \quad \frac{2}{T}\int_0^T \mathbf{R}(\widehat{u}) \times \cos(k\omega t) \, dt = \mathbf{0}, \quad \frac{2}{T}\int_0^T \mathbf{R}(\widehat{u}) \times \sin(k\omega t) \, dt = \mathbf{0},$$
$$(4)$$

with $k = 1, \ldots, H$, $T = 2\pi/\omega$ and $\mathbf{R}(\widehat{u}) = \mathbf{M}\ddot{\widehat{u}} + \mathbf{C}\dot{\widehat{u}} + \mathbf{K}\widehat{u} + \mathbf{F}_{nl}(\widehat{u}) - \mathbf{F}_{ex}(t)$. Equations (4) corresponds to a set of $N(2H + 1)$ algebraic equations with unknowns x and y.

2.3 Equations to Be Solved

In our application, $\mathbf{F}_{nl}(u) = k_{nl}u^3$ is polynomial and (4) corresponds to a polynomial system. In order to simplify the presentation and reduce the number of variables, we will only consider a single harmonic approximation of the periodic solution, i.e., $H = 1$ in (3). Moreover, as the nonlinearity is odd, no continuous component will be retained, i.e., $x^{(0)} = \mathbf{0}$ in (3). Under these hypothesis,

(4) corresponds to a system of $2N$ polynomial equations which can be written in the following form (dropping the harmonic index $^{(k)}$):

$$
\begin{aligned}
\alpha(\omega)x_i + \delta(\omega)y_i - \beta x_{i-1} - \beta x_{i+1} + \gamma x_i(x_i^2 + y_i^2) = f_i^c, & \quad i = 1, \ldots, N, \\
\alpha(\omega)y_i - \delta(\omega)x_i - \beta y_{i-1} - \beta y_{i+1} + \gamma y_i(x_i^2 + y_i^2) = f_i^s, & \quad i = 1, \ldots, N,
\end{aligned} \tag{5}
$$

where f_i^c (resp. f_i^s) denotes the amplitude of the excitation forces relative to the $\cos(\omega t)$ (resp. $\sin(\omega t)$) term, and with the following expression for the different coefficients:

$$
\alpha(\omega) = k + 2k_c - \omega^2 m, \quad \beta = k_c, \quad \gamma = \frac{3}{4}k_{nl}, \quad \delta(\omega) = \omega c.
$$

In our application, we are interested in forced and free solutions.

Forced Solutions. In the forced case ($f^c \neq 0$ or $f^s \neq 0$), the angular frequency ω is set by the excitation forces and (5) will be solved for x and y. Depending on the symmetry of the excitation forces, System (5) may present some invariance properties. We will choose $f_i^c = 1$, $f_i^s = 0$ for all $i = 1, \ldots, N$ so that System (5) will be invariant under the action of the dihedral group $\mathcal{D}_N$.

Free Solutions. In the free case, we aims at finding solutions of an unforced, undamped version of System (2), also called nonlinear normal modes (NNM) [11–13]. In order to simplify we will only search for solutions where all dof vibrate "in-phase" (monophase NNM [14]) by imposing $y_i = 0$ for all $i = 1, \ldots, N$, thus resulting in the following polynomial system with N equations:

$$
\alpha(\omega)x_i - \beta x_{i-1} - \beta x_{i+1} + \gamma x_i^3 = 0, \quad i = 1, \ldots, N. \tag{6}
$$

The angular frequency ω will be set to an arbitrary value and System (6) will be solved for x. Again (6) is invariant under the action of the dihedral group $\mathcal{D}_N = \mathcal{C}_n \times \mathcal{C}_2$. The system is also invariant under change of sign, characterized by the group $\mathcal{C}_2 = \{e, b \mid b^2 = e\}$ with $b(x) = -x$.

3   Solving Multivariate Polynomial Systems

In this section, we present the method used to solve symmetric polynomial systems. We describe the eigenvalue method and we show how to include symmetry of the system in order to reduce the number of solutions as proposed in [6], leading to the so called eigenvalue method with symmetry. Finally we propose an algorithm to summarize the process.

3.1   Notation and Gröbner Basis

We will denote by $\mathbb{C}[x]$ the ring of multivariate polynomials with complex coefficients in the variables $x = (x_1, \ldots, x_n)$. We will consider the graded reverse lexicographic order on $\mathbb{C}[x]$. We will denote by $\mathrm{LM}(f)$ and $\mathrm{LC}(f)$ the leading monomial and the leading coefficient of a polynomial f, we will denote by $\mathrm{LT}(f) = \mathrm{LC}(f)\mathrm{LM}(f)$ its leading term.

Given a multivariate polynomial system $P(x) = [p_1(x), \ldots, p_n(x)]$, with $p_j \in \mathbb{C}[x]$, we denote by $\mathcal{I} = \langle P \rangle = \langle p_1, \ldots, p_n \rangle$ the ideal of $\mathbb{C}[x]$ generated by the polynomial system P. We denote G a Gröbner basis of $\mathcal{I}$, and $\mathrm{NF}(\cdot)$ the normal form operator. The set G being a Gröbner basis, the monomials $\mathcal{B} = \{x^\alpha \mid x^\alpha \notin \langle \mathrm{LT}(G) \rangle\}$ form a basis of the algebra $A = \mathbb{C}[x]/\mathcal{I}$, defined as the quotient of $\mathbb{C}[x]$ by the ideal $\mathcal{I}$.

If the polynomial system $P(x) = 0$ has only a finite number of solutions (say D solutions), the ideal $\mathcal{I}$ is zero-dimensional, and it can be shown [6] that A is of finite dimension D.

3.2   Multiplication Matrices Method

Given a polynomial f in $\mathbb{C}[x]$, we consider the map $m_f : A \to A$, defined by $m_f(h) = fh$, for any h in A. Since A is a finite-dimensional algebra the map m_f can be represented by a matrix $\mathbf{M}_f$ relative to the basis $\mathcal{B}$. The matrix $\mathbf{M}_f$ is called *multiplication matrix* and it is characterized by the relation $f \, \mathcal{B} = \mathbf{M}_f \, \mathcal{B}$ (modulo $\mathcal{I}$), or equivalently:

$$f \, \mathcal{B}_i = \sum_{j=1}^{D} M_{i,j}^f \, \mathcal{B}_j \quad \mathrm{mod}(\mathcal{I}), \ i = 1, \ldots, D. \tag{7}$$

The coefficients of line i of the matrix $\mathbf{M}_f$ can be obtained by computing the normal form of each product $f \, \mathcal{B}_i$ and by expressing the results as a linear combination of elements of $\mathcal{B}$.

For particular choices of $f = x_p$, $p = 1, \ldots, n$, it can be shown that the eigenvalues of the multiplication matrices $\mathbf{M}_{x_p}$ are related to the zeros of the polynomial system. Indeed, substituting $f = x_p$ into (7), for any x, we have:

$$\left(\mathbf{M}_{x_p} - x_p \mathbf{I} \right) \mathcal{B}(x) = 0 \quad \mathrm{mod}(\mathcal{I}). \tag{8}$$

It follows that the vector $\left(\mathbf{M}_{x_p} - x_p \mathbf{I} \right) \mathcal{B}(x)$ can therefore be expressed as a combination of the polynomials in P. Now, let's suppose that x^* is a root of P. Then $p_i(x^*) = 0$ for all $i = 1, \ldots, n$, and (8) shows that x_p^* is an eigenvalue of $\mathbf{M}_{x_p}$ associated to the eigenvector $\mathcal{B}(x^*)$. Note that the eigenvector should be normalized so that its first component equals 1 (in order to match with the associated polynomials $\mathcal{B}_1(x) = 1$).

Going further, it can be shown [8] that the components of the roots are given by the eigenvalues of $\mathbf{M}_{x_p}$, $p = 1, \ldots, n$, associated with common eigenvectors $\mathcal{B}_k$.

We follow the method given in [4], considering only one multiplication matrix associated with a linear combination of the variables $f = \sum_{i=1}^{n} c_i \, x_i$, where c_i are rational numbers chosen such that the value of $f(x^{(k)})$ is different for each solution $x^{(k)}$, $k = 1, \ldots, D$. Generally, random choices for coefficients c_i are sufficient to ensure this properties almost surely [4]. The search for the roots of system P is then simply conducted by solving the eigenvalue problem $(\mathbf{M}_f - f\mathbf{I})\mathcal{B} = 0$, and by reading the solutions in the eigenvectors $\mathcal{B}_k = \mathcal{B}(x^{(k)})$, $k = 1, \ldots, D$.

3.3   Introducing Symmetry

Invariant Polynomial Systems. Due to the symmetry properties of the mechanical structure, the polynomial systems to be solved in our applications (see section 2.3) also possess a symmetric structure. Here we will consider that the polynomial system to be solved is equivariant under the action of a group $\mathcal{G}$, that is $P(g(x)) = g(P)(x)$, $\forall g \in \mathcal{G}$, where $g \in \mathcal{G}$ is a permutation operation defined by $g(x) = [x_{g(1)}, \ldots, x_{g(n)}]$. The set of invariant polynomial under $\mathcal{G}$ is denoted $\mathbb{C}[x]^{\mathcal{G}}$ and defined by: $\mathbb{C}[x]^{\mathcal{G}} = \{f \in \mathbb{C}[x] \mid f(g(x)) = f(x), \forall g \in \mathcal{G}\}$. We denote by $\mathcal{I}^{\mathcal{G}} = \mathcal{I} \cap \mathbb{C}[x]^{\mathcal{G}}$ the ideal invariant under the action of the group $\mathcal{G}$.

Quotient Decomposition. It can be shown that $\mathbb{C}[x]$ can be decomposed into a direct sum of isotypic components [6, 15], such that $\mathbb{C}[x] = V_1 \oplus V_2 \oplus \ldots \oplus V_K$, where the V_i's are the isotypic components (related to the K irreducible representations of group $\mathcal{G}$ [6]), and where the first component is the invariant ring itself: $V_1 = \mathbb{C}[x]^{\mathcal{G}}$. By defining $\mathcal{I}_i = \mathcal{I} \cap V_i$, the algebra $A = \mathbb{C}[x]/\mathcal{I}$ can be decomposed into a direct sum as follows [6]:

$$A = \mathbb{C}[x]^{\mathcal{G}}/\mathcal{I}^{\mathcal{G}} \oplus V_2/\mathcal{I}_2 \oplus \ldots \oplus V_K/\mathcal{I}_K \tag{9}$$

The space $\mathbb{C}[x]^{\mathcal{G}}$ can be decomposed into the following direct sum (Hironaka decomposition) [6]:

$$\mathbb{C}[x]^{\mathcal{G}} = \oplus_i S_i \, \mathbb{C}[\pi] = \mathbb{C}[\pi] \oplus S_2 \mathbb{C}[\pi] \oplus S_3 \mathbb{C}[\pi] \oplus \cdots \oplus S_p \mathbb{C}[\pi]$$

where $\pi = [\pi_1, \ldots, \pi_n]$ is the set of primary polynomial invariants related to $\mathcal{G}$, and $S_2, \ldots, S_n$ correspond to the secondary polynomial invariants related to $\mathcal{G}$. The primary polynomial invariants π can be found by using the Reynold projection operator [16]. In this work, we compute the primary and secondary invariants using the `invariant_ring` command of Singular [17].

Using the Primary Polynomial Invariants. In the following, the primary invariants will be used to find solutions of a symmetric polynomial system. Let's suppose that we can find the values of the primary invariants $\pi^{(k)} = \pi(x^{(k)})$ for each solution $x^{(k)}$, then by solving the following systems:

$$\pi(x) = \pi^{(k)}, \quad k = 1, \ldots, D_{\mathcal{G}},$$

for x by a Newton-like method, one can compute an unique occurrence of solution $x^{(k)}$ and the other can be generated by applying the group's actions on $x^{(k)}$, i.e., $g(x^{(k)}), \forall g \in \mathcal{G}$.

We will compute the values of the primary invariants $\pi^{(k)}$ for each solution $x^{(k)}$ with the multiplication matrix method. However, as shown in [6], the multiplication matrices related to the primary invariants are redundant as they contain the same eigenvalues several times. In a suited basis of A, it is even shown that the multiplication matrices associated to the primary invariants are block diagonal [6, Thm. 3], with each block containing the same eigenvalues [6, Prop. 8]. Thus, only the first diagonal block (related to the subspace $\mathbb{C}[\boldsymbol{x}]^{\mathcal{G}}/\mathcal{I}^{\mathcal{G}}$) is of interest to compute the values of the primary invariants.

All that is left to do here, is to find a basis $\mathcal{B}'$ of A that makes the multiplication matrices block diagonal. More precisely, it is sufficient to find a basis $\mathcal{B}^{\mathcal{G}} = [\mathcal{B}'_1, \ldots, \mathcal{B}'_{D_{\mathcal{G}}}]$ of $\mathbb{C}[\boldsymbol{x}]^{\mathcal{G}}/\mathcal{I}^{\mathcal{G}}$ in agreement with the direct sum decomposition in (9).

Construction of an Adapted Basis. The goal is to find a basis $\mathcal{B}^{\mathcal{G}}$ of $\mathbb{C}[\boldsymbol{x}]^{\mathcal{G}}/\mathcal{I}^{\mathcal{G}}$ (with $\#\mathcal{B}^{\mathcal{G}} = D_{\mathcal{G}}$) in agreement with the direct sum decomposition in (9), in order to construct the first block of a multiplication matrix. As in the previous section, the multiplication matrix will be related to a polynomial $f = \sum_{i=1}^{n} c_i \, \pi_i$, where c_i are rational coefficients chosen randomly.

The basis $\mathcal{B}^{\mathcal{G}}$ should only contains invariant polynomials, and their normal forms should be sufficient to express all remainders r in the division of f $\mathcal{B}_i^{\mathcal{G}}$ by $\mathcal{I}$ (i.e., $r = \sum_{j=1}^{D_{\mathcal{G}}} M_{i,j}^{\mathcal{G}} \, \mathrm{NF}(\mathcal{B}_j^{\mathcal{G}})$).

We suppose that a Gröbner basis G of $\mathcal{I}$ is known. Let NF the normal form operator for G. At start, we set $\mathcal{B}_1^{\mathcal{G}} = 1$.

The construction of the basis then goes as follows. For $\mathcal{B}_i^{\mathcal{G}}$ in $\mathcal{B}^{\mathcal{G}}$ we compute the normal form $r = \mathrm{NF}(f \ \mathcal{B}_i^{\mathcal{G}})$. Then, until the remainder r equals zero, we search if there exists $\mathcal{B}_j^{\mathcal{G}}$ in $\mathcal{B}^{\mathcal{G}}$ such that $\mathrm{LM}(\mathrm{NF}(\mathcal{B}_j^{\mathcal{G}})) = \mathrm{LM}(r)$:

- if such a $\mathcal{B}_j^{\mathcal{G}}$ exists, then we divide r by $\mathrm{NF}(\mathcal{B}_j^{\mathcal{G}})$: $r = q \, \mathrm{NF}(\mathcal{B}_j^{\mathcal{G}}) + h$ and we save the (numeric) matrix coefficient $M_{i,j}^{\mathcal{G}} = q$. Finally, we affect $r = h$, and search for a new divisor of $\mathrm{LT}(r)$.
- if not, we will create a new basis term $\mathcal{B}_k^{\mathcal{G}}$ whose leading monomial equals $\mathrm{LM}(r)$ by considering the Reynold projection of $\mathrm{LM}(r)$, ie: $\mathcal{B}_k^{\mathcal{G}} = \mathrm{Re}_{\mathrm{LM}(r)}$. However, it may happen that $\mathrm{LM}(\mathrm{NF}(\mathrm{Re}_{\mathrm{LM}(r)})) \neq \mathrm{LM}(r)$. In that case, we modify the Reynold projection by subtracting the high order term until $\mathrm{LM}(\mathrm{NF}(\mathrm{Re}_{\mathrm{LM}(r)})) = \mathrm{LM}(r)$. This is done by searching into the basis an element $\mathcal{B}_{j_0}^{\mathcal{G}}$ such that $\mathrm{LM}(\mathrm{NF}(\mathcal{B}_{j_0}^{\mathcal{G}})) = \mathrm{LT}(\mathrm{NF}(\mathrm{Re}_{\mathrm{LM}(r)}))$ and by modifying the Reynold projection : $\mathrm{Re}_{\mathrm{LM}(r)} = \mathrm{Re}_{\mathrm{LM}(r)} - c_{j_0}\mathcal{B}_{j_0}^{\mathcal{G}}$. Once the invariant is computed, we divide r by the new element.

This process is repeated until all products f $\mathcal{B}_i$, $i = 1, \ldots, D_{\mathcal{G}}$, have been computed. The basis construction is summarized in Algorithm 1.

Algorithm 1. Computation of a basis $\mathcal{B}^{\mathcal{G}}$ of the invariant space $\mathbb{C}[\boldsymbol{x}]^{\mathcal{G}}/\mathcal{I}$, and construction of the multiplication matrix of the invariant variable $f = \sum c_j \pi_j$

#PRELIMINARIES
compute a Gröbner basis G of P with the grevlex order
initialize $f = \sum_j c_j \pi_j$, $\mathcal{B}_1^{\mathcal{G}} = 1$, $n = 1$
#BASIS COMPUTATION
$j = 1$
while $j < n + 1$ **do**

compute the normal form $r = \mathrm{NF}(f\mathcal{B}^{\mathcal{G}}_j)$
while $r \neq 0$ **do**
    if $\mathrm{LM}(r) \in \mathrm{NF}(\mathcal{B}^{\mathcal{G}})$ **then**
        find k such that $\mathrm{LM}(r) = \mathrm{NF}(\mathcal{B}^{\mathcal{G}}_k)$
        reduce $r : r = q\mathcal{B}^{\mathcal{G}}_k + h$
        save $M_{j,k} = q$ and update $: r = h$
    else
        compute the Reynold projection $Re(\boldsymbol{x}) = Re_{\mathrm{LM}(r)}(\boldsymbol{x})$
        if $\mathrm{LM}(\mathrm{NF}(Re))=\mathrm{LM}(r)$ **then**
            affect $\mathcal{B}^{\mathcal{G}}_{n+1} = Re$
        else
            while $\mathrm{LM}(\mathrm{NF}(Re)) \neq \mathrm{LM}(r)$, reduce the Reynold projection: $Re = Re - c_k\mathcal{B}^{\mathcal{G}}_k$
            affect $\mathcal{B}^{\mathcal{G}}_{n+1} = Re$
        end if
        reduce the normal form $r : r = qRe + h$
        save $M_{j,n+1} = q$ and update $: n = n + 1, r = h$
    end if
end while
$j = j + 1$
end while
return the multiplication matrix $\boldsymbol{M}_f$ and the basis $\mathcal{B}^{\mathcal{G}}$

The termination of the algorithm is related to the finite dimension of $\mathbb{C}[\boldsymbol{x}]/\mathcal{I}$. The fact that the algorithm actually output a basis of the invariant space is related to the fact that the normal forms of the basis vectors have different leading terms (see also [6]).

4   Numerical Applications

In this section, we apply the eigenvalue method with symmetry to the system given in Section 2.3. The algorithm presented in the previous section has been implemented with the Maple 15 software. The numerical applications will be conducted for system with $N = 2, 4$ degrees of freedom. In the two cases, free and forced analysis are conducted. Solutions for a particular frequency are computed with the multiplication matrix method, and we give an overview of the system dynamics by applying continuation methods [3]. Finally, an NNM analysis is carried for $3 \leq N \leq 6$ in order to show the decrease in the number of solutions.

4.1   Simple Example with 2 Degrees of Freedom

As a first application, we study a system with $N = 2$ degrees of freedom. In this case, (2) reduces to the following dynamic system:

$$\begin{aligned} m\ddot{u}_1 + c\dot{u}_1 + (k + k_c)u_1 - k_cu_2 + k_{nl}u_1^3 = f_1(t), \\ m\ddot{u}_2 + c\dot{u}_2 + (k + k_c)u_2 - k_cu_1 + k_{nl}u_2^3 = f_2(t). \end{aligned} \tag{10}$$

The application of the HBM with only one harmonic ($u_i = x_i \cos(\omega t) + y_i \sin(\omega t)$) leads to the following system of polynomial equations:

$$
\begin{aligned}
\alpha x_1 - \beta x_2 + \delta y_1 + \gamma x_1(x_1^2 + y_1^2) &= f_c, \\
\alpha y_1 - \beta y_2 - \delta x_1 + \gamma y_1(x_1^2 + y_1^2) &= f_s, \\
\alpha x_2 - \beta x_1 + \delta y_2 + \gamma x_2(x_2^2 + y_2^2) &= f_c, \\
\alpha y_2 - \beta y_1 - \delta x_2 + \gamma y_2(x_2^2 + y_2^2) &= f_s,
\end{aligned}
\tag{11}
$$

with $\alpha = k + k_c - \omega^2 m$, $\beta = k_c$, $\gamma = \frac{3}{4}k_{nl}$ and $\delta = \omega c$. The frequency parameter will be set to $\omega = \frac{25}{10}$ (however the search for multiple solutions can be conducted for any values of ω), leading to the following numerical values:

$$
\alpha = \frac{-17}{4}, \; \beta = 1, \gamma = \frac{3}{4}, \; \delta = \frac{1}{10}, \; f_c = 1, \; f_s = 0.
\tag{12}
$$

Monophase NNM Analysis. We search for monophase NNM solutions of (11) (undamped, unforced). In this case, the system (6) reduces to the following:

$$
\begin{aligned}
\alpha x_1 - \beta x_2 + \gamma x_1^3 &= 0, \\
\alpha x_2 - \beta x_1 + \gamma x_2^3 &= 0.
\end{aligned}
\tag{13}
$$

We consider the order grevlex with $x_1 > x_2$. Since the leading term of each equation are co-prime, the system P is already in a Gröbner basis form. We computed a normal set and we show the algebra $A = \mathbb{C}[x]/\langle P \rangle$ is of dimension 9. The system (13) is invariant under permutation of variable and under change of sign. This invariance property corresponds to the group $\mathcal{G} = \mathcal{C}_2 \times \mathcal{C}_2$. All element $g \in \mathcal{G}$ can be represented by a matrix $\mathbf{M}_g = \mathbf{A}^{i_g}\mathbf{B}^{i_g}$ with

$$
\mathbf{A} = \begin{bmatrix} 0 & 1 \\ 1 & 0 \end{bmatrix}, \; \mathbf{B} = \begin{bmatrix} -1 & 0 \\ 0 & -1 \end{bmatrix}.
$$

Using `Singular`, we know that the primary invariant of $\mathcal{G}$ are $\pi_1 = x_1 x_2$ and $\pi_2 = \frac{1}{2}(x_1^2 + x_2^2)$. We set $f = \pi_1 + \frac{2}{3}\pi_2$, and we construct the multiplication matrix of f in an symmetry adapted basis of $A^{\mathcal{G}}$ using Algorithm 1. The basis $\mathcal{B}^{\mathcal{G}}$ and the multiplication matrix $\mathbf{M}_f$ are given by

$$
\mathcal{B}^{\mathcal{G}} = [1, \; \frac{1}{2}(x_1^2 + x_2^2), \; x_1 x_2, \; x_1^2 x_2^2], \quad \mathbf{M}_f = \begin{bmatrix} 0 & \frac{4}{3} & 1 & 0 \\ 0 & \frac{46}{39} & \frac{59}{9} & \frac{2}{3} \\ 0 & \frac{19}{9} & \frac{68}{9} & 1 \\ 0 & \frac{471}{27} & \frac{1187}{27} & \frac{68}{9} \end{bmatrix}.
$$

The computation of eigenvalues $\boldsymbol{\lambda} = f(x^*)$ and eigenvectors $\mathcal{B}^{\mathcal{G}}(x^*)$ of $\mathbf{M}_f$ gives (after normalization of the first component):

$$
\boldsymbol{\lambda} = \begin{bmatrix} 0 \\ 16.3333 \\ 2.4444 \\ 1.4444 \end{bmatrix}, \quad \mathcal{B}^{\mathcal{G}}(x^*) = \begin{bmatrix} 1.00 & 1.00 & 1.00 & 1.00 \\ 0 & 7.00 & 2.83 & 4.33 \\ 0 & 7.00 & -1.33 & -4.33 \\ 0 & 49.00 & 1.77 & 18.77 \end{bmatrix}.
$$

Here π_1 and π_2 belong to the invariant basis $\mathcal{B}^{\mathcal{G}}$ ($\pi_1 = \mathcal{B}_3^{\mathcal{G}}$ and $\pi_2 = \mathcal{B}_2^{\mathcal{G}}$), so that their values $\boldsymbol{\pi}(\boldsymbol{x}^*)$ can directly be read into the eigenvectors $\mathcal{B}^{\mathcal{G}}(\boldsymbol{x}^*)$ (at line 3 and line 2), leading to the 4 following systems of equations:

$$(\pi_1(\boldsymbol{x}), \ \pi_2(\boldsymbol{x})) \in \{ \ (0,0), \ (7,7), \ (-1.33, 2.83), \ (-4.33, 4.33) \ \} \quad (14)$$

The nonlinear system in (14) are solved by a Newton Raphson method. Four different solutions are obtained, see (15), and they are depicted on Fig.1. We verified that those solutions are actually solutions of $\boldsymbol{P}(\boldsymbol{x}) = \boldsymbol{0}$ by computing the values of $\|\boldsymbol{P}(\boldsymbol{x}^*)\|$ in following table for solution quality of (13) at $\omega = \frac{25}{10}$

solution	1	2	3	4
value $\|\boldsymbol{P}(\boldsymbol{x}^*)\|$	0.04	0.11	0.00	0.04
relative diff. from NR sol. (%)	x	0.23	0.00	0.32

To assess the quality of the real solutions, we also compared them to refined solutions obtained with a Newton algorithm applied on $\boldsymbol{P}$ with starting points $\boldsymbol{x}_0 = \boldsymbol{x}^*$. It is seen that solutions from the eigenvalue method are indeed very close to the actual roots of $\boldsymbol{P}$, as their relative differences lie below 0.5%. In any cases, a few Newton iterations should be applied to overcome the numerical error due to numerical rounding of rational numbers in the multiplication matrix.

$$(x_1, x_2) \in \{ \ (0,0), \ (-2.65, -2.65), \ (2.31, -0.58), \ (2.08, -2.08) \ \}. \quad (15)$$

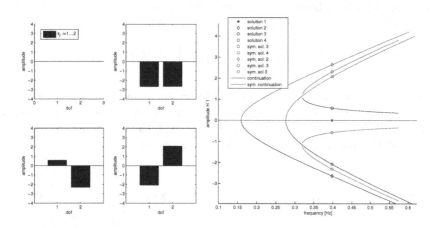

Fig. 1. Left:Form of the real solutions of system (13) found by the invariant multiplication matrix method. Right: Frequency continuation of the solutions obtained at $f = \frac{1}{2\pi}\frac{25}{10}$ and their symmetrics relative to the group's operations.

The application of the group actions generates 5 other solutions. At the end the total set of solutions contains 9 elements as indicated by the dimension of the quotient space. However, the use of symmetry decreased the size of the eigenvalue problem from 9 to 4, leading to only 4 solutions (one for each orbit of solutions).

In order to give an overview of the system dynamics, we use the four solutions in (15) as starting points for a continuation procedure on the parameter ω. The results are depicted on Fig. 1 and correspond to the monophase nonlinear normal modes of the systems. Four types of solution can be identified: the trivial solution (sol.1), an in-phase solution (sol. 2), an out-of-phase solution (sol. 4) and a localized solution (sol. 3) which corresponds to a bifurcation of the out-of-phase solution.

Forced Analysis. We now turn to the forced analysis of system (11). We compute a Gröbner basis G with 12 elements relatively to the grevlex order with $y_2 < y_1 < x_2 < x_1$. We compute $\dim(A) = 11$, thus the system has 11 solutions. The system is invariant under the action of $\mathcal{G} = \mathcal{C}_2 = \{e,\ a\ |\ a^2 = e\ \}$ with $a(x_1, y_1, x_2, y_2) = (x_2, y_2, x_1, y_1)$. The representation of $\mathcal{G}$ is chosen such that a is represented by $\mathbf{M}_a = \begin{bmatrix} \mathbf{0} & \mathbf{I}_2 \\ \mathbf{I}_2 & \mathbf{0} \end{bmatrix}$.

The primary invariant of $\mathcal{G}$ are given by $\pi_1 = \frac{1}{2}(x_1 + x_2)$, $\pi_2 = \frac{1}{2}(y_1 + y_2)$, $\pi_3 = x_1 x_2$ and $\pi_4 = y_1 y_2$; and the multiplication matrix is computed for $f = \pi_1 + \pi_2 + \pi_3 + \pi_4$. By using Algorithm 1 we compute a basis $\mathcal{B}^{\mathcal{G}}$ of $A^{\mathcal{G}}$ with 7 elements. All primary invariants are in $\mathcal{B}^{\mathcal{G}}$ except for π_3. Thus, the normal form of π_3 is computed and the result is expressed in terms of elements of $\mathcal{B}^{\mathcal{G}}$: $\pi_3 = \mathbf{c}^T \mathcal{B}^{\mathcal{G}}$. After solving the eigenvalue problem, the values of π_3 at the solutions point are given by $\pi_3(\mathbf{x}^*) = \mathbf{c}^T \mathcal{B}^{\mathcal{G}}(\mathbf{x}^*)$.

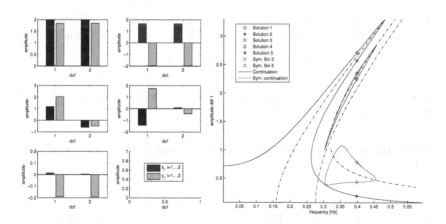

Fig. 2. left: Form of the real solutions of system (11) found by the invariant multiplication matrix method. Right: Frequency continuation of the solution obtained at $f = \frac{1}{2\pi} \frac{25}{10}$ and their symmetric relative to the group operation.

The solutions of $P(x) = 0$ are then evaluated by solving the 7 nonlinear systems $\pi = \mathcal{B}^{\mathcal{G}}(x^*)$: 7 solutions (5 real and 2 complex) are found by a Newton algorithm, and the form of the real solutions are depicted on Fig.2.

Assessment of the solution's quality for (11) at $\omega = \frac{25}{10}$ is given in the following table:

solution	1	2	3	4	5
value $\|P(x^*)\|$	0.00	0.00	0.00	0.00	0.02
relative diff. from NR sol. (%)	0.03	0.02	0.00	0.00	2.80

Note that solutions from the eigenvalue method are close to the actual roots of P, as their relative differences lie below 3%. To obtain the full set of solution, we apply the group actions and generate 4 more solutions, leading to a total of 11 solutions (7 real and 4 complex) as indicated by the dimension of the quotient space.

The application of the continuation procedure for the 5 real solutions from the invariant system (Fig.2) shows that 3 solutions belong to the principale resonance curve, and that 2 solutions belong to closed curves corresponding to a localized motion. The application of the group action generates another closed curve solution corresponding to the change of coordinates $(u_1, u_2) \to (u_2, u_1)$ in the dynamic system (10). All forced solutions are positioned around the backbone curves coming from the monophase NNM analysis.

4.2   Simple Example with 4 Degrees of Freedom

For $N = 4$, the application of the HBM with one harmonic on (2) leads to the following system:

$$\begin{aligned}
\alpha x_i - \beta x_{i+1} - \beta x_{i-1} + \delta y_i + \gamma x_i(x_i^2 + y_i^2) = f_i^c, & \quad i = 1, \dots, 4, \\
\alpha y_i - \beta y_{i+1} - \beta y_{i-1} - \delta x_i + \gamma y_i(x_i^2 + y_i^2) = f_i^s, & \quad i = 1, \dots, 4,
\end{aligned} \tag{16}$$

with $\alpha = k + 2k_c - \omega^2 m$, $\beta = k_c$, $\gamma = \frac{3}{4}k_{nl}$ and $\delta = \omega c$. In the NNM analysis, the frequency parameter will be set to $\omega = \frac{31}{10}$, leading to the following numerical values:

$$\alpha = \frac{-661}{100}, \ \beta = 1, \gamma = \frac{3}{4}, \ \delta = \frac{1}{10}, \ f_c = 1, \ f_s = 0.$$

In the forced analysis, the angular frequency will be set by $\omega = \frac{25}{10}$, leading to the numerical values in (12).

Monophase NNM Analysis. For the monophase analysis the system is the following:

$$\alpha x_i - \beta x_{i+1} - \beta x_{i-1} \gamma x_i^3 = 0, \quad i = 1, \dots, 4. \tag{17}$$

As in the previous example, the system is already in a gröbner basis form for the grevlex order, and we have $\dim(A) = 81$. The invariance group is taken as $\mathcal{G} = \mathcal{C}_4 \times \mathcal{C}_2$. The primary invariant of $\mathcal{G}$ are given by:

$$\begin{aligned}
\pi_1 &= x_1 x_3 + x_2 x_4, & \pi_2 &= x_1 x_2 + x_2 x_3 + x_3 x_4 + x_4 x_1, \\
\pi_3 &= x_1^2 + x_2^2 + x_3^2 + x_4^2, & \pi_4 &= x_1 x_2 x_3 x_4.
\end{aligned}$$

The application of Algorithm 1 leads to the construction of a basis $\mathcal{B}^{\mathcal{G}}$ with 14 elements. Following method exposed in the previous section, 14 real solutions are obtained by solving the invariant systems, and their forms are depicted in Fig.3. The assessment of the solutions quality for (17) at $\omega = \frac{31}{10}$ is given in the following table

solution	1	2	3	4	5	6	7	8	9	10	11	12	13	14
residual $\|P(x^*)\|$	0.58	0.68	0.00	0.23	0.00	0.97	0.00	0.00	0.00	0.00	0.00	0.00	0.00	0.00
rel. diff. from NR sol. (%)	x	0.74	0.00	0.90	0.01	2.11	0.01	0.01	0.01	0.01	0.01	0.00	0.00	0.00

showing that all solutions of the invariant systems are indeed solutions of $P(x) = 0$. The total set of solutions is generated by applying the group action leading to 81 solutions:

solution	1 2 3 4 5 6 7 8 9 10 11 12 13 14	total
occurence	1 2 8 2 8 8 4 8 8 8 4 4 8 8	81

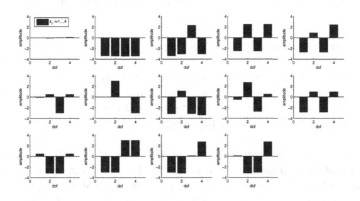

Fig. 3. Form of the real solutions of system (17) found by the invariant multiplication matrix method

Forced Analysis. We now turn to the forced analysis of system (16). First, the angular frequency parameter is set to $\omega = \frac{25}{10}$. In this case the computation of a Gröbner basis and a normal set for the grevlex order tells us that the quotient space A is of dimension 147. The invariance group is taken as $\mathcal{G} = \mathcal{C}_4 \times \mathcal{C}_2$ represented in $\mathbb{C}^8$ by the following matrices:

$$\mathbf{M}_r = \begin{bmatrix} 0 & I_2 & 0 & 0 \\ 0 & 0 & I_2 & 0 \\ 0 & 0 & 0 & I_2 \\ I_2 & 0 & 0 & 0 \end{bmatrix}, \quad \mathbf{M}_s = \begin{bmatrix} I_2 & 0 & 0 & 0 \\ 0 & 0 & 0 & I_2 \\ 0 & 0 & I_2 & 0 \\ 0 & I_2 & 0 & 0 \end{bmatrix}.$$

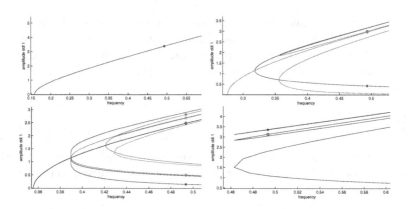

Fig. 4. Frequency continuation of the solution obtained at $f = \frac{1}{2\pi}\frac{31}{10}$ and their symmetric relative to the group operation (only positive amplitudes of the first dof are depicted). From top left to botom right: Mode 1 (solution 2); Mode 2 (solutions 7, 11, 12, 13, 14); Mode 3 (solutions 4, 5 ,6 ,9, 10); Disconnected solutions (solutions 3, 8).

The primary invariant of $\mathcal{G}$ are given by:

$$\pi_1 = y_1 + y_2 + y_3 + y_4, \quad \pi_2 = x_1 + x_2 + x_3 + x_4, \quad \pi_3 = y_1 y_3 + y_2 y_4,$$
$$\pi_4 = y_1 x_3 + y_3 x_1 + y_2 x_4 + y_4 x_2, \quad \pi_5 = x_1 x_3 + x_2 x_4,$$
$$\pi_6 = y_1 y_2 + y_2 y_3 + y_3 y_4 + y_4 y_1, \quad \pi_7 = x_1 x_2 x_3 x_4,$$
$$\pi_8 = x_1^2 x_2^2 + x_2^2 x_3^2 + x_3^2 x_4^2 + x_4^2 x_1^2 + y_1 y_2 y_3 y_4.$$

With Algorithm 1 we compute a basis $\mathcal{B}^{\mathcal{G}}$ with 33 elements, and the multiplication matrix associated to the polynomial $f = \sum_i c_i \pi_i$ is also of size 33. In this case all primary invariant are in the basis except for π_7, for which we compute its normal form and express it in term of elements of $\mathcal{B}^{\mathcal{G}}$ as $\pi_7 = c^T \mathcal{B}^{\mathcal{G}}$. The solutions of the eigenvalue problem then leads to 33 possible values (5 real and 28 complex) for the primary invariants. Finally, the solutions of the 5 real invariant systems lead to 5 real solutions of the polynomial system $P(x) = 0$ depicted on Fig. 5.

The application of the group's actions on the real solutions generates only two other solutions (i.e., the symmetric of solution 3 and 4). The frequency continuation of the solutions is depicted on Fig. 5. Again, three solutions belong to the principal resonance curve (corresponding to a motion shape on the first NNM), and two solutions belong to a closed curve solution corresponding to a motion shape on a bifurcation of the second NNM (i.e., a localized motion on only two dof corresponding to the monophase NNM solution 11 in Fig. 3).

4.3   NNM Analysis for $3 \leq N \leq 6$

In this last application, we consider the monophase NNM analysis of system (2) and we compare the results of Algorithm 1 with the eigenvalue method

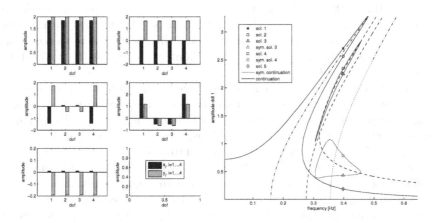

Fig. 5. Left: Form of the real solutions of system (16) found by the invariant multiplication matrix method at $\omega = \frac{25}{10}$. Right: Frequency continuation of the solution obtained at $f = \frac{1}{2\pi}\frac{25}{10}$ and their symmetric relative to the group operation. The backbone curve of NNM 1, NNM 2, NNM3 and a bifurcation of NNM 2 are also depicted.

without symmetry. The application of the harmonic balance method, leads to the polynomial system (6). In order to illustrate the reduction in the number of solution, Algorithm 1 is applied for N from 3 to 6. The invariance groupe is taken as $\mathcal{G} = \mathcal{D}_N = \mathcal{C}_N \times \mathcal{C}_2$, where $\mathcal{C}_2$ is related to the transformation $\boldsymbol{x} \to -\boldsymbol{x}$. Results are summarized in the following table:

N	$\dim(\mathbb{C}[\boldsymbol{x}]/\mathcal{I})$	$\dim(\mathbb{C}[\boldsymbol{x}]^\mathcal{G}/\mathcal{I}^\mathcal{G})$	reduction ratio
	(CPU time [s])	(CPU time [s])	
3	27 (0.4)	6 (0.3)	22.22 %
4	81 (2.4)	14 (0.6)	17.2 %
5	243 (21.3)	26 (5.12)	10.70 %
6	729 (196.7)	68 (151.3)	9.33 %

It can be seen that taking into account symmetry decreases the number of solutions down to 10% of the total number of solutions. This number should be even smaller if taking into account larger groups (i.e. $\mathcal{G} = \mathcal{C}_N \times \mathcal{C}_2 \times \mathcal{C}_2$).

From an algorithmic point of view, the methods (with or without symmetry) share several parts: (i) Groebner basis computation and (ii) normal set computation. In this simple exemple, the system is already in a Groebner basis form. However, for larger systems this part may be time consuming and the two methods equally suffer from this drawback.

From the previous table, it can be seen that the method with symmetry tends to be a bit faster to assemble the matrix (the computation of the multiplication matrix in the non symmetric case is done from an modified version of Algorithm 1). Since the matrix size is smaller, the resolution of the eigenvalue problem is faster in the method with symmetry. However, in this example, due to the small matrix size involved (≤ 1000), the resolution costs of the method with symmetry (eigenvalue computation + Newton methods) slightly overheads the costs of the eigenvalue computation in the method without symmetry (for $N = 6$ both times are around 5 seconds).

Finally, the overall computation costs of the two methods for the present example are at the same order of magnitude. However, they are several advantages in using the symmetric method. First, the solution set is much smaller and already sorted, which greatly simplifies the analysis. Second, we recall that the two methods are subjected to rounding-off errors, and that Newton iteration are needed to decrease this error. Since there is less solutions in the method with symmetry, the raffined solution are obtained faster and no sorting has to be done.

A limitation of the proposed method is related to the computation cost of the primary invariants. Indeed, the computation of primary invariants for the dihedral group $\mathcal{D}_N$ begin to be time consuming for $N > 6$. However, this computation can be considered as a preliminary computation since the primary invariants can be reused for any system having the same symmetry properties. Nevertheless, further investigations should be carried to compute the primary invariants for $\mathcal{D}_N$ for large N.

5   Discussion, Conclusion

This paper presents the application of the eigenvalue method with symmetry for solving polynomial systems arising in the vibrations study of nonlinear mechanical structures by the HBM. The system under consideration correspond to N duffing oscillators, linearly coupled. The application of the HBM with one harmonic on this system generates polynomial equations, which are invariant under some transformations (cyclic permutation, change of sign, ...).

The application of the eigenvalue method with symmetry for solving the invariant polynomial system shows that this method is well adapted for this kind of problem. Taking into account symmetry decreases the size of the multiplication matrix. Each solution corresponds to a unique orbit of solutions induced by the group's action. The solutions are very close to the actual solutions of the polynomial system, even in the presence of rounding-off errors.

The best results are obtained when searching for free solutions (NNM) of the dynamic system. In the forced case, the method is interesting when the spacial distribution of the excitation also presents symmetry properties. In the worst case scenario (symmetry breaking excitation) the system is not longer invariant, and the method no longer applicable.

Further applications to larger systems seems limited by several factors. The first drawback is related to Gröbner basis computation. For large number of variables, it can take a great amount of time even with the grevlex ordering. Second, it is not clear how to efficiently find primary invariants of large groups such a $\mathcal{D}_N$ or $\mathcal{D}_N \times \mathcal{C}_2$ for large N. However, the computation of the invariants is needed only once per invariance group as they can be reused for any subsequent computation on system having the same invariance properties.

Although this method has limitations, we have to recall that numerical methods, such has homotopie, are also subjected to limitations that restrict the size of the polynomial system to be solved. In this context, the fact that the eigenvalue method with symmetry automatically sorts the solutions is an improvement as it simplifies the analysis of the system.

References

1. Grolet, A., Thouverez, F.: Free and forced vibration analysis of nonlinear system with cyclic symmetry: application to a simplified model. Journal of Sound and Vibration 331, 2911–2928 (2012)
2. Sarrouy, E., Grolet, A., Thouverez, F.: Global and bifurcation analysis of a structure with cyclic symmetry. International Journal of Nonlinear Mechanics 46, 727–737 (2011)
3. Nayfey, A.H., Balanchandran, B.: Applied nonlinear dynamics. Wiley-Interscience (1995)
4. Sommese, A.J., Wampler, C.W.: The numerical solution of polynomials arising in engineering and science. World Scientific Publishing (2005)
5. Li, T.Y.: Solving polynomial systems with polyhedral homotopie. Taiwanese Journal of Mathematics 3, 251–279 (1999)
6. Corless, R.M., Gatermann, K., Kotsireas, I.: Using symmetries in the eigenvalue method for polynomial systems. Journal of Symbolic Computation 44, 1536–1550 (2009)
7. Yokoyama, K., Noro, M., Takeshima, T.: Solutions of systems of algebraic equations and linear maps on residue class rings. Journal of Symbolic Computation 14, 399–417 (1992)
8. Moller, H.M., Tenberg, R.: Multivariate polynomial system solving using intersections of eigenspaces. Journal of Symbolic Computation 32, 513–531 (2001)
9. Auzinger, W., Stetter, H.J.: An elimination algorithm for the computation of all zeros of a system of multivariate polynomial equations. In: Numerical Mathematics Singapore, pp. 11–30 (1988)
10. Groll, G., Ewins, D.J.: The harmonic balance method with arc-length continuation in rotor stator contact problems. Journal of Sound and Vibration 241(2), 223–233 (2001)
11. Kerschen, G., Peeters, M., Golinval, J.C., Vakakis, A.F.: Nonlinear normal modes, part i: A useful framework for the structural dynamicist. Mechanical System and Signal Processing 23, 170–194 (2009)
12. Peeters, M., Kerschen, G., Viguié, R., Sérandour, G., Golinval, J.C.: Nonlinear normal modes, part ii: toward a practical computation using continuation technique. Mechanical System and Signal Processing 23, 195–216 (2009)
13. Vakakis, A.F.: Normal mode and localiation in nonlinear systems. Wiley-Interscience (1996)
14. Peeters, M.: Toward a practical modal analysis of non linear vibrating structures using nonlinear normal modes. PhD thesis, University of Liège (2007)
15. Gatermann, K., Guyard, F.: An introduction to invariant and moduli. Journal of Symbolic Computation 28, 275–302 (1999)
16. Cox, D.A., Little, J., O'Shea, D.: Ideals, Varieties, and Algorithms: An Introduction to Computational Algebraic Geometry and Commutative Algebra, 3/e (Undergraduate Texts in Mathematics). Springer-Verlag New York, Inc. (2007)
17. Decker, W., Greuel, G.-M., Pfister, G., Schönemann, H.: SINGULAR 3-1-6 — A computer algebra system for polynomial computations (2012), http://www.singular.uni-kl.de

Symbolic-Numerical Solution of Boundary-Value Problems with Self-adjoint Second-Order Differential Equation Using the Finite Element Method with Interpolation Hermite Polynomials

Alexander A. Gusev[1], Ochbadrakh Chuluunbaatar[1,2], Sergue I. Vinitsky[1],
Vladimir L. Derbov[3], Andrzej Góźdź[4], Luong Le Hai[1,5],
and Vitaly A. Rostovtsev[1]

[1] Joint Institute for Nuclear Research, Dubna, Russia,
gooseff@jinr.ru
[2] National University of Mongolia, UlaanBaatar, Mongolia
[3] Saratov State University, Saratov, Russia
[4] Institute of Physics, Maria Curie-Skłodowska University, Lublin, Poland
[5] Belgorod State University, Belgorod, Russia

Abstract. We present a symbolic algorithm generating finite-element schemes with interpolating Hermite polynomials intended for solving the boundary-value problems with self-adjoint second-order differential equation and implemented in the Maple computer algebra system. Recurrence relations for the calculation in analytical form of the interpolating Hermite polynomials with nodes of arbitrary multiplicity are derived. The integrals of interpolating Hermite polynomials are used for constructing the stiffness and mass matrices and formulating a generalized algebraic eigenvalue problem. The algorithm is used to generate Fortran routines that allow solution of the generalized algebraic eigenvalue problem with matrices of large dimension. The efficiency of the programs generated in Maple and Fortran is demonstrated by the examples of exactly solvable quantum-mechanical problems with continuous and piecewise continuous potentials.

1 Introduction

The study of mathematical models that describe tunneling and channeling of composite quantum systems through multidimensional barriers, photo-ionization and photo-absorption in molecular, atomic, nuclear, and quantum-dimensional semiconductor systems, requires high-accuracy efficient algorithms and programs for solving boundary-value problems (BVPs) [7,5,8,9,13].

In this direction, using the variation-projection BVP formulation and finite element method (FEM) with Lagrange interpolation elements [12,2,1], the symbolic-numeric algorithms (SNAs) and programs have been elaborated [5,6,4]. This implementation of FEM using the interpolation Lagrange polynomials (ILPs) was such that it preserved only the continuity of the solution itself in

V.P. Gerdt et al. (Eds.): CASC Workshop 2014, LNCS 8660, pp. 138–154, 2014.

the course of its numerical approximation on a finite-element grid. However, in the above class of problems, particularly, in quantum-dimensional semiconductor systems, the continuity should be preserved not only for the solution (wave function) itself, but also for the probability current [2,10]. The required continuity of the solution derivatives can be preserved in FEM numerical approximation using the interpolation Hermite polynomials (IHPs) [3,11].

This motivated the aim of the present work, namely, the use of FEM with IHPs to elaborate SNAs implemented in Maple-Fortran for the solution of the BVPs with self-adjoint second order differential equation, and the analysis of the approximate numerical solutions in benchmark calculations.

In this paper, we present a symbolic algorithm implemented in Maple computer algebra system (CAS) that generates finite-element calculation schemes for solving BVPs for the self-adjoint second-order differential equation using interpolating Hermite polynomials. We derived recurrence relations for the calculation of the IHPs with nodes of arbitrary multiplicity. The stiffness and mass matrices are expressed via the integrals of products of the BVP coefficient functions, the IHPs and their derivatives. The result is used to formulate a generalized algebraic eigenvalue problem solved in Maple for matrices of small dimension. We use the symbolic algorithm to generate Fortran routines that allow the solution of the generalized algebraic eigenvalue problem with matrices of large dimension. We demonstrate the efficiency of the programs generated in Maple and Fortran for 100×100 and higher-order matrices, respectively, in benchmark calculations for exactly solvable quantum-mechanical problems with continuous and piecewise continuous potentials.

The paper is organized as follows. In Section 2, the formulation of BVPs and variational functional is presented. Section 3 describes the algorithm that generates algebraic problems using the finite element method with interpolation Hermite polynomials. In Section 4, the benchmark calculations are analysed. The obtained results and further development of SNA are discussed in Conclusion.

2   Formulation of BVPs

We consider a self-adjoint second-order differential equation with respect to the unknown solution $\Phi(z)$ in the region $z \in \Omega_z = (z^{\min}, z^{\max})$ [4]

$$(D - 2E)\,\Phi(z) = 0, \quad D = -\frac{1}{f_1(z)}\frac{\partial}{\partial z}f_2(z)\frac{\partial}{\partial z} + V(z). \tag{1}$$

If no additional restrictions are explicitly specified, we assume $f_1(z) > 0$, $f_2(z) > 0$, and $V(z)$ to be continuous functions that have derivatives up to the order of $\kappa^{\max} \geq 1$ in the domain $z \in \bar{\Omega}_z = [z^{\min}, z^{\max}]$. In quantum mechanics, Eq. (1) is actually the Schrödinger equation that describes a particle with the wave function $\Phi(z)$ and the energy E.

For a discrete-spectrum problem, the eigenfunctions $\Phi(z) = \Phi_m(z) \in \mathcal{H}_2^2$ in the Sobolev space $\mathcal{H}_2^2$ corresponding to the eigenvalues $E_1 < E_2 < \ldots < E_m <$

... are to satisfy the boundary condition of the first (I) and/or the second (II) and/or the third (III) kind at given values of parameters $\mathcal{R}(z^t)$

$$\text{(I)}: \quad \Phi_m(z^t) = 0, \quad t = \min \text{ and/or max}, \tag{2}$$

$$\text{(II)}: \quad f_1(z)\frac{d\Phi_m(z)}{dz}\bigg|_{z=z^t} = 0, \quad t = \min \text{ and/or max}, \tag{3}$$

$$\text{(III)}: \quad \frac{d\Phi_m(z)}{dz}\bigg|_{z=z^t} = \mathcal{R}(z^t)\Phi_m(z^t), \quad t = \min \text{ and/or max} \tag{4}$$

along with the normalization and orthogonality condition

$$\langle \Phi_m(z)|\Phi_{m'}(z)\rangle = \int_{z^{\min}}^{z^{\max}} f_1(z)(\Phi_m(z))^*\Phi_{m'}(z)dz = \delta_{mm'}. \tag{5}$$

The solution of the above BPVs can be reduced to the calculation of stationary points of a variational functional [12,6]

$$\Xi(\Phi, E, z^{\min}, z^{\max}) \equiv \int_{z^{\min}}^{z^{\max}} \Phi^*(z)\,(D - 2E)\,\Phi(z)dz = \Pi(\Phi, E, z^{\min}, z^{\max})$$

$$- f_2(z^{\max})\Phi^*(z^{\max})\mathcal{R}(z^{\max})\Phi(z^{\max}) + f_2(z^{\min})\Phi^*(z^{\min})\mathcal{R}(z^{\min})\Phi(z^{\min}), \tag{6}$$

where the symmetric functional $\Pi(\Phi, E, z^{\min}, z^{\max})$ is expressed as

$$\Pi(\Phi, E, z^{\min}, z^{\max}) = \int_{z^{\min}}^{z^{\max}} \left[f_2(z)\frac{d\Phi^*(z)}{dz}\frac{d\Phi(z)}{dz} + f_1(z)\Phi^*(z)V(z)\Phi(z) \tag{7} \right.$$

$$\left. - f_1(z)2E\Phi^*(z)\Phi(z) \right] dz.$$

Here $\mathcal{R}(z) \to \infty$ and $\mathcal{R}(z) = 0$ for discrete spectrum problem with BCs (I) and BCs (II), Eqs. (2) and (3), respectively.

3　FEM Generation of Algebraic Problems

High-accuracy computational schemes for solving the BVP (1)–(4) can be derived from the variational functional (6), (7) basing on the FEM. The general idea of the FEM in one-dimensional space is to divide the interval $[z^{\min}, z^{\max}]$ into many small domains referred to as elements. The size of the elements can be defined free enough to account for physical properties or qualitative behavior of the desired solutions, such as smoothness.

The interval $\Delta = [z^{\min}, z^{\max}]$ is covered by a set of n elements $\Delta_j = [z_j^{\min}, z_j^{\max} \equiv z_{j+1}^{\min}]$ in such a way that $\Delta = \bigcup_{j=1}^{n} \Delta_j$. Thus, we obtain the grid

$$\Omega^{h_j(z)}[z^{\min}, z^{\max}] = \{z^{\min} = z_1^{\min}, z_j^{\max} = z_j^{\min} + h_j, j = 1, \ldots, n - 1, \tag{8}$$

$$z_n^{\max} = z_n^{\min} + h_n = z^{\max}\},$$

where $z_j^{\min} \equiv z_{j-1}^{\max}$, $j = 2, \ldots, n$ are the mesh points, and the steps $h_j = z_j^{\max} - z_j^{\min}$ are the lengths of the elements Δ_j.

3.1 Interpolation Hermite Polynomials

In each element Δ_j we define the equidistant sub-grid $\Omega_j^{h_j(z)}[z_j^{\min}, z_j^{\max}] = \{z_{(j-1)p} = z_j^{\min}, z_{(j-1)p+r}, r = 1, \ldots, p-1, z_{jp} = z_j^{\max}\}$ with the nodal points $z_r \equiv z_{(j-1)p+r}$ determined by the formula

$$z_{(j-1)p+r} = ((p-r)z_j^{\min} + rz_j^{\max})/p, \quad r = 0, \ldots, p. \tag{9}$$

As a set of basis functions $\{N_l(z, z_j^{\min}, z_j^{\max})\}_{l=0}^{l^{\max}}$, $l^{\max} = \sum_{r=0}^{p} \kappa_r^{\max}$ we will use the IHPs $\{\{\varphi_r^\kappa(z)\}_{r=0}^{p}\}_{\kappa=0}^{\kappa_r^{\max}-1}$ in the nodes z_r, $r = 0, \ldots, p$ of the grid (9). The values of the functions $\varphi_r^\kappa(z)$ with their derivatives up to the order $(\kappa_r^{\max} - 1)$, i.e. $\kappa = 0, \ldots, \kappa_r^{\max} - 1$, where $\kappa_r^{\max}$ is referred to as the multiplicity of the node z_r, are determined by the expressions [3]

$$\varphi_r^\kappa(z_{r'}) = \delta_{rr'}\delta_{\kappa 0}, \quad \left.\frac{d^{\kappa'}\varphi_r^\kappa(z)}{dz^{\kappa'}}\right|_{z=z_{r'}} = \delta_{rr'}\delta_{\kappa\kappa'}. \tag{10}$$

To calculate the IHPs we introduce the auxiliary weight function

$$w_r(z) = \prod_{r'=0, r'\neq r}^{p} \left(\frac{z - z_{r'}}{z_r - z_{r'}}\right)^{\kappa_{r'}^{\max}}, \quad w_r(z_r) = 1. \tag{11}$$

The weight function derivatives can be presented as a product

$$\frac{d^\kappa w_r(z)}{dz^\kappa} = w_r(z)g_r^\kappa(z),$$

where the factor $g_r^\kappa(z)$ is calculated by means of the recurrence relations

$$g_r^\kappa(z) = \frac{dg_r^{\kappa-1}(z)}{dz} + g_r^1(z)g_r^{\kappa-1}(z), \tag{12}$$

with the initial conditions

$$g_r^0(z) = 1, \quad g_r^1(z) \equiv \frac{1}{w_r(z)}\frac{dw_r(z)}{dz} = \sum_{r'=0, r'\neq r}^{p} \frac{\kappa_{r'}^{\max}}{z - z_{r'}}.$$

We will seek for the IHPs $\varphi_r^\kappa(z)$ in the following form:

$$\varphi_r^\kappa(z) = w_r(z) \sum_{\kappa'=0}^{\kappa_r^{\max}-1} a_r^{\kappa,\kappa'}(z - z_r)^{\kappa'}. \tag{13}$$

Differentiating the function (13) by z at the point of z_r and using Eq. (11), we obtain

$$\left.\frac{d^{\kappa'}\varphi_r^\kappa(z)}{dz^{\kappa'}}\right|_{z=z_r} = \sum_{\kappa''=0}^{\kappa'} \frac{\kappa'!}{\kappa''!(\kappa'-\kappa'')!} g_r^{\kappa'-\kappa''}(z_r)a_r^{\kappa,\kappa''}\kappa''!. \tag{14}$$

Hence we arrive at the expression for the coefficients $a_r^{\kappa,\kappa'}$

$$a_r^{\kappa,\kappa'} = \left(\left. \frac{d^{\kappa'}\varphi_r^{\kappa}(z)}{dz^{\kappa'}} \right|_{z=z_r} - \sum_{\kappa''=0}^{\kappa'-1} \frac{\kappa'!}{\kappa''!(\kappa'-\kappa'')!} g_r^{\kappa'-\kappa''}(z_r) a_r^{\kappa,\kappa''} \kappa''! \right) / \kappa'!. \quad (15)$$

Taking Eq. (10) into account, we finally get:

$$a_r^{\kappa,\kappa'} = \begin{cases} 0, & \kappa' < \kappa, \\ 1/\kappa'!, & \kappa' = \kappa, \\ -\sum_{\kappa''=\kappa}^{\kappa'-1} \frac{1}{(\kappa'-\kappa'')!} g_r^{\kappa'-\kappa''}(z_r) a_r^{\kappa,\kappa''}, & \kappa' > \kappa. \end{cases}$$

Note that all degrees of interpolation Hermite polynomials $\varphi_r^{\kappa}(z)$ do not depend on κ and equal $p' = \sum_{r'=0}^{p} \kappa_{r'}^{\max} - 1$. Below we consider only the IHPs with the nodes of identical multiplicity $\kappa_r^{\max} = \kappa^{\max}$, $r = 0, \ldots, p$. In this case, the degree of the polynomials is equal to $p' = \kappa^{\max}(p+1) - 1$. We introduce the following notation for such polynomials:

$$N_{\kappa^{\max}r+\kappa}(z, z_j^{\min}, z_j^{\max}) = \varphi_r^{\kappa}(z), \quad r = 0, \ldots, p, \quad \kappa = 0, \ldots, \kappa^{\max} - 1. \quad (16)$$

These IHPs form a basis in the space of polynomials having the degree $p' = \kappa^{\max}(p+1) - 1$ in the element $z \in [z_j^{\min}, z_j^{\max}]$ that have continuous derivatives up to the order $\kappa^{\max} - 1$ at the boundary points $z_j^{\min}$ and $z_j^{\max}$ of the element $z \in [z_j^{\min}, z_j^{\max}]$. The IHPs at $\kappa^{\max} = 1, 2, 3$ and $p = 4$ are shown in Fig. 1. It is seen that the values of IHP $N_{\kappa^{\max}p+\kappa}(z, z_j^{\min}, z_j^{\max})$ and $N_{\kappa}(z, z_{j+1}^{\min}, z_{j+1}^{\max})$ (at $r = p$ and $r = 0$) and their derivatives up to the order $\kappa^{\max} - 1$ coincide at the mutual point $z_j^{\max} = z_{j+1}^{\min}$ of the adjacent elements. Moreover, the boundary points are nodes (zeros) of multiplicity $\kappa^{\max}$ of other IHPs, irrespective of the length of elements of $[z_j^{\min}, z_j^{\max}]$ and $[z_{j+1}^{\min}, z_{j+1}^{\max}]$. This allows construction of a basis of piecewise and polynomial functions having continuous derivatives to the order of $\kappa^{\max} - 1$ in any set $\Delta = \bigcup_{j=1}^{n} \Delta_j = [z_j^{\min}, z_j^{\max}]$ of elements $\Delta_j = [z_j^{\min}, z_j^{\max} \equiv z_{j+1}^{\min}]$. The **Algorithm 1** of the IHP construction is presented in Appendix A and implemented in the CAS Maple.

3.2   Generation of Algebraic Eigenvalue Problems

We consider a discrete representation of the solutions $\Phi(z)$ of the problem (1), (5), (4) reduced by means of the FEM to the variational functional (6), (7) on the finite-element grid,

$$\Omega_{h_j(z)}^{p}[z^{\min}, z^{\max}] = [z_0 = z^{\min}, z_l, l = 1, \ldots, np - 1, z_{np} = z^{\max}],$$

with the mesh points $z_l = z_{jp} = z_j^{\max} \equiv z_{j+1}^{\min}$ of the grid $\Omega^{h_j(z)}[z^{\min}, z^{\max}]$ determined by Eq. (8) and the nodal points $z_l = z_{(j-1)p+r}$, $r = 0, \ldots, p$ of the sub-grids $\Omega_j^{h_j(z)}[z_j^{\min}, z_j^{\max}]$, $j = 1, \ldots, n$, determined by Eq. (9). The solutions

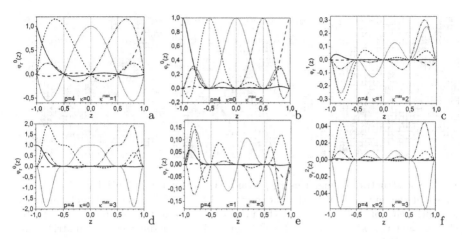

Fig. 1. The IHP coinciding at $\kappa^{\max} = 1$ with the ILP (a) and IHPs at $\kappa^{\max} = 2$ (b, c) and $\kappa^{\max} = 3$ (d, e, f). Here $p + 1 = 5$ is the number of nodes in the subinterval, $\Delta_j = [z_j^{\min} = -1, z_j^{\max} = 1]$. The grid nodes z_r are shown by vertical lines.

$\Phi^h(z) \approx \Phi(z)$ are sought for in the form of a finite sum over the basis of local functions $N_\mu^g(z)$ at each nodal point $z = z_k$ of the grid $\Omega_{h_j(z)}^p[z^{\min}, z^{\max}]$:

$$\Phi^h(z) = \sum_{\mu=0}^{L-1} \Phi_\mu^h N_\mu^g(z), \quad \Phi^h(z_l) = \Phi_{l\kappa^{\max}}^h, \quad \left.\frac{d^\kappa \Phi^h(z)}{dz^\kappa}\right|_{z=z_l} = \Phi_{l\kappa^{\max}+\kappa}^h \quad (17)$$

where $L = (pn+1)\kappa^{\max}$ is the number of local functions and Φ_μ^h at $\mu = l\kappa^{\max}+\kappa$ are the nodal values of the κth derivatives of the function $\Phi^h(z)$ (including the function $\Phi^h(z)$ itself for $\kappa = 0$) at the points z_l.

The local functions $N_\mu^g(z) \equiv N_{l\kappa^{\max}+\kappa}^g(z)$ are piecewise polynomials of the given order p', their derivative of the order κ at the node z_l equals one, and the derivative of the order $\kappa' \neq \kappa$ at this node equals zero, while the values of the function $N_\mu^g(z)$ with all its derivatives up to the order $(\kappa^{\max} - 1)$ equal zero at all other nodes $z_{l'} \neq z_l$ of the grid $\Omega_{h_j(z)}$, i.e., $\left.\frac{d^\kappa N_{l'\kappa^{\max}+\kappa'}}{dz^\kappa}\right|_{z=z_l} = \delta_{ll'}\delta_{\kappa\kappa'}$, $l = 0, \ldots, np$, $\kappa = 0, \ldots, \kappa^{\max} - 1$.

For the nodes z_l of the grid that do not coincide with the mesh points $z_j^{\max}$, i.e., at $l \neq jp$, $j = 1 \ldots n-1$, the polynomial N_μ^g at $\mu = ((j-1)p+r)\kappa^{\max} + \kappa$ has the form

$$N_{(p(j-1)+r)\kappa^{\max}+\kappa}^g = \begin{cases} N_{\kappa^{\max}r+\kappa}(z, z_j^{\min}, z_j^{\max}), & z \in \Delta_j; \\ 0, & z \notin \Delta_j, \end{cases} \quad (18)$$

i.e., it is defined as the IHP $N_{\kappa^{\max}r+\kappa}(z, z_j^{\min}, z_j^{\max})$ in the interval $z \in \Delta_j$ and zero otherwise. Since the points $z_j^{\min}$ and $z_j^{\max}$ are nodes of multiplicity $\kappa^{\max}$, such piecewise polynomial functions and their derivatives up to the order $\kappa^{\max}-1$

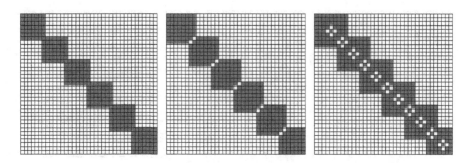

Fig. 2. The structure of matrices $B_{L_1 L_2}$ and $A_{L_1 L_2}$ for the potential $V(z) = 0$, the number of elements $n = 6$ in the entire interval $(z^{\min}, z^{\max})$, and different values of the multiplicity of nodes $\kappa^{\max}$ and the number of subintervals p. From left to right: $(\kappa^{\max}, p) = (1, 6)$, $(\kappa^{\max}, p) = (2, 3)$, $(\kappa^{\max}, p) = (3, 2)$. The dimensions of matrices are $L \times L$, $L = \kappa^{\max}(np + 1)$: 37×37, 38×38, 39×39.

are continuous in the entire interval Δ. In Fig. 1 such IHPs are plotted by dotted, short-dashed and dot-dashed lines.

For the nodal points of the grid z_l that coincide with one of the mesh points $z_j^{\max}$ belonging to two elements Δ_j and Δ_{j+1}, $j = 1 \ldots n - 1$, i.e., for $l = jp$, the polynomial, whose derivative of the order κ equals one at the node z_l, has the form

$$N_{p\kappa^{\max}j+\kappa}^g = \begin{cases} N_{\kappa^{\max}p+\kappa}(z, z_j^{\min}, z_j^{\max}), & z \in \Delta_j; \\ N_\kappa(z, z_{j+1}^{\min}, z_{j+1}^{\max}), & z \in \Delta_{j+1}; \\ 0, & z \notin \Delta_j \cup \Delta_{j+1}, \end{cases} \tag{19}$$

In other words, it is constructed by joining the polynomial $N_{p\kappa^{\max}+\kappa}(z, z_j^{\min}, z_j^{\max})$ defined in the element Δ_j with the polynomial $N_\kappa(z, z_{j+1}^{\min}, z_{j+1}^{\max})$ defined in the element Δ_{j+1}. This polynomial is also continuous with all its derivatives of the order $\kappa^{\max} - 1$ in the interval $z \in \Delta$. The corresponding IHPs are plotted in Fig. 1 by solid and long-dashed lines.

The substitution of the expansion (17) into the variational functional (6), (7) reduces the solution of the problem (1)–(5) to the solution of the generalized algebraic eigenvalue problem with respect to the desired set of eigenvalues E and eigenvectors $\boldsymbol{\Phi}^h = \{\Phi_\mu^h\}_{\mu=0}^{L-1}$:

$$(\tilde{\mathbf{A}} - 2E\,\mathbf{B})\boldsymbol{\Phi}^h = 0. \tag{20}$$

Here $\tilde{\mathbf{A}} = \mathbf{A} + \mathbf{M}_{\min} - \mathbf{M}_{\max}$ and $\mathbf{B}$ are symmetric $L \times L$ stiffness and mass matrices, $L = \kappa^{\max}(np+1)$, $\mathbf{M}_{\max}$ and $\mathbf{M}_{\min}$ are $L \times L$ matrices with zero elements except $M_{11} = f_2(z^{\min})R(z^{\min})$ and $M_{L+1-\kappa^{\max}, L+1-\kappa^{\max}} = f_2(z^{\max})R(z^{\max})$, respectively. The **Algorithm 2** that generates the local functions $N_\mu^g(z)$ defined by (18), (19) and the matrices $\mathbf{A}$ and $\mathbf{B}$ is described in Appendix B and implemented in the CAS Maple.

Table 1. Runge coefficients (24) for the eigenvalues (Runge Eigv) and the eigenfunction (Runge EigF) of the first three lower-energy states calculated for schemes with different $\kappa^{\max}$ and p up to order $p' = \kappa^{\max}(p+1) - 1 = 8$ at $h = 0.125$ for schemes with $p' = 7$, $p' = 8$, and at $h = 0.0625$ for the rest of the schemes. Theoretical estimates of Runge coefficient for the convergence of eigenvalues and eigenfunctions are $2p'$ and $(p' + 1)$, respectively. The execution time T_h (in seconds) for the mesh step $h = 1/32$ is presented in the last column.

$\kappa^{\max}$	p	p'	Runge Eigv			$2p'$	Runge EigF			$p'+1$	T_h
1	1	1	2.00	2.00	1.99	2	1.99	1.99	2.00	2	9.36
1	2	2	4.00	3.99	3.99	4	2.99	2.98	3.02	3	19.5
1	3	3	5.99	6.00	5.99	6	3.98	3.99	3.97	4	33.4
2	1	3	5.97	5.96	5.96	6	3.95	3.95	3.94	4	21.8
1	4	4	7.99	8.00	8.00	8	4.99	4.98	5.00	5	48.6
1	5	5	9.99	9.99	9.99	10	5.98	6.01	5.97	6	65.6
2	2	5	9.97	9.97	9.97	10	5.96	5.98	5.95	6	47.6
3	1	5	10.05	10.05	10.06	10	6.01	6.04	6.02	6	38.0
1	6	6	12.00	12.00	12.00	12	6.99	6.97	6.99	7	88.9
1	7	7	13.98	13.98	13.98	14	7.85	8.03	7.85	8	111.
2	3	7	13.88	13.87	13.87	14	7.77	7.95	7.77	8	82.3
4	1	7	13.59	13.58	13.57	14	7.61	7.57	7.59	8	59.6
1	8	8	16.13	16.00	15.99	16	9.00	8.82	9.09	9	139.
3	2	8	15.75	15.75	15.74	16	8.83	8.67	8.86	9	99.1

To solve equation (20) we have chosen the subspace iteration method [12,1] elaborated by Bathe [1] for the solution of large symmetric banded matrix eigenvalue problems. This method uses a skyline storage mode, which stores the components of the matrix column vectors within the nonzero band of the matrix and, therefore, is perfectly suitable for the banded FEM matrices. The procedure chooses a vector subspace of the full solution space and iterates upon the successive solutions in the subspace (for details, see [1]). Using the Rayleigh quotients for the eigenpairs, the iterations are repeated until the desired set of solutions in the iteration subspace converges to within the specified tolerance. Generally, 10–24 iterations are enough to converge the subspace to within the prescribed tolerance. If the matrix $\breve{\mathbf{A}}$ in Eq. (20) is not positive-definite, the problem (20) is replaced with the following problem: $\breve{\mathbf{A}}\boldsymbol{\Phi}^h = \breve{E}^h\,\mathbf{B}\boldsymbol{\Phi}^h$, $\breve{\mathbf{A}} = \tilde{\mathbf{A}} - \alpha\mathbf{B}$. The number α (the shift of the energy spectrum) is chosen such that the matrix $\breve{\mathbf{A}}$ is positive-definite. The eigenvector of this problem is the same, and $E^h = \breve{E}^h + \alpha$.

The theoretical estimate for the $\mathbf{H}^0$ norm of the difference between the exact solution $\Phi_m(z) \in \mathcal{H}_2^2$ and the numerical one $\Phi_m^h(z) \in \mathbf{H}^{\kappa^{\max}}$ has the order of

$$|E_m^h - E_m| \leq c_1\, h^{2p'}, \quad \left\|\Phi_m^h(z) - \Phi_m(z)\right\|_0 \leq c_2 h^{p'+1}, \tag{21}$$

where $h = \max_{1<j<n} h_j$ is the maximal step of the grid [12].

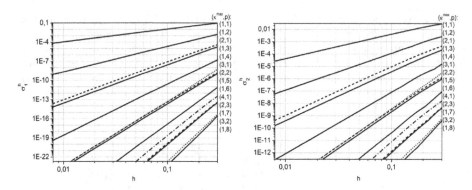

Fig. 3. Absolute errors $\sigma_1^h = |\varepsilon_1^{exact} - \varepsilon_1^h|$ and $\sigma_2^h = \max_{z \in \Omega^h(z)} |\chi_1^{exact}(z) - \chi_1^h(z)|$ for the ground state vs the grid step h calculated using approximation by IHPs with different $\kappa^{\max}$ and p

4   Benchmark Calculations

4.1   Modified Pöschl–Teller Potential

As an example, we consider the exactly solvable eigenvalue problem for Schrödinger equation in the units $\hbar = m = 1$:

$$\left(-\frac{d^2}{dz^2} + 2V(z) - 2E \right) \Phi(z) = 0, \tag{22}$$

with the modified Pöschl–Teller potential on the axis $z \in (-\infty, +\infty)$:

$$V(z) = -\frac{\alpha^2}{2} \frac{\lambda\,(\lambda - 1)}{(\cosh(\alpha\,z))^2}, \tag{23}$$

where $\alpha > 0$ and $\lambda > 0$ are real-value parameters. The parameters $\lambda = 11/2$ and $\alpha = 1$ were chosen such that the discrete spectrum problem for Eq. (22) with the potential (23) had five eigenvalues $2E_m = [-20.25, -12.25, -6.25, -2.25, -0.25]$ with the corresponding five eigenfunctions $\psi_m(x)$ known in the analytical form.

The numerical experiments using the finite-element grid $\Omega_{h_j(z)}^p[z^{\min} = -40,$ $z^{\max} = 40]$ demonstrated strict correspondence to the theoretical estimations (21) for eigenvalues and eigenfunctions. In particular, we calculated the Runge coefficients

$$\beta_l = \log_2 \left| \frac{\sigma_l^h - \sigma_l^{h/2}}{\sigma_l^{h/2} - \sigma_l^{h/4}} \right|, \quad l = 1, 2, \tag{24}$$

on three twice condensed grids with the absolute errors

$$\sigma_1^h = |E_m^{exact} - E_m^h|, \quad \sigma_2^h = \max_{z \in \Omega^h(z)} |\Phi_m^{exact}(z) - \Phi_m^h(z)| \tag{25}$$

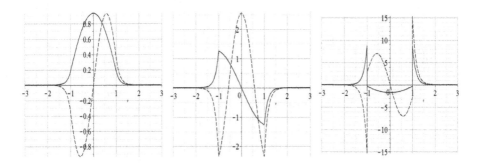

Fig. 4. The solutions and their first and second derivatives for the ground state (solid curves) and the first excited state (dashed curves) of the rectangular well potential problem

for the eigenvalues and eigenfunctions, respectively. From Eq. (25) we obtained the numerical estimations of the convergence order of the proposed numerical schemes, the theoretical estimates being $\beta_1 = 2p'$ and $\beta_2 = p' + 1$.

In Table 1, we show the Runge coefficients (24) for the eigenvalues (Runge Eigv) and the eigenfunction (Runge EigF) of the first three lower-energy states calculated for schemes with different $\kappa^{\max}$ and p up to order $p' = \kappa^{\max}(p+1)-1 = 8$. One can see that for the chosen $p' = 1 \div 8$, the numerical estimates of Runge coefficients lie within $2p' \pm 0.06$ for $p' = 1, \ldots, 6$ and $2p' \pm 0.56$ for $p' = 7, 8$ in the case of eigenvalues and within $(p' + 1) \pm 0.2$ in the case of eigenfunctions, which strongly corresponds to the theoretical error estimates (21). In Fig. 3, we show the dependence of absolute errors $\sigma_1^h = |\varepsilon_1^{exact} - \varepsilon_1^h|$ for eigenvalues and $\sigma_2^h = \max_{z \in \Omega^h(z)} |\chi_1^{exact}(z) - \chi_1^h(z)|$ for eigenfunctions of the ground state vs. the grid step h calculated using approximation by IHPs with different $\kappa^{\max}$ and p. In the double logarithmic scale, the errors lie on lines with different slopes that explicitly show the desirable order of approximation $p' = \kappa^{\max}(p + 1) - 1$ by IHPs with different $\kappa^{\max}$ and p.

For calculations, we used the program KANTBP 1.1 with the specified accuracy of $\sim 10^{-34}$ and the relative error tolerance of the eigenvalues $\epsilon_1 = 4 \cdot 10^{-34}$, implemented in Intel Fortran 77 on the computer 2 x Xeon 3.2 GHz, 4 GB RAM. The data type QUADRUPLE PRECISION provided 32 significant digits. The running time T_h for $h = 1/32 = 0.03125$ is presented in the last column of Table 1.

4.2   Rectangular Well Potential

For piecewise continuous potentials (or potentials with discontinuous derivatives), the approximation by IHPs does not converge to the desired solution with increasing number of nodes. Within the FEM approach, the following technique is used. Let the potential have the form $V(z) = \{V_i(z), z \in (\zeta_i^{\min}, \zeta_i^{\max})\}$, $\zeta_{i+1}^{\min} = \zeta_i^{\max}$, where $V_i(z)$ are $(p' + 1)$-times differentiable functions. The interval of the problem definition is divided into a set of subintervals $[z_j^{\min}, z_j^{\max}]$

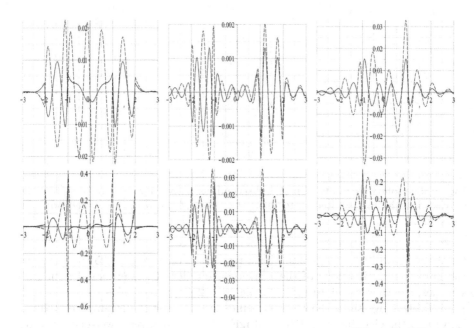

Fig. 5. The difference of numerical and exact eigenfunctions $D_{swp,0}^{\kappa^{\max},p} = \psi_0^{\kappa^{\max},p}(z) - \psi_0(z)$ (solid curves) and $D_{swp,1}^{\kappa^{\max},p} = \psi_1^{\kappa^{\max},p}(z) - \psi_1(z)$ (dashed curves) (upper panels) and their first derivatives (lower panels) for rectangular well potential for $n = 10$ elements in the interval $(-5,5)$ and different values of the multiplicity of nodes $\kappa^{\max}$ and the number of subinterval divisions p. >From left to right: $(\kappa^{\max}, p) = (1,3)$, $(\kappa^{\max}, p) = (2,1)$, $(\kappa^{\max}, p) = (3,1)$.

$(z_j^{\max} \equiv z_{j+1}^{\min})$, such that every point $\zeta_i^{\min}$, in which the second derivative of the solution is discontinuous, coincides with some boundary point $z_j^{\min}$.

Consider, e.g., the exactly solvable discrete-spectrum problem for Eq. (22) with the rectangular well potential $2V(z) = V_0$, if $|z| \le a$, and $2V(z) = 0$ otherwise. At $a = 1$, $2V_0 = -50$ the discrete-spectrum problem has five eigenfunctions (see Fig. 4), expressed in the analytical form via five eigenvalues $2E_m = [-48.109146, -42.474904, -33.232792, -20.714111, -5.965365]$.

Since the first two eigenfunctions rapidly decrease, it is sufficient to use the finite-element grid $\Omega_{h_j(z)}^p[z^{\min} = -5, z^{\max} = 5]$. The calculation error for the first two eigenvalues is presented in Table 2. It is seen that the scheme with $\kappa^{\max} = 1$ and $\kappa^{\max} = 2$ having the same order of accuracy $p' = 3$ and $p' = 5$ ($p' = \kappa^{\max}(p + 1) - 1$) yield nearly the same error (at $n = 20$, $h = 1/2$ the error is about 10^{-2} and $4 \cdot 10^{-6}$, respectively), while for $\kappa^{\max} = 3$, the error is much higher (about 10^{-2} at $n = 20$, $h = 1/2$). In Table 2, we show the Runge coefficients (24) for the eigenvalues of the first two lower-energy states calculated for schemes with different $\kappa^{\max}$ and p with order $p' = \kappa^{\max}(p + 1) - 1 = 3$ and $p' = \kappa^{\max}(p+1) - 1 = 5$. One can see that for the chosen $p' = 3, 5$, the numerical estimates of Runge coefficients lie within $2p' \pm 0.5$ for schemes with $\kappa^{\max} = 1, 2$

Table 2. The absolute errors $\sigma_1^h(E_0)$ and $\sigma_1^h(E_1)$ of eigenvalues of ground and first exited state for square well potential for $a = 1$ and $2V_0 = -50$. The Runge coefficient (Ru) from (24) for the eigenvalues at $h = 1/4$, $n = 40$ and its theoretical estimates $(2p')$ are given in last two columns.

(κ^{max}, p)	p'	$\sigma_1^{h=1}(E_0)$	$\sigma_1^{h=1/2}(E_0)$	$\sigma_1^{h=1/4}(E_0)$	$\sigma_1^{h=1/8}(E_0)$	$\sigma_1^{h=1/16}(E_0)$	Ru	$2p'$
(1,3)	3	1.93e-02	1.39e-03	4.44e-05	8.83e-07	1.48e-08	5.65	6
(2,1)	3	5.70e-02	3.15e-03	1.00e-04	2.21e-06	4.14e-08	5.50	6
(1,5)	5	2.47e-04	1.67e-06	3.82e-09	5.26e-12	2.22e-12	10.3	10
(2,2)	5	4.01e-04	2.59e-06	6.12e-09	8.59e-12	2.20e-13	9.51	10
(3,1)	5	1.48e-02	2.66e-03	3.51e-04	4.40e-05	5.50e-06	2.99	10
(κ^{max}, p)	p'	$\sigma_1^{h=1}(E_1)$	$\sigma_1^{h=1/2}(E_1)$	$\sigma_1^{h=1/4}(E_1)$	$\sigma_1^{h=1/8}(E_1)$	$\sigma_1^{h=1/16}(E_1)$	Ru	$2p'$
(1,3)	3	9.96e-02	4.38e-03	1.25e-04	2.40e-06	3.96e-08	5.70	6
(2,1)	3	2.92e-01	1.14e-02	3.08e-04	6.33e-06	1.14e-07	5.60	6
(1,5)	5	6.44e-04	3.75e-06	7.93e-09	1.04e-11	2.63e-12	9.99	10
(2,2)	5	9.40e-04	5.66e-06	1.27e-08	1.74e-11	2.06e-13	9.53	10
(3,1)	5	6.70e-02	1.07e-02	1.39e-03	1.74e-04	2.17e-05	3.01	10

which strongly corresponds to the theoretical error estimates (21). While the scheme with $\kappa^{max} = 3$, $p = 1$ of fifth order $p' = 5$ gives Runge coefficient $\beta_1 = 3$. Maximal discrepancies arise in the vicinities of discontinuity of the potential well (at $z = \pm 1$) because of a worse approximation of function with discontinuous second derivative by means of functions with continuous one.

It is due to the fact that the first derivative of the solution has a discontinuity at $z = \pm a$ displayed in Fig 4. To illustrate this fact, we display in Fig. 5 the discrepancies of eigenfunctions and their first derivatives. It is seen that the scheme with $\kappa^{max} = 2$, $p = 1$ provides better approximation for eigenfunctions among schemes of third order $p' = 3$. The scheme of fifth order $p' = 5$ with $\kappa^{max} = 3$, $p = 1$ leads to worse approximation in comparison with schemes of third order.

5   Conclusion

We presented the SNAs for solving the BVPs with self-adjoint second order differential equation using the FEM with interpolation Hermite polynomials. The proposed approach preserves the property of continuity of derivatives of the desired solutions. We demonstrated the efficiency of the programs generated in Maple and Fortran for 100×100 and greater-order matrices, respectively, in benchmark calculations for exactly solvable quantum-mechanical problems with continuous and piecewise continuous potentials. The analysis of approximate numerical solutions in benchmark calculations with smooth potentials shows that the order $p' = \kappa^{\max}(p+1) - 1$ of the elaborated FEM schemes strongly corresponds to the theoretical error estimates. Schemes of higher order p' allow high-accuracy results at larger step of the finite-element grid, provided that the derivative of the p'th order is a smooth function. Schemes with the fixed order p' have similar rate convergence, the execution time being smaller for greater $\kappa^{\max}$ due to smaller dimension of matrices used in the calculations. However, if the κth derivative of the desired solution has discontinuity points, i.e., for potentials having a discontinuous derivative of the order $\kappa - 2$, the schemes with $\kappa^{\max} \geq \kappa$ operate worse, because in this case, the solution having discontinuous κ^{th}th derivatives is approximated by functions having no such discontinuities.

In future, the elaborated calculation schemes, algorithms, and programs will be applied to the analysis of models of molecular, atomic, and nuclear systems, as well as to quantum-dimensional systems such as quantum dots, wires, and wells in bulk semiconductors, and smooth irregular wave-guide structures with piecewise continuous potentials.

The authors thank Professor V.P. Gerdt for collaboration. The work was partially supported by the Russian Foundation for Basic Research (RFBR) (grants No. 14-01-00420 and 13-01-00668) and the Bogoliubov–Infeld program.

A   Algorithm 1. Generation of IHPs

Input:
$z^{\min}, z^{\max}$, (formal parameters) the boundary points of the interval;
p is the number of subintervals: $p+1$ is the number of nodes of IHPs;
$\kappa^{\max}$ is the multiplicity of nodes;
$f_1(z)$ and $f_2(z)$ are coefficient functions from (1);
Output:
$N_{l_1}(z, z^{\min}, z^{\max})$ are IHPs, $l_1 = 0, \ldots, l_{\max}$, i.e. $l_{\max} + 1$ is number of IHPs;
$A_{l_1;l_2}(z^{\min}, z^{\max})$ and $B_{l_1;l_2}(z^{\min}, z^{\max})$ are auxiliary integrals;
Local:
$l_{\max} = \kappa^{\max}(p+1) - 1$ is largest index of IHPs, $l_{\max} + 1$ is number of IHPs;
z_r are nodes in subinterval;
$w_r(z)$ are weight functions;
$g_r^{\kappa}(z)$ are derivatives of order κ divided by weight function;
$a_r^{\kappa,\kappa'}$ are coefficients of expansion (13);

―――

1: generation of IHPs and calculation of integrals in the interval $[z^{\min}, z^{\max}]$
1.1.:   for $r:=0$ to p do
   $z_r = ((p-r)z^{\min} + rz^{\max})/p$;
   end for;
1.2.:   for $r:=0$ to p do
1.2.1:   auxiliary weight function
   $w_r(z) = \prod_{r'=0, r' \neq r}^{p} \left(\frac{z - z_{r'}}{z_r - z_{r'}}\right)^{\kappa^{\max}}$;
1.2.2:   recurrence relation for calculating the function $g_r^{\kappa}(z)$
   $g_r^0(z) = 1$;

$$g_r^1(z) = \sum_{r'=0, r' \neq r}^{p} \frac{\kappa^{\max}}{z - z_{r'}};$$

for $\kappa := 2$ to $\kappa^{\max} - 1$ do

$$g_r^\kappa(z) = \frac{dg_r^{\kappa-1}(z)}{dz} + g_r^1(z) g_r^{\kappa-1}(z);$$

end for;

1.2.3:   recurrence relation for calculation of coefficients $a_r^{\kappa, \kappa'}$

for $\kappa := 0$ to $\kappa^{\max} - 1$ do

$$a_r^{\kappa, \kappa} = 1/\kappa'!;$$

for $\kappa' := \kappa + 1$ to $\kappa^{\max} - 1$ do

$$a_r^{\kappa, \kappa'} = -\sum_{\kappa''=\kappa}^{\kappa'-1} \frac{1}{(\kappa' - \kappa'')!} g_r^{\kappa' - \kappa''}(z_r) a_r^{\kappa, \kappa''};$$

end for;

1.2.4:   calculation of IHP

$$N_{\kappa^{\max} r + \kappa}(z, z^{\min}, z^{\max}) \equiv \varphi_r^\kappa(z) = w_r(z) \sum_{\kappa'=\kappa}^{\kappa^{\max} - 1} a_r^{\kappa, \kappa'} (z - z_r)^{\kappa'};$$

end for;

end for;

$$l_{\max} = \kappa^{\max}(p + 1) - 1;$$

1.3:   calculation of the auxiliary integrals

for $l_1 := 0$ to $l_{\max}$ do

for $l_2 := l_1$ to $l_{\max}$ do

$$A_{l_1; l_2}(z^{\min}, z^{\max}) = \int_{z^{\min}}^{z^{\max}} f_2(z) \frac{dN_{l_1}(z, z^{\min}, z^{\max})}{dz} \frac{dN_{l_2}(z, z^{\min}, z^{\max})}{dz} dz;$$

$$B_{l_1; l_2}(z^{\min}, z^{\max}) = \int_{z^{\min}}^{z^{\max}} f_1(z) N_{l_1}(z, z^{\min}, z^{\max}) N_{l_2}(z, z^{\min}, z^{\max}) dz;$$

end for;

end for;

Remarks. 1. In commonly used coordinates, the integrals in Step 1.3. are calculated analytically. If $f_1(z)$ or $f_2(z)$ are such that these integrals cannot be calculated analytically, then one can apply the expansion over the interpolation polynomials.

2. The auxiliary integrals $A_{l_1; l_2}(z^{\min}, z^{\max})$ and $B_{l_1; l_2}(z^{\min}, z^{\max})$ are symmetric with respect to permutations of their indexes.

B   Algorithm 2: FEM Generation of Algebraic Eigenvalue Problem

Input:

n is the number of subintervals $\Delta_j = [z_j^{\min}, z_j^{\max} = z_j^{\min} + h_j]$;

$\Delta_j = [z_j^{\min}, z_j^{\max}]$ are sets of subintervals ($z_j^{\max} \equiv z_{j+1}^{\min}$);

p is the number of divisions of subintervals: $p + 1$ is the number of nodes of IHP;

$\kappa^{\max}$ is the multiplicity of nodes;

$N_{l_1}(z, z^{\min}, z^{\max})$ are IHP;

$A_{l_1; l_2}(z^{\min}, z^{\max})$ and $B_{l_1; l_2}(z^{\min}, z^{\max})$ are auxiliary integrals from the **Algorithm 1**;

$V(z)$ is coefficient function from (1);

Output:
z_l are nodes in the whole interval, $l = 0, \ldots, np$;
N_l^g are piecewise polynomials;
$A_{L_1 L_2}$ and $B_{L_1 L_2}$ are matrices of algebraic eigenvalue problem (20);
Local:
$l_{\max} = \kappa^{\max}(p+1) - 1$ where $l_{\max} + 1$ is number of IHP;
$L = \kappa^{\max}(np+1)$ is the dimension of the algebraic eigenvalue problem.

2.1. calculation of grid points
$\quad z_0 = z_1^{\min}$;
$\quad$ for $j := 1$ to n do
$\quad\quad$ for $r := 1$ to $p - 1$ do
$\quad\quad\quad z_{(j-1)p+r} = ((p - r)z_j^{\min} + rz_j^{\max})/p$;
$\quad\quad$ end for;
$\quad\quad z_{jp} = z_j^{\max}$;
$\quad$ end for;
2.2. calculation of piecewise polynomials
$\quad$ for $\kappa := 0$ to $\kappa^{\max} - 1$ do
$\quad N_\kappa^g = \{N_\kappa(z, z_1^{\min}, z_1^{\max}), z \in \Delta_1\}$;
$\quad\quad$ for $j := 1$ to n do
$\quad\quad\quad$ for $r := 1$ to $p - 1$ do
$\quad\quad\quad N_{((j-1)p+r)\kappa^{\max}+\kappa}^g = \{N_{\kappa^{\max}r+\kappa}(z, z_j^{\min}, z_j^{\max}), z \in \Delta_j; 0, z \notin \Delta_j\}$;
$\quad\quad\quad$ end for;
$\quad\quad\quad$ if $(j < n)$ then
$\quad\quad\quad N_{jp\kappa^{\max}+\kappa}^g := \{N_{\kappa^{\max}p+\kappa}(z, z_j^{\min}, z_j^{\max}), z \in \Delta_j;$
$\quad\quad\quad\quad\quad\quad\quad\quad N_\kappa(z, z_{j+1}^{\min}, z_{j+1}^{\max}), z \in \Delta_{j+1}; 0, z \notin \Delta_j \cup \Delta_{j+1}\}$;
$\quad\quad\quad$ else
$\quad\quad\quad N_{np\kappa^{\max}+\kappa}^g := \{N_{\kappa^{\max}p+\kappa}(z, z_n^{\min}, z_n^{\max}), z \in \Delta_n; 0, z \notin \Delta_n\}$;
$\quad\quad\quad$ end if;
$\quad\quad$ end for;
$\quad$ end for;
2.3. Generation of matrices **A** and **B**
$\quad$ for $j := 1$ to n do
$\quad\quad$ for $l_1 := 0$ to $l_{\max} - 1$ do
$\quad\quad L_1 = p\kappa^{\max}(j - 1) + l_1 + 1$;
$\quad\quad\quad$ for l_2 from l_1 to $l_{\max} - 1$ do
$\quad\quad\quad L_2 = p\kappa^{\max}(j - 1) + l_2 + 1$;
$\quad\quad\quad A_{L_1 L_2} = A_{L_1 L_2} + A_{l_1; l_2}(z_j^{\min}, z_j^{\max})$
$\quad\quad\quad\quad\quad + \int_{z_j^{\min}}^{z_j^{\max}} f_1(z)dz N_{L_1}(z, z_j^{\min}, z_j^{\max})V(z)N_{L_2}(z, z_j^{\min}, z_j^{\max})$;
$\quad\quad\quad B_{L_1 L_2} = B_{L_1 L_2} + B_{l_1; l_2}(z_j^{\min}, z_j^{\max})$;
$\quad\quad$ end for (j, l_1, l_2)

Remarks. 1. If the coefficients of the equation (1) are given in the tabular form, then we use the following matrix elements in Step 1.3 of Algorithm 1 and Step 2.3 of Algorithm 2:

$$\int_{z_j^{\min}}^{z_j^{\max}} f_1(z)dz N_{L_1}(z, z_j^{\min}, z_j^{\max}) V(z) N_{L_2}(z, z_j^{\min}, z_j^{\max})$$

$$= \sum_{r=0}^{p} \sum_{\kappa=0}^{\kappa_{\max}-1} V^{(\kappa)}(z_{(j-1)p+r}) V_{l_1;l_2;\kappa^{\max}r+\kappa}(z_j^{\min}, z_j^{\max})), \quad (26)$$

where $V_{l_1;l_2;l_3}(z^{\min}, z^{\max})$ are determined by integrals with IHPs

$$V_{l_1;l_2;l_3}(z_j^{\min}, z_j^{\max}) = \int_{z_j^{\min}}^{z_j^{\max}} f_1(z) N_{l_1}(z, z_j^{\min}, z_j^{\max})$$

$$\times N_{l_2}(z, z_j^{\min}, z_j^{\max}) N_{l_3}(z, z_j^{\min}, z_j^{\max})dz. \quad (27)$$

The obtained expression will be exact for polynomial potentials of the degree smaller than p'. Generally this decomposition leads to numerical eigenfunctions and eigenvalues with the accuracy of order about $p' + 1$. If the integrals in Step 1.3 of Algorithm 1 and Step 2.3 of Algorithm 2 cannot be calculated in the analytical form, then the Gauss integration rule [1,6] with $p' + 1$ nodes is applied and held the theoretical estimations (21).

2. Using the local coordinate $\eta \in [-1, 1]$ related to the absolute coordinate z as $z = z_j^{\min} + h_j(1+\eta)/2$, $\frac{dz}{d\eta} = h_j/2$, one should exploit the following expansions of the function and its first derivative

$$\hat{\Phi}(z) = \sum_{r=0}^{p} \sum_{\kappa=0}^{\kappa^{\max}-1} \hat{\Phi}_{\kappa^{\max}r+\kappa} N_{\kappa^{\max}r+\kappa}(\eta, -1, 1) \left(\frac{dz}{d\eta}\right)^{\kappa},$$

$$\frac{d\hat{\Phi}(z)}{dz} = \sum_{r=0}^{p} \sum_{\kappa=0}^{\kappa^{\max}-1} \hat{\Phi}_{\kappa^{\max}r+\kappa} \frac{dN_{\kappa^{\max}r+\kappa}(\eta, -1, 1)}{d\eta} \left(\frac{dz}{d\eta}\right)^{\kappa-1}.$$

3. The matrices $A_{L_1 L_2}$ and $B_{L_1 L_2}$ are symmetric, their dimension is $L \times L$, where $L = \kappa^{\max}(np + 1)$. They consist of n sub-matrices with the dimension $\kappa^{\max}(p + 1) \times \kappa^{\max}(p + 1)$. The intersections of these sub-matrices are blocks having the dimension $\kappa^{\max} \times \kappa^{\max}$. These blocks include elements that equal zero in both matrices $B_{L_1 L_2}$ and $A_{L_1 L_2}$ for $V(z) = 0$ and become nonzero in the matrix $A_{L_1 L_2}$, when $V(z) \neq 0$. The existence of such elements is a manifestation of the IHPs symmetry. The total number of elements in all these blocks is $(n(p^2 + 2p) + 1)(\kappa^{\max})^2$. Examples of banded matrix structures are shown in Fig. 2.

4. To impose the BC (III) in $z^{\min}$ one should apply $A_{11} = A_{11} + f_2(z^{\min})R(z^{\min})$, while to impose the BC (III) in $z^{\max}$ one should apply $A_{L+1-\kappa^{\max}, L+1-\kappa^{\max}} = A_{L+1-\kappa^{\max}, L+1-\kappa^{\max}} - f_2(z^{\max})R(z^{\max})$. To impose the BC (I) in $z^{\min}$ one should drop first row and first column, while to apply the BC (I) in $z^{\max}$ one should drop row and column with number $L + 1 - \kappa^{\max}$.

5. For small matrix dimensions ~ 100, the desired solution of the problem generated at Step 2.3 is performed using the built-in procedures of the Maple LinearAlgebra package. For large matrix dimensions $\sim 100 \div 1000000$, the subspace iteration method is used, implemented in the Fortran program SSPACE [1].

References

1. Bathe, K.J.: Finite Element Procedures in Engineering Analysis, Englewood Cliffs. Prentice Hall, New York (1982)
2. Becker, E.B., Carey, G.F., Tinsley Oden, J.: Finite elements. An introduction, vol. I. Prentice-Hall, Inc., Englewood Cliffs (1981)
3. Berezin, I.S., Zhidkov, N.P.: Computing Methods, vol. I. Pergamon Press, Oxford (1965)
4. Chuluunbaatar, O., Gusev, A.A., Vinitsky, S.I., Abrashkevich, A.G.: ODPEVP: A program for computing eigenvalues and eigenfunctions and their first derivatives with respect to the parameter of the parametric self-adjoined Sturm-Liouville problem. Comput. Phys. Commun. 180, 1358–1375 (2009)
5. Chuluunbaatar, O., Gusev, A.A., Gerdt, V.P., Kaschiev, M.S., Rostovtsev, V.A., Samoylov, V., Tupikova, T., Vinitsky, S.I.: A symbolic-numerical Algorithm for solving the eigenvalue problem for a hydrogen atom in the magnetic field: cylindrical coordinates. In: Ganzha, V.G., Mayr, E.W., Vorozhtsov, E.V. (eds.) CASC 2007. LNCS, vol. 4770, pp. 118–133. Springer, Heidelberg (2007)
6. Chuluunbaatar, O., et al.: KANTBP: A program for computing energy levels, reaction matrix and radial wave functions in the coupled-channel hyperspherical adiabatic approach. Comput. Phys. Commun. 177, 649–675 (2007)
7. Cwiok, S., et al.: Single-particle energies, wave functions, quadrupole moments and g-factors in an axially deformed Woods-Saxon potential with applications to the two-centre-type nuclear problems. Comput. Phys. Communications 46, 379–399 (1987)
8. Gusev, A.A., Chuluunbaatar, O., Gerdt, V.P., Rostovtsev, V.A., Vinitsky, S.I., Derbov, V.L., Serov, V.V.: Symbolic-numeric algorithms for computer analysis of spheroidal quantum dot models. In: Gerdt, V.P., Koepf, W., Mayr, E.W., Vorozhtsov, E.V. (eds.) CASC 2010. LNCS, vol. 6244, pp. 106–122. Springer, Heidelberg (2010)
9. Gusev, A.A., Vinitsky, S.I., Chuluunbaatar, O., Gerdt, V.P., Rostovtsev, V.A.: Symbolic-numerical algorithms to solve the quantum tunneling problem for a coupled pair of ions. In: Gerdt, V.P., Koepf, W., Mayr, E.W., Vorozhtsov, E.V. (eds.) CASC 2011. LNCS, vol. 6885, pp. 175–191. Springer, Heidelberg (2011)
10. Ramdas Ram-Mohan, L.: Finite Element and Boundary Element Aplications in Quantum Mechanics. Oxford University Press, New York (2002)
11. Samarski, A.A., Gulin, A.V.: Numerical methods, Nauka, Moscow (1989) (in Russian)
12. Strang, G., Fix, G.J.: An Analysis of the Finite Element Method. Prentice-Hall, Englewood Cliffs (1973)
13. Vinitsky, S., Gusev, A., Chuluunbaatar, O., Rostovtsev, V., Le Hai, L., Derbov, V., Krassovitskiy, P.: Symbolic-numerical algorithm for generating cluster eigenfunctions: quantum tunneling of clusters through repulsive barriers. In: Gerdt, V.P., Koepf, W., Mayr, E.W., Vorozhtsov, E.V. (eds.) CASC 2013. LNCS, vol. 8136, pp. 427–442. Springer, Heidelberg (2013)

Sporadic Examples of Directed Strongly Regular Graphs Obtained By Computer Algebra Experimentation

Štefan Gyürki[1,3] and Mikhail Klin[2,3]

[1] Institute of Information Engineering, Automation and Mathematics
Faculty of Chemical and Food Technology, Slovak University of Technology
Radlinského 9, 812 37 Bratislava, Slovak Republic
stefan.gyurki@stuba.sk
[2] Department of Mathematics,
Ben-Gurion University of the Negev
84105 Beer Sheva, Israel
klin@cs.bgu.ac.il
[3] Institute of Mathematics and Computer Science
Matej Bel University
974 11 Banská Bystrica, Slovak Republic

Abstract. We report about the results of the application of modern computer algebra tools for construction of directed strongly regular graphs. The suggested techniques are based on the investigation of non-commutative association schemes and Cayley graphs over non-Abelian groups. We demonstrate examples of directed strongly regular graphs for 28 different parameter sets, for which the existence of a corresponding digraph has not been known before.

1   Introduction

This paper is devoted to the computer algebra experimentation in the area of algebraic graph theory, the part of mathematics on the edge between graph theory, linear algebra, and group theory. The main objects of interest in algebraic graph theory are highly symmetric graphs, where level of symmetry might be measured both on group-theoretical and purely combinatorial levels. Two books [1] and [11] reflect impressive progress in this part of mathematics.

Nowadays computer algebra tools, and especially GAP (Groups, Algorithms, Programming - a System for Computational Discrete Algebra [9]), together with a few of its share packages, become an inalienable part of modern graph theory and combinatorics. A significant portion of striking combinatorial structures was discovered and analyzed with the aid of a computer. The main subject of interest in the presented text are directed strongly regular graphs (briefly DSRGs), a natural generalization of a classical (in algebraic graph theory) concept from simple to directed graphs. The concept of a DSRG was suggested and investigated by A. Duval in [5]. For a while it remained unnoticed, however, during last 15 years this class of structures is becoming more and more popular.

V.P. Gerdt et al. (Eds.): CASC Workshop 2014, LNCS 8660, pp. 155–170, 2014.
© Springer International Publishing Switzerland 2014

The initial concept of a strongly regular graph (briefly SRG) has a number of relatively independent origins of interest in such diverse areas like design of statistical experiments, finite geometries, applied permutation groups, and also complexity theory of algorithms. Indeed, it is well-known that SRGs are usually regarded as most sophisticated structures for the problems of isomorphism testing of graphs and determination of the automorphism group of graphs. The main combinatorial invariant of a SRG is its parameter set, in the sense of [2]. Typically, classification of SRGs is arranged for each parameter set separately. Similar situation is also observed for DSRGs. However, these structures appear even more frequently. For example, while there are 36 parameter sets for SRGs on up to 50 vertices, this number is 225 for DSRGs. On the other hand, the central problems of the identification of DSRGs and determination of their symmetry are on the same level of difficulty as it appears for the classical case of strongly regular graphs.

In this context, DSRGs provide, in comparison with SRGs much more wide training polygons for the experts in the complexity theory which allows more diversity (undirected versus directed) for investigated graphs.

The previous experience (earned, in particular, by M. Klin and his coauthors) shows that a clever use of computers helps to discover new examples of DSRGs and after that to reach an honest theoretical generalization of the detected structures. This line of activity stimulated the authors to join their efforts in a new attempt. At this stage, we are concentrating on the association schemes as possible origins of new DSRGs. Namely, we wish to consider any association scheme $\mathcal{M}$, for which a suitable union of classes provides a DSRG, preferably new, moreover, with a new parameter set.

The paper is organized as follows. In Section 2, the necessary basic notions are introduced. In Section 3, we describe our approach to the problem of finding new directed strongly regular graphs using computer algebra experimentation. In Section 4, the mentioned strategies are explained with enough rigorous details and the results of different approaches are reported. In Section 5, the results of a classical strategy using Cayley graphs are submitted. We conclude with a discussion and summary of new graphs, being discovered.

2   Preliminaries

Below we present brief account of most significant concepts exploited in the paper. We refer to [2] and [19] for more information.

2.1   General Concepts

A *simple graph* Γ is a pair (V, E), where V is a finite set of *vertices*, and E is a set of 2-subsets of V which are called *edges*.

A *directed graph* (briefly digraph) Γ is a pair (V, R) where V is the set of *vertices* and R is a binary relation on V, that is a subset of the set V^2 of all ordered pairs of elements in V. The pairs in R are called *directed arcs* or *darts*. The vertex set of Γ is denoted by $V(\Gamma)$ and the dart set is denoted by $R(\Gamma)$.

A *balanced incomplete block design* (BIBD) is a pair $(\mathcal{P}, \mathcal{B})$ where $\mathcal{P}$ is the point set of cardinality v, and $\mathcal{B}$ is a collection of b k-subsets of $\mathcal{P}$ (*blocks*) such that each element of $\mathcal{P}$ is contained in exactly r blocks and any 2-subset of V is contained in exactly λ blocks. The numbers v, b, r, k, and λ are *parameters* of the BIBD. From the parameters v, k, λ the remaining two are determined uniquely, therefore, we use just the triplet of parameters (v, k, λ) for a BIBD.

For any finite group H, the *group ring* $\mathbb{Z}H$ is defined as the set of all formal sums of elements of H, with coefficients from $\mathbb{Z}$. Let X denote a non-empty subset of H. The element $\sum_{x \in X} x$ in $\mathbb{Z}H$ is called a *simple quantity*, and it is denoted as $\underline{X}$. Suppose now that $e \notin X$, where e is the identity element of the group H. Then the digraph $\Gamma = \text{Cay}(H, X)$ with vertex set H and dart set $\{(x, y) : x, y \in H, yx^{-1} \in X\}$ is called the *Cayley digraph* over H with respect to X.

2.2 Strongly Regular Graphs

A graph Γ with adjacency matrix $A = A(\Gamma)$ is called *regular*, if there exists a positive integer k such that $AJ = JA = kJ$, where J is the all-one matrix. The number k is called *valency* of Γ. A simple regular graph with valency k is said to be *strongly regular* (SRG, for short) if there exist integers λ and μ such that for each edge $\{u, v\}$ the number of common neighbors of u and v is exactly λ; while for each non-edge $\{u, v\}$ the number of common neighbors of u and v is equal to μ. Previous condition can be rewritten equivalently into the equation $A^2 = kI + \lambda A + \mu(J - I - A)$ using the adjacency matrix of Γ. The quadruple (n, k, λ, μ) is called the *parameter set* of an SRG Γ.

2.3 Directed Strongly Regular Graphs

A possible generalization of the notion of SRGs for directed graphs was given by Duval [5]. While the family of SRGs has been well-studied in the algebraic graph theory cf. [2], the directed version has not received enough attention.

A *directed strongly regular graph* (DSRG) with parameters (n, k, t, λ, μ) is a regular directed graph on n vertices with valency k, such that every vertex is incident with t undirected edges, and the number of paths of length 2 directed from a vertex x to another vertex y is λ, if there is an arc from x to y, and μ otherwise. In particular, a DSRG with $t = k$ is an SRG, and a DSRG with $t = 0$ is a doubly regular tournament. Throughout the paper we consider only DSRGs satisfying $0 < t < k$, which are called *genuine* DSRGs.

The adjacency matrix $A = A(\Gamma)$ of a DSRG with parameters (n, k, t, λ, μ), satisfies $AJ = JA = kJ$ and $A^2 = tI + \lambda A + \mu(J - I - A)$.

Example 1. The smallest example of a DSRG is appearing on 6 vertices. Its parameter set is $(6, 2, 1, 0, 1)$ and it is depicted in Fig. 1.

Remark 1. In this paper, we are using for DSRG's 5-tuple of parameters in the order (n, k, t, λ, μ), however, in several other papers the order (n, k, μ, λ, t) is used.

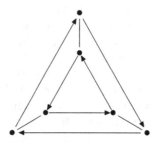

Fig. 1. The smallest genuine DSRG

Proposition 1 ([5]). *If Γ is a DSRG with parameter set (n, k, t, λ, μ) and adjacency matrix A, then the complementary graph $\bar{\Gamma}$ is a DSRG with parameter set $(n, \bar{k}, \bar{t}, \bar{\lambda}, \bar{\mu})$ with adjacency matrix $\bar{A} = J - I - A$, where*

$$\bar{k} = n - k + 1$$
$$\bar{t} = n - 2k + t - 1$$
$$\bar{\lambda} = n - 2k + \mu - 2$$
$$\bar{\mu} = n - 2k + \lambda.$$

Remark 2. Proposition 1 allows us to restrict our search for the DSRGs with $2k < n$, due to complementation, and clearly a discovery of a DSRG with new parameter set implies a discovery of a DSRG on the complementary parameter set. As a consequence, throughout the paper we display just the parameter sets satisfying $2k < n$.

For a directed graph Γ let Γ^T denote the digraph obtained by reversing all the darts in Γ. Then Γ^T is called the *reverse* of Γ. In other words, if A is the adjacency matrix of Γ, then A^T is the adjacency matrix of Γ^T.

The following proposition was observed by Ch. Pech, and presented in [18]:

Proposition 2 ([18]). *Let Γ be a DSRG. Then the graph Γ^T is also a DSRG with the same parameter set.*

We say that two DSRGs Γ_1 and Γ_2 are *equivalent*, if $\Gamma_1 \cong \Gamma_2$, or $\Gamma_1 \cong \Gamma_2^T$, or $\Gamma_1 \cong \bar{\Gamma}_2$, or $\Gamma_1 \cong \bar{\Gamma}_2^T$; otherwise they are called *non-equivalent*. (In other words, Γ_1 is equivalent to Γ_2 if and only if Γ_1 is isomorphic to Γ_2 or to a graph obtained from Γ_2 via reverse and complementation.) From our point of view the interesting DSRGs are those which are non-equivalent.

The parameters n, k, t, λ, μ are not independent. Relations to be satisfied for such parameter sets are usually called *feasibility conditions*. Most important and, in a sense, basic conditions are the following (for their proof see [5]):

$$k(k + \mu - \lambda) = t + (n - 1)\mu. \tag{1}$$

There exists a positive integer d such that:

$$d^2 = (\mu - \lambda)^2 + 4(t - \mu) \tag{2}$$

$$d \mid (2k - (\mu - \lambda)(n - 1)) \tag{3}$$

$$n - 1 \equiv \frac{2k - (\mu - \lambda)(n - 1)}{d} \pmod{2} \tag{4}$$

$$n - 1 \geq \left| \frac{2k - (\mu - \lambda)(n - 1)}{d} \right|. \tag{5}$$

Further:

$$0 \leq \lambda < t < k$$
$$0 < \mu \leq t < k$$
$$-2(k - t - 1) \leq \mu - \lambda \leq 2(k - t).$$

We have to mention that for a feasible parameter set it is not guaranteed that a DSRG with that parameter set does exist. A feasible parameter set for which at least one DSRG Γ exists is called *realizable*, otherwise *non-realizable*. The smallest example of a non-realizable parameter set is $(14, 5, 4, 1, 2)$, what was shown in [18].

2.4   Coherent Configurations and Association Schemes

Under a *color graph* Γ we mean an ordered pair $(V, \mathcal{R})$, where V is a set of vertices and $\mathcal{R}$ a partition of $V \times V$ into binary relations. The elements of $\mathcal{R}$ are called *colors*, and the number of colors is the *rank* of Γ. In other words, a color graph is an edge-colored complete directed graph with loops, whose arcs are colored by the same color if and only if they belong to the same binary relation.

A *coherent configuration* is a color graph $\mathcal{M} = (\Omega, \mathcal{R})$, $\mathcal{R} = \{R_i \mid i \in I\}$, such that the following axioms are satisfied:

(i) The diagonal relation $\Delta_\Omega = \{(x, x) \mid x \in \Omega\}$ is a union of relations $\cup_{i \in I'} R_i$, for a suitable subset $I' \subseteq I$.

(ii) For each $i \in I$ there exists $i' \in I$ such that $R_i^T = R_{i'}$, where $R_i^T = \{(y, x) \mid (x, y) \in R_i\}$ is the relation transposed to R_i.

(iii) For any $i, j, k \in I$, the number $p_{i,j}^k$ of elements $z \in \Omega$ such that $(x, z) \in R_i$ and $(z, y) \in R_j$ is a constant depending only on i, j, k, and independent of the choice of $(x, y) \in R_k$.

The numbers $p_{i,j}^k$ are called *intersection numbers*, or sometimes *structure constants* of $\mathcal{M}$. A coherent configuration $\mathcal{M}$ is called *commutative*, if for all $i, j, k \in I$ we have $p_{ij}^k = p_{ji}^k$, otherwise *non-commutative*.

Let (G, Ω) be a permutation group. G acts naturally on $\Omega \times \Omega$ by $(x, y)^g = (x^g, y^g)$. The orbits of this action are called *2-orbits* (or *orbitals*) of (G, Ω), and denoted by $2-\mathrm{Orb}(G, \Omega)$. It is easy to check that $(\Omega, 2-\mathrm{Orb}(G, \Omega))$ is a coherent configuration for every permutation group (G, Ω). The coherent configurations which appear in this manner are called *Schurian*, otherwise *non-Schurian*.

An *association scheme* $\mathcal{M} = (\Omega, \mathcal{R})$ is a *homogeneous* coherent configuration, i.e., where the diagonal relation Δ_Ω does belong to $\mathcal{R}$. Hence, a very important source of association schemes are transitive permutation groups, since their 2-orbits form a homogeneous coherent configuration.

Let $\mathcal{M}$ be a coherent configuration of rank r. To each relation R_i in $\mathcal{M}$ we can assign a 0–1-matrix A_i such that $(A_i)_{xy} = 1 \iff (x, y) \in R_i$. Then clearly $\sum_{i=1}^{r} A_i = J$ and $A_i A_j = \sum_{k=1}^{r} p_{ij}^k A_k$. Matrices $A_1, \ldots, A_r$ generate an algebra $\mathcal{W}$ over $\mathbb{C}$, which is called *coherent algebra* of rank r and degree n, and we write $\mathcal{W} = \langle A_1, \ldots, A_r \rangle$.

Let $\mathcal{W}_1, \mathcal{W}_2$ be two coherent algebras of order n. Then $\mathcal{W}_1 \cap \mathcal{W}_2$ is again a coherent algebra, therefore, there exists a unique minimal coherent algebra $\mathcal{W}$ containing a given set $\{M_1, \ldots, M_t\}$ of 0–1-matrices of order $n \times n$. This algebra is called *coherent closure* of $M_1, \ldots, M_t$ and it is denoted $\langle\langle M_1, \ldots, M_t \rangle\rangle$. In particular, to a DSRG Γ with adjacency matrix A, we can associate the coherent closure $\mathcal{W}(\Gamma) = \langle\langle A \rangle\rangle$.

To each coherent configuration $\mathcal{M}$, we can assign three groups: $\mathrm{Aut}(\mathcal{M})$, $\mathrm{CAut}(\mathcal{M})$ and $\mathrm{AAut}(\mathcal{M})$. The (combinatorial) *group of automorphisms* $\mathrm{Aut}(\mathcal{M})$ consists of the permutations $\phi : \Omega \to \Omega$ which preserve the relations, i.e., $R_i^\phi = R_i$ for all $R_i \in \mathcal{R}$. The *color automorphisms* preserve relations setwise, i.e., for $\phi : \Omega \to \Omega$ we have $\phi \in \mathrm{CAut}(\mathcal{M})$ if and only if for all $i \in I$ there exists $j \in I$ such that $R_i^\phi = R_j$. An *algebraic automorphism* is a bijection $\psi : \mathcal{R} \to \mathcal{R}$ which satisfies $p_{ij}^k = p_{i^\psi j^\psi}^{k^\psi}$. We refer to [19] for a discussion of these concepts.

Graphs and digraphs can be regarded as binary relations, while association schemes are collections of binary relations in the sense of our definition. Therefore, it is natural to ask:

Question 1. Assume $\mathcal{M}$ is an association scheme of order n. Can we obtain a DSRG on n vertices as a union of suitable classes in $\mathcal{M}$?

It turns out that there is no standard easy way to reply to the question.

A very important necessary condition posed for the initial association scheme $\mathcal{M}$ was given in [18]:

Theorem 1 ([18]). *Let Γ be a genuine directed strongly regular graph. Then the coherent closure $\mathcal{W}(\Gamma)$ is non-commutative, and its rank is at least 6.*

In other words, we have to consider non-commutative association schemes of rank at least 6, when we are searching for directed strongly regular graphs as unions of relations in a prescribed association scheme.

Significant part of our results was achieved following the strategy of creating suitable non-commutative association schemes and taking unions of their relations.

3 General Approach to the Computer Experimentation

3.1 Main Methodology

Assume that $\mathcal{M} = (\Omega, \mathcal{R})$ is an association scheme and $G = \mathrm{Aut}(\mathcal{M})$. Let r be the rank of $\mathcal{M}$, thus, $\mathcal{M}$ has $r - 1$ classes. In many cases below, G acts transitively on Ω. Moreover, $\mathcal{M}$ is the Schurian scheme obtained from permutation group (G, Ω), however, this restriction is not obligatory in the framework of the described approach.

Let Γ be a putative DSRG (with order n), which is obtained via union of suitable classes of $\mathcal{M}$. Then clearly one has to inspect 2^{r-1} possible unions.

First evident restriction is to look simultaneously for all possible parameter sets of DSRGs of order n; recall that this data is available at [3]. Typically, in this project, our attention was restricted only to the open parameter sets.

At the second step, one has to consider multisets of valencies of symmetric and antisymmetric classes in $\mathcal{M}$ and to find in advance which subsets of classes of $\mathcal{M}$ may in principle provide a mixed graph with prescribed pair of valencies $(t, k-t)$, respectively. Getting such a list is a simple case of the famous knapsack problem, however, we were using a very naive approach to provide all solutions.

In order to eliminate in further search duplicates of isomorphic graphs it might be helpful to work with the representatives of orbits on sets of relations in the action of the color group $\mathrm{CAut}(\mathcal{M})$. Sometimes, preliminary sorting with the aid of the action of the algebraic group $\mathrm{AAut}(\mathcal{M})$ might be also of help. Nevertheless, according to the gathered practical experience, in most of the cases the order of $\mathrm{AAut}(\mathcal{M})$ is relatively small, thus, we are facing exponential complexity in cases of schemes with relatively small valencies. Therefore, we decided to restrict our systematic attempts just to the association schemes of rank not larger than 25.

Finally, for each selected "suspicious" union of relations from $\mathcal{M}$ we have to check whether it is providing a DSRG or not. Here use of the known structure constants of $\mathcal{M}$ is very crucial: indeed, instead of inspection of the adjacency matrix $A(\Gamma)$ of a putative graph Γ we arrange calculations with the tensor of structure constants of $\mathcal{M}$.

The computational scheme outlined above is, in a sense, the ideal plan of activities, which were arranged in the course of computations. In many cases, we preferred to use ingredients of a brute-force approach, rather than to being involved in a more sophisticated programming. Since in many cases it was impossible to execute an exhaustive search, it was substituted by an ad hoc selection of simple "promising" subsets of candidates.

3.2   Computer Tools

We run all computations in the software GAP [9] with its share packages GRAPE [28] together with nauty [24] for computation with graphs; an unpublished package COCO-II [27] written by S. Reichard for computations with association schemes and coherent configurations; and the package SetOrbit [26] written by Ch. Pech and S. Reichard, and documented in [25], for finding representatives of orbits of group actions on sets of various size.

In addition, some ad hoc computational tricks were used from time to time, like to exploit a simple variation of the calculation of the coherent closure of an auxiliary graph, which is related to the putative DSRG Γ, as well as some helpful functions for the calculations with association schemes borrowed from the site [13].

3.3 Sources for Association Schemes

Recall that first open parameter set for a DSRG appears for order $n = 22$. With growing of n the fraction of open parameter sets is becoming more essential. This dictated our strategy in the selection of candidates for association schemes being considered. In what follows, we report only about successful attempts, resulted in discovery of graphs with open parameter sets. However, as a byproduct, many graphs with known parameter sets were also considered (their comparison with known ones remains as one of tasks for a more systematic approach in the future).

Roughly speaking, we distinguish a few different typical origins in our search:

- use of existing catalogues of association schemes;
- inspection of groups of automorphisms of some "famous" vertex-transitive graphs;
- consideration of incidence structures;
- investigation of Cayley graphs.

In the next section, we are paying reasonable attention to a more detailed discussion of each of these approaches.

4   Unions of Relations in Association Schemes

Here we consider several strategies for finding non-commutative association schemes which serve as input for searching new DSRGs.

4.1   Search Using Catalogue of Small Association Schemes

For executing our strategy it is enough to consider non-commutative association schemes of small order at the first stage. They are systematically arranged according to their order and rank in the catalogue of Hanaki and Miyamoto [13].

The number of new parameter sets, for which we succeed, using exactly this approach, is 12, see Table 2 in Summary. For several parameter sets, we have found a few non-equivalent DSRGs. Table 1 contains just the digraphs which are mutually non-equivalent. In this Table 1, we display sufficient portion of information for reconstructing discovered DSRGs using the catalogue of association schemes by Hanaki and Miyamoto.

Remark 3. We noticed that in [13] "class" of association schemes is used instead of their "rank". Clearly, the number of classes is less by one than the rank.

From Table 1 it is easy to observe that graph nr. 13 is a (spanning) subgraph of graphs nr. 25 and 26; while nr. 14 is a subgraph of nr. 19.

4.2   Actions of Group of Automorphisms of Graphs

Jørgensen in [17] and [15] announced the existence of a DSRG with the parameter set $(108, 10, 3, 0, 1)$. The author provided us the adjacency matrix of this new

Table 1. DSRGs from small association schemes

nr.	(n, k, t, λ, μ)	Union of relations	AS order	AS rank	nr.cat
1	$(30, 13, 11, 6, 5)$	$1, 2, 4, 6, 8$	30	11	184
2	$(36, 13, 7, 4, 5)$	$4, 5, 6, 9$	36	11	49
3	$(36, 13, 7, 4, 5)$	$1, 2, 3, 4, 8, 12, 14, 16$	36	20	28
4	$(36, 13, 7, 4, 5)$	$1, 2, 3, 4, 8, 12, 14, 17$	36	20	28
5	$(36, 13, 7, 4, 5)$	$1, 2, 3, 4, 5, 10, 13, 14$	36	20	30
6	$(36, 13, 7, 4, 5)$	$1, 2, 3, 4, 6, 8, 12, 18$	36	20	40
7	$(36, 13, 7, 4, 5)$	$1, 2, 3, 4, 6, 8, 13, 16$	36	20	40
8	$(36, 13, 11, 2, 6)$	$1, 3, 5, 7, 10$	36	13	57
9	$(45, 16, 8, 5, 6)$	$1, 3, 5, 8$	45	10	18
10	$(45, 16, 8, 5, 6)$	$2, 3, 5, 8$	45	10	18
11	$(50, 16, 10, 3, 6)$	$1, 2, 6, 9, 13$	50	14	9
12	$(50, 23, 13, 10, 11)$	$1, 2, 6, 8, 10, 12$	50	14	17
13	$(54, 8, 3, 2, 1)$	$1, 6, 8, 12$	54	18	103
14	$(54, 16, 12, 6, 4)$	$4, 5, 10, 12, 16$	54	18	109
15	$(54, 19, 9, 6, 7)$	$1, 3, 4, 6, 11, 14, 16$	54	18	111
16	$(54, 19, 9, 6, 7)$	$1, 3, 4, 6, 11, 14, 17$	54	18	111
17	$(54, 20, 16, 6, 8)$	$2, 3, 4, 5, 11, 13, 14$	54	18	109
18	$(54, 20, 16, 6, 8)$	$8, 9, 11, 13, 14$	54	18	109
19	$(54, 21, 17, 8, 8)$	$1, 2, 3, 4, 5, 10, 12, 16$	54	18	109
20	$(54, 21, 17, 8, 8)$	$1, 8, 9, 10, 12, 16$	54	18	109
21	$(54, 25, 14, 11, 12)$	$1, 2, 4, 6, 10, 12, 14, 16$	54	18	106
22	$(54, 25, 14, 11, 12)$	$1, 2, 4, 6, 10, 12, 15, 17$	54	18	106
23	$(54, 25, 14, 11, 12)$	$1, 2, 4, 6, 10, 13, 15, 16$	54	18	106
24	$(54, 25, 14, 11, 12)$	$1, 2, 4, 6, 11, 13, 15, 17$	54	18	106
25	$(54, 25, 14, 11, 12)$	$1, 2, 6, 8, 10, 12, 14, 16$	54	18	103
26	$(54, 25, 14, 11, 12)$	$1, 2, 6, 8, 10, 12, 15, 17$	54	18	103
27	$(54, 25, 14, 11, 12)$	$1, 2, 6, 8, 10, 13, 14, 17$	54	18	103
28	$(54, 25, 14, 11, 12)$	$1, 2, 6, 8, 11, 13, 14, 16$	54	18	103

digraph, and we managed to explain it in terms of unions of relations in the Schurian association scheme of the group of automorphism of the Pappus graph in the action on the ordered triples of its vertices. For more details see [12].

This successful attempt inspired us to go ahead in a similar spirit. In fact, we investigated actions of the group of automorphisms of several symmetric graphs on certain orbits of various k-sets and k-tuples. Usually, due to high time and space complexity, we took just $k \in \{2, 3, 4\}$. Restricting group action to an orbit we ensure that the resulted action is transitive on it, and from this action we create the Schurian association scheme. When it passes the test for being non-commutative, then there is sense to execute the search for DSRGs as unions of relations in these schemes. Once more, due to high time-complexity, we restricted ourselves just for the cases when the rank was not greater than 25 and the size of the orbit not greater than 110. Therefore, our search is far from being exhaustive. If one goes higher with the rank, then he could probably find new DSRGs.

One can find origins of this strategy in [8], Example 3.4. The authors took the lattice graph on 9 points and investigated an action of a subgroup of its group of automorphism on the edges. Our strategy is a slight generalization of it, since we do not consider only the pairs of two adjacent vertices, but also actions on any 2-sets, 3-sets, ordered pairs, ordered triplets of vertices and sometimes on 4-sets, ordered quadruples.

Using our strategy, we succeed in the following cases (the starting "famous" graphs are available via Internet, e.g. from the home page of A. Brouwer):

- from the Petersen graph we obtained a DSRG$(60, 13, 5, 2, 3)$;
- from the Shrikhande graph we get a DSRG$(48, 10, 6, 2, 2)$ and $(48, 13, 7, 2, 4)$;
- from the Heawood graph we obtained DSRG $(84, 31, 17, 12, 11)$, $(84, 29, 19, 6, 12)$ and $(84, 39, 27, 18, 18)$, see also Section 4.3;
- from the unique SRG$(21, 10, 3, 6)$ we obtained a DSRG$(105, 36, 16, 11, 13)$;

Explicit descriptions of these digraphs are shown in Appendix.

4.3   Actions of Group of Automorphisms of Combinatorial Designs

Let us now start from a block design $\mathcal{D} = (\mathcal{P}, \mathcal{B})$ with the point set $\mathcal{P}$ and block set $\mathcal{B}$, let $I = I(\mathcal{D})$ be the Levi graph of $\mathcal{D}$, that is the graph with vertex set $\mathcal{P} \cup \mathcal{B}$ and two vertices being adjacent if and only if the corresponding elements of $\mathcal{D}$ are incident. Clearly, $I(\mathcal{D})$ is a bipartite graph. The group $\mathrm{Aut}(I(\mathcal{D}))$ either coincides with the group $\mathrm{Aut}(\mathcal{D})$, or it is twice larger (the latter corresponds to the case when $\mathcal{D}$ is a symmetric self-dual design).

For a number of designs $\mathcal{D}$, we investigated the action of the group $G = \mathrm{Aut}(I(\mathcal{D}))$ on certain orbits of various k-subsets and k-tuples of vertices of I. The same limitations for values of k, order and rank of related Schurian association scheme (like in previous section) remain valid. The execution of search for DSRGs has been started, provided the appearing Schurian schemes were non-commutative.

Using this strategy we succeed in the following cases:

- Considering the unique $(7, 3, 1)$-design is equivalent of consideration of the Heawood graph in the previous section, since the Levi graph of the $(7, 3, 1)$-design is the Heawood graph;
- from the unique $(9, 3, 1)$-design we get DSRGs with parameter sets $(72, 20, 14, 4, 6)$, $(72, 21, 15, 6, 6)$, $(72, 22, 9, 6, 7)$;
- from the $(10, 4, 2)$-design with group of automorphisms of order 720 we get DSRGs with parameter sets $(60, 26, 20, 10, 12)$ and $(90, 28, 16, 10, 8)$;
- from the $(15, 3, 1)$-design which has group of automorphisms of order 288 we get DSRG with parameter set $(72, 26, 10, 8, 10)$;
- from the $(15, 7, 3)$-design with group of automorphisms of order 1152 we get a DSRG with parameter set $(72, 19, 11, 2, 6)$;
- from the $(25, 4, 1)$-design with group of automorphisms of order 504 we get a DSRG with parameter set $(63, 22, 10, 7, 8)$.

In all these cases, we refer to the description of block designs provided in [4].

Explicit descriptions of all digraphs constructed in this subsection are shown in Appendix.

Example 2. Consider the unique $(7, 3, 1)$-design F, that is the Fano plane. In this case, the group G of order 336 acts transitively on the vertex set of graph $I(F)$ of size 14. Let us consider configuration, which consists of two lines, their intersection point and another point in one line not belonging to the other line. Clearly, there are $\binom{7}{2} \cdot 4 = 84$ possibilities to select such a configuration. It is easy to see that both groups $\text{Aut}(F)$ and $G = \text{Aut}(I(F))$ (of order 168 and 336, respectively) act transitively on the set Ω of cardinality 84. The advantage of the group G is that the corresponding association scheme is of rank 25, that is on the edge of our computational possibilities. The remaining details relevant to the precise description of the resulted DSRG(84, 31, 17, 12, 11) are in Appendix.

It is worthy to notice that the incidence graph $I(F)$ is isomorphic to the Heawood graph considered above. In fact, the considerations from Heawood graph were fulfilled in advance (this took a few days of computational time) and it was exceptionally extended up to groups of rank 30, which lead to the discovery of DSRG(84, 29, 19, 6, 12) and (84, 39, 27, 18, 18).

Example 3. Consider the unique $(9, 3, 1)$-design. We identify its set of points with the set $\mathcal{P} = \{1, 2, \ldots, 9\}$ and its set of blocks with

$$\mathcal{B} = \{\{1, 2, 3\}, \{1, 4, 7\}, \{1, 5, 9\}, \{1, 6, 8\}, \{2, 4, 9\}, \{2, 5, 8\}, \{2, 6, 7\},$$

$$\{3, 4, 8\}, \{3, 5, 7\}, \{3, 6, 9\}, \{4, 5, 6\}, \{7, 8, 9\}\}.$$

The full group of automorphisms of $(\mathcal{P}, \mathcal{B})$ can be identified with the permutation group $G = \langle (1, 7, 3, 2, 6, 9, 4, 5), (4, 6, 5)(7, 8, 9) \rangle$ of order 432 and degree 9. (Of course, it could be also regarded as a permutation group of degree $|\mathcal{P}| + |\mathcal{B}| = 21$.) Let us consider the action of G on the 3-set $\{1, 2, 4\}$. There are $72 = 12 \cdot 3 \cdot 2$ possibilities (from geometric arguments) for such a selection. Denote by $\mathcal{O}$ the entire set of selected configurations. The automorphism group of $(\mathcal{P}, \mathcal{B})$ acts naturally on $\mathcal{O}$ as a permutation group $\tilde{G}$ of degree 72, rank 16. By choosing suitable subsets of 2-orbits of $\tilde{G}$, we get two non-equivalent DSRGs with parameter set $(72, 22, 9, 6, 7)$. For a representation of $\tilde{G}$ as a permutation group of degree 72 see the group called H_9 in Appendix.

5   New Sporadic Examples as Cayley Digraphs

In this section, we construct some new DSRGs of order 32 and 39 as Cayley digraphs. Among them, we obtain the first DSRG with parameter set $(39, 16, 12, 7, 6)$.

The following lemma is crucial for testing whether a Cayley digraph is DSRG.

Lemma 1 ([18],[14]). *The Cayley digraph* $\text{Cay}(G, X)$ *is DSRG with parameter set* (v, k, t, λ, μ) *if and only if the equation* $\underline{X} \cdot \underline{X} = t \cdot \underline{e} + \lambda \cdot \underline{X} + \mu \cdot (\underline{G} - \underline{e} - \underline{X})$ *holds in* $\mathbb{Z}G$.

5.1   Cayley Digraphs on 32 Vertices

We now show how to obtain new DSRGs for parameter sets $(32, 9, 6, 1, 3)$, $(32, 13, 9, 4, 6)$ and $(32, 14, 10, 6, 6)$.

Let us take the wreath product group $H = S_2 \wr \mathbb{Z}_4$ of order 32. (Here for wreath product we follow notation from [22], which is inherited from L.A. Kalužnin.) Each element $h \in H$ can be uniquely represented as $h = (g; k_1, k_2)$, where $g \in S_2$, $k_1, k_2 \in \mathbb{Z}_4$. Let x be a generator of $\mathbb{Z}_4$, and $\pi = (1\,2) \in S_2$. In order to shorten description we display just the triple $i\,j\,l$ instead of $(\pi^i; x^j, x^l)$.

Let us define six subsets of H:

$$X_1 = \{002, 011, 012, 032, 033, 100, 101, 102, 103\},$$
$$X_2 = \{010, 011, 030, 031, 033, 100, 102, 111, 113, 121, 123, 130, 132\},$$
$$X_3 = \{010, 021, 022, 023, 030, 100, 102, 111, 113, 121, 123, 130, 132\},$$
$$X_4 = \{002, 011, 012, 032, 033, 100, 101, 102, 103, 120, 121, 122, 123\},$$
$$X_5 = \{001, 003, 011, 012, 032, 033, 100, 101, 102, 103, 110, 111, 112, 113\},$$
$$X_6 = \{001, 003, 011, 012, 032, 033, 100, 101, 102, 103, 110, 112, 131, 133\}.$$

It is a routine-work to check using Lemma 1 that the following proposition holds:

Proposition 3. *The Cayley digraph* $\Gamma_i = \mathrm{Cay}(H, X_i)$ *is a DSRG with parameter set*

a) $(32, 9, 6, 1, 3)$, *for* $i = 1$;
b) $(32, 13, 9, 4, 6)$, *for* $i = 2, 3, 4$, *and*
c) $(32, 14, 10, 6, 6)$, *for* $i = 5, 6$.

According to [3], DSRGs with parameter sets mentioned in the previous proposition had been constructed in [10], [23].

Using computer algebra system GAP [9] along with the computer package GRAPE [11] in common with nauty [24] we have tested that all the digraphs constructed in Proposition 3 are pairwise non-equivalent and none of these DSRGs were obtained earlier.

5.2   Cayley Digraphs on 39 Vertices

In this subsection, we construct DSRGs for all feasible parameter sets on 39 vertices. Hence, we obtain also a DSRG for the parameter set $(39, 16, 12, 7, 6)$ for which such a graph has not been known at the time of writing this paper. All these graphs arise as Cayley digraphs over a metacyclic group of order 39. For more constructions from metacyclic groups, we refer the reader to [6].

Let us take the group G presented as

$$G = \langle a, b : a^{13} = b^3 = e, ba = a^9 b \rangle \le AGL(1, 13)$$

and let us define eight of its subsets:

$X_1 = \{a, a^5, a^8, a^{12}, b, a^2b, a^4b, a^3b^2, a^7b^2, a^{11}b^2\}$,

$X_2 = \{a, a^5, a^8, a^{12}, b, a^4b, a^7b, a^{11}b, a^2b^2, a^4b^2, a^8b^2, a^{11}b^2\}$,

$X_3 = \{a, a^5, a^8, a^{12}, b, a^4b, a^7b, a^{10}b, a^3b^2, a^4b^2, a^{10}b^2, a^{11}b^2\}$,

$X_4 = \{a^2, a^4, a^9, a^{11}, b, a^4b, a^6b, a^{11}b, a^3b^2, a^5b^2, a^9b^2, a^{11}b^2\}$,

$X_5 = X_2 \cup \{a^{12}b, a^3b^2\}$,

$X_6 = X_4 \cup \{a^2b, a^7b^2\}$,

$X_7 = \{a^2, a^4, a^9, a^{11}, a^2b, a^3b, a^6b, a^8b, a^{11}b, a^{12}b, b^2, a^3b^2, a^4b^2, a^6b^2, a^7b^2, a^{10}b^2\}$,

$X_8 = \{a^4, a^5, a^8, a^9, b, ab, a^4b, a^6b, a^8b, a^{10}b, b^2, a^2b^2, a^4b^2, a^6b^2, a^8b^2, a^{10}b^2\}$.

Proposition 4. *The Cayley digraph* $\Gamma_i = \text{Cay}(G, X_i)$ *is a DSRG with parameter set*

a) $(39, 10, 6, 1, 3)$, *for* $i = 1$;

b) $(39, 12, 4, 3, 4)$, *for* $i = 2, 3, 4$,

c) $(39, 14, 6, 5, 5)$, *for* $i = 5, 6$, *and*

d) $(39, 16, 12, 7, 6)$ *for* $i = 7, 8$.

Remark 4. The DSRG with parameter set $(39, 10, 6, 1, 3)$ is isomorphic to the one constructed in [23] and described using partial sum families. The two digraphs with parameters $(39, 16, 12, 7, 6)$ are non-equivalent.

Remark 5. In [16] the author states that it may happen that the so-called Krein parameter $q^\theta_{\theta\theta}$ is always non-negative if 0 and -1 are not eigenvalues of a DSRG Γ in consideration. The existence of a DSRG with parameters $(39, 10, 6, 1, 3)$ disproves it, since in this case $q^\theta_{\theta\theta} = -3/8$ (this fact was somehow not mentioned in [23]).

6   Conclusion and Summary

The main genre of this paper is computer algebra experimentation for the purposes of algebraic graph theory. Using techniques and ideas, which were before reflected in [6], [8], [7], [18] and [21], the author Š. Gyürki arranged a more systematical search for DSRGs, relying on the above described strategies.

We think that the approaches outlined above carry features of methodological innovations, though in a few cases they simply stem from careful analysis of previous successful computations done by M. Klin et al.

Our next goal was of a definite "sporting" interest: to present examples of new DSRGs for previously open parameter sets. Altogether we reached such a success for 28 new parameter sets, see Table 2 below.

Of course, the foremost goal at a computer algebra experimentation (cf. [20]) is to reach a successful theoretical generalization of the obtained new results. In the case of the sporadic examples of new DSRGs, this would mean to try to embed at least some of the new examples into new infinite classes of DSRGs. We are pleased to claim that this task was successfully fulfilled in the course of our project. In fact, already after the submission of the initial version of this paper we succeeded to generalize the presented digraph with parameter set $(32, 14, 10, 6, 6)$ to the infinite series of DSRGs with parameters $(2n^2, 4n - 2, 2n + 2, n + 2, 6)$. The corresponding paper is in preparation. Hence, we can finally claim that one more corollary of the reported project is creation of new (striking in the eyes of the authors) patterns of successful insight:

- to observe a short sequence of parameter sets with similar properties;
- to formulate a plausible conjecture about a possible putative infinite series of combinatorial structures;
- to prove this conjecture on purely theoretical level, that is finally, without the use of a computer.

Table 2 below provides a brief summary of our computer aided discoveries.

Table 2. Summary

n	k	t	λ	μ	ps	am	constructed in	n	k	t	λ	μ	ps	am	constructed in
30	13	11	6	5	Yes	1	Section 4.1	54	19	9	6	7	Yes	2	Section 4.1
32	9	6	1	3	No	1	Section 5.1	54	20	16	6	8	Yes	2	Section 4.1
32	13	9	4	6	No	3	Section 5.1	54	21	17	8	8	Yes	2	Section 4.1
32	14	10	6	6	No	2	Section 5.1	54	25	14	11	12	Yes	8	Section 4.1
36	13	7	4	5	Yes	6	Section 4.1	60	13	5	2	3	Yes	1	Section 4.2
36	13	11	2	6	Yes	1	Section 4.1	60	26	20	10	12	Yes	2	Section 4.3
39	10	6	1	3	No	0	Section 5.2	63	22	10	7	8	Yes	1	Section 4.3
39	12	4	3	4	No	3	Section 5.2	72	19	11	2	6	Yes	1	Section 4.3
39	14	6	5	5	No	2	Section 5.2	72	20	14	4	6	Yes	1	Section 4.3
39	16	12	7	6	Yes	2	Section 5.2	72	21	15	6	6	Yes	1	Section 4.3
45	16	8	5	6	Yes	2	Section 4.1	72	22	9	6	7	Yes	2	Section 4.3
48	10	6	2	2	Yes	1	Section 4.2	72	26	10	8	10	Yes	1	Section 4.3
48	13	7	2	4	Yes	1	Section 4.2	84	29	19	6	12	Yes	1	Section 4.2
50	16	10	3	6	Yes	1	Section 4.1	84	31	17	12	11	Yes	1	Section 4.2
50	23	13	10	11	Yes	1	Section 4.1	84	39	27	18	18	Yes	2	Section 4.2
54	8	3	2	1	Yes	1	Section 4.1	90	28	16	10	8	Yes	1	Section 4.3
54	16	12	6	4	Yes	1	Section 4.1	105	36	16	11	13	Yes	1	Section 4.2

Remark 6. Abbreviations used in Table 2: ps – Is parameter set new?; am – The amount of new constructed DSRGs.

Remark 7. A more detailed version of the paper contains also Appendix with all details regarding constructed DSRGs. It is available on request from Š. Gyürki and will finally appear on his home page.

Acknowledgements. The first author gratefully acknowledges the contribution of the Scientific Grant Agency of the Slovak Republic under the grant 1/1005/12.

This research was also supported by the Project: Mobility - enhancing research, science and education at the Matej Bel University, ITMS code: 26110230082, under the Operational Program Education cofinanced by the European Social Fund.

We thank L. Jørgensen for generous sharing with us of his preliminary results related to the DSRG on 108 vertices. A long-standing cooperation with Ch. Pech and S. Reichard in the use of computer algebra tools is appreciated.

We thank the reviewers for helpful constructive remarks and suggestions.

References

1. Biggs, N.: Algebraic Graph Theory, 2nd edn. Cambridge Mathematical Library. Cambridge University Press, Cambridge (1993); (1st ed. (1974))
2. Brouwer, A.E., Haemers, W.H.: Spectra of Graphs. Universitext. Springer, New York (2012)
3. Brouwer, A.E., Hobart, S.: Tables of directed strongly regular graphs (April 2014), http://homepages.cwi.nl/~aeb/
4. Colbourn, C.J., Dinitz, J.H.: The Handbook of Combinatorial Designs, 2nd edn. Chapman & Hall/CRC, Boca Raton (2007)
5. Duval, A.M.: A directed graph version of strongly regular graphs. J. Combin. Th. A 47, 71–100 (1988)
6. Duval, A.M., Iourinski, D.: Semidirect product constructions of directed strongly regular graphs. J. Combin. Th. A 104, 157–167 (2003)
7. Fiedler, F., Klin, M.H., Muzychuk, M.: Small vertex-transitive directed strongly regular graphs. Discrete Math. 255, 87–115 (2002)
8. Fiedler, F., Klin, M., Pech, C.: Directed strongly regular graphs as elements of coherent algebras. In: Denecke, K., Vogel, H.-J. (eds.) General Algebra and Discrete Mathematics: Proc. Conf. on General Algebra and Discrete Mathematics, Potsdam 1998, pp. 69–87. Shaker Verlag, Aachen (1999)
9. GAP – Groups, Algorithms, Programming – a System for Computational Discrete Algebra, http://www.gap-system.org
10. Godsil, C.D., Hobart, S.A., Martin, W.J.: Representations of directed strongly regular graphs. Europ. J. Combin. 28, 1980–1993 (2007)
11. Godsil, C.D., Royle, G.: Algebraic Graph Theory. Graduate Texts in Mathematics, vol. 207. Springer, New York (2001)
12. Gyürki, Š., Klin, M.: On a new directed strongly regular graph on 108 vertices constructed by Jørgensen, and graphs related to it (2013) (manuscript)
13. Hanaki, A., Miyamoto, I.: Catalogue of Small Association Schemes (2013), http://kissme.shinshu-u.ac.jp/as/ (accessed: November 2013)
14. Hobart, S.A., Shaw, T.J.: A note on a family of directed strongly regular graphs. Europ. J. Combin. 20, 819–820 (1999)
15. Jørgensen, L.K.: New mixed Moore graphs and directed strongly regular graphs, http://vbn.aau.dk/files/166247351/R_2013_13.pdf
16. Jørgensen, L.K.: Non-existence of directed strongly regular graphs. Discrete Math. 264, 111–126 (2003)

17. Jørgensen, L.K.: Variations and generalizations of Moore Graphs. In: The International Workshop on Optimal Networks Topologies 2012, Bandung (2012), Slides are available on http://people.math.aau.dk/~leif

18. Klin, M., Munemasa, A., Muzychuk, M., Zieschang, P.H.: Directed strongly regular graphs obtained from coherent algebras. Lin. Alg. Appl. 377, 83–109 (2004)

19. Klin, M., Muzychuk, M., Pech, C., Woldar, A., Zieschang, P.H.: Association schemes on 28 points as mergings of a half-homogeneous coherent configuration. European J. Combin. 28(7), 1994–2025 (2007)

20. Klin, M., Pech, C., Reichard, S., Woldar, A., Ziv-Av, M.: Examples of computer experimentation in algebraic combinatorics. Ars Math. Contemp. 3(2), 237–258 (2010)

21. Klin, M., Pech, C., Zieschang, P.H.: Flag algebras of block designs: I. Initial notions, Steiner 2-designs and generalized quadrangles. Preprint, MATH-AL-10-1998, Technische Universität Dresden (1998)

22. Klin, M., Pöschel, R., Rosenbaum, K.: Angewandte Algebra für Mathematiker und Informatiker, Einführung in gruppentheoretisch-kombinatorische Methoden, Berlin (Applied algebra for mathematicians and information scientists. Introduction to Group-theoretical Combinatorial Methods.) VEB Deutscher Verlag der Wissenschaften (1988) (German)

23. Martinez, L., Araluze, A.: New tools for construction of directed strongly regular digraphs: Difference digraphs and partial sum families. J. Combin. Th. B 100, 720–728 (2010)

24. McKay, B.D.: nauty user's guide, ver. 1.5, Technical Report TR-CS-90-02. Computer Science Department, Australian National Univ. (1990)

25. Pech, C., Reichard, S.: Enumerating set orbits. In: Klin, M., et al. (eds.) Algorithmic Algebraic Combinatorics and Gröbner Bases, pp. 137–150. Springer, Heidelberg (2009)

26. Pech, C., Reichard, S.: The SetOrbit package for GAP, http://www.math.tu-dresden.de/~pech

27. Reichard, S.: COCO II (personal communication)

28. Soicher, L.H.: GRAPE: A system for computing with graphs and groups. Groups and Computation, New Brunswick (1991); DIMACS Ser. Discrete Math. Theoret. Comput. Sci. 11, pp. 287–291, Amer. Math. Soc., Providence, RI (1993)

On the Parallelization of Subproduct Tree Techniques Targeting Many-Core Architectures

Sardar Anisul Haque, Farnam Mansouri, Marc Moreno Maza

University of Western Ontario, London, Ontario, Canada
{haque.sardar,mansouri.farnam}@gmail.com, moreno@csd.uwo.ca

Abstract. We propose parallel algorithms for operations on univariate polynomials (multi-point evaluation, interpolation) based on subproduct tree techniques and targeting many-core GPUs. On those architectures, we demonstrate the importance of adaptive algorithms, in particular the combination of parallel plain arithmetic and parallel FFT-based arithmetic. Experimental results illustrate the benefits of our algorithms.

1 Introduction

We investigate the use of Graphics Processing Units (GPUs) in the problems of evaluating and interpolating polynomials. Many-core GPU architectures were considered in [17] and [18] in the case of numerical computations, with the purpose of obtaining better support, in terms of accuracy and running times, for the development of polynomial system solvers.

Our motivation, in this work, is also to improve the performance of polynomial system solvers. However, we are targeting symbolic, thus exact, computations. In particular, we aim at providing GPU support for solvers of polynomial systems with coefficients in finite fields, such as the one presented in [14]. This case handles problems from cryptography and serves as a base case for the so-called modular methods [4], since those methods reduce computations with rational number coefficients to computations with finite field coefficients.

Finite fields allow the use of asymptotically fast algorithms for polynomial arithmetic, based on Fast Fourier Transforms (FFTs) or, more generally, subproduct tree techniques[1], which have the advantage of providing a more general setting than FFTs. More precisely, evaluation points do not need to be successive powers of a primitive root of unity. Evaluation and interpolation based on subproduct tree techniques have "essentially" (up to log factors) the same algebraic complexity estimates as their FFT-based counterparts. However, their implementation is known to be challenging.

In this work, we report on the first GPU implementation (using CUDA [16]) of subproduct tree techniques for multi-point evaluation and interpolation of univariate polynomials. The parallelization of those techniques raises the following challenges on hardware accelerators:

---

[1] Chapter 10 of [5] and the paper [1] contain overviews of those techniques.

V.P. Gerdt et al. (Eds.): CASC Workshop 2014, LNCS 8660, pp. 171–185, 2014.
© Springer International Publishing Switzerland 2014

1. The divide-and-conquer formulation of operations on subproduct-trees is not sufficient to provide enough parallelism and one must also parallelize the underlying polynomial arithmetic operations, in particular multiplication.
2. Algorithms based on FFT (such as subproduct tree techniques) are memory bound since the ratio of work to memory access is essentially constant, which makes those algorithms not well suited for multi-core architectures.
3. During the course of the execution of a subproduct tree operation (construction, evaluation, interpolation) the degrees of the involved polynomials vary greatly, thus so does the work load of the tasks, which makes those algorithms complex to implement on many-core GPUs.

The contributions of this work are summarized below. We propose parallel algorithms for performing subproduct tree construction, evaluation and interpolation. We also report on their implementation on many-core GPUs. See Sections 3, 5 and 6, respectively. We enhance the traditional algorithms for polynomial evaluation and interpolation based on subproduct tree techniques, by introducing the data-structure of a *subinverse tree*, which we use to implement both evaluation and interpolation, see Section 4. For subproduct tree operations targeting many-core GPUs, we demonstrate the importance of *adaptive algorithms*[2] That is, algorithms that adapt their behavior according to the available computing resources. In particular, we combine *parallel plain arithmetic* and *parallel fast arithmetic*. For the former we rely on [7] and, for the latter we extend the work of [13]. The span and parallelism overhead of our algorithm are measured considering the *many-core machine model* of [8]. The paper [15] briefly discusses the parallelization of FFT-based multi-point evaluation without considering parallelism overhead, adaptive algorithms nor reporting on an implementation.

To evaluate our implementation, we measure the effective memory bandwidth of our GPU code for parallel multi-point evaluation and interpolation on a card with a theoretical maximum memory bandwidth of 148 GB/S, our code reaches peaks at 64 GB/S. Since the arithmetic intensity of our algorithms is high, we believe that this is a promising result.

All implementation of subproduct tree techniques that we are aware of are serial only. This includes [3] for $GF(2)[x]$, the FLINT library[9] and the Modpn library [10]. Hence we compare our code against probably the best serial C code (the FLINT library) for the same operations. For sufficiently large input data and on NVIDIA Tesla C2050, our code outperforms its serial counterpart by a factor ranging between 20 to 30. Experimental data are provided in Section 7. Our code is freely available in source, under GPL license, as part of the project *CUDA Modular Polynomial* (CUMODP) whose web site is http://www.cumodp.org.

2   Background

We refer to [16] for notions related to GPU programming. We review below the notion of a subproduct tree and specify costs for the underlying polynomial

---

[2] A famous example of adaptive algorithm usage was for computing 2,700 billion decimal digits of π on a desktop computer by F. Bellard http://bellard.org/pi/.

arithmetic used in our implementation. Notations and hypotheses introduced in this section are used throughout this paper. Let $n = 2^k$ for some positive integer k and let $\mathbb{K}$ be a finite field. Let $u_0, \ldots, u_{n-1} \in \mathbb{K}$. Define $m_i = x - u_i$, for $0 \le i < n$. We assume that each $u_i \in \mathbb{K}$ can be stored in one machine word.

Subproduct Tree. The subproduct tree $M_n := \mathsf{SubproductTree}(u_0, \ldots, u_{n-1})$ is a complete binary tree of height $k = \log_2 n$. The j-th node of the i-th level of M_n is denoted by $M_{i,j}$, where $0 \le i \le k$ and $0 \le j < 2^{k-i}$, and is defined by $M_{i,j} = m_{j \cdot 2^i} \cdot m_{j \cdot 2^i + 1} \cdots m_{j \cdot 2^i + (2^i - 1)} = \prod_{0 \le \ell < 2^i} m_{j \cdot 2^i + \ell}$. Each $M_{i,j}$ can be defined recursively by $M_{0,j} = m_j$ and $M_{i+1,j} = M_{i,2j} \cdot M_{i,2j+1}$. The i-th level of M_n has 2^{k-i} polynomials with degree of 2^i. Since each element of $\mathbb{K}$ fits a machine word, storing the subproduct tree M_n requires at most $n \log_2 n + 3n - 1$ words.

Algorithm 1. $\mathsf{SubproductTree}(m_0, \ldots, m_{n-1})$

Input: $m_0 = (x - u_0), \ldots, m_{n-1} = (x - u_{n-1}) \in \mathbb{K}[x]$ with $u_i \in \mathbb{K}$, $n = 2^k$, $k \in \mathbb{N}$.
Output: The subproduct-tree M_n.
for $j = 0$ *to* $n - 1$ **do**
$\quad\lfloor\ M_{0,j} = m_j$;

for $i = 1$ *to* k **do**
$\quad$ **for** $j = 0$ *to* $2^{k-i} - 1$ **do**
$\quad\quad\lfloor\ M_{i,j} = M_{i-1,2j} M_{i-1,2j+1}$;

return M_n;

Multi-Point Evaluation and Interpolation. Given a univariate polynomial $f \in \mathbb{K}[x]$ of degree less than n, we define $\chi(f) = (f(u_0), \ldots, f(u_{n-1}))$. The map χ is called the *multi-point evaluation map* at $u_0, \ldots, u_{n-1}$. When $u_0, \ldots, u_{n-1}$ are pairwise distinct, then it realizes an isomorphism of $\mathbb{K}$-vector spaces $\mathbb{K}[x]/\langle m \rangle$ and $\mathbb{K}^n$, where $m = \prod_{0 \le i < n}(x - u_i)$. The inverse map χ^{-1} can be computed via Lagrange interpolation. Given values $v_0, \ldots, v_{n-1} \in \mathbb{K}$, the unique polynomial $f \in \mathbb{K}[x]$ of degree less than n which takes the value v_i at the point u_i for all $0 \le i < n$ is: $f = \sum_{i=0}^{n-1} v_i s_i m / (x - u_i)$ where $s_i = \prod_{i \ne j, 0 \le j < n} 1 / (u_i - u_j)$.

Complexity Measures. Since we are targeting GPU implementation, our parallel algorithms are analyzed using an appropriate model of computation introduced in [8]. The complexity measures are the *work* (i.e. algebraic complexity estimate) the *span* (i.e. running time on infinitely many processors) and the *parallelism overhead*. This latter is the total time for transferring data between the global memory and the local memories of the streaming multi-processors (SMs).

Plain Multiplication. The number of arithmetic operations for multiplying two polynomials with degree less than d using the *plain (schoolbook) multiplication* is $\mathsf{M}_{\mathrm{plain}}(d) = 2d^2 - 2d + 1$. In our GPU implementation, when d is small enough, each polynomial product is computed by a single thread-block and thus within the local memory of a single SM. In this case, we use $2d+2$ threads for one polynomial multiplication. Each thread copies one coefficient from global memory to the local memory. Each of these threads, except one, is responsible for computing one coefficient of the output polynomial and writes that coefficient back to global memory. So the span and parallelism overhead are $d+1$ and $2U$ respectively, where $1/U$ is the throughput measured in word per second, see [8].

FFT-Based Multiplication. The number of operations for multiplying two polynomials with degree less than d using *Cooley-Tukey's FFT* algorithms is $\mathsf{M}_{\mathrm{FFT}}(d) = 9/2\, d^{\triangleleft} \log_2(d^{\triangleleft}) + 4d^{\triangleleft}$ [11]. Here $d^{\triangleleft} = 2^{\lceil \log_2(2d-1) \rceil}$. In our GPU implementation, which relies on Stockham FFT algorithm, this number of operations becomes: $\mathsf{M}_{\mathrm{FFT}}(d) = 15d^{\triangleleft} \log_2(d^{\triangleleft}) + 2d^{\triangleleft}$, see [13]. The span and parallelism overhead of our FFT-based multiplication are $15d^{\triangleleft} + 2d^{\triangleleft}$ and $(36d^{\triangleleft} - 21)U$ respectively.

Polynomial Division. Given $a, b \in \mathbb{K}[x]$, with $\deg(a) \geq \deg(b)$ we denote by $\mathsf{Remainder}(a, b)$ the remainder in the *Euclidean division* of a by b. The number of arithmetic operations for computing $\mathsf{Remainder}(a, b)$, by plain division, is $(\deg(b) + 1)(\deg(a) - \deg(b) + 1)$. In our GPU implementation, we perform plain division for small degree polynomials, in which case a, b are stored into the local memory of an SM. For larger polynomials, we use an FFT-based algorithm to be discussed later. Returning to plain division, we use $\deg(b) + 1$ threads to implement this operation. Each thread reads one coefficient of b and at most $\lceil \frac{\deg(a)+1}{\deg(b)+1} \rceil$ coefficients of a from the global memory. For the output, at most $\deg(b)$ threads write the coefficients of the remainder to the global memory. The span and parallelism overhead are $2(\deg(a) - \deg(b) + 1)$ and $(2 + \lceil \frac{\deg(a)+1}{\deg(b)+1} \rceil)U$.

Reversal of a Polynomial. For $f \in \mathbb{K}[x]$ of degree $d > 0$ and for $e \geq d$, the *reversal* of order e of f is the polynomial denoted by $\mathsf{rev}_e(f)$ and defined as $\mathsf{rev}_e(f) = x^e f(1/x)$. In our implementation, we use one thread for each coefficient of the input and output. So the span and overhead are 1 and $2U$, respectively.

Inverse Modulo a Power of x. For $f \in \mathbb{K}[x]$, with $f(0) = 1$, and $\ell \in \mathbb{N}$ the *modular inverse* of f modulo x^ℓ is denoted by $\mathsf{Inverse}(f, \ell)$ and is uniquely defined by $\mathsf{Inverse}(f, \ell)\, f \equiv 1 \mod (x^\ell)$. One can compute $\mathsf{Inverse}(f, \ell)$ by Newton iteration, see [5, Chapter 10] for details in sequential time $O(\mathsf{M}_{\mathrm{FFT}}(\ell))$.

To help the reader following the complexity analysis presented in the sequel of this paper, a Maple worksheet can be found at http://cumodp.org/links.html. It provides estimates for space allocation, work (total of number of arithmetic operations), span (parallel running time) and parallelism overhead for constructing subproduct tree and sub-inverse tree (our proposed data structure). Recall that the parallelism overhead measures the time for transferring data between the device global memory and the SMs' shared memories. The estimates that we provide follow our CUDA implementation available at http://cumodp.org.

3 Subproduct Tree Construction

We propose an adaptive algorithm for constructing the subproduct tree $M_n :=$ SubproductTree$(u_0, \ldots, u_{n-1})$. We fix an integer H with $1 \leq H \leq k$. We call the following procedure an *adaptive algorithm for computing M_n with threshold H*:

1. for each level h, with $1 \leq h \leq H$, nodes are computed via plain multiplication,
2. for each level h, with $H + 1 \leq h \leq k$, nodes are computed via FFT-based multiplication.

This algorithm is adaptive in the sense that it takes into account the amount of available resources, as well as the input data size. Indeed, as specified in Section 2, each plain multiplication is performed by a single SM, while each FFT-based multiplication is computed by a kernel call, thus using several SMs. In fact, this kernel computes a number of FFT-based products concurrently.

Before analyzing this adaptive algorithm, we recall that, if the subproduct tree M_n is computed by means of a single multiplication algorithm, with multiplication time[3] M(n), Lemma 10.4 in [5] states that the total number of operations for constructing M_n is at most M$(n) \log_2 n$ operations in $\mathbb{K}$. We also note that the leading coefficient of each polynomial in M_n is one. Thus this coefficient does not need to be stored in memory. Moreover, this allows us to multiply two polynomials at level i, for $H + 1 \leq i \leq k - 1$, via FFTs of size 2^{i+1} (instead of 2^{i+2} with a naive approach that would ignore that leading coefficients are one).

Another implementation trick is the so-called *FFT doubling*. At a level $H + 2 \leq i \leq k$, for $0 \leq j \leq 2^{k-i} - 1$, consider how to compute $M_{i,j}$ from $M_{i-1,2j}$ and $M_{i-1,2j+1}$. Since the values of $M_{i-1,2j}$ and $M_{i-1,2j+1}$ at 2^{i-1} points have already been computed (via FFT), it is sufficient, in order to determine $M_{i,j}$, to evaluate $M_{i-1,2j}$ and $M_{i-1,2j+1}$ at 2^{i-1} additional points. To do this, we write $f \in \{M_{i-1,2j}, M_{i-1,2j+1}\}$ as $f = f_0 + x^{2^{i-2}} f_1$, with $\deg(f_0) < 2^{i-2}$, and evaluate each of f_0, f_1 at those 2^{i-1} additional points. While this trick brings savings in terms of work, it increases memory footprint, in particular parallelism overheads. Integrating this trick in our implementation is work in progress and, in the rest of this paper, the theoretical and experimental results do not rely on it.

Proposition 1. *The number of arithmetic operations of the adaptive algorithm for computing M_n with threshold H is*

$$n \left(\frac{15}{2} \log_2(n)^2 + \frac{19}{2} \log_2(n) + 2^H - \frac{15}{2} H^2 - \frac{17}{2} H - \frac{1}{2^H} \right).$$

Proposition 2. *The number of machine words required for storing M_n, with threshold H is given below*

$$n \left(\log_2(n) - H + 5 \right) + (-H - 2) \left(n + \frac{n}{2^{H+1}} \right) + 2nH \left(1 + \frac{1}{2^{H+2}} \right)$$

---

[3] This notion is defined in [5, Chapter 8]

Proposition 3. *Span and overhead for constructing M_n with threshold H using our adaptive method are* span_{M_n} *and* $\mathsf{overhead}_{M_n}$ *respectively, where*

$$\mathsf{span}_{M_n} = \frac{15}{2}\left(\log_2(n)+1\right)^2 - \frac{7}{2}\log_2(n) + 2^{H+1} - \frac{15}{2}\left(H+1\right)^2 + \frac{9}{2}H - 2$$

and

$$\mathsf{overhead}_{M_n} = \left(\left(18\left(\log_2(n)+1\right)^2 - 35\log_2(n) - 18\left(H+1\right)^2 + 35\,H\right) + 2\,H\right)U.$$

The proof of Propositions 1, 2 and 3 are based on the hypotheses stated in Section 2 and elementary calculations, which, to the interest of space, can be found in our MAPLE worksheet at http://cumodp.org/links.html.

Propositions 1 and 3 imply that for a fixed a H, the parallelism (ratio work to span) is in $\Theta(n)$ which is very satisfactory. We stress the fact that this result could be achieved because both our plain and FFT-based multiplications are parallelized. Observe also that, for a fixed n, parallelism overhead decreases as H increases: that is, plain multiplication suffers less parallelism overheads than FFT-based multiplication on GPUs.

It is natural to ask how to choose H so as to minimize work and span. Elementary calculations, using our MAPLE worksheet suggest $6 \le H \le 7$. However, in degrees 2^6 and 2^7, parallelism overhead is too high for FFT-based multiplication and, experimentally, the best choice appeared to be $H = 8$.

4   Subinverse Tree Construction

For $f \in \mathbb{K}[x]$ of degree less than n, evaluating f on the point set $\{u_0, \ldots, u_{n-1}\}$ is done by Algorithm 2 by calling TopDownTraverse$(f, k, 0, M_n, F)$. An array F of length n is passed to this procedure such that $F[i]$ receives $f(u_i)$ for $0 \le i \le n-1$. The function call Remainder$(f, M_{i,j})$ relies on plain division whenever $i < H$ holds, where H is the threshold of Section 3. Fast division is applied when polynomials are large enough and, actually, can not be stored within the local memory of a streaming multiprocessor.

Algorithm 2. TopDownTraverse(f, i, j, M_n, F)

Input: $f \in \mathbb{K}[x]$ with $\deg(f) < 2^i$, i and j are integers such that $0 \le i \le k$,
 $\qquad 0 \le j < 2^{k-i}$ and F is an array of length n.
if $i == 0$ **then**
$\quad\mid\quad F[j] = f;$
$\quad\mid\quad$ return;

$f_0 = $ Remainder$(f, M_{i-1, 2j});$
$f_1 = $ Remainder$(f, M_{i-1, 2j+1});$
TopDownTraverse$(f_0, i-1, 2j, M_n, F);$
TopDownTraverse$(f_1, i-1, 2j+1, M_n, F);$

Fast division requires computing $\mathsf{Inverse}(\mathsf{rev}_{2^i}(M_{i,j}), 2^i)$, for $H \leq i \leq k$ and $0 \leq j < 2^{k-i}$, see Chapter 9 in [5]. However, this latter calculation has, in principle, to be done via Newton iteration. As mentioned in Section 2, this latter provides little opportunities for concurrent execution. To overcome this performance issue, we introduce a strategy that relies on a new data structure called *subinverse tree*. In this section, we first define subinverse trees and describe their implementation. Then, we analyze the complexity of constructing a subinverse tree.

Definition 1. *For the subproduct tree $M_n := \mathsf{SubproductTree}(u_0, \ldots, u_{n-1})$, the subinverse tree associated with M_n, denoted by InvM_n, is a complete binary tree of the same format as M_n, defined as follows. For $0 \leq i \leq k$, for $0 \leq j < 2^{k-i}$, the j-th node of level i in InvM_n contains the univariate polynomial $\mathsf{InvM}_{i,j}$ of less than degree 2^i and defined by*

$$\mathsf{InvM}_{i,j}\, \mathsf{rev}_{2^i}(M_{i,j}) \equiv 1 \mod x^{2^i}.$$

Note that we do not store the polynomials of the subinverse tree InvM_n below level H. Indeed, for those levels, we rely on plain division for the function calls $\mathsf{Remainder}(f, M_{i,j})$ in Algorithm 2.

Proposition 4. *Let InvM_n be the subinverse tree associated with the subproduct tree M_n, with the threshold $H < k$. Then, the amount of space required for storing InvM_n, is $(k - H)n$.*

The following lemma is a simple observation from which we derive Proposition 5 and, thus, the principle of subinverse tree construction.

Lemma 1. *Let $\mathbb{R}$ be a commutative ring with identity element. Let $a, b, c \in \mathbb{R}[x]$ be univariate polynomials such that $c = a\,b$ and $a(0) = b(0) = 1$ hold. Let $d = \deg(c) + 1$. Then, we have $c(0) = 1$ and $\mathsf{Inverse}(c, d) \mod x^d$ can be computed from a and b as follows: $\mathsf{Inverse}(c, d) \equiv \mathsf{Inverse}(a, d) \cdot \mathsf{Inverse}(b, d) \mod x^d$.*

Proposition 5. *Let $\mathsf{InvM}_{i,j}$ be the jth polynomial (from left to right) of the subinverse tree at level i, where $0 < i < k$ and $0 \leq j < 2^{k-i}$. We have:*

$$\mathsf{InvM}_{i,j} \equiv \mathsf{Inverse}(\mathsf{rev}_{2^{i-1}}(M_{i-1,2j}), 2^i) \cdot \mathsf{Inverse}(\mathsf{rev}_{2^{i-1}}(M_{i-1,2j+1}), 2^i) \mod x^{2^i}$$

where $\mathsf{InvM}_{i,j} = \mathsf{Inverse}(\mathsf{rev}_{2^i}(M_{i,j}), 2^i)$ from Definition 1.

We observe that computing $\mathsf{InvM}_{i,j}$ requires $\mathsf{Inverse}(\mathsf{rev}_{2^{i-1}}(M_{i-1,2j}), 2^i)$ and $\mathsf{Inverse}(\mathsf{rev}_{2^{i-1}}(M_{i-1,2j+1}), 2^i)$. However, at level $i - 1$, the nodes $\mathsf{InvM}_{i-1,2j}$ and $\mathsf{InvM}_{i-1,2j+1}$ are $\mathsf{Inverse}(\mathsf{rev}_{2^{i-1}}(M_{i-1,2j}), 2^{i-1})$ and $\mathsf{Inverse}(\mathsf{rev}_{2^{i-1}}(M_{i-1,2j+1}), 2^{i-1})$ respectively. To take advantage of this observation, we call $\mathsf{OneStepNewtonIteration}(\mathsf{rev}_{2^{i-1}}(M_{i-1,2j}), \mathsf{InvM}_{i-1,2j}, i - 1)$ and $\mathsf{OneStepNewtonIteration}(\mathsf{rev}_{2^{i-1}}(M_{i-1,2j+1}), \mathsf{InvM}_{i-1,2j+1}, i - 1)$, see Algorithm 3, so as to obtain $\mathsf{Inverse}(M_{i-1,2j}, 2^i)$ and $\mathsf{Inverse}(M_{i-1,2j+1}, 2^i)$ respectively. Algorithm 3 performs a single iteration of *Newton iteration*'s algorithm. Finally, we perform one truncated polynomial multiplication, as stated in Proposition 5, to obtain $\mathsf{InvM}_{i,j}$. We apply this technique to compute all the polynomials of level i of the subinverse tree, for $H + 1 \le i \le k$.

Since we do not store the leading coefficients of the polynomials in the subproduct tree, our implementation relies on a modified version of Algorithm 3, namely Algorithm 4.

Algorithm 3. OneStepNewtonIteration(f, g, i)

Input: $f, g \in \mathbb{R}[x]$ such that $f(0) = 1$, where $\deg(g) =\le 2^i$ and $fg \equiv 1 \mod x^{2^i}$.
Output: $g^{\triangleleft} \in \mathbb{R}[x]$ such that $fg^{\triangleleft} \equiv 1 \mod x^{2^{i+1}}$.
$g^{\triangleleft} = (2g - fg^2) \mod x^{2^{i+1}}$;
return $g^{\triangleleft}$;

Let $f = \mathsf{rev}_{2^i}(M_{i,j})$ and $g = \mathsf{InvM}_{i,j}$. From Definition 1, we have $fg \equiv 1 \mod x^{2^i}$. Note that $\deg(fg) \le 2^{i+1} - 1$ holds. Let $e^{\triangleleft} = -fg + 1$. Thus $e^{\triangleleft}$ is a polynomial of degree at most $2^{i+1} - 1$. Moreover, from the definition of a subinverse tree, we know its least significant 2^i coefficients are zeros. Let $e = e^{\triangleleft}/x^{2^i}$. So $\deg(e) \le 2^i - 1$. In Algorithm 3, we have $g^{\triangleleft} \equiv g \mod x^{2^i}$. We can compute $g^{\triangleleft}$ from eg and g. The advantage of working with e instead of $e^{\triangleleft}$ is that the degree of $e^{\triangleleft}$ is twice the degree of e. In Algorithm 4, we compute e as $e = -\mathsf{rev}_{2^i}(M_{i,j} \cdot \mathsf{rev}_{2^i-1}(\mathsf{InvM}_{i,j}) - x^{2^{i+1}-1})$.

Algorithm 4. EfficientOneStep$(M_{i,j}^{\triangleleft}, \mathsf{InvM}_{i,j}, i)$

Input: $M_{i,j}^{\triangleleft} = M_{i,j} - x^{2^i}$, $\mathsf{InvM}_{i,j}$.
Output: g, such that $g \; \mathsf{rev}_{2^i}(M_{i,j}) \equiv 1 \mod x^{2^{i+1}}$.
$a = \mathsf{rev}_{2^i-1}(\mathsf{InvM}_{i,j})$;
$b = a - x^{2^i-1}$;
$c = \mathsf{convolution}(a, M_{i,j}^{\triangleleft}, 2^i)$;
$d = \mathsf{rev}_{2^i}(c + b)$;
$e = -d$;
$h = e \; \mathsf{InvM}_{i,j} \mod x^{2^i}$;
$g = hx^{2^i} + \mathsf{InvM}_{i,j}$;
return g;

The *Middle product* technique [6] is used in Algorithm 3 for computing c.

For a given i, with $H < i \le k$, and for $0 \le j < 2^{k-i}$, Algorithm 5 computes the polynomial $\mathsf{InvM}_{i,j}$. Algorithm 5 calls Algorithm 4 twice to increase the accuracy of $\mathsf{InvM}_{i-1,2j}$ and $\mathsf{InvM}_{i-1,2j+1}$ to x^{2^i}. Then it multiplies those latter polynomials and applies a mod operation. Algorithm 6 is the top level algorithm which creates the subinverse tree InvM_n using a bottom-up approach and calling Algorithm 5 for computing each node $\mathsf{InvM}_{i,j}$ for $H \le i \le k$ and $0 \le j < 2^{k-i}$.

Algorithm 5. InvPolyCompute(M_n,InvM, i, j)

Input: M_n and InvM are the subproduct tree and subinverse tree respectively.
Output: c such that $c \operatorname{rev}_{2^i}(M_{i,j}) \equiv 1 \bmod \ x^{2^i}$.
$M_{i-1,2j}^{\lessdot} = M_{i-1,2j} - x^{2^{i-1}}$;
$M_{i-1,2j+1}^{\lessdot} = M_{i-1,2j+1} - x^{2^{i-1}}$;
$a = \mathsf{EfficientOneStep}(M_{i-1,2j}^{\lessdot},\mathsf{InvM}_{i-1,2j}, i - 1)$;
$b = \mathsf{EfficientOneStep}(M_{i-1,2j+1}^{\lessdot},\mathsf{InvM}_{i-1,2j+1}, i - 1)$;
$c = ab \bmod \ x^{2^i}$;
return c;

Algorithm 6. SubinverseTree(M_n, H)

Input: M_n is the subproduct tree and $H \in \mathbb{N}$.
Output: the subinverse tree InvM$_n$
for $j = 0 \ldots 2^{k-H} - 1$ do
$\quad \lfloor$ InvM$_{H,j}$ = Inverse($M_{H,j}$, deg($M_{H,j}$));
for $i = (H + 1) \ldots k$ do
$\quad$ for $j = 0 \ldots 2^{k-i} - 1$ do
$\quad \quad \lfloor$ InvM$_{i,j}$ = InvPolyCompute(M_n,InvM$_{i,j}$);

return InvM$_n$;

Propositions 6 and 7 imply that for a fixed a H, the parallelism (ratio work to span) is in $\Theta(n)$ which is satisfactory.

Proposition 6. *For the subproduct tree M_n, with threshold H, the number of arithmetic operations for constructing the subinverse tree* InvM$_n$ *using Algorithm 6 is:*

$$n\left(10\left(3\log_2(n)^2 + \log_2(n) - 3H^2 - 7H - 4\right) + \frac{16\,4^{2^H}}{3 \cdot 2^H} + 2 - \frac{1}{3 \cdot 2^H} - \frac{2}{2^{H-2^H}}\right).$$

Proposition 7. *For the subproduct tree M_n with threshold H, the span and overhead of constructing the subinverse tree* InvM$_n$ *by Algorithm 6 are* span$_{\mathsf{InvM}_n}$ *and* overhead$_{\mathsf{InvM}_n}$ *respectively, where*

$$\mathsf{span}_{\mathsf{InvM}_n} = \frac{75}{2}\log_2(n)^2 - \frac{107}{2}\log_2(n) + 2 \cdot 4^H + 4 \cdot 2^H - \frac{75}{2}H^2 - \frac{43}{2}H + 14$$

and

$$\mathsf{overhead}_{\mathsf{InvM}_n} = U\left(90\log_2(n)^2 - 255\log_2(n) + 2^{H+1} - 90H^2 + 75H + 166\right).$$

5   Polynomial Evaluation

Algorithm 2 solves the multi-point evaluation problem using subproduct tree technique. To do so, we construct the subproduct tree M_n with threshold H

and the corresponding subinverse tree InvM_n. Then, we run Algorithm 2, which requires polynomial division. We implement both plain and fast division. For the latter, we rely on the subinverse tree, as described in Section 4

Proposition 8. *For the subproduct tree M_n with threshold H and its corresponding subinverse tree InvM_n, the number of arithmetic operations of Algorithm 2 is:*

$$30\,n\log_2(n)^2 + 106\,n\log_2(n) + n\,2^{H+1} - 30\,n\,H^2 - 46\,n\,H + 74\,n + 16\,\frac{n}{2^H} - 8.$$

In [12], the algebraic complexity estimate for performing multi-point evaluation (which only considers multiplication cost and ignores other coefficient operations) is $7M(n/2)\log_2(n) + O(M(n))$. Considering for $M(n)$ a multiplication time like the one based on Cooley-Tukey's algorithm (see Section 2) the running time estimate of [12] becomes similar to the estimate of Proposition 8. Since our primary goal is paralllelization, we view this comparison as satisfactory. Furthermore, Propositions 8 and 9 imply that for a fixed a H, the parallelism (ratio work to span) is in $\Theta(n)$ which is satisfactory as well.

Proposition 9. *Given a subproduct tree M_n with threshold H and the corresponding subinverse tree InvM_n, span and overhead of Algorithm 2 are $\mathsf{span}_{\mathsf{eva}}$ and $\mathsf{overhead}_{\mathsf{eva}}$ respectively, where*

$$\mathsf{span}_{\mathsf{eva}} = 15\log_2(n)^2 + 23\log_2(n) + 6 \times 2^H - 15\,H^2 - 22\,H - 2$$

and

$$\mathsf{overhead}_{\mathsf{eva}} = \left(36\log_2(n)^2 + 3\log_2(n) - 36\,H^2 + 2\,H\right)U.$$

6   Polynomial Interpolation

As recalled in Section 2, we rely on Lagrange interpolation. Our interpolation procedure, inspired by the recursive algorithm in [5, Chapter 10], relies on Algorithm 7 below, which proceeds in a bottom-up traversal fashion.

Algorithm 7 computes a binary tree such that the j-th node from the left at level i is a polynomial $I_{i,j}$ of degree $2^i - 1$, for $0 \le i \le k$, $0 \le j \le 2^{k-i} - 1$. The root $I_{k,0}$ is the desired polynomial. We use the same threshold H as for the construction of the subproducttree tree:

1. for each node $I_{i,j}$ where $1 \le i \le H$ and $0 \le j < 2^{k-i}$, we compute $I_{h,j}$ using plain multiplication.
2. for each node $I_{i,j}$, with $H + 1 \le i \le k$, we compute the $I_{i,j}$ using FFT-based multiplication.

In Theorem 10.10 in [5], the complexity estimate for the *Linear Combination* is $(\mathsf{M}(n) + O(n))\log(n)$. In Proposition 10, we present a more precise estimate.

Algorithm 7. LinearCombination($M_n, c_0, \ldots, c_{n-1}$)

Input: Precomputed subproduct tree M_n for the evaluation points $u_0, \ldots, u_{n-1}$,
   and $c_0, \ldots, c_{n-1} \in \mathbb{K}$, with $n = 2^k$ for $k \in \mathbb{N}$

Output: $\sum_{0 \leq i < n} c_i m/(x - u_i) \in \mathbb{K}[x]$, where $m = \prod_{0 \leq i < n}(x - u_i)$

for $j = 0$ to $n - 1$ do
 $\lfloor$ $I_{0,j} = c_j$;

for $i = 1$ to k do
  for $j = 0$ to $2^{k-i} - 1$ do
   $\lfloor$ $I_{i,j} = M_{i-1,2j} I_{i-1,2j+1} + M_{i-1,2j+1} I_{i-1,2j}$;

return $I_{k,0}$;

Proposition 10. *For the subproduct tree M_n with threshold H, the number of arithmetic operations Algorithm 7 is given below*

$$15\,n\log_2(n)^2 + 20\,n\log_2(n) + 11\,n + 13\,nH - 15\,nH^2 + n\,2^{H+1} - n\,2^{1-H}.$$

Proposition 11. *For the subproduct tree M_n with threshold H and the corresponding subinverse tree InvM_n, the span and overhead of Algorithm 7 are span_{lc} and $\mathsf{overhead}_{lc}$ respectively, where*

$$\mathsf{span}_{lc} = \frac{15}{2}\log_2(n)^2 + \frac{25}{2}\log_2(n) + 2^{H+1} - \frac{15}{2}H^2 - \frac{21}{2}H - 2$$

and

$$\mathsf{overhead}_{lc} = 18\log_2(n)^2 + \log_2(n) - 18\,H^2 + 4\,H.$$

Finally we use Algorithm 8 in which we first compute $c_0, \ldots, c_{n-1}$, and then we call Algorithm 7. Algorithm 8 is adapted from Algorithm 10.11 in [5].

Algorithm 8. FastInterpolation($u_0, \ldots, u_{n-1}, v_0, \ldots, v_{n-1}$)

Input: $u_0, \ldots, u_{n-1} \in \mathbb{K}$ such that $u_i - u_j$ is a unit for $i \neq j$, and $v_0, \ldots, v_{n-1} \in \mathbb{K}$,
   and $n = 2^k$ for $k \in \mathbb{N}$

Output: The unique polynomial $P \in \mathbb{K}[x]$ of degree less than n such that
   $P(u_i) = v_i$ for $0 \leq i < n$

$M_n := \mathsf{SubproductTree}(u_0, \ldots, u_{n-1})$;

Let m be the root of M_n;

Compute $m^\triangleleft(x)$ the derivative of m;

$\mathsf{InvM}_n := \mathsf{SubinverseTree}(M_n, H)$;

$\mathsf{TopDownTraverse}(m^\triangleleft(x), i, j, M_n, F)$;

return $\mathsf{LinearCombination}(M_n, v_0/F[0], \ldots, v_{n-1}/F[n-1])$;

From the different propositions of this paper, it follows that, for a fixed H, the parallelism (ratio work to span) of Algorithm 8 is in $\Theta(n)$ which is satisfactory.

7   Experimentation

The algorithms presented in this paper have been implemented in CUDA [16] as part of the CUMODP library. The FFT-based algorithms of this library are described [13,14] while those based on plain arithmetic are presented in [7]. As mentioned before, our FFT computations use Stockham algorithms which is known to be more appropriate for many-core GPUs thah the one of Cooley-Tukey. We focus on radix-10 FFTs [13] and rely on an optimized version of Montgomery's trick [12] for modular multipoint [4].

Table 1. Effective memory bandwidth (in GB/S). The input size is $n = 2^k$.

k	Evaluation	Interpolation
11	0.2554	0.3403
12	0.5596	0.7054
13	1.2947	1.6182
14	2.5838	3.1445
15	5.2702	6.3464
16	9.6193	11.4143
17	16.4358	18.7800
18	22.6172	26.7590
19	32.3230	38.7674
20	40.4644	49.0012
21	46.7343	57.0978
22	50.8830	62.4516
23	52.9413	64.2464

Table 2. Multiplication timings (in sec.) for polynomials of size 2^k: CUMODP vs FLINT

k	CUMODP (s)	FLINT (s)	Ratio
11	0.0019	0.002	1.029
12	0.0032	0.003	0.917
13	0.0023	0.008	3.441
14	0.0039	0.013	3.346
15	0.0032	0.023	7.216
16	0.0065	0.045	6.942
17	0.0084	0.088	10.475
18	0.0122	0.227	18.468
19	0.0198	0.471	23.738
20	0.0266	1.011	27.581
21	0.0718	2.086	29.037
22	0.1451	4.419	30.454
23	0.3043	9.043	29.717

We run our CUDA codes on a NVIDIA Tesla M2050 GPU card and we run the other codes on the same machine equipped with an Intel Xeon X5650 CPU at 2.67GHz. Our test cases use random points or random polynomials with coefficients in a prime field whose characteristic is a 30-bit prime number.

With Table 1 we evaluate the intrinsic quality of this implementation while with Tables 2, 3 and Figures 1, 2 we provide comparative benchmark results.

One of the major factors of performance in GPU applications is of *memory bandwidth*. For our implementation of multi-point evaluation and interpolation, this factor is presented for various input sizes in the Table 1. The maximum memory bandwidth for our GPU card is 148 GB/S. Since our code has a high arithmetic intensity, we believe that our experimental results are promising, while leaving room for improvement.

In Table 2, we compare two implementations of FFT-based polynomial multiplication. The first one is that the CUMODP library, presented [13]. The second one is from the FLINT library [9]. From the experimental data, it is clear that, our CUDA code for FFT-based multiplication outperforms its FLINT counterpart only in size larger than 2^{13}. Thus, we need to implement another

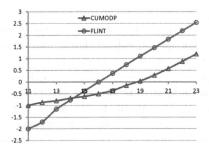
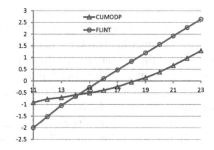

Fig. 1. Multi-point timings (using radix-10 log-scales on both axes): CU-MODP vs FLINT

Fig. 2. Interpolation timings (using radix-10 log-scales on both axes): CU-MODP vs FLINT

multiplication algorithm to have better performance in low-to-average degrees. This is work in progress.

In Table 3 we compare our implementation of multi-point polynomial evaluation and polynomial interpolation with that of the FLINT library. These timings are also available in the form of plots with Figures 1 and 2 where radix-10 log-scales are used on both axes.

We found that our implementation does not perform well until degree 2^{15}. In degree 2^{23}, we achieve a 21 times speedup factor w.r.t. FLINT, which is a satisfactory result. Nevertheless, we believe that by improving our multiplication routine for polynomials of degrees 2^9 to 2^{13}, we would have better performance in both polynomial evaluation and interpolation in these middle ranges.

Table 3. Multi-point evaluation and interpolation timings (in sec.) with input size 2^k: CUMODP vs FLINT

	Evaluation			Interpolation		
k	GPU (s)	FLINT (s)	Ratio	GPU (s)	FLINT (s)	Ratio
11	0.1012	0.01	0.0987	0.1202	0.01	0.0831
12	0.1361	0.02	0.1468	0.1671	0.03	0.1794
13	0.1580	0.07	0.4429	0.1963	0.09	0.4584
14	0.2034	0.17	0.8354	0.2548	0.22	0.8631
15	0.2415	0.41	1.6971	0.3073	0.53	1.7242
16	0.3126	0.99	3.1666	0.4026	1.26	3.1294
17	0.4285	2.33	5.4375	0.5677	2.94	5.1780
18	0.7106	5.43	7.6404	0.9034	6.81	7.5379
19	1.0936	12.63	11.5484	1.3931	15.85	11.3768
20	1.9412	29.2	15.0420	2.4363	36.61	15.0268
21	3.6927	67.18	18.1923	4.5965	83.98	18.2702
22	7.4855	153.07	20.4486	9.2940	191.32	20.5851
23	15.796	346.44	21.9321	19.6923	432.13	21.9441

8 Conclusion

We discussed fast multi-point evaluation and interpolation of univariate polynomials over a finite field on GPU architectures. We have combined algorithmic techniques like subproduct trees, subinverse trees, plain polynomial arithmetic, FFT-based polynomial arithmetic. Up to our knowledge, this is the first report on a parallel implementation of subproduct tree techniques. The source code of our algorithms is freely available in CUMODP-Library website http://cumodp.org/.

The experimental results are promising. Room for improvement, however, still exists, in particular for efficiently multiplying polynomials in the range of degrees from 2^9 to 2^{13}. Filling this gap is work in progress.

References

[1] Bernstein, D.J.: Fast multiplication and its applications. In: Buhler, J., Stevenhagen, P. (eds.) Algorithmic Number Theory: Lattices, Number Fields, Curves and Cryptography (2008)

[2] Bostan, A., Schost, É.: Polynomial evaluation and interpolation on special sets of points. J. Complexity 21(4), 420–446 (2005)

[3] Brent, R.P., Gaudry, P., Thomé, E., Zimmermann, P.: Faster multiplication in gf(2)[x]. In: van der Poorten, A.J., Stein, A. (eds.) ANTS-VIII 2008. LNCS, vol. 5011, pp. 153–166. Springer, Heidelberg (2008)

[4] Dahan, X., Moreno Maza, M., Schost, É., Wu, W., Xie, Y.: Lifting techniques for triangular decompositions. In: Proceedings of the 2005 International Symposium on Symbolic and Algebraic Computation, ISSAC 2005, pp. 108–115. ACM, New York (2005)

[5] Gathen, J., Gerhard, J.: Modern Computer Algebra. Cambridge University Press (1999)

[6] Hanrot, G., Quercia, M., Zimmermann, P.: The middle product algorithm i. Appl. Algebra Eng., Commun. Comput. 14(6), 415–438 (2004)

[7] Haque, S.A., Moreno Maza, M.: Plain polynomial arithmetic on GPU. In: J. of Physics: Conf. Series, vol. 385, IOP Publishing (2012)

[8] Haque, S.A., Moreno Maza, M., Xie, N.: A many-core machine model for designing algorithms with minimum parallelism overheads. CoRR, abs/1402.0264 (2014)

[9] Hart, W.B.: Fast Library for Number Theory: An Introduction. In: Fukuda, K., van der Hoeven, J., Joswig, M., Takayama, N. (eds.) ICMS 2010. LNCS, vol. 6327, pp. 88–91. Springer, Heidelberg (2010)

[10] Li, X., Moreno Maza, M., Rasheed, R. and Schost, É.: The Modpn library: Bringing fast polynomial arithmetic into MAPLE. J. Symb. Comput., 46(7), 841–858 (2011)

[11] Moreno Maza, M., Xie, Y.: Balanced dense polynomial multiplication on multi-cores. Int. J. Found. Comput. Sci. 22(5), 1035–1055 (2011)

[12] Montgomery, P.L.: An FFT Extension of the Elliptic Curve Method of Factorization. PhD thesis, University of California Los Angeles, USA (1992)

[13] Moreno Maza, M., Pan, W .: Fast polynomial arithmetic on a GPU. In: J. of Physics: Conference Series, vol. 256 (2010)

[14] Moreno Maza, M., Pan, W.: Solving bivariate polynomial systems on a GPU. In: J. of Physics: Conference Series, vol. 341 (2011)

[15] Murao, H., Fujise, T.: Towards an efficient implementation of a fast algorithm for multipoint polynomial evaluation and its parallel processing. In: Proc. of PASCO, pp. 24–30. ACM (1997)

[16] Nickolls, J., Buck, I., Garland, M., Skadron, K.: Scalable parallel programming with CUDA. Queue 6(2), 40–53 (2008)

[17] Tanaka, S., Chou, T., Yang, B.-Y., Cheng, C.-M., Sakurai, K.: Efficient parallel evaluation of multivariate quadratic polynomials on GPUs. In: Lee, D.H., Yung, M. (eds.) WISA 2012. LNCS, vol. 7690, pp. 28–42. Springer, Heidelberg (2012)

[18] Verschelde, J., Yoffe, G.: Evaluating polynomials in several variables and their derivatives on a GPU computing processor. In: Proc. of the 26th International Parallel and Distributed Processing Symposium Workshops & PhD Forum, IPDPSW 2012, pp. 1397–1405. IEEE Computer Society (2012)

Deterministically Computing Reduction Numbers of Polynomial Ideals

Amir Hashemi[1,2], Michael Schweinfurter[3], and Werner M. Seiler[3]

[1] Department of Mathematical Sciences, Isfahan University of Technology
Isfahan, 84156-83111, Iran
[2] School of Mathematics, Institute for Research in Fundamental Sciences (IPM),
Tehran, 19395-5746, Iran
Amir.Hashemi@cc.iut.ac.ir
[3] Institut für Mathematik, Universität Kassel
Heinrich-Plett-Straße 40, 34132 Kassel, Germany
{michael.schweinfurter, seiler}@mathematik.uni-kassel.de

Abstract. We discuss the problem of determining reduction numbers of a polynomial ideal $\mathcal{I}$ in n variables. We present two algorithms based on parametric computations. The first one determines the absolute reduction number of $\mathcal{I}$ and requires computations in a polynomial ring with $(n - \dim \mathcal{I}) \dim \mathcal{I}$ parameters and $n - \dim \mathcal{I}$ variables. The second one computes via a Gröbner system the set of all reduction numbers of the ideal $\mathcal{I}$ and thus in particular also its big reduction number. However, it requires computations in a ring with $n \dim \mathcal{I}$ parameters and n variables.

1 Introduction

One of the fundamental ideas behind Gröbner bases is the reduction of questions about general polynomial ideals to monomial ideals. In the context of determining invariants of an ideal like projective dimension or Castelnuovo-Mumford regularity, it is therefore interesting to know when these invariants possess the same values for an ideal and its leading ideal. It is well-known that in many instances the invariants of the leading ideal provide an upper bound for those of the polynomial ideal and that in generic position, i.e. when the leading ideal is the generic initial ideal, the values even coincide.

From an algorithmic point of view, it is not easy to work with the generic initial ideal. While it is comparatively easy to determine it with probabilistic method, there exists no simple test to verify that one has really obtained the generic initial ideal. However, relaxing the conditions on the leading ideal somewhat one can introduce generic positions which share many properties with the generic initial ideal and which are effectively checkable with deterministic algorithms. In [9], the authors showed that for many purposes it suffices to ensure that the leading ideal is quasi-stable (i.e. that the given ideal possesses a Pommaret basis [17,18]) in order to achieve that many invariants can be immediately read off the Pommaret basis.

V.P. Gerdt et al. (Eds.): CASC Workshop 2014, LNCS 8660, pp. 186–201, 2014.

Our article [9] was mainly concerned with invariants and concepts related to the minimal free resolution of the given ideal. In this work, we study the *reduction number*, an invariant which was introduced by Northcott and Rees [15] and which intuitively measures the complexity of computations in the associated factor ring. It is also related to some other invariants like the degree, the arithmetic degree and the Castelnuovo-Mumford regularity (see [3,20,22] for more details). Independently, Conca [4] and Trung [21] proved that the reduction number of an ideal is bounded by the one of its leading ideal (for an arbitrary term order) and Trung [20] showed that for the generic initial ideal (for the degree reverse lexicographic order) equality holds.

Trung [21] also presented an approach to the effective determination of various reduction numbers. However, his method is very expensive. We will show that it is indeed impossible to design a "simple" algorithm for reduction numbers where we mean by "simple" an approach based solely on the analysis of leading terms. Nevertheless, we will provide two alternative methods which we believe to be more efficient than the one presented by Trung. Our first method is based on directly adding the right number of sufficiently generic linear forms and yields the absolute reduction number. Our second method determines the whole set of possible reduction numbers (and thus in particular both the absolute and the big reduction number) using a Gröbner system.

Throughout this article, we will use the following notations. $\mathcal{P} = \Bbbk[x_1, \ldots, x_n]$ is an n-dimensional polynomial ring over some infinite field $\Bbbk$ with homogeneous maximal ideal $\mathfrak{m}$. If not stated otherwise, the term order will always be the degree reverse lexicographic order induced by $x_n \prec \cdots \prec x_1$. We assume that we are given a fixed homogeneous ideal $\mathcal{I} \trianglelefteq \mathcal{P}$ of dimension D and write for the corresponding factor ring $\mathcal{R} = \mathcal{P}/\mathcal{I}$. A non-singular matrix $A = (a_{ij}) \in \mathrm{GL}(n, \Bbbk)$ induces on $\mathcal{P}$ the linear change of coordinates $x \mapsto A \cdot x$ transforming the given ideal $\mathcal{I}$ into a new ideal $A \cdot \mathcal{I} \trianglelefteq \mathcal{P}$. Finally, given a term $t \in \mathcal{P}$, we denote by $w(t)$ the largest integer ℓ such that $x_\ell \mid t$.

The article is organised as follows. The next section collects some known facts about reduction numbers and generic initial ideals. Section 3 introduces some novel generalised notions of stability for monomial ideals. The following section extends the for us crucial notion of weak D-stability to polynomial ideals and presents a deterministic algorithm to transform any ideal into weakly D-stable position. After these preparations, we present in Section 5 an algorithm for computing the absolute reduction number. In the final section, we exploit Gröbner systems to compute the set of all possible reduction numbers.

2   Reduction Numbers and the Generic Initial Ideal

We recall some basic facts about reduction numbers. There exist several equivalent approaches to defining them; for our purposes the following one is particularly convenient. Let $y_1, \ldots, y_D \in \mathcal{P}_1$ be D linear forms defining a Noether normalisation of $\mathcal{R}$. Then the ideal $\mathcal{J} = \mathcal{I} + \langle y_1, \ldots, y_D \rangle$ is called a *minimal reduction* of $\mathcal{I}$ and the *reduction number* $r_{\mathcal{J}}(\mathcal{R})$ with respect to $\mathcal{J}$ is the largest non-vanishing degree in the factor ring $\mathcal{P}/\mathcal{J}$. We write for the set of all possible reduction numbers $\mathrm{rSet}(\mathcal{R}) = \{r_{\mathcal{J}}(\mathcal{R}) \mid \mathcal{J} \text{ minimal reduction of } \mathcal{I}\}$. The *(absolute) reduction number* $r(\mathcal{R})$ is the minimal element of $\mathrm{rSet}(\mathcal{R})$, the *big reduction number* $br(\mathcal{R})$ the maximal one. As already mentioned above, the former one appeared first in the work of Northcott and Rees [15]; the latter one was much later introduced by Vasconcelos [23].

While it is easy to construct some minimal reduction $\mathcal{J}$, the obvious key problem in computing $r(\mathcal{R})$ consists of identifying a $\mathcal{J}$ with $r_{\mathcal{J}}(\mathcal{R}) = r(\mathcal{R})$. In the sequel, we will use the following three results. The first one characterises all minimal reductions of a *monomial* ideal in Noether position. Any such ideal has a minimal generator of the form x_{n-D}^α. The second result relates for a strongly stable ideal (which is always in Noether position) $r(\mathcal{R})$ with the exponent α. The final result bounds for arbitrary ideals $r(\mathcal{R})$ by $r(\mathcal{P}/\operatorname{lt}\mathcal{I})$.

Lemma 1 ([3, Lemma 5]). *Let $\mathcal{I} \trianglelefteq \mathcal{P}$ be a monomial ideal such that the variables $x_{n-D+1}, \dots, x_n$ induce a minimal reduction. Then every minimal reduction is induced by linear forms*

$$y_i = x_{n-D+i} + \sum_{j=1}^{n-D} a_{ij}x_j, \quad a_{ij} \in \Bbbk. \tag{1}$$

Theorem 2 ([3, Thm. 11]). *Let $\mathcal{I} \trianglelefteq \mathcal{P}$ be a strongly stable monomial ideal. Then $\mathcal{I}$ has a minimal generator x_{n-D}^α and we have $r(\mathcal{R}) = r_{\mathcal{J}}(\mathcal{R}) = \alpha - 1$ for any minimal reduction $\mathcal{J}$ of $\mathcal{I}$.*

Theorem 3 ([4, Thm. 1.1], [21, Cor. 3.4]). *For any ideal $\mathcal{I} \trianglelefteq \mathcal{P}$ and any term order $\prec$, the inequality $r(\mathcal{R}) \le r(\mathcal{P}/\operatorname{lt}\mathcal{I})$ holds.*

Galligo [5] proved for a base field $\Bbbk$ of characteristic 0 that almost any linear coordinate transformation leads to the same leading ideal, the *generic initial ideal* $\operatorname{gin}\mathcal{I}$ (for more information see [7]). Bayer and Stillman [2] extended this result to positive characteristic. A for us important result of Trung asserts that for the generic initial ideal the inequality in Theorem 3 becomes an equality.

Theorem 4 (Galligo, [5], [2]). *There exists a nonempty Zariski open subset $\mathcal{U} \subseteq \mathrm{GL}(n, \Bbbk)$ such that $\operatorname{lt}(A \cdot \mathcal{I}) = \operatorname{lt}(A' \cdot \mathcal{I})$ for all matrices $A, A' \in \mathcal{U}$.*

Theorem 5 ([20, Thm. 4.3]). *For the degree reverse lexicographic order, we always find $r(\mathcal{R}) = r(\mathcal{P}/\operatorname{gin}\mathcal{I})$.*

3   Some Generalised Notions of Stability

Stable and strongly stable ideals form two important classes of monomial ideals. We introduce now generalisations of these concepts depending on an integer ℓ. In the context of determining reduction numbers, it will turn out that the case $\ell = D$ is of particular interest. Like for the classical stability notions, it is easy to see that it always suffices, if the defining property is satisfied by the minimal generators of the ideal.

Definition 6. *Let $0 \leq \ell < n$ be an integer. The monomial ideal $\mathcal{I}$ is ℓ-stable, if for every term $t \in \mathcal{I}$ with $w(t) \geq n - \ell$ and every $i < w(t)$ the term $x_i t / x_{w(t)}$ also lies in $\mathcal{I}$. For a weakly ℓ-stable ideal $\mathcal{I}$, the above condition must be satisfied only for all $i \leq n - \ell$. Finally, $\mathcal{I}$ is strongly ℓ-stable, if for every term $t \in \mathcal{I}$ with $w(t) \geq n - \ell$, every $j \geq n - \ell$ with $x_j \mid t$ and every $i < j$ the term $x_i t / x_j$ also lies in $\mathcal{I}$.*

Example 7. We consider first for $n = 6$ the ideal

$$\mathcal{I} = \langle x_1,\ x_4^2,\ x_3 x_4,\ x_2 x_4,\ x_2 x_3,\ x_4^2,\ x_5^3,\ x_4 x_5^2,\ x_3 x_5^2,\ x_2 x_5^2,\ x_3^2 x_5,\ x_3^3,\ x_5^2 x_6^2,$$
$$x_4 x_5 x_6^2,\ x_3 x_5 x_6^2,\ x_2 x_5 x_6^2,\ x_3^2 x_6^2,\ x_5 x_6^4,\ x_4 x_6^4,\ x_3 x_6^4,\ x_2 x_6^4,\ x_6^6 \rangle,$$

the leading ideal of the fifth Katsura ideal. As one can easily see that here $D = 0$, it suffices to check the defining property for the generators containing x_6 and it turns out that $\mathcal{I}$ is 0-stable. However, $\mathcal{I}$ is not stable, as for example $x_3 x_4 \in \mathcal{I}$ but $x_3^2 \notin \mathcal{I}$.

Consider now for $n = 5$ the monomial ideal

$$\mathcal{I} = \langle x_1^2,\ x_2^3,\ x_1 x_2^2,\ x_3^2 x_2^2,\ x_2 x_3^2 x_1,\ x_3^5,\ x_2 x_3^4,\ x_1 x_3^4,\ x_3^4 x_4^2,\ x_2 x_3^3 x_4^2,\ x_1 x_3^3 x_4^2,$$
$$x_3^3 x_4^4,\ x_3^2 x_2 x_4^4,\ x_1 x_3^2 x_4^4,\ x_3 x_2^2 x_4^4,\ x_3 x_2 x_1 x_4^4,\ x_1 x_2 x_3 x_4^3 x_5^2,\ x_1 x_3 x_4^6,\ x_2^2 x_4^6,$$
$$x_1 x_2 x_4^6,\ x_2^2 x_4^5 x_5^2,\ x_1 x_2 x_4^5 x_5^2,\ x_3 x_2^2 x_4^3 x_5^4,\ x_1 x_3 x_4^5 x_5^3,\ x_2 x_3^3 x_4^4 x_5^2,\ x_1 x_3^3 x_4^4 x_5^5,$$
$$x_1 x_2 x_4^4 x_6^6,\ x_1 x_4^6 x_5^5,\ x_2^2 x_4^4 x_5^7,\ x_2 x_3 x_4^5 x_5^7 \rangle.$$

Since here $D = 2$, we must check the defining property of a weakly D-stable ideal only for the terms containing x_3, x_4, x_5 and one readily verifies that $\mathcal{I}$ is weakly D-stable. However, it is not D-stable because $t = x_1 x_4^6 x_5^5 \in \mathcal{I}$ but $t x_4 / x_5 \notin \mathcal{I}$.

The generic initial ideal is always Borel-fixed, i. e. invariant under the natural action of the Borel group [2,6]. In general, it depends on the characteristic of the base field whether a given ideal is Borel-fixed. In characteristic zero, the Borel-fixed ideals are precisely the strongly stable ones. We provide now the analogous result for strong ℓ-stability.

Definition 8. *The* Borel group *is the subgroup $\mathcal{B} < \mathrm{GL}(n, \Bbbk)$ consisting of all lower triangular invertible $n \times n$ matrices. For any integer $0 \leq \ell < n$, we define the ℓ-Borel group as the subgroup $\mathcal{B}_\ell \leq \mathcal{B}$ consisting of all matrices $A \in \mathcal{B}$ such that for $i < n - \ell$ we have $a_{ii} = 1$ and $a_{ij} = 0$ for $i \neq j$.*

Proposition 9. *Assume that $\operatorname{char} \Bbbk = 0$. The monomial ideal $\mathcal{I} \trianglelefteq \mathcal{P}$ is strongly ℓ-stable, if and only if it is invariant under the ℓ-Borel group $\mathcal{B}_\ell$.*

Proof. Assume first that $\mathcal{I}$ is ℓ-stable and consider a generating set $\mathcal{H}$ of it. The transformation induced by an element $A = (a_{ij}) \in \mathcal{B}_\ell$ is of the form

$$
\begin{aligned}
x_i &\to x_i && \text{if } i < n - \ell, \\
x_i &\to a_{ii} x_i + \textstyle\sum_{j=n-l}^{i-1} a_{ij} x_j && \text{if } i \geq n - \ell.
\end{aligned}
\tag{2}
$$

One immediately sees that any generator $t \in \mathcal{H}$ with $w(t) < n - \ell$ remains unchanged under the action of A. If $w(t) \geq n - \ell$, then t is transformed into a polynomial $f_t = A \cdot t$. It follows again from (2) that any term in the support of f_t is obtained from t by applying a sequence of "elementary moves" of the form $s \to x_j s / x_k$ with $j < k$ where $x_k \mid s$. In this sequence we always have $k \geq n - \ell$ and thus the strong ℓ-stability of $\mathcal{I}$ implies that all appearing terms s lie in $\mathcal{I}$. Furthermore, t itself always lies in the support of f_t.

Consider now the elements t of $\mathcal{H}$ with $w(t) \geq n - \ell$ sorted reverse lexicographically. If t is the largest term among these, then $w(s) < w(t)$ for all $s \neq t$ appearing in the support of f_t. Thus they are multiples of elements of $\mathcal{H}$ which remain unchanged under the operation of A and can be eliminated. If t is the second largest term, then the support may in addition contain multiples of the largest term; otherwise we can apply the same argument. By iteration, we obtain that the whole ideal remains invariant.

For the converse, we need the assumption on the characteristic. If char $\Bbbk = 0$ (and thus no coefficient drops out when we transform a term), then we may revert the above arguments: if $\mathcal{I}$ is invariant under $\mathcal{B}_\ell$, then all terms appearing in the support of f_t must lie in $\mathcal{I}$ and hence $\mathcal{I}$ is strongly ℓ-stable. $\qquad\square$

In relation to our previous work [9], it is of interest to show that a D-stable ideal is automatically quasi-stable. The proof depends on the following characterisation of ℓ-stability which is of independent interest.

Proposition 10. *The monomial ideal $\mathcal{I} \trianglelefteq \mathcal{P}$ is ℓ-stable, if and only if it satisfies for all $0 \leq i \leq \ell$*

$$\langle \mathcal{I}, x_n, \ldots, x_{n-i+1} \rangle : x_{n-i} = \langle \mathcal{I}, x_n, \ldots, x_{n-i+1} \rangle : \mathfrak{m} \,. \tag{3}$$

Proof. Assume first that $\mathcal{I}$ is ℓ-stable and let t be a term such that $x_{n-i} t \in \langle \mathcal{I}, x_n, \ldots, x_{n-i+1} \rangle$ for some $i \leq \ell$. If $w(t) > n - i$, then $t \in \langle x_n, \ldots, x_{n-i+1} \rangle$ and nothing is to be proven. Otherwise we have $x_{n-i} t \in \mathcal{I}$ and $w(x_{n-i} t) = n - i \geq n - \ell$. Because of the ℓ-stability, this entails that $x_j t = x_j (x_{n-i} t) / x_{n-i} \in \mathcal{I}$ for all $j \leq n - \ell$. Hence $t \langle x_1, \ldots, x_{n-i} \rangle \subseteq \mathcal{I}$ implying $t \mathfrak{m} \subseteq \langle \mathcal{I}, x_n, \ldots, x_{n-i+1} \rangle$.

For the converse consider a term $t \in \mathcal{I}$ with $w(t) = n - i \geq n - \ell$. Because of (3), we have $t / x_{n-i} \in \mathcal{I} : x_{n-i} \subseteq \langle \mathcal{I}, x_n, \ldots, x_{n-i+1} \rangle : \mathfrak{m}$. Hence $x_j t / x_{n-i} \in \langle \mathcal{I}, x_n, \ldots, x_{n-i+1} \rangle$ for all $j \leq n$. If $j \leq n - i$, then $w(x_j t / x_{n-i}) \leq n - i$ and thus we must have $x_j t / x_{n-i} \in \mathcal{I}$ so that $\mathcal{I}$ is ℓ-stable. $\qquad\square$

Corollary 11. *A D-stable monomial ideal $\mathcal{I}$ is quasi-stable.*

Proof. According to the previous proposition, (3) holds for all $0 \leq i \leq D$. As a preparatory step, we claim that this fact implies that for these values of i also

$$\langle \mathcal{I}, x_n, \ldots, x_{n-i+1} \rangle : x_{n-i}^\infty = \langle \mathcal{I}, x_n, \ldots, x_{n-i+1} \rangle : \mathfrak{m}^\infty \,. \tag{4}$$

Indeed, if the term t lies in the ideal on the left hand side, then an integer s exists such that $x_{n-i}^s t \in \langle \mathcal{I}, x_n, \ldots, x_{n-i+1} \rangle$ and therefore

$$x_{n-i}^{s-1} t \in \langle \mathcal{I}, x_n, \ldots, x_{n-i+1} \rangle : x_{n-i} = \langle \mathcal{I}, x_n, \ldots, x_{n-i+1} \rangle : \mathfrak{m} \,.$$

Applying this argument a second time yields

$$x_{n-i}^{s-2}t \in \left(\langle \mathcal{I}, x_n, \ldots, x_{n-i+1} \rangle : \mathfrak{m} \right) : x_{n-i}$$
$$= \left(\langle \mathcal{I}, x_n, \ldots, x_{n-i+1} \rangle : x_{n-i} \right) : \mathfrak{m}$$
$$= \langle \mathcal{I}, x_n, \ldots, x_{n-i+1} \rangle : \mathfrak{m}^2 .$$

Thus we find by iteration that $t \in \langle \mathcal{I}, x_n, \ldots, x_{n-i+1} \rangle : \mathfrak{m}^s$ proving the claim.

It follows that x_{n-i} is not a zero divisor in $\mathcal{P}/(\langle \mathcal{I}, x_n, \ldots, x_{n-i+1} \rangle : \mathfrak{m}^\infty)$ for all $0 \le i < D$. Indeed, if $f \in \mathcal{P}$ satisfies $x_{n-i}f \in \langle \mathcal{I}, x_n, \ldots, x_{n-i+1} \rangle : \mathfrak{m}^\infty$, then an exponent s exists such that $x_{n-i}f\mathfrak{m}^s \subseteq \langle \mathcal{I}, x_n, \ldots, x_{n-i+1} \rangle$ and hence $x_{n-i}^{s+1}f \in \langle \mathcal{I}, x_n, \ldots, x_{n-i+1} \rangle$. But this implies $f \in \langle \mathcal{I}, x_n, \ldots, x_{n-i+1} \rangle : x_n^\infty = \langle \mathcal{I}, x_n, \ldots, x_{n-i+1} \rangle : \mathfrak{m}^\infty$. Now the assertion follows from [18, Prop. 4.4]. □

Example 12. Weak D-stability is not sufficient for quasi-stability, as one can see from the ideal $\langle x_1^2, x_1 x_3 \rangle$ where $n = 3$ and $D = 2$. One easily verifies that it is weakly D-stable but not quasi-stable. And for the same values of n and D the ideal $\langle x_1^3, x_1 x_2 \rangle$ shows that the converse of Corollary 11 does not hold, as it is quasi-stable but not (weakly) D-stable.

Remark 13. Assume that the monomial ideal $\mathcal{I}$ is weakly ℓ-stable for some ℓ and that $t = x_1^{\alpha_1} \cdots x_n^{\alpha_n} \in \mathcal{I}$. It follows immediately from Definition 6 that any term of the form $x_1^{\alpha_1+\beta_1} \cdots x_{n-\ell}^{\alpha_{n-\ell}+\beta_{n-\ell}}$ with $\beta_1 + \cdots + \beta_{n-\ell} = \alpha_{n-\ell+1} + \cdots + \alpha_n$ is then also contained in $\mathcal{I}$. If we introduce for $1 \le j \le \ell$ the homogeneous polynomials

$$g_j = \sum_{\beta_1^{(j)} + \cdots + \beta_{n-\ell}^{(j)} = \alpha_{n-\ell+j}} a_{\beta_1^{(j)}, \ldots, \beta_{n-\ell}^{(j)}}^{(j)} x_1^{\beta_1^{(j)}} \cdots x_{n-\ell}^{\beta_{n-\ell}^{(j)}}$$

with arbitrary coefficients $a_{\beta_1^{(j)}, \ldots, \beta_{n-\ell}^{(j)}}^{(j)} \in \Bbbk$, then it follows from the observation above that the polynomial

$$f_t = x_1^{\alpha_1} \cdots x_{n-\ell}^{\alpha_{n-\ell}} g_1 \cdots g_\ell$$

also lies in $\mathcal{I}$. Each term in its support is of the form $x_1^{\alpha_1+\beta_1} \cdots x_{n-\ell}^{\alpha_{n-\ell}+\beta_{n-\ell}}$ with $\beta_i = \beta_i^{(1)} + \cdots + \beta_i^{(\ell)}$ and by construction $\beta_1 + \cdots + \beta_{n-\ell} = \alpha_{n-\ell+1} + \cdots + \alpha_n$.

Proposition 14. *A weakly D-stable ideal $\mathcal{I}$ is always in Noether position.*

Proof. A D-dimensional monomial ideal is in Noether position, if and only if for all $1 \le j \le n - D$ a pure power $x_j^{e_j}$ is contained in $\mathcal{I}$. Assume first that there exists a term $t \in \mathcal{I} \cap \Bbbk[x_{n-D+1}, \ldots, x_n]$. Then Remark 13 immediately implies for $e = \deg t$ that $x_j^e \in \mathcal{I}$ for all $1 \le j \le n - D$ and we are done. If $\mathcal{I} \cap \Bbbk[x_{n-D+1}, \ldots, x_n] = \emptyset$, then the D-dimensional cone $1 \cdot \Bbbk[x_{n-D+1}, \ldots, x_n]$ lies completely in the complement of $\mathcal{I}$. Assume that for some $1 \le j \le n - D$ no power of x_j was contained in $\mathcal{I}$. Since $D = \dim \mathcal{I}$, it is not possible that the complement of $\mathcal{I}$ contains a $(D+1)$-dimensional cone. Thus we must have $\mathcal{I} \cap \Bbbk[x_j, x_{n-D+1}, \ldots, x_n] \ne \emptyset$. But if a term t of degree e lies in this intersection, then again by Remark 13 $x_j^e \in \mathcal{I}$ in contradiction to our assumption. □

The simple Algorithm 1 verifies whether a given monomial ideal is weakly D-stable without a priori knowledge of the dimension D of $\mathcal{I}$. For showing its correctness, we note that if $\mathcal{I}$ is weakly D-stable, then the number d computed in Line 2 equals D by Proposition 14 and by Definition 6 of weak D-stability we never get to Line 6. If $\mathcal{I}$ is not weakly D-stable, then $d \geq D$ (this estimate holds for any monomial ideal) and soon or later we will reach Line 6. The bit complexity of the algorithm is polynomial in kn, as one can easily see that the number of operations in the two **for**-loops is at most $k^2 n^3$.

Algorithm 1. WDS-Test: Test for weak D-stability

Input: minimal basis $G = \{m_1, \ldots, m_k\}$ of monomial ideal $\mathcal{I} \lhd \mathcal{P}$
Output: The answer to: is $\mathcal{I}$ weakly D-stable?
  1: $e := \max \{\deg(m_1), \ldots, \deg(m_k)\}$
  2: $d :=$ smallest ℓ such that $x_i^e \in \mathcal{I}$ for $i = 1, \ldots, n - \ell$
  3: **for all** $x_1^{e_1} \cdots x_h^{e_h} \in G$ with $h \geq n - d$ and $e_h > 0$ **do**
  4:     **for** $j = 1, \ldots, n - d$ **do**
  5:         **if** $x_1^{e_1} \cdots x_{h-1}^{e_{h-1}} x_h^{e_h - 1} x_j \notin \langle G \rangle$ **then**
  6:             **return** false
  7:         **end if**
  8:     **end for**
  9: **end for**
 10: **return** true

---

4   Weak D-Stability for Polynomial Ideals

In the previous section, we considered exclusively monomial ideals. All the notions introduced in Definition 6 can be straightforwardly extended to polynomial ideals by saying that an ideal $\mathcal{I}$ satisfies some form of stability, if its leading ideal $\mathrm{lt}\,\mathcal{I}$ satisfies this form of stability. Galligo's Theorem 4 immediately implies that after a generic change of coordinate $A \in \mathrm{GL}(n, \Bbbk)$ the transformed ideal $A \cdot \mathcal{I}$ possesses any stability property here considered. Thus in principle a random coordinate transformation (almost) always provides a "nice" leading ideal.

However, from a computational point of view, random transformations are rather unpleasant, as they destroy all sparsity typically present in ideal bases. It is therefore of great interest to see whether for some notion of stability it is possible to design a *deterministic* algorithm which yields a fairly sparse transformation A such that $A \cdot \mathcal{I}$ has the desired stability property. In a forthcoming work [1], we will study this question in depth and provide such an algorithm for many important stability notions. Here, we only present a variation of this algorithm for the case of weak D-stability. For lack of space, we omit the (non-trivial) termination proof which will be given in [1].

Algorithm 2 works by performing incrementally very sparse transformations where all variables except one remain unchanged and this one undergoes a transformation of the form $x_i \to x_i + a x_j$ where $j < i$ and $a \in \Bbbk \setminus \{0\}$ is a generic parameter. The pair (i, j) is chosen in such a way that each transformation leads to true progress towards a weakly D-stable position, if a does not take one of finitely many "bad" values. In practice, we always use the value $a = 1$. If this accidentally represents a "bad" value, then we will automatically perform the same transformation a second time which corresponds to $a = 2$. Obviously, after a finite number of iterations (which can be bounded via the degrees of the generators), we will reach a "good" value, since $\Bbbk$ is an infinite field.

Algorithm 2. WDS-TRAFO: Transformation to weakly D-stable position

---

Input: Gröbner basis G of homogeneous ideal $\mathcal{I} \trianglelefteq \mathcal{P}$
Output: a linear change of coordinates Ψ such that $\Psi(\mathcal{I})$ is weakly D-stable
 1: $D := \dim \mathcal{I}$; $\Psi := \mathrm{id}$
 2: **while** $\exists g \in G,\ 1 \le j \le n - D : i = w(\mathrm{lt}\, g) \ge n - D \wedge x_j\, \mathrm{lt}\,(g)/x_i \notin \langle \mathrm{lt}\, G \rangle$ **do**
 3: $\quad \psi := (x_i \mapsto x_i + x_j)$; $\Psi = \psi \circ \Psi$
 4: $\quad G := \mathrm{GR\ddot{O}BNERBASIS}\big(\psi(G)\big)$
 5: **end while**
 6: **return** Ψ

Algorithm 2 is not in an optimised form. In practice, if one finds more than one suitable pair (i, j), it appears natural to perform several transformations simultaneously, as each iteration of the **while** loop requires a Gröbner basis computation. Furthermore, one should take into account that the input for these computations is typically already fairly close to a Gröbner basis. Hence it is probably useful to apply some specialised algorithm exploiting this fact. A prototype implementation of Algorithm 2 in MAPLE can be found at http://amirhashemi. iut.ac.ir/softwares.

Example 15. We consider for $n = 3$ the ideal $\mathcal{I} = \langle x_1^3, x_2^2 x_3, x_2^3 \rangle$ with $D = 1$. This ideal is not weakly D-stable, since $x_1(x_2^2 x_3)/x_3 \notin \mathcal{I}$ and, according to Algorithm 2, we perform the change of coordinates $\psi_1 : x_3 \mapsto x_1 + x_3$. The transformed ideal $\mathcal{I}_1 = \psi_1(\mathcal{I})$ has the leading ideal $\langle x_1^3, x_1 x_2^2, x_2^3, x_2^2 x_3^3 \rangle$ and is also not D-stable, since $x_1(x_1 x_2^2)/x_2 \notin \mathrm{lt}\, \mathcal{I}_1$. Thus in the second iteration the **while** loop performs the change of coordinate $\psi_2 : x_2 \mapsto x_1 + x_2$. The leading ideal of the transformed ideal $\mathcal{I}_2 = \psi_2(\mathcal{I}_1)$ is by chance even the generic initial ideal $\mathrm{gin}\, \mathcal{I} = \langle x_1^3, x_1^2 x_2, x_1 x_2^2, x_2^4, x_1^2 x_3^3 \rangle$ and thus of course weakly D-stable.

5 Computing the Absolute Reduction Number

We consider first the case of a monomial ideal and extend Theorem 2 from strongly stable ideals to weakly D-stable ones. Our proof follows closely the arguments of the original proof by Bresinsky and Hoa [3].

Theorem 16. *Let $\mathcal{I} \trianglelefteq \mathcal{P}$ be a weakly D-stable monomial ideal. Then $\mathcal{I}$ has a minimal generator x_{n-D}^α and $r(\mathcal{R}) = r_{\mathcal{J}}(\mathcal{R}) = \alpha - 1$ for any minimal reduction $\mathcal{J}$ of $\mathcal{I}$.*

Proof. Since $\mathcal{I}$ is assumed to be weakly D-stable, $x_{n-D+1}, \ldots, x_n$ induce a minimal reduction by Proposition 14 and we can apply Lemma 1. Consider the D linear forms $y_i = x_{n-D+i} + a_{i,1}x_1 + \cdots + a_{i,n-D}x_{n-D}$ with $1 \leq i \leq D$ and arbitrary coefficients $a_{i,j} \in \Bbbk$ and set $\mathcal{J}_1 = \mathcal{I} + \langle y_1, \ldots, y_D \rangle$.

We claim that $r_{\mathcal{J}_1}(\mathcal{R}) = r_{\mathcal{J}_2}(\mathcal{R})$ where $\mathcal{J}_2 = \mathcal{I} + \langle x_{n-D+1}, \ldots, x_n \rangle$. It is enough to show the identity $\mathcal{I}_1 = \mathcal{I}_2$ where $\mathcal{P}/\mathcal{J}_1 \simeq \Bbbk[x_1, \ldots, x_{n-D}]/\mathcal{I}_1$ and $\mathcal{P}/\mathcal{J}_2 \simeq \Bbbk[x_1, \ldots, x_{n-D}]/\mathcal{I}_2$. One easily sees that $\mathcal{I}_2 = \mathcal{I} \cap \Bbbk[x_1, \ldots, x_{n-D}]$ and thus trivially $\mathcal{I}_2 \subseteq \mathcal{I}_1$. The converse inclusion $\mathcal{I}_1 \subseteq \mathcal{I}_2$ follows by Remark 13 which entails that for any term $t = x_1^{\alpha_1} \cdots x_n^{\alpha_n} \in \mathcal{I}$ the corresponding term

$$\tilde{t} = x_1^{\alpha_1} \cdots x_{n-D}^{\alpha_{n-D}} \prod_{j=1}^{D}(-a_{j,1}x_1 - \cdots - a_{j,n-D}x_{n-D})^{\alpha_{n-D+j}} \in \mathcal{I}_1$$

also lies in $\mathcal{I}$ and hence in $\mathcal{I}_2$.

Proposition 14 also implies that $\mathcal{I}$ has a minimal generator of the form x_{n-D}^{α} for some $\alpha \in \mathbb{N}$. Hence, $r_{\mathcal{J}_2}(\mathcal{R}) \geq \alpha - 1$. On the other hand, $x_{n-D}^{\alpha} \in \mathcal{I}$ implies by Remark 13 that any term $x_1^{\alpha_1} \cdots x_{n-D}^{\alpha_{n-D}}$ of degree α also belongs to $\mathcal{I}$ and thus $r_{\mathcal{J}_2}(\mathcal{R}) \leq \alpha - 1$. Therefore $r_{\mathcal{J}_2}(\mathcal{R}) = \alpha - 1$ proving the second assertion. $\square$

We have thus identified a class of monomial ideals, the weakly D-stable ideals, for which it is particularly simple to determine their reduction number. Given a polynomial ideal $\mathcal{I}$, we may use Algorithm 2 to render it weakly D-stable and obtain then immediately the reduction number of its leading ideal $\operatorname{lt}\mathcal{I}$. According to Theorem 3, this number gives us an upper bound for $r(\mathcal{R})$. We introduce now a more specialised class of ideals for which we can guarantee that $\mathcal{I}$ and $\operatorname{lt}\mathcal{I}$ have the same reduction number. We denote here for a monomial ideal $\mathcal{L}$ by $\deg_{x_k} \mathcal{L}$ the maximal x_k-degree of a minimal generator of $\mathcal{L}$.

Definition 17. *Let $0 \leq \ell < n$ be an integer. The homogeneous ideal $\mathcal{I} \trianglelefteq \mathcal{P}$ is weakly ℓ-minimal stable, if its leading ideal $\operatorname{lt}\mathcal{I}$ is weakly ℓ-stable and if for any linear change of coordinates $A \in \operatorname{GL}(n, \Bbbk)$ such that $\operatorname{lt}(A \cdot \mathcal{I})$ is still weakly ℓ-stable, we have $\deg_{x_{n-\ell}} \operatorname{lt}\mathcal{I} \leq \deg_{x_{n-\ell}} \operatorname{lt}(A \cdot \mathcal{I})$.*

Again it is easy to see that this is a generic notion, as any coordinate transformation A with $\operatorname{lt}(A \cdot \mathcal{I}) = \operatorname{gin}\mathcal{I}$ leads to a weakly ℓ-minimal stable position.

Example 18. Consider for $n = 3$ the ideal $\mathcal{I} = \langle x_1x_3, x_1x_2 + x_2^2, x_1^2 \rangle$ introduced by Green [7]. One finds that the leading ideal $\operatorname{lt}\mathcal{I} = \langle x_1^2, x_1x_2, x_1x_3, x_2^3, x_2^2x_3 \rangle$ is even strongly stable and thus of course weakly D-stable (with $D = 1$ here). However, $\mathcal{I}$ is not weakly D-minimal stable, as $\operatorname{gin}(\mathcal{I}) = \langle x_1^2, x_1x_2, x_2^2, x_1x_3^2 \rangle$ and thus has a lower degree in x_2.

Example 19. We consider for $n = 4$ the ideal

$$\mathcal{I} = \langle x_1x_4 - x_2x_3, \ x_2^3 - x_1x_3^2, \ x_2^2x_4 - x_1^3 \rangle;$$

it represents the special case $a = 2$, $b = 3$ of [3, Example 15]. Here $D = 2$ and the ideal $\mathcal{I}$ is not weakly D-stable. The following linear change of coordinates $\Psi : x_2 \mapsto x_1 + x_2, x_3 \mapsto x_1 + x_3$ transforms $\mathcal{I}$ into a weakly D-stable (in fact, even strongly stable) ideal $\mathcal{I}_1$ with leading ideal

$$\operatorname{lt}\mathcal{I}_1 = \langle x_1^2,\ x_1 x_2^2,\ x_2^3,\ x_1 x_2 x_3^2,\ x_1 x_3^3,\ x_2^2 x_3^3,\ x_2 x_3^4 \rangle \, .$$

Note that although this leading ideal is different from

$$\operatorname{gin}\mathcal{I} = \langle x_1^2,\ x_1 x_2^2,\ x_2^3,\ x_1 x_2 x_3^2,\ x_2^2 x_3^2,\ x_1 x_3^4,\ x_2 x_3^4 \rangle \, ,$$

both ideals have the same minimal generator x_2^3. Thus $\mathcal{I}_1$ is weakly D-minimal stable and we see that in this example the set of transformations leading to weakly D-minimal position is strictly larger than the one leading to the generic initial ideal.

Theorem 20. *Let $\mathcal{I} \trianglelefteq \mathcal{P}$ be a weakly D-minimal stable homogeneous ideal. Then $\operatorname{lt}\mathcal{I}$ has a minimal generator x_{n-D}^α and $r(\mathcal{R}) = r(\mathcal{P}/\operatorname{lt}\mathcal{I}) = \alpha - 1$.*

Proof. Since $\operatorname{lt}\mathcal{I}$ is weakly D-stable, it possesses by Proposition 14 a minimal generator x_{n-D}^α and thus $r(\mathcal{P}/\operatorname{lt}\mathcal{I}) = \alpha - 1$ by Proposition 16. As $\mathcal{I}$ is assumed to be weakly D-minimal stable, x_{n-D}^α must also be a minimal generator of $\operatorname{gin}\mathcal{I}$ and hence $r(\mathcal{R}) = r(\mathcal{P}/\operatorname{gin}\mathcal{I}) = \alpha - 1$ by Theorem 5.   □

Unfortunately, Theorem 20 is mainly of theoretical interest, as we are not able to provide a simple deterministic algorithm for the construction of a change of coordinates leading to be weakly D-minimal stable position. We present now Algorithm 3 for the computation of $r(\mathcal{R})$. Instead of a coordinate transformation, it is based on a parametric computation. The main point will be to keep the number of parameters as small as possible.

Algorithm 3. REDNUM: (Absolute) Reduction Number

Input: Gröbner basis G of a homogeneous ideal $\mathcal{I} \triangleleft \mathcal{P}$
Output: the absolute reduction number $r(\mathcal{R})$
1: $D := \dim \mathcal{I}$
2: $\tilde{G} := G$ with x_{n-D+i} replaced by $-\sum_{j=1}^{n-D} a_{ij} x_j$ for all $i > 0$
3: $\tilde{\mathcal{I}} := \langle \tilde{G} \rangle_{\tilde{\mathcal{P}}}$
4: $\mathcal{H} := \text{POMMARETBASIS}(\tilde{\mathcal{I}})$
5: **return** $\deg \mathcal{H} - 1$

The algorithm simply adds D linear forms y_i of the special form (1). The occuring coefficients a_{ij} are then considered as undetermined parameters. Replacing in the ideal $\mathcal{I}$ every variable x_{n-D+i} with $i > 0$ by $-\sum_{j=1}^{n-D} a_{ij} x_j$, we obtain a new homogeneous ideal $\tilde{\mathcal{I}}$ in the polynomial ring $\tilde{\mathcal{P}} = \Bbbk(a_{ij})[x_1, \ldots, x_{n-D}]$

over the field of rational functions in the $D(n - D)$ parameters a_{ij} and compute its Pommaret basis (see [17,18] and references therein).

Theorem 21. *Algorithm 3 correctly determines $r(\mathcal{R})$.*

Proof. We consider first the addition of D generic linear forms $z_i = \sum_{j=1}^{n} b_{ij}x_j$ to the ideal $\mathcal{I}$. This leads to an ideal $\hat{\mathcal{I}}$ in the polynomial ring $\hat{\mathcal{P}} = \Bbbk(b_{ij})[x_1, \ldots, x_n]$ depending on Dn parameters and n variables. It follows from the classical proof of the existence of a Noether normalisation (see e.g. [8, Thm. 3.4.1]) over an infinite field that $\hat{\mathcal{I}}$ is a zero-dimensional ideal (which thus possesses a finite Pommaret basis).

We now claim that the absolute reduction number $r(\mathcal{R})$ is one less than the Castelnuovo-Mumford regularity $\operatorname{reg}\hat{\mathcal{I}}$. According to [18, Cor. 9.5], $\operatorname{reg}\hat{\mathcal{I}}$ is given by the degree of the Pommaret basis of $\hat{\mathcal{I}}$, so that this claim implies that $r(\mathcal{R})$ can be read off the Pommaret basis of $\hat{\mathcal{I}}$. The correctness of the claim follows from a simple genericity argument.

We build recursively $\Bbbk(b_{ij})$-linear generating systems of the vector spaces $\hat{\mathcal{I}}_q$ for all degrees $q = 1, 2, \ldots$ by taking all elements of $\mathcal{H}$ of degree q and adding all products of elements of the previous generating system multiplied with a variable x_j. We collect the coefficients of the obtained generators in a matrix. Entering generic values for the parameters b_{ij} leads to the maximal possible rank of this matrix and thus to the lowest possible dimension of the complement of the degree q component of the corresponding specialisation of $\hat{\mathcal{I}}$. The absolute reduction number is the largest value of q for which we cannot achieve a zero-dimensional complement. Hence a generic choice of the parameters leads to the correct value of the absolute reduction number $r(\mathcal{R})$. Since computing over $\Bbbk(b_{ij})$ corresponds to the generic branch of the parametric computation and since for a zero-dimensional ideal $\operatorname{reg}\hat{\mathcal{I}}$ is the lowest degree q where $\hat{\mathcal{I}}_q = \hat{\mathcal{P}}_q$, we conclude that our claim is correct.

Now consider the $(D \times n)$-matrix (b_{ij}): if the determinant of the submatrix composed of the last D column does not vanish, then by a Gaussian elimination we obtain a set of linear forms y_i in the "reduced" triangular form (1) leading to the same ideal $\hat{\mathcal{I}}$. As the intersection of two Zariski open sets is again Zariski open, this observation proves that generically also the reduced ansatz (1) used in our algorithm yields the correct absolute reduction number. Because of the special form of this ansatz, we may solve the linear forms for the variables x_{n-D+i} and then perform the computations in the polynomial ring $\tilde{\mathcal{P}}$ depending only on $D(n - D)$ parameters and $n - D$ variables. $\square$

Remark 22. Since the Algorithms 2 and 3 are based on Gröbner or Pommaret bases and the worst case complexity of computing Gröbner bases is doubly exponential in the number of variables (as shown by Mayr and Meyer [12]), we conclude that the complexity of these algorithms is also doubly exponential in the number of variables.

Example 23. For $n = 4$, the homogenised Weispfenning94 ideal $\mathcal{I} \lhd \Bbbk[x_1, \ldots, x_4]$ is generated by the polynomials

$$f_1 = x_2^4 + x_1 x_2^2 x_3 + x_1^2 x_4^2 - 2x_1 x_2 x_4^2 + x_2^2 x_4^2 + x_3^2 x_4^2,$$
$$f_2 = x_1 x_2^4 + x_2 x_3^4 - 2x_1^2 x_2 x_4^2 - 3x_4^5,$$
$$f_3 = -x_1^3 x_2^2 + x_1 x_2 x_3^3 + x_2^4 x_4 + x_1 x_2^2 x_3 x_4 - 2x_1 x_2 x_4^3.$$

Here $D = 2$ and we replace x_4 by $-(a_{4,1} x_1 + a_{4,2} x_2)$ and x_3 by $-(a_{3,1} x_1 + a_{3,2} x_2)$ in $\mathcal{I}$ to obtain the new ideal $\tilde{\mathcal{I}} \lhd \Bbbk(a_{3,1}, a_{3,2}, a_{4,1}, a_{4,2})[x_1, x_2]$. We compute a Pommaret basis $\mathcal{H}$ of $\tilde{\mathcal{I}}$ and get as leading terms

$$\operatorname{lt} \mathcal{H} = \left\{ x_1^4, x_1^3 x_2^2, x_1^2 x_2^3, x_1 x_2^5, x_2^6 \right\}.$$

Therefore $r(\mathcal{R}) = 6 - 1 = 5$.

Our second example proves that there cannot exist a "simple" algorithm for computing the (absolute) reduction number. By "simple" we mean that the algorithm uses exclusively information obtained from the leading terms (like for instance Algorithm 2 to transform into weakly D-stable position).

Example 24. We consider again Example 18 of Green. It follows immediately from the above presented bases that here $r(\mathcal{R}) = 1 < 2 = r(\mathcal{P}/\operatorname{lt} \mathcal{I})$. Following Algorithm 3, we replace x_3 by $-(a_1 x_1 + a_2 x_2)$ in order to obtain the ideal $\tilde{\mathcal{I}}$. Then we compute a Pommaret basis $\mathcal{H}$ of $\tilde{\mathcal{I}}$ and get for the leading terms

$$\operatorname{lt} \mathcal{H} = \left\{ x_1^2, x_1 x_2, x_2^2 \right\}.$$

Hence our algorithm yields the correct result $r(\mathcal{R}) = 1$. Since $\mathcal{L} = \operatorname{lt} \mathcal{I}$ is in fact even strongly stable, we conclude that $\operatorname{gin} \mathcal{L} = \mathcal{L}$. Hence the leading terms of the generators of $\mathcal{I}$ cannot contain any information on how to transform $\mathcal{I}$ into a position such that the transformed ideal and its leading ideal share the same reduction number.

6   Big Reduction Numbers and Gröbner Systems

We present now an approach that is able to determine the whole reduction number set $\operatorname{rSet}(\mathcal{R})$ and thus in particular both the absolute and the big reduction number. Our method is based on the theory of Gröbner systems, a notion introduced by Weispfenning [24] who also provided a first algorithm for computing such systems. Subsequently, improvements and alternatives were presented by many authors [10,11,13,14,16]. Our calculations were done using a MAPLE implementation of the DISPGB algorithm of Montes which is available at http://amirhashemi.iut.ac.ir/softwares.

In the sequel, we denote by $\tilde{\mathcal{P}} = \mathcal{P}[\mathbf{a}] = \Bbbk[\mathbf{a}, \mathbf{x}]$ a *parametric* polynomial ring where $\mathbf{a} = a_1, \ldots, a_m$ represents the parameters and $\mathbf{x} = x_1, \ldots, x_n$ the variables. Let $\prec_{\mathbf{x}}$ (resp. $\prec_{\mathbf{a}}$) be a term order for the power products of the variables x_i (resp. the parameters a_i). Then we introduce the block elimination term order $\prec_{\mathbf{x}, \mathbf{a}}$ in the usual manner: for all $\alpha, \gamma \in \mathbb{N}_0^n$ and all $\beta, \delta \in \mathbb{N}_0^m$, we define $\mathbf{a}^\delta \mathbf{x}^\gamma \prec_{\mathbf{x}, \mathbf{a}} \mathbf{a}^\beta \mathbf{x}^\alpha$, if either $\mathbf{x}^\gamma \prec_{\mathbf{x}} \mathbf{x}^\alpha$ or $\mathbf{x}^\gamma = \mathbf{x}^\alpha$ and $\mathbf{a}^\delta \prec_{\mathbf{a}} \mathbf{a}^\beta$.

Definition 25. *A finite set of triples* $\left\{(\tilde{G}_i, N_i, W_i)\right\}_{i=1}^{\ell}$ *with finite sets* $\tilde{G}_i \subset \tilde{\mathcal{P}}$ *and* $N_i, W_i \subset \mathcal{Q} = \mathbb{k}[\mathbf{a}]$ *is a* Gröbner system *for a parametric ideal* $\tilde{\mathcal{I}} \trianglelefteq \tilde{\mathcal{P}}$ *with respect to the block order* $\prec_{\mathbf{x},\mathbf{a}}$, *if for every index* $1 \leq i \leq \ell$ *and every specialisation homomorphism* $\sigma : \mathcal{Q} \to \mathbb{k}$ *such that*

$$\text{(i) } \forall g \in N_i : \sigma(g) = 0, \qquad \text{(ii) } \forall h \in W_i : \sigma(h) \neq 0 \qquad (5)$$

$\sigma(\tilde{G}_i)$ *is a Gröbner basis of* $\sigma(\tilde{\mathcal{I}}) \trianglelefteq \mathcal{P}$ *with respect to the order* $\prec_{\mathbf{x}}$ *and if for any point* $a \in \mathbb{k}^m$ *an index* $1 \leq i \leq \ell$ *exists such that* $a \in \mathcal{V}(N_i) \setminus \mathcal{V}(W_i)$.

Thus a Gröbner systems yields a Gröbner basis for all possible values of the parameters $\mathbf{a}$. Weispfenning [24, Theorem 2.7] proved that every parametric ideal $\mathcal{I} \trianglelefteq \mathcal{S}$ possesses a Gröbner system, but in general the system is not unique. Basically every algorithm (in particular the DISPGB algorithm used by us) produces Gröbner systems such that given one specific triple $(\tilde{G}_i, N_i, W_i)$ all specialisations σ satisfying (5) yield the same leading terms $\mathrm{lt}\,\sigma(G_i)$ so that we can speak of a monomial ideal $\mathcal{L}_i \trianglelefteq \mathcal{P}$ determined by the conditions (N_i, W_i). In the sequel, we will always assume that a Gröbner system with this property is used. As a simple corollary, we find then that the reduction number set of an ideal $\mathcal{I} \trianglelefteq \mathcal{P}$ is always finite. Our proof also yields an explicit method for computing it.

Theorem 26. *Let* $\mathcal{I} \trianglelefteq \mathcal{P}$ *be a homogeneous ideal. Then its reduction number set* $\mathrm{rSet}(\mathcal{R})$ *is finite.*

Proof. By definition, any minimal reduction of $\mathcal{I}$ is induced by D linear forms

$$y_i = \sum_{j=1}^{n} a_{i,j} x_j, \qquad i = 1, \ldots, D \qquad (6)$$

with $a_{i,j} \in \mathbb{k}$ and minimality is equivalent to $\mathcal{J} = \mathcal{I} + \langle y_1, \ldots, y_D \rangle$ being a zero-dimensional ideal. Considering the coefficients $a_{i,j}$ as parameters, we may identify $\mathcal{J}$ with a parametric ideal $\tilde{\mathcal{I}} \trianglelefteq \tilde{\mathcal{P}}$. Let $\left\{(\tilde{G}_i, N_i, W_i)\right\}_{i=1}^{\ell}$ be a Gröbner system for $\tilde{\mathcal{I}}$. Without loss of generality, we may assume that for the first s triples the associated monomial ideals $\mathcal{L}_i$ are zero-dimensional, whereas all other triples lead to monomial ideals of positive dimension. Hence precisely the parameter values satisfying one of the conditions (N_i, W_i) with $1 \leq i \leq s$ define minimal reductions. If d_i is the highest degree such that $(\mathcal{L}_i)_{d_i} \neq \mathcal{P}_{d_i}$, then it follows that $\mathrm{rSet}(\mathcal{R}) = \{d_1, \ldots, d_s\}$. □

Remark 27. Any Gröbner system for a parametric ideal $\tilde{\mathcal{I}}$ contains one generic branch where the set N_i of equations is empty. Obviously, the corresponding leading ideal $\mathcal{L}_i$ must be the generic initial ideal $\mathrm{gin}\,\mathcal{I}$ and we have $d_i = r(\mathcal{R})$. This observation immediately yields an alternative proof of [21, Cor. 2.2]: for almost all minimal reductions $\mathcal{J}$ we find $r_{\mathcal{J}}(\mathcal{R}) = r(\mathcal{R})$.

Example 28. Let us consider again Green's Example 18 where $D = 1$. Hence we set $\tilde{\mathcal{I}} = \langle x_1^2, x_1 x_3, x_2^2 + x_1 x_2, a_1 x_1 + a_2 x_2 + a_3 x_3 \rangle$. The Gröbner system for $\tilde{\mathcal{I}}$ consists of 4 triples. For simplicity, we present in the following list for each branch as first entry only the corresponding leading ideal $\mathcal{L}_i$; the other two entries are the equations N_i and the inequations W_i, respectively.

$$
\begin{array}{lll}
\{x_1,\ x_2^2,\ x_3^2,\ x_2 x_3\} & \{\} & \{a_1,\ a_2,\ a_1 - a_2\} \\
\{x_1,\ x_2^2,\ x_3^2,\ x_2 x_3\} & \{a_1 - a_2\} & \{a_2\} \\
\{x_1,\ x_2^2,\ x_3^2\} & \{a_2\} & \{a_1\} \\
\{x_2,\ x_1^2,\ x_3^2,\ x_1 x_3\} & \{a_1\} & \{\}
\end{array}
$$

We observe that all four branches lead to zero-dimensional leading ideals and their reduction numbers are $1, 1, 2, 1$, respectively. Therefore, $\mathrm{rSet}(\mathcal{R}) = \{1, 2\}$ and $br(\mathcal{R}) = 2$.

Remark 29. For comparison, we briefly outline Trung's constructive characterisation [21] of the big reduction number of an ideal. He also takes D linear forms (6) with undetermined coefficients $a_{i,j}$ and proceeds with the ideal $\mathcal{J} = \mathcal{I} + \langle y_1, \ldots, y_D \rangle \trianglelefteq \mathcal{P}$ (note that he does not work in the parametric polynomial ring $\tilde{\mathcal{P}}$). Then he introduces the matrix M_d of the coefficients of the generators in a $\Bbbk$-linear basis of $\mathcal{J}_d$ (which are elements in $\mathcal{Q}$). Let $\mathcal{V}_d$ be the variety of the ideal generated in $\mathcal{Q}$ by all the minors of M_d of the size of the number of terms of degree d. Then, $br(\mathcal{R})$ is the largest d such that $\mathcal{V}_d \neq \mathcal{V}_{d+1}$ [21, Cor. 2.3].

Note, however, that a priori it is unclear how to detect that one has obtained the largest d with this property. Thus his approach becomes truely algorithmic only by combining it with another result of his, namely that $br(\mathcal{R}) + 1$ is bounded by the Castelnuovo-Mumford regularity $\mathrm{reg}(\mathcal{I})$ [19, Prop. 3.2]. Now one can check all degrees d until $\mathrm{reg}(\mathcal{I})$—which has to be computed first—and then finally decide on the value of $br(\mathcal{R})$. While the computation of a Gröbner system is surely a rather expensive operation, we strongly believe that it is much more efficient that the determination and subsequent analysis of large determinantal ideals. Furthermore, our approach yields directly all possible values for the reduction number, whereas Trung must consider one determinantal ideal after the other (of increasing size).

Finally, we note that Trung [21] proved that $br(\mathcal{R}) \leq br(\mathcal{P}/\mathrm{lt}\,\mathcal{I})$ if $\mathcal{R}$ is Cohen-Macaulay. He also claimed that generally one cannot compare $br(\mathcal{R})$ and $br(\mathcal{P}/\mathrm{lt}\,\mathcal{I})$. However, he did not provide a concrete example where the above inequality is violated—which we will do now.

Example 30. Consider for $n = 3$ the ideal

$$\mathcal{I} = \langle x_1^2 x_2 + x_1 x_2^2,\ x_2^3 + x_2^2 x_3,\ x_1 x_3^5, x_2^2 x_3^5,\ x_1^2 x_3 + x_1 x_2 x_3,\ x_1^3 - x_1 x_2^2 \rangle.$$

The given generators form already a Gröbner basis and thus $D = 1$. $\mathrm{lt}\,\mathcal{I}$ is quasi-stable and, using Pommaret bases, one easily shows that the depth of $\mathcal{R}$ is 0 and $\mathcal{R}$ is not Cohen-Macaulay. With $\mathcal{J} = \mathcal{I} + \langle x_1 + x_2 + x_3 \rangle$, a simple computation yields that $\mathrm{lt}\,\mathcal{J} = \langle x_1, x_2 x_3^2, x_2^2 x_3, x_2^3, x_3^7 \rangle$ and thus $br(\mathcal{R}) \geq r_{\mathcal{J}}(\mathcal{R}) = 6$. For showing that $br(\mathcal{R}) = 6$, we set $\tilde{\mathcal{I}} = \mathcal{I} + \langle a_1 x_1 + a_2 x_2 + a_3 x_3 \rangle \trianglelefteq \tilde{\mathcal{P}}$. The Gröbner system of this ideal shows that the reduction numbers of the zero-dimensional branches are $3, 5, 6$, respectively, and therefore $br(\mathcal{R}) = 6$. On the other hand, lt $\mathcal{I} = \langle x_1^2 x_3, x_2^3, x_1^2 x_2, x_1^3, x_1 x_3^5, x_2^2 x_3^5 \rangle$. We set $\tilde{\mathcal{I}}_1 = \text{lt}\,\mathcal{I} + \langle a_1 x_1 + a_2 x_s + a_3 x_3 \rangle$, and compute its Gröbner system. Only three branches are zero-dimensional and they all have as reduction number 3. This shows that $br(\mathcal{P}/\,\text{lt}\,\mathcal{I}) = 3 < br(\mathcal{R})$.

Acknowledgements. The first author greatly appreciates financial support by DAAD (German Academic Exchange Service) for a stay at Universität Kassel in summer 2013. He also would like to thank his coauthors for the invitation, hospitality, and support. The research of the first author was in part supported by a grant from IPM (No. 92550420).

References

1. Albert, M., Hashemi, A., Pytlik, P., Schweinfurter, M., Seiler, W.: Effective genericity for polynomial ideals (in preparation, 2014)
2. Bayer, D., Stillman, M.: A theorem on refining division orders by the reverse lexicographic orders. Duke J. Math. 55, 321–328 (1987)
3. Bresinsky, H., Hoa, L.: On the reduction number of some graded algebras. Proc. Amer. Math. Soc. 127, 1257–1263 (1999)
4. Conca, A.: Reduction numbers and initial ideals. Proc. Amer. Math. Soc. 131, 1015–1020 (2002)
5. Galligo, A.: A propos du théorème de préparation de Weierstrass. In: Norguet, F. (ed.) Fonctions de Plusieurs Variables Complexes. Lecture Notes in Mathematics, vol. 409, pp. 543–579. Springer, Berlin (1974)
6. Galligo, A.: Théorème de division et stabilité en géometrie analytique locale. Ann. Inst. Fourier 29(2), 107–184 (1979)
7. Green, M.: Generic initial ideals. In: Elias, J., Giral, J., Miró-Roig, R., Zarzuela, S. (eds.) Six Lectures on Commutative Algebra. Progress in Mathematics, vol. 166, pp. 119–186. Birkhäuser, Basel (1998)
8. Greuel, G.M., Pfister, G.: A SINGULAR Introduction to Commutative Algebra, 2nd edn. Springer, Berlin (2008)
9. Hashemi, A., Schweinfurter, M., Seiler, W.: Quasi-stability versus genericity. In: Gerdt, V.P., Koepf, W., Mayr, E.W., Vorozhtsov, E.V. (eds.) CASC 2012. LNCS, vol. 7442, pp. 172–184. Springer, Heidelberg (2012)
10. Kapur, D., Sun, Y., Wang, D.: A new algorithm for computing comprehensive Gröbner systems. In: Koepf, W. (ed.) Proc. ISSAC 2010, pp. 29–36. ACM Press (2010)
11. Kapur, D., Sun, Y., Wang, D.: An efficient algorithm for computing a comprehensive Gröbner system of a parametric polynomial system. J. Symb. Comput. 49, 27–44 (2013)
12. Mayr, E., Meyer, A.: The complexity of the word problems for commutative semigroups and polynomial ideals. Adv. Math. 46, 305–329 (1982)
13. Montes, A.: A new algorithm for discussing Gröbner bases with parameters. J. Symb. Comput. 33, 183–208 (2002)
14. Montes, A., Wibmer, M.: Gröbner bases for polynomial systems with parameters. J. Symb. Comput. 45, 1391–1425 (2010)

15. Northcott, D., Rees, D.: Reduction of ideals in local rings. Cambridge Philos. Soc. 50, 145–158 (1954)
16. Sato, Y., Suzuki, A.: A simple algorithm to compute comprehensive Gröbner bases using Gröbner bases. In: Dumas, J.-G (ed.) Proc. ISSAC 2006, pp. 326–331. ACM Press (2006)
17. Seiler, W.: A combinatorial approach to involution and δ-regularity I: Involutive bases in polynomial algebras of solvable type. Appl. Alg. Eng. Comm. Comp. 20, 207–259 (2009)
18. Seiler, W.: A combinatorial approach to involution and δ-regularity II: Structure analysis of polynomial modules with Pommaret bases. Appl. Alg. Eng. Comm. Comp. 20, 261–338 (2009)
19. Trung, N.: Reduction exponent and degree bounds for the defining equations of a graded ring. Proc. Amer. Math. Soc. 101, 229–236 (1987)
20. Trung, N.: Gröbner bases, local cohomology and reduction number. Proc. Amer. Math. Soc. 129, 9–18 (2001)
21. Trung, N.: Constructive characterization of the reduction numbers. Compos. Math. 137, 99–113 (2003)
22. Vasconcelos, W.: The reduction number of an algebra. Compos. Math. 106, 189–197 (1996)
23. Vasconcelos, W.: Reduction numbers of ideals. J. Alg. 216, 652–664 (1999)
24. Weispfenning, V.: Comprehensive Gröbner bases. J. Symb. Comp. 14, 1–29 (1992)

A Note on Global Newton Iteration Over Archimedean and Non-Archimedean Fields

Jonathan D. Hauenstein[1,*], Victor Y. Pan[2,**], and Agnes Szanto[1,***]

[1] North Carolina State University
[2] Lehman College - City University of New York

Abstract. In this paper, we study iterative methods on the coefficients of the *rational univariate representation (RUR)* of a given algebraic set, called a *global Newton iterations*. We compare two natural approaches to define locally quadratically convergent iterations: the first one involves Newton iteration applied to the approximate roots individually and then interpolation to find the RUR of these approximate roots; the second one considers the coefficients in the exact RUR as zeroes of a high dimensional map defined by polynomial reduction and applies Newton iteration on this map. We prove that over fields with a p-adic valuation these two approaches give the same iteration function. However, over fields equipped with the usual Archimedean absolute value they are not equivalent. In the latter case, we give explicitly the iteration function for both approaches. Finally, we analyze the parallel complexity of the different versions of the global Newton iteration, compare them, and demonstrate that they can be efficiently computed. The motivation for this study comes from the certification of approximate roots of overdetermined and singular polynomial systems via the recovery of an exact RUR from approximate numerical data.

1 Introduction

Let $F_1, \ldots, F_n \in \mathbb{K}[x_1, \ldots, x_n]$ be polynomials with coefficients from a field $\mathbb{K}$, $\mathcal{J} := \langle F_1, \ldots, F_n \rangle$ the ideal they generate, and assume that $\mathcal{J}$ is zero dimensional and radical. We consider two cases for the coefficient field $\mathbb{K}$:

Non-Archimedean Case: Let R be a principal ideal domain, $\mathbb{K}$ its field of fractions, and p an irreducible element in R. Then we can equip $\mathbb{K}$ with the p-adic valuation, which defines a non-Archimedean metric on vector spaces over $\mathbb{K}$.

Archimedean Case: In this case, $\mathbb{K}$ is a subfield of $\mathbb{C}$ and it is equipped with the usual absolute value. Then, the usual Euclidean norm defines an Archimedean metric on vector spaces over $\mathbb{K}$.

* Research was partly supported by NSF grant DMS-1262428 and DARPA Young Faculty Award.
** Research was partly supported by NSF grant CCF-1116736.
*** Research partly supported by NSF grant CCF-1217557.

V.P. Gerdt et al. (Eds.): CASC Workshop 2014, LNCS 8660, pp. 202–217, 2014.

The objective of this paper is to study iterative methods on the coefficients of the *rational univariate representation (RUR)* of a component of $\mathcal{J}$, and compare them in the Archimedean and the non-Archimedean cases. The RUR of a component of $\mathcal{J}$, originally defined in [32], is a simple representation of a subset of the common roots of $F_1, \ldots, F_n$, expressing the coordinates of these common roots as Lagrange interpolants at nodes which are given as the roots of a univariate polynomial (see definition below).

We study two natural approaches for iterations that are locally quadratically convergent to an exact RUR of a component of $\mathcal{J}$, based on Newton's method:

- To update an RUR, apply the usual $n \times n$ Newton iteration to each common root of the old RUR, and compute the updated RUR which defines these updated roots. In this approach we assume inductively that the common roots of the iterated RUR's are known exactly.
- Consider the map that takes an RUR and returns the reduced form of the input polynomials $F_1, \ldots, F_n$ modulo the RUR. Since an exact RUR of a component of $\mathcal{J}$ is a zero of this map, we apply Newton's method to this map.

Note that in the p-adic case the first iteration was studied in [14], where they gave the iteration function explicitly and analyzed its complexity in terms of straight-line programs, while the second approach was proposed in [37], without giving the iteration explicitly.

The main results of this paper are as follows. First, we prove that the above two approaches give the same iteration function in the p-adic valuation. Next, we show that in the Archimedean case the two iterations are not equivalent. In this case, we give the explicit iteration functions for both approaches, and show that they are also different from the iteration function presented in [14]. We illustrate the methods on an example involving mobility of spacial mechanisms. Finally, we analyze the parallel complexity of both approaches in the Archimedean case: for the first approach we use $n \times n$ Newton iterations independently for each root and an efficient parallel Vandermonde linear solver for Lagrange interpolation, while for the second approach we utilize a new efficient version of the algorithms of [30] to solve Toeplitz-like linear systems that uses a more efficient displacement representation with factor circulant matrices defined in [31, Example 4.4.2] rather than triangular Toeplitz matrices.

Note that because of the page restrictions of this submission, we could not include in this version most of the proofs and the detail of the algorithms used in our complexity bounds. A full version of this paper is uploaded on the archive server arxiv.org [17].

1.1 Related Work

The motivation to study numerical approximations of RUR's come from a work in progress in [2] to certify approximate roots of overdetermined and singular polynomial systems over $\mathbb{Q}$. For well-constrained non-singular systems, Smale's α-theory (see [6, Chapters 8 and 14]) gives a tool for the certification of approximate roots, as was explored and implemented in **alphaCertified** [18].

However, **alphaCertified** does not straightforwardly extend to overdetermined or singular systems: in [18] they propose to use universal lower bounds for the minimum of positive polynomials on a disk, such as in [22], but they conclude that such bounds are "too small to be practical". To overcome this difficulty, in [2] it is proposed to iteratively compute the exact RUR of a rational component from approximations of the roots, and then use the machinery of [18] to certify approximate roots of this RUR. While [2] is devoted to considerations about the global behavior of the iteration, this paper considers different choices of the iteration function and their parallel complexity.

The iterative algorithms that are in the core of this paper are the Archimedean adaptations of what is known as "global Newton iteration" or "multivariate Hensel lifting" or "Newton-Hensel lifting" in the computer algebra literature, where it is defined for the non-Archimedean case. . Various versions of Newton-Hensel lifting were applied in many applications within computer algebra, including univariate polynomial factorization [41,27], multivariate polynomial factorization [8,15,23], gcd of sparse multivariate polynomials [24], lexicographic and general Gröbner basis computation of zero dimensional algebraic sets [37,40], geometric resolution of equi-dimensional algebraic sets [12,13,19,14], Chow forms [21], and sparse interpolation [3]. As mentioned above, the most related to this paper are the [37,14].

Computing numerical approximation to symbolic objects in the Archimedean metric is not new either. There is a significant literature studying such hybrid symbolic-numeric algorithms, and without trying to give a complete bibliography, the following summarizes the papers that are the closest to our work.

Closest to our approach is the literature on finding the vanishing ideal of a finite point set given with limited precision. In [7] they give an algorithm such that given *one* approximate zero of a polynomial system, finds the RUR of the irreducible component containing the corresponding exact roots in randomized polynomial time. The algorithm in [7] applies the univariate results of [25] using lattice basis reduction. The main point of our approach in this paper and in [2] is that we assume to know *all* approximate roots of a rational component, so in this case we can compute the exact RUR much more efficiently, and in parallel.

The papers [7,35,29,1,20,9,10] use a more general approach than the one here, by computing border bases for a given set of approximate roots, which avoids defining a random primitive element as is done for RURs used in this paper. For general polynomial systems numerical computation of Gröbner bases was proposed for example in [33,34,28,36]. The focus of these papers is to find numerically stable support for the bases, which we assume to be given here by the primitive element.

2 Preliminaries

Let us start with recalling the notion of Roullier's *Rational Univariate Representation (RUR)*, originally defined in [32]. Instead of defining the RUR of an ideal $\mathcal{J}$, here we only define the *RUR of a component of $\mathcal{J}$*, which is a weaker notion. We follow here the notation in [2].

Let $\mathbb{K}$ be a field. Given $\mathbf{F} = (F_1, \ldots, F_n) \in \mathbb{K}[x_1, \ldots, x_n]$ for some n, and assume that the ideal $\mathcal{J} := \langle F_1, \ldots, F_n \rangle$ is radical and zero dimensional. Then the factor ring $\mathbb{K}[x_1, \ldots, x_n]/\mathcal{J}$ is a finite dimensional vector space over $\mathbb{K}$, and we denote

$$\delta := \dim_{\mathbb{K}} \mathbb{K}[x_1, \ldots, x_n]/\mathcal{J}.$$

Furthermore, for almost all $(\lambda_1, \ldots, \lambda_n) \in \mathbb{K}^n$ (except a Zariski closed subset), the linear combination

$$u(x_1, \ldots, x_n) := \lambda_1 x_1 + \cdots + \lambda_n x_n$$

is a *primitive element* of $\mathcal{J}$, i.e. the powers $1, u, u^2, \ldots, u^{\delta-1}$ form a linear basis for $\mathbb{K}[x_1, \ldots, x_n]/\mathcal{J}$ (c.f. [32]).

In the algorithms that follow, we compute an RUR that may not generate the ideal $\mathcal{J}$, nevertheless the polynomials $F_1, \ldots, F_n$ vanish modulo the RUR. In this case, the RUR will define a *component of* $\mathcal{J}$. We have the following definition:

Definition 1. *Let* $\mathcal{J} = \langle F_1, \ldots, F_n \rangle \subset \mathbb{K}[x_1, \ldots, x_n]$ *be as above. Let* $\lambda_1 x_1 + \cdots + \lambda_n x_n$ *be a primitive element of* $\mathcal{J}$. *We call the polynomials*

$$\lambda_1 x_1 + \cdots + \lambda_n x_n - T, \ q(T), \ v_1(T), \ldots, v_n(T) \tag{1}$$

a Rational Univariate Representation (RUR) of a component of $\mathcal{J}$ *if it satisfies the following properties:*

- $q(T) \in \mathbb{K}[T]$ *is a monic polynomial of degree* $d \leq \delta$,
- $\gcd_T(q(T), q'(T)) = 1$ *where* $q'(T) = \frac{\partial q(T)}{\partial T}$,
- $v_1(T), \ldots, v_n(T) \in \mathbb{K}[T]$ *are all degree at most* $d - 1$ *and satisfy*

$$\lambda_1 v_1(T) + \cdots + \lambda_n v_n(T) = T,$$

- *for all* $i = 1, \ldots, n$ *we have*

$$F_i(v_1(T), \ldots, v_n(T)) \equiv 0 \mod q(T).$$

Next, let us recall the relationship between the RUR of a component of $\mathcal{J}$ and its (exact) roots. Let

$$V(\mathcal{J}) = \{\xi_1, \ldots, \xi_\delta\} \subset \mathbb{C}^n$$

be the set of common roots of $\mathcal{J}$. Denote $\xi_i = (\xi_{i,1}, \ldots, \xi_{i,n})$ for $i = 1, \ldots, \delta$. Then for any n-tuple $(\lambda_1, \ldots \lambda_n) \in \mathbb{K}^n$ such that for $i, j = 1, \ldots, \delta$

$$\lambda_1 \xi_{i,1} + \cdots + \lambda_n \xi_{i,n} \neq \lambda_1 \xi_{j,1} + \cdots + \lambda_n \xi_{j,n} \quad \text{if } i \neq j,$$

$u = \lambda_1 x_1 + \cdots + \lambda_n x_n$ is a primitive element for $\mathcal{J}$. Since all roots of $\mathcal{J}$ are distinct, such primitive element exist, and can be computed from the roots $\{\xi_1, \ldots, \xi_\delta\}$, or using randomization. Fix such $(\lambda_1, \ldots \lambda_n) \in \mathbb{K}^n$. For $d \leq \delta$ let $\{\xi_1, \ldots, \xi_d\}$ be a subset of $V(\mathcal{J})$ and define

$$\mu_i := \lambda_1 \xi_{i,1} + \cdots + \lambda_n \xi_{i,n}, \quad i = 1, \ldots d. \tag{2}$$

The RUR of the component of $\mathcal{J}$ corresponding to the subset $\{\xi_1, \ldots, \xi_d\} \subset V(\mathcal{J})$ is defined by

$$q(T) := \prod_{i=1}^{d} (T - \mu_i), \tag{3}$$

and for each $j = 1, \ldots, n$, the polynomial $v_j(T)$ is the unique Lagrange interpolant of degree at most $d - 1$ satisfying

$$v_j(\mu_i) = \xi_{i,j} \quad \text{for } i = 1, \ldots, d. \tag{4}$$

Note that if $\mathcal{J}$ is defined by polynomials over $\mathbb{K}$ then the polynomials in the RUR of $\mathcal{J}$ have coefficients in $\mathbb{K}$. Since this may not be true for all subsets of $V(\mathcal{J})$, we call a subset $\{\xi_1, \ldots, \xi_d\} \subset V(\mathcal{J})$ a *rational component of $\mathcal{J}$* if the corresponding RUR has also coefficients in $\mathbb{K}$.

3   Global Newton Iteration

In this section we describe iterative methods that improves the accuracy of the coefficients of the RUR of a component of $\mathcal{J}$. We use a similar approach as in [14, Sec 4], but instead of a coefficient ring with the p-adic absolute value, we make adaptations to a coefficient field $\mathbb{K} \subseteq \mathbb{C}$ equipped with the usual absolute value. We start with recalling the definitions given in [14, Sec 4].

3.1   Non-Archimedean Global Newton iteration

First, we briefly describe the global Newton iteration defined in [14, Sec 4]. There the coefficient domain is the ring $\mathbb{Q}[t]$ and the non-Archimedean metric is defined by the irreducible element $t \in \mathbb{Q}[t]$. They consider a square system $\boldsymbol{F} = (F_1, \ldots, F_n)$ with $F_i \in \mathbb{Q}[t][x_1, \ldots, x_n]$. Let

$$u(x_1, \ldots, x_n) = \lambda_1 x_1 + \cdots + \lambda_n x_n = T$$

be a random primitive element for $\mathcal{J} = \langle F_1, \ldots, F_n \rangle$. Furthermore, define

$$I := \langle t^k \rangle \text{ for some } k.$$

In [14, Sec 4] they assume that some initial approximate RUR is given for a component of $\mathcal{J}$:

$$q(T), \, \mathbf{v}(T) := (v_1(T), \ldots, v_n(T)) \in \mathbb{Q}[t][T],$$

satisfying the following assumptions.

Assumption 2. Let $\boldsymbol{F}$, $u = \sum_{i=1}^{n} \lambda_i x_i$, I, $q(T)$ and $\mathbf{v}(T)$ be as above. Then

1. $q(T)$ is monic and has degree d,
2. $v_i(T)$ has degree at most $d - 1$,

3. $F(v(T)) \equiv 0 \mod q(T) \mod I$
4. $\lambda_1 v_1(T) + \cdots + \lambda_n v_n(T) = T \mod I$,
5. $J_F(v(T)) := \left[\frac{\partial F_i}{\partial x_j}(v(T)) \right]_{i,j=1}^n$ is invertible modulo $q(T)$ and I.

They define the following updates:

Definition 3. *Assume that F, $\mathcal{J}$, $u = \sum_{i=1}^n \lambda_i x_i$, I, $q(T)$ and $v(T)$ satisfy Assumption 2. Then in [14, Section 4] they define*

$$w(T) := v(T) - \left(J_F(v(T))^{-1} F(v(T)) \mod q(T) \right) \mod I^2,$$

$$\Delta(T) := \sum_{i=1}^n \lambda_i w_i(T) - T \mod I^2,$$

$$V(T) := w(T) - \left(\Delta(T) \cdot \frac{\partial w(T)}{\partial T} \mod q(T) \right) \mod I^2,$$

$$Q(T) := q(T) - \left(\Delta(T) \cdot \frac{\partial q(T)}{\partial T} \mod q(T) \right) \mod I^2.$$

In [14, Section 4] they prove the following:

Proposition 4. *Assume that F, u, $q(T)$, $v(T)$ and I satisfy Assumption 2 and let $w(T)$, $\Delta(T)$, $V(T)$, $Q(T)$ be as in Definition 3. Then*

(i) $v(T) \equiv w(T) \equiv V(T)$, $q(T) \equiv Q(T)$ and $\Delta(T) \equiv 0 \mod I$
(ii) $F(w(T)) \equiv 0 \mod q(T) \mod I^2$,
(iii) $\langle q(T), U - T - \Delta(T), x_1 - w_1(T), \ldots, x_n - w_n(T) \rangle = \langle Q(U), T - U - \Delta(U), x_1 - V_1(U), \ldots, x_n - V_n(U) \rangle \mod I^2$,
(iv) $F(V(T)) \equiv 0 \mod Q(T) \mod I^2$,
(v) $\lambda_1 V_1(T) + \cdots + \lambda_n V_n(T) = T \mod I^2$.

3.2 First Construction

Our first variation of Definition 3 will have the property that it agrees to the approximate RUR obtained from the approximate roots via Lagrange interpolation as was described in the Preliminaries. We give our definition over some general coefficient ring R, but later we will use $R = \mathbb{K} \subset \mathbb{C}$, or $\mathbb{Q}[t]/I^2$. We need the following assumptions:

Assumption 5. Let $F = (F_1, \ldots, F_n)$, $u = \lambda_1 x_1 + \cdots + \lambda_n x_n$, $q(T)$ and $v(T) = (v_1(T), \ldots, v_n(T))$ polynomials over some Euclidean domain R as above. We assume that

1. $q(T)$ is monic and has degree d,
2. $v_i(T)$ has degree at most $d - 1$,
3. $\frac{\partial q(T)}{\partial T}$ is invertible modulo $q(T)$,
4. $\lambda_1 v_1(T) + \cdots + \lambda_n v_n(T) = T$,
5. $J_F(v(T)) := \left[\frac{\partial F_i}{\partial x_j}(v(T)) \right]_{i,j=1}^n$ is invertible modulo $q(T)$.

Our first construction for the update is defined as follows:

Definition 6. *Assume that* $\boldsymbol{F}$, $u = \sum_{i=1}^{n} \lambda_i x_i$, $q(T)$ *and* $\mathbf{v}(T)$ *satisfy Assumption 5. Then we define*

$$\boldsymbol{w}(T) := \mathbf{v}(T) - \left(J_{\boldsymbol{F}}(\mathbf{v}(T))\right)^{-1} \boldsymbol{F}(\mathbf{v}(T)) \mod q(T)),$$

$$\Delta(T) := \sum_{i=1}^{n} \lambda_i w_i(T) - T \text{ so far the same as in Definition 3,} \tag{5}$$

$$\tilde{V}(T + \Delta(T)) := \boldsymbol{w}(T) \mod q(T), \tag{6}$$

$$\Delta\tilde{Q}(T + \Delta(T)) := -(T + \Delta(T))^d \mod q(T) \tag{7}$$

$$\tilde{Q}(T) := \Delta\tilde{Q}(T) + T^d \tag{8}$$

Note that in Definition 6 we define $\tilde{V}(T + \Delta(T))$ and not $\tilde{V}(T)$, but the coefficients of $\tilde{V}_i(T)$ can be obtained as solutions of linear systems. Similarly for the coefficients of $\Delta\tilde{Q}(T)$. In the next proposition we examine these linear systems and give conditions on the existence and uniqueness of their solutions.

Proposition 7. *The coefficients of the polynomials* $\tilde{V}_i(T)$ *in (6) for* $i = 1, \ldots, n$, *and the coefficients of the polynomial* $\Delta\tilde{Q}(T) = \tilde{Q}(T) - T^d$ *in (7) are the solutions of* $d \times d$ *linear systems with a common coefficient matrix that has columns which are the coefficient vectors of*

$$(T + \Delta(T))^j \mod q(T) \quad \text{for } j = 0, \ldots, d - 1.$$

This coefficient matrix is non-singular if and only if u *is a primitive element for the ideal* $\langle q(T), x_1 - w_1(T), \ldots, x_n - w_n(T) \rangle$.

Proof. See [17] for the proof. $\qquad\square$

The following proposition compares Definitions 3 and 6 in cases when the coefficient ring is $R = \mathbb{Q}[t]/I^2$ for $I = \langle t^k \rangle$ for some $k > 0$.

Proposition 8. *Assume that the conditions of Assumption 2 are satisfied and that* u *is a primitive element for* $\langle q(T), x_1 - w_1(T), \ldots, x_n - w_n(T) \rangle$ *as in Proposition 7. Let* $\boldsymbol{V}(T)$ *and* $Q(T)$ *be as in Definition 3 and* $\tilde{V}(T)$ *and* $\tilde{Q}(T)$ *be as in Definition 6 for* $R = \mathbb{Q}[t]/I^2$. *Then* $\boldsymbol{V}(T) \equiv \tilde{V}(T)$ *and* $Q(T) \equiv \tilde{Q}(T) \mod I^2$.

Proof. See [17] for the proof. $\qquad\square$

The next proposition connects the updated RUR defined in Definition 6 to the ones obtained by applying one step of Newton iteration on the approximate roots, as promised in the Introduction:

Proposition 9. *Let* $\boldsymbol{F} = (F_1, \ldots, F_n) \subset \mathbb{K}[x_1, \ldots, x_n]$ *as above. Assume that the polynomials* $u = \lambda_1 x_1 + \cdots + \lambda_n x_n$, $q(T)$, $\mathbf{v}(T) := (v_1(T), \ldots, v_n(T))$ *satisfy Assumption 5. Let* $\boldsymbol{z}_1, \ldots, \boldsymbol{z}_d \in \overline{\mathbb{K}}^n$ *be the exact roots of* $\langle q(T), x_1 - v_1(T), \ldots, x_n - v_n(T) \rangle \cap \mathbb{K}[x_1, \ldots, x_n]$, *where* $\overline{\mathbb{K}}$ *is the algebraic closure of* $\mathbb{K}$. *Let*

$$\tilde{\boldsymbol{z}}_i := \boldsymbol{z}_i - J_{\boldsymbol{F}}(\boldsymbol{z}_i)^{-1} \boldsymbol{F}(\boldsymbol{z}_i) \quad i = 1, \ldots, d$$

be one step of Newton iteration on these roots. Assume that $u(\tilde{z}_i) \neq u(\tilde{z}_j)$ for $i \neq j$. Then $\tilde{Q}(T), \tilde{V}_1(T), \ldots, \tilde{V}_n(T)$ defined in Definition 6 is the exact RUR of $\{\tilde{z}_1, \ldots, \tilde{z}_d\}$, with $\sum_{i=1}^n \lambda_i \tilde{V}_i(T) = T$.

Proof. See [17] for the proof. □

3.3 Second Construction

Our second variation of Definition 3 will have the property that it can be interpreted as an $(n+1)d$ dimensional Newton iteration as follows. Given $\boldsymbol{F} = (F_1, \ldots, F_n)$ and $u = \sum_{i=1}^n \lambda_i x_i$ in $R[x_1, \ldots, x_n]$ as before, we define the map

$$\Phi : R^{(n+1)d} \to R^{(n+1)d}$$

as the map of the coefficient vectors of the following degree $d-1$ polynomials:

$$\Phi : \begin{bmatrix} v_1(T) \\ \vdots \\ v_n(T) \\ \Delta q(T) \end{bmatrix} \mapsto \begin{bmatrix} F_1(\mathbf{v}(T)) \bmod q(T) \\ \vdots \\ F_n(\mathbf{v}(T)) \bmod q(T) \\ \sum_{i=1}^n \lambda_i v_i(T) - T \end{bmatrix}, \tag{9}$$

where

$$\Delta q(T) := q(T) - T^d.$$

If $u, q(T), v_1(T), \ldots, v_n(T)$ is an exact RUR of a component of $\langle \boldsymbol{F} \rangle$ then

$$\Phi(v_1(T), \ldots, v_n(T), \Delta q(T)) = 0.$$

So one can apply the $(n+1)d$ dimensional Newton iteration to locally converge to the coefficient vector of an exact RUR which is a zero of Φ. Note that below we will consider the map Φ as a map between

$$\Phi : (R[T]/\langle q(T) \rangle)^{n+1} \to (R[T]/\langle q(T) \rangle)^{n+1},$$

and note that $(R[T]/\langle q(T) \rangle)^{n+1}$ and $R^{(n+1)d}$ are isomorphic as vectors spaces when $R = \mathbb{K}$ a field. Moreover, as we will see below, the Newton iteration for Φ respects the algebra structure of $(R[T]/\langle q(T) \rangle)^{n+1}$ as well.

The first lemma gives the Jacobian matrix of Φ.

Lemma 10. *Let $\boldsymbol{F} = (F_1, \ldots, F_n)$, u, $q(T)$, $\mathbf{v}(T)$ and Φ be as above. For $i = 1, \ldots, n$ define $m_i(T)$ and $r_i(T)$ as the quotient and remainder in the division with remainder:*

$$F_i(\mathbf{v}(T)) = m_i(T)q(T) + r_i(T). \tag{10}$$

Then the Jacobian matrix of Φ defined in (9) and considered as a map on $(R[T]/\langle q(T) \rangle)^{n+1}$ is given by

$$J_\Phi(\mathbf{v}(T), \Delta q(T)) := \begin{array}{c} \overset{\displaystyle n \qquad\qquad 1}{\left[\begin{array}{ccc|c} & & & -m_1(T) \\ & J_{\boldsymbol{F}}(\mathbf{v}(T)) & & \vdots \\ & & & -m_n(T) \\ \hline \lambda_1 & \cdots & \lambda_n & 0 \end{array}\right]} \begin{array}{l} \\ n \\ \\ 1 \end{array} \end{array} \bmod q(T). \tag{11}$$

Proof. See [17] for the proof.   □

Next, we give explicitly the iteration function corresponding to the Newton iteration on Φ, using polynomial arithmetic modulo $q(T)$. We need the following:

Assumption 11. *Let* F, $u = \sum_{i=1}^{n} \lambda_i x_i$, $q(T)$ *and* $\mathbf{v}(T)$ *polynomials over some Euclidean domain* R *as above. In addition to the five conditions of Assumption 5, we also assume that*

6. $J_\Phi := J_\Phi(\mathbf{v}(T), \Delta q(T))$ *defined in (11) is invertible modulo* $q(T)$.

Definition 12. *Let* F, $u = \sum_{i=1}^{n} \lambda_i x_i$, $q(T)$ *and* $\mathbf{v}(T)$ *polynomials over* R *satisfying Assumption 11. Then we define*

$$\mathbf{w}(T) := \mathbf{v}(T) - \left(J_F(\mathbf{v}(T))\right)^{-1} F(\mathbf{v}(T)) \mod q(T),$$

$$\Delta(T) := \sum_{i=1}^{n} \lambda_i w_i(T) - T \text{ same as in Definitions 3 and 6}, \tag{12}$$

$$r(T) := F(\mathbf{v}(T)) \mod q(T) \tag{13}$$

$$U(T) := \frac{\partial \mathbf{v}(T)}{\partial T} - \left(J_F(\mathbf{v}(T))^{-1} \frac{\partial r(T)}{\partial T} \mod q(T)\right), \tag{14}$$

$$\Lambda(T) := \sum_{i=1}^{n} \lambda_i U_i(T) \text{ that we will show to be invertible mod } q(T) \tag{15}$$

$$\bar{\mathbf{V}}(T) := \mathbf{w}(T) - \left(\frac{\Delta(T)}{\Lambda(T)} U(T) \mod q(T)\right), \tag{16}$$

$$\bar{Q}(T) := q(T) - \left(\frac{\Delta(T)}{\Lambda(T)} \frac{\partial q(T)}{\partial T} \mod q(T)\right). \tag{17}$$

Remark 13. Note that if $R = \mathbb{Q}[t]$ and $I = \langle t^k \rangle$ for some $k \geq 1$ then

$$\frac{\Delta(T)}{\Lambda(T)} \equiv \Delta(T) \text{ and } U(T) \equiv \frac{\partial \mathbf{w}(T)}{\partial T} \mod q(T) \mod I^2,$$

thus our second construction is equivalent to the one in Definition 3. However, when our coefficient ring R is a field $\mathbb{K} \subset \mathbb{C}$, the polynomials in $U(T)$ are not the partial derivatives of the ones in $\mathbf{w}(T)$, so Definition 12 defines a different iteration from the one in Definition 3. Moreover, in general $\bar{\mathbf{V}}(T+\Delta(T)) \not\equiv \mathbf{w}(T)$ mod $q(T)$, this construction also differ from Definition 6.

The next proposition shows that $\bar{\mathbf{V}}(T)$ and $\bar{Q}(T)$ from Definition 12 are the Newton iterates for the function Φ.

Proposition 14. *Let* F, u, $q(T)$, $\mathbf{v}(T)$ *and* Φ *be such that Assumption 11 holds. Then* $\Lambda(T)$ *defined in (15) is invertible modulo* $q(T)$, *and thus* $\bar{\mathbf{V}}(T)$ *and* $\bar{Q}(T)$ *are well defined in Definition 12. Furthermore*

$$\begin{bmatrix} \bar{\mathbf{V}}(T) \\ \bar{Q}(T) - T^d \end{bmatrix} = \begin{bmatrix} \mathbf{v}(T) \\ q(T) - T^d \end{bmatrix} - J_\Phi^{-1} \cdot \begin{bmatrix} F(\mathbf{v}(T)) \\ \sum_{i=1}^{n} \lambda_i v_i(T) - T \end{bmatrix} \mod q(T), \tag{18}$$

where the vector on the right hand side is $\Phi(\mathbf{v}(T), q(T) - T^d)$. *Finally, we also have that* $\sum_{i=1}^{n} \lambda_i \bar{V}_i(T) = T$.

Proof. See [17] for the proof.   □

4   Example: A Cubic-Centered 12-Bar Linkage

To illustrate the application of these techniques, we compute an RUR for a rational component of a square system to prove that it solves an overdetermined system of equations arising from a 12-bar spherical linkage. The overdetermined polynomial system G consists of 17 quadratic and 2 linear polynomials in 18 variables for the linkage first described in [39], which is presented in [38, Fig. 3]. The trivial rotation of the cube is removed by placing the center of the cube at the origin and fixing two adjacent vertices, say P_7 and P_8, at $(-1, 1, -1)$ and $(-1, -1, -1)$, respectively. The 18 variables are the coordinates of the six remaining vertices of the cube, say $P_1, \dots, P_6$ with $P_i = (P_{ix}, P_{iy}, P_{iz})$. The 17 quadratic conditions force these free vertices to maintain their relative distances:

$$\|P_i - P_j\|^2 - 4 = 0, \; (i,j) \in \left\{ \begin{array}{c} (1,2), (3,4), (5,6), (1,5), (2,6), (3,7), \\ (4,8), (1,3), (2,4), (5,7), (6,8) \end{array} \right\}$$

$$\|P_i\|^2 - 3 = 0, \qquad i = 1, \dots, 6.$$

The irreducible components of these 17 quadratic polynomials was first described in [16, Table 1]. This decomposition shows that there is a unique irreducible surface S of degree 16, which is the current focus of study. In particular, the rational component is the 16 points arising from the intersection of S with the codimension two linear space defined by:

$$P_{3x} + P_{4x} + P_{2z} = 0,$$
$$P_{5x} - P_{6x} + P_{3y} + P_{3z} + 1 = 0.$$

In order to compute an RUR for this rational component, we consider the square polynomial system F consisting of these two linear equations and 16 quadratic equations obtained by adding the last quadratic above, i.e., $\|P_6\|^2 - 3$, to the other sixteen. Starting with a witness set for S, we used `Bertini` [4] to compute numerical approximations of the 16 points of interest. From these points, we observe that the variable P_{6z} is distinct, so we take the primitive element $u(P_1, \dots, P_6) = P_{6z} = T$. Next, we produce an initial guess for the monic univariate polynomial $q(T)$ of degree 16. Since $q(T)$ naturally has small integer coefficients, this polynomial was computed exactly:

$$q(T) = T^{16} + 20T^{15} + 210T^{14} + 1230T^{13} + 4212T^{12} + 4677T^{11} - 6886T^{10} - 21389T^9 + 58242T^8$$
$$- 45269T^7 - 6118T^6 + 58968T^5 - 103014T^4 + 119847T^3 - 91281T^2 + 40466T - 8291.$$

We produced an initial guess for the univariate polynomials $\mathbf{v}(T)$ of degree at most 15 via Lagrange interpolation using the computed numerical approximations. These are polynomials with rational coefficients having at most 5 digit numerators and denominators, which are presented in [17].

We refined the approximate RUR using Algorithm 15, and in 3 iterations (in roughly 1 second) we found the exact RUR. In the $\mathbf{v}(T)$ polynomials of the exact

RUR the numerators and denominators of the coefficients have at most 28 digits, for example (the full description of the exact RUR can be found in [17]):

$$\alpha = 1/32044717732213692797906558525$$
$$v_{3x}(T) = \alpha(-28811294935936308656103297T^{15} - 56469358709164889119641644T^{14}$$
$$- 5784420480830153173904226597T^{13} - 3227775460749576025678391459T^{12}$$
$$- 99098949465871882288837195827T^{11} - 3578358749346900975113448620T^{10}$$
$$+ 4426015108420575550019096058 9T^{9} + 8473157760188112871101856542 0T^{8}$$
$$- 1994911653787808024645153051 88T^{7} + 9572322983833968142397131457 8T^{6}$$
$$+ 7578713094175159648709310599 5T^{5} - 1825484700326150204205233749 37T^{4}$$
$$+ 2929594970035001755344520998 849T^{3} - 3296430424768572816050693148 89T^{2}$$
$$+ 1966741253646013620851198100 25T - 47452126308845628915580789974)$$

We checked $\boldsymbol{F}(\mathbf{v}(T)) \equiv 0 \mod q(T)$ and $\boldsymbol{G}(\mathbf{v}(T)) \equiv 0 \mod q(T)$ using exact arithmetics, meaning that we found an exact rational component of our zero dimensional overdetermined system.

5   Parallel Complexity

In this section, we study two parallel algorithms for our two constructions in Definitions 6 and 12, and analyze their parallel complexity. We express our complexity results as functions of the number of variables n and the number of roots d. In many applications, the number of roots d is large, possibly being an exponential function of n. Our goal is to demonstrate that we can efficiently distribute our computations to polynomially many processors in n and d so that the parallel computational time is polynomial in $\log(d)$ and n.

We present our analysis using the PRAM (Parallel Random Access Machine) arithmetic model of parallel computing [26], in which we more conveniently expose our complexity estimates, but we also cover them in terms of basic operations (like FFT), which are efficient under any reasonable model. $O_A(t, p)$ will denote the simultaneous upper bounds $O(t)$ on the parallel arithmetic time, and $O(p)$ on the number of processors involved.

5.1   Parallel Complexity of the First Construction from the Roots

By Proposition 9 the iterates of Definition 6 are the same as the Lagrange interpolants of the approximate roots obtained from one step local Newton iteration. We assume here that the coordinates of the approximate roots are given as floating point complex numbers, thus our base field is $\mathbb{K} := \mathbb{Q}(i)$. Below we give estimates on the parallel complexity of the following simple algorithm:

Algorithm 15. Computation of RUR from roots.

INPUT: *A primitive element* $u = \lambda_1 x_1 + \ldots + \lambda_n x_n$ *and approximate roots* $\boldsymbol{z}_1, \ldots, \boldsymbol{z}_d \in \mathbb{K}^n$. *We assume that the corresponding RUR* $u, q(T), \mathbf{v}(T)$ *satisfies Assumption 5 (but need not to be given explicitly).*
OUTPUT: *The updated RUR* $\tilde{Q}(T), \tilde{\boldsymbol{V}}(T)$ *defined in Definition 6, and its common roots* $\tilde{\boldsymbol{z}}_1, \ldots, \tilde{\boldsymbol{z}}_d \in \mathbb{K}^n$.

COMPUTATIONS:

1. *Compute* $\tilde{z}_i := z_i - J_F(z_i)^{-1} F(z_i)$ $i = 1, \ldots, d$
2. *Compute* $\tilde{Q}(T) := \prod_{i=1}^{d}(T - u(\tilde{z}_i))$
3. *Interpolate the polynomials* $\tilde{V}_1(T), \ldots, \tilde{V}_n(T)$ *such that* $\tilde{V}_j(u(\tilde{z}_i)) = \tilde{z}_{i,j}$
   for $i = 1, \ldots, d$ *and* $j = 1, \ldots n$.

We get the following complexity bounds for Algorithm 15:

Proposition 16. *Given* $u = \sum_{i=1}^{n} \lambda_i x_i$, *and* $z_1, \ldots, z_d \in \mathbb{K}^n$ *satisfying the assumptions of Algorithm 15. Then we can compute the polynomials* $\tilde{Q}(T), \tilde{V}(T)$ *of Definition 6 and the corresponding approximate roots* $\tilde{z}_1, \ldots, \tilde{z}_d \in \mathbb{K}^n$ *of* F *in two stages with respective costs*

$$O_A(\log^2(n), n^{\omega+1}) \text{ and } O_A(\log^2(d) \log^*(nd), nd/\log^*(d)).$$

Here $2 \leq \omega \leq 2.373$ *is the exponent of matrix multiplication and* $\log^*(d)$ *is the usual iterated logarithm function.*

Proof. See [17] for the proof.

5.2   Parallel Complexity of the Second Construction Using Modular Arithmetics

In this subsection, we analyze the parallel complexity of the iteration defined in Definition 12.

The bottleneck of the iteration defined in Definition 12 is the computation of the two modular inverses mod $q(T)$, especially if $d = \deg q(T)$ is large in comparison to n. We use Toeplitz-like linear solvers to compute modular inverses. In the full version of this paper [17] we describe an algorithm for the solution of a general Toeplitz-like linear system of equations, which refines and modifies the algorithm of [30] by using more efficient displacement representation with factor circulant matrices (see [31, Example 4.4.2]) rather than triangular Toeplitz matrices. Here we give a brief summary of the results without presenting the details.

To compute $p^{-1}(T) \mod q(T)$ for some $p(T) \in \mathbb{K}[T]$ of degree at most $d - 1$, relative prime to q, it is sufficient to compute polynomials s and t of degree at most $d - 1$ and $d - 2$ respectively, such that $sp + tq = 1$. If we define the Sylvester matrix $S := Sylv(p, q)$, then the coefficients of s and t comprise the vector $S^{-1}e$ where e is the coefficient vector of 1. We use the following algorithm to compute $S^{-1}e$ (see [30] or [17] for more details):

Algorithm 17. Structured LIN·SOLVE.

INPUT: S *an* $m \times m$ *nonsingular matrix and* e *a vector of size* m.
OUTPUT: *the vector* $S^{-1}e$.

COMPUTATIONS:

1. Let $A := I - \lambda S$ be the matrix polynomial, with I the identity matrix and λ a parameter, and compute the displacement generators of A (see [17])
2. Compute the displacement generators of $A^{-1} \bmod \lambda^m$ and $A^{-1}\mathbf{e} \bmod \lambda^m$ using Parametrized Newton's iteration as follows:
   (a) Let $X_0 := I$ and $p := \lceil \log_2(m) \rceil$.
   (b) For $i = 1, \ldots, p$ compute $X_i := X_{i-1}(2I - AX_{i-1}) \bmod \lambda^{2^i}$.
   (c) Output $A^{-1} \equiv X_p \bmod \lambda^m$ and $A^{-1}\mathbf{e} \equiv X_p\mathbf{e} \bmod \lambda^m$.
3. Compute the traces of the matrices S^i, $i = 1, 2, \ldots, m-1$, as the coefficients of the trace of the matrix $A^{-1} \bmod \lambda^m$.
4. Compute the coefficients $c_0, \ldots, c_{n-1}$ of the characteristic polynomial $\sum_{j=0}^{m} c_i \lambda^i = \det(\lambda I - S)$.
5. Note that $c_0 \neq 0$, write $c_n = 1$, and compute and output vector $S^{-1}\mathbf{e} = -\sum_{i=1}^{m}(c_i/c_0)S^{i-1}\mathbf{e}$.

The next proposition gives the parallel complexity of the computation of modular inverses using Algorithm 17:

Proposition 18. *Let $q(T) \in \mathbb{K}(T)$ be degree d and $p(T) \in \mathbb{K}(T)$ be degree at most $d - 1$ that is relatively prime to $q(T)$. Then we can compute $p^{-1}(T)$ mod $q(T)$ in parallel complexity $O_A(\log^2(d), d^2/\log(d))$.*

Proof. In [17] we prove that for an $m \times m$ matrix S of displacement rank r using the circulant representation of [31, Example 4.4.2], all matrices appearing in Algorithm 17 have circulant displacement rank at most $r+1$, and the asymptotic computational cost is dominated by the cost of the $\lceil \log(m) \rceil$ iterations within Step 2. Under this displacement representation, the ith step of Step 2(b) is equivalent to multiplying $O(r^2)$ bivariate polynomials of degree less than $2m$ in one variable and of degree at most 2^i in the other. These multiplications can be reduced essentially to performing $O(\log(m))$ two-dimensional discrete Fourier transforms at at most m^2 knots. Under the arithmetic PRAM model the above computations require at most $O(\log^2(m))$ time using $O(r^2m^2/\log(m))$ arithmetic processors (cf. [5, ch.4]). If $S = Sylv(p, q)$ then $m \leq 2d - 1$ and the displacement rank of S is at most 2, using either the factor circulant or the triangular Toeplitz model. This proves the claim. $\qquad \square$

Besides modular inverses, the computation of the polynomials in Definition 12 is dominated by the computation of the adjoint of the polynomial matrix $J_{\mathbf{F}}(\mathbf{v}(T))$ modulo $q(T)$. We can assume that all polynomials in the polynomial arithmetics involved as well as our input polynomials in $\mathbf{F}$ have degree at most $2d$. Then, according to [5, page 311], the parallel complexity of division with remainder using degrees at most $2d$ and d polynomials is $O_A(\log(d)\log^*(d), d/\log^*(d))$. Moreover, using [5, page 319], we can compute the adjoint (and the inverse) of an $n \times n$ scalar matrix in $O_A(\log^2(n), n^{\omega+1})$. Thus the adjoint of $J_{\mathbf{F}}(\mathbf{v}(T))$ modulo $q(T)$ can be computed in

$$O_A(\log(d)\log^2(n)\log^*(d), n^{\omega+1}d/\log^*(d)).$$

Combining all the above, we get the following proposition. The most significant difference between its complexity bounds and the ones in Proposition 16 is the extra d factor in the required number of processors.

Proposition 19. *Assume that we are given F, u, $q(T)$ and $\mathbf{v}(T)$ satisfying Assumption 11. Assume further that the polynomials in F have degree at most $2d$. Then we can compute the polynomials $\bar{Q}(T)$, $\bar{V}(T) = (\bar{V}_1(T), \ldots, \bar{V}_n(T))$ of Definition 12 with the cost $O_A(\log^2(n)\log^2(d)\log^*(d), n^{\omega+1}d^2/\log^*(d))$.*

References

1. Abbott, J., Fassino, C., Torrente, M.-L.: Stable border bases for ideals of points. J. Symbolic Comput. 43(12), 883–894 (2008)
2. Akoglu, T.A., Hauenstein, J.D., Szanto, A.: Certifying solutions to overdetermined and singular polynomial systems over Q. (2013) (manuscript)
3. Avendaño, M., Krick, T., Pacetti, A.: Newton-Hensel interpolation lifting. Found. Comput. Math. 6(1), 81–120 (2006)
4. Bates, D.J., Hauenstein, J.D., Sommese, A.J., Wampler, C.W.: Bertini: software for numerical algebraic geometry, `bertini.nd.edu`
5. Bini, D., Pan, V.Y.: Polynomial and matrix computations. Progress in Theoretical Computer Science, vol. 1. Birkhäuser Boston Inc., Boston (1994)
6. Blum, L., Cucker, F., Shub, M., Smale, S.: Complexity and real computation. Springer, New York (1998)
7. Castro, D., Pardo, L.M., Hägele, K., Morais, J.E.: Kronecker's and Newton's approaches to solving: a first comparison. J. Complexity 17(1), 212–303 (2001)
8. Chistov, A.L.: An algorithm of polynomial complexity for factoring polynomials, and determination of the components of a variety in a subexponential time. Zap. Nauchn. Sem. Leningrad. Otdel. Mat. Inst. Steklov. (LOMI) 137, 124–188 (1984)
9. Fassino, C.: Almost vanishing polynomials for sets of limited precision points. J. Symbolic Comput. 45(1), 19–37 (2010)
10. Fassino, C., Torrente, M.-L.: Simple varieties for limited precision points. Theoret. Comput. Sci. 479, 174–186 (2013)
11. Ferguson, H.R.P., Bailey, D.H., Arno, S.: Analysis of PSLQ, an integer relation finding algorithm. Math. Comp. 68(225), 351–369 (1999)
12. Giusti, M., Heintz, J., Hägele, K., Morais, J.E., Pardo, L.M., Montaña, J.L.: Lower bounds for Diophantine approximations. J. Pure Appl. Algebra 117/118, 277–317 (1997)
13. Giusti, M., Heintz, J., Morais, J.E., Morgenstern, J., Pardo, L.M.: Straight-line programs in geometric elimination theory. J. Pure Appl. Algebra 124, 101–146 (1998)
14. Giusti, M., Lecerf, G., Salvy, B.: A Gröbner free alternative for polynomial system solving. J. Complexity 17(1), 154–211 (2001)
15. Grigorév, D.Y.: Factoring polynomials over a finite field and solution of systems of algebraic equations. Zap. Nauchn. Sem. Leningrad. Otdel. Mat. Inst. Steklov. (LOMI) 137, 20–79 (1984)
16. Hauenstein, J.D.: Numerically computing real points on algebraic sets. Acta Appl. Math. 125(1), 105–119 (2013)

17. Hauenstein, J.D., Pan, V.Y., Szanto, A.: Global Newton Iteration over Archimedean and non-Archimedean Fields - Full Version. arXiv:1404.5525
18. Hauenstein, J.D., Sottile, F.: Algorithm 921: alphaCertified: certifying solutions to polynomial systems. ACM Trans. Math. Software 38(4), Art. ID 28, 20 (2012)
19. Heintz, J., Krick, T., Puddu, S., Sabia, J., Waissbein, A.: Deformation techniques for efficient polynomial equation solving. J. Complexity 16(1), 70–109 (2000)
20. Heldt, D., Kreuzer, M., Pokutta, S., Poulisse, H.: Approximate computation of zero-dimensional polynomial ideals. J. Symbolic Comput. 44(11), 1566–1591 (2009)
21. Jeronimo, G., Krick, T., Sabia, J., Sombra, M.: The computational complexity of the Chow form. Found. Comput. Math. 4(1), 41–117 (2004)
22. Jeronimo, G., Perrucci, D.: On the minimum of a positive polynomial over the standard simplex. J. Symbolic Comput. 45(4), 434–442 (2010)
23. Kaltofen, E.: Polynomial-time reductions from multivariate to bi- and univariate integral polynomial factorization. SIAM J. Comput. 14(2), 469–489 (1985)
24. Kaltofen, E.: Sparse Hensel lifting. In: Caviness, B.F. (ed.) GI-Fachtagung 1973. LNCS, vol. 204, pp. 4–17. Springer, Berlin (1985)
25. Kannan, R., Lenstra, A.K., Lovász, L.: Polynomial factorization and nonrandomness of bits of algebraic and some transcendental numbers. Math. Comp. 50(181), 235–250 (1988)
26. Karp, R., Ramachandran, V.: Parallel algorithms for shared-memory machines. In: Handbook of Theoretical Computer Science, vol. A, pp. 869–941. Elsevier (1990)
27. Lenstra, A.K., Lenstra Jr., H.W., Lovász, L.: Factoring polynomials with rational coefficients. Math. Ann. 261(4), 515–534 (1982)
28. Lichtblau, D.: Exact computation using approximate Gröbner bases. available in the Wolfram electronic library
29. Mourrain, B., Trébuchet, P.: Stable normal forms for polynomial system solving. Theoret. Comput. Sci. 409(2), 229–240 (2008)
30. Pan, V.Y.: Parametrization of Newton's iteration for computations with structured matrices and applications. Comput. Math. Appl. 24(3), 61–75 (1992)
31. Pan, V.Y.: Structured Matrices and Polynomials: Unified Superfast Algorithms. Birkhäuser/Springer, Boston/New York (2001)
32. Rouillier, F.: Solving zero-dimensional systems through the rational univariate representation. Journal of Applicable Algebra in Engineering, Communication and Computing 9(5), 433–461 (1999)
33. Shirayanagi, K.: An algorithm to compute floating point Groebner bases. In: Mathematical Computation with Maple V: Ideas and Applications, Ann Arbor, MI, pp. 95–106. Birkhäuser Boston, Boston (1993)
34. Shirayanagi, K.: Floating point Gröbner bases. Math. Comput. Simulation 42(4-6), 509–528 (1996); Symbolic Computation, New Trends and Developments (Lille, 1993)
35. Stetter, H.J.: Numerical polynomial algebra. Society for Industrial and Applied Mathematics (SIAM), Philadelphia (2004)
36. Traverso, C., Zanoni, A.: Numerical stability and stabilization of Groebner basis computation. In: Proceedings of the 2002 International Symposium on Symbolic and Algebraic Computation, pp. 262–269 (electronic). ACM, New York (2002)
37. Trinks, W.: On improving approximate results of buchberger's algorithm by newton's method. In: Caviness, B. (ed.) ISSAC 1985 and EUROCAL 1985. LNCS, vol. 204, pp. 608–612. Springer, Heidelberg (1985)

38. Wampler, C.W., Hauenstein, J.D., Sommese, A.J.: Mechanism mobility and a local dimension test. Mech. Mach. Theory 46(9), 1193–1206 (2011)
39. Wampler, C.W., Larson, B., Edrman, A.: A new mobility formula for spatial mechanisms. In: Proc. DETC/Mechanisms & Robotics Conf., September 4-7 (2007)
40. Winkler, F.: A p-adic approach to the computation of Gröbner bases. J. Symbolic Comput. 6(2-3), 287–304 (1988)
41. Zassenhaus, H.: On Hensel factorization. I. J. Number Theory 1, 291–311 (1969)

Invariant Manifolds in the Classic and Generalized Goryachev–Chaplygin Problem

Valentin Irtegov and Tatyana Titorenko

Institute for System Dynamics and Control Theory SB RAS,
134, Lermontov str., Irkutsk, 664033, Russia
irteg@icc.ru

Abstract. With the aid of computer algebra methods, we have conducted qualitative analysis of the phase space for the classic and generalized Goryachev–Chaplygin problem. In particular, we have found a series of new invariant manifolds of various dimension which possess some extremal property. Motions on a one-dimensional invariant manifold have been investigated. It was shown that these motions are asymptotically stable on this manifold, and one of equilibrium points on the manifold is a limit point for these motions.

Keywords: the Goryachev–Chaplygin problem, computer algebra, invariant manifolds, stability.

1 Introduction

In recent decades, when new methods for investigation of integrable systems of finite and infinite dimension have been developed (see, e.g., [9], [8] and references therein), an interest was resumed in the integrable problems of classic mechanics which are often a source of new ideas, analysis methods, and concrete applications. For example, the integrable problems of rigid body dynamics can be a base for constructing the integrable vector fields on sphere and ellipsoid [1]. In this way, one can obtain new interesting results even in the classic problems.

The topic of the present paper is qualitative analysis of the differential equations of motion for the classic and generalized Goryachev–Chaplygin problem with the aid of computer algebra (CA) tools. It is known, when initial data are arbitrary, the classic problem of motion of a rigid body with a fixed point in constant gravity field is integrated in quadratures in the Euler, Lagrange, and Kowalewski cases only. Under some restrictions imposed on initial data, new integrable systems appear, among which there exists the case found by Goryachev and Chaplygin [2], [4]. It takes place at the zeroth level of area integral.

The Goryachev–Chaplygin problem was studied in many works, only a small part of them is devoted to analysis of invariant manifolds (IMs), i.e., sets of nonzero-dimension composed of trajectories of motion equations. Finding and investigating IMs in the initial phase space of the problem, as a rule, requires bulky computations. In this work, we applied computer algebra system "Mathematica" as computing tool. We used a procedure for obtaining IMs together with

V.P. Gerdt et al. (Eds.): CASC Workshop 2014, LNCS 8660, pp. 218–229, 2014.

the first integrals of vector fields on these IMs [5]. The procedure is reduced to solving the stationary equations for a family of first integrals of the problem with respect to some part of phase variables and some part of parameters occurring in this family. Considering the first integrals, found in such a way, as IMs of a higher level, we can write down the equations of motion on these IMs and already solve the problem of finding IMs for these equations. The higher level IMs keep the property of invariance after "lifting up" them into the initial phase space. Using the above approach, we have found IMs for the classic and generalized Goryachev–Chaplygin problem, which are not described in the literature, and investigated their qualitative properties.

2   Formulation of the Problem

Let us consider the differential equations of the generalized Goryachev–Chaplygin problem:

$$\dot{M}_1 = (3M_3 + \lambda)M_2 - \frac{a\gamma_2}{\gamma_3^3}, \qquad \dot{\gamma}_1 = (4M_3 + \lambda)\gamma_2 - M_2\gamma_3,$$

$$\dot{M}_2 = -(3M_3 + \lambda)M_1 + \mu\gamma_3 + \frac{a\gamma_1}{\gamma_3^3}, \quad \dot{\gamma}_2 = -(4M_3 + \lambda)\gamma_1 + M_1\gamma_3,$$

$$\dot{M}_3 = -\mu\gamma_2, \qquad \dot{\gamma}_3 = M_2\gamma_1 - M_1\gamma_2. \qquad (1)$$

When $\lambda = 0, a = 0$, equations (1) describe the Goryachev–Chaplygin top, and the Goryachev–Chaplygin gyrostat when $\lambda \neq 0, a = 0$. In these cases, variables M_i ($i = 1, 2, 3$) are interpreted as the components of the kinetic moment vector, and γ_i ($i = 1, 2, 3$) as the direction cosines of the upward vertical; λ is the gyrostatic parameter, the parameter μ is proportional to the coordinate of the center of mass.

The case $\lambda \neq 0, a \neq 0$ corresponds to the generalized Goryachev–Chaplygin gyrostat. Some quantum analogue for this case has been constructed [6], and some quantum mechanical interpretation for the parameter a has been given therein.

The above equations have the following first integrals

$$2H = M_1^2 + M_2^2 + 4M_3^2 + 2\lambda M_3 + 2\mu\gamma_1 + \frac{a}{\gamma_3^2} = 2h, \qquad (2)$$

$$V_1 = M_1\gamma_1 + M_2\gamma_2 + M_3\gamma_3 = 0, \qquad (3)$$

$$V_2 = \gamma_1^2 + \gamma_2^2 + \gamma_3^2 = 1, \qquad (4)$$

$$F = (M_3 + \frac{\lambda}{2})(M_1^2 + M_2^2 + \frac{a}{\gamma_3^2}) - \mu M_1\gamma_3 = c_2. \qquad (5)$$

Note that the additional integral F for system (1) exists when the constant of the integral V_1 is equal to zero.

The goal of our work is to analyze the structure of the phase space of system (1), in particular, to find IMs, which possess some extremal properties, and to analyze them basing on these extremal properties. The class of such IMs is of interest by possibility to apply the 2nd Lyapunov method for the investigation of their stability.

3 Finding Invariant Manifolds

3.1 The Invariant Manifolds of the Generalized Goryachev–Chaplygin Problem

For qualitative analysis of the problem, in particular, for finding IMs of equations (1), we will use the Routh–Lyapunov method [7] and some of its extensions [5]. According to this method, the IMs of equations (1) can be obtained by solving the conditional extremum problem for the first integrals of these equations. For this purpose, a linear or nonlinear combination of the first integrals (a family of the first integrals) is constructed, and the necessary extremum conditions for this family with respect to phase variables are written. As a result, the problem of finding IMs for the differential equations is reduced to an algebraic problem.

Following the method chosen, we take the linear combination of first integrals (2)–(5)

$$K = \lambda_0 H - \lambda_1 V_1 - \frac{1}{2}\lambda_2 V_2 - \lambda_3 F, \tag{6}$$

and write down the necessary conditions for the integral K to have an extremum with respect to phase variables M_i, γ_i $(i = 1, 2, 3)$:

$$\partial K/\partial M_1 = \lambda_0 M_1 - \lambda_1 \gamma_1 - \lambda_3 (M_1 (2M_3 + \lambda) - \mu\gamma_3) = 0,$$
$$\partial K/\partial M_2 = \lambda_0 M_2 - \lambda_1 \gamma_2 - \lambda_3 M_2 (2M_3 + \lambda) = 0,$$
$$\partial K/\partial M_3 = \lambda_0 (4M_3 + \lambda) - \lambda_1 \gamma_3 - \lambda_3(M_1^2 + M_2^2 + \frac{a}{\gamma_3^2}) = 0,$$
$$\partial K/\partial\gamma_1 = \lambda_0\, \mu - \lambda_2\gamma_1 - \lambda_1 M_1 = 0, \quad \partial K/\partial\gamma_2 = -\lambda_2\gamma_2 - \lambda_1 M_2 = 0,$$
$$\partial K/\partial\gamma_3 = \lambda_3 \left(\frac{a\,(2M_3 + \lambda)}{\gamma_3^3} + \mu M_1\right) - \gamma_3\lambda_2 - \lambda_1 M_3 - \frac{a\lambda_0}{\gamma_3^3} = 0. \tag{7}$$

Stationary equations (7) are a system of rational equations with parameters $\lambda_i, \lambda, \mu, a$. We add expression (3) (the equation of connection) to these equations and, further, we will solve the problem of conditional extremum for the integral K.

The solutions of system (3), (7) in the case when its equations are dependent allow one to define the IMs of differential equations (1) which correspond to the family of the first integrals K. The dependence conditions of these equations can be obtained, e.g., as the conditions of vanishing system Jacobian. In some cases, it is more convenient to obtain both the desired dependence conditions and the IMs themselves by solving the stationary equations for a family of first integrals with respect to some part of phase variables and some part of family parameters. It is suitable, e.g., for the Goryachev–Chaplygin problem. There exist constraints on the constants of the first integrals and system (3), (7) is overdetermined.

We have transformed system (3), (7) into a polynomial one, and for the resulting system we have constructed a Gröbner basis (with "Mathematica" function *GroebnerBasis*), taking as unknowns $\gamma_1, \gamma_2, M_2, \lambda_0, \lambda_1, \lambda_2$. The lexicographic monomial ordering $\gamma_2 > \gamma_1 > M_2 > \lambda_2 > \lambda_1 > \lambda_0$ was used. The basis polynomials in the form of polynomial equations write

$$-\gamma_3^2 \lambda_0^3 + (M_3 + \lambda)\gamma_3^2 \lambda_3 \lambda_0^2 + ((2M_3 + \lambda)\gamma_3^2 M_3 + a)\lambda_3^2 \lambda_0 - (a\,(2M_3 + \lambda)$$
$$+\mu\,\gamma_3^3 M_1)\lambda_3^3 = 0, \quad \gamma_3 \lambda_3 \lambda_1 + (\lambda_0 - \lambda_3(2M_3 + \lambda))\lambda_0 = 0,$$
$$-\gamma_3^4 \lambda_3 \lambda_2 + (a\lambda_3 - \gamma_3^2 \lambda_0 M_3)(\lambda_3(2M_3 + \lambda) - \lambda_0) + \mu\,\gamma_3^3 \lambda_3^2 M_1 = 0, \qquad (8)$$
$$-\gamma_3^2 \lambda_3^2 M_2^2 + (\lambda_0^2 - \lambda_3^2 M_1^2 + 2\lambda_0 \lambda_3 M_3)\gamma_3^2 - a\lambda_3^2 = 0,$$
$$-(a\,(2M_3 + \lambda) + \mu\gamma_3^3 M_1)\,\lambda_3^2 M_1 \gamma_3 \gamma_1 + a(\lambda_0^2 - \lambda_3^2 M_1^2 + \lambda_0 \lambda_3 M_3)\,\gamma_3^2$$
$$+(\lambda_0 + \lambda_3 M_3)(\lambda_0 M_1 - \lambda_3 M_1(2M_3 + \lambda) + \mu\gamma_3 \lambda_3)\,\gamma_3^4 M_1 - a^2 \lambda_3^2 = 0,$$
$$-(a\,(2M_3 + \lambda) + \mu\,\gamma_3^3 M_1)\,\lambda_3^2 \gamma_2 + [(\lambda_0 + \lambda_3 M_3)(\lambda_0 - \lambda_3(\lambda + 2M_3))\,\gamma_3^2$$
$$-a\lambda_3^2]\,\gamma_3 M_2 = 0. \qquad (9)$$

Under the values of $\lambda_0, \lambda_1, \lambda_2$

$$\lambda_0 = \frac{1}{3}\lambda_3(M_3 + \lambda) - \frac{2^{1/3}\lambda_3^2\,[(M_3 + \lambda)^2\,\gamma_3^2 + 3((2M_3 + \lambda)\,\gamma_3^2 M_3 + a)]}{3\,\sigma_1}$$
$$-\frac{\sigma_1}{3\,2^{1/3}\,\gamma_3^2}, \qquad (10)$$
$$\lambda_1 = -\frac{\lambda_0 \rho_1}{\gamma_3 \lambda_3}, \quad \lambda_2 = \frac{\mu\lambda_3^2\gamma_3^3 M_1 - (a\lambda_3 - \lambda_0\gamma_3^2 M_3)\rho_1}{\lambda_3\gamma_3^4}, \qquad (11)$$

found from equations (8), equations (9) define the family of IMs of codimension 3 (λ_3 is the family parameter) of the initial differential equations. Here $\rho_1 = \lambda_0 - \lambda_3(2M_3 + \lambda)$, $\sigma_1 = \gamma_3 \lambda_3\,[(27\mu\gamma_3^3 M_1 - (5M_3 + 2\lambda)((M_3 + \lambda)(4M_3 + \lambda)\,\gamma_3^2 - 9a))\,\gamma_3 + \sqrt{\chi_1}]^{1/3}$.

It should be noted that excessive solutions can appear because of a transformation of the system of rational equations (3), (7) into a polynomial system. Hence, we need to verify all solutions obtained. For this purpose, we shall apply the IM definition: the derivative of IM expressions calculated due to the corresponding equations of motion must vanish on these expressions. In the case in question, it is more convenient to do the verification with the aid of maps of some atlas [3] on the elements of this IMs family, e.g., map 1

$$\gamma_1 = \frac{-a^2\lambda_3^2 + a\gamma_3^2(\lambda_0^2 - \lambda_3^2 M_1^2 + \lambda_0\lambda_3 M_3) + (\lambda_3 M_3 + \lambda_0)[\rho_1 M_1 + \mu\,\gamma_3\lambda_3]\,\gamma_3^4 M_1}{(a\,(2M_3 + \lambda) + \mu\,\gamma_3^3 M_1)\,\lambda_3^2\gamma_3 M_1},$$
$$\gamma_2 = \frac{[\,a\,\lambda_3^2 - \gamma_3^2(\lambda_3 M_3 + \lambda_0)\rho_1\,]\,\rho_2}{(a\,(2M_3 + \lambda) + \mu\gamma_3^3 M_1)\lambda_3^3}, \quad M_2 = -\frac{\rho_2}{\gamma_3\lambda_3} \qquad (12)$$

and map 2

$$\gamma_1 = \frac{-a^2\lambda_3^2 + a\gamma_3^2(\lambda_0^2 - \lambda_3^2 M_1^2 + \lambda_0\lambda_3 M_3) + (\lambda_3 M_3 + \lambda_0)[\rho_1 M_1 + \mu\,\gamma_3\lambda_3]\,\gamma_3^4 M_1}{(a\,(2M_3 + \lambda) + \mu\,\gamma_3^3 M_1)\,\lambda_3^2\gamma_3 M_1},$$
$$\gamma_2 = -\frac{[\,a\,\lambda_3^2 - \gamma_3^2(\lambda_3 M_3 + \lambda_0)\rho_1\,]\,\rho_2}{(a\,(2M_3 + \lambda) + \mu\gamma_3^3 M_1)\lambda_3^3}, \quad M_2 = \frac{\rho_2}{\gamma_3\lambda_3}. \qquad (13)$$

The maps were obtained from equations (9) by solving them with respect to variables γ_1, γ_2, M_2. Here $\rho_2 = \sqrt{\gamma_3^2(\lambda_0^2 - \lambda_3^2 M_1^2 + 2\lambda_0\lambda_3 M_3)} - a\lambda_3^2$, $\chi_1 = [(5M_3 + 2\lambda)((M_3+\lambda)(4M_3+\lambda)\gamma_3^2 - 9a)\gamma_3 - 27\mu\gamma_3^4 M_1]^2 - 4[\gamma_3^2(7M_3^2 + 5\lambda M_3 + \lambda^2) + 3a]^3$. For brevity, the expression for λ_0 (10) is not substituted into the above formulae.

The vector field on the elements of the family of IMs (9) in map 1 writes

$$\dot{M}_1 = -\left[\frac{3M_3 + \lambda}{\gamma_3\lambda_3} + \frac{a(a\lambda_3^2 - \gamma_3^2(\lambda_3 M_3 + \lambda_0)\rho_1)}{(a(\lambda + 2M_3) + \mu\gamma_3^3 M_1)\gamma_3^2\lambda_3^3}\right]\rho_2,$$

$$\dot{M}_3 = -\frac{(\mu\rho_1[a\lambda_3^2 - \gamma_3^2(\lambda_3 M_3 + \lambda_0)]\rho_2}{\lambda_3^3(a(2M_3 + \lambda) + \mu\gamma_3^3 M_1)},$$

$$\dot{\gamma}_3 = -\frac{[(\lambda_3 M_3 + \lambda_0)a\gamma_3^2\lambda_0 + (\lambda_3 M_3 + \lambda_0)\mu\gamma_3^5\lambda_3 M_1 - a^2\lambda_3^2]\rho_2}{(a(2M_3 + \lambda) + \mu\gamma_3^3 M_1)\gamma_3^2\lambda_3^3 M_1}. \qquad (14)$$

These equations were derived from equations (1) by eliminating variables γ_1, γ_2, M_2 from them with the aid of (12).

The expressions λ_0 (10), λ_2 (11) are the first integrals of differential equations (14). According to first integral definition, it can be verified by direct differentiation of these expressions by virtue of system (14).

The equations of the vector field on the elements of the family of IMs (9) in map 2 differ from equations (14) by signs before right-hand expressions: the minus sign should be instead of the plus sign, and vice versa. The relations λ_0 (10), λ_2 (11) are also the first integrals of these equations.

In order to take into account a constraint imposed on the constant of the integral V_2, it is sufficient to add this integral to equations (9). The resulting equations will define the family of IMs of codimension 4 for equations (1).

3.2   The Invariant Manifolds of the Classic Goryachev–Chaplygin Problem

Let us consider the problem of finding IMs for the motion equations of the Goryachev–Chaplygin gyrostat (when $a = 0$). In this case, stationary equations (7) become polynomial ones that considerably simplifies the problem.

Now we add the integral V_2 to equations (3) and (7). For the resulting system under the condition $a = 0$, we construct a Gröbner basis, taking as unknowns $M_1, M_2, \gamma_1, \gamma_3, \lambda_0, \lambda_1, \lambda_2$. The lexicographic monomial ordering $M_1 > M_2 > \gamma_3 > \gamma_1 > \lambda_2 > \lambda_1 > \lambda_0$ is used.

The basis polynomials are factorized, and they can be written in the form of 2 polynomial systems as follows.

The 1st subsystem:

$$\lambda_0 = 0, \ \lambda_1 = 0, \ \lambda_2 = 0, \qquad (15)$$

$$M_1 = 0, \ M_2 = 0, \ \gamma_3 = 0, \ \gamma_1^2 + \gamma_2^2 = 1. \qquad (16)$$

The 2nd subsystem:

$$\lambda_0^2 - \lambda_0\lambda_3\,(M_3 + \lambda) - \lambda_3^2\,[(2M_3 + \lambda)M_3 + \mu\,(\gamma_1 + 1)] = 0,$$
$$\lambda_1^2 + \mu\,\lambda_3\,(\lambda_0 M_3 - \lambda_3[\,(2M_3 + \lambda)M_3 + \mu\,(\gamma_1 + 1)]\,) = 0, \quad \lambda_2 + \mu\,\lambda_0 = 0, \quad (17)$$
$$\mu\,\lambda_3\,\gamma_2^2 + \mu\,\lambda_3\,\gamma_1\,(\gamma_1 + 1) + (\lambda_3\,(2M_3 + \lambda) - \lambda_0)M_3 = 0,$$
$$\mu\,\lambda_3\,\gamma_3 + \lambda_1 = 0, \quad \mu\,\lambda_3^2\,M_1 - \lambda_1\,(\lambda_3 M_3 + \lambda_0) = 0,$$
$$-\mu\,\lambda_3^2\,(\gamma_1 + 1)M_2 + \lambda_1\,(\lambda_3\,M_3 + \lambda_0)\,\gamma_2 = 0. \tag{18}$$

An analysis of the 1st subsystem has shown that equations (16) define IM of codimension 4 of the initial differential equations (where $a = 0$). The latter is verified by the IM definition. This IM possesses the extremal property: it delivers a stationary value to the integral F.

¿From a mechanical viewpoint, equations (16) describe pendulum-like oscillations of the gyrostat around a motionless axis of dynamic symmetry occupying a horizontal position. Indeed, the vector field on the IM writes

$$\dot{M_3} = -\mu\,\sqrt{1 - \gamma_2^2}, \quad \dot{\gamma_2} = (4M_3 + \lambda)\,\sqrt{1 - \gamma_2^2}. \tag{19}$$

These equations were derived from equations (1) by eliminating variables γ_1, γ_3, M_1, M_2 from them with the aid of (16). Equations (19) written in the Euler angles have the form: $\ddot{\varphi} + \mu\cos\varphi = 0$, $3\dot{\varphi} + \lambda = 0$.

An analysis of the 2nd subsystem has shown that under the values of $\lambda_0, \lambda_1, \lambda_2$

$$\lambda_0 = \frac{\lambda_3}{2}\,(M_3 + \lambda + \sqrt{(3M_3 + \lambda)^2 + 4\mu\,(\gamma_1 + 1)}), \quad \lambda_2 = -\mu\lambda_0,$$
$$\lambda_1 = \sqrt{\mu\,\lambda_3}\,\sqrt{\lambda_3\,[(2M_3 + \lambda)M_3 + \mu(\gamma_1 + 1)] - \lambda_0 M_3}, \tag{20}$$

found from equations (17), equations (18) define the family of IMs of codimension 4 of differential equations (1) (where $a = 0$). The latter is also verified by the IM definition with the aid of maps of some atlas on the elements of this IMs family, e.g., map 1

$$M_1 = \frac{(\lambda_3 M_3 + \lambda_0)\,\varrho_1}{\sqrt{\mu}\,\lambda_3^{3/2}}, \quad M_2 = \frac{(\lambda_3 M_3 + \lambda_0)\,\varrho_1\,\varrho_2}{\mu\,\lambda_3^2\,(\gamma_1 + 1)}, \quad \gamma_2 = \frac{\varrho_2}{\sqrt{\mu\lambda_3}}, \quad \gamma_3 = -\frac{\varrho_1}{\sqrt{\mu\lambda_3}} \tag{21}$$

and map 2

$$M_1 = \frac{(\lambda_3 M_3 + \lambda_0)\,\varrho_1}{\sqrt{\mu}\,\lambda_3^{3/2}}, \quad M_2 = -\frac{(\lambda_3 M_3 + \lambda_0)\,\varrho_1\,\varrho_2}{\mu\,\lambda_3^2\,(\gamma_1 + 1)}, \quad \gamma_2 = -\frac{\varrho_2}{\sqrt{\mu\lambda_3}}, \quad \gamma_3 = -\frac{\varrho_1}{\sqrt{\mu\lambda_3}}.$$

These maps were derived from equations (18) by solving them with respect to variables $M_1, M_2, \gamma_2, \gamma_3$. Here $\varrho_1 = \sqrt{\lambda_3\,((2M_3 + \lambda)M_3 + \mu\,(\gamma_1 + 1)) - \lambda_0 M_3}$, $\varrho_2 = \sqrt{\lambda_0 M_3 - \lambda_3\,[M_3(2M_3 + \lambda) + \mu\,(\gamma_1 + 1)\,\gamma_1]}$. Likewise above, for brevity, the expression for λ_0 was not substituted into the above formulae.

The vector field on the elements of the family of IMs (18) in map 1 writes

$$\dot{\gamma}_1 = \frac{\varrho_2}{\mu^{3/2} \lambda_3^{5/2} (\gamma_1 + 1)} \Big[M_3 (\lambda_3 M_3 + \lambda_0) (\lambda_3 (2M_3 + \lambda) - \lambda_0)$$

$$+ \mu \lambda_3 (\gamma_1 + 1) (\lambda_3 (5M_3 + \lambda) + \lambda_0) \Big], \quad \dot{M}_3 = -\frac{\sqrt{\mu}\, \varrho_2}{\sqrt{\lambda_3}}. \tag{22}$$

Expressions λ_0, λ_2 (20) are the first integrals of differential equations (22).

The equations of the vector field on the elements of the family of IMs (18) in map 2 differ from equations (22) by signs before right-handed expressions: the minus sign should be instead of the plus sign, and vice versa. The relations λ_0, λ_2 are also the first integrals of these equations.

So, we can assert.

Assertion 1. Solving the stationary equations of a family of first integrals with respect to some part of phase variables and some part of parameters occurring in this family allows one to obtain, as solutions of these equations, the equations of IMs for the initial equations of motion and the expressions for the parameters in the form of functions of phase variables. These functions are the first integrals of vector fields on the found IMs.

For the vector fields on the IMs we can again state the problem of finding IMs. Such IMs we will call second-level IMs.

3.3   Second-Level Invariant Manifolds

Now, let us consider the problem of obtaining the IMs for differential equations (14). These equations have first integrals λ_0 (10), λ_2 (11).

Within the Routh–Lyapunov method, any first integral defines a family of IMs of codimension-one for the corresponding differential equations. These IMs possess the extremal property since any first integral provides a stationary value to its square. Therefore, all the techniques for obtaining first integrals (including partial first integrals) can be regarded as procedures for finding IMs for a given system.

¿From aforesaid it follows that the integrals λ_0 (10), λ_2 (11) define the families of IMs for equations (14), i.e., they are second-level IMs.

The IMs found in such a way are stable in the Lyapunov sense. Consider, e.g., the family of IMs λ_0 (10). We introduce the deviation

$$y = \frac{1}{3} \lambda_3 (M_3 + \lambda) - \frac{2^{1/3} \lambda_3^2 [(M_3 + \lambda)^2 \gamma_3^2 + 3((2M_3 + \lambda) \gamma_3^2 M_3 + a)]}{3\,\sigma_1}$$

$$- \frac{\sigma_1}{3\, 2^{1/3} \gamma_3^2} - \lambda_0$$

of the perturbed motion from the unperturbed value of λ_0.

Considering y^2 as a Lyapunov's function, we can conclude on stability of the elements of the IMs family in question.

Second-level IMs can be "lifted up" into the initial phase space by a standard technique. To this end, in the case under consideration, we add equations (9) to equation (10) (or (11)). Next, we should add equation (4) to the latter equations in order to take into account the constraint imposed on the constant of the integral V_2. The resulting equations will define the family of one-dimensional IMs for equations (1) that is verified by the IM definition.

Analogously we can obtain IMs for differential equations (22). First integrals λ_0, λ_2 (20) will be the families of IMs for these equations. In this case, to "lift up" these families into the initial phase space, it is sufficient to add equations (18) to expression λ_0 (or λ_2). The resulting equations define the family of one-dimensional IMs for equations (1) (where $a = 0$) that is also verified by the IM definition. The equations of the given IMs family are presented below.

$$\frac{\lambda_3}{2}\left(M_3 + \lambda + \sqrt{(3M_3 + \lambda)^2 + 4\mu\,(\gamma_1 + 1)}\right) = \tilde{\lambda}_0,$$

$$\mu\,\lambda_3\,\gamma_2^2 + \mu\,\lambda_3\,\gamma_1\,(\gamma_1 + 1) + (\lambda_3\,(2M_3 + \lambda) - \lambda_0)M_3 = 0,$$

$$\mu\,\lambda_3\,\gamma_3 + \lambda_1 = 0, \quad \mu\,\lambda_3^2\,M_1 - \lambda_1\,(\lambda_3 M_3 + \lambda_0) = 0,$$

$$-\mu\,\lambda_3^2\,(\gamma_1 + 1)M_2 + \lambda_1\,(\lambda_3\,M_3 + \lambda_0)\,\gamma_2 = 0.$$

This family in one of maps written on its elements has the form

$$M_1 = \frac{\sqrt{\hat{\lambda}_0}\,(\lambda_3 M_3 + \hat{\lambda}_0)\sqrt{\hat{\lambda}_0 - \lambda_3(2M_3 + \lambda)}}{\sqrt{\mu}\,\lambda_3^2}, \quad M_2 = \frac{\sqrt{\hat{\lambda}_0}\,\varrho_3}{\sqrt{\mu}\,\lambda_3^2},$$

$$\gamma_1 = \frac{(\lambda_3 M_3 + \hat{\lambda}_0)\,(\hat{\lambda}_0 - \lambda_3(2M_3 + \lambda))}{\mu\lambda_3^2} - 1, \quad \gamma_2 = \frac{\sqrt{\hat{\lambda}_0 - \lambda_3\,(2M_3 + \lambda)}\,\varrho_3}{\mu\lambda_3^2},$$

$$\gamma_3 = -\frac{\sqrt{\hat{\lambda}_0}\,\sqrt{\hat{\lambda}_0 - \lambda_3\,(2M_3 + \lambda)}}{\sqrt{\mu}\lambda_3}.$$

Here $\varrho_3 = \sqrt{\mu\,\lambda_3^2\,(2\lambda_3 M_3 + \hat{\lambda}_0) - (\lambda_3 M_3 + \hat{\lambda}_0)^2(\hat{\lambda}_0 - \lambda_3(2M_3 + \lambda))}$, $\hat{\lambda}_0, \lambda_3$ are the family parameters.

4 About Motions on the Invariant Manifolds and Their Stability

Besides the problem of finding the IMs for equations (1), we used CA tools for a parametric analysis of these IMs and investigation of their stability. The analysis of the found IMs families for various values of parameters λ_i, λ, μ allowed us to isolate a one-dimensional IM, motions on which are asymptotically stable, and one of equilibrium points on the manifold is a limit point for these motions.

Let us consider one of the above families of IMs written in one of its maps:

$$\gamma_1 = \frac{\lambda_2 \left[2a\lambda_0\lambda_3^2 - \lambda_2\lambda_3^2\gamma_3^4 + \lambda_0^2(\lambda\lambda_3 - 3\lambda_0)\gamma_3^2\right]}{2\mu\,\lambda_0^4} + \frac{\mu\,\lambda_0}{\lambda_2},\ \gamma_2 = \frac{\hat{\rho}}{\lambda_0},$$

$$M_1 = \frac{\lambda_2\left[-2a\lambda_0\lambda_3^2 + \lambda_2\lambda_3^2\gamma_3^4 + \lambda_0^2(3\lambda_0 - \lambda\lambda_3)\gamma_3^2\right]}{2\mu\,\lambda_0^3\lambda_3\,\gamma_3},\ M_2 = -\frac{\hat{\rho}}{\lambda_3\,\gamma_3},$$

$$M_3 = \frac{1}{2}\left(\frac{\lambda_2\lambda_3\gamma_3^2}{\lambda_0^2} + \frac{\lambda_0}{\lambda_3} - \lambda\right). \tag{23}$$

Here

$$\hat{\rho} = \sqrt{\frac{(2\lambda_0^3 - \lambda\lambda_0^2\lambda_3 + \lambda_2\lambda_3^2\gamma_3^2)\,\gamma_3^2}{\lambda_0} + \frac{\lambda_2^2\left[2a\lambda_0\lambda_3^2 - \lambda_2\lambda_3^2\gamma_3^4 + \lambda_0^2(\lambda\lambda_3 - 3\lambda_0)\,\gamma_3^2\right]^2}{4\,\mu^2\,\tilde{\lambda}_0^6} - a\lambda_3^2},$$

$\lambda_0, \lambda_2, \lambda_3$ are the family parameters.

When $a = 0$, $3\lambda^2 = 16\,\mu$, equations (23) define the family of IMs for the motion equations of the Goryachev–Chaplygin gyrostat. Further, we shall show that under some restrictions imposed on parameters λ_i there exist asymptotically stable motions on the elements of the given IMs family.

When $\tilde{\lambda}_0 = -\lambda\lambda_3/6$, $\tilde{\lambda}_2 = \lambda^3\lambda_3/32$, the above equations have the form

$$\gamma_1 = -\frac{27\,\gamma_3^4}{8} + \frac{9\,\gamma_3^2}{2} - 1,\ \gamma_2 = -2\sqrt{2}\,(1 - \frac{9}{8}\,\gamma_3^2)^{3/2}\,\gamma_3,$$

$$M_1 = \frac{3\,\lambda}{4}\left(1 - \frac{3}{4}\,\gamma_3^2\right)\gamma_3,\ M_2 = -\frac{\sqrt{2}\,\lambda}{3}\,(1 - \frac{9}{8}\,\gamma_3^2)^{3/2},$$

$$M_3 = \frac{\lambda}{4}\left(\frac{9}{4}\,\gamma_3^2 - \frac{7}{3}\right), \tag{24}$$

and represent a submanifold of IMs family (23). The differential equation of the vector field on this IM writes

$$\dot{\gamma}_3 = \frac{\sqrt{2}\,\lambda}{3}\,(1 - \frac{9}{8}\,\gamma_3^2)^{3/2}, \tag{25}$$

and it can be integrated in elementary functions

$$\gamma_3(t) = \frac{2\sqrt{2}\,(\lambda t - 48\,C_1)}{3\sqrt{4 + (\lambda t - 48\,C_1)^2}}. \tag{26}$$

Here C_1 is a constant of integration.

Next, we set $\gamma_3(t_0) = \gamma_3^0$ in (26) and find C_1 as the function of γ_3^0:

$$C_1 = \frac{\lambda t_0}{48} - \frac{\gamma_3^0}{16\sqrt{2}\,\sqrt{1 - 9/8\,\gamma_3^{0^2}}}.$$

Taking into account the latter relation, we have

$$\gamma_3(t, \gamma_3^0) = \frac{2\,\sqrt{2}\,(3\gamma_3^0 + \sqrt{2}\,\lambda\,\sqrt{1 - 9/8\gamma_3^{0^2}}\,(t - t_0))}{3\sqrt{8\,(1 - 9/8\gamma_3^{0^2}) + \left(3\gamma_3^0 + \sqrt{2}\,\lambda\sqrt{1 - 9/8\gamma_3^{0^2}}(t - t_0)\right)^2}}. \tag{27}$$

One can see from (27) that functions $\gamma_3(t, \gamma_3^0)$ take real values when $|\gamma_3^0| \leq 2\sqrt{2}/3$. Thus, the motions on IM (24) are described by functions (27).

Let us investigate some properties of the functions under the above constraint on γ_3^0.

We conclude from (25) that there exist two equilibrium points on IM (24)

$$\gamma_3^0 = -\frac{2\sqrt{2}}{3}, \quad \gamma_3^0 = \frac{2\sqrt{2}}{3}. \tag{28}$$

With the aid of the "Mathematica" function "Limit", it is not difficult to show that all solutions originating within interval $|\gamma_3^0| < 2\sqrt{2}/3$ approach the 2nd equilibrium point of (28) when $t \to \infty$, i.e.

$$\lim_{t \to \infty} \gamma_3(t, \gamma_3^0) = \frac{2\sqrt{2}}{3}.$$

Next, we consider solution (27) as unperturbed one, and the solution

$$\tilde{\gamma}_3(t, \gamma_3^0) = \frac{2\sqrt{2}\,[3\,(\gamma_3^0 + \delta) + \sqrt{2}\,\lambda\,\sqrt{1 - 9/8\,(\gamma_3^0 + \delta)^2}\,(t - t_0)]}{3\,\sqrt{8\,(1 - 9/8\,(\gamma_3^0 + \delta)^2) + [3\,(\gamma_3^0 + \delta) + \sqrt{2}\,\lambda\,\sqrt{1 - 9/8\,(\gamma_3^0 + \delta)^2}\,(t - t_0)]^2}},$$

as perturbed one. Here $|\gamma_3^0 + \delta| \leq 2\sqrt{2}/3$, and δ is the arbitrary constant which characterizes a perturbation of the initial data γ_3^0.

As the limit of expression $\Delta = \tilde{\gamma}_3(t, \gamma_3^0) - \gamma_3(t, \gamma_3^0)$ is equal to zero as $t \to \infty$, and the derivative

$$\dot{\gamma}_3(t, \gamma_3^0) = \frac{8\sqrt{2}\,\lambda\,(8 - 9\gamma_3^{0^2})^{3/2}}{[3\,(\lambda(t - t_0)(\lambda(t - t_0)(8 - 9\gamma_3^{0^2}) + 12\gamma_3^0\sqrt{8 - 9\gamma_3^{0^2}}) + 32]^{3/2}}$$

of function (27) is a positive monotonically decreasing function $\forall\, t \geq t_0$ and $|\gamma_3^0| \leq 2\sqrt{2}/3$ then all solutions originating within the interval under consideration, as well as equilibrium point $\gamma_3^0 = 2\sqrt{2}/3$, are asymptotically stable.

4.1   About Stationary Solutions and Their Stability

Equilibrium points (28) correspond to the following solutions

$$\gamma_1 = \frac{1}{3},\ \gamma_2 = 0,\ \gamma_3 = -\frac{4}{3\sqrt{2}},\ M_1 = -\frac{\lambda}{3\sqrt{2}},\ M_2 = 0,\ M_3 = -\frac{\lambda}{12}; \tag{29}$$

$$\gamma_1 = \frac{1}{3},\ \gamma_2 = 0,\ \gamma_3 = \frac{4}{3\sqrt{2}},\ M_1 = \frac{\lambda}{3\sqrt{2}},\ M_2 = 0,\ M_3 = -\frac{\lambda}{12} \tag{30}$$

of differential equations (1) (where $a = 0$) in the initial phase space. We further investigate some properties of these solutions.

Solutions (29), (30) possess the extremal property: they deliver a stationary value to the integral K when $a = 0$, $\lambda_0 = -\lambda\lambda_3/6$, $\lambda_2 = \lambda^3\lambda_3/32$, $\lambda_1 = \sqrt{2}\lambda^2\lambda_3/8$ and $a = 0$, $\lambda_0 = -\lambda\lambda_3/6$, $\lambda_2 = \lambda^3\lambda_3/32$, $\lambda_1 = -\sqrt{2}\lambda^2\lambda_3/8$, respectively. Indeed, substitute expressions (29) (or (30)) and the values of the parameters corresponding to them, into stationary equations (7). The latter equations vanish.

The solutions in question are degenerate because the Jacobian of system (7) vanishes for these solutions. Mechanically, they define the permanent rotations of the gyrostat around the upward vertical when an axis of rotation in the body does not coincide with the principal inertia axes of the body.

Using integral K (6), we can investigate solutions (29), (30) for stability. For the equations of perturbed motion, the integral K in the vicinity of solution (29) will be

$$\Delta K = -\lambda_3 \left[\frac{\lambda^3 z_1^2}{64} + \frac{\lambda^3 z_2^2}{64} + \frac{\lambda^3 z_3^2}{64} + \frac{\sqrt{2}\lambda^2 z_1 z_4}{8} - \frac{3\lambda^2 z_3 z_4}{16} + \frac{1}{2}\lambda z_4^2 + \frac{\sqrt{2}\lambda^2 z_2 z_5}{8} \right.$$
$$\left. + \frac{1}{2}\lambda z_5^2 + \frac{\sqrt{2}\lambda^2 z_3 z_6}{8} - \frac{\sqrt{2}}{3}\lambda z_4 z_6 + \frac{1}{3}\lambda z_6^2 + (z_4^2 + z_5^2)z_6 \right]. \tag{31}$$

Here $z_1 = \gamma_1 - 1/3$, $z_2 = \gamma_2$, $z_3 = \gamma_3 + 4/(3\sqrt{2})$, $z_4 = M_1 + \lambda/(3\sqrt{2})$, $z_5 = M_2$, $z_6 = M_3 + \lambda/12$ are the deviations of perturbed motion from unperturbed one.

On the linear manifold

$$\delta H = \frac{3\lambda^2}{16}z_1 - \frac{\lambda z_4}{3\sqrt{2}} + \frac{2}{3}(\lambda + 3)z_6 = 0,$$

$$\delta V_1 = -\frac{1}{3}\left(\frac{\lambda z_1}{\sqrt{2}} + \frac{\lambda z_3}{4} - z_4 + \frac{4z_6}{\sqrt{2}}\right) = 0, \quad \delta V_2 = \frac{2}{3}(z_1 - 2\sqrt{2}\,z_3) = 0$$

the ΔK writes

$$2\Delta\tilde{K} = -\lambda\lambda_3 \left(\frac{27\,\lambda^2\,z_1^2}{32} + \xi^2\right),$$

where $\xi = \sqrt{2}\,\lambda\,z_2/8 + z_5$.

Since the above quadratic form is sign-definite for the variables appearing in it when $\lambda_3 \neq 0$ and $\lambda \neq 0$, these conditions are sufficient for the stability of solution (29) with respect to variables z_1, z_3, z_4, z_6, ξ. So, in the given case, we have proved stability with respect to a part of variables [10]. Analogous result has been obtained for solution (30).

5   Conclusion

With the aid of CA tools, we have conducted some analysis of the structure of the phase space for the classic and generalized Goryachev–Chaplygin problem. In particular, new families of IMs of various dimension have been found for these problems. Some of the IMs possess the extremal property: they deliver a stationary value to the problem first integrals.

A parametric analysis of the obtained IMs families allowed us to isolate a one-dimensional IM, on which there exist asymptotically stable motions, and one of equilibrium points on the manifold is a limit point for these motions. The equilibrium points on the IM correspond to permanent rotations in the initial phase space. The sufficient conditions of stability with respect to a part of the phase variables have been derived for the latter motions.

The presented results show that the approach used in this work for finding and for analysing IMs in combination with CA methods, in particular, Gröbner bases, allows one to solve such problems effectively.

The work was supported by the Presidium of the Russian Academy of Sciences, basic research program No. 17.1. This work was also partially financed by Grant Department of the President of the Russian Federation for the state support of leading scientific schools (grant No. 5007.2014.9).

References

1. Bolsinov, A.V., Kozlov, V.V., Fomenko, A.T.: The Maupertuis principle and geodesic flows on the sphere arising from integrable cases in the dynamics of a rigid body. Russian Math. Surveys. 50(3), 473–501 (1995)
2. Chaplygin, S.A.: A new particular solution of the problem of the rotation of a heavy rigid body about a fixed point. Trudy Otd. Fiz. Nauk Obshch. Lyubitelei Yestestvoznaniya 12(1), 1–4 (1904)
3. Godbillon, C.: Géometrie Différentielle et Mécanique Analytique, Collection Méthodes Hermann, Paris (1969)
4. Goryachev, D.N.: The motion of a heavy rigid body about a fixed point in the case when A=B=4C. Mat. Sbornik Kruzhka Lyubitelei Mat. Nauk. 21(3), 431–438 (1900)
5. Irtegov, V.D., Titorenko, T.N.: The invariant manifolds of systems with first integrals. J. Appl. Math. Mech. 73(4), 379–384 (2009)
6. Komarov, I.V., Kuznetsov, V.B.: The generalized Goryachev Chaplygin gyrostat in quantum mechanics. Differential Geometry, Lie Groups and Mechanics. Trans. of LOMI scientifc seminar USSR Acad. of Sciences 9, 134–141 (1987)
7. Lyapunov, A.M.: On Permanent Helical Motions of a Rigid Body in Fluid. Collected Works, vol. 1. USSR Acad. Sci., Moscow–Leningrad (1954)
8. Oden, M.: Rotating Tops: A Course of Integrable Systems. Udmurtiya univ., Izhevsk (1999)
9. Reyman, A.G., Semenov-Tian-Shansky, M.A.: Integrable Systems (Theoretic-group Approach). Institute of Computer Science, Izhevsk (2003)
10. Rumyantsev, V.V., Oziraner, A.S.: Motion Stability and Stabilization with Respect to Part of Variables. Nauka, Moscow (1987)

Coherence and Large-Scale Pattern Formation in Coupled Logistic-Map Lattices via Computer Algebra Systems

Maciej Janowicz and Arkadiusz Orłowski

Katedra Informatyki, Szkoła Główna Gospodarstwa Wiejskiego w Warszawie
ul. Nowoursynowska 159, 02-776 Warszawa, Poland
{maciej_janowicz,arkadiusz_orlowski}@sggw.pl

Abstract. Three quantitative measures of the spatiotemporal behavior of the coupled map lattices: reduced density matrix, reduced wave function, and an analog of particle number, have been introduced. Making extensive use of two computer algebra systems (Maxima and Mathematica) various properties of the above mentioned parameters have been thoroughly studied. Their behavior suggests that the logistic coupled-map lattices approach the states which resemble the condensed states of systems of Bose particles. In addition, pattern formation in two-dimensional coupled map lattices based on the logistic mapping has been investigated with respect to the non-linear parameter, the diffusion constant and initial as well as boundary conditions.

Keywords: coupled logistic-map lattices, Bose–Einstein condensation, pattern formation, computer algebra systems.

1 Introduction

Coupled map lattices (CMLs) [3,8] have long become a useful tool to investigate spatiotemporal behavior of extended and possibly chaotic dynamical systems [12,24,15,13]. It is so even though the most standard CML, that based on the coupling of logistic maps, is physically not particularly appealing as it is fairly remote from any model of natural phenomena. Other, more complicated CMLs, have found interesting applications in physical modeling. One should mention here CMLs developed to describe the Rayleigh–Benard convection [26], dynamics of boiling [25,4], formation and dynamics of clouds [27], crystal growth processes, and hydrodynamics of two-dimensional flows [14].

The most important characteristic quantities employed to study various types of CMLs include co-moving Lyapunov spectra, mutual information flow, spatiotemporal power spectra, Kolmogorov–Sinai entropy density, pattern entropy [14]. More recently, the detrended fluctuation analysis, structure function analysis, local dimensions, embedding dimension, and recurrence analysis have also been introduced for CMLs [17].

The purpose of this paper is to analyze the interesting features of the above-mentioned most standard coupled map lattices which resemble the characteristics

V.P. Gerdt et al. (Eds.): CASC Workshop 2014, LNCS 8660, pp. 230–241, 2014.

of the condensates of Bose particles as well as those associated with formation of patterns in two spatial dimensions. In particular, we investigate the formation of such patterns for relatively short times; their dependence on two parameters which define CML as well as the initial conditions is found numerically. Thus, the present work is very much in the spirit of classical papers [15,13,14]. We believe, however, that the subject is very far from being exhausted as it is quite easy to find interesting patterns not discovered in the above works. More importantly, we combine searching for interesting patterns with the introduction of three additional quantities with the help of which one can characterize the dynamics and statistical properties of CMLs. These are the reduced density matrix, the reduced wave function, and a quantity which is an analog of the number of particles. This is motivated, in part, by what we feel is the need to slightly deemphasize the connection of CMLs with finite-dimensional dynamical systems, and make their analysis similar to that of classical field theory, especially the Gross–Pitaevskii equation which is used in the physics of Bose–Einstein condensation [2,16]. Application of the classical field-theoretical methods in the physics of condensates have been described, e.g., in [5,6,22].

Many interesting patterns emerge in the system while it still exhibits a transient behavior as can be seen, e.g., in the plots of the "number of particles". We have not attempted here to reach the regime of stationary dynamics in each case. The problem for which times such a stationary regime becomes established is beyond the scope of this work. We are content with the transient regime as long as something interesting about the coherence properties and about the patterns can be observed. Let us notice that remarkable results on the transient behavior of extended systems with chaotic behavior have been obtained, e.g., in [23,9].

In addition, we observe that the condensate-like behavior has been reported in other systems which are not connected with many Bose particles. Of particular interest are the developments in the theory of complex networks [1,21]. Here, however, we explore the condensate-like behavior in the coupled map lattices.

The main body of this work is organized as follows. The mathematical model as well as the basic definitions of reduced density matrix and reduced wave function are introduced in Section 2. Section 3 provides a justification of our claim that the coupled map lattices based on logistic map exhibit properties which appear to be analogous to those of the Bose–Einstein condensates (BEC). The description of numerical results concerning pattern formation are contained in Section 4, while Section 5 comprises a few concluding remarks.

In performing the presented investigations we used extensively two computer algebra systems, namely Maxima and Mathematica. They provided excellent tools for both numerical simulations and graphical presentations. Due to the user-friendly environment as well as computational power they allowed to overcome the burden of testing many important special cases. They helped also to organize an enormous and complex set of interesting worth-to-study possibilities into a relatively coherent picture still staying in touch with quite nontrivial physical ideas that motivated this study.

2   The Model

Let us consider the classical field $\psi(x, y, t)$ defined on a two-dimensional spatial lattice. Its evolution in (dimensionless, discrete) time t is given by the following equation:

$$\psi(x, y, t+1) = (1 - 4d)f(\psi(x, y, t)) + d\,[f(\psi(x + 1, y, t)) + f(\psi(x - 1, y, t)) \\ + f(\psi(x, y + 1, t)) + f(\psi(x, y - 1, t))] \tag{1}$$

where the function f is given by:

$$f(\psi) = c\psi(1 - \psi), \tag{2}$$

and the parameters c and d are constant. The set of values taken by ψ is the closed interval $[0, 1]$. From the physical point of view the above diffusive model is rather bizarre, containing a field-dependent diffusion. There is no conserved quantity here which could play the role of energy or the number of excitations.

In the following the coefficient d will be called the "diffusion constant", and the coefficient c - the "non-linear parameter". It is assumed that ψ satisfies either the periodic boundary or Dirichlet (with $\psi = 0$) conditions on the borders of simulation box. The size of that box is $N \times N$. All our simulations have been performed with $N = 256$.

Let $\tilde{\psi}$ be the two-dimensional discrete Fourier transform of ψ,

$$\tilde{\psi}(m, n) = \sum_{x=0}^{N-1} \sum_{y=0}^{N-1} e^{2\pi i m x / N} e^{2\pi i n y / N} \psi(x, y), \tag{3}$$

Thus, $\tilde{\psi}$ may be interpreted as the momentum representation of the field ψ.

Below we investigate the relation between a CML described by Eq.(1) and a Bose–Einstein condensate. Therefore, let us invoke the basic characteristics of the latter which are so important that they actually form a part of its modern definition. These are [18,28]: (1) the presence of off-diagonal long-range order (ODLRO) and (2) The presence of one eigenvalue of the one-particle reduced density matrix which is much larger than all other eigenvalues.

Let us notice that the property (2) corresponds to the well-known intuitive definition of the Bose–Einstein condensate. Namely, taking into account that the one-particle reduced density matrix $\rho^{(1)}$ has the following decomposition in terms of eigenvalues λ_j and eigenvectors $|\phi_j\rangle$:

$$\rho^{(1)} = \sum_j \lambda_j |\phi_j\rangle\langle\phi_j|$$

we realize that if one of the eigenvalues is much larger than the rest, then the majority or at least a substantial fraction of particles is in the same single-particle quantum state.

In addition, for an idealized system of Bose particles with periodic boundary conditions and without external potential, the following signature of condensation is also to be noticed: (3) The population of the zero-momentum mode is much larger than population of all other modes.

The properties (1) and (2) acquire quantitative meaning only if the one-particle reduced density matrix is defined. However, as our model is purely classical, the definition of that density matrix is not self-evident. Here we make use of the classical-field approach to the theory of Bose–Einstein condensation [6,11] and define the quantities:

$$\bar{\rho}(x, x') = \langle \sum_{y=0}^{N-1} \psi(x, y)\psi(x', y) \rangle_t, \tag{4}$$

and

$$\rho(x, x') = \bar{\rho}(x, x') / \sum_x \bar{\rho}(x, x). \tag{5}$$

We shall call the quantity $\rho(x, x')$ the reduced density matrix of CML. The above definition in terms of an averaged quadratic form made of ψ seems quite natural, especially because ρ is a real symmetric, positive-definite matrix with the trace equal to 1. The sharp brackets $\langle \ldots \rangle_t$ denote the time averaging:

$$\langle (\ldots) \rangle_t = \frac{1}{T_s} \sum_{t=T-T_s}^{T} (\ldots),$$

where T is the total simulation time and T_s is the averaging time. In our numerical experiments, T has been equal to 3000, and T_s has been chosen to be equal to 300.

We can provide the quantitative meaning to the concept of off-diagonal long-range order (ODLRO) by saying that it is present in the system if

$$\rho(x_1 + x, x_1 - x)$$

does not go to zero with increasing x [28]. If this is the case, the system possesses the basic property (1) of Bose–Einstein condensates.

Let W be the largest eigenvalue of ρ. We will say that CML is in a "condensed state" if W is significantly larger that all other eigenvalues of ρ. If this is the case, the system possesses property (2) of the Bose–Einstein condensates. The corresponding eigenvector, $F(x)$, will be called the reduced wave function of the (condensed part of) coupled map lattice.

One thing which still requires explanation is that the above definition of the reduced density matrix involves not only temporal, but also spatial averaging over y. This is performed just for technical convenience, namely, to avoid dealing with too large matrices. Strictly speaking, we are allowed to assess the presence or absence of ODLRO only in one (x) direction. But that direction is arbitrary, as we might equally well consider averaging over x without any qualitative change in the results.

In the classical field theory the quantity $\tilde{\psi}^{\star}\tilde{\psi}$ represents the particle density in momentum space; in the corresponding quantum theory, upon the raising of ψ, $\psi^{\star}$ to the status of operators, $\tilde{\psi}^{\star}\tilde{\psi}$ would be called the particle number operator. Analogously, we introduce the number P which – just for the purpose of the present article – will be called the "particle number", and is defined as:

$$P = \sum_{m=-N/2}^{N/2-1} \sum_{n=-N/2}^{N/2-1} |\tilde{\psi}(m,n)|^2. \tag{6}$$

All the above definitions are modelled after the corresponding definitions in the non-relativistic classical field theory.

3   Similarity to Bose-Condensed Systems

We have performed our numerical experiment with six values of the non-linear parameter c ($3.5+0.1 \cdot i$, $i = 0, 1, ..., 5$), twenty five values of the diffusion constant d ($0.01 \cdot j$, $j = 1, 2, .., 25$), four different initial conditions, and two different boundary conditions. The boundary conditions have been chosen as periodic ones, the latter with $\psi = 0$ at all boundaries. To save some space, the tables below contain the results for d being multiples of 0.05, but the results for other d do not differ qualitatively from those reported below. The following initial conditions have been investigated. The first – type (A) – initial conditions are such that $\psi(x,y,t)$ is "excited" only at a single point at $t = 0$: $\psi(N/2, N/2, 0) = 0.5$, and $\psi(x,y,0)$ is equal to zero at all other (x,y). Type (B) initial conditions are such that $\psi(x,y,t)$ has initially two non-vanishing values: $\psi(N/4, N/4, 0) = \psi(3N/4, 3N/4, 0) = 0.5$. By type (C) initial conditions we mean those with $\psi(x,y,0)$ being a Gaussian function, $\psi(x,y,0) = 0.5\exp(-0.01((x - N/2)^2 + (y - N/2)^2))$. In type (D) initial conditions, the Gaussian has been replaced with a sine function, $\psi(x,y,0) = 0.5\sin(10x/(N - 1))$. How the largest eigenvalue of the time-averaged reduced density matrix depends on c and d for various types of initial conditions is presented in Tables 1-4.

There are several interesting observations which can be made in connection with Tables 1–4. Firstly, with exception of the case $d = 0.25$ and arbitrary c for type (A) initial conditions, the system exhibits one eigenvalue of the reduced

Table 1. Largest eigenvalue of the reduced density matrix. Periodic boundary conditions and type (A) initial conditions.

$d\backslash c$	3.5	3.6	3.7	3.8	3.9	4.0
0.05	0.920	0.909	0.905	0.908	0.905	0.902
0.10	0.929	0.911	0.905	0.914	0.904	0.902
0.15	0.948	0.917	0.929	0.945	0.907	0.904
0.20	0.999	0.996	0.986	0.912	0.908	0.905
0.25	0.499	0.496	0.493	0.483	0.454	0.453

Table 2. Largest eigenvalue of the reduced density matrix. Periodic boundary conditions and type (B) initial conditions.

$d\backslash c$	3.5	3.6	3.7	3.8	3.9	4.0
0.05	0.921	0.909	0.906	0.909	0.905	0.902
0.10	0.940	0.918	0.902	0.923	0.904	0.902
0.15	0.945	0.910	0.914	0.954	0.907	0.904
0.20	0.912	0.911	0.898	0.899	0.908	0.905
0.25	0.457	0.457	0.459	0.481	0.455	0.453

Table 3. Largest eigenvalue of the reduced density matrix. Periodic boundary conditions and type (C) initial conditions.

$d\backslash c$	3.5	3.6	3.7	3.8	3.9	4.0
0.05	0.925	0.911	0.909	0.907	0.902	0.902
0.10	0.938	0.920	0.901	0.906	0.904	0.902
0.15	0.953	0.928	0.910	0.915	0.906	0.905
0.20	0.923	0.913	0.923	0.893	0.907	0.905
0.25	0.910	0.903	0.888	0.909	0.906	0.904

Table 4. Largest eigenvalue of the reduced density matrix. Periodic boundary conditions and type (D) initial conditions.

$d\backslash c$	3.5	3.6	3.7	3.8	3.9	4.0
0.05	0.927	0.913	0.901	0.899	0.893	0.858
0.10	0.934	0.924	0.911	0.904	0.896	0.887
0.15	0.940	0.934	0.919	0.918	0.901	0.882
0.20	0.940	0.932	0.925	0.924	0.902	0.881
0.25	0.934	0.931	0.925	0.921	0.919	0.880

density matrix which is much larger than all other eigenvalues for all other values of c and d and both types of initial conditions. This is one of the most important features of the Bose-condensed matter, as explained in Section 2. Our system clearly has the property (2) of BEC. Secondly, for the case $d = 0.25$ and types (A) and (B) initial conditions, the largest eigenvalue is slightly lower than 0.5. We have checked that, for each c, there are *two* almost equal eigenvalues which are much larger than all other eigenvalues. The presence of two such eigenvalues of the reduced density matrix also has its analog in the physics of Bose–Einstein condensation; it is characteristic for the so-called quasi-condensates [19,20,11]. Further, it seems there are certain regularities in the c and d dependence of the maximal eigenvalue. In most (but not all) cases, the value of W appears to decrease with growing c for given d. In all cases except of $d = 0.25$, W has had the largest value for c equal to 3.5, that is, below the threshold of chaos for a single logistic map. Let us recall that the value 3.56995... corresponds to the onset of chaos for a single map as it is at the end of period-doubling bifurcations, see, e.g., [7].

4 Large-Scale Pattern Formation

We have observed the following general rules in the process of pattern formation in our system. Firstly, the patterns are incomparably better developed (much better visible) for any "structured" initial conditions (like those considered in this work) than in the case of random initial conditions. The initial inhomogeneities (or "seeds") serve the building of large structures much better than fully random conditions, which is fairly intuitive. The patterns are best developed for smaller values of the non-linear parameter and intermediate values of the diffusion constant.

Fig. 1. Grayscale shaded contour graphics representing the values of the field ψ after 3000 time steps for $d = 0.05$, periodic boundary conditions, and three values of c for type (A) initial conditions; (a)$c = 3.5$, (b)$c = 3.6$, (c)$c = 3.7$. Brighter regions are those with higher values of ψ.

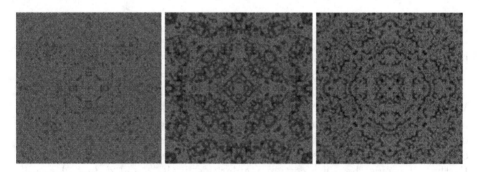

Fig. 2. Grayscale shaded contour graphics representing the values of the field ψ after 3000 time steps for $d = 0.25$, periodic boundary conditions and three values of c and type (A) initial conditions; (a) $c = 3.8$, (b) $c = 3.9$, (c) $c = 4.0$. Brighter regions are those with higher values of ψ

In Figs. 1–6, we show shaded-contour plots representing the values of the field $\psi(x, y)$ after 3000 time steps for periodic boundary conditions and types (A-C) initial conditions. There are no figures for type (D) (sinusoidal) boundary conditions because they are quite uninteresting, displaying merely the stripes corresponding to the sinusoidal initial "excitation".

Naturally, the large structures visible in Figs. 1–6 reflect, to some extent, the symmetry of the simulation box. More interesting observation is that the change from periodic ($c = 3.5$) to chaotic ($c = 3.6$) regime – as defined for individual maps – does not lead, in the case of very slow diffusion, to any spectacular change of the pattern.

The most characteristic feature of the fast-diffusion (i.e., large d) case is the disappearance of the large-scale structures for any type of initial conditions. However, somewhat more pronounced grainy structures reappear for $c = 3.9$.

Fig. 3. Grayscale shaded contour graphics representing the values of the field ψ after 3000 time steps for $d = 0.05$, periodic boundary conditions, three values of c, type (B) initial conditions; (a) c = 3.5, (b) c = 3.6, (c) c = 3.7. Brighter regions are those with higher values of ψ

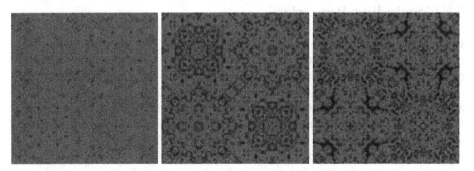

Fig. 4. Grayscale shaded contour graphics representing the values of the field ψ after 3000 time steps for $d = 0.25$, periodic boundary conditions, three values of c and type (B) initial conditions; (a) $c = 3.8$, (b) $c = 3.9$, (c) $c = 4.0$. Brighter regions are those with higher values of ψ

Fig. 5. Grayscale shaded contour graphics representing the values of the field ψ after 3000 time steps for $d = 0.05$, periodic boundary conditions, three values of c, type (C) initial conditions; (a) c $= 3.5$, (b) c $= 3.6$, (c) c $= 3.7$. Brighter regions are those with higher values of ψ

Fig. 6. Grayscale shaded contour graphics representing the values of the field ψ after 3000 time steps for $d = 0.25$, periodic boundary conditions, three values of c and type (C) initial conditions; (a) $c = 3.8$, (b) $c = 3.9$, (c) $c = 4.0$. Brighter regions are those with higher values of ψ

5   Concluding Remarks

Perhaps the most interesting of the various features of the considered system of coupled map lattices is that it appears to be "condensed" if the most standard measures of the classical field theory of Bose condensates are applied. That is, for a majority of parameter values we have observed that a gap between the largest eigenvalue of the reduced density matrix and the rest has been developed. Only for $d = 0.25$ we have observed not a single, but rather two eigenvalues which are much larger than all remaining ones. The latter fact might be of independent interest, as it seems to correspond with the so-called "quasi-condensates". Secondly, the prominent characteristic of the system is the presence of large-scale patterns for smaller values of the "diffusion constant" d, $d \leq 0.2$ and not too large values of the non-linear parameter, $c \leq 3.8$. Thirdly, a very strong dependence of both the presence and qualitative features of the patterns on the initial conditions is to be noticed. The latter fact should be a warning against restricting oneself to one type of initial conditions - namely, the purely random ones - which is very often met in the literature. The most interesting facts can be overlooked this way. Interestingly, the strong dependence of patterns on the initial conditions takes place even in the non-chaotic regime of the parameter c. Fourthly, for very slow diffusion ($d = 0.05$) we have found that the "number of particles" - defined in a natural way - is an approximate constant of motion for sufficiently large times (because the period-2 oscillations have very small amplitude). If the system exhibits period-2 or period-4 oscillations, the number of particle fluctuates around two (or four) values, as if there were two (four) different systems.

We have, in addition, performed similar numerical experiments with another version of the logistic map, reaching similar conclusions [10]. The same statement seems to be valid in the case of standard (rather than logistic) map employed as a basis for the coupled map lattice. However, we have only very preliminary results in that case.

Finally, we would like to observe that the domination of zeroth mode in the momentum space suggests that a kind of Bogoliubov approximation could be applicable. This might lead to an efficient analytical approach to the dynamics of CML.

We plan to develop further our attempt of using classical field-theoretical concepts in coupled map lattices. Work is in progress of their using in the case of three-dimensional CMLs based on logistic maps as well as other physically more appealing discrete systems.

This paper benefits a lot from extensive use of various features and possibilities offered by two important representatives of computer algebra systems: Maxima and Mathematica. Both of them have turned out to become comprehensive environments for symbolic, numerical and graphical work. It is their numerical and graphical capabilities which have been especially important in this work.

We would also like to mention that preliminary results exist for a quantum version of coupled logistic maps. In the calculation of quantum expectation values of operators defined in terms of ψ and its conjugate momentum the capabilities of Mathematica and Maxima pertaining to the operations with patterns, large amounts of symbols, and functional programming, become critical. We would like, however, to report those results elsewhere, when they become mature.

Acknowledgments. One of us (MJ) is grateful to Mariusz Gajda and Emilia Witkowska for offering several helpful discussions.

References

1. Bianconi, G., Barabasi, A.-L.: Bose–Einstein condensation in complex networks. Phys. Rev. Lett. 86, 5632–5635 (2001)
2. Dalfovo, F., Giorgini, S., Pitaevskii, L.P., Stringari, S.: Theory of Bose–Einstein condensation in trapped gases. Rev. Mod. Phys. 71, 463–512 (1999)

3. Chazottes, J.R., Fernandez, B.: Dynamics of Coupled Map Lattices and Related Spatially Extended Systems. Springer, New York (2005)
4. Ghoshdastidar, P.S., Chakraborty, I.: A coupled map lattice model of flow boiling in a horizontal tube. J. Heat Transfer 129, 1737–1741 (2007)
5. Góral, K., Gajda, M., Rzążewski, K.: Multi-mode description of an interacting Bose–Einstein condensate. Opt. Express 8, 92–98 (2001)
6. Góral, K., Gajda, M., Rzążewski, K.: Thermodynamics of an interacting trapped Bose–Einstein gas in the classical field approximation. Phys. Rev. A 66, 051602(R) (4 pages) (2002)
7. Hale, J., Koçak, H.: Dynamics and Bifurcations. Springer, Berlin (1991)
8. Ilachinski, A.: Cellular Automata. A Discrete Universe. World Scientific, Singapore (2001)
9. Janosi, I.M., Flepp, L., Tel, T.: Exploring transient chaos in an NMR-laser experiment. Phys. Rev. Lett. 73, 529–532 (1994)
10. Janowicz, M., Orłowski, A.: Coherence properties of coupled chaotic map lattices. Acta Phys. Polon. A 120, A-114–A-118 (2011)
11. Kadio, D., Gajda, M., Rzążewski, K.: Phase fluctuations of a Bose–Einstein condensate in low-dimensional geometry. Phys. Rev. A 72, 013607 (9 pages) (2005)
12. Kaneko, K.: Period-doubling of kink-antikink patterns, quasi-periodicity in antiferro-like structures and spatial intermittency in coupled map lattices – Toward a prelude to a "Field Theory of Chaos". Prog. Theor. Phys. 72, 480–486 (1984)
13. Kaneko, K.: Pattern dynamics in spatiotemporal chaos. Physica D 34, 1–41 (1989)
14. Kaneko, K.: Simulating physics with coupled map lattices – Pattern dynamics, information flow, and thermodynamics of spatiotemporal chaos. In: Kawasaki, K., Onuki, A., Suzuki, M. (eds.) Pattern Dynamics, Information Flow, and Thermodynamics of Spatiotemporal Chaos, pp. 1–52. World Scientific, Singapore (1990)
15. Kapral, R.: Pattern formation in two-dimensional arrays of coupled, discrete-time Oscillators. Phys. Rev. A 31, 3868–3879 (1985)
16. Leggett, A.: Bose–Einstein condensation in the alkali gases: some fundamental concepts. Rev. Mod. Phys. 73, 307–356 (2001)
17. Muruganandam, P., Francisco, F., de Menezes, M., Ferreira, F.F.: Low dimensional behavior in three-dimensional coupled map lattices. Chaos, Solitons and Fractals 41, 997–1004 (2009)
18. Penrose, O., Onsager, L.: Bose–Einstein condensation and liquid helium. Phys. Rev. 104, 576–584 (1956)
19. Petrov, D.S., Shlyapnikov, G.V., Walraven, J.T.M.: Regimes of quantum degeneracy in trapped 1D gases. Phys. Rev. Lett. 85, 3745–3749 (2000)
20. Petrov, D.S., Shlyapnikov, G.V., Walraven, J.T.M.: Phase-fluctuating 3D Bose–Einstein condensates in elongated traps. Phys. Rev. Lett. 87, 050404 (4 pages) (2001)
21. Reka, A., Barabasi, A.-L.: Statistical mechanics of complex networks. Rev. Mod. Phys. 74, 47–97 (2002)
22. Schmidt, H., Góral, K., Floegel, F., Gajda, M., Rzążewski, K.: Probing the classical field approximation – thermodynamics and decaying vortices. J. Opt. B: Quantum Semiclassical Opt. 5, S96 (2003)
23. Sinha, S.: Transient 1/f Noise. Phys. Rev. E 53, 4509–4513 (1996)
24. Waller, I., Kapral, R.: Spatial and temporal structure in systems of coupled nonlinear oscillators. Phys. Rev. A 30, 2047–2055 (1984)

25. Yanagita, T.: Coupled map lattice model for boiling. Phys. Lett. A 165, 405–408 (1992)

26. Yanagita, T., Kaneko, K.: Rayleigh–Benard convection: Pattern, chaos, spatiotemporal chaos and turbulence. Physica D 82, 288–313 (1995)

27. Yanagita, T., Kaneko, K.: Modeling and characterization of cloud dynamics. Phys. Rev. Lett. 78, 4297–4300 (1997)

28. Yang, C.N.: Concept of off-diagonal long-range order and the quantum phases of liquid He and of superconductors. Rev. Mod. Phys. 34, 694–704 (1962)

On the Computation of the Determinant of a Generalized Vandermonde Matrix

Takuya Kitamoto

Faculty of Education, Yamaguchi University
kitamoto@yamaguchi-u.ac.jp

Abstract. "Vandermonde" matrix is a matrix whose (i, j)th entry is in the form of x_i^j. The matrix has a lot of applications in many fields such as signal processing and polynomial interpolations. This paper generalizes the matrix, and let its (i, j) entry be $f_j(x_i)$ where $f_j(x)$ is a polynomial of x. We present an efficient algorithm to compute the determinant of the generalized Vandermonde matrix. The algorithm is composed of two sub-algorithms: the one that depends on given polynomials $f_j(x)$ and the one that does not. The latter algorithm (the one does not depend on $f_j(x)$) can be performed beforehand, and the former (the one that depends on $f_j(x)$) is mainly composed of the computation of determinants of numerical matrices. Determinants of the generalized Vandermonde matrices can be used, for example, to compute the optimal H_∞ and H_2 norm of a system achievable by a static feedback controller (for details, see [18],[19]).

1 Introduction

Vandermonde matrix, named after Alexandre-Thóphile Vandermonde, is a matrix with the form

$$\begin{bmatrix} 1 & x_1 & \cdots & x_1^{n-1} \\ 1 & x_2 & \cdots & x_2^{n-1} \\ \vdots & \vdots & \vdots & \vdots \\ 1 & x_n & \cdots & x_n^{n-1} \end{bmatrix}.$$

The matrix is well-known in linear algebra, and cited in every standard textbook with various applications, such as FFT in signal processing ([1],p.183), polynomial interpolations ([2],p.114) and so on. The matrix is also known to have close relationships with Frobenius formula in representation theory ([3]). The determinant of the matrix is known to be $\prod_{i<j}(x_i - x_j)$, and efficient algorithms to compute its inverse is being investigated ([20]).

This paper first generalizes the Vandermonde matrix, and presents an efficient algorithm to compute the determinant of the generalized Vandermonde matrix. In references [4] - [6], some generalizations of the matrix in the form of

$$\begin{bmatrix} x_1^{k_1} & x_1^{k_2} & \cdots & x_1^{k_n} \\ x_2^{k_1} & x_2^{k_2} & \cdots & x_2^{k_n} \\ \vdots & \vdots & \vdots & \vdots \\ x_n^{k_1} & x_n^{k_2} & \cdots & x_n^{k_n} \end{bmatrix},$$

V.P. Gerdt et al. (Eds.): CASC Workshop 2014, LNCS 8660, pp. 242–255, 2014.

are considered, where k_i are integers satisfying $0 \leq k_1 \leq \cdots \leq k_n$. This paper considers further generalization, and let its (i, j)th entry to be a polynomial $f_j(x_i)$, not monomial of x_i. This paper proposes an efficient algorithm to compute the determinant of the generalized Vandermonde matrix.

Computations of the determinant of a matrix has a long history, and a lot of research and efficient algorithms has been reported (see, for example, [7] - [10]). Among them, many papers and researches focused on the determinant computation of a *polynomial* matrix (for example, [11] - [13]), and its implementations has been reported (for example, [14] - [17]).

Although we can apply these algorithms to the generalized Vandermonde matrix, we focus on a special form of the generalized Vandermonde matrix in the paper, and present an algorithm that exploits the structure.

Applications of such generalized Vandermonde matrix appear in the references [18],[19], where the determinant of the matrix is required to compute the solution of algebraic Riccati equation (in the references, algorithms to compute the optimal H_2, H_∞ norm are proposed). When the size n of the matrix is large, it is practically difficult to compute the determinant, and an efficient algorithm to compute the determinant is required. The algorithm in this paper can be applied to solve such problem.

This paper is composed as follows: In Section 2, we give notations. Then, in Section 3, we present an algorithm to compute the determinant of the generalized Vandermonde matrix, which is the main contribution of the paper. In Section 4, we analyze the computational complexities of our algorithm, and give a bound of the complexities. In Section 5, we present the result of numerical experiments, and lastly, in Section 6, we conclude.

2 Notations

The following notations are applied in the paper:

N : The set of natural numbers
R : The set of real numbers
Z : The set of integers
K^n : n-dimensional vector with its elements in K
$[\, r_1, \, \cdots, \, r_n \,]$: n-dimensional vector with elements $r_1, \cdots, r_n$
$\{\, r_1, \, \cdots, \, r_n \,\}$: Set with elements $r_1, \, \cdots, \, r_n$.
$|A|$: The number of elements of set A
$_kC_l$: Binomial coefficient, i.e., $\frac{k!}{l!(k-l)!}$.
deg$_x(f(x))$: The degree of polynomial $f(x)$ with respect to x.
tdeg$_x(f(x_1, \cdots, x_n))$: The total degree of polynomial $f(x)$ with respect to x_j $(j = 1, \cdots, n)$, i.e., $\sum_{j=1}^n \deg_{x_j}(f(x_1, \cdots, x_n))$.

To simplify the notation, we denote n-dimensional vector by lower case bold letter, e.g., **e** and **i** denote vector

$$\mathbf{e} = [\, e_1, \, \cdots, \, e_n \,], \quad \mathbf{i} = [\, i_1, \, \cdots, \, i_n \,].$$

By s_i $(i = 1, \cdots, n)$, we denote elementary symmetric polynomials of total degree i, i.e.,

$$s_1 = x_1 + \cdots + x_n, \quad \cdots, \quad s_n = x_1 \cdots x_n.$$

3   Computation of the Determinant

3.1   Problem Formulation

Let $f_j(x) \in \mathbf{R}[x]$ $(j = 1, \cdots, n)$ be

$$f_j(x) \stackrel{\text{def}}{=} w_{0,j} + w_{1,j}x + \cdots + w_{u_j,j}x^{u_j} \quad (w_{i,j} \in \mathbf{R}) \tag{1}$$

and consider the following matrix M:

$$M = \begin{bmatrix} f_1(x_1) & f_2(x_1) & \cdots & f_n(x_1) \\ f_1(x_2) & f_2(x_2) & \cdots & f_n(x_2) \\ \vdots & \vdots & \vdots & \vdots \\ f_1(x_n) & f_2(x_n) & \cdots & f_n(x_n) \end{bmatrix} \tag{2}$$

In the following, we denote the maximum of u_j $(\in \mathbf{Z})$ in (1) by $\bar{u}$, i.e. $\bar{u} = \max_{r=1,\cdots,n} u_r$. We also denote n-dimensional vector $[\, \bar{w}_{i,1}, \cdots, \bar{w}_{i,n} \,]$ $(0 \le i \le \bar{u})$ by $\mathbf{w}_i$, where $\bar{w}_{i,j}$ is defined by

$$\bar{w}_{i,j} = \begin{cases} w_{i,j}, & \text{when } i \le u_j, \\ 0, & \text{otherwise.} \end{cases} \tag{3}$$

Note that with the above $\mathbf{w}_i$, matrix M can be written as

$$
M = \begin{bmatrix} \bar{w}_{0,1} + \bar{w}_{1,1}x_1 + \cdots + \bar{w}_{\bar{u},1}x_1^{\bar{u}} & \cdots & \bar{w}_{0,n} + \bar{w}_{1,n}x_1 + \cdots + \bar{w}_{\bar{u},n}x_1^{\bar{u}} \\ \bar{w}_{0,1} + \bar{w}_{1,1}x_2 + \cdots + \bar{w}_{\bar{u},1}x_2^{\bar{u}} & \cdots & \bar{w}_{0,n} + \bar{w}_{1,n}x_2 + \cdots + \bar{w}_{\bar{u},n}x_2^{\bar{u}} \\ \vdots & \vdots & \vdots \\ \bar{w}_{0,1} + \bar{w}_{1,1}x_n + \cdots + \bar{w}_{\bar{u},1}x_n^{\bar{u}} & \cdots & \bar{w}_{0,n} + \bar{w}_{1,n}x_n + \cdots + \bar{w}_{\bar{u},n}x_n^{\bar{u}} \end{bmatrix}
$$

$$
= \begin{bmatrix} [\, \bar{w}_{0,1}, \cdots, \bar{w}_{0,n}\,] + [\, \bar{w}_{1,1}, \cdots, \bar{w}_{1,n}\,]x_1 + \cdots + [\, \bar{w}_{\bar{u},1}, \cdots, \bar{w}_{\bar{u},n}\,]x_1^{\bar{u}} \\ [\, \bar{w}_{0,1}, \cdots, \bar{w}_{0,n}\,] + [\, \bar{w}_{1,1}, \cdots, \bar{w}_{1,n}\,]x_2 + \cdots + [\, \bar{w}_{\bar{u},1}, \cdots, \bar{w}_{\bar{u},n}\,]x_2^{\bar{u}} \\ \vdots \\ [\, \bar{w}_{0,1}, \cdots, \bar{w}_{0,n}\,] + [\, \bar{w}_{1,1}, \cdots, \bar{w}_{1,n}\,]x_n + \cdots + [\, \bar{w}_{\bar{u},1}, \cdots, \bar{w}_{\bar{u},n}\,]x_n^{\bar{u}} \end{bmatrix}
$$

$$
= \begin{bmatrix} \mathbf{w}_0 + \mathbf{w}_1 x_1 + \cdots + \mathbf{w}_{\bar{u}}x_1^{\bar{u}} \\ \mathbf{w}_0 + \mathbf{w}_1 x_2 + \cdots + \mathbf{w}_{\bar{u}}x_2^{\bar{u}} \\ \vdots \\ \mathbf{w}_0 + \mathbf{w}_1 x_n + \cdots + \mathbf{w}_{\bar{u}}x_n^{\bar{u}} \end{bmatrix}. \tag{4}
$$

When $f_j(x) = x^{j-1}$, matrix M in (2) is the Vandermonde matrix, and the matrix can be viewed as a generalization of the Vandermonde matrix. From the form of matrix M, it is easy to see that $\text{Det}(M)$ is an alternating polynomial in x_i $(i = 1, \cdots, n)$, and $\text{Det}(M)$ can be written as

$$\text{Det}(M) = g(x_1, \cdots, x_n) \prod_{i > j} (x_i - x_j), \tag{5}$$

where $g(x_1, \cdots, x_n)$ is a symmetric polynomial in x_i $(i = 1, \cdots, n)$. This implies that there exists a polynomial $h(s_1, \cdots, s_n)$ in $s_1, \cdots, s_n$ satisfying

$$h(s_1, \cdots, s_n) = g(x_1, \cdots, x_n) = \frac{\text{Det}(M)}{\prod_{i > j} (x_i - x_j)}. \tag{6}$$

Given $g(x_1, \cdots, x_n)$, $\text{Det}(M)$ can be easily computed with (5), and we present an algorithm to compute $h(s_1, \cdots, s_n)$ in (6).

3.2 Algorithm

Given degree $u_j \in \mathbf{Z}$ of polynomial $f_j(x)$ in (1), let Ψ be the set of vectors defined by

$$\Psi \stackrel{\text{def}}{=} \left\{ [\, i_1, \cdots, i_n \,] \in \mathbf{Z}^n \mid 0 \le i_1 < \cdots < i_n \le \bar{u}, \sum_{j=1}^{n} i_j \le \sum_{j=1}^{n} u_j \right\} \tag{7}$$

(recall that $\bar{u} = \max_{j=1,\cdots,n} u_j$). For each element $\mathbf{i}$ $(= [\, i_1, \cdots, i_n \,])$ of Ψ, we define $\Delta^{(\mathbf{i})}$ as the set composed of vector $\mathbf{j}$ $(= [\, j_1, \cdots, j_n \,])$ that is a permutation of $[\, i_1, \cdots, i_n \,]$, i.e.,

$$\Delta^{(\mathbf{i})} \stackrel{\text{def}}{=} \{ [\, j_1, \cdots, j_n \,] \in \mathbf{Z}^n \mid \{j_1, \cdots, j_n\} = \{i_1, \cdots, i_n\} \}. \tag{8}$$

We denote the permutation symbol of $(j_1, \cdots, j_n)$ by $\sigma(j_1, \cdots, j_n)$, i.e.,

$$\sigma(j_1, \cdots, j_n) \stackrel{\text{def}}{=} (-1)^{\tau(j_1, \cdots, j_n)}, \tag{9}$$

where $\tau(j_1, \cdots, j_n)$ denotes the number of permutation inversions in permutation $(j_1, \cdots, j_n)$. For example, when $n = 2$, $\bar{u} = 2$, we have $\Psi = \{[0, 1], [0, 2], [1, 2]\}$. For $\mathbf{i} = [1, 2]$, $\Delta^{(\mathbf{i})}$ is given by $\Delta^{(\mathbf{i})} = \{[1, 2], [2, 1]\}$, and for each $\mathbf{j} \in \Delta^{(\mathbf{i})}$, $\sigma(\mathbf{j})$ is given as $\sigma([1, 2]) = 1$, $\sigma([2, 1]) = -1$.

Our algorithm is based on the following two theorems:

Theorem 1. *Given a matrix M in the form of (2), let Ψ be the set in (7). For each element $\mathbf{i} = [\, i_1, \cdots, i_n \,]$ of Ψ, let $p^{(\mathbf{i})}(x_1, \cdots, x_n)$ be the polynomial in $x_1, \cdots, x_n$ defined by*

$$p^{(\mathbf{i})}(x_1, \cdots, x_n) \stackrel{\text{def}}{=} \sum_{[j_1, \cdots, j_n] \in \Delta^{(\mathbf{i})}} \sigma(j_1, \cdots, j_n) x_1^{j_1} \cdots x_n^{j_n}, \tag{10}$$

where $\Delta^{(i)}$ and $\sigma(j_1, \cdots, j_n)$ are functions defined by (8) and (9), respectively. Let $n \times n$ matrix $\mathbf{W}^{(i)}$ be defined by

$$\mathbf{W}^{(i)} \stackrel{def}{=} \begin{bmatrix} \mathbf{w}_{i_1} \\ \mathbf{w}_{i_2} \\ \vdots \\ \mathbf{w}_{i_n} \end{bmatrix} = \begin{bmatrix} \bar{w}_{i_1,1} & \bar{w}_{i_1,2} & \cdots & \bar{w}_{i_1,n} \\ \bar{w}_{i_2,1} & \bar{w}_{i_2,2} & \cdots & \bar{w}_{i_2,n} \\ \vdots & \vdots & \vdots & \vdots \\ \bar{w}_{i_n,1} & \bar{w}_{i_n,2} & \cdots & \bar{w}_{i_n,n} \end{bmatrix}, \tag{11}$$

and let $\phi^{(i)}$ be its determinant, i.e.,

$$\phi^{(i)} \stackrel{def}{=} Det(\mathbf{W}^{(i)}), \tag{12}$$

where $\bar{w}_{i,j}$ are real numbers defined by (3). Then, we have

$$Det(M) = \sum_{i \in \Psi} \phi^{(i)} p^{(i)}(x_1, \cdots, x_n). \tag{13}$$

Proof.

From (4), matrix M can be written as

$$M = \begin{bmatrix} \mathbf{w}_0 + \mathbf{w}_1 x_1 + \cdots + \mathbf{w}_{\bar{u}} x_1^{\bar{u}} \\ \mathbf{w}_0 + \mathbf{w}_1 x_2 + \cdots + \mathbf{w}_{\bar{u}} x_2^{\bar{u}} \\ \vdots \\ \mathbf{w}_0 + \mathbf{w}_1 x_n + \cdots + \mathbf{w}_{\bar{u}} x_n^{\bar{u}} \end{bmatrix}. \tag{14}$$

From the properties of the determinant, we see that

$$Det(M) = Det \left(\begin{bmatrix} \mathbf{w}_0 + \mathbf{w}_1 x_1 + \cdots + \mathbf{w}_{\bar{u}-1} x_1^{\bar{u}-1} \\ \mathbf{w}_0 + \mathbf{w}_1 x_2 + \cdots + \mathbf{w}_{\bar{u}} x_2^{\bar{u}} \\ \vdots \\ \mathbf{w}_0 + \mathbf{w}_1 x_n + \cdots + \mathbf{w}_{\bar{u}} x_n^{\bar{u}} \end{bmatrix} \right) +$$
$$Det \left(\begin{bmatrix} \mathbf{w}_{\bar{u}} \\ \mathbf{w}_0 + \mathbf{w}_1 x_2 + \cdots + \mathbf{w}_{\bar{u}} x_2^{\bar{u}} \\ \vdots \\ \mathbf{w}_0 + \mathbf{w}_1 x_n + \cdots + \mathbf{w}_{\bar{u}} x_n^{\bar{u}} \end{bmatrix} \right) x_1^{\bar{u}}.$$

Using the above formula repeatedly, we obtain

$$Det(M) = \sum_{j_1, \cdots, j_n \in \{0, \cdots, \bar{u}\}} Det \left(\begin{bmatrix} \mathbf{w}_{j_1} \\ \vdots \\ \mathbf{w}_{j_n} \end{bmatrix} \right) x_1^{j_1} \cdots x_n^{j_n}$$
$$= \sum_{j_1, \cdots, j_n \in \{0, \cdots, \bar{u}\}} \phi^{(j)} x_1^{j_1} \cdots x_n^{j_n} \tag{15}$$

where $\mathbf{j} = [\, j_1, \cdots, j_n \,]$. If $j_k = j_l$ for some $k, l \in \{1, \cdots, n\}$ $(k \neq l)$, then matrix $\mathbf{W}^{(\mathbf{j})}$ in (11) is singular and $\phi^{(\mathbf{j})} = 0$. Thus, (15) can be written as

$$\mathrm{Det}(M) = \sum_{j_1, \cdots, j_n \in \{0, \cdots, \bar{u}\}, \; j_k \neq j_l \, (k \neq l)} \phi^{(\mathbf{j})} x_1^{j_1} \cdots x_n^{j_n}. \tag{16}$$

Given integers $j_1, \cdots, j_n$ satisfying $j_k \in \{0, \cdots, \bar{u}\}$ $(k = 1, \cdots, n)$, $j_k \neq j_l$ $(k \neq l)$, let $\mathbf{i} = [\, i_1, \cdots, i_n \,]$ be the permutation of $[\, j_1, \cdots, j_n \,]$ satisfying $0 \leq i_1 < \cdots < i_n \leq \bar{u}$. From the properties of the determinant, we obtain

$$\phi^{(\mathbf{j})} = \sigma(j_1, \cdots, j_n)\phi^{(\mathbf{i})}. \tag{17}$$

Substituting (17) into (16), we see that

$$\mathrm{Det}(M) = \sum_{j_1, \cdots, j_n \in \{0, \cdots, \bar{u}\}, \; j_k \neq j_l \, (k \neq l)} \left(\phi^{(\mathbf{i})} \sigma(j_1, \cdots, j_n) x_1^{j_1} \cdots x_n^{j_n} \right).$$

Collecting the above summation with respect to $\phi^{(\mathbf{i})}$, we obtain

$$\mathrm{Det}(M) = \sum_{\mathbf{i} \in \hat{\Psi}} \phi^{(\mathbf{i})} p^{(\mathbf{i})}(x_1, \cdots, x_n), \tag{18}$$

where $\hat{\Psi}$ is defined by

$$\hat{\Psi} = \{\, [\, i_1, \cdots, i_n \,] \in \mathbf{Z}^n \mid 0 \leq i_1 < \cdots < i_n \leq \bar{u} \,\}. \tag{19}$$

Now, let $\mathbf{i} = [\, i_1, \cdots, i_n \,]$ be an element of $\hat{\Psi}$. We will show that

$$\left(\sum_{r=1}^{n} i_r > \sum_{r=1}^{n} u_r \right) \Rightarrow \phi^{(\mathbf{i})} p^{(\mathbf{i})}(x_1, \cdots, x_n) = 0. \tag{20}$$

Suppose on the contrary that there exists element $\mathbf{i}$ of $\hat{\Psi}$ such that

$$\left(\sum_{r=1}^{n} i_r > \sum_{r=1}^{n} u_r \right) \text{ and } \left(\phi^{(\mathbf{i})} p^{(\mathbf{i})}(x_1, \cdots, x_n) \neq 0 \right). \tag{21}$$

Since we have $\mathrm{tdeg}_x \left(\phi^{(\mathbf{i})} p^{(\mathbf{i})}(x_1, \cdots, x_n) \right) = \sum_{r=1}^{n} i_r$, inequality $\left(\phi^{(\mathbf{i})} p^{(\mathbf{i})}(x_1, \cdots, x_n) \neq 0 \right)$ implies

$$\mathrm{tdeg}_x \,(\text{Right-hand side of (18)}) \geq \sum_{r=1}^{n} i_r. \tag{22}$$

On the other hand, from the form of matrix M and the property of the determinant, we see that

$$\mathrm{tdeg}_x \,(\mathrm{Det}(M)) \leq \sum_{r=1, \cdots, n} \mathrm{tdeg}_x \,(f_r(x)) = \sum_{r=1, \cdots, n} u_r \tag{23}$$

which implies

$$\text{tdeg}_x (\text{Left-hand side of (18)}) \leq \sum_{r=1}^{n} u_r. \tag{24}$$

Inequalities (22) and (24) imply that $(\sum_{r=1}^{n} i_r \leq \sum_{r=1}^{n} u_r)$, which clearly contradicts (21). Hence, (20) is proved. Therefore, (18) can be written as

$$\text{Det}(M) = \sum_{i \in \hat{\Psi}} \phi^{(i)} p^{(i)}(x_1, \cdots, x_n)$$

$$= \sum_{i \in \hat{\Psi}, \ \sum_{r=1}^{n} i_r \leq \sum_{r=1}^{n} u_r} \phi^{(i)} p^{(i)}(x_1, \cdots, x_n)$$

$$= \sum_{i \in \Psi} \phi^{(i)} p^{(i)}(x_1, \cdots, x_n), \tag{25}$$

which proves the theorem.

Theorem 2. *Polynomial $p^{(i)}(x_1, \cdots, x_n)$ in (10) is an alternating polynomial in $x_1, \cdots, x_n$ and can be written as*

$$p^{(i)}(x_1, \cdots, x_n) = \left(\prod_{i>j} (x_i - x_j) \right) \sum_{e} \gamma^{(i,e)} s_1^{e_1} \cdots s_n^{e_n} \quad (\gamma^{(i,e)} \in \mathbf{R}). \tag{26}$$

Proof

It is enough to show that

$$p^{(i)}(x_1, \cdots, x_k, \cdots, x_l, \cdots, x_n) = -p^{(i)}(x_1, \cdots, x_l, \cdots, x_k, \cdots, x_n) \ (k \neq l) \tag{27}$$

To simplify the notation, we assume that $k = 1$, $l = 2$ and show that

$$p^{(i)}(x_1, x_2, x_3, \cdots, x_n) = -p^{(i)}(x_2, x_1, x_3, \cdots, x_n). \tag{28}$$

From the definition (8) of of $\Delta^{(i)}$, we see

$$[j_1, \ j_2, \ j_3, \ \cdots, \ j_n] \in \Delta^{(i)} \Leftrightarrow [j_2, \ j_1, \ j_3, \ \cdots, \ j_n] \in \Delta^{(i)}. \tag{29}$$

Thus, $p^{(i)}(x_1, \cdots, x_n)$ can be written as

$$p^{(i)}(x_1, \cdots, x_n) = \sum_{j \in \Delta^{(i)}} \sigma(j_1, \cdots, j_n) x_1^{j_1} \cdots x_n^{j_n}$$

$$= \sum_{j \in \Delta^{(i)}, \ j_1 < j_2} \left\{ \sigma(j_1, j_2, j_3, \cdots, j_n) x_1^{j_1} x_2^{j_2} x_3^{j_3} \cdots x_n^{j_n} + \right.$$

$$\left. \sigma(j_2, j_1, j_3, \cdots, j_n) x_1^{j_2} x_2^{j_1} x_3^{j_3} \cdots x_n^{j_n} \right\}$$

$$= \sum_{j \in \Delta^{(i)}, \ j_1 < j_2} \left\{ \sigma(j_1, j_2, j_3, \cdots, j_n) \left(x_1^{j_1} x_2^{j_2} - x_1^{j_2} x_2^{j_1} \right) x_3^{j_3} \cdots x_n^{j_n} \right\},$$

from which (28) is clear. This completes the proof.

From the above two theorems, we see that $h(s_1, \cdots, s_n)$ can be written as

$$h(s_1, \cdots, s_n) = \frac{\text{Det}(M)}{\prod_{i>j}(x_i - x_j)}$$

$$= \sum_{\mathbf{i} \in \Psi} \phi^{(\mathbf{i})} \left(\sum_{\mathbf{e}} \gamma^{(\mathbf{i}, \mathbf{e})} s_1^{e_1} \cdots s_n^{e_n} \right). \tag{30}$$

Note that $p^{(\mathbf{i})}(x_1, \cdots, x_n)$ in (10) is independent of $\mathbf{W}^{(\mathbf{i})}$, hence $f_j(x)$. This implies that $\left(\sum_{\mathbf{e}} \gamma^{(\mathbf{i}, \mathbf{e})} s_1^{e_1} \cdots s_n^{e_n} \right) = \frac{p^{(\mathbf{i})}(x_1, \cdots, x_n)}{\prod_{i>j}(x_i - x_j)}$ in (30) is also independent of $f_j(x)$. Thus, this part can be computed beforehand, and once we have computed it for sufficiently many $\mathbf{i}$ and put it into a table, we need not to compute it again. The computation of the part can be performed, for example, by the following algorithm with Groebner basis:

Algorithm 1. Computation of $\left(\sum_{\mathbf{e}} \gamma^{(\mathbf{i}, \mathbf{e})} s_1^{e_1} \cdots s_n^{e_n} \right)$

Input : $\mathbf{i} \in \mathbf{Z}^n$
Output : $\sum_{\mathbf{e}} \gamma^{(\mathbf{i}, \mathbf{e})} s_1^{e_1} \cdots s_n^{e_n}$ $\left(= \frac{p^{(\mathbf{i})}(x_1, \cdots, x_n)}{\prod_{i>j}(x_i - x_j)} \right)$

$\langle 1 \rangle$ Let $\xi(x_1, \ldots, x_n) = \frac{p^{(\mathbf{i})}(x_1, \cdots, x_n)}{\prod_{i<j}(x_i - x_j)}$.
$\langle 2 \rangle$ Compute Groebner basis of a set of polynomials

$$\{z - \xi(x_1, \cdots, x_n), x_1 + \cdots + x_n - s_1, \cdots, x_1 \cdots x_n - s_n\}$$

with lexicographic ordering $x_1, \cdots, x_n \succ z, s_1, \cdots, s_n$ and express $\xi(x_1, \ldots, x_n)$ as a polynomial in $s_1, \cdots, s_n$.

Having computed $\sum_{\mathbf{e}} \gamma^{(\mathbf{i}, \mathbf{e})} s_1^{e_1} \cdots s_n^{e_n}$ for sufficiently many $\mathbf{i}$ with **Algorithm 1**, we can use the following algorithm to compute $h(s_1, \cdots, s_n)$.

Algorithm 2. Computation of $h(s_1, \cdots, s_n)$

Input : $f_j(x) = w_{0,j} + \cdots + w_{u_j,j} x^{u_j}$ $(j = 1, \cdots, n)$
Output : $h(s_1, \cdots, s_n) = \sum_{\mathbf{i} \in \Psi} \phi^{(\mathbf{i})} \times \left(\sum_{\mathbf{e}} \gamma^{(\mathbf{i}, \mathbf{e})} s_1^{e_1} \cdots s_n^{e_n} \right)$

$\langle 1 \rangle$ Compute Ψ in (7).
$\langle 2 \rangle$ Compute $\phi^{(\mathbf{i})}$ for each $\mathbf{i}$ in Ψ.
$\langle 3 \rangle$ Output $\sum_{\mathbf{i} \in \Psi} \phi^{(\mathbf{i})} \left(\sum_{\mathbf{e}} \gamma^{(\mathbf{i}, \mathbf{e})} s_1^{e_1} \cdots s_n^{e_n} \right)$.

3.3 Numerical Example

Let $n = 2$ and

$$f_1(x) = 1 - x + 2x^2 + x^3, \quad f_2(x) = 2 - 3x - x^2 + 2x^3. \tag{31}$$

We will compute $h(s_1, s_2)$ with **Algorithm 2**.

⟨1⟩ Since we have $u_1 = u_2 = \bar{u} = 3$, $\varPsi$ is given by

$$\varPsi = \{ \ [0,1], \ [0,2], \ [0,3], \ [1,2], \ [1,3], \ [2,3] \ \}. \tag{32}$$

For each $\mathbf{i}$ in (32), $\sum_{\mathbf{e}} \gamma^{(\mathbf{i},\mathbf{e})} s_1^{e_1} \cdots s_n^{e_n}$ computed by Algorithm 1 is given by

$$1, \ s_1, \ s_1^2 - s_2, \ s_2, \ s_1 s_2, \ s_2^2,$$

respectively.

⟨2⟩ $\phi^{(\mathbf{i})}$ for each $\mathbf{i}$ in $\varPsi$ is given by

$$\phi^{([0,1])} = \begin{vmatrix} \bar{w}_{0,1} & \bar{w}_{0,2} \\ \bar{w}_{1,1} & \bar{w}_{1,2} \end{vmatrix} = \begin{vmatrix} w_{0,1} & w_{0,2} \\ w_{1,1} & w_{1,2} \end{vmatrix} = \begin{vmatrix} 1 & 2 \\ -1 & -3 \end{vmatrix} = -1,$$

$$\phi^{([0,2])} = \begin{vmatrix} \bar{w}_{0,1} & \bar{w}_{0,2} \\ \bar{w}_{2,1} & \bar{w}_{2,2} \end{vmatrix} = \begin{vmatrix} w_{0,1} & w_{0,2} \\ w_{2,1} & w_{2,2} \end{vmatrix} = \begin{vmatrix} 1 & 2 \\ 2 & -1 \end{vmatrix} = -5,$$

$$\phi^{([0,3])} = \begin{vmatrix} \bar{w}_{0,1} & \bar{w}_{0,2} \\ \bar{w}_{3,1} & \bar{w}_{3,2} \end{vmatrix} = \begin{vmatrix} w_{0,1} & w_{0,2} \\ w_{3,1} & w_{3,2} \end{vmatrix} = \begin{vmatrix} 1 & 2 \\ 1 & 2 \end{vmatrix} = 0,$$

$$\phi^{([1,2])} = \begin{vmatrix} \bar{w}_{1,1} & \bar{w}_{1,2} \\ \bar{w}_{2,1} & \bar{w}_{2,2} \end{vmatrix} = \begin{vmatrix} w_{1,1} & w_{1,2} \\ w_{2,1} & w_{2,2} \end{vmatrix} = \begin{vmatrix} -1 & -3 \\ 2 & -1 \end{vmatrix} = 7,$$

$$\phi^{([1,3])} = \begin{vmatrix} \bar{w}_{1,1} & \bar{w}_{1,2} \\ \bar{w}_{3,1} & \bar{w}_{3,2} \end{vmatrix} = \begin{vmatrix} w_{1,1} & w_{1,2} \\ w_{3,1} & w_{3,2} \end{vmatrix} = \begin{vmatrix} -1 & -3 \\ 1 & 2 \end{vmatrix} = 1,$$

$$\phi^{([2,3])} = \begin{vmatrix} \bar{w}_{2,1} & \bar{w}_{2,2} \\ \bar{w}_{3,1} & \bar{w}_{3,2} \end{vmatrix} = \begin{vmatrix} w_{2,1} & w_{2,2} \\ w_{3,1} & w_{3,2} \end{vmatrix} = \begin{vmatrix} 2 & -1 \\ 1 & 2 \end{vmatrix} = 5.$$

⟨3⟩ $h(s_1, \cdots, s_n)$ is computed to be

$$\begin{aligned} h(s_1, \cdots, s_n) &= (-1) \cdot 1 + (-5)s_1 + 0(s_1^2 - s_2) + 7s_2 + 1 \cdot s_1 s_2 + 5s_2^2 \\ &= -1 - 5s_1 + 7s_2 + s_1 s_2 + 5s_2^2 \\ &= -1 - 5x_1 - 5x_2 + 7x_1 x_2 + x_1^2 x_2 + x_1 x_2^2 + 5x_1^2 x_2^2. \end{aligned}$$

As you can see from the above example, in general, $h(s_1, \cdots, s_n)$ expressed as a polynomial in $s_1, \cdots, s_n$ has less terms than $h(s_1, \cdots, s_n)$ expressed in $x_1, \cdots, x_n$. In other words, $h(s_1, \cdots, s_n)$ in $s_1, \cdots, s_n$ is more sparse than $h(s_1, \cdots, s_n)$ in $x_1, \cdots, x_n$, which contributes the efficiency of **Algorithm 2**. The gap increases as n and $\bar{u}$ increase. For example, when $n = 5$ and $\bar{u} = 6$, $h(s_1, \cdots, s_n)$ in $s_1, \cdots, s_n$ has at most 21 terms, while $h(s_1, \cdots, s_n)$ in $x_1, \cdots, x_n$ has 243 terms in general.

3.4   Conditions for Det(M) = 0

In this subsection, we present the following simple criteria for Det(M) = 0.

Theorem 3. *Given a matrix M in the form of (2), suppose that*

$$\bar{u} < n - 1, \tag{33}$$

where $\bar{u} = \max_{j=1,\cdots,n} u_j$. Then, we have Det($M$) = 0.

Proof. From (13), it is enough to show that set Ψ defined by (7) has no elements. Suppose on the contrary that there exists an element $\mathbf{i} = [\ i_1, \ \cdots, \ i_n\ ]$ of Ψ. Then, from (7), we see $0 \leq i_1 < i_2 < \cdots < i_n \leq \bar{u}$, which implies that

$$n - 1 \leq i_n \leq \bar{u}. \tag{34}$$

The above inequality clearly contradicts (33). This completes the proof.

4   Complexity Analysis

This section discusses the computational complexities of **Algorithm 2**. Computational complexities in steps $\langle 1 \rangle$ and $\langle 3 \rangle$ of **Algorithm 2** are negligible, and we only have to analyze the complexity of step $\langle 2 \rangle$, where we compute the determinant of an $n \times n$ numerical matrix $|\Psi|$ times. Hence, the computational complexity $c(n, \mathbf{u})$ of step $\langle 2 \rangle$ is given by

$$c(n, \mathbf{u}) = d(n)|\Psi|, \tag{35}$$

where $d(n)$ is the complexity required to compute the determinant of an $n \times n$ numerical matrix. From the definition (7) of Ψ, it is easy to see that Ψ can be written as

$$\Psi = \Psi_1 \cap \Psi_2, \tag{36}$$

where Ψ_1 and Ψ_2 are the set defined by

$$\Psi_1 \overset{\text{def}}{=} \{\ [\ i_1, \ \cdots, \ i_n\ ] \in \mathbf{Z}^n \mid 0 \leq i_1 < \cdots < i_n \leq \bar{u}\}, \tag{37}$$

$$\Psi_2 \overset{\text{def}}{=} \left\{ [\ i_1, \ \cdots, \ i_n\ ] \in \mathbf{Z}^n \mid \sum_{j=1}^{n} i_j \leq \sum_{j=1}^{n} u_j \right\}. \tag{38}$$

From (37), we have $|\Psi_1| = {}_{\bar{u}+1}C_n$. This and (36) imply that

$$|\Psi| \leq |\Psi_1| = {}_{\bar{u}+1}C_n. \tag{39}$$

For the equality of (39), we have the following lemma:

Lemma 1. *We have $|\Psi| = |\Psi_1|$ if and only if*

$$n\bar{u} - \frac{n(n-1)}{2} \leq \sum_{j=1}^{n} u_j. \tag{40}$$

Proof.
   Since $|\Psi| = |\Psi_1| \Leftrightarrow \Psi_1 \subseteq \Psi_2$, it is enough to show

$$n\bar{u} - \frac{n(n-1)}{2} \leq \sum_{j=1}^{n} u_j \Leftrightarrow \Psi_1 \subseteq \Psi_2. \tag{41}$$

Proof for $\Rightarrow$ of (41)
Suppose

$$n\bar{u} - \frac{n(n-1)}{2} \leq \sum_{j=1}^{n} u_j, \tag{42}$$

and let $\mathbf{i} \in \mathbf{Z}^n$ be an element of Ψ_1. From (37), we see $i_j \leq \bar{u} - n + j$, which implies

$$\sum_{j=1}^{n} i_j \leq \sum_{j=1}^{n} (\bar{u} - n + j) = n\bar{u} - \frac{n(n-1)}{2}. \tag{43}$$

This and (42) imply $\sum_{j=1}^{n} i_j \leq \sum_{j=1}^{n} u_j$, hence $\mathbf{i} \in \Psi_2$. Therefore $\Psi_1 \subseteq \Psi_2$, and $\Rightarrow$ of (41) is proved.
Proof for $\Leftarrow$ of (41)
Suppose $\Psi_1 \subseteq \Psi_2$, and let $\mathbf{i} = [\,\bar{u} - n + 1, \cdots, \bar{u} - 1, \bar{u}\,]\ (\in \mathbf{Z}^n)$. Since $\mathbf{i} \in \Psi_1$, we obtain $\mathbf{i} \in \Psi_2$, which implies

$$\sum_{j=1}^{n} u_j \geq \sum_{j=1}^{n} i_j = \sum_{j=1}^{n} (\bar{u} - n + j) = n\bar{u} - \frac{n(n-1)}{2}.$$

Therefore, $\Leftarrow$ of (41) is proved. This completes the proof.

The above lemma and (39) imply the following theorem:

Theorem 4. *The computational complexity $c(n, \mathbf{u})$ of step $\langle 2 \rangle$ in* **Algorithm 2** *satisfies*

$$c(n, \mathbf{u}) \leq d(n)\,(_{\bar{u}+1}C_n) \tag{44}$$

where equality holds if and only if condition (40) is satisfied.

Roughly speaking, equality condition (40) is satisfied when u_j (the degree of $f_j(x)$) $(j = 1, \cdots, n)$ are not so different (for example when the degree of $f_j(x)$ are the same i.e., $\bar{u} = u_1 = \cdots = u_n$). In such cases, (44) gives the exact computational complexity.

With the fraction-free Gaussian elimination method, the complexity $d(n)$ in terms of ring operations is given by $d(n) = O(n^3)$, which implies that

$$c(n, \mathbf{u}) = O\left(\frac{n^3(\bar{u} + 1)!}{n!(\bar{u} - n + 1)!}\right)$$

when condition (40) is satisfied. We note that algorithms for the determinant computation with better computational complexity ($d(n) = O(n^{2.698})$) have already reported (for details, refer to [7],[8]), and we can apply the algorithms to compute $\phi^{(\mathbf{i})}$ in step $\langle 2 \rangle$. However, such asymptotically fast algorithms will be effective only for a large n, (say $n > 100$), and it would not help to improve the efficiency of our algorithm (our algorithm is impractical for such a large n).

Table 1. Computation time (in milli seconds)

$n\backslash\bar{u}$	3	4	5	6	7	8
3	1.26	3.1	5.28	10.3	38.38	45.24
4	1.26	2.76	8.14	20.86	44.94	91.44
5	×	2.82	5.32	15.92	43.04	108.62
6	×	×	2.74	8.14	23.74	68.04
7	×	×	×	4.98	10.9	35.28
8	×	×	×	×	8.42	19.02
9	×	×	×	×	×	14.04

Table 2. Computation time (in milli seconds)

$n\backslash\bar{u}$	3	4	5	6	7	8
3	0.64	0.	0.94	2.52	9.38	9.36
4	5.3	6.84	13.76	17.48	25.58	36.22
5	×	27.16	34.62	49.6	71.48	103.58
6	×	×	139.78	100.5	141.02	207.5
7	×	×	×	220.58	278.9	388.4
8	×	×	×	×	682.64	876.12
9	×	×	×	×	×	3325.7

5   Numerical Experiments

We performed numerical experiments to confirm the efficiency of **Algorithm 2**. We performed the experiments on the machine with Core i7-2640M 2.80 GHz processor and 8 Gbyte memory, using computer algebra system *Mathematica* 7.0.

In the experiments, $f_j(x) = \alpha_{0,j} + \alpha_{1,j}x + \cdots + \alpha_{\bar{u},j}x^{\bar{u}}$ $(j = 1, \cdots, n)$ in (1) are randomly generated with the conditions $(\alpha_{i,j} \in \mathbf{Z}, |\alpha_{i,j}| \leq 10)$, and $h(s_1, \cdots, s_n)$ are computed with **Algorithm 2**. More concretely, coefficients $\alpha_{i,j}$ were randomly generated with *Mathematica* command `Random[Integer,-10,10]`. Computations of $\phi^{(i)}$ (the determinant of numerical matrix $\mathbf{W}^{(i)}$) in step $\langle 2 \rangle$ of **Algorithm 2** are performed by *Mathematica* built-in function `Det` (according to [21], `Det` uses modular methods and row reduction, constructing a result using the Chinese remainder theorem).

Table 1 shows the computation time (average of 10 trials) where $u_j = \bar{u}$ $(j = 1, \cdots, n)$ ($\bar{u}$ is a given natural number). Each row shows the time where n (the size of the matrix) is fixed and $\bar{u}$ (the degree of the polynomial) is changed between $3 \sim 8$, and each column shows the time where $\bar{u}$ is fixed and n is changed between $3 \sim 9$. Symbol × denotes the case where $\mathrm{Det}(M) = 0$ from

Theorem 3. It is clear from Table 1 that when n is fixed, the computation time grows as $\bar{u}$ increases. On the other hand, when $\bar{u}$ is fixed, the computation time may not grow as n increases. The reason for this is as follows: When we increase

n, although the size of matrix M increase, $|\Psi|$ (the number of elements of set Ψ) decreases, which can decrease total computational complexities. For example, $\bar{u} = n - 1$, we have $|\Psi| = 1$ (in this case, $\mathrm{Det}(M)$ is a constant number).

Table 2 shows the computation time to compute $h(s_1, \cdots, s_n)$ with *Mathematica* built-in command Det i.e., we computed the determinant of matrix M with *Mathematica* command Det [M]. Comparing Table 1 and 2, we see that when $n \geq 5$, the computation time of Table 1 is better than that of Table 2 with just one exception (the case with $n = 5, \bar{u} = 8$). This indicates that our algorithm is advantageous when the size of given matrix is large.

6   Conclusion

In this paper, we generalize the Vandermonde matrix and proposed an efficient algorithm to compute the determinant of the generalized Vandermonde matrix. Proposed algorithm is composed of two sub-algorithm, **Algorithm 1** ($f_j(x)$ independent algorithm) and **Algorithm 2** ($f_j(x)$ dependent algorithm). **Algorithm 1** can be performed independently of input polynomials $f_j(x)$, and once we have performed the algorithm for sufficiently many $\mathbf{i} \in \mathbf{Z}^n$, the algorithm need not to be performed any more. **Algorithm 2** is mainly composed of the computations of the determinant of numerical matrices, and can be performed efficiently. We analyze the computational complexity of **Algorithm 2**, and present a bound of the complexity that gives us the exact complexity in many cases.

Acknowledgment. This work was supported by JSPS Grant-in-Aid for Scientific Research (C) KAKENHI 23540139. The author is grateful to anonymous reviewers for the comments on the paper which certainly lead to improve quality and clarity of the paper.

References

1. Golub, G.H., Loan, C.F.V.: Matrix Computations. Johns Hopkins Univ. Johns Hopkins Univ. Press (1986)
2. Hoffman, K.M., Kunze, R.: Linear Algebra, 2nd edn. Prentice Hall, Englewood Cliffs (1971)
3. Vinberg, E.B.: Frobenius formula. Encyclopaedia of Mathematics (2014), http://eom.springer.de/f/f041780.htm (accessed April 12, 2014)
4. Rowland, T.: Generalized vandermonde matrix. Wolfram MathWorld, http://mathworld.wolfram.com/GeneralizedVandermondeMatrix.html (accessed April 12, 2014)
5. Kalman, D.: The generalized Vandermonde matrix. Mathematics Magazine 57(1), 15–21 (1984)
6. Gohberg, I., Kaashoek, M.A., Lerer, L., Rodman, L.: Common multiples and common divisors of matrix polynomials, II. Vandermonde and resultant matrices. Linear and Multilinear Algebra 12(3), 159–203 (1982)

7. Kaltofen, E., Villard, G.: On the complexity of computing determinants. In: Proc. of Fifth Asian Symposium on Computer Mathematics, ASCM 2001, Matsuyama, Japan, pp. 13–27 (2001)
8. Kaltofen, E., Villard, G.: On the complexity of computing determinants. Computational Complexity 30(3-4), 91–130 (2004)
9. Krattenthaler, C.: Advanced determinant calculus, Séminaire Lotharingien Combin. 42, B42q (1999)
10. Krattenthaler, C.: Advanced determinant calculus: a complement. Linear Algebra Appl. 411, 68–166 (2005)
11. Gentleman, W.M., Johnson, S.C.: Analysis of Algorithms, A Case Study: Determinants of Matrices with Polynomial Entries. Journal ACM Transactions on Mathematical Software 2(3), 232–241 (1976)
12. Horrowitz, E., Sahni, S.: On Computing the Exact Determinant of Matrices with Polynomial Entries. Journal of the ACM 22(1), 33–50 (1975)
13. Sasaki, T., Murao, H.: Efficient Gaussian Elimination Method for Symbolic Determinants and Linear Systems. Journal ACM Transactions on Mathematical Software 8(3), 277–289 (1982)
14. Michael, T.: McClellan. A Comparison of Algorithms for the Exact Solution of Linear Equations. Journal ACM Transactions on Mathematical Software 3(2), 147–158 (1977)
15. Martin, L.: Griss. An efficient sparse minor expansion algorithm. In: Proc. of ACM 1976, pp. 429–434 (1976)
16. Marco, A., Martinez, J.: Parallel computation of determinants of matrices with polynomial entries. Journal of Symbolic Computation 37(6), 749–760 (2004)
17. Timothy, S., Freeman, G.M.: Imirzian, Erich Kaltofen and Lakshman Yagati. Dagwood: a system for manipulating polynomials given by straight-line programs. ACM Transactions on Mathematical Software 14(3), 218–240 (1988)
18. Kitamoto, T., Yamaguchi, T.: The optimal H_∞ norm of a parametric system achievable using a static feedback controller. The IEICE Trans. Funda. E90-A(11), 2496–2509 (2007)
19. Kitamoto, T., Yamaguchi, T.: On the parametric LQ control problem. The IEICE Trans. Funda. J91-A(3), 349–359 (2008)
20. Neagoe, V.: Inversion of the van der monde matrix. IEEE Signal Processing Letters 3(4), 119–120 (1996)
21. MATHEMATICA TUTORIAL, Some Notes on Internal Implementation, Exact Numerical Linear Algebra,
http://reference.wolfram.com/mathematica/tutorial/
SomeNotesOnInternalImplementation.en.html (accessed May 31, 2014)

Towards Conflict-Driven Learning for Virtual Substitution

Konstantin Korovin[1], Marek Koša[2], and Thomas Sturm[2]

[1] The University of Manchester, UK
korovin@cs.man.ac.uk
[2] Max-Planck-Institut für Informatik, 66123 Saarbrücken, Germany
{mkosta,sturm}@mpi-inf.mpg.de

Abstract. We consider satisfiability modulo theory-solving for linear real arithmetic. Inspired by related work for the Fourier–Motzkin method, we combine virtual substitution with learning strategies. For the first time, we present virtual substitution—including our learning strategies—as a formal calculus. We prove soundness and completeness for that calculus. Some standard linear programming benchmarks computed with an experimental implementation of our calculus show that the integration of learning techniques into virtual substitution gives rise to considerable speedups. Our implementation is open-source and freely available.

1 Introduction

Recently there has been considerable progress in real satisfiability modulo theory-solving (SMT) triggered by the idea to adopt from Boolean satisfiability-solving (SAT) conflict analysis and learning techniques [6,13,9,10,7,12,1,2]. On the other hand, during the past twenty years there has been a number of successful applications of real quantifier elimination methods, of which real SMT-solving is a special case [5,18,19,17,20,21,16].

Unfortunately, the SMT-solving community and the symbolic computation community working on quantifier elimination have been quite disconnected. The underlying frameworks are not really compatible, which makes it hard to even recognize that ideas from the respective other side are valuable. We would like to contribute to closing this gap by presenting as a calculus in the style of abstract DPLL [15] a special case of one successful approach to real quantifier elimination, viz. virtual substitution. In this paper we restrict ourselves to feasibility checking for systems of linear constraints. On that basis, we integrate conflict analysis and learning techniques in the spirit of the SMT ideas mentioned above.

Applying a very general technique like real quantifier elimination to that very special fragment can not compete with dedicated simplex-based methods. However, our approach has a considerable potential for generalizations, in particular to non-linear real arithmetic.

The plan of the paper is as follows: Section 2 provides a quick introduction into the concept of virtual substitution for readers not familiar with that theory. In Section 3 we formalize virtual substitution for the special case considered

V.P. Gerdt et al. (Eds.): CASC Workshop 2014, LNCS 8660, pp. 256–270, 2014.

here as a basic calculus where learning is only used to avoid cycles. This basic calculus essentially corresponds to a straightforward recursive implementation of the method. In Section 4 we proceed to an enhanced calculus featuring a learning technique based on linear algebra. Technically, that enhanced calculus is obtained by exchanging only one rule in the basic calculus so that soundness, completeness, and complexity results for the basic calculus can be mostly reused. Finally, in Section 5 we discuss computational experiments, comment on related work, and mention possible future work.

2   A Quick Introduction to Virtual Substitution

We consider formulas over the language $L = (0, 1, +, -, \geq)$ with the usual semantics over the real numbers. Given a quantifier-free L-formula Q an *elimination set* for Q and x is a finite nonempty set E of abstract elimination terms such that

$$\mathbb{R} \models \exists x[Q] \longleftrightarrow \bigvee_{e \in E} Q[x /\!\!/ e].$$

Here we are using a *virtual substitution* $[x /\!\!/ e]$, which is defined to map atomic formulas to quantifier-free formulas rather than terms to terms. In practice, this is combined with powerful simplification techniques for intermediate and final results.

In this paper we consider the special case where E contains only *linear elimination terms* of the form t/b, where t is an L-term and $b \in \mathbb{N} \setminus \{0\}$. Then virtual substitution can be defined as formally substituting t/b for x in a suitable extension language of L containing division and rewriting the result as an L-formula by dropping the positive denominator b. Although it is one of the principal strengths of virtual substitution methods that they can inherently deal with arbitrary Boolean combinations, we restrict ourselves here to conjunctions of atomic formulas as input.

Lemma 1. *Consider a formula* $F = I_1 \geq 0 \wedge \cdots \wedge I_n \geq 0$. *Let* x *be a variable occurring in* F. *Then an elimination set* E *for* F *and* x *can be computed as follows: Turn each inequality in* F *that contains* x *into an equation. Then formally solve the equation with respect to* x, *and add the solution to* E. $\quad\square$

As an example consider

$$F = -2x_1 - x_2 + 5 \geq 0 \wedge x_1 + x_2 + 5 \geq 0 \wedge -x_1 + x_2 + 3 \geq 0.$$

By Lemma 1, an elimination set for F and x_1 is given by

$$E = \{(-x_2 + 5)/2, -x_2 - 5, x_2 + 3\}.$$

$$\exists x_1[F] \longleftrightarrow F[x_1 /\!\!/ (-x_2 + 5)/2] \vee F[x_1 /\!\!/ -x_2 - 5] \vee F[x_1 /\!\!/ x_2 + 3]$$
$$= (0 \geq 0 \wedge x_2 + 15 \geq 0 \wedge 3x_2 + 1 \geq 0) \vee$$
$$(x_2 + 15 \geq 0 \wedge 0 \geq 0 \wedge 2x_2 + 8 \geq 0) \vee$$
$$(-3x_2 - 1 \geq 0 \wedge 2x_2 + 8 \geq 0 \wedge 0 \geq 0)$$
$$\longleftrightarrow x_2 + 4 \geq 0.$$

In practice, there are many optimizations, which lead to smaller elimination sets. For instance, when there are both lower and upper bounds on the considered variable, only one of these has to be taken into account, i.e., in our example $E = \{-x_2 - 5\}$ would actually be sufficient. For our calculi to be presented in the next sections, we are exclusively going to use the elimination sets according to Lemma 1. It is going to be crucial that each elimination term originates from exactly one constraint.

Consider the elimination of several variables, say, $\exists x_2 \exists x_1 F$. Eliminating x_1 as above we obtain $\exists x_2 \bigvee_{e \in E} F[x_1 /\!\!/ e]$. Before eliminating x_2 we can move $\exists x_2$ inside the disjunction and eliminate x_2 independently within each disjunct, which our calculi are implicitly going to do. This idea reduces the asymptotic worst-case complexity of the procedure from doubly exponential to singly exponential deterministic time in the input word length [22]. Recall that the Fourier–Motzkin method [14], in contrast, is doubly exponential [3, Section 4-4], [23], and that the simplex method is singly exponential in the worst-case as well [8], although it is known to perform much better than this bound on practical input.

For a more thorough introduction into the virtual substitution framework, we refer the reader to [11,5].

3   A Basic Calculus

In this section we introduce a conflict-driven calculus for deciding satisfiability of systems

$$F = I_1 \geq 0 \wedge \cdots \wedge I_n \geq 0$$

of linear inequalities. By $\mathrm{var}(F)$ we denote the finite set of variables occurring in F. Without loss of generality we assume that $\mathrm{var}(F) \neq \emptyset$. Our calculus is based on the virtual substitution method, which we combine with conflict analysis and learning. In this section we are going to present a basic version with a primitive concept of learning, which leads to exhaustive enumeration of all test terms in the sense of [22,11]. That calculus will serve as a basis for an enhanced calculus with stronger learning techniques, which we are going to present in the next section.

3.1   States

The states of our calculus are either $\top$ (sat), $\bot$ (unsat), or triplets (F, S, L). F is the input system, which will not be modified by any calculus rule.

S is a stack $\langle (x_1, \nu_1), \ldots, (x_k, \nu_k) \rangle$, growing to the right. Given a stack $S = \langle (x_1, \nu_1), \ldots, (x_k, \nu_k) \rangle$, we denote $\langle (x_1, \nu_1), \ldots, (x_{k+1}, \nu_{k+1}) \rangle$ by $S \,|\, (x_{k+1}, \nu_{k+1})$. For $i \in \{1, \ldots, k-1\}$, ν_i is a pair (t_i, J_i), where $J_i \in \{I_1, \ldots, I_n\}$ and

$$t_i = -\frac{1}{b} \left(\sum_{x \in V} a_x x + a_0 \right)$$

is derived from $J_i[x_1 /\!\!/ t_1] \ldots [x_{i-1} /\!\!/ t_{i-1}]$, which equals $\sum_{x \in V} a_x x + b x_i + a_0 \geq 0$ with $V = \mathrm{var}(F) \setminus \{x_1, \ldots, x_i\}$. In other words, t_i is the formal solution of $J_i = 0$

with respect to x_i subject to choices for $x_1, \ldots, x_{i-1}$ based on S. We call t_i an *elimination term*. Since for given $\langle (x_1, \nu_1), \ldots, (x_{i-1}, \nu_{i-1}) \rangle$ the elimination term t_i is uniquely determined by J_i and x_i, we allow ourselves the convenient notation $x_i \leftarrow t_i (J_i)$ instead of $(x_i, (J_i, x_i))$. The last stack element ν_k is either a pair as described above, or "?" or "$\perp$." In the last two cases, we also write $x_i \leftarrow$? and $x_i \leftarrow \perp$, respectively.

Finally, we have a set L of *lemmas*, each of which is a disjunction of negated equations $\sum_{x \in \mathrm{var}(F)} a_x x \neq 0$, where $a_x \in \mathbb{Z}$.

For a given system F of linear inequalities, the initial state of our calculus is $(F, \langle \rangle, \emptyset)$.

3.2 Rules

Before discussing the rules of our calculus we need some definitions. A quantifier-free formula Q is *trivially inconsistent* if it is ground and equivalent to "false."

Let Q be a quantifier-free formula. Given $S = \langle x_1 \leftarrow t_1(J_1), \ldots, x_k \leftarrow t_k(J_k) \rangle$ we define the successive—in contrast to simultaneous—virtual substitution of S into Q as

$$Q/S = Q[x_1 /\!/ t_1] \ldots [x_k /\!/ t_k].$$

Here $[x_i /\!/ t_i]$ denotes the virtual substitution of t_i for x_i in the sense of [22,11]. While virtual substitution in general maps atomic formulas to quantifier-free formulas, it is easy to see that for our linear inequalities one generally obtains atomic formulas. Specifically, for $\varrho \in \{=, >, \geq, \neq\}$ we have

$$(c_1 x_1 + J \varrho 0) \left[x_1 /\!/ b^{-1} K \right] = (c_1 K + bJ \varrho 0),$$

where $J = c_2 x_2 + \cdots + c_m x_m$, $K = a_2 x_2 + \cdots + a_m x_m$, $c_i, a_i \in \mathbb{Z}$, and $b \in \mathbb{N} \setminus \{0\}$. Our definition naturally generalizes to sets of quantifier-free formulas, in particular to L. Furthermore, we observe that the very special situation of our linear constraints allows to define virtual substitution even for terms:

$$(c_1 x_1 + J) \left[x_1 /\!/ b^{-1} K \right] = c_1 K + bJ.$$

Note that this differs from the standard definition of term substitution by dropping a positive integer denominator b.

Our first two rules decide for the next variable to be eliminated and assign to it one possible elimination term, respectively:

DECIDE :

$\quad (F, S, L) \;\vdash\; (F, S \,|\, x_{k+1} \leftarrow ?, L)$

$\qquad$ where S does not contain "?" or "$\perp$"

$\quad$ **if** F/S is not trivially inconsistent, and $x_{k+1} \in \mathrm{var}(F/S)$.

SUBSTITUTE :

$\quad (F, S \,|\, x_k \leftarrow ?, L) \;\vdash\; (F, S \,|\, x_k \leftarrow \mathrm{eterm}(F, S, L, x_k), L)$

Given F, S, and $x \in \mathrm{var}(F/S)$, we denote the elimination set described in Lemma 1 by $E(F/S, x)$. Recall that every elimination originates from a single constraint. For our purposes here we are going to assume that $E(F/S)$ actually contains pairs (t, J), where t is an elimination term originating from J/S. Formally we write:

$$\mathbb{R} \models \exists x [F/S] \longleftrightarrow \bigvee_{(t,J) \in E(F/S,x)} F/S \,|\, x \leftarrow t\,(J). \tag{1}$$

We call an elimination term (t, J) L-admissible, if $L/(S \,|\, x \leftarrow t\,(J))$ is not trivially inconsistent. The elimination term function $\mathrm{eterm}(F, S, L, x)$ enumerates all L-admissible elements of $E(F/S, x)$. In the end it returns "$\perp$."

The following two rules handle the situation that our trial substitutions have led to an inconsistency. We learn just enough not to repeat our unlucky decisions in the future. Afterwards, we successively remove elements from the top of S until F/S is not trivially inconsistent anymore.

LEAF CONFLICT :

$$(F, S, L) \;\vdash\; \left(F, S, L \cup \{\textstyle\bigvee_{i=1}^{k} J_i \neq 0\}\right)$$

   where $S = \langle x_1 \leftarrow t_1\,(J_1), \ldots, x_k \leftarrow t_k\,(J_k)\rangle$, $k \geq 1$

   if F/S is trivially inconsistent, and L/S is not trivially inconsistent.

LEAF BACKTRACK :

$$(F, S \,|\, x_i \leftarrow t_i\,(J_i) \,|\, \ldots \,|\, x_k \leftarrow t_k\,(J_k), L) \;\vdash\; (F, S \,|\, x_i \leftarrow ?, L)$$

   if $L/S \,|\, x_i \leftarrow t_i\,(J_i)$ is triv. inconsistent, and L/S is not triv. inconsistent.

It is not hard to see that based on our limited learning in LEAF CONFLICT we leaf-backtrack exactly one step. This is going to be improved with our enhanced calculus in the next section.

The following two rules are concerned with the situation that the enumeration of some elimination set $E(F/S, x_k)$ has ended with eterm delivering "$\perp$" in SUBSTITUTE. Similarly to LEAF CONFLICT we learn in INNER CONFLICT not to return to the particular subproblem to eliminate x_k from F/S. Afterwards, INNER BACKTRACK can backtrack exactly one step:

INNER CONFLICT :

$$(F, S \,|\, x_k \leftarrow \perp, L) \;\vdash\; \left(F, S \,|\, x_k \leftarrow \perp, L \cup \{\textstyle\bigvee_{i=1}^{k-1} J_i \neq 0\}\right)$$

   where $S = \langle x_1 \leftarrow t_1\,(J_1), \ldots, x_{k-1} \leftarrow t_{k-1}\,(J_{k-1})\rangle$

   if L/S is not trivially inconsistent.

INNER BACKTRACK :

$$(F, S \,|\, x_{k-1} \leftarrow t_{k-1}\,(J_{k-1}) \,|\, x_k \leftarrow \perp, L) \;\vdash\; (F, S \,|\, x_{k-1} \leftarrow ?, L)$$

   if $L/S \,|\, x_{k-1} \leftarrow t_{k-1}\,(J_{k-1})$ is trivially inconsistent.

Finally, we fail when the elimination set for the first-chosen variable is exhausted. We succeed when F/S becomes ground and equivalent to "true:"

FAIL :

$$(F, \langle x_1 \leftarrow \bot \rangle, L) \vdash \bot$$

SUCCEED :

$$(F, S, L) \vdash \top$$

if $\mathrm{var}(F/S) = \emptyset$, and F/S is equivalent to "true."

To conclude the discussion of our basic calculus we would like to point out that it is deterministic in the following sense: Every reachable state (F, S, L) matches the premise of exactly one of the rules.

3.3   Soundness

Lemma 2 (Invariants of the Calculus). *Consider*

$$(F, \langle \rangle, \emptyset) \vdash^n (F, S' \mid x_k \leftarrow \nu_k, L'),$$

where $S' = \langle x_1 \leftarrow t_1\,(J_1), \ldots, x_{k-1} \leftarrow t_{k-1}\,(J_{k-1}) \rangle$.

(i) If $\nu_k = (t_k, J_k)$, *then for all* $l \in \{1, \ldots, k\}$ *the following holds:*

$$\mathbb{R} \models \exists [F/\langle x_1 \leftarrow t_1\,(J_1), \ldots, x_l \leftarrow t_l\,(J_l) \rangle] \longleftrightarrow \exists \left[F \wedge \bigwedge_{i=1}^{l} J_i = 0 \right].$$

(ii) For $\nu_k = ?$ *or* $\nu_k = \bot$ *the equivalence in (i) holds for all* $l \in \{1, \ldots, k-1\}$.

Proof. We simultaneously prove (i) and (ii) by induction on n.

The stack of the initial state is empty, and since $\bigwedge \emptyset$ is defined as "true," we obtain $\mathbb{R} \models \exists [F] \longleftrightarrow \exists [F]$. This proves both (i) and (ii) for $n = 0$.

Consider now $(F, \langle \rangle, \emptyset) \vdash^n (F, S, L) \vdash (F, S' \mid x_k \leftarrow \nu_k, L')$. Assume that both (i) and (ii) hold for (F, S, L). We show by case distinction on the rule applied in the last derivation step that $(F, S' \mid x_k \leftarrow \nu_k, L')$ satisfies both (i) and (ii).

DECIDE yields $S' = S$, and $\nu_k = ?$ so that we are in case (ii), which holds by the induction hypothesis.

With SUBSTITUTE we have $S = S' \mid x_k \leftarrow ?$ and either $\nu_k = (t_k, J_k)$ or $\nu_k = \bot$. In the first case we see that

$$\mathbb{R} \models \exists \big[[F/S'] / \langle x_k \leftarrow t_k\,(J_k) \rangle \big] \longleftrightarrow \exists \big[[F \wedge \bigwedge_{i=1}^{k-1} J_i = 0] \wedge J_k = 0 \big].$$

In the second case we can directly apply the induction hypothesis.

LEAF CONFLICT and INNER CONFLICT both yield $S' \mid x_k \leftarrow \nu_k = S$, and we apply the induction hypothesis.

With LEAF BACKTRACK and INNER BACKTRACK we obtain $S = S'T$ for some possibly empty stack T. Furthermore, $\nu_k = ?$, i.e., we are in case (ii). The fact that S' is a prefix of S allows us to apply the induction hypothesis.

Finally, FAIL and SUCCEED do not match our considered derivation.   □

Lemma 3. *Consider* $(F, \langle \rangle, \emptyset) \vdash^n (F, S', L')$. *Then the following hold:*

(i) $\mathbb{R} \models F \longrightarrow \bigwedge L'$.
(ii) If $S' = S'_1 \,|\, x_k \leftarrow \bot$, *then* $\mathbb{R} \models \neg \exists [F/S'_1]$, *i.e.,* F/S'_1 *is unsatisfiable.*

Proof. To start with, we remark that FAIL and SUCCEED rules do not match our situation. We simultaneously prove (i) and (ii) by induction on n.

(i) For $n = 0$, since $\bigwedge \emptyset$ is defined as "true," we obtain $\mathbb{R} \models F \longrightarrow \bigwedge \emptyset$. Consider now $(F, \langle \rangle, \emptyset) \vdash^n (F, S, L) \vdash (F, S', L')$ and assume that both (i) and (ii) hold for n.

With DECIDE, LEAF BACKTRACK, INNER BACKTRACK, or SUBSTITUTE we have $L' = L$.

With LEAF CONFLICT we have $S' = S = \langle x_1 \leftarrow t_1(J_1), \ldots, x_k \leftarrow t_k(J_k) \rangle$ for $k \geq 1$. By definition of the rule, we know that F/S' is trivially inconsistent, in particular F/S' is unsatisfiable. Lemma 2(i) implies that $F \wedge \bigwedge_{i=1}^{k} J_i = 0$ is unsatisfiable. We equivalently transform

$$\mathbb{R} \models \neg \exists \left[F \wedge \bigwedge_{i=1}^{k} J_i = 0 \right] \longleftrightarrow \forall \left[F \longrightarrow \bigvee_{i=1}^{k} J_i \neq 0 \right],$$

i.e., F implies also the lemma newly learned in LEAF CONFLICT.

With INNER CONFLICT we have $S' = S = S'_1 \,|\, x_k \leftarrow \bot$, where $S'_1 = \langle x_1 \leftarrow t_1(J_1), \ldots, x_{k-1} \leftarrow t_{k-1}(J_{k-1}) \rangle$. By the induction hypothesis for (ii), F/S'_1 is unsatisfiable. By Lemma 2(ii) it follows that also $F \wedge \bigwedge_{i=1}^{k-1} J_i = 0$ is unsatisfiable, and we proceed in analogy to the previous case.

(ii) For $n = 0$, we observe that the stack of the initial state is empty. Again, consider $(F, \langle \rangle, \emptyset) \vdash^n (F, S, L) \vdash (F, S', L')$ and assume that both (i) and (ii) hold for n.

With DECIDE, LEAF BACKTRACK, INNER BACKTRACK, or LEAF CONFLICT the top element of S' is not of the form $x_k \leftarrow \bot$.

With SUBSTITUTE assume that $S' = S'_1 \,|\, x_k \leftarrow \bot$. According to (1), we know that

$$\mathbb{R} \models \exists x_k [F/S'_1] \longleftrightarrow \bigvee_{(t_k, J_k) \in E(F/S'_1, x_k)} F/S'_1 \,|\, x_k \leftarrow t_k(J_k).$$

On the other hand, by (i), $\mathbb{R} \models F \longrightarrow \bigwedge L$, in particular

$$\mathbb{R} \models [F/S'_1 \,|\, x_k \leftarrow t_k(J_k)] \longrightarrow \bigwedge L/S'_1 \,|\, x_k \leftarrow t_k(J_k).$$

Inspection of SUBSTITUTE shows that $\mathrm{eterm}(F, S'_1, L, x_k) = \bot$, which means that $L/S'_1 \,|\, x_k \leftarrow t_k(J_k)$ is equivalent to "false" for all $(t_k, J_k) \in E(F/S'_1, x_k)$. Together $\mathbb{R} \models \exists x_k [F/S'_1] \longrightarrow$ false, i.e., F/S'_1 is unsatisfiable.

With INNER CONFLICT we have $S' = S$, and we can directly apply the induction hypothesis. $\qquad \square$

Theorem 4 (Soundness)

(i) If $(F, \langle\rangle, \emptyset) \vdash^* \perp$, then F is unsatisfiable.
(ii) If $(F, \langle\rangle, \emptyset) \vdash^* \top$, then F is satisfiable.

Proof (i) Since $\perp$ is reachable only by FAIL we are in the following situation:

$$(F, \langle\rangle, \emptyset) \vdash^* (F, \langle x_1 \leftarrow \perp\rangle, L) \vdash \perp.$$

Now choose $S_1' = \langle\rangle$ in Lemma 3(ii).

(ii) Since $\top$ is reachable only by SUCCEED we are in a situation:

$$(F, \langle\rangle, \emptyset) \vdash^* (F, S, L) \vdash \top,$$

where $S = \langle x_1 \leftarrow t_1(J_1), \ldots, x_k \leftarrow t_k(J_k)\rangle$, and F/S does not contain variables and is equivalent to "true." Denote by E_i the set $E(F/\langle x_1 \leftarrow t_1(J_1), \ldots, x_{i-1} \leftarrow t_{i-1}(J_{i-1})\rangle, x_i)$. According to (1), it is easy to see that $\exists x_1 \ldots \exists x_k F$ is equivalent to

$$\bigvee_{(s_k, M_k) \in E_k} \cdots \bigvee_{(s_1, M_1) \in E_1} F/\langle x_1 \leftarrow s_1(M_1), \ldots, x_k \leftarrow s_k(M_k)\rangle.$$

Furthermore, in exactly one of the above disjuncts we have $(s_1, M_1) = (t_1, J_1)$, $\ldots$, $(s_k, M_k) = (t_k, J_k)$, i.e., our stack S is substituted into F. Since F/S is equivalent to "true," the entire disjunction is equivalent to "true." Hence, F is satisfiable. □

3.4 Completeness

Our goal is to assign to each state $\mathcal{S} = (F, S, L)$ of a derivation with our calculus a weight $(n, k, h, l) \in \mathbb{N}^4$, which lexicographically decreases with each derivation step. For the definition of the weight we proceed in three stages:

1. The input constraints F in a given state $\mathcal{S}$ determine a finite abstract F-*tree* $\mathcal{T}$. Each node of $\mathcal{T}$ is either a *variable node* containing a variable chosen by DECIDE or a *term node* containing an elimination term or the unique root node, which is considered a term node, as well.
2. Using L in $\mathcal{S}$, each term node is labeled either *active* or *inactive*. Using S in $\mathcal{S}$ exactly one node of $\mathcal{T}$ is *selected*. If the selected node is a variable node, then it is additionally labeled "?" or "$\perp$."
3. From the labeled tree $\mathcal{T}$ we can finally determine a weight for our state $\mathcal{S}$.

We are now going to make precise these three steps. In the following we denote by τ_i the elimination term (t_i, J_i). Firstly, $\mathcal{T}$ is uniquely described by giving for each node N the path from the unique root node to N: The empty path $()$ describes the root node. Given a path $(x_1, \tau_1, \ldots, x_k, \tau_k)$ to a term node, we obtain a variable node $(x_1, \tau_1, \ldots, x_k, \tau_k, x_{k+1})$ for each variable x_{k+1} occurring in $F/\langle x_1 \leftarrow \tau_1, \ldots, x_k \leftarrow \tau_k\rangle$. Given a path $(x_1, \tau_1, \ldots, x_k, \tau_k, x_{k+1})$ to a variable node, we obtain a term node with path $(x_1, \tau_1, \ldots, x_k, \tau_k, x_{k+1}, \tau_{k+1})$ for each elimination term $\tau_{k+1} \in E(F/\langle x_1 \leftarrow \tau_1, \ldots, x_k \leftarrow \tau_k\rangle, x_{k+1})$.

Secondly, a term node $(x_1, \tau_1, \ldots, x_k, \tau_k)$ is inactive if $L/\langle x_1 \leftarrow \tau_1, \ldots, x_k \leftarrow \tau_k \rangle$ is trivially inconsistent; otherwise it is active. The selected node essentially corresponds to the stack S: If $S = \langle x_1 \leftarrow \tau_1, \ldots, x_k \leftarrow ? \rangle$ or $S = \langle x_1 \leftarrow \tau_1, \ldots, x_k \leftarrow \bot \rangle$, then we select the variable node $(x_1, \tau_1, \ldots, x_{k-1}, \tau_{k-1}, x_k)$ and label it "?" or "$\bot$," respectively. If $S = \langle x_1 \leftarrow \tau_1, \ldots, x_k \leftarrow \tau_k \rangle$, then we select the term node $(x_1, \tau_1, \ldots, x_k, \tau_k)$.

Finally, the weight of $\mathcal{T}$ is $(n, k, h, l) \in \mathbb{N}^4$, where

- n is the number of active nodes,
- k is the number of inactive nodes on the path from the root to the selected node,
- h is the depth of $\mathcal{T}$ minus the length of the path from the root to the selected node,
- $l = 1$ if the selected node is labeled "?," otherwise $l = 0$.

Lemma 5. *Consider states (F, S, L) and (F, S', L') with weights (n, k, h, l) and (n', k', h', l'), respectively:*

$$(F, S, L) \vdash (F, S', L') \implies (n, k, h, l) >_{lex} (n', k', h', l').$$

Proof. DECIDE yields $n = n'$, $k = k'$, $h > h'$, $l < l'$. SUBSTITUTE depends on the return value of eterm; in either case $n = n'$, $k = k'$, $h \geq h'$, $l > l'$. LEAF CONFLICT and INNER CONFLICT yield $n > n'$, $k = k'$, $h = h'$, $l = l'$. LEAF BACKTRACK and INNER BACKTRACK yield $n = n'$, $k > k'$, $h < h'$, $l < l'$. Finally, notice that FAIL and SUCCEED cannot produce (F, S', L').   □

At this point it is clear that our calculus performs only sound derivations and always terminates. For our main result we have to make sure that for all reachable states different from "$\bot$" and "$\top$" there is always at least one calculus rule applicable. We had already mentioned in Subsection 3.2 that in fact exactly one rule is applicable. For the sake of completeness, we state this formally once more.

Lemma 6. *Consider $(F, \langle\rangle, \emptyset) \vdash^* (F, S, L)$. There is one and only calculus rule and at least one state S such that $(F, S, L) \vdash S$.*   □

Theorem 7 (Completeness). *Given a system F of linear inequalities, every derivation beginning in the initial state $(F, \langle\rangle, \emptyset)$ terminates either in state "$\bot$" or in state "$\top$."*

Proof. By Lemma 5 every derivation starting in the initial state terminates. Assume for a contradiction that the final state is neither "$\bot$" nor "$\top$," then Lemma 6 admits another derivation step, a contradiction.   □

3.5   Complexity

From earlier complexity results on the virtual substitution method [22] it is clear that there is a singly exponential upper bound on the worst-case complexity of the calculus bounding the number of substitutions as well as the overall number of derivation steps. This complexity bound persists when extending the approach to strict inequalities, which is straightforward. It is noteworthy that this bound is one exponential step smaller than the known lower bound for the Fourier–Motzkin method.

4   An Enhanced Calculus

Recall from Subsection 3.2 our definition of virtual substitution into terms, which differs from regular substitution by eliminating positive denominators. Similar to regular substitution, that virtual substitution can be expressed via linear combinations:

Lemma 8. *Consider a stack* $S = \langle x_1 \leftarrow t_1 (J_1), \ldots, x_k \leftarrow t_k (J_k) \rangle$ *and a linear term* K*. Then one can compute* $\alpha_1, \ldots, \alpha_k \in \mathbb{Q}, b \in \mathbb{N} \setminus \{0\}$ *such that*

$$K/S = b \sum_{i=1}^{k} \alpha_i J_i + bK. \tag{2}$$

Proof. Let $K' = K[x_1/t_1] \ldots [x_k/t_k]$ be the result of *regular substitution* of stack S into K. It is clear that there exist $\alpha_1, \ldots, \alpha_k \in \mathbb{Q}$ such that $K' = K + \sum_{i=1}^{k} \alpha_i J_i$. On the other hand, there exists $b \in \mathbb{N} \setminus \{0\}$ such that $K/S = bK'$. It follows that (2) has a solution. Compute $b := \frac{K/S}{K'}$. Using this fixed b consider $f = 0$, where f is the recursive polynomial

$$bK + b \sum_{i=1}^{k} \alpha_i J_i - K/S \in \mathbb{Z}[\alpha_1, \ldots, \alpha_k][\mathrm{var}(F)].$$

Solve the linear system $p_1 = 0, \ldots, p_{|\mathrm{var}(F)|} = 0, q = 0$, where $p_1, \ldots, p_{|\mathrm{var}(F)|} \in \mathbb{Z}[\alpha_1, \ldots, \alpha_k]$ are the coefficients of the variables in f, and $q \in \mathbb{Z}[\alpha_1, \ldots, \alpha_k]$ is the constant term of f. □

We call $(K \geq 0) \in F$ a *conflicting inequality* with respect to S if $(K \geq 0)/S$ is trivially inconsistent.

Lemma 9. *Consider a state* (F, S, L)*, where* $S = \langle x_1 \leftarrow t_1 (J_1), \ldots, x_k \leftarrow t_k (J_k) \rangle$*, and let* $(K \geq 0) \in F$ *be a conflicting inequality with respect to* S*. Compute* $\alpha_1, \ldots, \alpha_k \in \mathbb{Q}$ *as in Lemma 8. Then*

$$\mathbb{R} \models F \longrightarrow \bigvee_{\substack{i=1 \\ \alpha_i < 0}}^{k} J_i \neq 0.$$

Proof. Assume that F holds. Using the fact that $K \geq 0$ is a conflicting inequality with respect to S and Lemma 8 we can compute $\alpha_1, \ldots, \alpha_k \in \mathbb{Q}$ such that

$$-1 = \mathrm{sgn}(K/S) = \mathrm{sgn}\left(\sum_{i=1}^{k} \alpha_i J_i + K\right) = \mathrm{sgn}\left(\sum_{\substack{i=1 \\ \alpha_i \geq 0}}^{k} \alpha_i J_i + \sum_{\substack{i=1 \\ \alpha_i < 0}}^{k} \alpha_i J_i + K\right).$$

On the other hand, it is clear that F implies non-negative linear combinations of its constraints, i.e., $\sum_{i=1,\alpha_i \geq 0}^{k} \alpha_i J_i + K \geq 0$. Assume for a contradiction that $\bigwedge_{i=1,\alpha_i<0}^{k} J_i = 0$, which implies $\sum_{i=1,\alpha_i<0}^{k} \alpha_i J_i \geq 0$. This leads to the contradiction $\sum_{i=1,\alpha_i<0}^{k} \alpha_i J_i + \sum_{i=1,\alpha_i \geq 0}^{k} \alpha_i J_i + K \geq 0$. □

Given a conflicting inequality $(K \geq 0) \in F$ with respect to some stack S, the following new rule uses a function lincomb computing $\alpha_1, \ldots, \alpha_k \in \mathbb{Q}$ as in Lemma 8 in order to learn the corresponding disjunction from Lemma 9:

ANALYZE CONFLICT :

$\quad (F, S, L) \vdash (F, S, L \cup \{\bigvee_{i=1,\alpha_i<0}^{k} J_i \neq 0\})$

$\qquad$ where $S = \langle x_1 \leftarrow t_1(J_1), \ldots, x_k \leftarrow t_k(J_k)\rangle$

$\qquad$ and $(\alpha_1, \ldots, \alpha_k) = \text{lincomb}(S, K \geq 0)$.

$\quad$ **if** F contains a conflicting inequality $K \geq 0$ with respect to S.

Our enhanced calculus replaces LEAF CONFLICT in favor of ANALYZE CONFLICT. Since for any choice of $\alpha_1, \ldots, \alpha_k \in \mathbb{Q}$ we have

$$\mathbb{R} \models \bigvee_{\substack{i=1 \\ \alpha_i<0}}^{k} J_i \neq 0 \longrightarrow \bigvee_{i=1}^{k} J_i \neq 0,$$

the lemmas learned now are at least as strong as the ones learned with the basic calculus.

It is quite clear that one could even learn the stronger lemma $\bigvee_{\alpha_i<0} J_i > 0$. However, computational experiments indicate that this does not perform significantly better than the version described above.

4.1 Soundness

Inspection of Lemma 2 and its proof shows that that lemma remains valid also for ANALYZE CONFLICT instead of LEAF CONFLICT. From Lemma 9 it immediately follows that Lemma 3(i) remains valid. Finally, Lemma 3(ii) remains valid, because as with LEAF CONFLICT the premise of the assertion is false. On these grounds the Soundness Theorem 4 for the basic calculus remains valid for our enhanced calculus.

4.2 Completeness

For the termination of our enhanced calculus, we construct weights for the states in the same way as in Subsection 3.4. Then Lemma 5 remains valid. For proving this we have to supplement that in the case of an application of our new ANALYZE CONFLICT rule we obtain $n > n'$, $k = k'$, $h = h'$, $l = l'$. Finally, Lemma 6 remains valid. On these grounds, the Completeness Theorem 7 remains valid.

4.3 Complexity

It is clear that the single exponential upper worst-case bound on the time complexity discussed for the basic calculus in Subsection 3.5 holds also for the enhanced calculus. It is open whether there is a polynomial upper bound, but we do not believe so.

5 Computational Experiments and Conclusions

We have implemented both calculi in the Reduce package Redlog [4]. Both Redlog and Reduce are freely available on SourceForge.[1]

In our implementation we use the selection of either upper or lower bounds, which we have briefly discussed after the example in Section 2. This is supplemented with the virtual substitution of either ∞ or $-\infty$, respectively. From a learning point of view, these substitutions can be essentially ignored. Thus, on the practical side, we are already a bit closer to the *virtual* substitution than in our theory developed here.

For getting an impression of the performance of our enhanced calculus in contrast to the basic calculus we have computed some of the smaller examples from Netlib:[2] afiro (28 constraints, 32 variables), blend (75 constrains, 83 variables), kb2 (44 constrains, 41 variables), sc50a (51 constrains, 48 variables), sc50b (51 constrains, 48 variables), sc105 (106 constrains, 103 variables). In addition, we have computed kmc, a Klee–Minty cube of dimension 50 [8].

The results are collected in Table 1. The columns "basic" and "enhanced" refer to our respective calculi. The columns "rlqe" refer to Redlog's implementation of quantifier elimination by virtual substitution, which accepts way more general input than our examples considered here. For each example we give the number of performed substitutions and the computation times for the various methods.

All our examples are originally linear optimization problems, while in their present form our calculi are limited to feasibility. Under "original" in Table 1 we collect data for the original constraint sets ignoring the target functions. All these examples are in fact feasible so that there is a certain risk that we accidentally find feasible points very early.

This observation motivates the additional computation of two variants for each of these examples: As a first variant, we add a constraint $c \leq \lceil m \rceil$, where c is the objective function and $m \in \mathbb{Q} \setminus \mathbb{Z}$ is its known minimum subject to the constraints. The idea is that this makes the problem "just feasible." The corresponding results are collected in Table 1 under "feasible." Similarly, we add $c \leq \lfloor m \rfloor$ to render the problem infeasible but "almost feasible." The corresponding results are collected in Table 1 under "infeasible."

We observe that the number of substitutions performed with the enhanced calculus is in many cases orders of magnitude smaller than the number of substitutions performed by the basic calculus. The computation times with the

---

[1] http://reduce-algebra.sourceforge.net/

[2] http://www.netlib.org/lp/data/

Table 1. Netlib examples computed by our implementation. For each example the first line gives the number of performed substitutions, and the second line gives the CPU running times. All computations have been performed on a 2.4 GHz Intel Xeon E5-4640 running Debian Linux 64 bit.

	original			feasible			infeasible		
	basic	enhanced	rlqe	basic	enhanced	rlqe	basic	enhanced	rlqe
afiro	333,355 14 s	136 52 ms	267 42 ms	909,315 37 s	183 70 ms	546 86 ms	34,382,742 23 min	183 70 ms	574 92 ms
blend	83 103 ms	83 103 ms	1,592 706 ms	n/a >8 h	n/a >8 h	n/a >8 h	n/a >8 h	n/a >8 h	n/a >8 h
kb2	43 17 ms	43 19 ms	6,965 7 s	n/a >8 h	n/a >8 h	n/a >8 h	n/a >8 h	n/a >8 h	n/a >8 h
kmc	50 18 ms	50 19 ms	50 18 ms	50 18 ms	50 18 ms	50 18 ms	50 19 ms	50 20 ms	48 18 ms
sc50a	86 16 ms	70 49 ms	7,535 2 s	166,894 7 s	600 673 ms	24,700 7 s	15,064,009 10 min	568 594 ms	56,668 17 s
sc50b	48 14 ms	48 14 ms	1,561 353 ms	49 14 ms	49 16 ms	602 161 ms	216,952 13 s	49 18 ms	2,223 545 ms
sc105	n/a >8 h	4,450 39 s	1,785,226 21 min	n/a >8 h	6,701 1 min	n/a >8 h	n/a >8 h	5,427 44 s	n/a >8 h

enhanced calculus are often dramatically shorter and never significantly longer. The few slightly longer computation times can be explained by an overhead caused by solving systems of linear equations for finding the α_i in ANALYZE CONFLICT.

Although our calculus can not directly compete with simplex-based methods for solving systems of linear inequalities, it gives significant improvements over basic virtual substitution. Since virtual substitution can handle parameters and quantifier alternations [22,11], we believe our calculus has a great potential in such extensions.

Concerning the general approach and the role of possible parameters, virtual substitution is somewhat similar to the Fourier–Motzkin method. However, recall from Subsection 3.5 that the worst case time complexity for our approach is one exponential step better than that of the Fourier–Motzkin method.

We are confident that generalization of our calculi to strict inequalities and negated equations is quite straightforward. We think that the treatment of conjunctive systems of polynomial inequalities of arbitrary degrees is a challenging but realistic next step. From a theoretical point of view, generalization of virtual substitution to arbitrary degrees is well-understood [24]. From a practical point of view, there are robust implementations in Redlog for degree two, which have been successfully applied to numerous problems from science and engineering during the past twenty years [5,18,19,17,20,21,16].

Acknowledgments. This research was supported in part by the German Transregional Collaborative Research Center SFB/TR 14 AVACS and by the ANR/DFG project SMArT. The project arose out of discussions at Dagstuhl Seminar 13411, *Deduction and Arithmetic*, held in October 2013.

References

1. Abraham, E., Loup, U., Corzilius, F., Sturm, T.: A lazy SMT-solver for a non-linear subset of real algebra. In: Proceedings of the SMT 2010 (2010)
2. Corzilius, F., Ábrahám, E.: Virtual substitution for SMT-solving. In: Owe, O., Steffen, M., Telle, J.A. (eds.) FCT 2011. LNCS, vol. 6914, pp. 360–371. Springer, Heidelberg (2011)
3. Dantzig, G.B.: Linear Programming and Extensions. Princeton University Press (1963)
4. Dolzmann, A., Sturm, T.: Redlog: Computer algebra meets computer logic. ACM SIGSAM Bulletin 31(2), 2–9 (1997)
5. Dolzmann, A., Sturm, T., Weispfenning, V.: Real quantifier elimination in practice. In: Algorithmic Algebra and Number Theory, pp. 221–247. Springer (1998)
6. Fränzle, M., Herde, C., Teige, T., Ratschan, S., Schubert, T.: Efficient solving of large non-linear arithmetic constraint systems with complex Boolean structure. JSAT 1, 209–236 (2007)
7. Jovanović, D., de Moura, L.: Solving non-linear arithmetic. In: Gramlich, B., Miller, D., Sattler, U. (eds.) IJCAR 2012. LNCS (LNAI), vol. 7364, pp. 339–354. Springer, Heidelberg (2012)

8. Klee, V., Minty, G.: How good is the simplex algorithm? In: Proceedings of the Third Symposium on Inequalities, pp. 159–175. Academic Press (1972)

9. Korovin, K., Tsiskaridze, N., Voronkov, A.: Conflict resolution. In: Gent, I.P. (ed.) CP 2009. LNCS, vol. 5732, pp. 509–523. Springer, Heidelberg (2009)

10. Korovin, K., Voronkov, A.: Solving systems of linear inequalities by bound propagation. In: Bjørner, N., Sofronie-Stokkermans, V. (eds.) CADE 2011. LNCS, vol. 6803, pp. 369–383. Springer, Heidelberg (2011)

11. Loos, R., Weispfenning, V.: Applying linear quantifier elimination. The Computer Journal 36(5), 450–462 (1993)

12. Loup, U., Scheibler, K., Corzilius, F., Ábrahám, E., Becker, B.: A symbiosis of interval constraint propagation and cylindrical algebraic decomposition. In: Bonacina, M.P. (ed.) CADE 2013. LNCS, vol. 7898, pp. 193–207. Springer, Heidelberg (2013)

13. McMillan, K.L., Kuehlmann, A., Sagiv, M.: Generalizing DPLL to richer logics. In: Bouajjani, A., Maler, O. (eds.) CAV 2009. LNCS, vol. 5643, pp. 462–476. Springer, Heidelberg (2009)

14. Motzkin, T.S.: Beiträge zur Theorie der linearen Ungleichungen. Doctoral dissertation, Universität Zürich (1936)

15. Nieuwenhuis, R., Oliveras, A., Tinelli, C.: Solving SAT and SAT modulo theories: From an abstract Davis–Putnam–Logemann–Loveland procedure to DPLL(T). Journal of the ACM 53(6), 937–977 (2006)

16. Platzer, A., Quesel, J.D., Rümmer, P.: Real world verification. In: Schmidt, R.A. (ed.) CADE 2009. LNCS, vol. 5663, pp. 485–501. Springer, Heidelberg (2009)

17. Sofronie-Stokkermans, V.: Hierarchical and modular reasoning in complex theories: The case of local theory extensions. In: Konev, B., Wolter, F. (eds.) FroCos 2007. LNCS (LNAI), vol. 4720, pp. 47–71. Springer, Heidelberg (2007)

18. Sturm, T.: Real Quantifier Elimination in Geometry. Doctoral dissertation, Universität Passau, Germany (1999)

19. Sturm, T., Tiwari, A.: Verification and synthesis using real quantifier elimination. In: Proceedings of the ISSAC 2011, pp. 329–336. ACM Press (2011)

20. Sturm, T., Weber, A., Abdel-Rahman, E.O., El Kahoui, M.: Investigating algebraic and logical algorithms to solve Hopf bifurcation problems in algebraic biology. Math. Comput. Sci. 2(3), 493–515 (2009)

21. Weber, A., Sturm, T., Abdel-Rahman, E.O.: Algorithmic global criteria for excluding oscillations. Bull. Math. Biol. 73(4), 899–916 (2011)

22. Weispfenning, V.: The complexity of linear problems in fields. J. Symb. Comput. 5(1&2), 3–27 (1988)

23. Weispfenning, V.: Parametric linear and quadratic optimization by elimination. Technical Report MIP-9404, Universität Passau, Germany (1994)

24. Weispfenning, V.: Quantifier elimination for real algebra—the quadratic case and beyond. Appl. Algebr. Eng. Comm. 8(2), 85–101 (1997)

Sharpness in Trajectory Estimation for Planar Four-points Piecewise-Quadratic Interpolation

Ryszard Kozera[1], Lyle Noakes[2], and Piotr Szmielew[1,3]

[1] Warsaw University of Life Sciences - SGGW
Faculty of Applied Informatics and Mathematics
Nowoursynowska str. 159, 02-776 Warsaw, Poland
[2] Department of Mathematics and Statistics
The University of Western Australia
35 Stirling Highway, Crawley W.A. 6009, Perth, Australia
[3] University of Warsaw
Institute of Philosophy
Krakowskie Przedmiecie str. 3, 00-927 Warsaw, Poland
{ryszard_kozera,piotr_szmielew}@sggw.pl,
lyle.noakes@maths.uwa.edu.au

Abstract. This paper discusses the problem of fitting non-parametric planar data $Q_m = \{q_i\}_{i=0}^m$ with four-points piecewise-quadratic interpolant to estimate an unknown convex curve γ in Euclidean space E^2 sampled more-or-less uniformly. The derivation of the interpolant involves non-trivial algebraic and symbolic computations. As it turns out, exclusive symbolic computations with *Wolfram Mathematica 9* are unable to explicitly construct the interpolant in question. The alternative solution involves human and computer interaction. The theoretical asymptotic analysis concerning this interpolation scheme as already demonstrated yields quartic orders of convergence for trajectory estimation. This paper verifies in affirmative the sharpness of the above asymptotics via numerical tests and independently via analytic proof based on symbolic computations. Finally, we prove the necessity of admitting more-or-less uniformity and strict convexity to attain at least quartic order of convergence for trajectory approximation. In case of violating strict convexity of γ we propose a corrected interpolant $\bar{Q}$ which preserves quartic order of convergence.

1 Introduction

The sampled ordered planar data points $Q_m = \{q_i\}_{i=0}^m$ with $\gamma(t_i) = q_i \in E^2$ define parametric data $(\{t_i\}_{i=0}^m, Q_m)$. Here curve $\gamma : [0, T] \to E^2$ with $t_0 = 0$ and $t_m = T$. On the other hand Q_m represents the non-parametric data if $\{t_i\}_{i=0}^m$ are not given. Under such condition the unknown knots $\{t_i\}_{i=0}^m$ must be first somehow approximated by properly guessed $\{\hat{t}_i\}_{i=0}^m$. The latter permits to apply specific interpolation scheme $\hat{\gamma} : [0, \hat{T}] \to E^2$, with $\hat{t}_0 = 0$ and $\hat{t}_m = \hat{T}$. Note that in order to compare γ with $\hat{\gamma}$ a reparameterization $\psi : [0, T] \to [0, \hat{T}]$ needs to be also determined.

V.P. Gerdt et al. (Eds.): CASC Workshop 2014, LNCS 8660, pp. 271–285, 2014.

It is also required that $t_i < t_{i+1}$ and $q_i \neq q_{i+1}$, $q_i \neq q_{i+2}$, $q_i \neq q_{i+3}$. In addition, the curve γ is assumed to be regular (i.e. $\gamma' \neq \mathbf{0}$) and strictly convex of class C^4.

In order to estimate the unknown curve γ with an arbitrary interpolant $\tilde{\gamma}$: $[0, T] \to E^2$ it is necessary to assume that knots $\{t_i\}_{i=0}^m \in V_G^m$, i.e., that the following *admissibility condition* is satisfied:

$$V_G^m = \{\{t_i\}_{i=1}^m : \lim_{m \to \infty} \delta_m = 0, \text{ where } \delta_m = \max_{0 \leq i \leq m-1} (t_{i+1} - t_i)\}. \quad (1)$$

From now on, unless required, subscript m in δ_m is omitted. This paper discusses one particular subfamily $V_{mol}^m \subset V_G^m$, namely the so-called *more-or-less uniform samplings* (see, e.g., [1] or [2]) defined as:

$$\kappa\delta \leq t_{i+1} - t_i \leq \delta, \quad (2)$$

where $\kappa \in (0, 1]$ (κ depends here on sampling), for all $i \in [0, m]$ and m arbitrary.

Specific examples for interpolating real life reduced data in *computer graphics* (light-source motion estimation or image rendering), *computer vision* (image segmentation or video compression), *geometry* (trajectory, curvature of area estimation) or in *engineering and physics* (fast particles' motion estimation) can be found among all in [3], [4] and [5].

2   Problem Formulation and Motivation

We introduce now a formal definition of convergence orders (for $n = 2$).

Definition 1. *Consider the family $F_\delta : [0, T] \to E^2$ (in our case $F_\delta = (\hat{\gamma} \circ \psi - \gamma)(t)$). We say that $F_\delta = O(\delta^\eta)$ if $\|F_\delta\| = O(\delta^\eta)$ (where $\|\cdot\|$ denotes the Euclidean norm). The latter can be reformulated to: $\exists_{K>0} \exists_{\delta_0} \|F_\delta\| \leq K\delta^\eta$, for all $\delta \in (0, \delta_0)$ and $t \in [0, T]$.*

The following result is established in [1] (see also [6]):

Theorem 1. *Let regular $\gamma \in C^4$ be strictly convex and be sampled more-or-less uniformly (i.e., $\{t_i\}_{i=0}^m \in V_{mol}^m$) with t_i unknown. Then there is a piecewise-quadratic $Q : [0, \hat{T}] \to E^2$ calculable in terms of Q_m, the sequence of guessed knots $\{\hat{t}_i\}_{i=0}^m$ and a piecewise-C^∞ reparameterization $\psi : [0, T] \to [0, \hat{T}]$ with*

$$Q \circ \psi = \gamma + O(\delta^4), \quad (3)$$

where ψ is a reparameterization [7] defined as piecewise-cubic Lagrange interpolant. Note that according to previous notation we choose particular $\hat{\gamma} = Q$.

Recall that a regular curve γ in E^2 is strictly convex when either $\forall_{t \in [0,T]}(K(t) > 0)$ or $\forall_{t \in [0,T]}(K(t) < 0)$, where

$$K(t) = \frac{\det(\gamma'(t), \gamma''(t))}{\|\gamma'(t)\|^3}, \quad (4)$$

is the curvature. Then since $[0, T]$ is compact and $K(t) \in C^0$, the curvature $K(t)$ is in fact separated from 0.

Note that quartic order for *length estimation* $d(Q) - d(\gamma) = O(\delta^4)$ (proved in [1]) is experimentally verified to be sharp in [1], [8], [9]. However, the sharpness tests have been so-far not conducted for the *trajectory estimation* which asymptotics is determined by formula (3).

This paper will numerically and analytically (via symbolic computations) confirm the sharpness of (3). Recall, that by sharpness we understand the existence of at least one curve $\gamma \in C^4$ (strictly convex and regular) specifically sampled according to more-or-less uniformity (2) which yields exactly quartic order of convergence in trajectory approximation.

In addition, the numerical tests performed in this paper justify the necessity of admitting strict convexity and more-or-less uniformity in Th. 1. Namely, if either curve γ is not sampled more-or-less uniformly or is not convex (with no inflection points) there is a considerable deceleration in asymptotic order (3) including undesirable effect of divergence. Even worse, if inflection points are admitted the four-point quadratic interpolant discussed in this paper cannot be constructed, thus rendering the entire scheme useless.

Finally, for non-convex curves a new interpolant $\bar{Q}$ (see Section 4) is proposed. More specifically in the neighborhood of inflection points the original interpolant Q is substituted by cumulative chord piecewise-cubics which preserve sharply quartic order of convergence in trajectory estimation (see [10]). Experiments confirm that $\bar{Q} \circ \psi = \gamma + O(\delta^4)$ holds.

The next section introduces the construction of discussed herein four-point quadratic interpolant Q (with the aid of symbolic calculation).

2.1   Quadratics Interpolating Planar Quadruples of Points

Consider Q_m sampled more-or-less uniformly and suppose that m is a positive integer multiple of 3. For a given quadruple of sampling points $Q_m^{i,4} = (q_i, q_{i+1}, q_{i+2}, q_{i+3})$, define the quadratic $Q^i : [0, \beta_i] \to E^2$

$$Q^i(s) = a_0^i + a_1^i s + a_2^i s^2 \qquad (5)$$

satisfying

$$Q^i(0) = q_i, \quad Q^i(1) = q_{i+1}, \quad Q^i(\alpha_i) = q_{i+2}, \quad Q^i(\beta_i) = q_{i+3}, \qquad (6)$$

where $0 \le i \le m - 3$, $a_0^i, a_1^i, a_2^i \in E^2$ and $1 < \alpha_i < \beta_i$. For simplicity the subscripts in α_i, β_i, a_0^i, a_1^i and a_2^i, unless necessary, are omitted.

Evidently as $a_0 = q_i$ and $a_2 = q_{i+1} - a_0 - a_1$ equations (5) and (6) give two vector equations (both in E^2):

$$a_1 \alpha + (p_1 - a_1)\alpha^2 = p_\alpha \quad \text{and} \quad a_1 \beta + (p_1 - a_1)\beta^2 = p_\beta, \qquad (7)$$

where $(p_1, p_\alpha, p_\beta) = (q_{i+1} - q_i, q_{i+2} - q_i, q_{i+3} - q_i)$.

Consequently (7) represents four quadratic scalar equations with four scalar unknowns a_{11}, a_{12}, α and β (here $a_1 = (a_{11}, a_{12})$). First both $\hat{t}_{i+2} = \alpha$ and

$\hat{t}_{i+3} = \beta$ estimating the unknown parameters t_{i+3} and t_{i+4} are found. Note that we may safely assume (as done in (6)) that $\hat{t}_i = 0$ and $\hat{t}_{i+1} = 1$ which can be achieved upon a simple normalization step. In order to compute α and β let us introduce the following:

$$c = -det(p_\alpha, p_\beta), \quad d = -det(p_\beta, p_1)/c, \quad e = -det(p_\alpha, p_1)/c,$$

where $c, d, e \neq 0$ by strict convexity. Define next:

$$\rho_1 = \sqrt{e(1 + d - e)/d} \quad \text{and} \quad \rho_2 = \sqrt{d(1 + d - e)/e}, \tag{8}$$

with real roots in (8) again by strict convexity.

The following lemma is proved by symbolic computations (see [1] or [9]):

Lemma 1. *If γ is a planar curve of class C^3 (strictly convex) and is sampled more-or-less uniformly, then system (7) has two solutions in (α, β):*

$$\alpha_\pm = \frac{(1 \pm \rho_1)}{e - d} \quad \text{and} \quad \beta_\pm = \frac{(1 \pm \rho_2)}{e - d},$$

provided ρ_1, ρ_2 are real and $e - d \neq 0$. Moreover for $(\alpha_\pm, \beta_\pm)$ we have:

$$a_1 = \frac{p_{\alpha_\pm} - \alpha_\pm^2 p_1}{\alpha_\pm - \alpha_\pm^2} \quad \text{and} \quad a_2 = \frac{\alpha_\pm p_1 - p_{\alpha_\pm}}{\alpha_\pm - \alpha_\pm^2}.$$

It can be shown (see [1] or [8]) that more-or-less uniformity and convexity assures $\rho_1, \rho_2 \in \mathbb{R}$, $c, d, e \neq 0$, $e \neq d$ and $1 < \alpha_+ < \beta_+$.

We will now outline the proof of Lemma 1 with the aid of symbolic calculations performed in Mathematica.

Solving (7) gives two equations with the constraints on a_1:

$$a_1 = (\alpha^2 p_1 - p_\alpha)/((-1 + \alpha)\alpha),$$
$$a_1 = (\beta^2 p_1 - p_\beta)/((-1 + \beta)\beta).$$

Next substituting a_1 into (7) gives:

$$\frac{\beta\left(\alpha^2 p_1 - \alpha\beta p_1 + (\beta - 1)p_\alpha\right)}{(\alpha - 1)\alpha} = p_\beta.$$

Furthermore we have:

$$p_\beta = \frac{\beta\left(\alpha^2 p_1 - \alpha\beta p_1 + (\beta - 1)p_\alpha\right)}{(\alpha - 1)\alpha},$$
$$p_\beta(\alpha - 1)\alpha = \beta(\alpha^2 p_1 - \alpha\beta p_1 + (\beta - 1)p_\alpha),$$
$$p_\beta(\alpha - 1)\alpha = \beta\alpha^2 p_1 - \alpha\beta^2 p_1 + (\beta^2 - \beta)p_\alpha,$$
$$\alpha\beta(\alpha p_1 - \beta p_1) = (\beta - \beta^2)p_\alpha + p_\beta(\alpha^2 - \alpha),$$
$$\alpha\beta(\alpha - \beta)p_1 = (\beta - \beta^2)p_\alpha - p_\beta(\alpha - \alpha^2). \tag{9}$$

For two vectors $p_\alpha = (p_{\alpha 1}, p_{\alpha 2})$, $p_\beta = (p_{\beta 1}, p_{\beta 2})$ consider their respective orthogonal counterparts: $p_\alpha^\perp = (-p_{\alpha 2}, p_{\alpha 1})$, $p_\beta^\perp = (-p_{\beta 2}, p_{\beta 1})$. Taking the dot product $\langle \cdot, \cdot \rangle$ of (9) first with $p_\beta^\perp$ and then with $p_\alpha^\perp$ results in:

$$
\begin{aligned}
\alpha\beta(\alpha - \beta)\langle p_1, p_\beta^\perp \rangle &= (\beta - \beta^2)\langle p_\alpha, p_\beta^\perp \rangle, \\
\alpha\beta(\alpha - \beta)\langle p_1, p_\alpha^\perp \rangle &= -(\alpha - \alpha^2)\langle p_\beta, p_\alpha^\perp \rangle.
\end{aligned}
\tag{10}
$$

Since by strict convexity $\mathrm{span}\{p_\beta^\perp, p_\alpha^\perp\} = E^2$ holds asymptotically, formulas (9) and (10) are equivalent. Furthermore

$$
\frac{\alpha(\alpha - \beta)\langle p_1, p_\beta^\perp \rangle}{\langle p_\alpha, p_\beta^\perp \rangle} = 1 - \beta \quad \text{and} \quad \frac{\beta(\alpha - \beta)\langle p_1, p_\alpha^\perp \rangle}{-\langle p_\beta, p_\alpha^\perp \rangle} = \alpha - 1.
$$

Consequently the latter yields:

$$
c = -\langle p_\beta, p_\alpha^\perp \rangle = \langle p_\alpha, p_\beta^\perp \rangle,
\tag{11}
$$

which in turn renders:

$$
d = \frac{-\langle p_1, p_\beta^\perp \rangle}{c} = \frac{-\langle p_1, p_\beta^\perp \rangle}{\langle p_\alpha, p_\beta^\perp \rangle} \quad \text{and} \quad e = \frac{-\langle p_1, p_\alpha^\perp \rangle}{c} = \frac{\langle p_1, p_\alpha^\perp \rangle}{\langle p_\beta, p_\alpha^\perp \rangle}.
\tag{12}
$$

Combining (10) with (11) and (12) yields two scalar equations in α and β:

$$
\begin{aligned}
\alpha(\alpha - \beta)d &= \beta - 1, \\
\beta(\alpha - \beta)e &= \alpha - 1.
\end{aligned}
\tag{13}
$$

Applying *Mathematica* (to compute α and β) to (13) subject to the constraint:

$$
1 < \alpha < \beta
\tag{14}
$$

results in endless calculations with no explicit formulas for α and β. Both *Solve* and *Simplify* functions give the following three pairs of possible solutions:

$$
(\alpha = 1, \beta = 1),
$$

$$
\left(\alpha = -\frac{\sqrt{de(d - e + 1)} + d}{d(d - e)}, \beta = \frac{\sqrt{de(d - e + 1)} + e}{e(e - d)} \right),
$$

$$
\left(\alpha = \frac{\sqrt{de(d - e + 1)} - d}{d(d - e)}, \beta = \frac{e - \sqrt{de(d - e + 1)}}{e(e - d)} \right).
$$

The first pair is obviously discarded (due to (14)). On the other hand only the second pair α, β satisfies constraint (14). Therefore our interpolation scheme defined by (5), (6) and (14) is explicitly determined.

In the next section we will *numerically* and independently *symbolically* verify the sharpness of Th. 1 and prove the need for strict convexity and more-or-less uniformity (2).

3   Experiments

Our tests are performed in *Mathematica* 9.0.1 using Intel Core2Duo 2.4 GHz processor with 16 GiB of RAM and on PL-Grid infrastructure [11].

Since $T = \Sigma_{i=1}^m (t_{i+1}-t_i) \leq m\delta$ the following holds $m^{-\eta} = O(\delta^\eta)$, for arbitrary $\eta > 0$ mentioned in Definition 1 (see also [13]). Therefore, for the verification of any asymptotics expressed in terms of $O(\delta^\eta)$ it is sufficient to examine the claims of Th. 1 in terms of $O(1/m^\eta)$ asymptotics.

Recall that for a parametric smooth planar curve $\gamma : [0, T] \to E^2$ (with $[0, T]$ compact) and m varying between $m_{min} \leq m \leq m_{max}$ the i-th component of the error for γ estimation by Q^i is defined as follows:

$$E_m^i = \sup_{t\in[t_i,t_{i+2}]} \|(Q^i \circ \psi_i)(t) - \gamma(t)\| = \max_{t\in[t_i,t_{i+2}]} \|(Q^i \circ \psi_i)(t) - \gamma(t)\|. \qquad (15)$$

The maximal value E_m for each $m = 3k$ is found by using *Mathematica* numerical optimization function: *NMaximize* [12]. From the set of *absolute errors* $\{E_m\}_{m=m_{min}}^{m=m_{max}}$ the numerical estimate of η is calculated using a linear regression applied to the collection of points $(\log(m), -\log(E_m))$ (where $m_{min} \leq m \leq m_{max}$). The *Mathematica's* built-in function *LinearModelFit* renders the estimated coefficient η from the computed regression line $y(x) = \eta x + b$.

3.1   Curves and Sampling

In this subsection, testing curves and more-or-uniform samplings are introduced.

a) Curves: The first example refers to the specific curves used in our experimentation.

Example 1. (*i*) Define strictly convex (here with $K(t) < 0$ - see (4)) two planar curves: first *a semicircle*

$$\gamma_{sc}(t) = (\cos(\pi(1 - t)), \sin(\pi(1 - t))) \subset E^2, \text{ for } t \in [0, 1],$$

and then *a planar spiral* (see Fig. 1)

$$\gamma_{spl}(t) = ((6\pi - t)\cos(t), (6\pi - t)\sin(t)) \subset E^2, \text{ for } t \in [0, 5\pi].$$

(*ii*) Let us now introduce two non-strictly convex curves. First we consider the curve without inflection point (4) (i.e., with $K(t) \geq 0$) $\gamma_{pol4}(t) = (t, t^4)$ for $t \in [-1, 1]$ (see Fig. 2a), where $K(0) = 0$ and $K(t) > 0$ for $t \neq 0$. Then one admits the curve with inflection points (i.e., with $K(t)$ varying its signs) $\gamma_{pol3}(t) = (t, t^3)$ over $[-1, 1]$. The curve γ_{pol} has visibly inflection point at $t = 0$ (plotted in Fig. 2b). Note that γ_{pol4} and γ_{pol3} have different domain range than $[0, T]$. This, however, can be easily achieved by simple affine mapping.

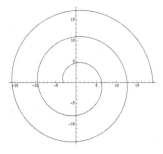

Fig. 1. A strictly convex spiral γ_{spl}

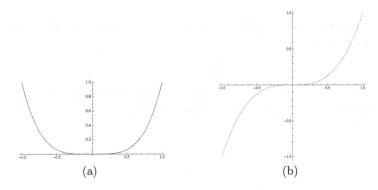

(a)   (b)

Fig. 2. Non-strictly convex curves: a) γ_{pol4} b) γ_{pol3}, without or with inflection points

□

b) Samplings: The next example includes different more-or-less uniform samplings.

Example 2. Both curves from Example 1i sampled uniformly, i.e., with $t_i = \frac{iT}{m}$ presented in Fig. 3. The uniform sampling is evidently more-or-less uniform with $\kappa = 1$ (see (2)).

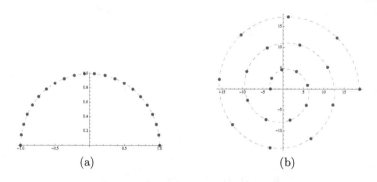

Fig. 3. Strictly convex curves: a) γ_{sc} and b) γ_{spl} sampled uniformly with $m = 21$

$\square$

Example 3. (*i*) Our tests use different more-or-less uniform samplings. The first one is defined as follows:

$$t_i = \frac{i}{m} + \frac{(-1)^{i+1}}{3m}, \tag{16}$$

with $\kappa = \frac{1}{5}$ in (2). Both curves determined in Example 1i sampled according to (16) (modulo rescaling if needed) are presented in Fig. 4.

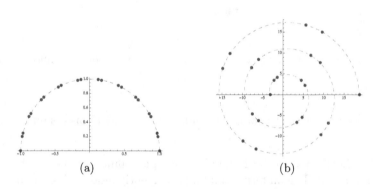

Fig. 4. Curves a) γ_{sc} b) γ_{spl} sampled according to (16) with $m = 21$

(*ii*) Another more-or-less uniform sampling applied in our experiments is:

$$t_i = \begin{cases} \frac{i}{m}, & \text{if } i \text{ is even,} \\ \frac{i}{m} + \frac{1}{2m} & \text{if } i = 4k + 1, \\ \frac{i}{m} - \frac{1}{2m} & \text{if } i = 4k + 3, \end{cases} \tag{17}$$

with $\kappa = \frac{1}{3}$ in (2). Similarly both curves γ_{sc}, γ_{spl} sampled according to (17) are presented in Fig. 5.

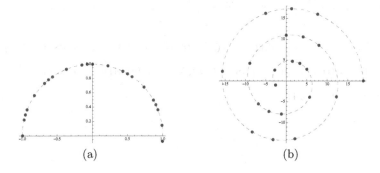

Fig. 5. Curves a) γ_{sc} b) γ_{spl} sampled according to (17) with $m = 21$

□

3.2   Analytical Construction of Interpolant

The construction of interpolant Q^i for each consecutive quadruple of points $(q_i, q_{i+1}, q_{i+2}, q_{i+3})$ is explicitly determined by Lemma 1. However, the symbolic computational burden increases heavily with m getting larger. Even for small m this issue is noticeable as indicated below.

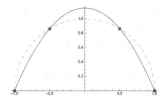

Fig. 6. γ_{sc} interpolated using 4-points quadratic with uniform sampling and $m = 3$

Case $m = 3$: We construct now the interpolant Q^0 based on single quadruple of interpolation knots $\{t_i\}_{i=0}^{m=3}$ generated from uniform sampling $\left(0, \frac{1}{3}, \frac{2}{3}, 1\right)$. The corresponding sampling points (q_0, q_1, q_2, q_3) obtained from the curve γ_{sc} are $\left((-1, 0), \left(-\frac{1}{2}, \frac{\sqrt{3}}{2}\right), \left(\frac{1}{2}, \frac{\sqrt{3}}{2}\right), (1, 0)\right)$ and $Q_{3,sc}^0(s) = \left(\frac{s-2}{2}, \frac{4s-s^2}{2\sqrt{3}}\right)$ with $s \in [0, 4]$. The derivation of the interpolant $Q_{3,sc}^0$ in *Mathematica* takes around $0.000191s$.

Case $m = 6$: Consider now interpolation scheme (5), (6) based on two four-point segments. The corresponding uniform knots are $\left(0, \frac{1}{6}, \frac{1}{3}, \frac{1}{2}, \frac{2}{3}, \frac{5}{6}, 1\right)$. The respective interpolation points are:

$$\left((-1, 0), \left(-\frac{\sqrt{3}}{2}, \frac{1}{2}\right), \left(-\frac{1}{2}, \frac{\sqrt{3}}{2}\right), (0, 1), \left(\frac{1}{2}, \frac{\sqrt{3}}{2}\right), \left(\frac{\sqrt{3}}{2}, \frac{1}{2}\right), (1, 0)\right).$$

The interpolants $Q^0_{6,sc}$, $Q^3_{6,sc} : [0, 2 + \frac{2}{\sqrt{3}}] \to E^2$ are determined by:

$$\hat{Q}^0_{6,sc}(s) = (-1 + 0.0490381s + 0.0849365s^2, 0.584936s - 0.0849365s^2),$$
$$\hat{Q}^3_{6,sc}(s) = (0.584936s - 0.0849365s^2, 1 - 0.0490381s - 0.0849365s^2),$$

evaluated over $[0, 3.1547]$.

Fig. 7. γ_{sc} interpolated using 4-points quadratic with uniform sampling and $m = 6$

3.3 Sharpness of Th. 1 via Symbolic Computation

The sharpness of trajectory estimation from Th. 1 is proved now analytically with the aid of symbolic computations.

Proof. Define part of the circle $\gamma_c : [0, 1] \to E^2$, where $\gamma_c(t) = (\cos(t), \sin(t))$ (which is a strictly convex curve). For sharpness it suffices to consider only one segment $I_0 = [0, 3\bar{\delta}]$. In doing so let $t_0 = 0$, $t_1 = \bar{\delta}$, $t_2 = \frac{3}{2}\bar{\delta}$ and $t_3 = 3\bar{\delta}$. Over all of the remaining segments the distribution of sampling is similar. Note that here $\delta = \frac{3}{2}\bar{\delta}$ (see (1)) and more-or-less uniformity (2) holds with $\kappa = \frac{1}{3}$. We examine now $f_0(t) = (\hat{Q}^0_{m,c} \circ \psi_0)(t) - \gamma_c(t)$ over I_0, where $\psi_0 : [0, 3\bar{\delta}] \to [0, \beta]$ is a cubic satisfying $\psi_0(0) = 0$, $\psi_0(\bar{\delta}) = 1$, $\psi_0(\frac{3}{2}\bar{\delta}) = \alpha$ and $\psi_0(3\bar{\delta}) = \beta$. The exact asymptotics of $\sup_{t \in [0,3\bar{\delta}]} f_0(t) = \max_{t \in [0,3\bar{\delta}]} f_0(t)$ needs to be determined. To prove sharpness of (3) it is sufficient to justify it over at least one subinterval i.e. over $[0, \bar{\delta}]$, $[\bar{\delta}, \frac{3}{2}\bar{\delta}]$ or $[\frac{3}{2}\bar{\delta}, 3\bar{\delta}]$, respectively. We prove more by selecting all of them. Moreover, for the latter it is also sufficient to choose special points $\bar{t}$ from each subinterval, selected here as $\bar{t} = \frac{t_i + t_{i+1}}{2}$, with $i = 0, 1, 2$. The function f_0 symbolically reduced with the aid of Mathematica's *Simplify* and Taylor expansion *Series* yields

$$f_0\left(\frac{\bar{\delta}}{2}\right) = \left(-\frac{5\bar{\delta}^4}{64} + O\left(\bar{\delta}^6\right), -\frac{43\bar{\delta}^5}{384} + O\left(\bar{\delta}^6\right)\right), \quad \text{over } [0, \bar{\delta}],$$

$$f_0\left(\frac{5\bar{\delta}}{4}\right) = \left(\frac{35\bar{\delta}^4}{2048} + O\left(\bar{\delta}^6\right), \frac{581\bar{\delta}^5}{24576} + O\left(\bar{\delta}^6\right)\right), \quad \text{over } [\bar{\delta}, 3\bar{\delta}/2],$$

$$f_0\left(\frac{9\bar{\delta}}{4}\right) = \left(-\frac{405\bar{\delta}^4}{2048} + O\left(\bar{\delta}^6\right), -\frac{2133\bar{\delta}^5}{8192} + O\left(\bar{\delta}^6\right)\right), \quad \text{over } [3\bar{\delta}/2, 3\bar{\delta}].$$

Evidently as $\delta = \frac{3}{2}\bar{\delta}$ the latter result on each segment $[0, \bar{\delta}]$, $[\bar{\delta}, \frac{3}{2}\bar{\delta}]$ and $[\frac{3}{2}\bar{\delta}, 3\bar{\delta}]$ independently proves sharpness of Th. 1 by symbolic computation. $\square$

3.4   Sharpness of Th. 1 via Numerical Computations

The numerical tests applied to the curves and samplings introduced in subsection 3.1 are presented in Table 1 (recall that η is introduced in Definition 1). They all confirm numerically the sharpness of asymptotics estimate established by Th. 1. The tests are conducted with the aid of (15), for $m \in \{99, \ldots, 120\}$.

Table 1.

Curve	Sampling	By Th. 1 $\eta = 4$
γ_{spl}	uniform	4.039
γ_{sc}	uniform	4.037
γ_{spl}	(16)	3.995
γ_{sc}	(16)	4.037
γ_{spl}	(17)	4.021
γ_{sc}	(17)	4.037

The next subsection justifies the necessity of imposing constraints such as more-or-less uniformity and strict convexity stipulated by Th. 1.

3.5   Counterexamples

The first example illustrates the impact of more-or-less uniformity assumption in Th. 1 on the convergence rate from (3).

Example 4. (*i*) Consider the following sampling (see e.g. [9]):

$$t_0 = 0, \quad t_i = \frac{(\sqrt{m} - 1)(i - 1)}{(m - 1)\sqrt{m}} + \frac{1}{\sqrt{m}} \quad \text{(for } i \in \{1, 2, \ldots, m\}). \quad (18)$$

Since $t_1 - t_0 = 1/\sqrt{m} = \delta$ and $t_{i+1} - t_i = \frac{1}{(1+\sqrt{m})\sqrt{m}} = \delta \cdot \frac{1}{1+\sqrt{m}}$ there is no κ satisfying (2), as $\lim_{m \to \infty} \frac{1}{1+\sqrt{m}} = 0$. Thus sampling (18) is not more-or-less uniform. On the other hand as here $\delta = 1/\sqrt{m}$ this sampling is still admissible (1). The plot of interpolation points sampled according to (18) is presented in Fig. 8, for $m = 21$ and γ_{sc} and γ_{spl}.

Linear regression applied to either γ_{spl} or γ_{sc} (for $m \in \{99, 102, \ldots, 120\}$) sampled according to (18) yields the following asymptotic estimates for trajectory approximation with four-point quadratic scheme (5):

$$\eta_{spl} = 2.2676 \quad \text{and} \quad \eta_{sc} = 2.1613. \quad (19)$$

The computed estimates from (19) indicate significant deceleration in asymptotics determined by (3).

(*ii*) We admit now another non-more-or-less uniform sampling:

$$t_0 = 0, \quad t_1 = \frac{1}{2m} - \frac{1}{m^2}, \quad t_2 = \frac{1}{2m}, \quad t_i = \frac{i}{m} \quad \text{(for } i \in \{3, 4, \ldots, m\}). \quad (20)$$

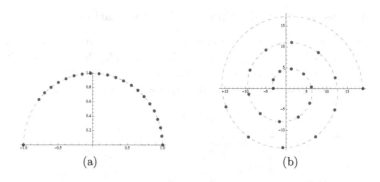

Fig. 8. Curves a) γ_{sc} and b) γ_{spl} sampled according to (18) with $m = 21$

Linear regression applied to either γ_{spl} or γ_{sc} (for $m \in \{99, 102, \ldots, 120\}$) sampled according to (20) yields the following non-decelerated asymptotic estimates for trajectory approximation with four-point quadratic (5):

$$\eta_{spl} = 4.047, \quad \text{and} \quad \eta_{sc} = 4.049. \tag{21}$$

This time both computed estimates from (21) comply with the asymptotics established by Th. 1.

Visibly for some non-more-or-less uniform samplings there is a duality in either reaching or breaking quartic convergence order.

□

The next example discusses the necessity of strict convexity in Th. 1.

Example 5. (*i*) First we admit the curve γ_{pol4} (see Example 1ii) without inflection point. Though our interpolant (5) is still constructable, Table 2 shows either a significant slowdown in convergence rate from (3) or admits a possible divergence (in case of computed estimates negative). This is related to the fact that $K(t)$ is not separated from zero.

Table 2.

Curve	Sampling	η
γ_{pol4}	uniform	−2.484
γ_{pol4}	(16)	−5.512
γ_{pol4}	(17)	0.4642

(*ii*) Consider now γ_{pol3} (see Example 1ii) in the two-side neighborhood of inflection point $t = 0$. This time the interpolation scheme (5) may not even be constructable. Indeed for sampling $t_0 = -1/m$, $t_1 = 0$, $t_2 = 1/m$ and $t_3 = 2/m$ both vectors p_1 and p_α (see (7)) are always co-linear thus yielding $e = 0$ which makes formula (8) incomputable.

□

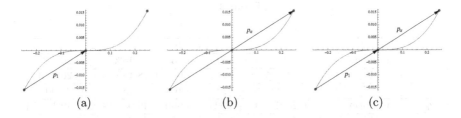

Fig. 9. Colinear (see c)) vectors a) p_1 and b) p_α on curve γ_{pol}

4   Four-Points Quadratics with Inflection Points

Recall cumulative chord cubic $\hat{\gamma}_3^i$ (see [10]) defined over each consecutive quadruple of points $(q_i, q_{i+1}, q_{i+2}, q_{i+3})$ for $\hat{t} \in [\hat{t}_i, \hat{t}_{i+3}]$:

$$\hat{\gamma}_3^i(\hat{t}) = \hat{\gamma}_3^i[\hat{t}_i] + \hat{\gamma}_3^i[\hat{t}_i, \hat{t}_{i+1}](\hat{t} - \hat{t}_i) + \hat{\gamma}_3^i[\hat{t}_i, \hat{t}_{i+1}, \hat{t}_{i+2}](\hat{t} - \hat{t}_i)(\hat{t} - \hat{t}_{i+1})$$
$$+ \hat{\gamma}_3^i[\hat{t}_i, \hat{t}_{i+1}, \hat{t}_{i+2}, \hat{t}_{i+3}](\hat{t} - \hat{t}_i)(\hat{t} - \hat{t}_{i+1})(\hat{t} - \hat{t}_{i+2}),$$

where $f[x_0, x_1, \ldots, x_k]$ denote respective divided differences (see e.g. [7]). Here:

$$\hat{t}_i = 0,$$
$$\hat{t}_{i+1} = \|q_{i+1} - q_i\|,$$
$$\hat{t}_{i+2} = \|q_{i+2} - q_{i+1}\| + \hat{t}_{i+1},$$
$$\hat{t}_{i+3} = \|q_{i+3} - q_{i+2}\| + \hat{t}_{i+2}.$$

Cumulative chord cubics $\hat{\gamma}_3$ (a sum-track of $\hat{\gamma}_3^i$) approximate trajectory of γ at least to order four which matches the same orders as if $\{t_i\}_{i=0}^m$ are given (see [10]). Recall that the unknown curvature $K(t)$ for applying the interpolant Q needs to be separated from zero. However, the proof of Th. 1 (see [1]) shows that curvature $K_Q(\hat{t}) = \det(Q'(\hat{t}), Q''(\hat{t}))/\|Q'(\hat{t})\|^3$ (computed from reduced data) is separated from zero for more-or-less uniformly sampled regular curve γ. Therefore, taking into account the latter we can set up an arbitrary zero curvature buffer zones $\varepsilon_0 > 0$. Namely we apply interpolant Q if $|K_Q(\hat{t})| > \varepsilon_0$ and otherwise cumulative chord piecewise $\hat{\gamma}_3^i$ which is not constrained by inflection points. Such new corrected interpolation scheme is denoted by $\bar{Q}_{\varepsilon_0}$ (we omit subscript ε_0).

Table 3 contains the results for asymptotic estimates in trajectory approximation by using corrected interpolant $\bar{Q}$. Note that ω used in the last column header represents the ratio of Q-segments against all segments.

Table 3.

Curve	Sampling	ε_0	m	η	ω
γ_{pol3}	uniform	1	$99 \leq m \leq 120$	4.138	$18.18\% \leq \omega \leq 20.00\%$
γ_{pol3}	(16)	1	$99 \leq m \leq 120$	4.098	$18.18\% \leq \omega \leq 20.00\%$
γ_{pol3}	(17)	1	$99 \leq m \leq 120$	4.104	$18.18\% \leq \omega \leq 20.00\%$
γ_{pol3}	uniform	0.1	$2001 \leq m \leq 2541$	3.960	$49.03\% \leq \omega \leq 49.11\%$
γ_{pol3}	(16)	0.1	$999 \leq m \leq 1041$	4.114	$48.95\% \leq \omega \leq 48.99\%$
γ_{pol3}	(17)	0.1	$2001 \leq m \leq 2541$	3.936	$49.11\% \leq \omega \leq 49.18\%$

5   Conclusion

In this paper, we confirm numerically and analytically (via symbolic computations) the sharpness of Th. 1. The symbolic computation is also used to construct four-point quadratic interpolant Q.

Additionally, the necessity of admitting more-or-less uniformity is also discussed. Our experiments clearly demonstrate that certain non-more-or-less uniform samplings yield decelerated quartic convergence rates in trajectory estimation upon using (5). On the other hand, as shown, there are some non-more-or-less uniform samplings which still preserve quartic order of convergence established by Th. 1. It remains an open question for which non-more-or-less uniform samplings the above duality holds. In case of slower rates, another question for examining real decelerated asymptotics arises.

Lastly the assumption of strict convexity in Th. 1 is also addressed. First, the tests on non-strictly convex curves (with inflection points excluded) render either significant slowdown or divergence in convergence rate (3). However, when inflection points are admitted four-point quadratic interpolant Q may not be constructable. Thus, to alleviate this deficiency we propose a new interpolation scheme $\bar{Q}$ insensitive to inflection points. Equally as both Q and $\hat{\gamma}_3$ over each respective segment yield quartic order of trajectory estimation the new interpolant $\bar{Q}$ preserves the same asymptotics. The latter is also confirmed here experimentally. We anticipate advantage in using Q over applying $\hat{\gamma}_3$ for estimating the curvatures.

More discussion on applications (including real data examples - see [3]) and theory of non-reduced data interpolation can be found in [2], [9], [4], [14], [15] or [16]. In particular different parameterizations $\{\hat{t}\}_{i=0}^m$ of the unknown interpolation knots $\{t\}_{i=0}^m$ are discussed, e.g., in [5], [6], [17] or [18].

Acknowledgement. This research was supported in part by PL-Grid Infrastructure [11].

References

1. Noakes, L., Kozera, R.: More-or-less-uniform sampling and lengths of curves. Quarterly of Applied Mathematics 3(61), 475–484 (2003)
2. Kozera, R., Noakes, L.: Piecewise-quadratics and exponential parameterization for reduced data. Applied Mathematics and Computation 221, 620–638 (2013)
3. Janik, M., Kozera, R., Kozioł, P.: Reduced data for curve modeling - applications in graphics, computer vision and physics. Advances in Science and Technology 7(18), 28–35 (2013)
4. Piegl, L., Tiller, W.: The NURBS Book. Springer, Heidelberg (1997)
5. Kvasov, B.I.: Methods of Shape-Preserving Spline Approximation. World Scientific Publishing Company, Singapore (2000)
6. Mørken, K., Scherer, K.: A general framework for high-accuracy parametric interpolation. Mathematics of Computation 66(217), 237–260 (1997)
7. De Boor, C.: A Practical Guide to Splines. Springer, Heidelberg (2001)
8. Noakes, L., Kozera, R.: Interpolating sporadic data. In: Heyden, A., Sparr, G., Nielsen, M., Johansen, P. (eds.) ECCV 2002, Part II. LNCS, vol. 2351, pp. 613–625. Springer, Heidelberg (2002)
9. Kozera, R.: Curve modeling via interpolation based on multidimensional reduced data. Studia Informatica 25(4B-61), 1–140 (2004)
10. Noakes, L., Kozera, R.: Cumulative Chord Piecewise Quadratics and Piecewise Cubics. Geometric Properties for Incomplete Data. In: Klette, R., Kozera, R., Noakes, L. (eds.) Computational Imaging and Vision, pp. 59–76. Springer (2006)
11. PL-Grid Infrastructure, http://www.plgrid.pl/en
12. Wolfram Mathematica 9, Documentation Center,
   http://reference.wolfram.com/mathematica/guide/Mathematica.html
13. Kozera, R., Noakes, L., Szmielew, P.: Trajectory estimation for Exponential Parameterization and Different Samplings. In: Saeed, K., Chaki, R., Cortesi, A., Wierzchoń, S. (eds.) CISIM 2013. LNCS, vol. 8104, pp. 430–441. Springer, Heidelberg (2013)
14. Kozera, R., Noakes, L., Szmielew, P.: Length Estimation for Exponential Parameterization and ϵ-Uniform Samplings. In: Huang, F., Sugimoto, A. (eds.) PSIVT 2013 Workshops. LNCS, vol. 8334, pp. 33–46. Springer, Heidelberg (2014)
15. Farin, G.: Curves and Surfaces for Computer Aided Geometric Design, 3rd edn. Academic Press, San Diego (1993)
16. Epstein, M.P.: On the influence of parameterization in parametric interpolation. SIAM J. Numer. Anal. 13, 261–268 (1976)
17. Koci, L.M., Simoncelli, A.C., Della Vecchia, B.: Blending parameterization of polynomial and spline interpolants, Facta Universitatis (NIŠ). Series Mathematics and Informatics 5, 95–107 (1990)
18. Lee, E.T.Y.: Choosing nodes in parametric curve interpolation. Computer-Aided Design 21(6), 363–370 (1987)

Scheme for Numerical Investigation of Movable Singularities of the Complex Valued Solutions of Ordinary Differential Equations

Radosław Antoni Kycia

University of Warsaw
Faculty of Mathematics, Informatics and Mechanics
Banacha 2, Warsaw, 02-097, Poland
rkycia@mimuw.edu.pl

Abstract. We present structure of integration scheme suitable for ordinary differential equations in some bounded region of the complex plane. The program which bases on these ideas can help to obtain qualitative information about the structure of singularities of solutions in the complex plane. It was tested on two representative examples.

Keywords: movable singularities, ordinary differential equations, numerical integration.

1 Introduction

Explicit integration of ordinary differential equations (ODEs) by means of elementary functions is generally impossible, even for simple equations. Therefore, approximate methods for constructing solutions like the expansion in a convergent power series are usually used. These power series have generally some finite radius of convergence in the complex plane due to the existence of singularities at the boundary of their balls of convergence.

ODEs have two types of singularities [I, Hi, D]. First type is called a fixed singularity. They are the singularities of the coefficients of the equation. This property makes that they are easy to localize. Second type of singularities is connected with the solutions of equations. The positions of theses singularities in the complex plane change as the initial data or the parameters of the equation varies, and therefore, they are called movable singularities. Usually, the solution is not known in a closed form, consequently, there is no knowledge of functional dependence of the solution on the initial data or the parameters of the equation. As a result, there is no general method to determine the positions of these singularities. The existence of movable singularities is the property of nonlinear equations, i.e., linear equations have only fixed singularities [I, Hi, D].

The simplest example of the equation that possesses this type of a singularity is

$$\frac{dy(x)}{dx} + y(x)^2 = 0, \tag{1}$$

V.P. Gerdt et al. (Eds.): CASC Workshop 2014, LNCS 8660, pp. 286–301, 2014.

which has the solution

$$y(x) = \frac{-1}{x - C},\qquad(2)$$

where C is determined by the initial data. The pole of the solution is located at the point $x = C$ and changes its position when C is varied.

Not only the existence but often the type of a singularity (pole, branch point, essential singularity) is important because it has profound implication on the global structure of solutions [C, G]. Painlevé formulated very restrictive definition of the solution of ODE [C], namely, it must be unambiguous mapping of the Riemann sphere onto itself, i.e., a single-valued function. To make multivalued prescription for a function the method of uniformization can be used. This can be done in two ways, either by defining the cuts on the complex plane which cannot be crossed if a given function would have to stay single-valued or by using equivalent approach that focuses on the definition of an appropriate Riemann surface [AF]. Success in applying uniformization relies on the ability to provide correct definition of cuts. Loosely speaking, one has to know how and where to place them in the complex plane. For the solutions which possess movable branch points there is no method to make them single-valued, and therefore, they are not 'solutions' in the sense of Painlevé. Such singularities which are obstacles to the uniformization procedure are called critical points. As a conclusion, the well-behaved solutions are such that they possess only poles as movable singularities. This feature of solutions is called the Painlevé property. There is also even more important issue — connection between the Painlevé property and the existence of algebraic first integrals [G]. These two subjects show that the knowledge of the location and the types of movable singularities is important as well as any tool that could provide some information about them.

Some hints about the location of movable singularities can be obtained using numerical methods. There are many approaches to this subject. If there is no need to know the location of singularities then the method which bases on numerical calculation of the coefficients of the local expansion of a solution and then estimation of the value of the radius of convergence of the series that defines this solution is used [AF, MA]. The method gives the distance from the expansion point to the nearest singularity, however, without any hint about the direction to this singularity. Sometimes the Padé approximation is applied to the truncated series solution [PTWF, FW]. This approximation allows one to make the analytic continuation of the solution from the real line to the complex plane. The zeros of the denominator of the approximation suggest where singularities are located. There is, however, a problem with interpretation of these results because of the fact that the zeros of numerator and denominator usually not cancel out exactly due to a small discrepancy between their positions in the complex plane which is a peculiar property of the Padé approximation. This behavior can produce artificial singularities. The simplest, the most obvious and direct method of finding singularities is to integrate the equation from the point at which the initial data are prescribed along some path in the complex plane and end when some condition which suggests the vicinity of a singularity, e.g., large norm of the solution, is met or when integration procedure leaves the domain in which the solution has to be analysed.

If the paths are dense enough in the examined region then this method can visualize the structure of singularities and hopefully helps to formulate some quantitative predictions. Sometimes using additional knowledge of the properties of the equation the methods described above can be combined to obtain efficient algorithm, see [FW] and the references therein.

Our aim is moderate – we want to describe structure and provide simple implementation of the program and library in C++ and Mathematica language [MATHEMATICA] which can be used to obtain qualitative information about the structure of singularities of the solutions of general ODEs in some domains in the complex plane using method of integration along paths that fill 'densely' this domain. The results obtained by the use of this program can help to develop intuition about the nature of solutions and indicate the way of how to approach the problem from the qualitative point of view, see [N, WS].

In the next section, we provide detailed definition of the problem and the method of solution using object-oriented approach in C++ language. Then in the following section, we will describe some results that can be obtained using the program. These examples will be used to test our program. Finally we discuss issues connected with further development of the software: parallelization, the use of additional numerical integration schemes and singularity detection methods. At the end we describe main points of implementation of the algorithm in Mathematica which will lead us to the reformulation of the problem in terms of functional programming.

2   The Definition of the Problem and Its Solution

We consider ODEs which can be written in the form of (usually nonautonomous) systems of first order differential equations

$$\frac{d\boldsymbol{y}(x)}{dx} = \boldsymbol{f}(\boldsymbol{y}; x), \quad \boldsymbol{y}(x) : x \in \mathbb{C} \to \mathbb{C}^n. \tag{3}$$

Function $\boldsymbol{f}$ is a vector of complex valued functions. To impose uniqueness of the solution it is usually assumed that f fulfils the Lipschitz condition with respect to $\boldsymbol{y}$ [Hi], however, this condition can be relaxed if one is interested in more general cases. The number n is called the rank of ODE. In order to define this problem properly initial conditions (IC) have to be provided. Their formulation will be discussed below. We ask about the nature of the solutions in the complex plane with emphasis on the location and structure of possible movable singularities. We start to present the way of obtaining algorithmic solution of this problem by describing the structure of C++ library which allows us to integrate (3).

The basic data types used in the program are complex numbers and vectors of complex numbers which are defined by

```
typedef complex< double >  cmplx;
typedef vector < cmplx >  cvector;
```

The most obvious implementation of the equation bases on the function derivs() that calculates the vector of the RHS of (3) and has to be supplied by the user, see Chap. 17 of [PTWF]. This function is wrapped in the class which is shown in Fig. 1.

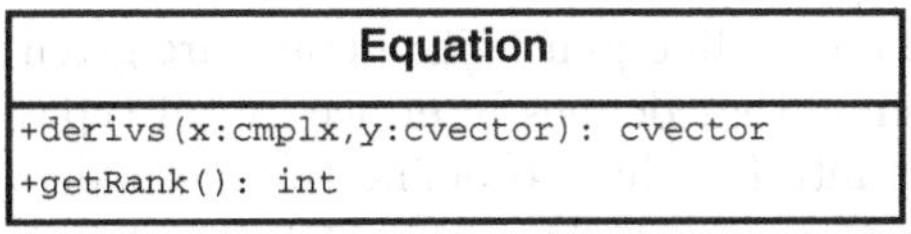

```
┌─────────────────────────────────────────────────┐
│                   Equation                      │
├─────────────────────────────────────────────────┤
│ +derivs(x:cmplx,y:cvector): cvector             │
│ +getRank(): int                                 │
└─────────────────────────────────────────────────┘
```

Fig. 1. The class Equation contains the method derivs(...), which calculates the RHS of (3) and the method getRank() that returns the rank of the equation. Both methods have to be supplied by the user.

The next part of properly defined initial value problem for ODEs are appropriate initial conditions (IC). They allow to start integration from the point x_0 at which $y(x_0) = y_0$, where y_0 is a constant vector. Sometimes, numerical integration cannot start from the point where IC were prescribed, e.g., due to the existence of a fixed singularity there. In these cases, some shift of IC to a new regular point has to be performed. This is usually achieved by the use of the local analytic expansion (if it exists) around this point. If a truncated Taylor series is used for this shift, then some control of the truncation error should be imposed. Then the starting point of integration can be chosen in a punctured disc with some small radius and the center at the expansion point. From this point of view, we can distinguish two cases:

- Initial conditions are defined at the fixed point x_0: $y(x_0) = y_0$.
- Initial conditions are defined in a punctured disc around x_0, i.e., at every x: $0 < |x_0 - x| < \epsilon$, for some small positive real ϵ.

This functionality is realized in the library by the method initialize(cmplx x0, Equation * eq), see Fig. 2. Only one set of global initial conditions should

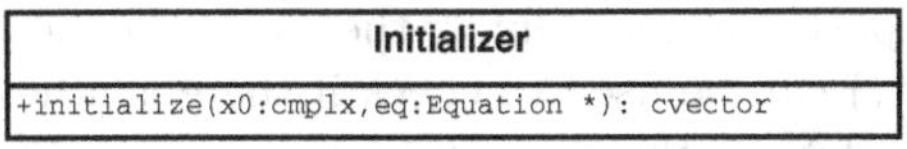

```
┌─────────────────────────────────────────────────┐
│                   Initializer                   │
├─────────────────────────────────────────────────┤
│ +initialize(x0:cmplx,eq:Equation *): cvector    │
└─────────────────────────────────────────────────┘
```

Fig. 2. Initializer class contains the method initialize(...), which calculates IC at x_0. The method has to be supplied by the user.

be imposed. All the other conditions at different points can be obtained by propagation of the global IC along some paths that do not cross singularities. This approach guarantees that all initial conditions will be consistent.

Numerical integration of ODE in the complex plane is performed along some path. To exclude pathological situations it is assumed that the path is a smooth piecewise curve that can be approximated by line segments of the given length $h > 0$, the more accurate the smaller is the value of h. Using numerical methods we will be integrating the equation along these line segments that build an approximation of the path.

The interface that realizes functionality of the line-segment curve is similar to the iterator design pattern [GHJV]. It also works similar to the stream that returns position of the next line joint. The details are given in Fig. 3. The class Path is an abstract class that defines interface for all types of paths. The class that implements this interface has to define the method that returns the first point of the path: getBeginning(), the method which can be used to check if there is the next element of the path: hasNext() and the method that returns this next element: getNext(). First type of two implemented paths is a semiline path built with a segments of the length h, which is realized by the class SemilinePath. It defines the path $x(t) = xbegin + (t + shift) \cdot e^{i\phi}$, where $xbegin$ is a complex point that defines the beginning point of the path, the real parameter t defines the position on the path, and $shift$ is an offset of t. getBeginning() returns $x(0)$ and kth call of getNext() returns $x(t_k)$, where $t_k = t_{k-1} + h$, $t_0 = 0$ and $k \in \mathbb{N}_{>0}$. The class SpiralPath implements a spiral path according to the formula $x(t) = (xbegin + (at + b)e^{i \cdot dir \cdot t})e^{i\phi}$. As before, $xbegin$ is a starting point in the complex plane, a, b are real parameters of the spiral, dir is a real parameter that determines the direction of wrapping of the spiral, and ϕ is some fixed angle of rotation of the spiral around $xbegin$.

We are interested in curves that cover 'densely' some domain in the complex plane. These curves have to have the same starting point at which initial conditions are specified. Therefore, it is natural to restrict our attention to the simply connected areas. Some additional restriction on the shape of the area has to be imposed if specific types of curves are used, e.g., for semilines, the domain has to be a star-shaped region, i.e., every point from the area can be connected with the central point x_0 along a segment line. The template class SimplyConnectedDomain parameterized by the variable pathType aggregates all paths of the given type along which integration will be performed. Before the integration procedure starts the object of this class has to be filled with paths using fill() method. Then ith path can be extracted by the method getPath(i). The number of paths stored in the object can be obtained using getNPaths() method. The class can also return paths sequentially by the use of the method getNextPath(). The method hasNext() can be used to check if there is a next path.

The class Domain defines the domain of integration. In this version of the program, it defines rectangular region in the complex plane with $zleft$ as the upper left corner and $zright$ as the lower right corner. These parameters have to be passed to the constructor of this class. The test if the point z is in the domain can be done with the help of isInDomain(z) method, which returns true if z is in the domain or false otherwise. Domain class can easily be altered to realize different shapes of the region of integration.

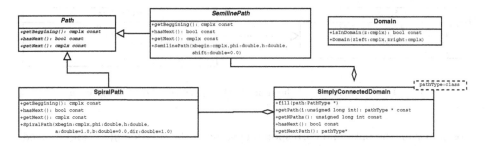

Fig. 3. The figure presents all the classes that define the topology of the region of integration

The last part of the library is a numerical integrator. It has to posses the following features:

– Stepper capability — every step defined by the complex number h can be divided into small steps, and integration along these steps can be independently but sequentially performed. The method of division can be regulated by the accuracy requirements and/or the constraints on the time of computation, see Chap. 17.2.3 of [PTWF].

– Monitor capability — approach to a possible singularity should be detected. There are various methods. The simplest approach which is used in our implementation is the test if the norm $||\boldsymbol{y}||_\infty = \max_{i\in\{1,...,n\}}\{|y_1|,\dots,|y_n|\}$ of the solution $\boldsymbol{y}$ is greater than a fixed large number.

The general scheme of integrator/stepping method is presented in Fig. 4. It is based on general ideas outlined in Chap. 17 of [PTWF]. The Stepper interface defines three methods that every class that realizes this interface (perform integration along a line segment) have to implement. The method makeStep() tries to make the step from x to $x + h$ with initial data y using internal integrator. It returns logical true if the step was successful or false if proximity of a singularity was detected – in current implementation, it is tested if the norm of the solution is greater than the value defined in the variable $solAbsMax$. The position x at which integration stopped can be obtained by calling getCurrentX() method and the vector of values of $y(x)$ at this position using getCurrentY() method. Currently, this interface is implemented by the stepper FixedRK4Stepper which divides the step h into small steps of length $hmax$ and makes fixed step integration along these small segments using classical fourth-order Runge–Kutta method [Bu]. New stepping algorithms will be implemented in the next versions of the library.

The solution along a segment path is stored as the STL C++ vector of tuples $\{x, \boldsymbol{y}(x)\}$, therefore, the solution in the domain is the collection of these solutions along different paths. The classes that store solution along one path and along all paths are presented in Fig. 5. PathSolution class was designed to store intermediate values x and $y(x)$ during integration along the path. The current

Fig. 4. The structure of the module that performs ODE integration

position x on the path and corresponding value of the solution $y(x)$ can be saved
in internal structure of the object of the class using fill() method. The class can
also save the results into file - method saveToFile(). When integration along a
single path is finished it is convenient to aggregate the objects of PathSolution
in the object of the class DomainSolution. This object can store and write all
paths into the file. When memory constraints are not a problem then collecting
all the results and then saving them to disk when integration is finished is a
faster approach than saving to disk every solution along a path right after it is
generated.

Fig. 5. The structure of the classes that store the solution

The final module of the program is the class that 'maps' initial-value problem
– the equation with IC – to the solution along a path using the selected method
of numerical integration inside prescribed domain. The structure of this class is
presented in Fig. 6. Mapper class have to be initialized by Equation object, IC
object, Domain constraints, and the object of the class which realizes Stepper
interface. Method makeStep() propagates initial conditions along a path and
returns pointer to the object of PathSolution that stores the solution along the
path.

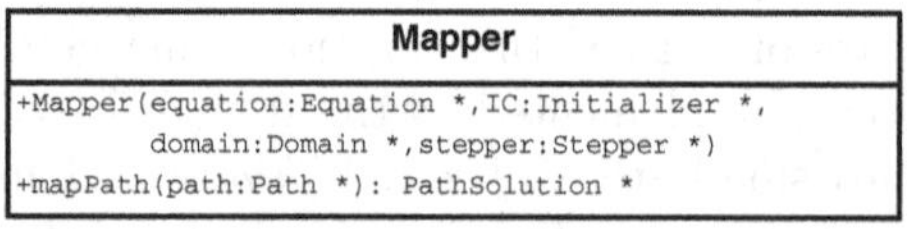

Mapper
+Mapper(equation:Equation *,IC:Initializer *, domain:Domain *,stepper:Stepper *)
+mapPath(path:Path *): PathSolution *

Fig. 6. Mapper class realize integration of ODE with IC along path. It stops integration
when the conditions that suggest singularity proximity are fulfilled.

The general scheme of generating of the solution of ODE in the complex plane
is represented by the following pseudo-code which is almost literally implemented
in the program:

```
Initialize Mapper using Equation, Initializer, Domain, and
specific stepper;
Prepare paths and store them in SimplyConnectedDomain object;
Initialize DomainSolution object;
For every path in SimplyConnectedDomain object:
{ Integrate IC along path using Mapper object;
  Store resulting PathSolution object in DomainSolution object; }
{ Integrate IC along path using Mapper object;
  Store resulting PathSolution object in DomainSolution object; }
Save data stored in DomainSolution to disk;
```

Generally, the computing time depends linearly on the number of paths. Some speed-up can be achieved by parallelization. This issue will be discussed later.

The run of the program generates the text file which stores the values of x and corresponding values of $y(x)$. Assuming that the paths cover densely the area, interpolation can be used to construct an approximate solution from these data. In current implementation[1], the set of gnuplot scripts [GNUPLOT] was created to visualize the solution. These scripts prepare various plots of the solution and save them to disk. They allow gnuplot to interpret the data from the text file generated by the program. They can also be run using GNU make [GNUMAKE] program. All commands start from 'make ' and then the option follows, e.g, 'make run'. Below there is the list of options:

- run – compile and run the program. Should be executed before generating plots.
- run-full – compile and run the program and then generate plot.
- absPlot, absContourPlot,abs2ContourPlot – every option creates a plot of the absolute value of the solution in the complex plane and saves it to disk.
- phasePlot, phaseContourPlot, phase2ContourPlot – every option creates a plot of the phase of the solution in the complex plane and saves it to disk.
- absPhasePlot – create the plot of the modulus of a solution with with color map that reflects the phase of the solution and save it to disk. The idea of this plot is based on [WS].
- areaOfConvergence – create the plot of the paths of integration in the complex plane and save it to disk.
- animate – create an animated gif file and open it in a default browser.
- Generate-doc – generate documentation from the code in html and TeX formats using Doxygen [DOXYGEN].
- clean – clean the directory from compilation and output files.

The program can be compiled and run on every operating system, however, the set of scripts which automatize compilation, plotting and generation of documentation are specific to the Linux operating systems. Nevertheless, they can be altered to run under other operating systems.

In the next section, we will provide two examples of application of this program.

---

[1] The program is available online at [KW].

3   Examples

In this section, we describe the application of the program to real-world problems. The structure of singularities (or lack thereof) of selected equations is well known, and, therefore, they can serve as test problems and as examples of the use of the program.

The simple fixed step size classical (fourth-order) Runge–Kutta method [Bu] was used as a first-order approximation to a more detailed study of the structure of singularities using more sophisticated integration methods. There is no obstacles in using adaptive time stepping by step doubling approach or in using embedded schemes, however, we only wanted to present the success of this approach. The simple Runge–Kutta method is a good first choice for such kind of problems as it was expressed in Chap. 17.3 of [PTWF]: 'That method does an excellent job of feeling its way through rocky or discontinuous terrain. It is also an excellent choice for a quick-and-dirty, low accuracy solution of a set of equations.' Therefore, as a method for qualitative analysis this method seems to be suitable.

The first example will be the Emden–Fowler equation.

3.1   The Emden–Fowler Equation

The equation has the following form

$$\frac{d^2y(x)}{dx^2} + \frac{\alpha}{x}\frac{dy(x)}{dx} + x^n y(x)^p = 0, \tag{4}$$

where $\alpha > 1$ is a real number and $n > -2$, $p > 1$ are integer constants. The equation has many applications [D], e.g., for $n = 2$, it is the famous Lane–Emden equation used in astrophysics [Hu]. The equation has fixed singularities at $x = 0$ and $x = \infty$.

There exists an analytic solution (convergent power series) for initial data $y(0) = c$, $y'(0) = 0$, where c is an arbitrary complex constant. It is defined in a punctured disc with the center at $x = 0$ even though it is singular point of the equation. This solution can be derived by introducing formal ansatz $y(x) = \sum_{k=0}^{\infty} a_k x^k$ into (4). Then the unique recurrence for the a_k coefficients can be obtained [Hu, KF]. The series has the following form

$$y(x) = c - \frac{c^p}{(n+2)(n+1+\alpha)} x^{n+2} + O(x^{n+3}), \tag{5}$$

where n is the parameter from (4). It can be proved that the series is convergent [KF]. However, it occurs that the series has finite radius of convergence due to the existence of movable singularities located at the rays connecting the origin with all $n + 2$ roots of -1, for details see [KF].

In numerical approach, integration cannot start from the singularity at $x = 0$, therefore, initial data have to be shifted slightly away from the origin by the use of (5) for x such that $0 < |x| < \epsilon$ for small ϵ that is much less that the radius of convergence of (5). The series can be truncated at arbitrary term if the accuracy is not important or it can be summed until prescribed accuracy is obtained. This truncated series has to be coded into Initializer::initialize() method. The equation has to be written in the form of a first-order system, which we give here for the reader's convenience ($' = \frac{d}{dx}$)

$$\begin{cases} u'(x) = v(x) \\ v'(x) = -\frac{\alpha}{x}v(x) - x^n u(x)^p, \end{cases} \tag{6}$$

where new variable $v(x)$ was introduced. This system has to be coded into Equation::derivs() method, and the rank of the equation in Equation::SetRank() method should be set to 2. The range of the domain of integration, number of paths, initial point of paths, and integration step also should be adjusted to the problem by editing main.cxx file. Then the command 'make run-full' produces the following results. Figure 7 presents the absolute value of the solution for $n = 1$. Similar pictures for other values of n can be found in [KF]. Movable singularities in the figure are located on the semilines connecting the origin with all three roots of $(-1)^{1/3}$. One can note that there are cuts (discontinuity in phase) that emanate from the singularities. The same assertion can be obtained using spiral paths instead of semilines. The equation (4) does not possess the Painlevé property, and the singularities are critical points, i.e., they are not poles.

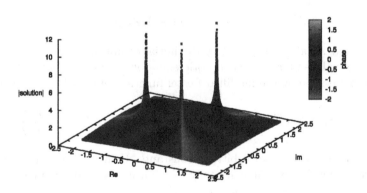

Fig. 7. Figure presents a plot of the modulus of the solution with colors that indicate the phase of it. It was plotted with points that sample the solution equidistantly along the paths of integration with distance between them $h = 0.01$. It is the solution of (4) with $\alpha = 2$, $p = 5$, $n = 1$ and with IC around $x = 0$ given by (5) with $c = 1.5$. The integration was performed along 5000 semilines that emanate from the origin and are parameterized by the polar angle from $[0; 2\pi)$ equidistantly. The starting point x_0 for the lines fulfils $|x_0| = 10^{-4}$. Animations are available online at [KW].

3.2   Electrochemical Reaction Equation

In this example, the analysis of linear equation will be shown. The equation

$$y''(x) - (s + x + \frac{1}{4}x^4)y(x) = 0, \tag{7}$$

where s is a real parameter assumed hereafter as a positive number, was obtained in [Bi] as an intermediate step in analysis of electrochemical reactions in transient experiments at channel and tubular electrodes. Similar equation was studied by Edward Charles Titchmarsch, see Chap. 5 of [Hi].

As it was mentioned above, the linear equation has no movable singularities, however, it will be shown that the output of the program can generate singularity-like results if not correctly interpreted.

The equation has the analytic solution around $x = 0$ in the form

$$y(x) = \sum_{k=0}^{\infty} a_k x^k$$

$$
\begin{array}{l}
a_0, a_1 \quad \text{arbitrary} \\
a_2 = -\frac{s}{2}a_0 \\
a_3 = -\frac{1}{6}(sa_1 + a_0) \\
\dots
\end{array}
\tag{8}
$$

The asymptotics at small x, i.e., the solution of the equation $y''(x) - (s+x)y(x) = 0$ is a combination of the Airy functions [OLBC]

$$y_0(x) = A_1 AiryAi(x + s) + A_2 AiryBi(x + s), \tag{9}$$

where A_1 and A_2 are determined by the initial conditions.

For large values of x, the asymptotic equation $y''(x) - \frac{1}{4}x^4y(x) = 0$ has the solution expressible by the modified Bessel functions [OLBC]

$$y_\infty(x) = B_1\sqrt{x}BesselI_{1/6}\left(\frac{x^3}{6}\right) + B_2\sqrt{x}BesselI_{-1/6}\left(\frac{x^3}{6}\right), \tag{10}$$

where B_1 and B_2 are also arbitrary constants. Using the asymptotics of the Bessel functions [OLBC] it can easily be shown that the dominant term for $|x| \to \infty$ behaves as $\frac{\exp(\frac{x^3}{6})}{\sqrt{x}}$, i.e., its modulus grows exponentially fast except of six directions in the complex plane. However, the solution is singular only at infinity. This can give a false imagination that there are only specific directions along which the solution is bounded, and other directions give singularity of a solution at some finite distance from the origin. It is due to the fact that the modulus of the solution grows fast, and the method used to indicate possible singularity in the program checks only if the modulus of the solution is bounded by some large but finite number. To remove this ambiguity in interpretation the better algorithm of indication of the existence of singularities is needed as it will be discussed in the next section.

Figure 8 presents the solution. Comparing this figure with Fig. 5.3 of [Hi] it is evident that the perturbation of the Titchmarsch equation by adding x to the coefficient of $y(x)$ is irrelevant, i.e., the leading term x^4 as $|x| \to \infty$ is dominant. Without x term (7) is invariant under reflection $x \to -x$ – this broken symmetry of the plot of the solution can be noticed from Fig. 8.

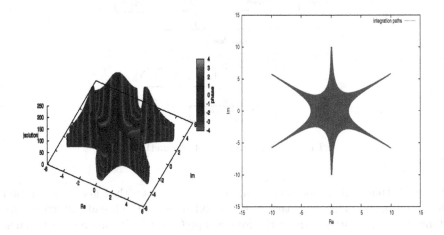

Fig. 8. Left figure presents the solution of (7) with IC around $x = 0$ given by (8) with $y(0) = 1$, $y'(0) = 0$. The right figure shows the paths along which integration was performed. The paths are line segments that emanate from the origin. The ending point of the path is located at the point at which the absolute value of the solution crossed the prescribed bound $\|y\|_\infty \leq 10^5$. The structure and values of parameters of the paths are the same as in the previous examples. Animations are available online at [KW].

For further examples, we refer the reader to the papers [K] and [KF] where the program was used as a tool that helps to resolve the structure of movable singularities of specific ODEs.

In the next section, generalization and improvement of the method will be proposed.

4   General Discussion and Prospects for Future

This section contains description of the ways of how to generalize the program and prospect for future development of it.

The first issue is the parallelization. The sequential methods of numerical integration like the Runge–Kutta methods that use the result of the previous step to derive the next one are almost impossible to parallelize. However, when one considers integration along paths then every integration is independent from one another, therefore, the easiest and the most obvious solution is to create a pool

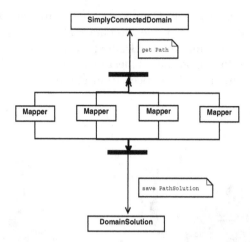

Fig. 9. Parallelization of the method

of threads that will do the same task as Mapper class – they generate solution along the paths. This approach is described in Fig. 9. Threads which realize Mapper functionality get the paths from SimplyConnectedDomain object. When integration along a path is finished then the solution along the path is saved in DomainSolution object. The access to the set of paths and saving solution have to be implemented as critical sections. This functionality was implemented and tested. The parallel version of the program is available online at [KW].

The next issue is related to the development of more efficient integration procedures, which can localize the positions of singularities with the precision higher than classical fourth-order Runge–Kutta method or any other standard method. In ideal situation, if during the integration along a path, the solution starts to become 'singular' then the method should try to determine the type of a singularity and handle it appropriately. One of the approach would be fitting some polynomial or logarithmic function to the solution along the path to determine the rate of divergence. In another approach, the singularity could be encircled by the path. When one turn around the singular point is performed then the jump in the solution is tested to see whether there is a branch point or a pole.

Finally, the issue of implementation of the algorithm using functional language approach can be considered as well. The building blocks of the method are quite general and can be encountered in many similar problems that require numerical integration of ODEs along a path. We can translate these concepts from an object-oriented approach to a functional one [ASuSu]:

- AreaQ – it is a predicate which determines if a point is in some area;
- Path – it is a function that returns points of a path;
- SolveAlongPath – it solves an equation with IC along a path and stops when appropriate condition that detects possible proximity of a singularity is met;

- DomainSolution – it is a higher order function which maps paths onto solutions using given integrator, equation, initial conditions, and area constraints;
- ShowDomainSolution – it plots the solution in a domain.

Using these ideas, the set of simple functions for Mathematica [MATHEMATICA] Computer Algebra System was created. The notebook with this library and examples of use can be downloaded from [KW]. The library in Mathematica is shorter and more compact than in C++. It is based on the following idea. If a path is given by an analytic prescription $p(t)$, e.g., line or spiral path, where t is a parameter along the path, then (3) can be pulled back on the path $x = p(t)$, which gives

$$\begin{cases} \frac{dy}{dt} = p'(t)\boldsymbol{f}(x(t), \boldsymbol{y}(t)) \\ \quad x'(t) = p'(t), \end{cases} \tag{11}$$

where $' = \frac{d}{dt}$ and where the last equation for the path was added. This transformation of the equation is automatically performed in our implementation. The system (11) with corresponding initial conditions can now be integrated with respect to t using Mathematica numerical ODE solver. For more details see notebook on [KW].

Implementation in the Mathematica language also explains how the algorithm can be rewritten in other languages which support functional approach like LISP, Scheme, MathML or Haskell to mention a few of them.

5 Conclusions

It was shown how general idea of obtaining qualitative information about the structure of the solutions of equations in the complex plane by integrating numerically equations along paths can be implemented in efficient way using an object-oriented approach. Then it was explained on some real-world examples how the program/library can be used. The examples served as a validation of the method and illustrate some difficulties in using this approach. They also show that this method is simple and powerful first approach in examination of the structure of movable singularities of ODEs. It was also discussed how the method can be augmented and improved. Implementation of the algorithm for Mathematica Computer Algebra System was also described.

Acknowledgment. RK is supported by the Warsaw Center of Mathematics and Computer Science from the funds of the Polish Leading National Research Centre (KNOW). Some sections were developed thanks to support of Polish National Science Centre grant UMO-2011/01/M/ST2/04126. The author is grateful to Galina Filipuk(MIM UW, PL), Tadeusz Chmaj(IFJ, PL) and Rod Halburd(UCL, UK) for enlightening discussions. The author also thanks the Referees for a number of helpful suggestions for improvement in the article.

References

[ASuSu]	Abelson, H., Sussman, G.J., Sussman, J.: Structure and Interpretation of Computer Programs, 2nd edn. MIT Press and McGraw-Hill (1996)
[AF]	Ablowitz, M.J., Fokas, A.S.: Complex Variables: Introduction and Applications, 2nd edn. Cambridge University Press (2003)
[Bi]	Bieniasz, L.K.: Automatic solution of the Singh and Dutt integral equations for channel or tubular electrodes, by the adaptive Huber method. J. Electroanal. Chem. 693, 95–104 (2013)
[Bu]	Butcher, J.C.: Numerical Methods for Ordinary Differential Equations, 2nd edn. John Wiley & Sons, Inc. (2008)
[C]	Conte, R. (ed.): The Painlevé Property. One Century Later. CRM Series in Mathematical Physics. Springer (1999)
[D]	Davis, H.T.: Introduction to Nonlinear Differential and Integral Equations. Dover Publications (2010)
[DOXYGEN]	Doxygen Web page, http://www.stack.nl/%7edimitri/doxygen/
[FW]	Fornberga, B., Weideman, J.A.C.: A numerical methodology for the Painlevé equations. J. Comput. Phys. 230, 5957–5973 (2011)
[GHJV]	Gamma, E., Helm, R., Johnson, R., Vlissides, J.: Design Patterns: Elements of Reusable Object-oriented Software, 1st edn. Addison–Wesley Professional (1994)
[GNUPLOT]	GNUPlot Web page, http://gnuplot.info/
[G]	Goriely, A.: Integrability and Nonintegrability of Dynamical Systems. Advanced Series on Nonlinear Dynamics. World Scientific (2001)
[Hi]	Hille, E.: Ordinary Differential Equations in the Complex Domain. Dover Publications (1997)
[Hu]	Hunter, C.: Series solutions for polytropes and the isothermal sphere. Mon. Not. R. Astron. Soc. 328, 839–847 (2001)
[I]	Ince, E.L.: Ordinary Differential Equations. Dover New York (1956)
[KF]	Kycia, R.A., Filipuk, G.: On the singularities of the Emden-Fowler type equations. In: Proc. ISAAC 2013 Conference (2013) (to appear)
[K]	Kycia, R.A.: On movable singularities of self-similar solutions of semilinear wave equations. In: Proc. from the Conference, "On Formal and Analytic Solutions of Differential and Difference Equations II", vol. 97, pp. 59–72. Banach Center Publ. (2012)
[KW]	Kycia, R.A.: Web page, http://www.mimuw.edu.pl/%7erkycia/
[MATHEMATICA]	Wolfram Mathematica Web page, http://www.wolfram.com/mathematica/
[GNUMAKE]	GNU make project Web page, https://www.gnu.org/software/make/
[MA]	Mohan, C., Al-Bayaty, A.R.: Power series solutions of the Lane–Emden equation. Astophysics and Space Science 73, 227–239 (1980)
[N]	Needham, T.: Visual Complex Analysis. Oxford University Press (1999)

[OLBC]   Olver, F.W.J., Lozier, D.W., Boisvert, R.F., Clark, C.W.: NIST
              Handbook of Mathematical Functions. Cambridge University
              Press (2010)
[PTWF]   Press, W.H., Teukolsky, S.A., Wetterling, W.T., Flannery, B.P.:
              Numerical Recipes: The Art of Scientific Computing, 3rd edn.
              Cambridge University Press (2007)
[WS]   Wegert, E., Semmler, G.: Phase plots of complex functions: A jour-
              ney in illustration. Notices Amer. Math. Soc. 58, 768–780 (2011)

Generalized Mass-Action Systems and Positive Solutions of Polynomial Equations with Real and Symbolic Exponents
(*Invited Talk*)

Stefan Müller and Georg Regensburger

Johann Radon Institute for Computational and Applied Mathematics (RICAM),
Austrian Academy of Sciences, Linz, Austria
{stefan.mueller,georg.regensburger}@ricam.oeaw.ac.at

Abstract. Dynamical systems arising from chemical reaction networks with mass action kinetics are the subject of chemical reaction network theory (CRNT). In particular, this theory provides statements about uniqueness, existence, and stability of positive steady states for all rate constants and initial conditions. In terms of the corresponding polynomial equations, the results guarantee uniqueness and existence of positive solutions for all positive parameters.

We address a recent extension of CRNT, called generalized mass-action systems, where reaction rates are allowed to be power-laws in the concentrations. In particular, the (real) kinetic orders can differ from the (integer) stoichiometric coefficients. As with mass-action kinetics, complex balancing equilibria are determined by the graph Laplacian of the underlying network and can be characterized by binomial equations and parametrized by monomials. In algebraic terms, we focus on a constructive characterization of positive solutions of polynomial equations with real and symbolic exponents.

Uniqueness and existence for all rate constants and initial conditions additionally depend on sign vectors of the stoichiometric and kinetic-order subspaces. This leads to a generalization of Birch's theorem, which is robust with respect to certain perturbations in the exponents. In this context, we discuss the occurrence of multiple complex balancing equilibria.

We illustrate our results by a running example and provide a MAPLE worksheet with implementations of all algorithmic methods.

Keywords: Chemical reaction network theory, generalized mass-action systems, generalized polynomial equations, symbolic exponents, positive solutions, binomial equations, Birch's theorem, oriented matroids, multistationarity.

1 Introduction

In this work, we focus on dynamical systems arising from (bio-)chemical reaction networks with *generalized* mass-action kinetics and positive solutions of the corresponding systems of generalized polynomial equations.

V.P. Gerdt et al. (Eds.): CASC Workshop 2014, LNCS 8660, pp. 302–323, 2014.

In chemical reaction network theory, as initiated by Horn, Jackson, and Feinberg in the 1970s [15,33,32], several fundamental results are based on the assumption of mass action kinetics (MAK). Consider the reaction

$$1\,A + 1\,B \to C \tag{1}$$

involving the reactant species A, B and the product C, where we explicitly state the stoichiometric coefficients of the reactants. The left- and right-hand sides of a reaction, in this case A+B and C, are called (stoichiometric) complexes. Let

$$[A] = [A](t)$$

denote the concentration of species A at time t, and analogously for B and C. Assuming MAK, the rate at which the reaction occurs is given by

$$v = k\,[A]^1[B]^1$$

with rate constant $k > 0$. In other words, the reaction rate is a monomial in the reactant concentrations [A] and [B] with the stoichiometric coefficients as exponents. Within a network involving additional species and reactions, the above reaction contributes to the dynamics of the species concentrations as

$$\frac{d}{dt}\begin{pmatrix}[A]\\[B]\\[C]\\[D]\\\vdots\end{pmatrix} = k\,[A][B]\begin{pmatrix}-1\\-1\\1\\0\\\vdots\end{pmatrix} + \cdots$$

In many applications, the reaction network is given, but the values of the rate constants are unknown. Surprisingly, there are results on existence, uniqueness, and stability of steady states that do not depend on the rate constants. See, for example, the lecture notes [16] and the surveys [17,19,30].

However, the validity of MAK is limited; it only holds for elementary reactions in homogeneous and dilute solutions. For biochemical reaction networks in intracellular environments, the rate law has to be modified. In previous work [40], we allowed generalized mass-action kinetics (GMAK) where reaction rates are power-laws in the concentrations. In particular, the exponents need not coincide with the stoichiometric coefficients and need not be integers. For example, the rate at which reaction (1) occurs may be given by

$$v = k\,[A]^a[B]^b$$

with kinetic orders $a, b > 0$. Formally, we specify the rate of a reaction by associating (here indicated by dots) with the reactant complex a kinetic complex, which determines the exponents in the generalized monomial:

$$A + B \;\to C$$

$$\vdots$$

$$aA + bB$$

Before we give the definition of generalized mass action systems, we introduce a running example, which will be used to motivate and illustrate general statements. Throughout the paper, we focus on algorithmic aspects of the theoretical results. Additionally, we provide a MAPLE worksheet[1] with implementations of all algorithms applied to the running example. For other applications of computer algebra to chemical reaction networks, we refer to [7,14,36,45].

Notation. We denote the strictly positive real numbers by $\mathbb{R}_>$. We define $e^x \in \mathbb{R}_>^n$ for $x \in \mathbb{R}^n$ component-wise, that is, $(e^x)_i = e^{x_i}$; analogously, $\ln(x) \in \mathbb{R}^n$ for $x \in \mathbb{R}_>^n$ and $x^{-1} \in \mathbb{R}^n$ for $x \in \mathbb{R}^n$ with $x_i \neq 0$. For $x, y \in \mathbb{R}^n$, we denote the component-wise (or Hadamard) product by $x \circ y \in \mathbb{R}^n$, that is, $(x \circ y)_i = x_i y_i$; for $x \in \mathbb{R}_>^n$ and $y \in \mathbb{R}^n$, we define $x^y \in \mathbb{R}_>$ as $\prod_{i=1}^n x_i^{y_i}$.

Given a matrix $B \in \mathbb{R}^{n \times m}$, we denote by $b^1, \ldots, b^m$ its column vectors and by $b_1, \ldots, b_n$ its row vectors. For $x \in \mathbb{R}_>^n$, we define $x^B \in \mathbb{R}_>^m$ as

$$(x^B)_j = x^{b^j} = \prod_{i=1}^n x_i^{b_{ij}}$$

for $j = 1, \ldots, m$. As a consequence,

$$\ln(x^B) = B^T \ln x.$$

Finally, we identify a matrix $B \in \mathbb{R}^{n \times m}$ with the corresponding linear map $B \colon \mathbb{R}^m \to \mathbb{R}^n$ and write $\operatorname{im}(B)$ and $\ker(B)$ for the respective vector subspaces.

2   Running Example

We consider a reaction network based on the weighted directed graph

$$1 \underset{k_{21}}{\overset{k_{12}}{\rightleftarrows}} 2 \qquad 4 \underset{k_{54}}{\overset{k_{45}}{\rightleftarrows}} 5 \qquad (2)$$

with k_{31}, k_{23}, and 3 forming the remaining structure.

with 5 vertices, 6 edges and corresponding positive weights. Clearly, the edges represent reactions and the weights are rate constants. We assume that the network contains 4 species A, B, C, D and associate with each vertex a (stoichiometric) complex, that is, a formal sum of species:

$$A + B \rightleftarrows C \qquad A \rightleftarrows D$$

$$2A$$

[1] The worksheet is available at http://gregensburger.com/software/GMAK.zip.

In order to specify the reaction rates, e.g., $v_{12} = k_{12}[\mathrm{A}]^{\frac{1}{2}}[\mathrm{B}]^{\frac{3}{2}}$, we additionally associate a kinetic complex with each source vertex:

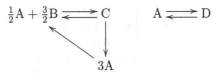

$$\tfrac{1}{2}\mathrm{A} + \tfrac{3}{2}\mathrm{B} \rightleftharpoons \mathrm{C} \qquad \mathrm{A} \rightleftharpoons \mathrm{D}$$

$$3\mathrm{A}$$

Writing

$$x = (x_1, x_2, x_3, x_4)^T$$

for the concentrations of species A, B, C, D, the dynamics of the generalized mass action system is given by

$$\frac{d}{dt}\begin{pmatrix} x_1 \\ x_2 \\ x_3 \\ x_4 \end{pmatrix} = \begin{pmatrix} -1 & 1 & 2 & -1 & -1 & 1 \\ -1 & 1 & 0 & 1 & 0 & 0 \\ 1 & -1 & -1 & 0 & 0 & 0 \\ 0 & 0 & 0 & 0 & 1 & -1 \end{pmatrix} \begin{pmatrix} k_{12}\,(x_1)^{\frac{1}{2}}(x_2)^{\frac{3}{2}} \\ k_{21}\,x_3 \\ k_{23}\,x_3 \\ k_{31}\,(x_1)^3 \\ k_{45}\,x_1 \\ k_{54}\,x_4 \end{pmatrix} = N\,v(x), \qquad (3)$$

where we fix an order on the edges, $E = \big((1,2),(2,1),(2,3),(3,1),(4,5),(5,4)\big)$, and introduce the stoichiometric matrix N and the vector of reaction rates $v(x)$.

We further decompose the system. Writing the stoichiometric and kinetic complexes as column vectors of the matrices

$$Y = \begin{pmatrix} 1 & 0 & 2 & 1 & 0 \\ 1 & 0 & 0 & 0 & 0 \\ 0 & 1 & 0 & 0 & 0 \\ 0 & 0 & 0 & 0 & 1 \end{pmatrix} \quad \text{and} \quad \tilde{Y} = \begin{pmatrix} \tfrac{1}{2} & 0 & 3 & 1 & 0 \\ \tfrac{3}{2} & 0 & 0 & 0 & 0 \\ 0 & 1 & 0 & 0 & 0 \\ 0 & 0 & 0 & 0 & 1 \end{pmatrix}$$

and using the incidence matrix of the graph (2),

$$I_E = \begin{pmatrix} -1 & 1 & 0 & 1 & 0 & 0 \\ 1 & -1 & -1 & 0 & 0 & 0 \\ 0 & 0 & 1 & -1 & 0 & 0 \\ 0 & 0 & 0 & 0 & -1 & 1 \\ 0 & 0 & 0 & 0 & 1 & -1 \end{pmatrix},$$

we can write the stoichiometric matrix as

$$N = Y\,I_E.$$

The vector of reaction rates $v(x)$ can also be decomposed by introducing a diagonal matrix

$$\Delta_k = \mathrm{diag}(k_{12}, k_{21}, k_{23}, k_{31}, k_{45}, k_{54})$$

containing the rate constants, a matrix indicating the source vertex of each reaction,

$$I_s = \begin{pmatrix} 1 & 0 & 0 & 0 & 0 & 0 \\ 0 & 1 & 1 & 0 & 0 & 0 \\ 0 & 0 & 0 & 1 & 0 & 0 \\ 0 & 0 & 0 & 0 & 1 & 0 \\ 0 & 0 & 0 & 0 & 0 & 1 \end{pmatrix},$$

and the vector of monomials determined by the kinetic complexes,

$$x^{\tilde{Y}} = \begin{pmatrix} (x_1)^{\frac{1}{2}}(x_2)^{\frac{3}{2}} \\ x_3 \\ (x_1)^3 \\ x_1 \\ x_4 \end{pmatrix}.$$

Then,

$$v(x) = \Delta_k\, I_s^T\, x^{\tilde{Y}},$$

and we can write

$$\frac{\mathrm{d}x}{\mathrm{d}t} = N\,v(x) = Y\,I_E\,\Delta_k\,I_s^T\,x^{\tilde{Y}}.$$

Note that the matrix

$$A_k = I_E\,\Delta_k\,I_s^T = \begin{pmatrix} -k_{12} & k_{21} & k_{31} & 0 & 0 \\ k_{12} & -(k_{21}+k_{23}) & 0 & 0 & 0 \\ 0 & k_{23} & -k_{31} & 0 & 0 \\ 0 & 0 & 0 & -k_{45} & k_{54} \\ 0 & 0 & 0 & k_{45} & -k_{54} \end{pmatrix} \quad (4)$$

depends only on the weighted digraph, while Y and $x^{\tilde{Y}}$ are determined by the stoichiometric and kinetic complexes. The resulting decomposition

$$\frac{\mathrm{d}x}{\mathrm{d}t} = Y\,A_k\,x^{\tilde{Y}}$$

is due to [33], where A_k is called kinetic matrix and the stoichiometric and kinetic complexes are equal, that is, $Y = \tilde{Y}$. The interpretation of A_k as a weighted graph Laplacian was introduced in [24] and used in [12,47,31,37,34], in particular, in connection with the matrix-tree theorem.

3 Generalized Mass Action Systems

We consider *directed graphs* $G = (V, E)$ given by a finite set of *vertices*

$$V = \{1, \ldots, m\}$$

and a finite set of *edges* $E \subseteq V \times V$. We often denote an edge $e = (i, j) \in E$ by $i \to j$ to emphasize that it is directed from the *source* i to the *target* j. Further, we write

$$V_s = \{i \mid i \to j \in E\}$$

for the set of source vertices that appear as a source of some edge.

Definition 1. *A* generalized chemical reaction network $(G, y, \tilde{y})$ *is given by a digraph* $G = (V, E)$ *without self-loops, and two functions*

$$y \colon V \to \mathbb{R}^n \quad and \quad \tilde{y} \colon V_s \to \mathbb{R}^n$$

assigning to each vertex a (stoichiometric) complex *and to each source a* kinetic complex.

We note that this definition differs from [40]. On the one hand, kinetic complexes were assigned also to non-source vertices, on the other hand, all (stoichiometric) complexes had to be different, and analogously the kinetic complexes.

Definition 2. *A* generalized mass action system $(G_k, y, \tilde{y})$ *is a generalized chemical reaction network* $(G, y, \tilde{y})$, *where edges* $(i, j) \in E$ *are labeled with* rate constants $k_{ij} \in \mathbb{R}_>$.

The contribution of reaction $i \to j \in E$ to the dynamics of the species concentrations $x \in \mathbb{R}^n$ is proportional to the *reaction vector* $y(j) - y(i) \in \mathbb{R}^n$. Assuming generalized mass action kinetics, the rate of the reaction is determined by the source kinetic complex $\tilde{y}(i)$ and the positive rate constant k_{ij}:

$$v_{i \to j}(x) = k_{ij}\, x^{\tilde{y}(i)}.$$

The ordinary differential equation associated with a generalized mass action system is defined as

$$\frac{dx}{dt} = \sum_{i \to j \in E} k_{ij}\, x^{\tilde{y}(i)} \big(y(j) - y(i)\big).$$

The change over time lies in the *stoichiometric subspace*

$$S = \operatorname{span}\{y(j) - y(i) \in \mathbb{R}^n \mid i \to j \in E\},$$

which suggests the definition of a (positive) *stoichiometric compatibility class* $(c' + S) \cap \mathbb{R}^n_>$ with $c' \in \mathbb{R}^n_>$.

In case every vertex is a source, that is, $V_s = V$, we introduce also the *kinetic-order subspace*

$$\tilde{S} = \operatorname{span}\{\tilde{y}(j) - \tilde{y}(i) \in \mathbb{R}^n \mid i \to j \in E\}.$$

In order to decompose the right-hand side of the ODE system, we define the matrices $Y \in \mathbb{R}^{n \times m}$ as $y^j = y(j)$ and $\tilde{Y} \in \mathbb{R}^{n \times m}$ as $\tilde{y}^j = \tilde{y}(j)$ for $j \in V_s$ and $\tilde{y}^j = 0$ otherwise (see also the remark below). Further, we introduce the weighted *graph Laplacian* $A_k \in \mathbb{R}^{m \times m}$: $(A_k)_{ij} = k_{ji}$ if $j \to i \in E$, $(A_k)_{ii} = -\sum_{i \to j \in E} k_{ij}$, and $(A_k)_{ij} = 0$ otherwise. We obtain:

$$\frac{dx}{dt} = Y A_k \, x^{\tilde{Y}}.$$

Note that $\tilde{y}^j$ can be chosen arbitrarily for $j \notin V_s$, since in this case $(A_k)^j = 0$ and hence $(A_k)^j x^{\tilde{y}^j} = 0$.

Steady states of the ODE satisfying $x \in \mathbb{R}^n_>$ and $A_k \, x^{\tilde{Y}} = 0$ are called *complex balancing equilibria*. We denote the corresponding set by

$$Z_k = \{x \in \mathbb{R}^n_> \mid A_k \, x^{\tilde{Y}} = 0\}.$$

Finally, the *(stoichiometric) deficiency* is defined as

$$\delta = m - l - s,$$

where m is the number of vertices, l is the number of connected components, and $s = \dim S$ is the dimension of the stoichiometric subspace.

Using $S = \mathrm{im}(Y I_E)$, where I_E is the incidence matrix of the graph (for a fixed order on E), we obtain the equivalent definition

$$\delta = \dim(\ker(Y) \cap \mathrm{im}(I_E)),$$

see for example [34]. Further, note that $\mathrm{im}(A_k) \subseteq \mathrm{im}(I_E)$. Now, if $\delta = 0$, then $\ker(Y) \cap \mathrm{im}(A_k) \subseteq \ker(Y) \cap \mathrm{im}(I_E) = \{0\}$, and there are no $x \in \mathbb{R}^n_>$ such that $Y A_k \, x^{\tilde{Y}} = 0$, but $A_k \, x^{\tilde{Y}} \neq 0$. In other words, if $\delta = 0$, there are no steady states other than complex balancing equilibria.

4   Graph Laplacian

A basis for the kernel of A_k in (4) is given by

$$(k_{31} k_{21} + k_{31} k_{23}, k_{12} k_{31}, k_{23} k_{12}, 0, 0)^T \quad \text{and} \quad (0, 0, 0, k_{54}, k_{45})^T.$$

Obviously, the support of the vectors coincides with the connected components of the graph. In general, this holds for the strongly connected components without outgoing edges.

Let $G_k = (V, E, k)$ be a weighted digraph without self-loops and A_k its graph Laplacian. Further, let l be the number of connected components (aka linkage classes) and $T_1, \ldots, T_t \subseteq V$ be the sets of vertices within the strongly connected components without outgoing edges (aka terminal strong linkage classes). Clearly, $t \geq l$. A fundamental result of CRNT [21] states that there exist linearly independent $\chi^1, \ldots, \chi^t \in \mathbb{R}^n_{\geq}$, where $\chi^\lambda_\mu > 0$ if $\mu \in T_\lambda$ and $\chi^\lambda_\mu = 0$ otherwise, such that $\ker(A_k) = \mathrm{span}\{\chi^1, \ldots, \chi^t\}$.

In fact, the non-zero entries in the basis vectors can be computed using the matrix-tree theorem:

$$\chi^\lambda_\mu = K_\mu, \quad \lambda \in \{1, \ldots, t\}$$

with *tree constants*

$$K_\mu = \sum_{T \in \mathcal{S}_\mu} \prod_{i \to j \in T} k_{ij}, \quad \mu \in \{1, \ldots, m\},$$

where $\mathcal{S}_\mu$ is the set of directed spanning trees (for the respective strongly connected component without outgoing edges) rooted at vertex μ; see [31,37,34]. We refer to [8] for further details and references on the graph Laplacian and a combinatorial proof of the matrix-tree theorem following [49].

If there exists $\psi \in \mathbb{R}^m_>$ with $A_k \psi = 0$, then every vertex resides in a strongly connected component without outgoing edges, that is, every connected component is strongly connected. In this case, the underlying unweighted digraph is called *weakly reversible*. Now, let $(G, y, \tilde{y})$ be a generalized chemical reaction network. If there exist rate constants k such that the generalized mass action system $(G_k, y, \tilde{y})$ admits a complex balancing equilibrium $x \in \mathbb{R}^n_>$, that is, $A_k x^Y = 0$, then G is weakly reversible.

5 Binomial Equations for Complex Balancing Equilibria

For a weakly reversible digraph, we know from the previous section that a basis for $\ker(A_k)$, parametrized by the weights, is given in terms of the l connected components and the m tree constants.

In our example, where $l = 2$ and $m = 5$, basis vectors of $\ker(A_k)$ are given by

$$(K_1, K_2, K_3, 0, 0)^T \quad \text{and} \quad (0, 0, 0, K_4, K_5)^T$$

with tree constants

$$(K_1, K_2, K_3, K_4, K_5) = (k_{31}\,k_{21} + k_{31}\,k_{23}, k_{12}\,k_{31}, k_{23}\,k_{12}, k_{54}, k_{45}).$$

Due to their special structure, we immediately find "binomial" basis vectors for the orthogonal complement $\ker(A_k)^\perp$,

$$(-K_2, K_1, 0, 0, 0)^T, \quad (0, -K_3, K_2, 0, 0)^T, \quad \text{and} \quad (0, 0, 0, -K_5, K_4)^T,$$

which are again determined by the connected components and tree constants. These vectors form a basis since they are linearly independent and

$$\dim \ker(A_k)^\perp = m - \dim \ker(A_k) = m - l = 5 - 2 = 3.$$

In our example, a complex balancing equilibrium $x \in \mathbb{R}^4_>$ with $\psi = x^{\tilde{Y}}$ and hence $A_k \psi = 0$, can equivalently be described as a positive solution of the binomial equations

$$\begin{pmatrix} -K_2 & K_1 & 0 & 0 & 0 \\ 0 & -K_3 & K_2 & 0 & 0 \\ 0 & 0 & 0 & -K_5 & K_4 \end{pmatrix} \psi = 0.$$

In other words, $\psi \in \ker(A_k)$ is equivalent to $\psi \perp \ker(A_k)^\perp$ or a basis thereof. Explicitly, we have $\psi = x^{\tilde{Y}} = ((x_1)^{\frac{1}{2}}(x_2)^{\frac{3}{2}}, x_3, (x_1)^3, x_1, x_4)^T$ and

$$K_1 x_3 - K_2 (x_1)^{\frac{1}{2}}(x_2)^{\frac{3}{2}} = 0, \quad K_2 (x_1)^3 - K_3 x_3 = 0, \quad K_4 x_4 - K_5 x_1 = 0. \quad (5)$$

Clearly, these considerations generalize to arbitrary weakly reversible digraphs: Based on the (strongly) connected components, we can characterize complex balancing equilibria by $m - l$ binomial equations with tree constants as coefficients.

Proposition 1. *Let A_k be the graph Laplacian of a weakly reversible digraph with positive weights and m vertices ordered within l connected components,*

$$L_\lambda = (i_\mu^\lambda)_{\mu=1,\ldots,m_\lambda} \quad \text{for } \lambda = 1,\ldots,l, \quad \text{where } \textstyle\sum_{\lambda=1}^l m_\lambda = m.$$

Let $\tilde{Y} \in \mathbb{R}^{n \times m}$ and

$$Z_k = \{x \in \mathbb{R}_>^n \mid A_k\, x^{\tilde{Y}} = 0\}.$$

Then,

$$Z_k = \{x \in \mathbb{R}_>^n \mid K_i\, x^{\tilde{y}^j} - K_j\, x^{\tilde{y}^i} = 0, \ (i,j) \in \mathcal{E}\}$$

where

$$\mathcal{E} = \{(i_\mu^\lambda, i_{\mu+1}^\lambda) \mid \lambda = 1,\ldots,l; \ \mu = 1,\ldots,m_\lambda - 1\}.$$

Note that the actual binomial equations depend on the order of the vertices within the connected components, but the zero set does not.

6 Binomial Equations with Real and Symbolic Exponents

In this section, we collect basic facts about positive real solutions of binomial equations with real exponents. We present the results in full generality, in particular, not restricted to complex balancing equilibria, and emphasize algorithmic aspects. Moreover, by reducing computations to linear algebra, we outline the treatment of symbolic exponents.

In an algebraic perspective, one usually considers solutions of binomial equations with integer exponents. We refer to [13] for an introduction including algorithmic aspects and an extensive list of references. An algorithm with polynomial complexity for computing solutions with non-zero or positive coordinates of parametric binomial systems is presented in [29]. For recent algorithmic methods for binomial equations and monomial parametrizations, see [1]. Toric geometry and computer algebra was introduced to the study of mass action systems in [25,27,26] and further developed in [12]. So-called *toric steady states* are solutions of binomial equations arising from polynomial dynamical systems [42].

In chemical reaction networks, it is natural to consider real exponents: kinetic orders, measured by experiments, need not be integers. Also in S-systems [46,48], defined by binomial power-laws, the exponents are real numbers identified from data. We note that binomial equations are implicit in the original works on chemical reaction networks [33,32].

In the following, we consider binomial equations

$$\alpha_i \, x^{a^i} - \beta_i \, x^{b^i} = 0 \quad \text{for } i = 1, \ldots, r$$

for $x \in \mathbb{R}^n_>$, where $a^i, b^i \in \mathbb{R}^n$ and $\alpha_i, \beta_i \in \mathbb{R}_>$. Clearly, x is a solution iff

$$x^{a^i - b^i} = \frac{\beta_i}{\alpha_i} \quad \text{for } i = 1, \ldots, r.$$

By introducing the exponent matrix $M \in \mathbb{R}^{n \times r}$, whose ith column is the vector $a^i - b^i$, and the vectors $\alpha, \beta \in \mathbb{R}^r_>$ with entries α_i and β_i, respectively, we can rewrite the above equation system as

$$x^M = \frac{\beta}{\alpha}.$$

More generally, we are interested for which $\gamma \in \mathbb{R}^r_>$ the equations

$$x^M = \gamma$$

have a positive solution. Taking the logarithm, we obtain the equivalent linear equations

$$M^T \ln x = \ln \gamma, \tag{6}$$

which reduces the problem to linear algebra.

In the rest of this section, we fix a matrix $M \in \mathbb{R}^{n \times r}$ and write

$$Z_{M,\gamma} = \{x \in \mathbb{R}^n_> \mid x^M = \gamma\}$$

for the set of all positive solutions with right-hand side $\gamma \in \mathbb{R}^r_>$.

Proposition 2. *The following statements hold:*

$$Z_{M,\gamma} \neq \emptyset \quad \text{for all } \gamma \in \mathbb{R}^r_> \quad \text{iff} \quad \ker(M) = \{0\}.$$

If $\ker(M) \neq \{0\}$, then

$$Z_{M,\gamma} \neq \emptyset \quad \text{for } \gamma \in \mathbb{R}^r_> \quad \text{iff} \quad \gamma^C = 1,$$

where $C \in \mathbb{R}^{r \times p}$ with $\text{im}(C) = \ker(M)$ and $\ker(C) = \{0\}$.

Proof. Using (6), $x^M = \gamma$ is equivalent to

$$\ln \gamma \in \text{im}(M^T) = \ker(M)^\perp.$$

Hence, $Z_{M,\gamma} \neq \emptyset$ for all $\gamma \in \mathbb{R}^r_>$ iff $\ker(M) = \{0\}$. If $\ker(M) \neq \{0\}$, then

$$\ln \gamma \in \ker(M)^\perp = \text{im}(C)^\perp \quad \Leftrightarrow \quad C^T \ln \gamma = 0 \quad \Leftrightarrow \quad \gamma^C = 1.$$

$\square$

Computing an explicit positive solution $x^* \in Z_{M,\gamma}$ (if it exists) in terms of γ is equivalent to computing a particular solution for the linear equations (6). For this, we use an arbitrary generalized inverse H of M^T, that is, a matrix $H \in \mathbb{R}^{n \times r}$ such that

$$M^T H M^T = M^T.$$

We refer to [4] for details on generalized inverses.

Proposition 3. *Let $\gamma \in \mathbb{R}^r_>$ such that $\ln \gamma \in \operatorname{im}(M^T)$. Let $H \in \mathbb{R}^{n \times r}$ be a generalized inverse of M^T. Then,*

$$x^* = \gamma^{H^T} \in Z_{M,\gamma}.$$

Proof. By assumption, $\ln \gamma = M^T z$ for some $z \in \mathbb{R}^n$. Then,

$$M^T \ln x^* = M^T H \ln \gamma = M^T H M^T z = M^T z = \ln \gamma$$

and hence $x^* \in Z_{M,\gamma}$ as claimed.   □

Given one positive solution $x^* \in Z_{M,\gamma}$, we have a generalized monomial parametrization for the set of all positive solutions.

Proposition 4. *Let $x^* \in Z_{M,\gamma}$. Then,*

$$Z_{M,\gamma} = \{x^* \circ e^v \mid v \in \operatorname{im}(M)^\perp\}.$$

If $\operatorname{im}(M)^\perp \neq \{0\}$, then

$$Z_{M,\gamma} = \{x^* \circ \xi^{B^T} \mid \xi \in \mathbb{R}^q_>\},$$

where $B \in \mathbb{R}^{n \times q}$ with $\operatorname{im}(B) = \operatorname{im}(M)^\perp$ and $\ker(B) = \{0\}$.

Proof. The first equality follows from (6): $x \in Z_{M,\gamma}$ iff $v = \ln x - \ln x^* \in \ker(M^T) = \operatorname{im}(M)^\perp$, that is, $x = x^* \circ e^v$ with $v \in \operatorname{im}(M)^\perp$.

Since the columns of B form a basis for $\operatorname{im}(M)^\perp$, we can write $v \in \operatorname{im}(M)^\perp$ uniquely as $v = B t$ for some $t \in \mathbb{R}^q$. By introducing $\xi = e^t \in \mathbb{R}^q_>$, we obtain

$$(e^v)_i = e^{v_i} = e^{\sum_j b_{ij} t_j} = \prod_j \xi_j^{b_{ij}} = \xi^{b_i} = (\xi^{B^T})_i,$$

that is, $e^v = \xi^{B^T}$.   □

Note that the conditions for the existence of positive solutions and the parametrization of all positive solutions, respectively, depend only on the vector subspaces $\ker(M)$ and $\operatorname{im}(M)^\perp = \ker(M^T)$.

Summing up, we have seen that computing positive solutions for binomial equations reduces to linear algebra involving the exponent matrix M. The matrices C, H and B from Propositions 2, 3, and 4 can be computed effectively if $M \in \mathbb{Q}^{n \times r}$ and C, B can be chosen to have only integer entries.

Moreover, the linear algebra approach to binomial equations allows to deal algorithmically with indeterminate (symbolic) exponents. We can use computer algebra methods for matrices with symbolic entries like Turing factoring (generalized PLU decomposition) [10] and its implementation [11]. Based on these methods, we can compute explicit monomial parametrizations with symbolic exponents for generic entries and investigate conditions for special cases. See Section 8 for an example.

7 Kinetic Deficiency

Applying the results from the previous section, we rewrite the binomial equations (5) from our example,

$$K_1 x_3 - K_2 (x_1)^{\frac{1}{2}} (x_2)^{\frac{3}{2}} = 0, \quad K_2 (x_1)^3 - K_3 x_3 = 0, \quad K_4 x_4 - K_5 x_1 = 0,$$

as

$$x^M = \kappa_k,$$

where

$$M = \begin{pmatrix} -\frac{1}{2} & 3 & -1 \\ -\frac{3}{2} & 0 & 0 \\ 1 & -1 & 0 \\ 0 & 0 & 1 \end{pmatrix} \tag{7}$$

and

$$\kappa_k = (K_2/K_1, K_3/K_2, K_5/K_4)^T,$$

which depends on the weights k via the tree constants K.

Recall that the binomial equations depend on the basis vectors for $\ker(A_k)^\perp$ which are determined by the relation $\mathcal{E} = \{(1,2), (2,3), (4,5)\}$. To specify the resulting exponent matrix M and the right-hand side κ_k, we have fixed an order on the relation. By abuse of notation, we write

$$\mathcal{E} = ((1,2), (2,3), (4,5)).$$

Hence, $M = \tilde{Y} I_\mathcal{E}$ with

$$I_\mathcal{E} = \begin{pmatrix} -1 & 0 & 0 \\ 1 & -1 & 0 \\ 0 & 1 & 0 \\ 0 & 0 & -1 \\ 0 & 0 & 1 \end{pmatrix}. \tag{8}$$

In general, for a weakly reversible digraph with m vertices and l connected components, let $\mathcal{E}$ be a relation as in Proposition 1 with fixed order. We denote by $I_\mathcal{E} \in \mathbb{R}^{m \times (m-l)}$ the matrix with columns

$$e^j - e^i \quad \text{for } (i,j) \in \mathcal{E},$$

where e^i denotes the ith standard basis vector in $\mathbb{R}^m$. Clearly, the columns of $I_\mathcal{E}$ are linearly independent and hence $\dim \mathrm{im}(I_\mathcal{E}) = m - l$. To rewrite the binomial equations in Proposition 1, we define the exponent matrix $M \in \mathbb{R}^{n \times (m-l)}$ as

$$M = \tilde{Y} I_\mathcal{E},$$

the right-hand side $\kappa_k \in \mathbb{R}^{m-l}_>$ as

$$(\kappa_k)_{(i,j)} = K_j/K_i \quad \text{for } (i,j) \in \mathcal{E}, \tag{9}$$

and obtain

$$Z_k = \{x \in \mathbb{R}^n_> \mid x^M = \kappa_k\}.$$

We note that the actual matrix M depends on $\mathcal{E}$, but $\operatorname{im}(M)$ does not. This can be seen using the following fact.

Proposition 5. *Let $G = (V, E)$ be a digraph with m vertices and l connected components. Let $I_E \in \mathbb{R}^{m \times |E|}$ denote its incidence matrix (for fixed order on E), and let $I_\mathcal{E} \in \mathbb{R}^{m \times (m-l)}$ be as defined above. Then,*

$$\operatorname{im}(I_\mathcal{E}) = \operatorname{im}(I_E).$$

Proof. From graph theory (see for example [35]) and the argument above, we know that $\dim \operatorname{im}(I_E) = \dim \operatorname{im}(I_\mathcal{E}) = m - l$. It remains to show that $\operatorname{im}(I_E) \subseteq \operatorname{im}(I_\mathcal{E})$. We consider the column $e^j - e^i$ of I_E corresponding to the edge $(i, j) \in E$. Clearly, i and j are in the same connected component L_λ, in particular, $i = i^\lambda_{\mu(i)}$ and $j = i^\lambda_{\mu(j)}$, where we assume $\mu(i) < \mu(j)$. Then,

$$e^j - e^i = \sum_{\mu = \mu(i), \ldots, \mu(j)-1} e^{i^\lambda_{\mu+1}} - e^{i^\lambda_\mu},$$

where $e^{i^\lambda_{\mu+1}} - e^{i^\lambda_\mu}$ are columns of $I_\mathcal{E}$ corresponding to pairs $(i^\lambda_\mu, i^\lambda_{\mu+1})$ in $\mathcal{E}$. $\square$

Now, we see that $\operatorname{im}(M)$ equals the kinetic-order subspace $\tilde{S}$:

$$\operatorname{im}(M) = \operatorname{im}(\tilde{Y} I_\mathcal{E}) = \operatorname{im}(\tilde{Y} I_E) = \tilde{S}.$$

Finally, we recall that the number of independent conditions on κ_k for the existence of a positive solution of $x^M = \kappa_k$ is given by $\dim \ker(M)$, cf. Proposition 2. Observing $M \in \mathbb{R}^{n \times (m-l)}$, we obtain

$$\dim \ker(M) = m - l - \dim \operatorname{im}(M) = m - l - \dim \tilde{S}. \tag{10}$$

Hence, for a digraph with m vertices and l connected components, we define the *kinetic deficiency* as

$$\tilde{\delta} = m - l - \tilde{s},$$

where $\tilde{s} = \dim \tilde{S}$ denotes the dimension of the kinetic-order subspace.

8   Computing Complex Balancing Equilibria

Combining the results from the previous sections, we obtain the following constructive characterization of complex balancing equilibria in terms of quotients of tree constants.

Theorem 1. *Let A_k be the graph Laplacian of a weakly reversible digraph with positive weights, m vertices, and l connected components. Let $\tilde{Y} \in \mathbb{R}^{n \times m}$ be the matrix of kinetic complexes, $\tilde{s} = \dim \tilde{S}$ the dimension of the kinetic-order* *subspace, and* $\tilde{\delta} = m - l - \tilde{s}$ *the kinetic deficiency. Further, let* $M \in \mathbb{R}^{n \times (m-l)}$ *and* $\kappa_k \in \mathbb{R}_>^{m-l}$ *such that*

$$Z_k = \{x \in \mathbb{R}_>^n \mid A_k \, x^{\tilde{Y}} = 0\} = \{x \in \mathbb{R}_>^n \mid x^M = \kappa_k\}.$$

Then, the following statements hold:

(a) $Z_k \neq \emptyset$ *for all* k *iff* $\tilde{\delta} = 0$.

(b) *If* $\tilde{\delta} > 0$, *then*

$$Z_k \neq \emptyset \quad \text{iff} \quad (\kappa_k)^C = 1,$$

where $C \in \mathbb{R}^{(m-l) \times \tilde{\delta}}$ *with* $\operatorname{im}(C) = \ker(M)$ *and* $\ker(C) = \{0\}$.

(c) *If* $Z_k \neq \emptyset$, *then*

$$x^* = (\kappa_k)^{H^T} \in Z_k,$$

where $H \in \mathbb{R}^{n \times (m-l)}$ *is a generalized inverse of* M^T.

(d) *If* $x^* \in Z_k$ *and* $\tilde{s} < n$, *then*

$$Z_k = \{x^* \circ \xi^{B^T} \mid \xi \in \mathbb{R}_>^{n-\tilde{s}}\},$$

where $B \in \mathbb{R}^{n \times (n-\tilde{s})}$ *with* $\operatorname{im}(B) = \tilde{S}^\perp$ *and* $\ker(B) = \{0\}$.

Proof. By Propositions 2, 3, and 4. In fact, it remains to prove one implication in (a). Assume $Z_k \neq \emptyset$ for all k, that is, there exists a solution to $x^M = \kappa_k$ for all k. By Lemma 1 below, for all $\gamma \in \mathbb{R}_>^{m-l}$, there exists k such that $\kappa_k = \gamma$. Hence, there exists a solution to $x^M = \gamma$ for all γ. Using (10) and Proposition 2, we obtain $\tilde{\delta} = \dim \ker(M) = 0$. $\qquad\square$

Lemma 1. *Let* A_k *be the graph Laplacian of a weakly reversible digraph with positive weights,* m *vertices, and* l *connected components, and let* $\kappa_k \in \mathbb{R}_>^{m-l}$ *be the vector of quotients of tree constants defined in* (9). *For all* $\gamma \in \mathbb{R}_>^{m-l}$, *there exists* k *such that* $\kappa_k = \gamma$.

Proof. First, we show that every positive vector $\psi \in \mathbb{R}_>^m$ solves $A_k \psi = 0$ for some weights k. Indeed, for given k, the vector of tree constants $K \in \mathbb{R}_>^m$ solves $A_k K = 0$, and by choosing $k_{ij}^* = k_{ij} \frac{K_i}{\psi_i}$, one obtains

$$(A_{k^*} \psi)_i = \sum_{j=1}^m (A_{k^*})_{ij} \, \psi_j = \sum_{j \to i \in E} k_{ji}^* \, \psi_j - \sum_{i \to j \in E} k_{ij}^* \, \psi_i$$

$$= \sum_{j \to i \in E} k_{ji} \, K_j - \sum_{i \to j \in E} k_{ij} \, K_i = \sum_{j=1}^m (A_k)_{ij} \, K_j = (A_k K)_i = 0$$

for all $i = 1, \ldots, m$, that is, $A_{k^*} \, \psi = 0$.

Let $\mathcal{E}$ be a relation as in Proposition 1 with the obvious order. Using basis vectors of $\ker(A_k)$ having tree constants as entries, we find that

$$\frac{\psi_j}{\psi_i} = \frac{K_j}{K_i} = (\kappa_k)_{(i,j)} \quad \text{for all } (i,j) \in \mathcal{E}.$$

By choosing the entries of $\psi \in \mathbb{R}^m_>$ in the obvious order, every $\gamma \in \mathbb{R}^{m-l}_>$ can be attained by κ_k for some k. $\square$

Remark 1. Theorem 1 is constructive in the following sense:

- To test if the digraph G is weakly reversible, we compute the connected and the strongly connected components and check whether they are equal.
- The tree constants are computed in terms of the weights k, using (fraction-free) Gaussian elimination on the sub-matrices of A_k determined by the (strongly) connected components.
- Given the kinetic complexes $\tilde{Y} \in \mathbb{Q}^{n \times m}$ and the (strongly) connected components of the digraph, we compute a matrix M and a vector κ_k as introduced in Section 7.
- All matrices involved are computed by linear algebra from the exponent matrix M. This can also be done algorithmically if the kinetic complexes $\tilde{Y}$ and hence M contain indeterminate (symbolic) entries; see the end of Section 6.

In our example, $\tilde{\delta} = 5 - 2 - 3 = 0$ and a monomial parametrization of all complex balancing equilibria is given by

$$\left((\kappa_3)^{-1}, (\kappa_1)^{-\frac{2}{3}} (\kappa_2)^{-\frac{2}{3}} (\kappa_3)^{-\frac{5}{3}}, \kappa_2^{-1} (\kappa_3)^{-3}, 1 \right)^T \circ (\xi^3, \xi^5, \xi^9, \xi^3)^T,$$

where

$$\kappa \equiv \kappa_k = \left(\frac{k_{12}}{k_{21} + k_{23}}, \frac{k_{23}}{k_{31}}, \frac{k_{45}}{k_{54}} \right)^T$$

and $\xi \in \mathbb{R}_>$.

To conclude, we associate with each vertex of the graph a kinetic complex possibly containing symbolic coefficients, thereby specifying monomials with symbolic exponents:

$$aA + bB \rightleftarrows C \qquad A \rightleftarrows D \tag{11}$$
$$\searrow \qquad \downarrow$$
$$cA$$

In this setting, a monomial parametrization with symbolic exponents of all complex balancing equilibria is given by

$$\left((\kappa_3)^{-1}, (\kappa_1)^{-\frac{1}{b}} (\kappa_2)^{-\frac{1}{b}} (\kappa_3)^{\frac{a-c}{b}}, (\kappa_2)^{-1} (\kappa_3)^{-c}, 1 \right)^T \circ (\xi^b, \xi^{c-a}, \xi^{bc}, \xi^b)^T,$$

which is valid for non-zero $a, b, c \in \mathbb{R}$.

9 Generalized Birch's Theorem

Since the dynamics of generalized mass-action systems is confined to cosets of the stoichiometric subspace, we are interested in uniqueness and existence of complex balancing equilibria in every positive stoichiometric compatibility class.

Let G_k be a weakly reversible digraph with positive weights, m vertices and l connected components. For fixed rate constants k, a complex balancing equilibrium $x^* \in \mathbb{R}^n_>$ of the mass-action system $(G_k, y, \tilde{y})$ solves $A_k\, x^{\tilde{Y}} = 0$, where $A_k \in \mathbb{R}^{m \times m}$ is the graph Laplacian and $\tilde{Y} \in \mathbb{R}^{n \times m}$ is the matrix of kinetic complexes. Equivalently, it solves $x^M = \kappa_k$, where the columns of $M \in \mathbb{R}^{n \times (m-l)}$ are differences of kinetic complexes and the entries of $\kappa_k \in \mathbb{R}^{m-l}_>$ are quotients of the tree constants K, which depend on the weights k. In other words,

$$Z_k = \{x \in \mathbb{R}^n_> \mid A_k\, x^{\tilde{Y}} = 0\}$$
$$= \{x \in \mathbb{R}^n_> \mid x^M = \kappa_k\}.$$

Given a complex balancing equilibrium $x^* \in \mathbb{R}^n_>$, we further know that

$$Z_k = \{x^* \circ e^v \mid v \in \operatorname{im}(M)^\perp\}$$
$$= \{x^* \circ \xi^{B^T} \mid \xi \in \mathbb{R}^{\tilde{d}}_>\},$$

where the second equality holds if $\operatorname{im}(M)^\perp \neq \{0\}$ and $B \in \mathbb{R}^{n \times \tilde{d}}$ is defined as $\operatorname{im}(B) = \operatorname{im}(M)^\perp$ and $\ker(B) = \{0\}$.

For simplicity, we write $\tilde{W} = B^T \in \mathbb{R}^{\tilde{d} \times n}$ such that $\tilde{S} = \operatorname{im}(M) = \operatorname{im}(B)^\perp = \operatorname{im}(\tilde{W}^T)^\perp = \ker(\tilde{W})$. Analogously, we introduce a matrix $W \in \mathbb{R}^{d \times n}$ with full rank d such that $S = \ker(W)$.

If the intersection of the set of complex balancing equilibria with some compatibility class,

$$Z_k \cap (x' + S),$$

is non-empty, then there exist $\xi \in \mathbb{R}^{\tilde{d}}_>$ and $u \in S$ such that

$$x^* \circ \xi^{\tilde{W}} = x' + u.$$

Multiplication by W yields

$$W\,(x^* \circ \xi^{\tilde{W}}) = W\,x'$$

such that existence and uniqueness of complex balancing equilibria in every stoichiometric compatibility class are equivalent to surjectivity and injectivity of the generalized polynomial map

$$f_{x^*} \colon \mathbb{R}^{\tilde{d}}_> \to C^\circ \subseteq \mathbb{R}^d \tag{12}$$
$$\xi \mapsto W\,(x^* \circ \xi^{\tilde{W}}) = \sum_{i=1}^n x^*_i\, \xi^{\tilde{w}^i} w^i,$$

where C° is the interior of the polyhedral cone

$$C = \{W x' \in \mathbb{R}^d \mid x' \in \mathbb{R}^n_\geq\} = \left\{\sum_{i=1}^n x'_i\, w^i \in \mathbb{R}^d \mid x' \in \mathbb{R}^n_\geq\right\}.$$

In mass-action systems, where $S = \tilde{S}$ and hence $W = \tilde{W}$, one version [23] of Birch's theorem [5] states that f_{x^*} is a real analytic isomorphism of $\mathbb{R}^d_>$ onto C° for all $x^* \in \mathbb{R}^n_>$. We refer to [28, Sect. 5] for a recent overview on the use of Birch's theorem in CRNT and to [41] for the version used in algebraic statistics. Interestingly, Martin W. Birch's seminal paper on maximum likelihood methods for log-linear models was part of a PhD thesis at the University of Glasgow that was never submitted [22].

Recently, we have generalized Birch's theorem to $W \neq \tilde{W}$, cf. [40, Proposition 3.9]. To formulate the result, we define the sign vector $\sigma(x) \in \{-, 0, +\}^n$ of a vector $x \in \mathbb{R}^n$ by applying the sign function component-wise, and we write $\sigma(S) = \{\sigma(x) \mid x \in S\}$ for a subset $S \subseteq \mathbb{R}^n$.

Theorem 2. *Let $W \in \mathbb{R}^{d \times n}$, $\tilde{W} \in \mathbb{R}^{\tilde{d} \times n}$ and $S = \ker(W)$, $\tilde{S} = \ker(\tilde{W})$. If $\sigma(S) = \sigma(\tilde{S})$ and $(+, \ldots, +)^T \in \sigma(S^\perp)$, then the generalized polynomial map f_{x^*} in (12) is a real analytic isomorphism of $\mathbb{R}^d_>$ onto C° for all $x^* \in \mathbb{R}^n_>$.*

If $\tilde{\delta} = 0$, there exists a complex balancing equilibrium for all rate constants k, by Theorem 1. If further the generalized polynomial map f_{x^*} is surjective and injective for all x^*, then, by Theorem 2, there exists a unique steady state in every positive stoichiometric compatibility class for all k.

To illustrate the result, we consider the minimal (weakly) reversible weighted digraph

$$1 \underset{k_{21}}{\overset{k_{12}}{\rightleftarrows}} 2,$$

and associate with each vertex a (stoichiometric) complex

$$A + B \rightleftarrows C$$

as well as a kinetic complex

$$aA + bB \rightleftarrows C,$$

where $a, b > 0$. We find $S = \operatorname{im}(-1, -1, 1)^T$ and $\tilde{S} = \operatorname{im}(-a, -b, 1)^T$ and choose

$$W = \begin{pmatrix} 1 & 0 & 1 \\ 0 & 1 & 1 \end{pmatrix} \quad \text{and} \quad \tilde{W} = \begin{pmatrix} 1 & 0 & a \\ 0 & 1 & b \end{pmatrix}$$

such that $S = \ker(W)$ and $\tilde{S} = \ker(\tilde{W})$. Clearly, our generalization of Birch's theorem applies since

$$\sigma(S) = \left\{ \begin{pmatrix} - \\ - \\ + \end{pmatrix}, \begin{pmatrix} + \\ + \\ - \end{pmatrix}, \begin{pmatrix} 0 \\ 0 \\ 0 \end{pmatrix} \right\} = \sigma(\tilde{S})$$

and $(1, 1, 2)^T \in S^\perp$. Hence, there exists a unique solution $\xi \in \mathbb{R}^2_>$ for the system of generalized polynomial equations

$$x_1^* \xi_1 \begin{pmatrix} 1 \\ 0 \end{pmatrix} + x_2^* \xi_2 \begin{pmatrix} 0 \\ 1 \end{pmatrix} + x_3^* (\xi_1)^a (\xi_2)^b \begin{pmatrix} 1 \\ 1 \end{pmatrix} = \begin{pmatrix} y_1 \\ y_2 \end{pmatrix}$$

for all right-hand-sides $y \in C^\circ = \mathbb{R}^2_{>}$, all parameters $x^* \in \mathbb{R}^3_{>}$, and all exponents $a, b > 0$. Note that Birch's theorem guarantees the existence of a unique solution only for $a = b = 1$.

In terms of the generalized mass-action system above, we have the following result: Since $\tilde{\delta} = 2 - 1 - 1 = 0$, there exists a unique complex balancing equilibrium in every positive stoichiometric compatibility class for all $k_{12}, k_{21} > 0$ and all kinetic orders $a, b > 0$. Since $\delta = 2 - 1 - 1 = 0$, there are no other steady states.

10 Sign Vectors and Oriented Matroids

The characterization of surjectivity and injectivity of generalized polynomial maps involves sign vectors of real linear subspaces, which are basic examples of oriented matroids. (Whereas a matroid abstracts the notion of linear independence, an oriented matroid additionally captures orientation.)

The theory of oriented matroids provides a common framework to study combinatorial properties of various geometric objects, including point configurations, hyperplane arrangements, convex polyhedra, and directed graphs. See [2], [50, Chapters 6 and 7], and [44] for an introduction and overview, and [6] for a comprehensive study.

There are several sets of sign vectors associated with a linear subspace which satisfy the axiom systems for (co-)vectors, (co-)circuits, or chirotopes of oriented matroids. (In fact, there are non-realizable oriented matroids that do not arise from linear subspaces.)

For algorithmic purposes, the characterization of oriented matroids in terms of basis orientations is most useful. The chirotope of a matrix $W \in \mathbb{R}^{d \times n}$ (with rank d) is defined as the map

$$\chi_W : \{1, \ldots, n\}^d \to \{-, 0, +\}$$
$$(i_1, \ldots, i_d) \mapsto \operatorname{sign}(\det(w^{i_1}, \ldots, w^{i_d})),$$

which records for each d-tuple of vectors whether it forms a positively oriented basis of $\mathbb{R}^d$, a negatively oriented basis, or not a basis. Hence, chirotopes can be used to test algorithmically if the sign vectors of two subspaces are equal by comparing determinants of maximal minors.

More generally, the realization space of matrices defining the same oriented matroid as $W \in \mathbb{R}^{d \times n}$ (with rank d) is described by the semi-algebraic set

$$\mathcal{R}(W) = \{A \in \mathbb{R}^{d \times n} \mid \operatorname{sign}(\det(a^{i_1}, \ldots, a^{i_d})) =$$
$$\operatorname{sign}(\det(w^{i_1}, \ldots, w^{i_d})), \ 1 \leq i_1 < \cdots < i_d \leq n\}.$$

Mnëv's universality theorem [38] theorem states that already for oriented matroids with rank $d = 3$, the realization space can be "arbitrarily complicated"; see [6] for a precise statement and [3] for semi-algebraic sets and algorithms.

Concerning software, the C++ package TOPCOM [43] allows to compute efficiently chirotopes with rational arithmetic and generate all cocircuits (covectors with minimal support). There is also an interface to the open source computer algebra system SAGE.

In our running example, we have $\tilde{S} = \mathrm{im}(\tilde{Y} I_\mathcal{E}) = \mathrm{im}(M)$ with M as in (7). Analogously, $S = \mathrm{im}(Y I_\mathcal{E}) = \mathrm{im}(\mathcal{N})$ with

$$\mathcal{N} = \begin{pmatrix} -1 & 2 & -1 \\ -1 & 0 & 0 \\ 1 & -1 & 0 \\ 0 & 0 & 1 \end{pmatrix}. \tag{13}$$

To check the sign vector condition $\sigma(S) = \sigma(\tilde{S})$, we compare the chirotopes of $\mathcal{N}^T$ and M^T. Computing the signs of the four maximal minors of $\mathcal{N}^T$, we see that its chirotope is given by

$$\chi_{\mathcal{N}^T}(1,2,3) = -, \quad \chi_{\mathcal{N}^T}(1,2,4) = +, \quad \chi_{\mathcal{N}^T}(1,3,4) = -, \quad \chi_{\mathcal{N}^T}(2,3,4) = +.$$

Analogously, we compute the chirotope of M^T and verify $\chi_{\mathcal{N}^T} = \chi_{M^T}$. Clearly, the other sign vector condition $(+, \ldots, +)^T \in \sigma(S^\perp)$ also holds, for example, $(1,1,2,1)^T \in S^\perp$.

Since $\tilde{\delta} = 0$, we know from Theorems 1 and 2 that there exists a unique complex balancing equilibrium in every positive stoichiometric compatibility class for all rate constants k. Moreover, since $\delta = 5 - 2 - 3 = 0$, we know that there are no steady states other than complex balancing equilibria for the ODE (3).

In the setting of symbolic exponents (11), the exponent matrix amounts to

$$M = \begin{pmatrix} -a & c & -1 \\ -b & 0 & 0 \\ 1 & -1 & 0 \\ 0 & 0 & 1 \end{pmatrix} \tag{14}$$

and the chirotope of M^T (in the same order as above) is given by

$$-\mathrm{sign}(b), \quad \mathrm{sign}(bc), \quad \mathrm{sign}(a-c), \quad \mathrm{sign}(b)$$

for $a, b, c \neq 0$. Hence, there exists a unique steady state in every positive stoichiometric compatibility class for all rate constants and all exponents with $a, b, c > 0$ and $a < c$.

11 Multistationarity

A (generalized) chemical reaction network $(G, y, \tilde{y})$ has *the capacity for multistationarity* if there exist rate constants k such that the generalized mass action system $(G_k, y, \tilde{y})$ admits more than one steady state in some stoichiometric compatibility class.

In mass-action systems, every stoichiometric compatibility class contains at most one complex balancing equilibrium. However, in generalized mass action systems, multiple steady states of this type are possible [40, Proposition 3.2].

Proposition 6. *Let $(G, y, \tilde{y})$ be a generalized chemical reaction network. If G is weakly reversible and $\sigma(S) \cap \sigma(\tilde{S}^\perp) \neq \{0\}$, then $(G, y, \tilde{y})$ has the capacity for multiple complex balancing equilibria.*

Analogously, multiple toric steady states are possible (for networks with mass-action kinetics) if the sign vectors of two subspaces intersect non-trivially [9,42]. For deficiency one networks (with mass-action kinetics), the capacity for multi-stationarity is also characterized by sign conditions [18,20].

For precluding multistationarity, injectivity of the right-hand side of the dynamical system on cosets of the stoichiometric subspace is sufficient. In [39], we characterize injectivity of generalized polynomial maps on cosets of the stoichiometric subspace in terms of sign vectors. There, we also give a survey on injectivity criteria and discuss algorithms to check sign vector conditions.

For the last time, we return to our example, in particular, to the setting of symbolic kinetic complexes. Considering the matrix M in (14), a matrix B with $\text{im}(B) = \text{im}(M)^\perp = \tilde{S}^\perp$ is given by

$$B = (b, c - a, b\,c, b)^T$$

for $a, b, c \neq 0$. Hence, for $a, b, c > 0$ and $a > c$, we have $(+, -, +, +)^T \in \sigma(\tilde{S}^\perp)$.

On the other hand, considering the matrix $\mathcal{N}$ in (13) with $\text{im}(\mathcal{N}) = S$, we also have $(+, -, +, +)^T \in \sigma(S)$, and hence $\sigma(S) \cap \sigma(\tilde{S}^\perp) \neq \{0\}$. By Proposition 6, if the inequalities $a, b, c > 0$ and $a > c$ hold, then there exist rate constants k that admit more than one complex balancing equilibrium in some stoichiometric compatibility class.

References

1. Adrovic, D., Verschelde, J.: A polyhedral method to compute all affine solution sets of sparse polynomial systems (2013), http://arxiv.org/abs/1310.4128, arXiv:1310.4128 [cs.SC]
2. Bachem, A., Kern, W.: Linear programming duality. Springer, Berlin (1992)
3. Basu, S., Pollack, R., Roy, M.F.: Algorithms in real algebraic geometry, 2nd edn. Springer, Berlin (2006)
4. Ben-Israel, A., Greville, T.N.E.: Generalized inverses, 2nd edn. Springer, New York (2003)
5. Birch, M.W.: Maximum likelihood in three-way contingency tables. J. Roy. Statist. Soc. Ser. B 25, 220–233 (1963)
6. Björner, A., Las Vergnas, M., Sturmfels, B., White, N., Ziegler, G.M.: Oriented matroids, 2nd edn. Cambridge University Press, Cambridge (1999)
7. Boulier, F., Lemaire, F., Petitot, M., Sedoglavic, A.: Chemical reaction systems, computer algebra and systems biology. In: Gerdt, V.P., Koepf, W., Mayr, E.W., Vorozhtsov, E.V. (eds.) CASC 2011. LNCS, vol. 6885, pp. 73–87. Springer, Heidelberg (2011)
8. Brualdi, R.A., Ryser, H.J.: Combinatorial matrix theory. Cambridge University Press, Cambridge (1991)
9. Conradi, C., Flockerzi, D., Raisch, J.: Multistationarity in the activation of a MAPK: parametrizing the relevant region in parameter space. Math. Biosci. 211, 105–131 (2008)

10. Corless, R.M., Jeffrey, D.J.: The turing factorization of a rectangular matrix. SIGSAM Bull. 31, 20–30 (1997)
11. Corless, R.M., Jeffrey, D.J.: Linear Algebra in Maple. In: CRC Handbook of Linear Algebra, 2nd edn. Chapman and Hall/CRC (2013)
12. Craciun, G., Dickenstein, A., Shiu, A., Sturmfels, B.: Toric dynamical systems. J. Symbolic Comput. 44, 1551–1565 (2009)
13. Dickenstein, A.: A world of binomials. In: Foundations of Computational Mathematics, Hong Kong, pp. 42–67. Cambridge Univ. Press, Cambridge (2009)
14. Errami, H., Seiler, W.M., Eiswirth, M., Weber, A.: Computing Hopf bifurcations in chemical reaction networks using reaction coordinates. In: Gerdt, V.P., Koepf, W., Mayr, E.W., Vorozhtsov, E.V. (eds.) CASC 2012. LNCS, vol. 7442, pp. 84–97. Springer, Heidelberg (2012)
15. Feinberg, M.: Complex balancing in general kinetic systems. Arch. Rational Mech. Anal. 49, 187–194 (1972)
16. Feinberg, M.: Lectures on chemical reaction networks (1979), http://crnt.engineering.osu.edu/LecturesOnReactionNetworks
17. Feinberg, M.: Chemical reaction network structure and the stability of complex isothermal reactors–I. The deficiency zero and deficiency one theorems. Chem. Eng. Sci. 42, 2229–2268 (1987)
18. Feinberg, M.: Chemical reaction network structure and the stability of complex isothermal reactors–II. Multiple steady states for networks of deficiency one. Chem. Eng. Sci. 43, 1–25 (1988)
19. Feinberg, M.: The existence and uniqueness of steady states for a class of chemical reaction networks. Arch. Rational Mech. Anal. 132, 311–370 (1995)
20. Feinberg, M.: Multiple steady states for chemical reaction networks of deficiency one. Arch. Rational Mech. Anal. 132, 371–406 (1995)
21. Feinberg, M., Horn, F.J.M.: Chemical mechanism structure and the coincidence of the stoichiometric and kinetic subspaces. Arch. Rational Mech. Anal. 66, 83–97 (1977)
22. Fienberg, S.E.: Introduction to Birch (1963) Maximum likelihood in three-way contingency tables. In: Kotz, S., Johnson, N.L. (eds.) Breakthroughs in statistics, vol. II, pp. 453–461. Springer, New York (1992)
23. Fulton, W.: Introduction to toric varieties. Princeton University Press, Princeton (1993)
24. Gatermann, K., Wolfrum, M.: Bernstein's second theorem and Viro's method for sparse polynomial systems in chemistry. Adv. in Appl. Math. 34, 252–294 (2005)
25. Gatermann, K.: Counting stable solutions of sparse polynomial systems in chemistry. In: Symbolic Computation: Solving Equations in Algebra, Geometry, and Engineering, pp. 53–69. Amer. Math. Soc., Providence (2001)
26. Gatermann, K., Eiswirth, M., Sensse, A.: Toric ideals and graph theory to analyze Hopf bifurcations in mass action systems. J. Symbolic Comput. 40, 1361–1382 (2005)
27. Gatermann, K., Huber, B.: A family of sparse polynomial systems arising in chemical reaction systems. J. Symbolic Comput. 33, 275–305 (2002)
28. Gopalkrishnan, M., Miller, E., Shiu, A.: A Geometric Approach to the Global Attractor Conjecture. SIAM J. Appl. Dyn. Syst. 13, 758–797 (2014)
29. Grigoriev, D., Weber, A.: Complexity of solving systems with few independent monomials and applications to mass-action kinetics. In: Gerdt, V.P., Koepf, W., Mayr, E.W., Vorozhtsov, E.V. (eds.) CASC 2012. LNCS, vol. 7442, pp. 143–154. Springer, Heidelberg (2012)

30. Gunawardena, J.: Chemical reaction network theory for in-silico biologists (2003), http://vcp.med.harvard.edu/papers/crnt.pdf
31. Gunawardena, J.: A linear framework for time-scale separation in nonlinear biochemical systems. PLoS ONE 7, e36321 (2012)
32. Horn, F.: Necessary and sufficient conditions for complex balancing in chemical kinetics. Arch. Rational Mech. Anal. 49, 172–186 (1972)
33. Horn, F., Jackson, R.: General mass action kinetics. Arch. Rational Mech. Anal. 47, 81–116 (1972)
34. Johnston, M.D.: Translated Chemical Reaction Networks. Bull. Math. Biol. 76, 1081–1116 (2014)
35. Jungnickel, D.: Graphs, networks and algorithms, 4th edn. Springer, Heidelberg (2013)
36. Lemaire, F., Ürgüplü, A.: MABSys: Modeling and analysis of biological systems. In: Horimoto, K., Nakatsui, M., Popov, N. (eds.) ANB 2010. LNCS, vol. 6479, pp. 57–75. Springer, Heidelberg (2012)
37. Mirzaev, I., Gunawardena, J.: Laplacian dynamics on general graphs. Bull. Math. Biol. 75, 2118–2149 (2013)
38. Mnëv, N.E.: The universality theorems on the classification problem of configuration varieties and convex polytopes varieties. In: Topology and geometry—Rohlin Seminar. Lecture Notes in Math., vol. 1346, pp. 527–543. Springer, Berlin (1988)
39. Müller, S., Feliu, E., Regensburger, G., Conradi, C., Shiu, A., Dickenstein, A.: Sign conditions for injectivity of generalized polynomial maps with applications to chemical reaction networks and real algebraic geometry (2013) (submitted), http://arxiv.org/abs/1311.5493, arXiv:1311.5493 [math.AG]
40. Müller, S., Regensburger, G.: Generalized mass action systems: Complex balancing equilibria and sign vectors of the stoichiometric and kinetic-order subspaces. SIAM J. Appl. Math. 72, 1926–1947 (2012)
41. Pachter, L., Sturmfels, B.: Statistics. In: Algebraic statistics for computational biology, pp. 3–42. Cambridge Univ. Press, New York (2005)
42. Pérez Millán, M., Dickenstein, A., Shiu, A., Conradi, C.: Chemical reaction systems with toric steady states. Bull. Math. Biol. 74, 1027–1065 (2012)
43. Rambau, J.: TOPCOM: triangulations of point configurations and oriented matroids. In: Mathematical Software (Beijing 2002), pp. 330–340. World Sci. Publ, River Edge (2002)
44. Richter-Gebert, J., Ziegler, G.M.: Oriented matroids. In: Handbook of Discrete and Computational Geometry, pp. 111–132. CRC, Boca Raton (1997)
45. Samal, S.S., Errami, H., Weber, A.: PoCaB: A software infrastructure to explore algebraic methods for bio-chemical reaction networks. In: Gerdt, V.P., Koepf, W., Mayr, E.W., Vorozhtsov, E.V. (eds.) CASC 2012. LNCS, vol. 7442, pp. 294–307. Springer, Heidelberg (2012)
46. Savageau, M.A.: Biochemical systems analysis: II. The steady state solutions for an n-pool system using a power-law approximation. J. Theor. Biol. 25, 370–379 (1969)
47. Thomson, M., Gunawardena, J.: The rational parameterisation theorem for multisite post-translational modification systems. J. Theoret. Biol. 261, 626–636 (2009)
48. Voit, E.O.: Biochemical systems theory: A review. In: ISRN Biomath. 2013, 897658 (2013)
49. Zeilberger, D.: A combinatorial approach to matrix algebra. Discrete Math. 56, 61–72 (1985)
50. Ziegler, G.M.: Lectures on polytopes. Springer, New York (1995)

Lie Symmetry Analysis for Cosserat Rods

Dominik L. Michels[1], Dmitry A. Lyakhov[2], Vladimir P. Gerdt[3],
Gerrit A. Sobottka[4], and Andreas G. Weber[4]

[1] Department of Computing and Mathematical Sciences,
California Institute of Technology, 1200 E. California Blvd., MC 305-16, Pasadena,
CA 91125-2100, USA
dominik@caltech.edu

[2] Radiation Gaseous Dynamics Lab, A. V. Luikov Heat and Mass Transfer Institute
of the National Academy of Sciences of Belarus, P. Brovka St 15,
220072 Minsk, Belarus
lyakhovda@bsu.by

[3] Group of Algebraic and Quantum Computations,
Joint Institute for Nuclear Research, Joliot-Curie 6, 141980 Dubna,
Moscow Region, Russia
gerdt@jinr.ru

[4] Multimedia, Simulation and Virtual Reality Group,
Institute of Computer Science II, University of Bonn, Friedrich-Ebert-Allee 144,
53113 Bonn, Germany
{sobottka,weber}@cs.uni-bonn.de

Abstract. We consider a subsystem of the Special Cosserat Theory of
Rods and construct an explicit form of its solution that depends on three
arbitrary functions in (s, t) and three arbitrary function in t. Assuming
analyticity of the arbitrary functions in a domain under consideration,
we prove that the obtained solution is analytic and general. The Special Cosserat Theory of Rods describes the dynamic equilibrium of 1-
dimensional continua, i.e. slender structures like fibers, by means of a
system of partial differential equations.

Keywords: Cosserat Rods, General Solution, Janet Basis, Kirchhoff
Rods, Lie Symmetry Method.

1 Introduction

The Lie symmetry analysis of differential equations has become a powerful and
universal approach to obtain group-invariant solutions and to perform their classification (cf. [10] and references therein). Sophus Lie himself considered groups
of point and contact transformations to integrate systems of partial differential
equations (PDEs). His key idea was to obtain first infinitesimal generators of one-
parameter symmetry subgroups and then to construct the full symmetry group.
The study of symmetries of differential equations allows one to gain insight into
the structure of the problem they describe. In particular, the existence of Lie
symmetries means that one can find a decomposition of the differential equation
system into a transformed system of reduced order and a set of integrators that

V.P. Gerdt et al. (Eds.): CASC Workshop 2014, LNCS 8660, pp. 324–334, 2014.

in turn can be applied to develop more efficient numerical integration schemes for the governing differential equations.

In our contribution we focus on an equation subsystem of the Special Cosserat Theory of Rods (cf. [1]), a system of coupled partial differential equations, that govern the spatiotemporal evolution of the physical process of deformation of an one-dimensional continuum, the Cosserat rod (e.g. a fiber). In paper [15] the Lie symmetry analysis was applied to study symmetric properties of DNA modelled as a super-long elastic round rod. As a result, nontrivial infinitesimal symmetries of the dynamical Hamiltonian equations of the rod were detected and the related conserved quantities were derived. We consider here another model of the rod and apply the Lie symmetries to the subsystem of the governing system of partial differential equations. With assistance of computer algebra the Lie symmetry approach allowed us to construct a closed form of general analytical solution to the subsystem under consideration. Our motivation to do this research is based on the fact that the deformation modes of a Cosserat rod like bending, twisting, shearing, and extension typically evolve on different time scales which renders the problem inherently stiff (cf. [7]) and demands for appropriate methods for the numerical treatment of the governing PDE system. Knowledge of its structural properties can directly lead to more efficient solution methods.

1.1   Specific Contributions

We use computer algebra systems (specifically MAPLE, which provides sophisticated packages for the analysis of Lie symmetries in PDEs) in order to find Lie symmetries for proper systems and define the conditions under which they exist. In this regard our specific contributions are as follows.

⋄ We study a subsystem of the Special Cosserat Theory of Rods (cf. [1]) of the form

$$\partial_t \kappa(s,t) = \partial_s \omega(s,t) + \omega(s,t) \times \kappa(s,t),$$

by performing a Lie group analysis.
⋄ We construct an explicit form of the solution of the subsystem that depends on three arbitrary functions in (s,t) and three arbitrary functions in t.
⋄ We prove that the obtained solution is analytic and general.

2   Special Cosserat Theory of Rods

In this section we give a recap of the Special Cosserat Theory of Rods. Fibers can approximately be considered as one-dimensional continua that undergo bending, twisting, shearing, and longitudinal dilation deformation. Following [1], we consider the Euclidian 3-space $\mathbb{E}^3$ to be the abtract 3-dimensional inner product space. Its elements are denoted by lower-case, boldface, italic symbols. Let $\mathbb{R}^3$ be the set of triples of real numbers. Its elements are denoted by lower-case, boldface, sans-serif letters.

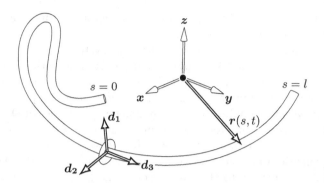

Fig. 1. The vector set $\{d_k\}$ forms a right-handed orthonormal basis at each point of the centerline. The directors d_1 and d_2 span the local material cross-section, whereas d_3 is perpendicular to the material cross-section. Note that in the presence of shear deformations d_3 is not tangent to the centerline of the fiber.

The motion of a special Cosserat Rod is given by

$$(s,t) \mapsto (r(s,t), d_1(s,t), d_2(s,t)), \tag{1}$$

where $r(s,t)$ is the centerline of the rod. It is furnished with a set of so-called orthonormal directors $\{d_1(s,t), d_2(s,t), d_3(s,t)\}$. $\{d_k\}$ is a right-handed orthonormal basis in $\mathbb{E}^3$, with $d_3 := d_1 \times d_2$. The directors d_1 and d_2 span the cross-section plane, see Fig. (1). The deformation of the rod is obtained if its motion defined by Eq. (1) is related to some reference configuration $\{r^\circ(s,t), d_1^\circ(s,t), d_2^\circ(s,t)\}$.

Further, there exist vector-valued functions κ and ω such that the directors evolve according to the kinematic relations

$$\partial_s d_k = \kappa \times d_k,$$
$$\partial_t d_k = \omega \times d_k,$$

where κ is the Darboux and ω the twist vector. Their components are given with respect to the orthonormal basis, i.e. $\kappa = \sum_{k=1}^{3} \kappa_k d_k$ and $\omega = \sum_{k=1}^{3} \omega_k d_k$.

The linear strains of the rod are given by $\nu = \sum_{k=1}^{3} \nu_k d_k = \partial_s r$ and the velocity of a cross-section material plane by $v = \partial_t r$. The triples $(\kappa_1, \kappa_2, \kappa_3)$, $(\omega_1, \omega_2, \omega_3)$, (ν_1, ν_2, ν_3), and (v_1, v_2, v_3) are denoted by κ, ω, ν, and v respectively. In particular, $\kappa := (\kappa_1, \kappa_2, \kappa_3)$ and $\nu := (\nu_1, \nu_2, \nu_3)$ are the strain variables that uniquely determine the motion of the rod described by Eq. (1) at every instant in time t (except for a rigid body motion). Their components have a physical meaning: they describe the bending of the rod with respect to the two major axes of the cross section (κ_1, κ_2), the torsion (κ_3), shear (ν_1, ν_2) and extension (ν_3). Moreover, since $\partial_t \partial_s d_k = \partial_s \partial_t d_k$ we obtain the compatibility equation

$$\partial_t \kappa = \partial_s \omega + \omega \times \kappa.$$

In the same sense we have

$$\partial_t \nu = \partial_s v.$$

2.1   Equations of Motion

The equations of motion for the rod read

$$\partial_s \boldsymbol{n} + \boldsymbol{f} = \rho A \partial_t \boldsymbol{v} + \rho \left(I_1 \partial_{tt} \boldsymbol{d}_1 + I_2 \partial_{tt} \boldsymbol{d}_2 \right),$$
$$\partial_s \boldsymbol{m} + \boldsymbol{\nu} \times \boldsymbol{n} + \boldsymbol{l} = \rho \left(I_1 \boldsymbol{d}_1 + I_2 \boldsymbol{d}_2 \right) \times \partial_t \boldsymbol{v} + \partial_t \left(\rho \boldsymbol{J} \boldsymbol{\omega} \right),$$

where $\boldsymbol{n} = \sum_{k=1}^3 n_k \boldsymbol{d}_k$ and $\boldsymbol{m} = \sum_{k=1}^3 m_k \boldsymbol{d}_k$ are the internal stresses and $\boldsymbol{f}$ and $\boldsymbol{l}$ are the external forces and torques acting on the rod, ϱ the linear density. I_1 and I_2 are the first mass moments of inertia of cross section per unit length and $\boldsymbol{J}$ is the inertia tensor of cross section per unit length. Further, we define $\mathsf{n} := (n_1, n_2, n_3)$ and $\mathsf{m} := (m_1, m_2, m_3)$. The shear forces are given by n_1 and n_2, the tension by $\boldsymbol{n} \cdot \boldsymbol{\nu} / \|\boldsymbol{\nu}\|$, bending moments by m_1 and m_2, and the twisting moment by m_3.

2.2   Constitutive Relations

In order to relate the internal stresses $\boldsymbol{n}$ and $\boldsymbol{m}$ to the kinematic quantities $\boldsymbol{\nu}$ and $\boldsymbol{\kappa}$ we introduce constitutive equations of the form

$$\mathsf{n}(s,t) = \hat{\mathsf{n}} \left(\boldsymbol{\kappa}(s,t), \boldsymbol{\nu}(s,t), s \right),$$
$$\mathsf{m}(s,t) = \hat{\mathsf{m}} \left(\boldsymbol{\kappa}(s,t), \boldsymbol{\nu}(s,t), s \right).$$

For fixed s, the common domain $\mathcal{V}(s)$ of these constitutive functions is a subset of $(\boldsymbol{\kappa}, \boldsymbol{\nu})$ describing orientation preserving deformations. $\mathcal{V}(s)$ consists at least of all $(\boldsymbol{\kappa}, \boldsymbol{\nu})$ that satisfy $\nu_3 = \boldsymbol{\nu} \cdot \boldsymbol{d}_3 > 0$.

The rod is called hyper-elastic, if there exists a strain-energy-function $W : \{(\boldsymbol{\kappa}, \boldsymbol{\nu} \in \mathcal{V})\} \to \mathbb{R}$ such that

$$\hat{\mathsf{n}} \left(\boldsymbol{\kappa}, \boldsymbol{\nu}, s \right) = \partial W \left(\boldsymbol{\kappa}, \boldsymbol{\nu}, s \right) / \partial \boldsymbol{\nu},$$
$$\hat{\mathsf{m}} \left(\boldsymbol{\kappa}, \boldsymbol{\nu}, s \right) = \partial W \left(\boldsymbol{\kappa}, \boldsymbol{\nu}, s \right) / \partial \boldsymbol{\kappa}.$$

The rod is called viscoelastic of strain-rate type of complexity 1 if there exist functions such that

$$\mathsf{n}(s,t) = \hat{\mathsf{n}} \left(\boldsymbol{\kappa}(s,t), \boldsymbol{\nu}(s,t), \partial_t \boldsymbol{\kappa}(s,t), \partial_t \boldsymbol{\nu}(s,t), s \right), \tag{2}$$
$$\mathsf{m}(s,t) = \hat{\mathsf{m}} \left(\boldsymbol{\kappa}(s,t), \boldsymbol{\nu}(s,t), \partial_t \boldsymbol{\kappa}(s,t), \partial_t \boldsymbol{\nu}(s,t), s \right). \tag{3}$$

For $\partial_t \boldsymbol{\kappa}(s,t) = \boldsymbol{0}$ and $\partial_t \boldsymbol{\nu}(s,t) = \boldsymbol{0}$, Eq. (2) and Eq. (3) become the so called equilibrium response functions and describe elastic behavior.

2.3   Material Laws

The constitutive laws for elastic material behavior become

$$\hat{\mathsf{n}}(s,t) = \left(GA \left(\nu_1 - \nu_1^\circ \right), GA \left(\nu_2 - \nu_2^\circ \right), EA \left(\nu_3 - \nu_3^\circ \right) \right),$$

with the initial strain vector field $\boldsymbol{\nu}^\circ(s)$, Young's modulus E, cross-section area A, and

$$\hat{\mathbf{m}}(s,t) = \left(E_b I_1 \left(\kappa_1 - \kappa_1^\circ\right), E_b I_2 \left(\kappa_2 - \kappa_2^\circ\right), G I_3 \left(\kappa_3 - \kappa_3^\circ\right)\right),$$

with the initial bending and torsion vector field $\boldsymbol{\kappa}^\circ(s)$, Young's modulus E_b of bending, and shear modulus G. The area moments of inertia are again denoted by I_1 and I_2, the polar moment of inertia with I_3.

Since bending stiffnesses $E_b I_1$, $E_b I_2$, and torsional stiffness $G_b I_3$ change with the fourth power of the fiber diameter they are usually orders of magnitude smaller than the tensile stiffness EA and shearing stiffness GA. This renders the problem of fiber simulation based on the Special Theory of Cosserat Rods inherently "stiff".

Kirchhoff Rods. We allude to the fact that in the classical theory of Kirchhoff the rod can undergo neither shear nor extension. This is accommodated by setting the linear strains to $\boldsymbol{\nu} := (\nu_1, \nu_2, \nu_3) = (0, 0, 1)$, (local coordinates). Geometrically this means, that the angle between the director $\mathbf{d}_3$ and the tangent to the centerline, $\partial_s \mathbf{r}$, always remains zero (no shear) and that the tangent to the centerline always has unit length (no elongation).

2.4 System of Governing Equations

The full system of partial differential equations governing the deformation of an elastic rod is thus given by the following first order system,

$$\partial_t \mathbf{d}_k = \boldsymbol{\omega} \times \mathbf{d}_k, \tag{4}$$

$$\partial_t \boldsymbol{\kappa} = \partial_s \boldsymbol{\omega} + \boldsymbol{\omega} \times \boldsymbol{\kappa}, \tag{5}$$

$$\partial_t \boldsymbol{\nu} = \partial_s \boldsymbol{v},$$

$$\partial_t \left(\rho \mathbf{J} \boldsymbol{\omega}\right) = \partial_s \left(\hat{m}_k(\boldsymbol{\kappa}, \boldsymbol{\nu}) \mathbf{d}_k\right) + \boldsymbol{\nu} \times \hat{n}_k(\boldsymbol{\kappa}, \boldsymbol{\nu}) \mathbf{d}_k,$$

$$\rho A \partial_t \boldsymbol{v} = \partial_s \left(\hat{n}_k(\boldsymbol{\kappa}, \boldsymbol{\nu}) \mathbf{d}_k\right).$$

If $(\hat{\mathbf{n}}, \hat{\mathbf{m}})$ satisfy the monotonicity condition, i.e. the matrix

$$\begin{bmatrix} \partial\hat{\mathbf{m}}/\partial\boldsymbol{\kappa} & \partial\hat{\mathbf{m}}/\partial\boldsymbol{\nu} \\ \partial\hat{\mathbf{n}}/\partial\boldsymbol{\kappa} & \partial\hat{\mathbf{n}}/\partial\boldsymbol{\nu} \end{bmatrix},$$

is positive-definite, then this system is hyperbolic. It can be written in the form of a conservation law

$$\partial_t \Phi(\zeta) = \partial_s \Psi(\zeta) + \Theta(\zeta),$$

with $\zeta = (\mathbf{d}_k, \boldsymbol{\kappa}, \boldsymbol{\nu}, \boldsymbol{\omega}, \boldsymbol{v})$. This system can be decoupled from the Kinematic Eq. (4) by decomposing it with respect to the basis $\{\mathbf{d}_k\}$ which yields

$$\partial_t \boldsymbol{\kappa} = \partial_s \boldsymbol{\omega} + \boldsymbol{\omega} \times \boldsymbol{\kappa}, \tag{6}$$

$$\partial_t \boldsymbol{\nu} = \partial_s \boldsymbol{v} + \boldsymbol{\kappa} \times \boldsymbol{v} - \boldsymbol{\omega} \times \boldsymbol{\nu},$$

$$\partial_t \left(\rho \mathbf{J} \boldsymbol{\omega}\right) = \partial_s \hat{\mathbf{m}} + \boldsymbol{\kappa} \times \hat{\mathbf{m}} + \boldsymbol{\nu} \times \hat{\mathbf{n}} - \boldsymbol{\omega} \times \left(\rho \mathbf{J} \boldsymbol{\omega}\right),$$

$$\rho A \partial_t \boldsymbol{v} = \partial_s \hat{\mathbf{n}} + \boldsymbol{\kappa} \times \hat{\mathbf{n}} - \boldsymbol{\omega} \times \left(\rho A \boldsymbol{v}\right). \tag{7}$$

If external forces (e.g. gravity) are to be considered as well, they have to be added to the right-hand side of Eq. (7) after transforming them into the local basis. For this purpose the Kinematic Eq. (4) has to be solved additionally.

3   Lie Symmetry Analysis

3.1   Problem Setting

As a first step in the direction of a better understanding of the Cosserat rod we begin our study with a subsystem of the Special Cosserat Theory of Rods. Precisely, we apply the classical symmetry method invented by Sophus Lie (cf. [10]) to the equation system given by Eq. (6),

$$\boldsymbol{F} = \boldsymbol{0}, \quad \boldsymbol{F} := \partial_s \boldsymbol{\omega} - \partial_t \boldsymbol{\kappa} + \boldsymbol{\omega} \times \boldsymbol{\kappa}, \tag{8}$$

which corresponds to Eq. (5) or Eq. (6).

3.2   Infinitesimal Criterion of Invariance

We apply the classical Lie-point symmetry method [10] to the quasi-linear first-order PDE system Eq. (8) in two independent variables s, t and six dependent variables that are the components of the vectors $\boldsymbol{\omega}$ and $\boldsymbol{\kappa}$.

A transformation of Eq. (8)

$$\begin{aligned} s' &= s'(s, t, \boldsymbol{\omega}(s,t), \boldsymbol{\kappa}(s,t)), \quad \boldsymbol{\omega}' = \boldsymbol{\omega}'(s, t, \boldsymbol{\omega}(s,t), \boldsymbol{\kappa}(s,t)), \\ t' &= t'(s, t, \boldsymbol{\omega}(s,t), \boldsymbol{\kappa}(s,t)), \quad \boldsymbol{\kappa}' = \boldsymbol{\kappa}'(s, t, \boldsymbol{\omega}(s,t), \boldsymbol{\kappa}(s,t)), \end{aligned} \tag{9}$$

that maps solutions to solutions is called Lie(-point) symmetry of Eq. (8). The vector field

$$X := \xi^1 \partial_s + \xi^2 \partial_t + \sum_{i=1}^{3} \left(\theta^i \partial_{\omega_i} + \vartheta^i \partial_{\kappa_i} \right), \tag{10}$$

is an infinitesimal generator of a one-parameter ($x \in \mathbb{R}$) Lie group of point transformations if its flow $\exp(x\,X)$ is a Lie(-point) symmetry. The coefficients $\xi^1, \xi^2, \theta^i, \vartheta^j$ ($i, j = 1, 2, 3$) in Eq. (10) are functions in independent and dependent variables, and below we shall use the vector notation $\boldsymbol{\theta} := \{\theta^1, \theta^2, \theta^3\}$, $\boldsymbol{\vartheta} := \{\vartheta^1, \vartheta^2, \vartheta^3\}$.

The equality

$$X^{(pr)} \boldsymbol{F} \mid_{\boldsymbol{F}=0} = 0, \tag{11}$$

is the *infinitesimal criterion of invariance* of Eq. (8) under a one-parameter Lie group of point transformations. Here $X^{(pr)}$ stands for the prolonged infinitesimal symmetry generator that, in addition to those in Eq. (10), contains extra terms caused by the presence of the first-order partial derivatives in Eq. (8). These extra terms are easily computed taking certain derivatives of the coefficient functions in the generator. The subscript in Eq. (11) indicates that equality $X^{(pr)} \boldsymbol{F} = 0$ must hold under condition $\boldsymbol{F} = 0$. Equality Eq. (11) implies an overdetermined determining system of linear PDEs for the coefficient functions of generator Eq. (10). Since the equations in Eq. (8) are quasi-linear and solved with respect to partial derivatives, generation of their determining equations is algorithmically straightforward.

3.3 Solving Determining Equations

To obtain the determining equations we use the Maple package DESOLV [6] and its routine *gendef*. It outputs 42 first-order PDEs. Generally, a reliable and powerful algorithmic way to solve a system determining equations is its transformation to a canonical involutive form or to a Gröbner basis form and then solving such canonical system (cf. [8]). The package DESOLV has the built-in routine *icde* that implements the Standard Form algorithm [12] for completion to involution. We prefer, however, completion to a Janet basis, the canonical involutive form based on Janet division (cf. [13] for theory of involution for algebraic and differential systems and references therein) and respectively, the Maple package JANET [5] that computes a Janet basis for a differential ideal generated by linear differential polynomials. We have two arguments in favor of this preference: (i) according to our benchmarking, package JANET is substantially faster than the routine *icde* in DESOLV; (ii) given a linear system of PDEs in the Janet involutive form and a set of its analytic solutions one can algorithmically check whether this set contains all analytic solutions (cf. [9,13]).

The involutive form of determining equations for Eq. (11) computed by package JANET contains 86 linear PDEs. To solve these equations we applied the routine *pdsolv* of the package DESOLV that exploits a number of heuristic algorithms for integration of linear PDEs including some advanced algorithms [14] oriented to integration of overdetermined systems of polynomially-nonlinear differential equations. It outputs a solution that depends on five arbitrary functions in the independent variables s, t. Two of these functions are unnecessary for us and can be omitted since they appear in the solution as shifts in the independent variables in accordance to the fact that the equations in Eq. (8) are autonomous. Taking this into account, the obtained solution can be presented as

$$\xi^1 = \xi^2 = 0, \quad \boldsymbol{\theta} = \hat{\boldsymbol{A}}\boldsymbol{\omega} + \partial_t\boldsymbol{p}, \quad \boldsymbol{\vartheta} = \hat{\boldsymbol{A}}\boldsymbol{\kappa} + \partial_s\boldsymbol{p}, \tag{12}$$

where

$$\hat{\boldsymbol{A}}(s,t) = \begin{bmatrix} 0 & -c(s,t) & b(s,t) \\ c(s,t) & 0 & -a(s,t) \\ -b(s,t) & a(s,t) & 0 \end{bmatrix}, \quad \boldsymbol{p}(s,t) := \begin{bmatrix} a(s,t) \\ b(s,t) \\ c(s,t) \end{bmatrix}, \tag{13}$$

and $a(s,t), b(s,t), c(s,t)$ are arbitrary smooth functions. Hereafter, we shall assume analyticity of these functions.

3.4 Lie Group of Point Symmetry Transformations

The routine *pdesolv* of DESOLV may not always find all solutions of the input differential system. For our purpose, however, the arbitrariness given by Eq. (12)

and Eq. (13) is sufficient for the construction of a general solution to Eq. (8). It is essential that the routine outputs a solution with maximally possible arbitrariness in the number of arbitrary functions depending on the both independent variables (s, t). This fact can be detected from the structure of differential dimensional polynomial [9]. It is easily computed by the routine *DifferentialSystemDimensionPolynomial* of the Maple package DIFFERENTIALTHOMAS [3] that takes the Janet basis form of the determining system as an input. In the case being considered the differential dimensional polynomial is given by

$$\frac{5}{2}l^2 + \frac{21}{2}l + 11 = 5\binom{l+2}{l} + 3\binom{l+1}{l} + 3.$$

The first term of this expression shows that the general analytic solution contains 5 arbitrary functions in 2 variables. If one applies the built-in Maple command *pdsolve*, then its output solution set also contains 5 arbitrary functions in two variables. However, this solution depends nonlocally (via some integrals) on the arbitrary functions, and is rather cumbersome.

Having obtained the structure Eq. (12) and Eq. (13) of the infinitesimal symmetry generator Eq. (10), we compute now the one-parameter Lie-point symmetry group of transformations Eq. (9) it generates. The symmetry group is given as solution (Lie's first fundamental theorem [10]) of the following system of two trivially solvable scalar ordinary differential equations

$$d_x s' = 0, \quad s'(0) = s \quad \implies \quad s' = s,$$
$$d_x t' = 0, \quad t'(0) = t \quad \implies \quad t' = t,$$

and two vector ones

$$d_x \boldsymbol{\omega}' = \hat{\boldsymbol{A}}\boldsymbol{\omega}' + \partial_t \boldsymbol{p}, \quad \boldsymbol{\omega}'(0) = \boldsymbol{\omega}, \tag{14}$$

$$d_x \boldsymbol{\kappa}' = \hat{\boldsymbol{A}}\boldsymbol{\kappa}' + \partial_s \boldsymbol{p}, \quad \boldsymbol{\kappa}'(0) = \boldsymbol{\kappa}. \tag{15}$$

Solution to Eq. (14) and Eq. (15) is computable with the Maple command *dsolve*. But the output of this command is awkward and it is not easy to obtain in compact form. Such compact form solution can be obtained by hand computation as follows.

It is not difficult to see (cf. [4]), that Eq. (14) is satisfied by the expression

$$\boldsymbol{\omega}' = \exp(x\hat{\boldsymbol{A}})\boldsymbol{\omega} + \exp(x\hat{\boldsymbol{A}}) \int_0^x \exp(-y\hat{\boldsymbol{A}})\, dy\, \partial_t \boldsymbol{p}. \tag{16}$$

Since this system satisfies the conditions of the classical existence and uniqueness theorem for systems of ordinary differential equations [11], it follows that Eq. (16) is the solution to Eq. (14).

Furthermore, application of the Cayley-Hamilton theorem [2] to the matrix $\hat{\boldsymbol{A}}$ in Eq. (13) gives

$$\hat{\boldsymbol{A}}^3 = -p^2 \hat{\boldsymbol{A}}, \quad p := |\boldsymbol{p}| = \sqrt{a^2(s, t) + b^2(s, t) + c^2(s, t)}. \tag{17}$$

By means of relation Eq. (17) expression Eq. (16) is transformed to

$$\omega' = \left(\hat{I} + \frac{\sin(px)\,\hat{A}}{p} + \frac{(1 - \cos(px))\,\hat{A}^2}{p^2} \right) \omega +$$
$$+ \left(\frac{px\hat{A}^2 + p^3\,x\hat{I} - \sin(px)\,\hat{A}^2 - \cos(px)\,p\hat{A} + p\hat{A}}{p^3} \right) \partial_t p, \quad (18)$$

where $\hat{I}$ is the 3×3 identity matrix. Formulae Eq. (13) and Eq. (18) show that without loss of generality the arbitrary vector p and matrix $\hat{A}$ for $x \neq 0$ can be rescaled to absorb the group parameter x. It is equivalent to putting $x := 1$. In so doing, transformation Eq. (18) can be rewritten in terms of arbitrary vector-function p as follows,

$$\omega' = \omega - \frac{\sin(p)}{p}\,p \times \omega + \frac{1 - \cos(p)}{p^2}\left(p\,(p\,\omega) - p^2\,\omega\right) + \partial_t p +$$
$$+ \frac{p - \sin(p)}{p^3}\left(p\,(p\,\partial_t p) - p^2\,\partial_t p\right) - \frac{1 - \cos(p)}{p^2}\,p \times \partial_t p. \quad (19)$$

Respectively, solution to Eq. (15) reads

$$\kappa' = \kappa - \frac{\sin(p)}{p}\,p \times \kappa + \frac{1 - \cos(p)}{p^2}\left(p\,(p\,\kappa) - p^2\,\kappa\right) + \partial_s p +$$
$$+ \frac{p - \sin(p)}{p^3}\left(p\,(p\,\partial_s p) - p^2\,\partial_s p\right) - \frac{1 - \cos(p)}{p^2}\,p \times \partial_s p. \quad (20)$$

Thus, transformations Eq. (19) and Eq. (20) are Lie-point symmetries of Eq. (8) for an arbitrary vector p in Eq. (13). We also verified directly with Maple that if ω, κ are solutions to Eq. (8), then ω', κ' are also solutions for any vector-function p.

3.5   General Solution

Consider now the special case of transformations Eq. (19) and Eq. (20) when

$$\omega := f(t), \quad f(t) = \{f_1(t), f_2(t), f_3(t)\}, \quad \kappa := 0, \quad (21)$$

where $f_1(t), f_2(t), f_3(t)$ are arbitrary analytic functions, and denote the image of Eq. (21) under the transformations by ω and κ:

$$\omega = f(t) - \frac{\sin(p)}{p}\,p \times f(t) + \frac{1 - \cos(p)}{p^2}\left(p\,(p\,f(t)) - p^2\,f(t)\right) + \partial_t p +$$
$$+ \frac{p - \sin(p)}{p^3}\left(p\,(p\,\partial_t p) - p^2\,\partial_t p\right) - \frac{1 - \cos(p)}{p^2}\,p \times \partial_t p, \quad (22)$$
$$\kappa = \partial_s p + \frac{p - \sin(p)}{p^3}\left(p\,(p\,\partial_s p) - p^2\,\partial_s p\right) - \frac{1 - \cos(p)}{p^2}\,p \times \partial_s p.$$

Now we can formulate and prove the main theoretical result of the present paper.

Proposition 1. *The vector-functions ω and κ expressed by formulae Eq. (22) in terms of the matrix- and vector-functions Eq. (13), Eq. (21) whose components are arbitrary analytic functions provide a solution to equations Eq. (8) that is general.*

Proof. Obviously, Eq. (21) is a solution. It follows that expressions Eq. (22) make up a solution. We show first that one can choose arbitrary functions $a(s,t)$, $b(s,t)$, and $c(s,t)$ to obtain any vector-function κ as the left-hand side of the second equality in Eq. (22). The partial derivatives $\partial_s a$, $\partial_s b$, $\partial_s c$ of a, b, c with respect to s appear linearly in Eq. (22). Direct computation with Maple of the Jacobian matrix $J_\kappa(\partial_s a, \partial_s b, \partial_s c)$ and its determinant gives the following compact expression:

$$\det\left(J_\kappa(\partial_s a, \partial_s b, \partial_s c)\right) = \frac{2(\cos(p) - 1)}{p^2} .$$

It is clear that one can always choose the *initial values* $a(0,t)$, $b(0,t)$, and $c(0,t)$ of the arbitrary functions to provide nonzero values of the Jacobian determinant, and hence to solve locally the last vector equality in Eq. (22) with respect to $\partial_s a$, $\partial_s b$, $\partial_s c$. Thereby, the equation can be brought into a first-order differential system solved with respect to the partial derivative and with analytic right-hand sides. Then, by the classical Cauchy-Kovalevskaya theorem (cf. [13]) the initial value (Cauchy) problem for the solved system has a unique analytic solution.

We now turn our attention to the first equation in Eq. (22) and show that having decided upon $a(s,t)$, $b(s,t)$, and $c(s,t)$, one can choose functions $f_1(t)$, $f_2(t)$, and $f_3(t)$ to obtain arbitrary analytic $\omega_1(0,t)$, $\omega_2(0,t)$, and $\omega_3(0,t)$. The part of Eq. (22) linear in $f(t)$ is identically equal to $\exp(\hat{A})f(t)$ (cf. Eq. (16)). Since the matrix $\hat{A}$ given by Eq. (13) is skew-symmetric, $\det(\exp(\hat{A})) = 1$ and Eq. (22) is solvable with respect to the analytic vector-function $f(t)$. It implies that the system Eq. (8), being in *normal* or *Cauchy-Kovalevskaya form*, admits unique analytic solutions for $\omega_1(s,t)$, $\omega_2(s,t)$, and $\omega_3(s,t)$. ☐

4 Conclusion and Future Work

In this contribution, we have studied a subsystem of the Special Cosserat Theory of Rods by performing a Lie group analysis. To be more specific, by applying modern computer algebra methods, algorithms, and software we have constructed an explicit form of solution to Eq. (8) that depends on three arbitrary functions in (s,t) and three arbitrary functions in t. Assuming analyticity of the arbitrary functions in a domain under consideration, we have proved that the obtained solution is analytic and general. This is a step towards generating detailed knowledge about the structure of the PDE system that governs the spatiotemporal evolution of the Cosserat rod.

In our future work these results will be used to develop algorithms based on combinations of numerical and analytical treatments of the governing equations to overcome the typical problems resulting from the system's stiffness. This approach would allow for larger step sizes compared to pure numerical solvers and at the same time combines efficiency and accuracy without sacrificing one for another.

Acknowledgements. The contribution of the third author (V.P.G.) was partially supported by the grant 13-01-00668 from the Russian Foundation for Basic Research. We thank Markus Lange-Hegermann and Paul Mueller for useful remarks. The authors are grateful to the reviewers' valuable comments that improved the manuscript.

References

1. Antman, S.S.: Nonlinear Problems of Elasticity. Appl. Math. Sci., vol. 107. Springer, New York (1995)
2. Atiyah, M.F., MacDonald, I.G.: Introduction to Commutative Algebra. Addison-Wesley Pub. Co., London (1969)
3. Bächler, T., Gerdt, V.P., Lange-Hegermann, M., Robertz, D.: Algorithmic Thomas Decomposition of algebraic and differential systems. J. Symb. Comput. 47, 1233–1266 (2012)
4. Bellman, R.: Introduction to Matrix Analysis, 2nd edn. Society for Industrial and Applied Mathematics, Philadelphia (1997)
5. Blinkov, Y.A., Cid, C.F., Gerdt, V.P., Plesken, W., Robertz, D.: The Maple package Janet: II. Linear partial differential equations. In: Ganzha, V.G., Mayr, E.W., Vorozhtsov, E.V. (eds.) Proceedings of the 6th Workshop on Computer Algebra in Scientific Computing / CASC 2003, Institut für Informatik, Technische Universität München, Garching, pp. 41–54 (2003)
6. Carminati, J., Vu, K.: Symbolic Computation and Differential Equations: Lie Symmetries. J. Symb. Comput. 29, 95–116 (2000)
7. Curtiss, C.F., Hirschfelder, J.O.: Integration of Stiff Equations. Proceedings of the National Academy of Sciences of the United States of America 38(3), 235–243 (1952)
8. Hereman, W.: Review of symbolic software for Lie symmetry analysis. In: Ibragimov, N.H. (ed.) CRS Handbook of Lie Group Analysis of Differential Equations, ch. 13. New Trends in Theoretical Developments and Computational Methods, vol. 3, pp. 367–413. CRS Press, Boca Raton (1996)
9. Lange-Hegermann, M.: The Differential Dimension Polynomial for Chatacterizable Differential Ideals. arXiv:1401.5959 (2014)
10. Oliveri, F.: Lie Symmetries of Differential Eqautions: Classical Results and Recent Contributions. Symmetry 2, 658–706 (2010)
11. Pontryagin, L.S.: Ordinary Differential Equations. Addison-Wesley Pub. Co., London (1962)
12. Reid, G.: Algorithms for reducing a system of PDEs to standard form, determining the dimension of its solution space and calculating its Taylor series solution. Eur. J. Appl. Math. 2, 293–318 (1991)
13. Seiler, W.M.: Involution - The Formal Theory of Differential Equations and Its Application in Computer Algebra. Algorithms and Computation in Mathematics, vol. 24. Springer, Heidelberg (2010)
14. Wolf, T.: The Symbolic Integration of Exact PDEs. J. Symb. Comput. 30, 619–629 (2000)
15. Zhao, W.-J., Weng, Y.-Q., Fu, J.-L.: Lie Symmetries and Conserved Quantities for Super-Long Elastic Slender Rod. Chinese Phys. Lett. 24, 2773–2776 (2007)

Real Polynomial Root-Finding by Means of Matrix and Polynomial Iterations

Victor Y. Pan

Departments of Mathematics and Computer Science
Lehman College and the Graduate Center of the City University of New York
Bronx, NY 10468 USA
victor.pan@lehman.cuny.edu
http://comet.lehman.cuny.edu/vpan/

Abstract. Frequently one seeks approximation to all r real roots of a polynomial of degree n with real coefficients, which also has nonreal roots. We split a polynomial into two factors, one of which has degree r and has r real roots. We approximate them at a low cost, and then decrease the arithmetic time of the known algorithms for this popular problem by roughly a factor of n/k, if k iterations prepare splitting. k is a small integer unless some nonreal roots lie close to the real axis, but even if there nonreal roots near the real axis, we substantially accelerate the known algorithms. We also propose a dual algorithm, operating with the associated structured matrices. At the price of minor increase of the arithmetic time, it facilitates numerical implementation. Our analysis and tests demonstrate the efficiency of our approach.

Keywords: polynomials, real roots, matrices, matrix sign iteration, companion matrix, real eigenvalues, Frobenius algebra, square root iteration, root squaring.

1 Introduction

In some applications, e.g., to algebraic and geometric optimization, one seeks real roots of a univariate polynomial

$$p(x) = \sum_{i=0}^{n} p_i x^i = p_n \prod_{j=1}^{n} (x - x_j), \quad p_n \neq 0, \tag{1}$$

of degree n that has real coefficients, r real roots $x_1, \ldots, x_r$, and $s = (n - r)/2$ pairs of nonreal complex conjugate roots $x_{r+1}, \ldots, x_n$ (typically $r \ll n$). This is a well studied subject (see [13, Chapter 15], [21], [25], and the bibliography therein), but we propose new efficient algorithms by extending and combining the techniques of [23] and [20]. We combine the two known low cost steps, recalled in Section 2, that is, splitting a polynomial into two factors whose two sets of roots are isolated from one another, and real root-finding when all n roots of the input polynomial are real. Namely, our iterative processes split out the factor

V.P. Gerdt et al. (Eds.): CASC Workshop 2014, LNCS 8660, pp. 335–349, 2014.
© Springer International Publishing Switzerland 2014

$s(x) = \prod_{j=1}^{r}(x - x_j)$ of degree r that shares with the input polynomial $p(x)$ all its real roots, and as soon as this factor has been computed, we readily approximate its r roots at a low computational cost. As a result, we yield the solution at the arithmetic cost $O(kn \log(n))$, provided that k iterations prepare splitting of the factor. Our iterative algorithms converge exponentially fast (with quadratic or cubic rates), and so $k = O(b+d)$, assuming the tolerance 2^{-b} to the errors of the output approximation and the minimal distance 2^{-d} of the nonreal roots from the real axis. Usually this bound on k is not large, except for the inputs having nonreal roots that lie very close to the real axis. In Remark 4, we discuss some techniques for handling even such harder inputs. According to our preliminary considerations (cf., e.g., Remark 8) and the test results, our algorithms can be implemented with a reasonably bounded precision of computing, but we leave the formal study of this subject and of the Boolean complexity of our algorithms as a challenge for further research.

We devise dual iterations with polynomials generated from the input polynomial $p(x)$ of (1) and with matrices generated from the companion matrix of this polynomial. In the latter case, we seek real eigenvalues of this matrix, extend the matrix sign classical iteration toward this goal, and employ the known results and techniques in this well developed area. Dealing with matrices one can engage efficient packages of subroutines available for numerical matrix computations with the IEEE standard double precision. The highly structured companion matrix generates the Frobenius matrix algebra, in which one can perform FFT-based computations in nearly linear time, that is, as fast as the similar operations with polynomials. In some cases, we take advantage of combining the power of operating with matrices and polynomials (see Remark 13). Finding their deeper synergistic combinations is another natural research challenge.

We present a number of promising algorithms. Algorithms 2 and 5 have the lowest estimated arithmetic cost $O(kn \log(n))$, which increases to $O(kn \log^2(n)) + c(n, r)$ for Algorithms 3 and 4. Here $c(n, r)$ is the overhead due to randomization for Algorithm 3 and to computing approximate polynomial GCDs for Algorithm 4. We include these algorithms since they use some promising techniques and since Algorithm 3 showed superior numerical stability in our tests.

We engage, extend, and combine the number of efficient methods available for complex polynomial root-finding, particularly the ones of [23] and [20], but we also propose new techniques and employ some old methods in novel and nontrivial ways. E.g., our Algorithm 2 streamlines and substantially modifies [23, Algorithm 9.1] by avoiding the stage of root-squaring and the application of the Cayley map, and similar comments apply to our adjustment of the matrix sign classical iteration to real eigen-solving. Most of the techniques of Algorithm 3 are implicit in [20, Section 5], but we specify the algorithm in some detail, include initial scaling, substantially modify the recovery of the eigenvalues, and combine it with Algorithm 2. Algorithms 4 and 5 are new, in spite of some links to Algorithms 2 and 3 and hence to [20, Section 5] and [23, Section 9]. Our interplay with matrix and polynomial computations to the benefit of both subjects (this idea can be traced back to [14] and [2]) as well as our exploitation of the complex plane geometry and of various transforms of the variable can be of independent interest. Our simple recipe for real root-finding by means of combining the root radii algorithm with Newton's iteration in Algorithm 1 works for a large class of inputs, and even the extension of our approach to the approximation of real eigenvalues of a real matrix can be of some potential interest.

Hereafter "ops" stands for "arithmetic operations", "lc(p)" stands for "the leading coefficient of $p(x)$". $D(X, r) = \{x : |x - X| \le r\}$ and $C(X, r) = \{x : |x - X| = r\}$ denote a disc and a circle on the complex plane, respectively. We write $\| \sum_i v_i x^i \|_q = (\sum_i |v_i|^q)^{1/q}$ for $q = 1, 2$ and $\| \sum_i v_i x^i \|_\infty = \max_i |v_i|$. A function is in $\tilde{O}(f(bc))$ if it is in $O(f(bc))$ up to polylogarithmic factors in b and c. agcd(u, v) denotes the *approximate greatest common divisor* of two polynomials $u(x)$ and $v(x)$ (see [1] on definitions and algorithms).

2   Basic Results for Polynomials

Next we present some building blocks for our root-finders. Besides the two cited results, used as the main blocks of our algorithms (that is, inexpensive splitting of a polynomial into two factors and fast real root-finding for a polynomial that has only real roots) we recall scaling, shifting, inverting and squaring the roots, their mapping from the real axis or from a real line interval into a fixed circle and back, and the approximation of the absolute values of all roots $|x_1|, \ldots, |x_n|$. All these operations can be performed at a low arithmetic cost as well.

Theorem 1. *(Root Radii Approximation, cf. [24], [13, Section 15.4], [5].) Assume a polynomial $p(x)$ of (1) and two scalars $c > 0$ and d. Define the n root radii $r_j = |x_{k_j}|$ for $j = 1, \ldots, n$ and $r_1 \ge r_2 \ge \cdots \ge r_n$, so that all roots lie in the disc $D(0, r_1)$. Then approximations $\tilde{r}_j$ such that $\tilde{r}_j \le r_j \le (1 + c/n^d)\tilde{r}_j$ for $j = 1, \ldots, n$ can be computed by using $O(n \log^2(n))$ ops.*

Theorem 2. *(Root Inversion, Shift and Scaling, cf. [17].) Given a polynomial $p(x)$ of (1) and two scalars a and b, we can compute the coefficients of the polynomial $q(x) = p(ax + b)$ by using $O(n \log(n))$ ops. We need only $2n - 1$ ops if $b = 0$. Reversing a polynomial inverts all its roots, involving no ops, because $p_{\text{rev}}(x) = x^n p(1/x) = \sum_{i=0}^n p_i x^{n-i} = p_n \prod_{j=1}^n (1 - xx_j)$.*

By combining Theorems 1 and 2 we can move the roots of a polynomial into a fixed disc, e.g., $D(0, 1) = \{x : |x| \le 1\}$.

Theorem 3. *(Root Squaring, cf. [10].) (i) Assume a monic polynomial $p(x)$ of (1), $p_n = 1$. Then the map $q(x) = (-1)^n p(\sqrt{x})p(-\sqrt{x})$ squares the roots, that is, $q(x) = \prod_{j=1}^n (x - x_j^2)$, and (ii) one can evaluate $p(x)$ at the k-th roots of unity for $k > 2n$ and then interpolate to $q(x)$ by using $O(n \log(n))$ ops.*

Theorem 4. *(The Cayley Maps, cf. [9].) The maps $y = (x - \sqrt{-1})/(x + \sqrt{-1})$ and $x = \sqrt{-1}(y + 1)/(y - 1)$ send the real axis $\{x : x$ is real$\}$ into the unit circle $C(0, 1) = \{x : |x| = 1\}$, and vice versa.*

Theorem 5. (Möbius Map.) (i) The maps $y = \frac{1}{2}(x+1/x)$ and $x = y \pm \sqrt{y^2 - 1}$ send the unit circle $C(0, 1)$ into the real line interval $[-1, 1] = \{y : \Im y = 0, -1 \le y \le 1\}$, and vice versa. (ii) Write $y = \frac{1}{2}(x + 1/x)$ and $y_j = \frac{1}{2}(x_j + 1/x_j)$, $j = 1, \ldots, n$. Then $q(y) = p(x)p(1/x) = q_n \prod_{j=1}^n (y - y_j)$ (cf. [3, eq. (14)]). (iii) Given a polynomial $p(x)$ of (1) one can interpolate to the polynomial $q(y) = p(x)p(1/x) = q_n \prod_{j=1}^n (y - y_j)$ by using $O(n \log(n))$ ops.

Proof. Follow [3, Section 2]. Apply the algorithms of [16] to interpolate to the polynomial $q(y)$ from its values at the Chebyshev knots at the cost $O(n \log(n))$.

Theorem 6. (Error Bounds of the Möbius Iteration.) Fix a complex $x = x^{(0)}$ and define the iterations

$$x^{(h+1)} = \frac{1}{2}(x^{(h)} + (x^{(h)})^{-1}) \text{ and } \gamma = \sqrt{-1} \text{ for } h = 0, 1, \ldots, \tag{2}$$

$$x^{(h+1)} = \frac{1}{2}(x^{(h)} - (x^{(h)})^{-1}) \text{ and } \gamma = 1 \text{ for } h = 0, 1, \ldots \tag{3}$$

If $x^{(0)}\gamma$ is real, then $x^{(h)}\gamma$ are real for all h. Otherwise $|x^{(h)} - \text{sign}(x)\sqrt{-1}/\gamma| \le \frac{2\tau^{2^h}}{1-\tau^{2^h}}$ for $\tau = |\frac{x-\text{sign}(x)}{x+\text{sign}(x)}|$ and $h = 0, 1, \ldots$

Proof. Under (2), for $\gamma = \sqrt{-1}$, the bound is from [3, page 500]). It is readily extended to the case of (3), for $\gamma = 1$.

Theorem 7. (Root-finding Where All Roots Are Real). The modified Laguerre algorithm of [8] converges to all roots of a polynomial $p(x)$ of (1) right from the start with superlinear convergence rate and uses $O(n)$ ops per iteration. Consequently the algorithm approximates all n roots within $\epsilon = 1/2^b$ by using $O(\log(b))$ iteration loops, performing $\tilde{O}(n \log(b))$ ops overall. This cost bound is optimal and is also supported by the alternative algorithms of [6] and [4].

Algorithm 1. (Real Root-finding via Root Radii Approximation.)

1. Compute approximations $\tilde{r}_1, \ldots, \tilde{r}_n$ to the root radii of a polynomial $p(x)$ of (1) (see Theorem 1). (This defines $2n$ candidates points $\pm \tilde{r}_1, \ldots, \pm \tilde{r}_n$ for the approximation of the r real roots $x_1, \ldots, x_r$.)

2. Evaluate the polynomial at these $2n$ points, at a low arithmetic and Boolean cost, to exclude a number of extraneous candidates.

3. Apply Newton's iteration $x^{(h+1)} = x^{(h)} - p(x^{(h)})/p'(x^{(h)})$, $h = 0, 1, \ldots$ concurrently at the remaining candidate points. (Its single concurrent step or a few steps, performed at a low arithmetic and Boolean cost (cf. [22]), should exclude the other extraneous candidates and refine the remaining approximations to the real roots, as long as these roots are well isolated from the nonreal roots.)

Theorem 8. (Splitting a Polynomial into Two Factors Over a Circle, cf. [24] or [13, Chapter 15].) Suppose a polynomial $t(x)$ of degree n has r roots in the disc $D(0, \rho)$ and $n - r$ roots outside the disc $D(0, R)$ for $R/\rho \ge 1 + 1/n$. Let $\epsilon = 1/2^b$ for $b \ge n$. Then we can compute two polynomials $\tilde{f}$ and $\tilde{g}$ such that

$||p - \tilde{f}\tilde{g}||_q \leq \epsilon||p||_q$ for $q = 1, 2$ or ∞, the polynomial $\tilde{f}$ of degree r has r roots inside the circle $C(0,1)$, and the polynomial $\tilde{g}$ of degree $n - r$ has $n - r$ roots outside the circle. The algorithm performs $O((\log^2(n) + \log(b))n\log(n))$ ops (that is, $O(n\log^3(n))$ ops for $\log(b) = O(\log^2(n))$), with a precision of $O(b)$ bits.

Remark 1. (Increasing Isolation by Means of Repeated Squaring.) Let the assumptions of Theorem 8 hold, except that $R/\rho = 1 + c/n^d < 1 + 1/n$, for two positive constants c and d. Then the map of Theorem 3 squares the ratio R/ρ. So $d = O(\log(n))$ applications of this map (using $O(n\log^2(n))$ ops overall) increase the ratio above $1 + 1/n$, which supports the application of Theorem 8.

3   Root-Finding as Eigen-Solving and Basic Results for Matrix Computations

3.1   Companion Matrix, Its Maps, and Maps of Its Eigenvalues

$$C_p = \begin{pmatrix} 0 & & & & -p_0/p_n \\ 1 & \ddots & & & -p_1/p_n \\ & \ddots & \ddots & & \vdots \\ & & \ddots & 0 & -p_{n-2}/p_n \\ & & & 1 & -p_{n-1}/p_n \end{pmatrix}$$

denotes the *companion matrix* of a polynomial $p(x)$ of (1). $p(x) = c_{C_p}(x) = \det(xI_n - C_p)$ is its *characteristic polynomial*. Its roots form the *spectrum* of C_p, and so our problem can be restated as the problem of real eigen-solving for the companion matrix C_p. Next we recall that operations with this matrix are as inexpensive as with polynomials and restate the maps for the variable x of the polynomials in terms of maps of the matrix C_p, playing the role of this variable.

Theorem 9. *(The Cost of Computations in the Frobenius Matrix Algebra, cf. [7].) The companion matrix $C_p \in \mathbb{C}^{n \times n}$ of a polynomial $p(x)$ of (1) generates the Frobenius matrix algebra $\mathcal{A}_p$. One needs $O(n)$ ops for addition, $O(n\log(n))$ ops for multiplication, and $O(n\log^2(n))$ ops for inversion in this algebra. One needs $O(n\log(n))$ ops to multiply a matrix in this algebra by a vector.*

3.2   Some Fundamental Matrix Computations

To study the eigen-solving for C_p, next we recall some fundamentals of matrix computations. In the next subsection we focus on the basic properties of eigenvalues and eigenspaces of matrices, that we use in our algorithms.

$M^T = (m_{ji})_{i,j=1}^{n,m}$ is the transpose of a matrix $M = (m_{ij})_{i,j=1}^{m,n}$. M^H is its Hermitian transpose. $I = I_n = (\mathbf{e}_1 \mid \mathbf{e}_2 \mid \ldots \mid \mathbf{e}_n)$ is the $n \times n$ identity matrix whose columns are the n coordinate vectors $\mathbf{e}_1, \mathbf{e}_2, \ldots, \mathbf{e}_n$. $\mathrm{diag}(b_j)_{j=1}^s = \mathrm{diag}(b_1, \ldots, b_s)$ is the $s \times s$ diagonal matrix with the diagonal entries $b_1, \ldots, b_s$.

A matrix Q is *unitary* if $Q^H Q = I$ or $QQ^H = I$. Let $(Q, R) = (Q(M), R(M))$ for an $m \times n$ matrix M of rank n denote a unique pair of unitary $m \times n$ matrix Q and upper triangular $n \times n$ matrix R such that $M = QR$ and all diagonal entries of the matrix R are positive [9, Theorem 5.2.2].

M^+ is the Moore–Penrose pseudo inverse of M [9, Section 5.5.4]. An $n \times m$ matrix $X = M^{(I)}$ is a left (resp. right) inverse of an $m \times n$ matrix M if $XM = I_n$ (resp. if $MY = I_m$). $M^{(I)} = M^+$ for a matrix M of full rank. $M^{(I)} = M^H$ for an orthogonal matrix M. $M^{(I)} = M^{-1}$ for a nonsingular matrix M.

$\mathcal{R}(M)$ is the range of a matrix M, that is the linear space generated by its columns. A matrix of full column rank is a *matrix basis* of its range.

3.3   Eigenspaces and Eigenvalues

Definition 1. $\mathcal{S}$ is the invariant subspace *of a square matrix M if $M\mathcal{S} = \{M\mathbf{v} : \mathbf{v} \in \mathcal{S}\} \subseteq \mathcal{S}$. A scalar λ is an* eigenvalue *of a matrix M associated with an* eigenvector $\mathbf{v}$ *if $M\mathbf{v} = \lambda\mathbf{v}$. All eigenvectors associated with an eigenvalue λ of M form an eigenspace $\mathcal{S}(M, \lambda)$, which is an invariant space. Its dimension d is the* geometric multiplicity *of λ. The eigenvalue is simple if $d = 1$. The set $\Lambda(M)$ of all eigenvalues of the matrix M is called its* spectrum.

Our next goal is to limit eigen-solving for the matrix C_p to the study of its invariant space of dimension r associated with the r real eigenvalues. The following theorem is basic for this step.

Theorem 10. *(Decreasing the Eigenproblem Size to the Dimension of an Invariant Space, cf. [26, Section 2.1].) Let $U \in \mathbb{C}^{n \times r}$, $\mathcal{R}(U) = \mathcal{U}$, and $M \in \mathbb{C}^{n \times n}$. Then $\mathcal{U}$ is an invariant space of M if and only if there exists a matrix $L \in \mathbb{C}^{k \times k}$ such that $MU = UL$ or equivalently $L = U^{(I)} MU$. The matrix L is unique (that is independent of the choice of the left inverse $U^{(I)}$) if U is a matrix basis for the space $\mathcal{U}$. Hence $MU\mathbf{v} = \lambda U\mathbf{v}$ if $L\mathbf{v} = \lambda\mathbf{v}$, $\Lambda(L) \subseteq \Lambda(M)$, and if U is an orthogonal matrix, then $L = U^H MU$.*

To facilitate the computation of the desired invariant space of C_p, we reduce the task to the case of an appropriate matrix function, for which the solution is simpler, but we still solve our problem, because, by virtue of the following theorem, a matrix function shares its invariant spaces with the matrix C_p.

Theorem 11. *(Reduction of the Eigenproblem for a Matrix to That for a Matrix Function.) Suppose M is a square matrix, a rational function $f(\lambda)$ is defined on its spectrum, and $M\mathbf{v} = \lambda\mathbf{v}$. Then (i) $f(M)\mathbf{v} = f(\lambda)\mathbf{v}$. (ii) Let $\mathcal{U}$ be the eigenspace of the matrix $f(M)$ associated with its eigenvalue μ. Then this is an invariant space of the matrix M generated by its eigenspaces associated with all its eigenvalues λ such that $f(\lambda) = \mu$. (iii) The space $\mathcal{U}$ is associated with a single eigenvalue of M if μ is a simple eigenvalue of $f(M)$.*

We readily verify part (i), which implies parts (ii) and (iii).

Suppose we have computed a matrix basis $U \in \mathbb{C}^{n \times r}$ for an invariant space $\mathcal{U}$ of a matrix function $f(M)$ of an $n \times n$ matrix M. By virtue of Theorem 11 this is a matrix basis of an invariant space of the matrix M. We can first compute a left inverse $U^{(I)}$ or the orthogonalization $Q = Q(U)$ and then approximate the eigenvalues of M associated with this eigenspace as the eigenvalues of the $r \times r$ matrix $L = U^{(I)}MU = Q^H MQ$ (cf. Theorem 10). Empirically the QR algorithm uses $O(r^3)$ ops at the latter stage.

Given an approximation $\tilde{\mu}$ to a simple eigenvalue of a matrix function $f(M)$, we can compute an approximation $\tilde{u}$ to an eigenvector u of the matrix $f(M)$ associated with this eigenvalue, recall from Theorem 11 that this is also an eigenvector of the matrix M, associated with its simple eigenvalue, and approximate this eigenvalue by the Rayleigh Quotient $\frac{\tilde{u}^T M \tilde{u}}{\tilde{u}^T \tilde{u}}$.

3.4   Some Maps in the Frobenius Matrix Algebra

For a polynomial $p(x)$ of (1) and a rational function $f(x)$ defined on the set $\{x_i\}_{i=1}^n$ of its roots, the rational matrix function $f(C_p)$ has spectrum $\Lambda(f(C_p)) = \{f(x_i)\}_{i=1}^n$, by virtue of Theorem 11. In particular, the maps

$$C_p \to C_p^{-1}, \; C_p \to aC_p + bI, \; C_p \to C_p^2, \; C_p \to \frac{C_p + C_p^{-1}}{2}, \; \text{and } C_p \to \frac{C_p - C_p^{-1}}{2}$$

induce the maps of the eigenvalues of the matrix C_p, and thus induce the maps of the roots of the characteristic polynomial $p(x)$ given by the equations

$$y = 1/x, \; y = ax + b, \; y = x^2, \; y = 0.5(x + 1/x), \text{ and } y = 0.5(x - 1/x),$$

respectively. By using the reduction modulo $p(x)$, define the five dual maps

$$y = (1/x) \mod p(x), \; y = ax + b \mod p(x), \; y = x^2 \mod p(x),$$
$$y = 0.5(x + 1/x) \mod p(x), \text{ and } y = 0.5(x - 1/x) \mod p(x),$$

where $y = y(x)$ denote polynomials. Apply the two latter maps recursively, to define two iterations with polynomials modulo $p(x)$ as follows, $y_0 = x$, $y_{h+1} = 0.5(y_h + 1/y_h) \mod p(x)$ (cf. (3)) and $y_0 = x$, $y_{h+1} = 0.5(y_h - 1/y_h) \mod p(x)$, $h = 0, 1, \ldots$. More generally, define the iteration $y_0 = x$, $y_{h+1} = ay_h + b/y_h \mod p(x)$, $h = 0, 1, \ldots$, for any pair of scalars a and b. Here $y_h = y_h(x)$ are the characteristic polynomials of the matrices $M_0 = C_p$, $M_{h+1} = 0.5(M_h \pm M_h^{-1})$ and $M_0 = C_p$, $M_{h+1} = aM_h + bM_h^{-1}$, $h = 0, 1, \ldots$, respectively.

4   Real Root-Finders

4.1   Möbius Iteration

Theorem 6 implies that right from the start of iteration (3) the values $x^{(h)}$ converge to $\pm\sqrt{-1}$ exponentially fast unless the initial value $x^{(0)}$ is real, in which case all iterates $x^{(h)}$ are real. It follows that right from the start the values $y^{(h)} = (x^{(h)})^2 + 1$ converge to 0 exponentially fast unless $x^{(0)}$ is real, in which

case all values $y^{(h)}$ are real and exceed 1. Write $q_h(y) = \prod_{j=1}^{n}(y - (x_j^{(h)})^2 - 1)$
for $h = 1, 2, \ldots$ and $u_h(y) = \prod_{j=1}^{r}(y - (x_j^{(h)})^2 - 1)$. The roots of the polynomials $q_h(y)$ and $u_h(y)$ are the images of all roots and of the real roots of the polynomial $p(x)$ of (1), respectively, produced by the composition of the maps (3) and $y^{(h)} = (x^{(h)})^2 + 1$. Therefore $q_h(y) \approx y^{2s}u_h(y)$ for large integers h where the polynomial $u_h(y)$ has degree r and has exactly r real roots, all of them exceeding 1. Hence for sufficiently large integers h, we can closely approximate the polynomial $y^{2s}u_h(y)$ simply by the sum of the $r + 1$ leading terms of the polynomial $q_h(y)$. To verify that the $2s$ trailing coefficients nearly vanish, we need just $2s$ comparisons. The above argument shows correctness of the following algorithm.

Algorithm 2. Möbius iteration for real root-finding.
INPUT: *two integers n and r, $0 < r < n$, and the coefficients of a polynomial $p(x)$ of equation (1) where $p(0) \neq 0$.*
OUTPUT: *approximations to the real roots $x_1, \ldots, x_r$ of $p(x)$.*
INITIALIZATION: *Write $p_0(x) = p(-x\sqrt{-1})$.*
COMPUTATIONS:

1. *Recursively compute the polynomials $p_{h+1}(y) = p_h(x)p_h(1/x)$ for $y = (x + 1/x)/2$ and $h = 0, 1, \ldots$ (Part (ii) of Theorem 5 and Theorem 6 define the images of the real and nonreal roots of the polynomial $p(x)$ for all h.)*
2. *Periodically, at some selected Stages k, compute the polynomials*

$$t_h(y) = (-1)^n q_k(\sqrt{y+1}) q_h(-\sqrt{y+1})$$

   where $q_k(z) = p_k(z)/\mathrm{lc}(p_k)$ (cf. Theorems 2 and 3). When the integer k becomes large enough, so that $2s$ trailing coefficients of the polynomial $q_k(x)$ nearly vanish, approximate the factor $v_k(x)$ of the polynomial $t_k(x)$ that has r real roots on the ray $\{x : x \leq -1\}$ (see above).
3. *Apply one of the algorithms of [6], [4], and [8] (cf. Theorem 7) to approximate the r roots $z_1, \ldots, z_r$ of the polynomial $v_k(x)$.*
4. *Extend the descending process from [15], [18] and [3] to recover approximations to the r roots $-x_i\sqrt{-1}$, $i = 1, \ldots, r$, of the polynomial $p_0(x) = p(-x\sqrt{-1})$. First approximate $2r$ candidates for r roots of the polynomial $q_k(y)$ lying on the imaginary axis and select r of them on which the polynomial $q_k(y)$ nearly vanishes. Similarly define from these r roots $2r$ candidates for approximating the r roots of $p_{k-1}(x)$ lying on the imaginary axis. Recursively descend down to the r roots of $p_0(x)$ lying on the imaginary axis. This process is not ambiguous because only r roots of the polynomial $p_h(x)$ lie on that axis for each h, by virtue of Theorem 6.*
5. *Having approximated the r roots $-x_i\sqrt{-1}$, $i = 1, \ldots, r$, output the approximations to the real roots $x_1, \ldots, x_r$ of the polynomial $p(x)$.*

Like lifting Stage 1, descending Stage 4 involves order of $kn\log(n)$ ops, which also bounds the overall cost of performing the algorithm.

Remark 2. (Countering Degeneracy.) If $p(0) = p_0 = \cdots = p_m = 0 \neq p_{m+1}$, then we should output the real root $x_0 = 0$ of multiplicity m and apply the algorithm to the polynomial $p(x)/x^m$ to approximate the other real roots. Alternatively we can apply the algorithm to the polynomial $q(x) = p(x - s)$ for a shift value s such that $q(0) \neq 0$. With probability 1, this holds for Gaussian random variable s, but alternatively we can approximate the root radii of the polynomial $p(x)$ (cf. Theorem 1) to find a shift scalar s such that $q(x)$ has no roots near 0 as well.

Remark 3. (Saving the Recursive Steps of Stage 1.) We would decrease the parameter k of the cost estimate, if we approximate the factor $v_k(x)$ of the polynomial $t_k(x)$ for a smaller integer k. Theorem 8 enables us to do this (at a reasonable cost) if its assumptions are satisfied for $t(x) = t_k(x)$. We can verify if the assumptions hold by applying the root radii algorithm of Theorem 1. For a fixed k this requires $O(n \log^2(n))$ ops), so even the verification for all integers k in the range is not costly, unless the integer k is large, but we can periodically test just selected integers k, by applying binary search.

Remark 4. (Handling the Nearly Real Roots.) The integer parameter k and the overall cost of performing the algorithm are large if $2^{-d} = \min_{j=r+1}^{n} |\Im(x_j)|$ is small. To counter this deficiency, we can split out a factor $v_{k,+}(x)$ of the polynomial $p(x)$ having a degree $r_+ > r$ and having r_+ real and nearly real roots such that the other nonreal roots lie sufficiently far from the real axis. Indeed our convergence analysis and the techniques for splitting out the factor $v_k(x)$ can be readily extended to splitting out the factor $v_{k,+}(x)$. Having this factor approximated, we can tentatively apply to it the modified Laguerre algorithm of [8], expecting fast convergence to the r_+ roots of the polynomial $v_{k,+}(x)$ if all its roots lie on or sufficiently close to the real axis.

Remark 5. (The Number of Real Roots.) We assume that we are given the number r of the real roots (e.g., computed by means of non-costly techniques of computer algebra if the roots are distinct and simple), but we can compute this number as by-product of Stage 2, and similarly for our other algorithms. Moreover with a proper try-and-test policy we can apply our algorithm for at most $2 + 2\lceil \log(r) \rceil$ tentative choices of integers k in the range $[0, 2k - 1]$ to detect r.

Remark 6. The known upper bounds on the condition numbers of the roots of the computed polynomials $p_k(y)$ grow exponentially as k grows large (cf. [3, Section 3]). So, unless these bounds are overly pessimistic, Algorithm 2 is prone to numerical stability problems already for moderately large integers k.

4.2   Adjusted Matrix Sign Iteration

To avoid the latter potential deficiency, we replace the polynomial iteration at Stages 1 and 2 by the dual *matrix sign* classical iteration

$$Z_h = 0.5(Z_h + Z_h^{-1}) \text{ for } h = 0, 1, \ldots \qquad (4)$$

It maps the eigenvalues of the matrix Z_0 according to (2). Therefore, by virtue of part (ii) of Theorem 5, Stage 1 of Algorithm 2 maps the characteristic polynomials of the above matrices Z_h. Unlike the case of the latter map, working with matrices enables two minor implications: (i) we recover the desired real eigenvalues of the matrix C_p by means of our recipes of Section 3, without recursive descending, and (ii) we avoid scaling by $\sqrt{-1}$ and just slightly modify the iteration to keep the computations in the field of real numbers.

Algorithm 3. Matrix sign iterations modified for real eigen-solving.
INPUT AND OUTPUT *as in Algorithm 2, except that FAILURE can be output with a probability close to 0.*
COMPUTATIONS:

1. *Write $Y_0 = C_p$ and recursively compute the matrices*

$$Y_{h+1} = 0.5(Y_h - Y_h^{-1}) \text{ for } h = 0, 1, \dots. \tag{5}$$

   (For sufficiently large integers h, the $2s$ eigenvalues of the matrix Y_h lie near the points $\pm\sqrt{-1}$, whereas the r other eigenvalues are real, by virtue of Theorem 6.)
2. *Fix a sufficiently large integer k and compute the matrix $Y = Y_k^2 + I_n$. The map $Y_0 = C_p \to Y$ sends all nonreal eigenvalues of C_p to a small neighborhood of the origin 0 and sends all real eigenvalues of C_p into the ray $\{x : x \geq 1\}$.*
3. *Apply the randomized algorithms of [12] to compute the numerical rank of the matrix Y. Suppose it equals r. (Otherwise go back to Stage 1.) Generate a standard Gaussian random $n \times r$ matrix G and compute the matrices $H = YQ(G)$ and $Q = Q(H)$. (The analysis of preprocessing with Gaussian random multipliers in [12, Section 4], [19, Section 5.3] shows that, with a probability close to 1, the columns of the matrix Q closely approximate an orthogonal basis of the invariant space of the matrix Y associated with its r absolutely largest eigenvalues, which are the images of the real eigenvalues of the matrix C_p. Having this approximation is equivalent to having a small upper bound on the residual norm $\|Y - QQ^H Y\|$ [12], [19].) Verify the latter bound. In the unlikely case where the verification is failed, output FAILURE and stop the computations.*
4. *Otherwise compute and output approximations to the r eigenvalues of the $r \times r$ matrix $L = Q^H C_p Q$. They approximate the real roots of the polynomial $p(x)$. (Indeed, by virtue of Theorem 11, Q is a matrix basis for the invariant space of the matrix C_p associated with its r real eigenvalues. Therefore, by virtue of Theorem 10, the matrices C_p and L share these eigenvalues.)*

Stages 1 and 2 involve $O(kn \log^2(n))$ ops by virtue of Theorem 9. This exceeds the estimate for Algorithm 2 by a factor of $\log(n)$. Stage 3 adds $O(nr^2)$ ops and the cost a_{rn} of generating $n \times r$ standard Gaussian random matrix. The cost bounds are $O(nr^2)$ at Stage 4 and $O((kn \log^2(n) + nr^2) + a_{rn}$ overall.

Remark 7. (Counting Real Eigenvalues.) If the number of real eigenvalues is not given, we can apply binary search to compute it as the numerical rank of the matrices $Y_k^2 + I$ when this rank stabilizes.

Remark 8. (Avoiding Numerical Problems.) The images of nonreal eigenvalues of the matrix C_p converge to $\pm\sqrt{-1}$ in the recursive process of the algorithm. So the process involves ill conditioned matrices if and only if the images of some real eigenvalues of C_p lie close to 0. We can detect that this has occurred if it is hard to invert the matrix Y_h of (5) or by computing the smallest singular value of that matrix (e.g., by applying the Lanczos efficient, cf. [9, Proposition 9.1.4]). As soon as we detect an ill conditioned matrix Y_h, we would shift it (and hence shift its eigenvalues) by adding the matrix sI for a reasonably small real scalar s, which we can select by applying Theorem 1, heuristic, or randomization.

Remark 9. (Acceleration by Using Random Circulant Multiplier.) We can decrease the cost of performing Stage 3 to $a_{n+r} + O(n\log(n))$ by replacing an $n \times r$ standard Gaussian random multiplier by the product $\Omega C P$ where Ω and C are $n \times n$ matrices, Ω is the matrix of the discrete Fourier transform, C is a random circulant matrix, and P is an $n \times l$ random permutation matrix, for a sufficiently large l of order $r\log(r)$. (See [12, Section 11], [19, Section 6] for the analysis and for supporting probability and cost estimates. They are only slightly less favorable than in the case of a Gaussian random multiplier.) The overall arithmetic cost bound would change into $O(kn\log^2(n) + nr^2) + a_{r+n}$.

Remark 10. (Acceleration by Means of Scaling.) We can dramatically accelerate the initial convergence of Algorithm 3 by applying *determinantal scaling* (cf. [11]), that is, by computing the matrix Y_1 as follows, $Y_1 = 0.5(\nu Y_0 - (\nu Y_0)^{-1})$ for $\nu = 1/|\det(Y_0)|^{1/n} = |p_n^{(k)}/p_0^{(k)}|$, $Y_0 = C_{p^{(k)}}$, and $p^{(k)}(x) = \sum_{i=0}^n p_i^{(k)} x^i$.

Remark 11. (Hybrid Matrix and Polynomial Algorithms.) Can we modify Algorithm 3 to keep its advantages but to decrease the computational cost of its Stage 1 to the level $kn\log(n)$ of Algorithm 2? Yes, if all or almost all nonreal roots of the polynomial $p(x)$ lie not too far from the points $\pm\sqrt{-1}$, namely in the discs $D(\pm\sqrt{-1}, 1/2)$. Indeed in this case both iterations $Y_{h+1} = 0.5(Y_h^3 + 3Y_h)$ and $Y_{h+1} = -0.125(3Y_h^5 + 10Y_h^3 + 15Y_h)$ for $h = 0, 1, \ldots$ use $O(n\log(n))$ ops per loop. Right from the start they send the nonreal roots lying in these discs to the two points $\pm\sqrt{-1}$ with quadratic and cubic convergence rates, respectively (extend the proof of [3, Proposition 4.1]), while keeping the real roots real. This suggests the following policy. Heuristically or by applying Theorem 1 choose a proper integer h and run Algorithm 2 until all or almost all nonreal roots of $p(x)$ are moved into the discs $D(\pm\sqrt{-1}, 1/2)$. Then apply one of the two latter inversion-free variants of Algorithm 3 to the polynomial $q_h(x)$ produced by Algorithm 2. Descend from the output roots to the real roots of the polynomial $p(x)$. The hybrid algorithm combines the benefits of both Algorithms 2 and 3 when the above integer h is not large.

4.3   Adjusted Modular Square Root Iteration

The polynomial version of Algorithm 3 is known as the square root iteration. It mimics Algorithm 3, but replaces all rational functions in the matrix C_p by the same rational functions in the variable x, and then reduces all these functions modulo the input polynomial $p(x)$. The reduction does not affect the values of the functions at the roots of $p(x)$, and so these values are precisely the eigenvalues of the rational matrix functions involved in Algorithm 3.

Algorithm 4. Square root modular iteration modified for real root-finding.
INPUT AND OUTPUT *as in Algorithm 2.*
COMPUTATIONS:

1. *Write $y_0 = x$ and $Y_0 = C_p$ and (cf. (5)) compute the polynomials*

$$y_{h+1} = (y_h - y_h^{-1})/2 \quad \mod p(x). \tag{6}$$

2. *Periodically, for selected integers k, compute the polynomials $t_k = y_k^2 + 1$ mod $p(x)$ and $g_k(x) = \mathrm{agcd}(p, t_k)$.*
3. *If $\deg(g_k)) = n - r = 2s$, compute the polynomial $v_k \approx p(x)/g_k(x)$ of degree r. Otherwise continue the iteration of Stage 1.*
4. *Apply one of the algorithms of [6], [4], and [8] (cf. Theorem 7) to approximate the r roots $y_1, \ldots, y_r$ of the polynomial v_k. Output these approximations.*

By virtue of our comments preceding this algorithm, the values of the polynomials t_k at the roots of $p(x)$ equal to the images of the eigenvalues of the matrix C_p in Algorithm 3. Hence the values of the polynomials t_k at the nonreal roots converge to 0 as $k \to \infty$, whereas their values at the real roots stay far from 0. Therefore, for sufficiently large integers k, $\mathrm{agcd}(p, t_k)$ turn into the polynomial $\prod_{j=r+1}^{n}(x - x_j)$. This implies correctness of the algorithm. Its asymptotic computational cost is $O(kn \log^2(n))$ plus the cost of computing $\mathrm{agcd}(p, t_k)$ and choosing the integer k (see our next remark).

Remark 12. Compared to Algorithm 3, the latter algorithm reduces real root-finding essentially to the computation of $\mathrm{agcd}(p, t_k)$, but the complexity of this computation is not easy to estimate [1]. Moreover, the following example exhibits serious problems of numerical stability for this algorithm and apparently for the similar algorithms of [7] and [3]. Consider the case where $r = 0$. Then the polynomial $t(x)$ has degree at most $n - 1$, and its values at the n nonreal roots of the polynomial $p(x)$ are close to 0. This can only occur if $\|t_k\| \approx 0$.

Remark 13. We can concurrently perform Stages 1 of both Algorithms 3 and 4. The information about numerical rank at Stage 3 of Algorithm 3 can be a guiding rule for the choice of the integer parameter k and computing the polynomials t_k, g_k and v_k of Algorithm 4. Having the polynomial v_k available, Algorithm 4 produces the approximations to the real roots more readily than Algorithm 3 does this at its Stage 4.

5   Cayley Map and Root-Squaring

The following algorithm is somewhat similar to Algorithm 2, but employs repeated squaring of the roots instead of mapping them into their square roots.

Algorithm 5. Real root-finding by means of repeated squaring.

Assume a polynomial $p(x)$ of (1) with $p(0) \neq \pm\sqrt{-1}$ and proceed as follows.

1. Compute the polynomial $q(x) = p((x + \sqrt{-1})(x - \sqrt{-1})^{-1}) = \sum_{i=0}^{n} q_i x^i$. (This is the Cayley map, cf. Theorem 4. It moves the real axis, in particular, the real roots of $p(x)$, onto the unit circle $C(0,1)$.)

2. Write $q_0(x) = q(x)/q_n$, choose a sufficiently large integer k, and apply the k squaring steps of Theorem 3, $q_{h+1}(x) = (-1)^n q_h(\sqrt{x})q_h(-\sqrt{x})$ for $h = 1, \ldots, k-1$. (These steps keep the images of the real roots of $p(x)$ on the circle $C(0,1)$ for any k, while sending the images of every other root of $p(x)$ toward either the origin or the infinity.)

3. For a sufficiently large integer k, the polynomial $q_k(x)$ approximates the polynomial $x^s u_k(x)$ where the polynomial $u_k(x) = \sum_{i=0}^{r} u_i x^i$ has all its roots lying on the unit circle $C(0,1)$. Extract the approximation to this polynomial $u_k(x)$ from the coefficients of the polynomial $q_k(x)$.

4. Compute the polynomial $v_k(x) = \sqrt{-1}(u_k(x) + 1)(u_k(x) - 1)^{-1}$. (This is the inverse Cayley map. It sends the images of the real roots of the polynomial $p(x)$ from the unit circle $C(0,1)$ back to the real line.)

6. Apply one of the algorithms of [6], [4], and [8] to approximate the r real roots $z_1, \ldots, z_r$ of the polynomial $v_k(x)$ (cf. Theorem 7).

7. Apply the Cayley map $w_j = (z_j + \sqrt{-1})(z_j - \sqrt{-1})^{-1}$ for $j = 1, \ldots, r$ to extend Stage 6 to approximating the r roots $w_1, \ldots, w_r$ of the polynomials $u_k(x)$ and $y_k(x) = x^s u_k(x)$ lying on the unit circle $C(0,1)$.

8. Apply the descending process (similar to the ones of [15], [18], and of our Algorithm 2) to approximate the r roots $x_1^{(h)}, \ldots, x_r^{(h)}$ of the polynomials $q_h(x)$ lying on the unit circle $C(0,1)$ for $h = k-1, \ldots, 0$.

9. Apply the inverse Cayley map to approximate the r real roots $x_j = (x_j^{(0)} + \sqrt{-1})(x_j^{(0)} - \sqrt{-1})^{-1}$ of the polynomials $p(x)$.

Our analysis of Algorithm 2 (including its complexity estimates and the comments and recipes in Remarks 2–6) can be extended to Algorithm 5. The straightforward matrix version of this numerical algorithm, however, fails because high matrix powers have small numerical rank. Indeed their columns lie near the invariant space associated with the absolutely largest eigenvalues, and as a rule, this space has a small dimension. A more tricky modification, based on binomial factorization, promises to produce a working matrix iteration. We postpone its presentation.

6 Numerical Tests

Two series of numerical tests have been performed in the Graduate Center of the City University of New York by Ivan Retamoso and Liang Zhao. In both series, they tested Algorithm 3, without using the techniques of Remark 3, that is, in much weakened form. Still the test results are quite encouraging.

In the first series of tests, Algorithm 3 has been applied to one of the Mignotte benchmark polynomials, namely to $p(x) = x^n + (100x - 1)^3$. It is known that this polynomial has three ill conditioned roots clustered about 0.01 and has $n-3$ well conditioned roots. In the tests, Algorithm 3 has output the roots within the error less than 10^{-6} by using 9 iterations for $n = 32$ and $n = 64$ and by using 11 iterations for $n = 128$ and $n = 256$.

In the second series of tests they randomly generated polynomials $p(x)$ of degree $n = 50, 100, 150, 200, 250$ as the product $p(x) = f_1(x)f_2(x)$. They generated the polynomials $f_1(x)$ and $f_2(x)$ where $f_1(x) = \prod_{j=1}^{r}(x - x_j)$, $f_2(x) = \sum_{i=0}^{n-r} a_i x^i$, and x_i and a_j were i.i.d. standard Gaussian random variables, for $j = 1, \ldots, r$, $i = 0, \ldots, n - r$, and $r = 4, 8, 12, 16$. Hence the polynomial $p(x) = f_1(x)f_2(x)$ had at least r real roots. Then Algorithm 3 (performed with double precision) was applied to 100 randomly generated polynomials $p(x)$ for each pair of n and r, and the output data were recorded, namely, the numbers of iterations and the maximum difference of the output values of the roots from their values produced by MATLAB root-finding function "roots()". The test results were similar to the case of the Mignotte polynomials (see the Journal version of the paper).

Acknowledgement. I am grateful to NSF, for the support under Grant CCF 1116736, and to the reviewers, for their thoughtful and valuable comments.

References

1. Bini, D.A., Boito, P.: A fast algorithm for approximate polynomial GCD based on structured matrix computations. In: Operator Theory: Advances and Applications, vol. 199, pp. 155–173. Birkhäuser Verlag, Basel (2010)
2. Bini, D., Pan, V.Y.: Polynomial and Matrix Computations. Fundamental Algorithms, vol. 1. Birkhäuser, Boston (1994)
3. Bini, D., Pan, V.Y.: Graeffe's, Chebyshev, and Cardinal's processes for splitting a polynomial into factors. J. Complexity 12, 492–511 (1996)
4. Bini, D., Pan, V.Y.: Computing matrix eigenvalues and polynomial zeros where the output is real. SIAM J. on Computing 27(4), 1099–1115 (1998); (Also in Proc. of SODA 1991)
5. Bini, D.A., Robol, L.: Solving secular and polynomial equations: A multiprecision algorithm. J. Computational and Applied Mathematics (in press)
6. Ben-Or, M., Tiwari, P.: Simple algorithms for approximating all roots of a polynomial with real roots. J. Complexity 6(4), 417–442 (1990)
7. Cardinal, J.P.: On two iterative methods for approximating the roots of a polynomial. Lectures in Applied Mathematics 32, 165–188 (1996)
8. Du, Q., Jin, M., Li, T.Y., Zeng, Z.: The quasi-Laguerre iteration. Math. Comput. 66(217), 345–361 (1997)
9. Golub, G.H., Van Loan, C.F.: Matrix Computations, 3rd edn. The Johns Hopkins University Press, Baltimore (1996)
10. Householder, A.S.: Dandelin, Lobachevskii, or Graeffe. Amer. Math. Monthly 66, 464–466 (1959)
11. Higham, N.J.: Functions of Matrices. SIAM, Philadelphia (2008)
12. Halko, N., Martinsson, P.G., Tropp, J.A.: Finding structure with randomness: probabilistic algorithms for constructing approximate matrix decompositions. SIAM Review 53(2), 217–288 (2011)

13. McNamee, J.M., Pan, V.Y.: Numerical Methods for Roots of Polynomials, Part 2, XXII + 718 pages. Elsevier (2013)
14. Pan, V.Y.: Complexity of computations with matrices and polynomials. SIAM Review 34(2), 225–262 (1992)
15. Pan, V.Y.: Optimal (up to polylog factors) sequential and parallel algorithms for approximating complex polynomial zeros. In: Proc. 27th Ann. ACM Symp. on Theory of Computing, pp. 741–750. ACM Press, New York (1995)
16. Pan, V.Y.: New fast algorithms for polynomial interpolation and evaluation on the Chebyshev node set. Computers Math. Appls. 35(3), 125–129 (1998)
17. Pan, V.Y.: Structured Matrices and Polynomials: Unified Superfast Algorithms, Birkhäuser, Boston. Springer, New York (2001)
18. Pan, V.Y.: Univariate polynomials: nearly optimal algorithms for factorization and rootfinding. J. Symb. Computations 33(5), 253–267 (2002); Proc. version in ISSAC 2001, pp. 253–267, ACM Press, New York (2001)
19. Pan, V.Y., Qian, G., Yan, X.: Supporting GENP and Low-rank Approximation with Random Multipliers. Technical Report TR 2014008, PhD Program in Computer Science. Graduate Center, CUNY (2014), http://www.cs.gc.cuny.edu/tr/techreport.php?id=472
20. Pan, V.Y., Qian, G., Zheng, A.: Real and complex polynomial root-finding via eigen-solving and randomization. In: Gerdt, V.P., Koepf, W., Mayr, E.W., Vorozhtsov, E.V. (eds.) CASC 2012. LNCS, vol. 7442, pp. 283–293. Springer, Heidelberg (2012)
21. Pan, V.Y., Tsigaridas, E.P.: On the Boolean Complexity of the Real Root Refinement. Tech. Report, INRIA (2013), http://hal.inria.fr/hal-00960896; Proc. version in: M. Kauers (ed.) Proc. Intern. Symposium on Symbolic and Algebraic Computation (ISSAC 2013), pp. 299–306, Boston, MA, June 2013. ACM Press, New York (2013)
22. Pan, V.Y., Tsigaridas, E.P.: Nearly optimal computations with structured matrices. In: SNC 2014. ACM Press, New York (2014); Also April 18, 2014, arXiv:1404.4768 [math.NA] and, http://hal.inria.fr/hal-00980591
23. Pan, V.Y., Zheng, A.: New progress in real and complex Ppolynomial root-finding. Computers Math. Applics. 61(5), 1305–1334 (2011)
24. Schönhage, A.: The Fundamental Theorem of Algebra in Terms of Computational Complexity. Math. Department, Univ. Tübingen, Germany (1982)
25. Sagraloff, M., Mehlhorn, K.: Computing Real Roots of Real Polynomials, CoRR, abstract 1308.4088 (2013)
26. Watkins, D.S.: The Matrix Eigenvalue Problem: GR and Krylov Subspace Methods. SIAM, Philadelphia (2007)

On Testing Uniqueness of Analytic Solutions of PDE with Boundary Conditions

Sergey V. Paramonov[*]

Moscow State University, Moscow 119991, Russia
`s.v.paramonov@yandex.ru`

Abstract. We consider linear partial differential equations with polynomial coefficients and prove algorithmic undecidability of the following problem: to test whether a given equation of considered form has no more than one solution that is analytic on a domain and that satisfies some fixed boundary conditions. It is assumed that a polynomial which vanishes at each point of the domain boundary is known.

1 Introduction

We will consider linear differential operators of the form

$$L = \sum_{n \in S} a_n(x_1, \ldots, x_m) D_1^{n_1} \ldots D_m^{n_m}, \tag{1}$$

where S is a finite subset of $\mathbb{Z}_{\geq 0}^m$, $x_1, \ldots, x_m$ are independent variables, $a_n \in \mathbb{Z}[x_1, \ldots, x_m]$, $D_i = \frac{\partial}{\partial x_i}$. The set of such operators will be denoted by $\mathbb{Z}[D, x]$. Also we will use the following notation (here S is also a finite subset of $\mathbb{Z}_{\geq 0}^m$):

$$\delta_i = x_i \frac{\partial}{\partial x_i},$$

$$\mathbb{Z}[D] = \{\sum_{n \in S} a_n D_1^{n_1} \ldots D_m^{n_m}, \quad a_n \in \mathbb{Z}\},$$

$$\mathbb{Z}[\delta] = \{\sum_{n \in S} a_n \delta_1^{n_1} \ldots \delta_m^{n_m}, \quad a_n \in \mathbb{Z}\}.$$

Note that $\mathbb{Z}[D] \subset \mathbb{Z}[D, x]$ and $\mathbb{Z}[\delta] \subset \mathbb{Z}[D, x]$.

Consider the problems of testing the existence of solutions in the form of polynomials, rational functions, formal Laurent and power series for an equation $L(f) = 0$, $L \in \mathbb{Z}[D, x]$. It was studied in [1,6,7], where algorithmic undecidability of these problems was proved (Laurent series for several variables are considered in the form defined in [2], [3]). In the earlier paper of Denef and Lipshitz [4], the same problem was considered for inhomogeneous equations of form $L(f) = 1$. Proofs are based on the following facts (their proofs can be found, for example, in [4]):

---

[*] Supported by RFBR grant 13-01-00182-a.

V.P. Gerdt et al. (Eds.): CASC Workshop 2014, LNCS 8660, pp. 350–356, 2014.

1°. The problem of testing existence of a solution in the form of monomial $x_1^{n_1} \ldots x_m^{n_m}$ for a given equation $L(f) = 0$, where $L \in \mathbb{Z}[\delta]$, is equivalent to the problem of testing the existence of an integer solution for an arbitrary Diophantine equation (and hence is undecidable — see [5]).

2°. If the equation $L(f) = 0$, where $L \in \mathbb{Z}[\delta]$, has a non-zero solution that is the formal sum of monomials $f = \sum\limits_{n \in \mathbb{Z}^m} a_n x_1^{n_1} \ldots x_m^{n_m}$ then it also has a monomial solution $x_1^{n_1} \ldots x_m^{n_m}$, $n \in \mathbb{Z}^m$.

In this paper, we consider the problem of testing the existence of non-zero analytic solutions satisfying zero boundary conditions for a given linear differential equation $L(f) = 0$, $L \in \mathbb{Z}[D, x]$, and prove algorithmic undecidability of this problem.

2 Problem ZC

Let K be the field $\mathbb{C}$ of complex numbers or the field $\mathbb{R}$ of real numbers, U be an open domain in K^m, and $\overline{U}$ be the closure of U. If $f(x_1, \ldots, x_m)$ is a function defined on U then set

$$\overline{f}(x_1, \ldots, x_m) = \begin{cases} f(x_1, \ldots, x_m), & \text{if } (x_1, \ldots, x_m) \in U, \\ 0, & \text{if } (x_1, \ldots, x_m) \in \overline{U} \setminus U, \end{cases}$$

The function $\overline{f}(x_1, \ldots, x_m)$ is defined on $\overline{U}$.

For $\alpha \in \mathbb{Z}_{\geqslant 0}^m$, we denote by $f_\alpha(x_1, \ldots, x_m)$ the partial derivative of the function $f(x_1, \ldots, x_m)$:

$$f_\alpha(x_1, \ldots, x_m) = \frac{\partial^{|\alpha|} f(x_1, \ldots, x_m)}{\partial^{\alpha_1} x_1 \ldots \partial^{\alpha_m} x_m},$$

where $|\alpha| = \alpha_1 + \cdots + \alpha_m$.

Assume that $(0, \ldots, 0) \in U$; we say that U is *compatible* with a non-zero polynomial $q(x_1, \ldots, x_m) \in \mathbb{Z}[x_1, \ldots, x_m]$ if $q(x_1, \ldots, x_m) = 0$ at any point of the boundary of U. For example, the open m-dimensional ball of radius 1 with the center at the origin is compatible with the polynomial $q(x_1, \ldots, x_m) = x_1^2 + \cdots + x_m^2 - 1$, and the open square in $\mathbb{R}^2$ having the vertices $(-1, -1)$, $(-1, 1)$, $(1, -1)$, $(1, 1)$ is compatible with the polynomial $q(x_1, x_2) = (x_1 + 1)(x_2 + 1)(x_1 - 1)(x_2 - 1)$.

Let A be a finite subset of $\mathbb{Z}_{\geqslant 0}^m$. We say that a function $f(x_1, \ldots, x_m)$ satisfies zero boundary conditions with orders $\alpha \in A$ on U, if for any $\alpha \in A$ the function $\overline{f}_\alpha(x_1, \ldots, x_m)$ is continuous in $\overline{U}$.

Let U be a domain compatible with $q(x_1, \ldots, x_m) \in \mathbb{Z}[x_1, \ldots, x_m]$ and containing $(0, \ldots, 0)$. We consider the following problem:

Problem ZC (zero condition). For given:

1) $m \in \mathbb{Z}_{>0}$ (the number of independent variables),
2) a non-empty finite set $A \subset \mathbb{Z}_{\geqslant 0}^m$,

3) a non-zero polynomial $q(x_1, \ldots, x_m) \in \mathbb{Z}[x_1, \ldots, x_m]$ and, maybe, some additional information about U,
4) a differential operator $L \in \mathbb{Z}[D, x]$,

we study the algorithmic decidability of the existence of a non-zero solution $f(x_1, \ldots, x_m)$ of the equation $L(f) = 0$ such that
   (a) $f(x_1, \ldots, x_m)$ is analytic on U,
   (b) $\overline{f_\alpha}(x_1, \ldots, x_m)$ is continuous on $\overline{U}$ for any $\alpha \in A$.
(Function $\overline{f_\alpha}(x_1, \ldots, x_m)$ is continuous on U by the analyticity of $f(x_1, \ldots, x_m)$, and thus, the additional continuity of $\overline{f_\alpha}(x_1, \ldots, x_m)$ on the boundary of the domain is required.)

Domain U may not be uniquely determined by $q(x_1, \ldots, x_m)$. We do not specify exactly how the algorithm obtains the information about the U because below it does not matter for us.

In the examples given below, we can see that a non-zero analytic solution satisfying the zero boundary conditions exists in some cases and does not exist in some other cases.

Example 1. *Let* $m \in \mathbb{Z}_{>0}$, U *be the ball of radius 1 in* $\mathbb{R}^m$ *with the center at the origin,* $q(x_1, \ldots, x_m) = x_1^2 + \cdots + x_m^2 - 1$, $A = \{(0, \ldots, 0)\}$, $L = D_1^2 + \cdots + D_m^2$. *Then we obtain the Dirichlet problem:*

$$\frac{\partial^2 y(x_1, \ldots, x_m)}{\partial x_1^2} + \cdots + \frac{\partial^2 y(x_1, \ldots, x_m)}{\partial x_m^2} = 0$$

$$y(x_1, \ldots, x_m)|_{x_1^2 + \cdots + x_m^2 = 1} = 0.$$

It is well known that this problem with zero boundary conditions has no non-zero solutions. Thus the answer is "no".

Example 2. *Let* $m = 2$, U *be the halfplane in* $\mathbb{R}^2$ *bounded by line* $x_1 = 1$ *and containing the origin,* $q(x_1, x_2) = x_1 - 1$, $A = \{(0, 0), (1, 1)\}$ *and*

$$L = D_1^2 - x_1 D_1 - (x_1 + x_2)D_1 D_2 + 2 = 0.$$

We obtain the following problem: for the equation

$$\frac{\partial^2 y(x_1, x_2)}{\partial x_1^2} - x_1 \frac{\partial y(x_1, x_2)}{\partial x_1} - (x_1 + x_2)\frac{\partial^2 y(x_1, x_2)}{\partial x_1 \partial x_2} + 2y(x_1, x_2) = 0$$

test existence of a non-zero solution that is analytic in U *and satisfies the conditions*

$$y(x_1, x_2)|_{x_1=1} = 0, \qquad \frac{\partial^2 y(x_1, x_2)}{\partial x_1 \partial x_2}\bigg|_{x_1=1} = 0.$$

The answer is "yes", since $y(x_1, x_2) = x_1^2 - 1$ *is a solution.*

3 Undecidability of Problem ZC

To prove that problem ZC is algorithmically undecidable, we establish first the
following lemma:

Lemma 1. *Let $m \in \mathbb{Z}_{>0}$, A be a finite subset $\mathbb{Z}_{\geq 0}^m$ and U be a domain in
K^m compatible with the polynomial $q(x_1, \ldots, x_m) \in \mathbb{Z}[x_1, \ldots, x_m]$. Then the
Diophantine equation*

$$C(n_1, \ldots, n_m) = 0, \qquad C \in \mathbb{Z}[n_1, \ldots, n_m] \qquad (2)$$

*has a solution $(z_1, \ldots, z_m) \in \mathbb{Z}^m$ if and only if there exists a vector $(u_1, \ldots, u_m)$,
$u_i = \pm 1$, such that the differential equation $L(f) = 0$ with*

$$L = C(u_1 \delta_1, \ldots, u_m \delta_m) \frac{1}{q^{s+1}(x_1, \ldots, x_m)}, \qquad s = \max\{|\alpha| : \alpha \in A\},$$

has a non-zero solution which is analytic on U and satisfies zero boundary conditions with orders $\alpha \in A$.

Proof. First we prove that the existence of a solution for Eqn. (2) implies the
existence of the vector $(u_1, \ldots, u_m)$. Let the equation $C(n_1, \ldots, n_m) = 0$ have a
solution $(z_1, \ldots, z_m) \in \mathbb{Z}^m$. Set

$$u_i = \begin{cases} 1, & \text{if } z_i \geqslant 0, \\ -1, & \text{if } z_i < 0. \end{cases}$$

If $T(n_1, \ldots, n_m) = C(u_1 n_1, \ldots, u_m n_m)$, then $T(\delta_1, \ldots, \delta_m) = C(u_1 \delta_1, \ldots, u_m \delta_m)$.
The Diophantine equation $T(n_1, \ldots, n_m) = 0$ has a solution $(u_1 z_1, \ldots, u_m z_m) \in
\mathbb{Z}_{\geqslant 0}^m$, therefore, the differential equation $T(\delta_1, \ldots, \delta_m)(y) = 0$ has monomial
solution $y(x_1, \ldots, x_m) = x_1^{u_1 z_1} \ldots x_m^{u_m z_m}$ (see [1]). Hence if

$$L = T(\delta_1, \ldots, \delta_m) \frac{1}{q^{s+1}(x_1, \ldots, x_m)} = C(u_1 \delta_1, \ldots, u_m \delta_m) \frac{1}{q^{s+1}(x_1, \ldots, x_m)},$$

then the equation $L(f) = 0$ has the solution

$$f(x_1, \ldots, x_m) = x_1^{u_1 z_1} \ldots x_m^{u_m z_m} q^{s+1}(x_1, \ldots, x_m),$$

that is analytic in U (because it is polynomial) and satisfies zero boundary
conditions with orders $\alpha \in A$ (because all its partial derivatives up to order
$s = \max\{|\alpha| : \alpha \in A\}$ contain the factor $q(x_1, \ldots, x_m)$ and, therefore, vanish on
the boundary of U).

Now we prove that the existence of vector $(u_1, \ldots, u_m)$, $u_i = \pm 1$, implies the
existence of a suitable solution of equation (2). Assume that there exists such
vector $(u_1, \ldots, u_m)$, $u_i = \pm 1$ that if

$$L = C(u_1 \delta_1, \ldots, u_m \delta_m) \frac{1}{q^{s+1}(x_1, \ldots, x_m)}, \qquad s = \max\{|\alpha| : \alpha \in A\},$$

then the differential equation $L(f) = 0$ has non-zero solution $f(x_1, \ldots, x_m)$ that is analytic on U. Since $(0, \ldots, 0) \in U$, $f(x_1, \ldots, x_m)$ has a power series expansion at zero: $f(x_1, \ldots, x_m) = \sum_{n \in \mathbb{Z}_{\geqslant 0}^m} a_n x_1^{n_1} \ldots x_m^{n_m}$.

Consider the function $g(x_1, \ldots, x_m) = \frac{f(x_1, \ldots, x_m)}{q^{s+1}(x_1, \ldots, x_m)}$. The quotient of two formal power series (in our case, the second power series is a polynomial) can be represented as formal Laurent series. There are different definitions of Laurent series of several variables (see [2], [7]); we can use, for example, the iterative approach: $g(x_1, \ldots, x_m) \in K((x_1)) \ldots ((x_m))$. So $g(x_1, \ldots, x_m)$ can be represented as a non-zero sum of the monomials $g(x_1, \ldots, x_m) = \sum_{n \in \mathbb{Z}^m} b_n x_1^{n_1} \ldots x_m^{n_m}$. Note that $g(x_1, \ldots, x_m)$ is a solution of the equation

$$T(\delta_1, \ldots, \delta_m)(y) = 0, \quad T(\delta_1, \ldots, \delta_m) = C(u_1\delta_1, \ldots, u_m\delta_m).$$

Using the statement $2°$ from the introduction, we can see that this equation also has some monomial solution $y(x_1, \ldots, x_m) = x_1^{z_1} \ldots x_m^{z_m}$, where $(z_1, \ldots, z_m) \in \mathbb{Z}^m$. Hence the equation $T(n_1, \ldots, n_m) = 0$ has an integer solution $(z_1, \ldots, z_m)$ (see [1]). And since $T(n_1, \ldots, n_m) = C(u_1 n_1, \ldots, u_m n_m)$, the equation $C(n_1, \ldots, n_m) = 0$ has integer solution $(u_1 z_1, \ldots, u_m z_m)$. $\square$

Theorem 1. *Problem ZC is algorithmically undecidable.*

Proof. Suppose that there exists an algorithm solving problem ZC. Then we show how one can test whether an arbitrary Diophantine equation $C(n_1, \ldots, n_m) = 0$ has an integer solution using this algorithm.

Let $C \in \mathbb{Z}[n_1, \ldots, n_m]$ and U be the open m-dimensional ball of radius 1 with the center at the origin, $q(x_1, \ldots, x_m) = x_1^2 + x_2^2 + \cdots + x_m^2 - 1$, $A = \{(0, \ldots, 0)\}$. By Lemma 1, the equation $C(n_1, \ldots, n_m) = 0$ has an integer solution if and only if for some $(u_1, \ldots, u_m)$, $u_i = \pm 1$, the equation

$$C(u_1\delta_1, \ldots, u_m\delta_m) \frac{1}{q(x_1, \ldots, x_m)}(y) = 0 \tag{3}$$

has a non-zero solution y, that is analytic on U and satisfies zero boundary conditions with orders $\alpha \in A$. We can try 2^m possible vectors $(u_1, \ldots, u_m)$ and for each of them reduce equation (3) to the form $L(y) = 0$ ($L \in \mathbb{Z}[D, x]$). For each of these equations we test existence of an analytic solutions satisfying corresponding boundary conditions. In this way we establish the presence or absence of integer solutions of the equation $C(n_1, \ldots, n_m) = 0$. However, the problem of recognition of the existence of integer solutions for an arbitrary Diophantine equation is algorithmically undecidable (Davis–Putnam–Robinson–Matiyasevich theorem, see [5]), and, therefore, problem ZC is also algorithmically undecidable. $\square$

Remark 1. *In Item 3 of the formulation of problem ZC, additional information related to the domain U is mentioned. Our proof of Theorem 1 shows that problem ZC is algorithmically undecidable regardless of this possible additional information.*

4   Undecidability of Problem ZC with Fixed Order of Equation and Number of Variables

In the proof of statement 1° from the introduction, on which the proof of Theorem 1 is based, a mapping of Diophantine equations to differential equations, in which the order of equation and the number of variables stay the same, is constructed. Currently it is proved that there is no algorithm of testing existence of integer solutions for Diophantine equations of degree greater than or equal to four and also for Diophantine equations with eleven or more variables (see [8], [9]).

Remark 2. *Currently, for the third-degree equation and equations with from two to ten variables, the undecidability of this problem is neither proved nor disproved. Note that for first- and second-degree equations this problem is decidable — see [8].*

Returning in this regard to the problem of ZC, we can present a stronger version of Theorem 1 for some cases:

Theorem 2. *(i) If for a natural number n the problem of testing existence of integer solutions for nth degree Diophantine equations is algorithmically undecidable, then problem ZC with fixed order n of the differential operator L is also algorithmically undecidable.*

(ii) If for a natural number m, the problem of testing existence of integer solutions for Diophantine equations in m variables is algorithmically undecidable, then problem ZC with fixed number of variables m, a polynomial $q(x_1, \ldots, x_m)$ and a domain U compatible with it, is also algorithmically undecidable.

Proof. The proof of this statements is analogous to the proof of Theorem 1. Let $A = \{(0, \ldots, 0)\}$.

(i) By Lemma 1, the nth degree Diophantine equation $C(z_1, \ldots, z_m) = 0$ has an integer solution if and only if for some $(u_1, \ldots, u_m)$, $u_i = \pm 1$, the nth order differential equation

$$C(u_1\delta_1, \ldots, u_m\delta_m)\frac{1}{q(x_1, \ldots, x_m)}(y) = 0$$

has a non-zero solution y, that is analytic on some domain U containing 0 and satisfies zero boundary conditions with orders $\alpha \in A$ in this domain. Therefore, assuming the existence of an algorithm for solving the problem ZC with differential equations of order n, we obtain a contradiction because we can recognize the existence of solutions of an arbitrary nth degree Diophantine equation using this algorithm, but by the statement, this problem is algorithmically unsolvable.

(ii) The proof is analogous to the proof of (i). Instead of Diophantine and differential equations of degree/order n we consider Diophantine and differential equations with m variables, respectively. □

5   Uniqueness of Analytic Solutions

Problem ZC is connected with the question of uniqueness of analytic solutions for the case of inhomogeneous equation with non-zero boundary conditions: let there be given the domain U and the equation $L(f(x_1, \ldots, x_m)) = b(x_1, \ldots, x_m)$, $L \in \mathbb{Z}[D, x]$ with the set of boundary conditions of the form

$$f_\alpha(x_1, \ldots, x_m)\big|_{(x_1, \ldots, x_m) \in \overline{U} \setminus U} = \mu(x_1, \ldots, x_m), \quad \alpha \in \mathbb{Z}_{\geqslant 0}^m$$

(here the right-hand $\mu(x_1, \ldots, x_m)$ depends on α) and assume that some analytic solution satisfying these conditions is found. This solution is unique if and only if the answer to the question of problem ZC for the homogeneous equation $L(f) = 0$ with zero boundary conditions is negative (i.e., there is no non-zero solution). Hence the problem of testing uniqueness of analytic solutions, as well as the problem ZC, is algorithmically undecidable.

Acknowledgements. The author is grateful to his scientific adviser S. A. Abramov for the problem statement, attention to this study, and helpful suggestions.

References

1. Abramov, S.A., Petkovšek, M.: On polynomial solutions of linear partial differential and (q-)difference equations. In: Gerdt, V.P., Koepf, W., Mayr, E.W., Vorozhtsov, E.V. (eds.) CASC 2012. LNCS, vol. 7442, pp. 1–11. Springer, Heidelberg (2012)
2. Aparicio Monforte, A., Kauers, M.: Formal Laurent Series in Several Variables. Expositiones Mathematicae, pp. 350–367 (2012)
3. Aroca, F., Cano, J.M., Richard-Jung, F.: Power series solutions for non-linear PDE's. In: Proc. ISSAC 2003, pp. 15–22. ACM Press (2003)
4. Denef, J., Lipshitz, L.: Power series solutions of algebraic differential equations. Math. Ann. 267, 213–238 (1984)
5. Matiyasevich, Y.V.: Hilbert's Tenth Problem. MIT Press, Cambrige (1993)
6. Paramonov, S.V.: On rational solutions of linear partial differential or difference equations. Programming and Computer Software (2), 57–60 (2013)
7. Paramonov, S.V.: Checking existence of solutions of partial differential equations in the fields of Laurent series. Programming and Computer Software (2), 58–62 (2014)
8. Pheidas, T., Zahidi, K.: Undecidability of existential theories of rings and fields: A survey. Contemporary Mathematics 270, 49–106 (2000)
9. Zhi-Wei, S.: Reduction of unknowns in Diophantine representations. Science China Mathematics 35(3), 257–269 (1992)

Continuous Problems:
Optimality, Complexity, Tractability
(*Invited Talk*)

Leszek Plaskota

Institute of Applied Mathematics and Mechanics, University of Warsaw
Banacha 2, 02-097 Warsaw, Poland

Abstract. *Information-based complexity (IBC)* is a branch of computational complexity that studies continuous problems for which available information is partial, noisy, and priced. We present basic ideas of IBC and give some important results on optimal algorithms, complexity, and tractability of such problems. The focus is on numerical integration of univariate and multivariate functions.

1 Introduction

Since a digital computer is able to store and manipulate with finitely many real numbers, most computational problems of continuous mathematics can only be solved approximately using incomplete information. A branch of computational mathematics that studies the inherent difficulty of continuous problems for which available information is partial, noisy, and priced, is called *information-based complexity (IBC* in short). IBC emerged some 35 years ago as a consequence of the need for a mathematical theory to study aspects of computations related to continuous problems. Since then IBC developed in different directions, see, e.g., the monographs [36] [34] [18] [35] [38] [13] [25] [30].

Examples of computational problems of continuous mathematics include, e.g., numerical integration, function approximation, different optimization problems, or differential/integral equations. IBC seeks for algorithms that solve such problems not only efficiently but also optimally. The minimal cost of solving the problem within ε is its *ε-complexity*. The complexity depends on the *setting*. For instance, we may have the *worst case* setting with respect to a given function class, the *average case* setting with respect to a given probability measure, or *Monte Carlo* setting where one allows randomized algorithms. In the *asymptotic* setting, which is often considered in numerical analysis, one studies how the successive approximations converge to the true solution for each individual function from a class as the computational cost increases to infinity.

Problems that are defined on functions of many or even infinitely many variables play an important role in IBC. In physical or chemical applications, the number of variables can be millions. Such problems often suffer from the *curse of dimensionality*. This means that the ε-complexity grows exponentially fast as the number d of variables increases to ∞. For a long time, this notion had

V.P. Gerdt et al. (Eds.): CASC Workshop 2014, LNCS 8660, pp. 357–372, 2014.

been used informally. A systematic theoretical study of multivariate problems started only some 20 years ago when *tractability* was formally defined for continuous problems [39]. How to deal with the curse or, if possible, how to vanquish the curse, is a fundamental theoretical and practical question of contemporary computational mathematics. Some important questions have been answered only recently. The three volume monograph [20] [21] [22] is the present state of the art of this subject.

The purpose of this short survey is to present basic ideas of IBC and give some sample, but important results on optimal algorithms, complexity, and tractability that were obtained within this theory. The focus is on the worst case and randomized settings for the numerical integration which is one of the most important problems of numerical analysis, see e.g., [3] [32] for an account of various standard quadrature formulas.

Although *adaptive* quadrature formulas are frequently used in numerical or symbolic packages, see, e.g., [6], their behavior has not been satisfactorily explained. A common knowledge is that adaptive quadratures do well for rapidly varying functions. We show that IBC provides theoretical tools for answering this question; namely, we identify classes of piecewise smooth functions for which adaption helps a lot. Note that functions of this kind regularly appear in applications. Examples are shock computations or image representation.

Multivariate integration is one of the problems for which the curse of dimensionality frequently occurs in the worst case setting for many function classes. For a long time, switching to (non-deterministic) Monte Carlo methods seemed to be the only rescue. Only recently, it turned out that the (deterministic) quasi-Monte Carlo methods are able to break the curse as well for some classes of functions. This was first discovered empirically based on some finance applications [23], and then confirmed by theoretical studies in the IBC framework, see again [21] for a summary on this topic.

The basic ingredients of IBC such as the solution operator, information and algorithm are presented in Section 2. In Section 3 we discuss two fundamental questions of IBC, which are: existence of optimal linear algorithms, and when adaptive information is better than nonadaptive information. Section 4 is devoted to numerical integration in one variable. We show that for globally smooth functions adaption does not help, while for piecewise smooth functions it does help. Next we consider multivariate integration and show the curse of dimensionality for r-smooth functions in the worst case setting. Then we vanquish the curse by either switching to the randomized setting or changing the function space. In the latter case, we show a strong relation to *discrepancy*.

2   Basics of IBC

Many computational problems can be viewed as approximation of the values of an operator

$$S : F \to G,$$

where F is a linear space and G is a normed space with a norm $\| \cdot \|$. We usually think of F as a space of functions $f : D \to \mathbb{R}$ where the domain D is a measurable subset of $\mathbb{R}^d$. Examples include, e.g., zero finding, optimization, or function approximation. In this paper, we mainly consider *numerical integration*, in which case F is a linear space of integrable functions, $G = \mathbb{R}$, and $S = \text{Int}$ is defined as

$$\text{Int}(f) = \int_D f(x)\, dx. \tag{1}$$

Note that numerical integration is a special case of a problem represented by a *linear functional* on F.

We assume that our prior knowledge about f is that

$$f \in F_0$$

where F_0 is a subset of F. For instance, if F is equipped with a norm then F_0 can be the unit ball of F. During the computational process we can gain more *information* $y = [y_1, y_2, \ldots, y_n] \in \mathbb{R}^n$ about f by computation/observation of some linear functionals at f. For instance, if F is a function space then such information can be given by values $y_i = f(x_i)$ at finitely many points x_i of the domain. If F equipped with an inner product $\langle \cdot, \cdot \rangle_F$, then we may have $y_i = \langle f, \xi_i \rangle_F$ for some $\xi_i \in F$. In general, we distinguish *nonadaptive* and *adaptive* information. Nonadaptive information about f is given as

$$y_i = L_i(f), \qquad 1 \le i \le n,$$

where L_i are linear functionals from a class Λ of admissible functionals. In adaptive information, the successive functionals L_i are selected based on the information collected earlier. That is,

$$y_1 = L_1(f),$$
$$y_2 = L_2(f; y_1),$$
$$\cdots$$
$$y_n = L_n(f; y_1, y_2, \ldots, y_{n-1}).$$

We stress that, in adaptive information, the number $n = n(y)$ of information pieces can also be chosen adaptively. That is, in each step we make a decision whether we have enough information or want more information about f. In any case, $y = N(f)$ where

$$N : F \to Y$$

is an *information operator* and Y is the set of all its possible values. Clearly, if information is nonadaptive then $N : F \to \mathbb{R}^n$ is a linear mapping.

Having computed information y about f, an approximation to $S(f)$ is provided by an *algorithm* φ which is any mapping

$$\varphi : Y \to G.$$

(At this point we make a rather idealistic assumption that φ can be an arbitrary mapping.) Thus S is approximated by a composition

$$A = \varphi \circ N$$

that maps F to G.

Since information is usually *not* a one-to-one operator, we have to deal with an inevitable error of approximation. The error of an algorithm φ using information N can be measured in different ways, depending on the *setting*. In the *worst case* setting considered in this paper we have

$$e^{\mathrm{wor}}(\varphi, N) = \sup_{f \in F_0} \|S(f) - \varphi(N(f))\|.$$

This is the most conservative choice; if the error is at most ε then we are sure that $\|S(f) - \varphi(N(f))\| \le \varepsilon$ for each individual function $f \in F_0$.

Our aim is to approximate S within the error ε using as little functional evaluations as possible. Denoting by $\mathrm{card}(N)$ the maximum length $n = n(y)$ of information $y = N(f)$ for $f \in F_0$, we define the (worst case) ε-*complexity* of our problem as

$$\mathrm{comp}^{\mathrm{wor}}(F_0, \varepsilon) = \min\{n : \exists N \, \exists \varphi \text{ s.t. } \mathrm{card}(N) \le n, \, e^{\mathrm{wor}}(\varphi, N) \le \varepsilon\}.$$

Remark 1. The quantity $\mathrm{comp}^{\mathrm{wor}}(F_0, \varepsilon)$ is often called information ε-complexity. For many problems the information cost $\mathrm{card}(N)$ dominates the combinatory cost of φ. Therefore, for simplicity, we only consider the information cost.

3   Some General Results

For given information $N : F \to Y$, an algorithm φ^* is called *optimal* if it minimizes the error with respect to all possible algorithms using N, i.e.,

$$e^{\mathrm{wor}}(\varphi^*, N) = \inf_{\varphi : Y \to G} e^{\mathrm{wor}}(\varphi, N).$$

In the worst case setting, there is a nice interpretation of optimal algorithms and their errors. Indeed, observe that the error of φ can be equivalently written as

$$e^{\mathrm{wor}}(\varphi, N) = \sup_{y \in Y} \left(\sup_{g \in B(y)} \|g - \varphi(y)\| \right)$$

where $B(y) = \{S(f) : f \in F_0, N(f) = y\} \subset G$ is the set of all possible solutions corresponding to information y. Hence the minimal error that can be achieved by algorithms that use information N is the maximal (Chebyshev) radius of all the sets $B(y)$ for $y \in Y$. Recall that the radius $r(B)$ of a set $B \subset G$ is defined as the minimal radius of a ball containing B,

$$r(B) = \inf_{g_1 \in G} \sup_{g_2 \in B} \|g_2 - g_1\|.$$

Hence, if each $B(y)$ has a center $c_y \in G$, i.e., $r(B(y)) = \sup_{g_2 \in B(y)} \|g_2 - c_y\|$, then $\varphi(y) = c_y$ is an optimal algorithm. For this reason, the minimal error is called the *radius of information*, and denoted $r^{\mathrm{wor}}(N)$.

As a simple illustration, consider uniform approximation in the class F_0 of functions $f : [0,1] \to \mathbb{R}$ satisfying the Lipschitz condition,

$$|f(x) - f(y)| \le |x - y| \qquad \text{for all} \quad 0 \le x, y \le 1.$$

That is, $S(f) = f$ and $G = C([0, 1])$. Suppose the information operator is given as

$$N(f) = [f(x_1), f(x_2), \ldots, f(x_n)]$$

where $0 \leq x_1 < x_2 < \cdots \leq x_n \leq 1$. Then, for given information $y = [y_1, \ldots, y_n]$ about f, the center of $B(y)$ is $c_y = (f^- + f^+)/2$, where f^-, f^+ are the lower and upper envelopes,

$$f^-(x) = \min_{1 \leq i \leq n} y_i + |x - x_i|, \qquad f^+(x) = \max_{1 \leq i \leq n} y_i - |x - x_i|.$$

Moreover,

$$r^{\mathrm{wor}}(N) = \max\{x_1, (1 - x_n), (x_i - x_{i-1})/2, 2 \leq i \leq n\}.$$

Note that in this case the piecewise linear approximation,

$$\varphi(y) = \begin{cases} y_1, & 0 \leq x \leq x_1, \\ y_{i-1}\left(\frac{x-x_i}{x_{i-1}-x_i}\right) + y_i\left(\frac{x-x_{i-1}}{x_i-x_{i-1}}\right), & x_{i-1} < x \leq x_i, \ 2 \leq i \leq n, \\ y_n, & x_n < x \leq 1, \end{cases}$$

is also an optimal algorithm.

From the point of view of computational practice it is important to know whether the optimal algorithms have a relatively simple structure, e.g., are linear or affine. This question can be successfully answered when the problem S is a linear functional, e.g., $S = \mathrm{Int}$. Note that the widely used *quadratures*

$$Q_n(f) = \sum_{i=1}^{n} a_i f(x_i) \tag{2}$$

with fixed nodes x_i are linear algorithms.

Theorem 1 (see [31] [33]). *Suppose that S is a linear functional and information N is a linear mapping. If the set F_0 is convex then there exists an affine algorithm that is optimal. If, in addition, F_0 is symmetric about the zero then the optimal affine algorithm is linear.*

The importance of this theorem is clear. It says that the search for optimal algorithms can be restricted to affine (or linear) algorithms whose combinatory cost is proportional to $\mathrm{card}(N)$.

We add that Theorem 1 does not hold in general if S is linear, but not a functional. It does hold however when, e.g., S is a linear operator with domain F being a Hilbert space, and F_0 is the unit ball, see, e.g., [35, Sect. 5.5].

In the proof of Theorem 1 one uses the following convenient formula for the radius of linear information N with respect to convex classes F_0; namely,

$$r^{\mathrm{wor}}(N) = \sup_{h \in \mathrm{bal}(F_0) \cap \ker(N)} S(h) \tag{3}$$

where $\mathrm{bal}(F_0) = \{ (f_1 - f_2)/2 : f_1, f_2 \in F_0 \}$. This formula also plays a crucial role in the proof of the following result on the power of nonadaptive information.

Suppose that F_0 is convex and symmetric about an $f^* \in F$, i.e., if $f \in F_0$ then $2f^* - f \in F_0$. For a given adaptive information N^{ada} that uses functionals $L_i(\cdot; y_1, \ldots, y_{i-1})$ we set $y^* = N^{\mathrm{ada}}(f^*)$ and define nonadaptive information N^{non} as

$$N^{\mathrm{non}}(f) = [L_1(f), L_2(f; y_1^*), L_3(f; y_1^*, y_2^*), \ldots, L_n(f; y_1^*, \ldots, y_{n-1}^*)]$$

with $n = n(y^*)$. This means that for the choice of L_k we 'pretend' that we earlier saw $y_i^* = L_i(f^*; y_1^*, \ldots, y_{i-1}^*)$ instead of $y_i = L_i(f; y_1^*, \ldots, y_{i-1}^*)$ for $i = 1, 2, \ldots, k - 1$.

Theorem 2 (see [2]). *We have*

$$r^{\mathrm{wor}}(N^{\mathrm{non}}) \leq r^{\mathrm{wor}}(N^{\mathrm{ada}}).$$

Thus adaption does not help for approximating linear functionals S over convex and symmetric classes F_0.

Indeed, in this case the formula (3) reads

$$r^{\mathrm{wor}}(N^{\mathrm{non}}) = \sup\{ S(f) : f \in F_0, N^{\mathrm{non}}(f) = y^* \} - S(f^*),$$

which is the radius of the set of solutions $S(f)$ for $f \in F_0$ with $N^{\mathrm{non}}(f) = y^*$. By the definition of N^{non}, this coincides with the set of solutions for $f \in F_0$ with $N^{\mathrm{ada}}(f) = y^*$. Hence $r^{\mathrm{wor}}(N^{\mathrm{ada}})$ cannot be smaller than $r^{\mathrm{wor}}(N^{\mathrm{non}})$.

Theorem 2 generalizes to the case of linear operators S, where adaptive information can be at most twice better than nonadaptive information, see, e.g., [35, Sect. 5.2]

Remark 2. Theorems 1 and 2 have their counterparts in the average case setting with respect to Gaussian measures, and generalizations to the case of noisy information, see, e.g., [37] [5] [19], and the monographs [35] [25].

4   Univariate Integration

In this section, we consider the problem of numerical integration (1) of univariate functions. We assume that the class Λ of permissible information functionals consists of function evaluations, i.e., $L \in \Lambda$ iff there is $t \in D$ such that $L(f) = f(t) \ \forall f \in F$.

4.1   Smooth Functions

Suppose that F is the space of r times continuously differentiable functions $f : [0, 1] \to \mathbb{R}$ equipped with the norm

$$\|f\|_r = \max \left(\|f\|_\infty, \|f^{(r)}\|_\infty \right),$$

where $\|g\|_\infty = \max_{0 \le x \le 1} |g(x)|$, and $F_0 = W_r$ is the unit ball of F. Since the assumptions of Theorems 1 and 2 are satisfied, adaption does not help in the worst case setting, and quadratures (2) are optimal algorithms. For the class W_r, optimal quadratures are in general known up to constant factors.

Theorem 3 (see [36] [18]). *There are a_r and A_r such that*

$$a_r \, \varepsilon^{-1/r} \le \mathrm{comp}^{\mathrm{wor}}(W_r, \varepsilon) \le A_r \, \varepsilon^{-1/r}, \qquad 0 < \varepsilon < 1.$$

The upper complexity bound is attained by composite quadratures that are based on simple rules of order r and equidistant sampling.

To give a flavor of IBC proof techniques, we show the lower bound. To that end, we use the well known adversary argument. Suppose that an algorithm φ uses nonadaptive information $N(f) = [f(x_1), \dots, f(x_n)]$ with $0 \le x_1 < \dots < x_n \le 1$. In addition, we set $x_0 = x_n - 1$ and $x_{n+1} = 1 + x_1$. Then we choose $\Psi : \mathbb{R} \to \mathbb{R}$ to be any r times continuously differentiable function satisfying $\Psi|_{[0,1]} \in W_r$, $\Psi(x) = 0$ for $x \notin (0,1)$, and $\mathrm{Int}(\Psi) = a > 0$, and define

$$f = \sum_{i=0}^{n} \psi_i, \quad \text{where} \quad \psi_i(x) = h_i^r \, \Psi((x - x_i)/h_i), \quad h_i = x_{i+1} - x_i.$$

We have $\pm f \in W_r$ and $N(\pm f) = 0$. Hence $\mathrm{Int}(f)$ and $\mathrm{Int}(-f)$ are approximated by the same number $\varphi(0)$, and the error of φ is at least

$$(\mathrm{Int}(f) - \mathrm{Int}(-f))/2 = \mathrm{Int}(f) = a \sum_{i=1}^{n} h_i^{r+1} \ge a \, n^{-r}.$$

Since adaption does not help, $\mathrm{comp}^{\mathrm{wor}}(W_r, \varepsilon) \ge (a/\varepsilon)^{1/r}$, as claimed.

Theorem 3 can be a bit surprising since adaptive quadratures are quite popular in computational practice and are regularly used in symbolic and numerical packages such as Mathematica or Matlab. As an example, consider the probably most popular standard adaptive Simpson quadrature ASQ, first published in algorithm form in [12]. It can be conveniently written as a recursive function.

```
0    function ASQ(a, b, f, ε);
1      Q1 := Simpson(a, b, f);
2      Q2 := Simpson(a, (a+b)/2, f) + Simpson((a+b)/2, b, f);
3      if |Q1 − Q2| ≤ 15 ε then return Q2 else
4        return ASQ(a, (a+b)/2, f, ε/2) + ASQ((a+b)/2, b, f, ε/2)
```

Here a and b are the end-points of the interval and ε is the error demand.

A rough justification of ASQ is as follows. Let $I_i(f)$ be the integral of f in the ith subinterval of length h_i, and $Q_i^1(f)$, $Q_i^2(f)$ be the three- and five-point Simpson rules for this interval. Then $I_i(f) - Q_i^1(f) = -h_i^5 f^{(4)}(\xi_i)/2880$ and

$I_i(f) - Q_i^2(f) = -(h_i/2)^5 f^{(4)}(\eta_i)/2880$ for some ξ_i, η_i in the subinterval. Hence for 'small' h_i we should have

$$Q_i^1(f) - Q_i^2(f) = (I_i(f) - Q_i^2(f)) - (I_i(f) - Q_i^1(f)) \approx 15\,(I_i(f) - Q_i^2(f)),$$

and the overall error of approximating $\int_0^1 f(x)\,dx$ should be bounded as

$$\sum_i |I_i(f) - Q_i^2(f)| \approx \frac{1}{15}\sum_i |Q_i^1(f) - Q_i^2(f)| \le \sum_i \varepsilon\, h_i = \varepsilon.$$

ASQ is supposed to work well for functions $f \in C^4([0,1])$. However, it fails in many cases. If, for instance, $f(x) = \prod_{i=0}^4 (x - i/4)^2$ then ASQ returns zero independently of how small ε is. It is generally a difficult problem to identify a class F_0 of functions for which ASQ gives a correct answer using fewer function evaluations than the usual nonadaptive composite Simpson rule, see, e.g., [14] [15]. However, some results can be shown when analyzing the asymptotic behavior of the error for each individual f, see [26].

Specifically, assume in addition to $f \in C^4([0,1])$ that $f^{(4)}$ does not change sign, say,

$$f^{(4)}(x) \ge 0 \qquad \forall x. \tag{4}$$

Then, for sufficiently small ε (depending on f) adaptive Simpson quadrature returns an ε-approximation at cost proportional to $(\varepsilon\, L^{\mathrm{ASQ}}(f))^{-1/4}$ where

$$L^{\mathrm{ASQ}}(f) = \left(\int_0^1 (f^{(4)}(x))^{1/4}\,dx \right)^4.$$

This should be compared with the corresponding result for the (nonadaptive) composite Simpson rule using equidistant sampling, where the ε-approximation is asymptotically attained at cost proportional to $(\varepsilon\, L^{\mathrm{NSQ}}(f))^{-1/4}$ with

$$L^{\mathrm{NSQ}}(f) = \int_0^1 f^{(4)}(x)\,dx.$$

Thus in both cases the cost of obtaining an ε-approximation increases at the same rate $\varepsilon^{-1/4}$, as $\varepsilon \to 0$. However, we always have $L^{\mathrm{ASQ}}(f) \le L^{\mathrm{NSQ}}(f)$, and the ratio $L^{\mathrm{NSQ}}(f)/L^{\mathrm{ASQ}}(f)$ can be arbitrarily large for functions f in our class. For such functions the adaptive quadrature is asymptotically much better than the nonadaptive quadrature. The superiority of adaptive quadratures is even more striking when $f^{(4)}$ has an end-point singularity, e.g., $f(x) = \sqrt{x}$ and $[a,b] = [0,1]$. Then $L^{\mathrm{ASQ}}(f) < \infty$ while $L^{\mathrm{NSQ}}(f) = \infty$, and the nonadaptive procedure even loses the rate $\varepsilon^{-1/4}$.

Those asymptotic results may suggest that adaption can help in the worst case setting if we narrow the class W_r to, say, functions with nonnegative rth derivative. This is however not the case as such class is still convex and symmetric (about $f^*(x) = \frac{1}{2}x^r/r!$) and Theorem 2 applies.

Remark 3. It turns out that the adaptive subdivision strategy used by ASQ is *not* optimal for functions satisfying (4). The best strategy relies on keeping the local error in each subinterval at the same level. This strategy produces a subdivision such that the resulting composite Simpson quadrature asymptotically returns an ε-approximation at cost proportional to $(\varepsilon\, L^{\mathrm{OSQ}}(f))^{-1/4}$ where

$$L^{\mathrm{OSQ}}(f) = \left(\int_0^1 (f^{(4)}(x))^{1/5}\, \mathrm{d}x \right)^5,$$

see again [26] for details.

More examples of function classes for which adaption does help can be found in [19].

4.2   Singular Functions

Adaptive quadratures tend to sample denser in regions where the underlying function rapidly changes. Extreme examples of such functions are functions with singularities. Then we are able to rigorously show that adaptive quadratures are much superior to nonadaptive quadratures even in the worst case setting.

We consider a class $\widehat{W}_r$ of functions $f : [0,1] \to \mathbb{R}$ that are r-smooth and periodic except for one unknown *singular* point. Specifically, $f \in \widehat{W}_r$ iff there are $s = s_f \in [0,1)$ and a function $g = g_f \in W_r$ such that

$$f(x) = \begin{cases} g_f(x - s_f + 1), & 0 \le x < s_f, \\ g_f(x - s_f), & s_f \le x \le 1. \end{cases}$$

Hence $f \in \widehat{W}_r$ can be viewed as a function $g \in W_r$ with the argument shifted by s_f. Note that the class $\widehat{W}_r$ is symmetric about zero, but *not* convex. We have the following theorem.

Theorem 4 (see [27]). *Let $r \ge 2$.*

(i) *For any nonadaptive quadrature Q_n^{non} that uses n function evaluations we have*

$$e^{\mathrm{wor}}(Q_n^{\mathrm{non}}) \ge (n+1)^{-1}.$$

(ii) *There are adaptive quadratures Q_n^{ada}, each using no more than n function evaluations, for which*

$$e^{\mathrm{wor}}(Q_n^{\mathrm{ada}}) \le C_r\, n^{-r},$$

with some C_r independent of n.

The proof of (i) uses the already known adversary argument. That is, suppose that Q_n^{non} uses points $0 \le x_1 < x_2 < \cdots < x_n < 1$. We select k such that $x_{k+1} - x_k \ge 1/n$ (where $x_{n+1} = 1 + x_1$) and set $h = x_{k+1} - x_k - \delta$ with 'small' δ and $c = 2/(1+h)$. Finally, we define two functions, f generated by $g_f(x) = cx - 1$ and $s_f = x_k + \delta$, and $\tilde{f}$ generated by $g_{\tilde{f}}(x) = c(x + h) - 1$ and $s_{\tilde{f}} = x_{k+1}$ (or

$s_{\tilde{f}} = x_1$ if $k = n$). The functions f and $\tilde{f}$ share the same information, and therefore their integrals are approximated by the same value. Hence the error is at least

$$|\mathrm{Int}(f) - \mathrm{Int}(\tilde{f})|/2 = h(1 - ch/2) = h/(1 + h).$$

Since h can be arbitrarily close to $1/n$, (i) follows.

The construction of the adaptive quadrature Q_n^{ada} is quite simple. An adaptive mechanism is here used to detect the singularity. Specifically, for an initial uniform grid $t_i = ih$ (with $h \approx 1/n$) we compute the divided differences $d_i = f[t_i, t_{i+1}, \ldots, t_{i+r}]$ corresponding to all $r + 1$ successive points. Then similar procedure is repeated in the subinterval $[t_{i^*}, t_{i^*+r}]$, where $i^* = \arg\max_i |d_i|$, but with the mesh-size $h/2$. After such $r \log_2(1/h)$ bisection-like steps we obtain a critical subinterval $[u, v]$ of length h^{r+1}. Finally, in $[u, v]$ the integral is approximated by zero, while outside the usual composite quadrature of order r is applied using the original grid of size h.

It is clear that the error of Q_n^{ada} is of order h^r if the singular point s_f is in $(u, v]$. The correctness of Q_n^{ada} in case $s_f \notin [u, v]$ is justified by the following theoretical argument. Denote the discontinuity jumps of f and its derivatives by

$$\Delta_f^{(j)} = f^{(j)}(s_f^+) - f^{(j)}(s_f^-) = g_f^{(j)}(0) - g_f^{(j)}(1), \qquad 0 \le j \le r - 1.$$

Lemma 1 (see [29]). *There exists M_r with the following property.*
Suppose $f \in \widehat{W}_r$ with $s_f \in (t_k, t_{k+1}]$. If

$$|f[t_i, t_{i+1}, \ldots, t_{i+r}]| \le B \qquad \text{for all} \quad k + 1 - r \le i \le k,$$

then

$$|\Delta_f^{(j)}| \le M_r\left(B + \frac{1}{r!}\|f^{(r)}\|_\infty\right) h^{r-j}, \qquad 0 \le j \le r - 1.$$

Now, if $s_f \notin (u, v)$ then all the divided differences $|d_i| \le \|f^{(r)}\|_\infty/r!$, and by Lemma 1 the jumps $|\Delta_f^{(j)}| \le \frac{2M_r}{r!}\|f^{(r)}\|_\infty h^{r-j}$. It follows that this is enough for the error to be of order h^r even when the singularity is ignored and the composite rule with the mesh-size h is applied.

Summarizing this subsection we can say that singularities do not hurt. The worst case ε-complexity in the class $\widehat{W}_r$ is still of order $\varepsilon^{-1/r}$. However, one has to use adaptive methods to obtain the best error convergence.

Remark 4. Let us narrow the class $\widehat{W}_r$ to continuous functions, i.e., assume that $g_f(0) = g_f(1)$ (or $\Delta_f^{(0)} = 0$). It turns out that then for any regularity $r \ge 2$ one can obtain the worst case error of order n^{-2} using nonadaptive quadratures. Hence adaption does not help for $r = 2$. For $r \ge 3$ adaptive quadratures are better; however, the optimal quadratures are rather 'quasi-adaptive' since they use only about r adaptive points, independently of n, see [28] for details.

5   Multivariate Integration and Tractability

We now switch to d-variate integration,

$$\text{Int}_d(f) = \int_{[0,1]^d} f(x_1, x_2, \ldots, x_d) \, dx_1 \, dx_2 \ldots dx_d.$$

As before, we assume that Λ consists of function evaluations.

The problem is analyzed from the point of view of *tractability*. Roughly speaking, a problem is tractable if it can be solved in 'reasonable' time with respect to d and the inverse of ε. Formally we have different notions of tractability. For instance, a problem is *polynomially tractable* iff there are nonnegative p, q, and C such that

$$\text{comp}^{\text{wor}}(F_0, \varepsilon) \le C \, d^q \, \varepsilon^{-p} \qquad \text{for all } d \text{ and } \varepsilon \in (0, 1).$$

If, in addition, $q = 0$ then the problem is *strongly polynomially tractable*. We say that a problem suffers from the *curse of dimensionality* iff there are positive c, ε_0, and γ such that

$$\text{comp}^{\text{wor}}(F_0, \varepsilon) \ge c \, (1 + \gamma)^d \qquad \text{for all } \varepsilon \le \varepsilon_0 \text{ and infinitely many } d.$$

See [20] for other notions of tractability.

5.1   Smooth Functions

Consider the following generalization of the class of univariate functions of Subsection 4.1 to a class of d-variate functions. The space F consists of functions $f : [0,1]^d \to \mathbb{R}$ for which all the partial derivatives of order r exist and are continuous. The norm in F is

$$\|f\|_{d,r} = \max\left(\|f\|_\infty, \|D^\alpha f\|_\infty, |\alpha| = r \right),$$

where $\alpha = (\alpha_1, \ldots, \alpha_d)$, $|\alpha| = \sum_{i=1}^d \alpha_i$, and

$$D^\alpha f = \frac{\partial^{|\alpha|} f}{\partial x_1^{\alpha_1} \cdots \partial x_d^{\alpha_d}}.$$

Finally, the class $F_0 = W_{d,r}$ is the unit ball of F. Note that $W_{1,r} = W_r$. Since the error of the zero algorithms is 1, we also have that $\text{comp}^{\text{wor}}(W_{d,r}, \varepsilon) = 0$ for all d and $\varepsilon \ge 1$.

Theorem 5 (see [1] [18]). *There are $a_{d,r}$ and $A_{d,r}$ such that for all d and $0 < \varepsilon < 1$*

$$a_{d,r} \, \varepsilon^{-d/r} \le \text{comp}^{\text{wor}}(W_{d,r}, \varepsilon) \le A_{d,r} \, \varepsilon^{-d/r}.$$

The optimal quadratures are tensor products of univariate composite quadratures of order r.

Observe that this result is devastating for large d. For instance, if $d = 10^2$ (which is not too large in applications) and $r = 2$ then even for a moderate value of $\varepsilon = 10^{-2}$ we have $\varepsilon^{-d/r} = 10^{100}$, and the problem is practically unsolvable.

Since the exponent d/r at $1/\varepsilon$ in Theorem 5 can be arbitrarily large, the multivariate integration in the class $W_{d,r}$ is *not* polynomially tractable. However, based on that theorem, we cannot claim the curse of dimensionality as we do not know how $a_{d,r}$ and $A_{d,r}$ depend on d. The problem had been open for many years. The curse was shown in [33] for $r = 1$, and only recently in [10] for arbitrary r.

Theorem 6 (see [10]). *We have*

$$\mathrm{comp}^{\mathrm{wor}}(W_{d,r}, \varepsilon) \geq c_r \, (1 - \varepsilon) \, d^{d/(2r+3)} \qquad \textit{for all } d \textit{ and } \varepsilon \in (0,1),$$

where $c_r \in (0,1]$ depends only on r. Thus the multivariate integration in the class $W_{d,r}$ suffers from the curse of dimensionality with the super exponential lower bound in d.

Does Theorem 6 mean that it is practically impossible to integrate d-variate functions with large d? Not quite; we can break the curse by either switching from the worst case to another setting, or by changing the function class F_0.

5.2   Monte Carlo

In this subsection we consider the *randomized (Monte Carlo) setting* in which the information and/or algorithms are chosen at random. Formally, we have a family of (deterministic) information operators and algorithms $\{(N_\omega, \varphi_\omega)_{\omega \in \Omega}\}$ parameterized by a random variable $\omega \in \Omega$. In the computational process ω is randomly chosen and then the approximation $\varphi_\omega(N_\omega(f))$ to $S(f)$ is produced. Obviously, information N_ω can be adaptive or nonadaptive. In the randomized setting, the error is defined as

$$e^{\mathrm{rand}}\big((\varphi_\omega, N_\omega)_{\omega \in \Omega}\big) \;=\; \sup_{f \in F_0} \sqrt{\mathbb{E}\,|\,\mathrm{Int}_d(f) - \varphi_\omega(N_\omega(f))\,|^2}$$

where $\mathbb{E}$ is the expectation with respect to ω. An example is provided by the standard *Monte Carlo* method [16],

$$\mathrm{MC}_{d,n}(f) \;=\; \frac{1}{n} \sum_{i=1}^{n} f(t_i)$$

where t_i are independent and uniformly distributed random variables from $[0,1]^d$.

It is well known that

$$\mathbb{E}\,|\,\mathrm{Int}_d(f) - \mathrm{MC}_{d,n}(f)\,|^2 = \frac{\sigma^2(f)}{n}, \qquad \sigma^2(f) = \mathrm{Int}_d(f^2) - (\mathrm{Int}_d f)^2.$$

Thus for the class $W_{d,r}$ we have $e^{\mathrm{rand}}(\mathrm{MC}_n) \leq n^{-1/2}$, which immediately implies

$$\mathrm{comp}^{\mathrm{rand}}(W_{d,r}, \varepsilon) \leq \varepsilon^{-2}$$

and strong polynomial tractability with the exponent $p = 2$.

It is worthwhile to mention that for fixed d the exponent at $1/\varepsilon$ can be slightly improved. Indeed, we have the following result.

Theorem 7 (see [1] [18] [7]). *There are $a'_{d,r}$ and $A'_{d,r}$ such that*

$$a'_{d,r}\, \varepsilon^{-2/(1+2r/d)} \leq \mathrm{comp}^{\mathrm{rand}}(W_{d,r}, \varepsilon) \leq A'_{d,r}\, \varepsilon^{-2/(1+2r/d)}, \qquad 0 < \varepsilon < 1.$$

We add that the bound of Theorem 7 is achieved by a variant of variance reduction given as

$$\mathrm{MC}^*_{d,n}(f) = \mathrm{Int}(A_{d,n}f) + \mathrm{MC}_{d,n}(f - A_{d,n}f)$$

where $A_{d,n}f$ is the d-tensor product of piecewise polynomial interpolation of order r for univariate functions.

Observe that Theorem 7 implies that $p = 2$ is the best exponent of the strong polynomial tractability.

Remark 5. Since the methods $\mathrm{MC}_{d,n}$ and $\mathrm{MC}^*_{d,n}$ use nonadaptive information and linear algorithms, adaption does not help for integration in the randomized setting with respect to the class $W_{d,r}$. It is however not known whether Theorems 1 and 2 hold true in general in the randomized setting for convex and symmetric classes F_0.

5.3   Quasi-Monte Carlo

Quasi-Monte Carlo methods are deterministic counterparts of the (randomized) Monte Carlo $\mathrm{MC}_{d,n}$. These are methods of the form

$$\mathrm{QMC}_n(f) = \frac{1}{n} \sum_{i=1}^{n} f(t_i),$$

where the points t_i are chosen *deterministically*. We assume that f belongs to the space F of functions $f : [0,1]^d \to \mathbb{R}$ that are once continuously differentiable with respect to each variable, and $f(x_1, \ldots, x_d) = 0$ whenever at least one x_i is 1. The norm in F is given as

$$\|f\|^*_d = \|D^{(1,\ldots,1)}f\|_\infty = \sup_{x \in [0,1]^d} \left| \frac{\partial^d f}{\partial x_1 \cdots \partial x_d}(x) \right|.$$

Let $F_0 = W^*_d$ be the unit ball of F.

The error of $\mathrm{QMC}_{d,n}$ in the class W^*_d is closely related to the notion of discrepancy. This is a quantitative measure of the uniformity of the distribution of points in the unit cube. Formally, the *(star) discrepancy* of a point set $\{t_i\}_{i=1}^n \subset [0,1]^d$ is defined as

$$\mathrm{disc}^*(t_1, t_2, \ldots, t_n) = \sup_{x \in [0,1]^d} \left| \mathrm{Vol}([0,x)) - \frac{1}{n} \sum_{i=1}^{n} \mathbf{1}_{[0,x)}(t_i) \right|.$$

Here $[0,x) = [0, x_1) \times \cdots \times [0, x_d)$ is the d-dimensional rectangle anchored at 0.

From the well known Hlavka and Zaremba's identity [9] [40] it follows that

$$| \operatorname{Int}_d(f) - \operatorname{QMC}_{d,n}(f) | \leq \operatorname{disc}^\star(t_1, \ldots, t_n) \, \|f\|_d^\star, \qquad \forall f \in W_d^\star.$$

This implies that the worst case error of $\operatorname{QMC}_{d,n}$ in the class $W_d^\star$ is completely determined by the discrepancy of the points used,

$$e^{\operatorname{wor}}(\operatorname{QMC}_{d,n}) = \operatorname{disc}^\star(t_1, t_2, \ldots, t_n).$$

How to select the points $\{t_i\}_{i=1}^n$ to minimize the discrepancy is a well known open problem. The best known bound

$$\operatorname{disc}^\star(t_1, t_2, \ldots, t_n) \leq c_d \frac{(\ln n)^{d-1}}{n}$$

is provided, e.g., by Hammersley points or digital nets, see the monographs [17] [4]. It had long been believed that the presence of the factor $(\ln n)^{d-1}$ in the error formula of $\operatorname{disc}^\star$ makes the quasi-Monte Carlo methods applicable only for small dimensions d. However, numerical experiments [24] [23] done in the mid-1990s for some finance problems showed applicability of quasi-Monte Carlo even for $d = 360$. Since then this method has been successfully used to solve many other high dimensional problems. A couple of years later the polynomial tractability of quasi-Monte Carlo was shown using rigorous arguments.

Theorem 8 (see [11]). *There are points $\{t_i^*\}_{i=1}^n$ such that*

$$\operatorname{disc}^\star(t_1^*, t_2^*, \ldots, t_n^*) \leq C \, d^{1/2} \, n^{-1/2},$$

for some absolute constant C.

The proof uses, in particular, deep results from the theory of empirical processes, and is *not* constructive. Taking the inverse of the star discrepancy we conclude that the integration problem is in the class $W_d^\star$ polynomially tractable,

$$\operatorname{comp}^{\operatorname{wor}}(W_d^\star, \varepsilon) \leq C^2 \, d \, \varepsilon^{-2}.$$

We add that there is also known a lower bound $\operatorname{comp}^{\operatorname{wor}}(W_d^\star, \varepsilon) \geq c \, d \, \varepsilon^{-1}$ showing that we do not have strong polynomial tractability, see [8].

Acknowledgments. This research was supported by the National Science Centre, Poland, based on the decision DEC-2013/09/B/ST1/04275.

References

1. Bakhvalov, N.S.: On the approximate computation of multiple integrals. Vestnik MGU 4, 3–18 (1959) (in Russian)
2. Bakhvalov, N.S.: On the optimality of linear methods for operator approximation in convex classes. Comput. Math. Math. Phys. 11, 244–249 (1971)

3. Davis, P.J., Rabinowitz, P.: Methods of Numerical Integration, 2nd edn. Academic Press, Orlando (1984)
4. Dick, J., Pillichshammer, F.: Digital Nets and Sequences: Discrepancy Theory and Quasi-Monte Carlo Integration. Cambridge University Press, Cambridge (2010)
5. Donoho, D.L.: Statistical estimation and optimal recovery. Annals of Statistics 22, 238–270 (1994)
6. Gander, W., Gautschi, W.: Adaptive quadrature - revisited. BIT 40, 84–101 (2000)
7. Heinrich, S.: Random approximation in numerical analysis. In: Berstadt, et al. (eds.) Proc. of the Functional Analysis Conf., Essen 1991, pp. 123–171. Marcel Dekker, New York (1993)
8. Hinrichs, A.: Covering numbers, Vapnik-Cervonenkis classes and bounds for the star discrepancy. J. Complexity 20, 477–483 (2004)
9. Hlavka, E.: Über die Diskrepanz mehrdimensionaler Folgen mod 1. Math. Z. 77, 273–284
10. Hinrichs, A., Novak, E., Ullrich, M., Woźniakowski, H.: The curse of dimensionality for numerical integration of smooth functions. Math. Comp., http://dx.doi.org/10.1090/S0025-5718-2014-02855-X
11. Heinrich, S., Novak, E., Wasilkowski, G.W., Woźniakowski, H.: The inverse of the star discrepancy depends linearly on the dimension. Acta Arith 96, 279–302 (2001)
12. McKeeman, W.M.: Algorithm 145: Adaptive numerical integration by Simpson's rule. Commun. ACM 5, 604 (1962)
13. Kowalski, M., Sikorski, K., Stenger, F.: Selected Topics in Approximation and Computation. Oxford University Press, New York (1995)
14. Lyness, J.N.: Notes on the adaptive Simpson quadrature routine. J. Assoc. Comput. Mach. 16, 483–495 (1969)
15. Malcolm, M.A., Simpson, R.B.: Local versus global strategies for adaptive quadrature. ACM Trans. Math. Software 1, 129–146 (1975)
16. Metropolis, N., Ulam, S.: The Monte Carlo method. J. Amer. Statist. Assoc. 44, 335–341 (1949)
17. Niederreiter, H.: Random Number Generation and Quasi-Monte Carlo Methods. In: CBMS-NSF Regional Conf. Ser. in Appl. Math., vol. 63, SIAM, Philadelphia (1994)
18. Novak, E.: Deterministic and Stochastic Error Bounds in Numerical Analysis. Lecture Notes in Math., vol. 1349. Springer, Berlin (1988)
19. Novak, E.: On the power of adaption. J. Complexity 12, 199–237 (1996)
20. Novak, E., Woźniakowski, H.: Tractability of Multivariate Problems. Volume I: Linear Information. EMS Tracts in Math. 6 (2008)
21. Novak, E., Woźniakowski, H.: Tractability of Multivariate Problems. Volume II: Standard Information for Functionals. EMS Tracts in Math. 12 (2010)
22. Novak, E., Woźniakowski, H.: Tractability of Multivariate Problems. Volume III: Standard Information for Operators. EMS Tracts in Math. 18 (2012)
23. Papageorgiou, A.F., Traub, J.F.: Faster evaluation of multidimensional integrals. Comp. Phys. 11, 574–578 (1997)
24. Paskov, S., Traub, J.F.: Faster valuation of financial derivatives. J. Portfolio Management 22, 113–120 (1995)
25. Plaskota, L.: Noisy Information and Computational Complexity. Cambridge University Press, Cambridge (1996)
26. Plaskota, L.: Automatic integration using asymptotically optimal adaptive Simpson quadrature (submitted)
27. Plaskota, L., Wasilkowski, G.W.: Adaption allows efficient integration of functions with unknown singularities. Numerische Math. 102, 123–144 (2005)

28. Plaskota, L., Wasilkowski, G.W.: Uniform approximation of piecewise r-smooth and globally continuous functions. SIAM J. Numer. Analysis 47, 762–785 (2009)
29. Plaskota, L., Wasilkowski, G.W., Zhao, Y.: The power of adaption for approximating functions with singularities. Math. Comp. 77, 2309–2338 (2008)
30. Ritter, K.: Average Case Analysis of Numerical Problems. Lecture Notes in Math., vol. 1733. Springer, Berlin (2000)
31. Smolyak, S.A.: On optimal recovery of functions and functionals of them, PhD thesis, Moscow State Univ. (1965) (in Russian)
32. Sobolev, S.L., Vaskevich, V.L.: The Theory of Cubature Formulas. Kluwer Academic Publishers, Dordrecht (1997)
33. Sukharev, A.G.: On the existence of optimal affine methods for approximating linear functionals. J. Complexity 2, 317–322 (1986)
34. Traub, J.F., Wasilkowski, G.W., Woźniakowski, H.: Information, Uncertainty, Complexity. Addison-Wesley, Reading (1983)
35. Traub, J.F., Wasilkowski, G.W., Woźniakowski, H.: Information-Based Complexity. Academic Press, New York (1988)
36. Traub, J.F., Woźniakowski, H.: A General Theory of Optimal Algorithms. Academic Press, New York (1980)
37. Wasilkowski, G.W.: Information of varying cardinality. J. Complexity 2, 204–228 (1986)
38. Werschulz, A.G.: The Computational Complexity of Differential and Integral Equations: an Information-Based Approach. Oxford University Press, New York (1991)
39. Woźniakowski, H.: Tractability and strong tractability of linear multivariate problems. J. Complexity 10, 96–128 (1994)
40. Zaremba, K.S.: Some applications of multidimensional integration by parts. Ann. Polon. Math. 21, 85–96 (1968)

On Integrability of Evolutionary Equations in the Restricted Three-Body Problem with Variable Masses

Alexander N. Prokopenya[1,2], Mukhtar Zh. Minglibayev[3,4], and Baglan A. Beketauov[3]

[1] Warsaw University of Life Sciences – SGGW
Nowoursynowska str. 159, 02-776 Warsaw, Poland
alexander_prokopenya@sggw.pl
[2] Collegium Mazovia Innovative Higher School
Sokolowska str. 161, 08-110 Siedlce, Poland
[3] Al-Farabi Kazakh National University
al-Farabi ave. 71, Almaty, 050038 Kazakhstan
minglibayev@mail.ru
[4] Fessenkov Astrophysical Institute
Observatoriya 23, Almaty, 050020 Kazakhstan
Beketauov_Baglan@mail.ru

Abstract. The satellite version of the restricted three-body problem formulated on the basis of classical Gylden–Meshcherskii problem is considered. Motion of the point P_2 of infinitesimal mass about the point P_0 is described in the first approximation in terms of the osculating elements of the aperiodic quasi-conical motion, and an influence of the point P_1 gravity on this motion is analyzed. Long-term evolution of the orbital elements is determined by the differential equations written in the Hill approximation and averaged over the mean anomalies of points P_1 and P_2. Integrability of the evolutionary equations is analyzed, and the laws of mass variation have been found for which the evolutionary equations are integrable. All relevant symbolic calculations and visualizations are done with the computer algebra system Mathematica.

1 Introduction

The restricted three-body problem is a well-known model of celestial mechanics, having a lot of applications (see [9]). In the simplest case, it is assumed that two massive points P_0, P_1 move in the Keplerian orbits about their common center of mass, while the third point P_2 of negligible mass does not influence on their motion and moves in the gravitational field generated by P_0, P_1. This problem is not integrable, and so the perturbation theory is usually applied to the analysis of the point P_2 motion, and quite cumbersome symbolic calculations are involved. As a general solution of the two-body problem is known, one can consider in the first approximation that the point P_2 moves around the point P_0, for example, as a satellite and its Keplerian orbit is disturbed by the gravity of point P_1. Such a model has been used successfully in the study of satellite motion in the

V.P. Gerdt et al. (Eds.): CASC Workshop 2014, LNCS 8660, pp. 373–387, 2014.
© Springer International Publishing Switzerland 2014

system Earth–Moon or Sun–planet [4,5]. It was shown that doubly averaged equations of motion determining the evolution of satellite orbit may become integrable. The corresponding general solution may be found in analytic form, and it enables investigation of main qualitative features of the orbit parameters (see, for example, [10]).

If masses of points P_0 and P_1 vary with time as it takes place in case of a binary star, losing the mass due to the corpuscular and photon radiation, the problem becomes much more complicated because a general solution of the corresponding two-body problem cannot be found in an analytical form (see [2,1,6]). Actually, using the relative coordinate system with origin at point P_0, one can write equation of the point P_1 motion in the form

$$\frac{d^2\boldsymbol{R}_1}{dt^2} = -G(m_0(t) + m_1(t))\frac{\boldsymbol{R}_1}{R_1^3} , \tag{1}$$

where $\boldsymbol{R}_1$ is a radius-vector of point P_1, $R_1 = |\boldsymbol{R}_1|$, and G is the constant of gravitation. Equation (1) is known as the classical Gylden–Meshcherskii problem (see [2]), and its general solution can be found in symbolic form only for special cases. In the present paper, we assume that the masses $m_0(t)$ and $m_1(t)$ of points P_0 and P_1, respectively, vary isotropically with different rates, but their total mass reduces according to the joint Meshcherskii law

$$\frac{m_{00} + m_{10}}{m_0(t) + m_1(t)} = \sqrt{At^2 + 2Bt + C} \equiv v(t), \tag{2}$$

where $m_{00} = m_0(t_0)$, $m_{10} = m_1(t_0)$, t_0 is an initial instant of time, and parameters A, B, C are chosen in a way to satisfy the condition $v(t_0) = 1$ and $v(t)$ to be an increasing function for $t > t_0$. Then equation (1) is reduced to ordinary equation of Keplerian motion for constant masses by means of variables transformation (see [2])

$$\boldsymbol{R}_1(t) = v(t)\boldsymbol{R}(\tau) , \quad \frac{dt}{v^2(t)} = d\tau , \tag{3}$$

where $\boldsymbol{R}(\tau) = (X, Y, Z)$ is a new radius-vector, and τ is a new independent variable (new "time"). A uniform motion in a circle of radius a_1 situated in the coordinate plane XOY is a particular case of Keplerian motion and is given by

$$X(\tau) = a_1 \cos M_1(\tau), \quad Y(\tau) = a_1 \sin M_1(\tau), \quad Z(\tau) = 0, \tag{4}$$

where $M_1(\tau) = \omega_1\tau$, and angular velocity ω_1 is

$$\omega_1 = \left(AC - B^2 + \frac{K}{a_1^3}\right)^{1/2} , \quad K = G(m_{00} + m_{10}) .$$

Assuming that motion of point P_1 is determined by equations (3)-(4), we consider here the satellite version of the restricted three-body problem when the point P_2 moves around point P_0, being perturbed by the gravity of point P_1.

We use the Hill approximation [3], when a distance between points P_0 and P_1 is considered to be much greater than distance between P_0 and P_2. The main purpose of this paper is to find a class of functions $m_0(t)$, $m_1(t)$, for which the evolutionary equations, describing the secular perturbations of point P_2 trajectory, become integrable, and to obtain the corresponding solutions in analytic form. The relevant cumbersome symbolic calculation and visualization of the results are done with the computer algebra system Mathematica [11].

The paper is organized as follows. In section 2, we obtain the equations of point P_2 motion in the framework of the Hill approximation, considering an aperiodic motion on quasi-conical section as the unperturbed motion. Doubly averaging the equations of motion, we obtain the differential equations determining the long-term evolution of the orbital parameters. Then we look in section 3 for the solutions of the evolutionary equations in analytic form and analyze the conditions under which such solutions exist and describe a quasi-elliptic motion of point P_2. Finally, in section 4 we determine the mass variation laws for which a general solution of the evolutionary equations can be found in analytical form. And we conclude in section 5.

2 Evolutionary Equations

Assume that position of point P_2 in the relative coordinate system with origin at point P_0 is given by the radius-vector $\boldsymbol{R}_2$. Then equations of its motion are given by (see [6])

$$\frac{d^2 \boldsymbol{R}_2}{dt^2} = -Gm_0(t)\frac{\boldsymbol{R}_2}{R_2^3} - Gm_1(t)\frac{\boldsymbol{R}_1}{R_1^3} + Gm_1(t)\frac{\boldsymbol{R}_1 - \boldsymbol{R}_2}{R_{12}^3} \ , \tag{5}$$

where $R_2 = |\boldsymbol{R}_2|$, $R_{12} = |\boldsymbol{R}_1 - \boldsymbol{R}_2|$. Applying the scale transformation of spatial coordinates and time defined in (3), we reduce equation (5) to the form

$$\frac{d^2 \boldsymbol{r}}{d\tau^2} = -(AC - B^2)\boldsymbol{r} - Gm_0(t)v(t)\frac{\boldsymbol{r}}{r^3} + Gm_1(t)v(t)\left(\frac{\boldsymbol{R} - \boldsymbol{r}}{\varDelta_{12}^3} - \frac{\boldsymbol{R}}{R^3}\right) \ , \tag{6}$$

where $\boldsymbol{r}(\tau) = \boldsymbol{R}_2(t)/v(t)$ is a new radius-vector of point P_2, and $\varDelta_{12} = |\boldsymbol{R} - \boldsymbol{r}|$.

Note that masses $m_0(t)$ and $m_1(t)$ in (6) are arbitrary non-increasing functions satisfying the condition (2). It is convenient to represent them in the form

$$m_j(t) = \frac{m_{j0}}{v(t)\gamma_j(\tau)} \ , \quad (j = 0, 1) \ , \tag{7}$$

where the functions $\gamma_j(\tau)$ are constrained by the condition

$$\frac{m_{00}}{\gamma_0(\tau)} + \frac{m_{10}}{\gamma_1(\tau)} = m_{00} + m_{10} \ , \tag{8}$$

that follows from (2). Then equation (6) takes the form

$$\frac{d^2 \boldsymbol{r}}{d\tau^2} = -(AC - B^2)\boldsymbol{r} - \frac{Gm_{00}}{\gamma_0(\tau)}\frac{\boldsymbol{r}}{r^3} + \frac{Gm_{10}}{\gamma_1(\tau)}\left(\frac{\boldsymbol{R} - \boldsymbol{r}}{\varDelta_{12}^3} - \frac{\boldsymbol{R}}{R^3}\right) \ . \tag{9}$$

In case of $\gamma_0 = 1$, when each of the masses m_0 and m_1 decreases with time according to the joint Meshcherskii law (see (2), (7)–(8)), equation (9) reduces to the restricted three-body problem with constant masses. Note that appearance of a linear term $(AC - B^2)\boldsymbol{r}$ in the right-hand side of (9) does not destroy its integrability for $m_1 = 0$, although it can be integrated only in quadratures. So it is convenient to analyse the corresponding evolutionary equations under an assumption that in the first approximation, point P_2 moves around point P_0 on Keplerian orbit but its orbital parameters are disturbed by the gravity of point P_1 and by additional force being a linear function of $\boldsymbol{r}$. One can show that the differential equations determining evolution of the orbital parameters can then be integrated in analytic form.

To analyse a general case and to find other functions $\gamma_0(\tau)$ for which the evolutionary equations are integrable, one can apply similar approach, exploiting integrability of the differential equation

$$\frac{d^2 \boldsymbol{r}}{d\tau^2} = \frac{\ddot{\gamma}_0}{\gamma_0}\boldsymbol{r} - \frac{Gm_{00}}{\gamma_0(\tau)}\frac{\boldsymbol{r}}{r^3} , \tag{10}$$

where $\ddot{\gamma}_0 \equiv d^2\gamma_0/d\tau^2$. Note that $\gamma_0(\tau)$ in (10) is an arbitrary twice continuously differentiable function and this equation determines an aperiodic motion of a point on quasi-conical section (see [6,8]). The corresponding solution $\boldsymbol{r} = (x, y, z)$ can be represented in the form

$$\begin{aligned}
x = {} & \gamma_0 a((\cos E - e)(\cos\omega\cos\Omega - \sin\omega\sin\Omega\cos i) - \\
& -\sqrt{1 - e^2}(\sin\omega\cos\Omega\sin E + \cos\omega\sin\Omega\sin E\cos i)) , \\
y = {} & \gamma_0 a((\cos E - e)(\cos\omega\sin\Omega + \sin\omega\cos\Omega\cos i) - \\
& -\sqrt{1 - e^2}(\sin\omega\sin\Omega\sin E - \cos\omega\cos\Omega\sin E\cos i)) , \\
z = {} & \gamma_0 a((\cos E - e)\sin\omega\sin i + \sqrt{1 - e^2}\cos\omega\sin E\sin i) ,
\end{aligned} \tag{11}$$

where the constants a, e, i, Ω, ω are the analogues of orbital elements known from the classical two-body problem with constant masses (see, for example, [7]), and the eccentric anomaly E is determined by the equation

$$E - e\sin E = M = \frac{\sqrt{K_0}}{a^{3/2}}(\Phi(\tau) - \Phi(\tau_0)) . \tag{12}$$

The function $\Phi(\tau)$ in (12) is given by

$$\Phi(\tau) = \int_{\tau_0}^{\tau} \frac{d\tau}{\gamma_0^2(\tau)}, \quad K_0 = Gm_{00} .$$

In case of $\gamma_0 = 1$ equation (10) reduces to a pure Keplerian problem with constant masses when the variable M becomes a linear function of time known as the mean anomaly and the constant τ_0 is the time of perihelion passage (see [6,7]). Note that orbital parameters a, e, i, Ω, ω, and τ_0 are determined from the initial conditions of motion, and expressions (11), (12) determine an exact solution of the two-body problem (10) for any given function $\gamma_0(\tau)$ satisfying the conditions above.

As equation (9) does not contain a linear term being proportional to the second derivative of the function γ_0, one can add and subtract the corresponding term and rewrite the equation in the form

$$\frac{d^2 r}{d\tau^2} = \frac{\ddot{\gamma}_0}{\gamma_0} r - \frac{Gm_{00}}{\gamma_0(\tau)} \frac{r}{r^3} - \left[(AC - B^2) r + \frac{\ddot{\gamma}_0}{\gamma_0} r - \frac{Gm_{10}}{\gamma_1(\tau)} \left(\frac{R - r}{\Delta_{12}^3} - \frac{R}{R^3} \right) \right] . \quad (13)$$

Then its solution can be sought in the form (11) under the condition that the orbital parameters are functions of time. Such approach is known as a method of variation of constants and is widely used in the theory of differential equations.

To derive the differential equations determining the time evolution of orbital parameters in the simplest form, it is convenient to rewrite equation (13) in the Hamiltonian form and to change to the special set of canonical variables known as Delaunay's variables (see [6,8]). Three pairs of the corresponding canonical conjugate coordinates and momenta (l, L), (g, G) and (h, H) are related to the analogues of the Keplerian orbital elements as

$$l = M, \ L = \sqrt{K_0 a}, \ g = \omega, \ G = L\sqrt{1 - e^2}, \ h = \Omega, \ H = G \cos i . \quad (14)$$

The Hamiltonian function in the Delaunay variables may be written in the form

$$\mathcal{H} = -\frac{K_0^2}{2\gamma_0^2 L^2} + \frac{1}{2} \left(AC - B^2 + \frac{\ddot{\gamma}_0}{\gamma_0} \right) (x^2 + y^2 + z^2) - V, \quad (15)$$

where the function V is given by

$$V = \frac{K_1}{\gamma_1(\tau)} \left(\frac{1}{\Delta_{12}} - \frac{xX + yY + zZ}{R^3} \right) , \ K_1 = Gm_{10} ,$$

and components of vectors R and r are given by (4) and (11), respectively.

Assuming further the ratio of the distances r and R to be small ($r/R \ll 1$), one can expand the function V into a power series in terms of r/R and keep only the main term of the expansion in the Hamiltonian (15). It means that we consider the problem in the Hill approximation [3]. Then the Hamiltonian takes the form

$$\mathcal{H} = -\frac{K_0^2}{2\gamma_0^2 L^2} + \frac{1}{2} \left(AC - B^2 + \frac{\ddot{\gamma}_0}{\gamma_0} + \frac{K_1}{\gamma_1 a_1^3} \right) (x^2 + y^2 + z^2) -$$
$$- \frac{3K_1}{2\gamma_1 a_1^3} \left(x^2 \cos^2 M_1 + y^2 \sin^2 M_1 + xy \sin(2M_1) \right) . \quad (16)$$

As we are interested in the secular evolution of the point P_2 orbit under an influence of massive point P_1, one may disregard the short-period perturbations of orbital elements by means of averaging of the Hamiltonian (16) over the mean anomalies of points P_1 and P_2 (see [5]). The averaged Hamiltonian is determined as (see (12))

$$\bar{\mathcal{H}} = \frac{1}{4\pi^2} \int_0^{2\pi} \int_0^{2\pi} \mathcal{H} dM dM_1 = \frac{1}{4\pi^2} \int_0^{2\pi} \int_0^{2\pi} \mathcal{H}(1 - e \cos E) dE dM_1,$$

and is given by

$$\bar{\mathcal{H}} = -\frac{K_0^2}{2\gamma_0^2 L^2} + \frac{\gamma_0^2 L^4}{4K_0^2}\left(AC - B^2 + \frac{\ddot{\gamma}_0}{\gamma_0} + \frac{K_1}{\gamma_1 a_1^3}\right)\left(5 - \frac{3G^2}{L^2}\right) - \qquad (17)$$

$$-\frac{3K_1\gamma_0^2 L^4}{16\gamma_1 a_1^3 K_0^2}\left(2 + \frac{2H^2}{G^2} + \left(1 - \frac{G^2}{L^2}\right)\left(3 + \frac{3H^2}{G^2} + 5\cos(2g)\left(1 - \frac{H^2}{G^2}\right)\right)\right) ,$$

where the relationships (14) have been taken into account.

Obviously, the averaged Hamiltonian (17) does not depend on the mean anomaly $M \equiv l$ and so its canonical conjugate variable L is constant. The first term in the right-hand side of expression (17), depending only on L, influences on the time evolution of mean anomaly M but doesn't influence on other orbital parameters of point P_2. Therefore, if the rest terms of the Hamiltonian (17) contained the same multiplier γ_0^2/γ_1, depending on time, it would be possible to reduce the differential equations, determining the secular evolution of orbital parameters g, h, G, H, to the autonomous case by means of the scale transformation of time. We shall show later that such autonomous differential equations may be integrated. So let us consider a class of functions $\gamma_0(\tau)$ satisfying the condition

$$\frac{\ddot{\gamma}_0}{\gamma_0} + AC - B^2 = -\alpha\frac{K_1}{\gamma_1 a_1^3} , \qquad (18)$$

where α is a parameter. Then the Hamiltonian (17) can be rewritten as

$$\bar{\mathcal{H}} = -\frac{K_0^2}{2\gamma_0^2 L^2} + \frac{K_1\gamma_0^2 L^4}{4\gamma_1 a_1^3 K_0^2}\left[(1-\alpha)\left(5 - \frac{3G^2}{L^2}\right) - \frac{3}{2}\left(1 + \frac{H^2}{G^2}\right) - \right.$$

$$\left. - \frac{3}{4}\left(1 - \frac{G^2}{L^2}\right)\left(3 + \frac{3H^2}{G^2} + 5\cos(2g)\left(1 - \frac{H^2}{G^2}\right)\right)\right] . \qquad (19)$$

Differential equations for orbital parameters g, h, G, H are obtained in the standard Hamiltonian form as

$$\frac{dg}{d\tau} = \frac{\partial\bar{\mathcal{H}}}{\partial G} , \quad \frac{dG}{d\tau} = -\frac{\partial\bar{\mathcal{H}}}{\partial g} , \quad \frac{dh}{d\tau} = \frac{\partial\bar{\mathcal{H}}}{\partial H} , \quad \frac{dH}{d\tau} = -\frac{\partial\bar{\mathcal{H}}}{\partial h} . \qquad (20)$$

Substituting expression (19) into (20) and taking into account (14), after quite standard symbolic calculations we obtain differential equations determining the secular evolution of the orbital parameters in the form

$$\frac{dz}{dn} = 20z\sqrt{1-z}\sin^2 i\sin(2\omega) , \qquad (21)$$

$$\frac{di}{dn} = -\frac{10z}{\sqrt{1-z}}\sin i\cos i\sin(2\omega) , \qquad (22)$$

$$\frac{d\omega}{dn} = \frac{4}{\sqrt{1-z}} \left(5\cos^2 i \sin^2 \omega + (1-z)(2\alpha + 2 - 5\sin^2 \omega) \right) , \tag{23}$$

$$\frac{d\Omega}{dn} = -\frac{4\cos i}{\sqrt{1-z}} \left(1 - z + 5z\sin^2 \omega \right) , \tag{24}$$

where $z = e^2$, and n is a new dimensionless independent variable determined by the equation

$$dn = \frac{3K_1\gamma_0^2(\tau)a^2}{16\gamma_1(\tau)a_1^3\sqrt{K_0 a}} \, d\tau . \tag{25}$$

Note that the system of differential equations (21)-(24) looks similarly to the corresponding equations describing evolution of satellites of Uranus (see [10]). But due to dependence of the points masses on time equation (23) contains additional term $2\alpha(1-z)$ in the parentheses in the right-hand side and additional parameter α. Therefore, the system behaviour and its analysis should be more complicated, although it can be investigated in a similar way as in [10].

3   Integration of the Evolutionary Equations

Direct symbolic calculation shows that the system of three equations (21)-(23) has two independent integrals of motion

$$(1 - z) \cos^2 i = c_1 = const, \tag{26}$$

$$z\left(\frac{2}{5}N - \sin^2 i \sin^2 \omega \right) = c_2 = const, \tag{27}$$

where $N = 1+\alpha$ is a new parameter. This enables us to eliminate two variables in the system (21)-(23) and to reduce it to an ordinary differential equation with respect to the function $z(n)$ that can be integrated. As determination of the function $\Omega(n)$ reduces then to simple integrating the right-hand side of equation (24) we focus here on analysis of system (21), (26), (27) and will discuss solving the equations (22), (23) only if the corresponding solutions cannot be obtained from the integrals (26), (27).

Note that in case of quasi-elliptic motion of point P_2 eccentricity of its orbit should be less than 1 or $0 \le z < 1$. Hence, the first integral c_1 must belong to the interval $0 \le c_1 \le 1$. Consequently, for given c_1, expression (26) restricts possible values of z to the interval $0 \le z \le 1 - c_1$. Eliminating the variable i in the system (26)-(27), we obtain

$$\sin^2 \omega = \frac{(1 - z)(2Nz - 5c_2)}{5z(1 - z - c_1)} . \tag{28}$$

Then the condition $0 \le \sin^2 \omega \le 1$ gives two inequalities

$$2Nz - 5c_2 \ge 0 , \quad (5 - 2N)z^2 - z(5 - 2N - 5c_1 - 5c_2) - 5c_2 \le 0 . \tag{29}$$

Applying the Mathematica built-in function *Reduce* to the system (29) combined with inequalities $0 \leq z \leq 1 - c_1$, $0 \leq c_1 \leq 1$ and separating the results with the function *LogicalExpand*, one can get a long list of different solutions, determining possible values of the integral c_2 and the variable z, corresponding to quasi-elliptic motion of point P_2. Depending on the value of parameter N, one can separate three different cases which are considered below.

3.1   Case $N = \frac{5}{2}$

In this case, the system (29) reduces to the following inequalities

$$0 \leq c_2 \leq z \leq \frac{c_2}{c_1 + c_2} \ , \quad c_1 + c_2 \leq 1 \ , \quad 0 \leq c_1 \leq 1 \ . \tag{30}$$

Therefore, the domain of possible values of the integrals c_1, c_2 in the plane Oc_1c_2 is a triangle bounded by the lines $c_1 = 0$, $c_2 = 0$, $c_1 + c_2 = 1$.

Using expressions (26), (28), we eliminate the variables i and ω in (21) and obtain the following differential equation

$$\frac{dz}{dn} = 40 \, \mathrm{sgn}(\sin(2\omega_0)) \sqrt{(z - c_2)(c_2 - z(c_1 + c_2))} \ , \tag{31}$$

where the function $\mathrm{sgn}(x)$ determines a sign of $\sin(2\omega_0)$ at the initial instant of time $(\omega_0 = \omega(t_0))$. This equation is easily integrated, and its solution is given by

$$z = c_2 + \frac{c_2(1 - c_1 - c_2)}{c_1 + c_2} \sin^2 \left(20 \, \mathrm{sgn}(\sin(2\omega_0)) \sqrt{c_1 + c_2} \, n + \varphi_0 \right) \ , \tag{32}$$

where

$$\varphi_0 = \arcsin \sqrt{\frac{(z_0 - c_2)(c_1 + c_2)}{c_2(1 - c_1 - c_2)}} \ , \quad z_0 = z(0) \ .$$

Expression (32) shows that $z(n)$ is an oscillating function, and its values belong to the interval (30). The function $i(n) = \arccos \sqrt{c_1/(1 - z)}$ also oscillates, and an interval of its values is determined by inequality

$$\frac{c_1}{1 - c_2} \leq \cos^2 i \leq c_1 + c_2 \ .$$

One can readily check that the function $\omega(n)$ increases with time because its derivative (see (23)) is positive under the conditions (30), while its values are determined by the expression (28).

One should note that in case of $c_2 = 0$ there exists additional stationary solution of equation (21) that cannot be obtained as a limit case of (32). Actually, an equality $c_2 = 0$ takes place either in case of $z = 0$ or in case of $\sin^2 i = 1$ and $\sin^2 \omega = 1$ (see (27)). The second case implies $c_1 = 0$, and the corresponding solution is given by

$$0 \leq z = const < 1 \ , \quad i = \frac{\pi}{2} \ , \quad \omega = \frac{\pi}{2} \quad \text{or} \quad \omega = \frac{3\pi}{2} \ . \tag{33}$$

Solution (33) describes motion of point P_2 on elliptic orbit in a plane that is perpendicular to the orbital plane of point P_1. Note that in case of constant masses, such motion always results in collision of points P_0 and P_2 (see [4]).

3.2 Case $N > \frac{5}{2}$

Analysis of inequalities (29) shows that the domain of possible values of the integrals c_1 and c_2 in the plane Oc_1c_2 is a triangle determined by inequalities

$$c_1 \geq 0 \ , \quad c_2 \geq 0 \ , \quad c_1 \leq 1 - \frac{5c_2}{2N} \ . \tag{34}$$

On its boundary $c_2 = 0$, equations (21), (22) have only a stationary solution $z = 0$, $\cos^2 i = c_1$, while equation (23) takes the form

$$\frac{d\omega}{dn} = 4\left(2N - 5(1 - c_1)\sin^2\omega\right) \ ,$$

and is integrated in terms of elementary functions, the result is easily found with the Mathematica built-in function *DSolve*, for example.

On the other border $c_1 = 1 - 5c_2/(2N)$, inequalities (29) can be written in the form

$$z - \frac{5c_2}{2N} \geq 0 \ , \quad 2N - z(2N - 5) \leq 0 \ .$$

One can readily see that inside of the interval $z \in [0,1]$ there is only one point $z = 5c_2/(2N)$ satisfying these inequalities. Therefore, equations (21) and (22) have only a stationary solution $z = 5c_2/(2N) = 1 - c_1$, $\cos^2 i = 1$, and equation (23), taking the form

$$\frac{d\omega}{dn} = \frac{4}{\sqrt{c_1}}\left(2Nc_1 + 5(1 - c_1)\sin^2\omega\right) \ ,$$

is again integrable in terms of elementary functions.

On the third border $c_1 = 0$, when $0 \leq c_2 < 2N/5$ and $\cos^2 i = 0$, inequalities (29) can be written in the form

$$z - \frac{5c_2}{2N} \geq 0 \ , \quad (1 - z)\left(5c_2 - z(2N - 5)\right) \geq 0 \ . \tag{35}$$

Therefore, the variable z belongs to the interval

$$\frac{5c_2}{2N} \leq z < \frac{5c_2}{2N - 5} < 1 \ ,$$

for $0 \leq c_2 < 2N/5 - 1$, and

$$\frac{5c_2}{2N} \leq z \leq 1 \ ,$$

for $2N/5 - 1 \leq c_2 < 2N/5$. In this case, solution of equation (21) becomes more complicated, although it may be integrated in terms of elliptic functions. As the method applied is similar to the case when values of the integrals c_1, c_2 belong to the domain inside of the triangle (34) in the plane Oc_1c_2, let us consider such general case.

Using expressions (26), (28) and eliminating the variables i and ω in equation (21), one can rewrite it in the form

$$\frac{dz}{dn} = 8\mathrm{sgn}(\sin(2\omega_0))\sqrt{Q(z)} \,, \tag{36}$$

where the third-degree polynomial $Q(z)$ is given by

$$Q(z) = (2Nz - 5c_2)(5c_2 + z(5 - 2N - 5c_1 - 5c_2) - z^2(5 - 2N)) \,, \tag{37}$$

and it is assumed that the variable z takes only such values, for which $Q(z) \geq 0$. Solving the equation $Q(z) = 0$, we obtain in general three different roots

$$z_{1,2} = \frac{1}{2}\left[1 + \frac{5(c_1 + c_2)}{2N - 5} \pm \left(\left(1 + \frac{5(c_1 + c_2)}{2N - 5} \right)^2 - \frac{20c_2}{2N - 5} \right)^{1/2} \right] ,$$

$$z_3 = \frac{5c_2}{2N} \,. \tag{38}$$

Analysis of expressions (37), (38) shows that inside of the domain (34) two roots z_2, z_3 of polynomial $Q(z)$ belong to the interval $[0, 1]$, and $Q(z) \geq 0$ if $z_3 \leq z \leq z_2 < 1$, while the third root $z_1 \geq 1$. Then $Q(z)$ may be represented in the form

$$Q(z) = 2N(2N - 5)(z_1 - z)(z_2 - z)(z - z_3)$$

and equation (36) may be integrated in the elliptic quadrature. Its solution is

$$8 \, \mathrm{sgn}(\sin(2\omega_0))\sqrt{2N(2N - 5)} \, n = \int_{z_0}^{z} \frac{dz}{\sqrt{(z_1 - z)(z_2 - z)(z - z_3)}} \,. \tag{39}$$

An integral in the right-hand side of (39) is calculated in terms of the elliptic functions and the solution may be represented as

$$z(u) = z_3 + (z_2 - z_3)\mathrm{sn}^2 u \,, \tag{40}$$

where

$$u = 4\mathrm{sgn}(\sin(2\omega_0))\sqrt{2N(2N - 5)(z_1 - z_3)}n + u_0 \,, \quad u_0 = F(\varphi_0, \kappa^2) \,,$$

$$\kappa^2 = \frac{z_2 - z_3}{z_1 - z_3} < 1 \,, \quad \sin^2 \varphi_0 = \frac{z_0 - z_3}{z_2 - z_3} \,, \quad z_0 = z(0) \,.$$

Here $\mathrm{sn}u$ and $F(\varphi_0, \kappa^2)$ are the Jacobi elliptic sine and the incomplete elliptic integral of the first kind, respectively.

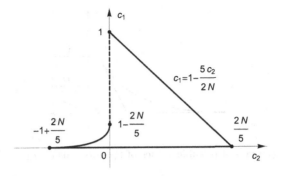

Fig. 1. Domain of possible values of the integrals c_1, c_2 for $0 \leq N < 5/2$

3.3   Case $N < \frac{5}{2}$

In case of $N \geq 0$ possible values of integrals c_1, c_2 must belong to the domain shown in Fig. 1, which are bounded by the lines

$$c_1 = 0 \ , \quad c_2 = 0 \ , \quad c_1 = 1 - \frac{5c_2}{2N} \ , \tag{41}$$

and the curve

$$c_1 = 1 - c_2 - \frac{2N}{5} - 2\sqrt{(-c_2)\left(1 - \frac{2N}{5}\right)} \ . \tag{42}$$

On the line $c_2 = 0$, polynomial $Q(z)$ has three roots, namely, $z_1 = 1 - c_1/(1 - 2N/5)$ and $z_{2,3} = 0$. The root z_1 is negative for $1 - 2N/5 < c_1 \leq 1$ (the corresponding points are shown in Fig. 1 as a dashed bold line), and equation (21) has the only solution $z = 0$. But for $0 \leq c_1 < 1 - 2N/5$, the root z_1 becomes smaller than 1, and the polynomial (37), taking a form

$$Q(z) = 2Nz^2(5 - 2N - 5c_1 - z(5 - 2N)) \ , \tag{43}$$

is non-negative for $0 \leq z \leq z_1$.

Substituting the polynomial (43) into equation (36), one can readily see that the differential equation is integrated in terms of elementary functions and its solution is determined by the equation

$$\ln \frac{\sqrt{a} - \sqrt{a - bz}}{\sqrt{a} + \sqrt{a - bz}} = 8\sqrt{10Na} \ \text{sgn}(\sin(2\omega_0))n + B_0 \ , \tag{44}$$

where

$$B_0 = \ln \frac{\sqrt{a} - \sqrt{a - bz_0}}{\sqrt{a} + \sqrt{a - bz_0}} \ , \quad a = 1 - c_1 - 2N/5 \ , \quad b = 1 - 2N/5 \ .$$

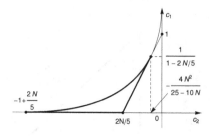

Fig. 2. Domain of possible values of the integrals c_1, c_2 for $N \le 0$

On the line $c_1 = 1 - 5c_2/(2N)$, polynomial $Q(z)$ takes the form

$$Q(z) = (2Nz - 5c_2)^2 \left(z \left(1 - \frac{5}{2N} \right) - 1 \right) ,$$

and has three zeros

$$z_1 = \frac{1}{1 - \frac{5}{2N}}, \quad z_{2,3} = \frac{5c_2}{2N}.$$

For $N \ge 0$ we have $c_2 > 0$ (see Fig. 1), and the root z_1 is negative, so equation (21) has only a stationary solution $z = z_{2,3} = 1 - c_1$. But for $N < 0$, when the line $c_1 = 1 - 5c_2/(2N)$ touches the curve (42) (see Fig. 2) and parameters c_1, c_2 must satisfy the conditions $0 \le c_1 \le 1/(1 - 2N/5)$, $2N/5 \le c_2 \le -4N^2/(25 - 10N)$, we obtain $0 < z_1 \le z_{2,3} \le 1$ and polynomial (37) is non-negative for $z \in [z_1, z_{2,3}]$. Then equation (36) is integrated in terms of elementary functions similar to the previous case (see (43), (44)).

On the curve (42), the polynomial $Q(z)$ takes the form

$$Q(z) = 2N(5 - 2N) \left(z - \frac{5c_2}{2N} \right) \left(z - \sqrt{\frac{(-c_2)}{1 - 2N/5}} \right)^2 ,$$

where we have taken into account that $c_2 < 0$ and $N < 5/2$. The corresponding roots are given by

$$z_{1,2} = \sqrt{\frac{(-c_2)}{1 - 2N/5}} , \quad z_3 = \frac{5c_2}{2N} .$$

One can readily check that for $0 \le N < 5/2$, we have $0 \le z_{1,2} \le 1$, $z_3 < 0$ and, hence, equation (21) has only a stationary solution $z = z_{1,2}$. This solution remains also for $N < 0$ and $-1 + 2N/5 \le c_2 < 2N/5$ when the root z_3 becomes greater than 1. But for $N < 0$ and $2N/5 \le c_2 < -4N^2/(25 - 10N)$ we obtain

$$\frac{1}{1 - 5/(2N)} < z_{1,2} < z_3 \le 1 .$$

Polynomial $Q(z)$ is non-negative for $z \in [z_{1,2}, z_3]$, and equation (21) is integrated in terms of elementary functions similarly to the cases above.

On the last boundary $c_1 = 0$, we have $\cos^2 i = 0$, and polynomial $Q(z)$ taking the form

$$Q(z) = (2Nz - 5c_2)(1 - z)(5c_2 + z(5 - 2N)) \,,$$

has three different roots

$$z_1 = 1, \quad z_2 = \frac{5c_2}{2N - 5}, \quad z_3 = \frac{5c_2}{2N} \,.$$

If $0 \leq N < 5/2$ and $c_2 \geq 0$ then these roots satisfy the inequalities

$$z_2 \leq 0 \leq z_3 \leq z_1 = 1 \,,$$

and the polynomial $Q(z) \geq 0$ for $z \in [z_3, z_1]$. If c_2 becomes negative then we obtain

$$z_3 \leq 0 \leq z_2 \leq z_1 = 1 \,,$$

and $Q(z) \geq 0$ for $z \in [z_2, z_1]$. Finally, for $N < 0$ and $(-1 + 2N/5) \leq c_2 \leq 2N/5$ the corresponding inequalities become

$$0 \leq z_2 \leq z_1 = 1 \leq z_3 \,,$$

and again $Q(z) \geq 0$ for $z \in [z_2, z_1]$. In all three cases one of the roots is outside the interval $[0, 1]$ and two other roots z_j, z_k are inside it, while $Q(z) \geq 0$ for $z \in [z_j, z_k]$. Then equation (36) is integrated in elliptic quadratures and its solution looks similarly to the expression (39).

Analysis of expressions (38) shows that at the internal points of the domains shown in Fig. 1, 2, we have similar situation, when three roots of the polynomial $Q(z)$ are different and only two of them belong to the interval $[0, 1]$. In all such cases, equation (21) is reduced to the form (36), and the result of its integration is expressed in terms of the elliptic functions with some permutation of the roots z_1, z_2, z_3.

4   Mass Variations

As we have seen above, the evolutionary equations are integrable in terms of elementary and elliptic functions if the functions $\gamma_0(\tau)$, $\gamma_1(\tau)$ satisfy equation (18). Taking into account condition (8), we can rewrite (18) in the form

$$\frac{d^2\gamma_0}{d\tau^2} + \left(AC - B^2 + \alpha\frac{K}{a_1^3}\right)\gamma_0(\tau) = \alpha\frac{K_0}{a_1^3} \,. \tag{45}$$

One can readily see that equation (45) is integrable, and its solution satisfying the condition $\gamma_0(0) = 1$ is given by

$$\gamma_0(\tau) = \alpha\frac{K_0}{\sigma^2 a_1^3} + \left(1 - \alpha\frac{K_0}{\sigma^2 a_1^3}\right)\cos(\sigma\tau) + \Phi\sin(\sigma\tau) \,, \tag{46}$$

where Φ is an arbitrary constant, and

$$\sigma^2 = AC - B^2 + \alpha \frac{K}{a_1^3} \, .$$

Taking into account equation (8), one can represent differential equation (25), determining the variable n, in the form

$$\frac{dn}{d\tau} = \frac{3a^{3/2}}{16a_1^3 K_0^{1/2}} (K\gamma_0^2(\tau) - K_0\gamma_0(\tau)) \, ,$$

where the function $\gamma_0(\tau)$ is given by (46). Obviously, this equation is easily integrated, and an explicit expression for the function $n(\tau)$ together with different explicit and implicit solutions $z(n)$ found in previous section gives a complete solution of the evolutionary equations in the considered restricted three-body problem with variable masses.

5   Conclusion

We have considered the satellite version of the restricted three-body problem formulated on the basis of the classical Gylden–Meshcherskii problem. We have obtained the evolutionary equations of the massless point P_2, describing a long-term evolution of its orbital elements, in the Hill approximation, and investigated their integrability. It was shown that the evolutionary equations are integrable in terms of the elementary and elliptic functions if masses of points P_0, P_1 vary isotropically with different rates determined by the expressions (2), (8), (46). Solutions of these equations describe quasi-elliptic motion of the point P_2 if initial conditions of motion are chosen in such a way that two integrals of motion c_1, c_2 belong to the domains shown in Fig. 1, 2. All relevant symbolic calculations and visualizations are done with the computer algebra system Mathematica.

Acknowledgement. This work was supported in part by the grant 0688/GF of the scientific-technical programs and projects of the Committee of Science of the Republic of Kazakhstan, 2012–2014.

References

1. Bekov, A.A., Omarov, T.B.: The theory of orbits in non-stationary stellar systems. Astronomy and Astrophysics Transactions 22, 145–153 (2003)
2. Berkovič, L.M.: Gylden–Meščerski problem. Celestial Mechanics 24, 407–429 (1981)
3. Hill, G.W.: Researches in the lunar theory. American J. Mathematics 1, 129–147 (1878)
4. Lidov, M.L.: The evolution of orbits of artificial satellites of planets under the action of gravitational perturbations of external bodies. Planetary and Space Science 9(10), 719–759 (1962)

5. Lidov, M.L., Vashkov'yak, M.A.: On quasi-satellite orbits in a restricted elliptic three-body problem. Astronomy Letters 20(5), 676–690 (1994)
6. Minglibayev, M.Z.: Dinamika gravitiruyushchikh tel s peremennymi massami i razmerami (Dymanics of Gravitating Bodies of Variable Masses and Sizes). LAP Lambert Academic Publ. (2012)
7. Morbidelli, A.: Modern Celestial Mechanics. Aspects of Solar System Dynamics. Taylor & Francis Inc., New York (2002)
8. Prokopenya, A.N., Minglibayev, M.Z., Mayemerova, G.M.: Symbolic calculations in studying the problem of three bodies with variable masses. Programming and Computer Software 40(2), 79–85 (2014)
9. Szebehely, V.: Theory of orbits. The restricted problem of three bodies. Academic Press, New York (1967)
10. Vashkov'yak, M.A.: Evolution of orbits of distant satellites of Uranus. Astronomy Letters 25(7), 476–481 (1999)
11. Wolfram, S.: The Mathematica Book, 4th edn. Wolfram Media/Cambridge University Press (1999)

Factoring Sparse Bivariate Polynomials Using the Priority Queue

Fatima K. Abu Salem[1,*], Khalil El-Harake[1], and Karl Gemayel[2]

[1] Computer Science Department, American University of Beirut, Lebanon
{fatima.abusalem,kme07}@aub.edu.lb
[2] School of Computational Science and Engineering,
Georgia Institute of Technology, U.S.,
karl@gatech.edu

Abstract. We revisit the polytope method for factoring sparse bivariate polynomials over finite fields, and address the bottleneck arising from solving the Hensel lifting equations using the sparse distributed polynomial representation. We revise the analysis when polynomials are represented as such, which reveals how performing the polynomial multiplications and ensuing additions in separate (serialised) phases causes the Hensel lifting phase to suffer from poor work, space, and I/O complexity, and hinges on the size of the intermediary output, as size is defined in the sparse distributed representation. We propose to overlap all polynomial arithmetic in one Hensel lifting step using a MAX priority queue. The overlapping approach adapts not only to the growth in the degree of the input polynomial but also to irregularities in the sparsity of intermediary output. It also results in evading expression swell and reducing the overall work and space complexity by an order of magnitude. When the priority queue is implemented as a cache-oblivious data structure, the overlapping approach achieves an order of magnitude improvement in I/O over the serialised approach, even when the latter is using cache efficient structures to assist in polynomial multiplications and additions. We present empirical results for the polytope method using a max-heap implementation of the global priority queue, which demonstrate extremely superior performance, and specifically against Magma, for sufficiently sparse input polynomials of very high degrees.

Keywords: Algorithms and Data Structures, Performance evaluation, Data Locality, Bivariate Polynomial Factorisation, Hensel Lifting, Newton Polytope.

1 Introduction

We address the fundamental problem of factoring polynomials over finite fields, which is integral to many routines in algebra and number theory. In many

---

* This work is supported by the Lebanese National Council on Scientific Research and the University Research Board of the American University of Beirut.

V.P. Gerdt et al. (Eds.): CASC Workshop 2014, LNCS 8660, pp. 388–402, 2014.

instances, the multivariate case is reduced to bivariate polynomial factorisation. Several leading factoring algorithms rely on Hensel lifting techniques that deliver fast algorithms in practice, but these algorithms of a "classical" flavour are designed for generic input [6,10]. To date, we do not know of an efficient algorithm dedicated for factoring sparse polynomials. The polytope method of [1] is based on ideas from polyhedral geometry and is intended to factor sparse polynomials more efficiently, by exploiting the structure of sufficiently sparse input polynomials whose polytopes consist of a few edges. However, the inner workings of Hensel lifting remain oblivious to fluctuations in the sparsity of intermediary polynomial output, which is a consequence of analysing and designing algorithms in the dense model for polynomial representation. In contrast, the input size in the sparse distributed representation consists of the number of non-zero terms of the input polynomial rather than its degree, which captures the fluctuations in sparsity throughout the factorisation process. More significantly, we note that deriving the cost analysis using this representation is more consistent for the purpose of standard benchmarking: Magma, Maple, Mathematica, and Singular, all use the sparse distributed polynomial representation by default.

Our contribution is a thorough analysis in the sparse distributed model, followed by a "data-structure"-centric improvement, for the Hensel lifting phase, given sparse bivariate polynomials over finite fields. The implications extend to any factoring algorithm that employs some form of Hensel lifting, and not only to the polytope method, despite that we emphasise the latter method when the input polynomial is sparse. In Section 3, we use the sparse distributed representation to transform the analysis of the costs associated with the Hensel lifting phase when the polynomial multiplications followed by addition of resulting products are performed in separate (serialised) phases, like in [1,6,10]. We derive that the asymptotic performance in work, space, and I/O is critically affected not only by the degree of the input polynomial, but also by the following factors: (i) the sparsity of each polynomial multiplication, and (ii) the sparsity of the resulting polynomial products to be merged into a final summand. We further show that even with advanced additive (merging) data structures like the cache aware tournament tree or the cache oblivious k-merger, the asymptotic performance of the serialised version in all three metrics is still poor. In Section 4, we re-engineer the Hensel lifting phase by overlapping all arithmetic using a MAX priority queue, which generalises the approach of [12,14] for a single polynomial multiplication. The analysis of this approach is examined using two implementations of the priority queue, consisting of a generic max-heap as well as a cache-oblivious data structure. We derive orders of magnitude reduction in work, space, and I/O when the overlapping approach is examined against all possible enhancements of the serialised version. Whereas the serialised approach is not able to fully exploit a cache-efficient data structure, our approach now becomes entirely cache-oblivious when a cache-oblivious priority queue is used. In Section 5, we report on our benchmarks. A modest implementation of the polytope method using the overlapping approach concludes significantly faster than with the serialised version. We use this observation to validate that the polytope method can now handle sparse polynomials significantly better, as it is no longer affected by the fluctuating sparsity of intermediary polynomial products. Our modest implementation also outperforms as Magma, up to polynomial degrees equal to 60000. More notably, Magma fails to process input instances of degrees 10,000 and higher.

2   Background

As a preamble, we follow the exposition in [12,14] to describe the sparse distributed representation, and simplify it further to suit univariate polynomials:

Definition 1. *Let* $w(x) \in \mathbb{F}[x]$ *denote a univariate polynomial with coefficients from* $\mathbb{F}$, *and let* $\#w$ *denote the number of non-zero terms of* $w(x)$. *A sparse distributed representation of* $w(x)$ *is obtained by writing* $w(x) = \sum_{i=1}^{\#w} w_i$ *such that* $w_i = a_i X_i$, *where* $a_i \neq 0$, *and* $X_1 > \ldots > X_{\#w}$ *according to some monomial ordering. We refer to each* X_i *as a monomial and* $a_i X_i$ *as a term.*

When choosing the lexicographical monomial ordering in the univariate case, we get a natural mapping to an order on exponents.

For a thorough review of various factoring algorithms that use Hensel Lifting techniques, we refer the reader to [18,19,20,21]. Recent methods that have a classical flavour appear in [6,10], and are not dedicated for sparse input. In [1] a method that generalises Hensel lifting is able to exploit a relationship between the input polynomial and its Newton polytope. The method has a "practical" flavour and is able to efficiently factor sparse input polynomials whose Newton polytopes contain only a few vertices, which tend to have a few Minkowski decompositions.

Thereafter we consider bivariate polynomials whose polytopes in the plane are referred to as polygons. Both the classical and polytope methods work by specialising one of the two variables of f by setting the other remaining variable to zero. The resulting univariate polynomials are factored and their factors lifted up to a suitable level using a form of Hensel lifting. Below, we summarise the workings of the two classes of algorithms in [1,10].

2.1   Classical Hensel Lifting

Let $\mathbb{F}$ denote a finite field of characteristic p, and consider a polynomial $f \in \mathbb{F}[x,y]$ with total degree n. We wish to obtain a polynomial factorisation of f into two factors g and h such that $f = gh$ and $g, h \in \mathbb{F}[x,y]$. Let $r = \deg g$ and $s = \deg h$. Write $f = \sum_{k=0}^{n} f_k y^k$ where $f_k \in \mathbb{F}[x]$ and $\deg(f_k) = n - k$, and determine $g = \sum_{k=0}^{r} g_k y^k$ or $h = \sum_{k=0}^{s} h_k y^k$ by revealing the univariate polynomials $\{g_k\}_{k=0}^{r}$ or $\{h_k\}_{k=0}^{s}$. To begin with, we require all boundary univariate factorisations $f_0 = g_0 h_0$ where g_0 and h_0 are coprime. Assume w.l.o.g. that $r \leq s$, and so $1 \leq r \leq \lfloor n/2 \rfloor$. Starting from a given pair (g_0, h_0), and for

$k = 1, \ldots, r$, the two polynomials g_k and h_k are determined using the following Hensel lifting equation:

$$g_0 h_k + h_0 g_k = f_k - \sum_{i=1}^{k-1} g_i h_{k-i}. \tag{1}$$

When $\deg(g_k) \leq r - k$ the pair (g_k, h_k) is unique and one can continue lifting. If the lifting concludes with the r'th step one returns whether $g = \sum_{k=0}^{r} g_k y^k$ divides f. Performing Hensel lifting using at least one boundary factorisation returns a monic factor g of f. This version for bivariate factorisation is one of several fastest generic algorithms that use Hensel lifting. It possesses exponential run-time in the worst-case if one has to eventually attempt all boundary factorisations. However, the method is known to be fast in practice. Indeed, the average case run-time is shown to be $O(n^4)$ using standard polynomial arithmetic [10]. Bostan et al.'s fast factoring algorithm in [6] is based on multi-moduli computation for univariate polynomials used in the Hensel lifting phase and achieves an $\tilde{O}(n^{\omega+1})$ running time algorithm, where ω denotes the matrix multiplication exponent ($2 \leq \omega \leq 3$). All of the work estimates given above are derived using the dense model for polynomial representation

2.2   Factoring Bivariate Polynomials Using Polygons

Consider all integer pairs (α, β) such that $x^\alpha y^\beta$ is a non-zero monomial of f. Then the set of all such pairs constitutes the *support vector* of f. The Newton (integral) polygon $N(f)$ is defined to be the convex hull in $\mathbb{Z}^2$ of all points in the support vector of f. One identifies suitable subsets $\{\Delta_i\}$ of edges belonging to $N(f)$, and specialises terms of f along each edge $\delta_j^{(i)} \in \Delta_i$, by setting one of the variables of f to be zero. Those specialisations are derived from the nonzero terms of f whose exponents make up integral points falling on each $\delta_j^{(i)}$, and we label them as $f_0^{\delta_j}$. The polytope method requires that for at least one Δ_i, the associated edge polynomials $f_0^{\delta_j}$ are squarefree, for all $\delta_j \in \Delta_i$. One can then begin lifting using the boundary factorisations of $f_0^{\delta_j} = g_0^{\delta_j} h_0^{\delta_j}$, for all $\delta_j \in \Delta_i$. As in above, given one boundary factorisation, one can then determine the associated $\{g_k\}$'s and $\{h_k\}$'s that satisfy the Hensel lifting equation

$$g_0^{\delta_j} h_k^{\delta_j} + h_0^{\delta_j} g_k^{\delta_j} = f_k^{\delta_j} - \sum_{j=1}^{k-1} g_j^{\delta_j} h_{k-j}^{\delta_j} \tag{2}$$

for $k = 1, \ldots, \min(\deg(g_0), \deg(h_0))$. By the proper transformations, the boundary specialisations and all the ensuing polynomials are treated as Laurent polynomials in one variable. For any coprime edges factorisation of f relative to Δ_i, there exists at most one full factorisation of f which extends it. Note that the standard Hensel lifting is a special instance of the polytope method, in the case when $N(f)$ consists of the triangle $(0, n)$, $(n, 0)$, and $(0, 0)$. Here, the lifting is initiated from the horizontal side defined by $(0, 0)$ and $(n, 0)$.

In addition to promising to perform well in practice given sparse input polynomials, this algorithm has the added advantage that it can help factor families of polynomials which possess the same Newton polygon. Still, the work complexity required by one full lifting round is given by $O(n^4)$ assuming standard polynomial arithmetic, and the dense model for polynomial representation.

In [2], the polytope method is implemented using a sparse polynomial representation, and the work and space become dependent on the number t of non-zero terms comprising the input polynomial. Particularly, for $t < d^{3/4}$ and using fast polynomial arithmetic over finite fields, the adaptation in [2] reduces the amount of work per one full step of Hensel lifting to $O(t^\lambda n^2 + t^{2\lambda} d\, L(n) + t^{4\lambda} n)$, for some $1/2 \le \lambda < 1$, and $L(n) = \log n \log \log n$. However, a downside of the approach in [2] is that it imposes strict assumptions on the number of non-zero terms arising in the intermediary polynomials produced during the Hensel lifting phase, and finally on the number of non-zero terms belonging to the output factors. The present work digresses significantly from these assumptions.

3   Revised Analysis in Sparse Distributed Represenation

In this section we revise the analysis of the main Hensel lifting Eq. (1) or its extension to polygons in Eq. (2), when polynomials are in sparse distributed representation.

In our work analysis we exclude the cost to perform coefficient arithmetic in the base field, and report only on monomial comparisons based on their degrees. This is because the cost to perform coefficient arithmetic is independent of the sparsity of the input or intermediary output. Our assessment of the space complexity targets the "working space", which corresponds to the amount of memory required to process the intermediary output. This exposes the amount of expression swell, and directly affects the work complexity, as opposed to the space required to store the input or output. To assess the I/O complexity, we proceed in line with existing models of computation [3,13]. We reason about a two-level memory hierarchy, featuring a primary and a secondary level of memory. Those two levels can be cache versus internal memory, or internal memory versus external memory. Thereafter, we shall distinguish them using the terms in-core versus out-of-core memory. In-core memory is of size M, and is organised using cache lines (disk blocks), respectively, each consisting of B consecutive words. All words in a single line are transferred together between in-core and out-of-core memory in one round (I/O operation) referred to as a cache miss (disk block transfer). Given a one-dimensional layout of n records in memory, the number of I/O's required to read them consecutively into in-core memory as $\Theta(n/B)$.

A single instance of Equations (1) or (2) has the input and output requirements stated below, and all previous treatments of Hensel lifting assume more or less the following serial flow of computation where Step 4 begins after Steps 1-3 have concluded:

Require: An integer $k \in \{1, \ldots, r\}$. Two sets of univariate polynomials over $\mathbb{F}$, $\{g_i\}_{i=1}^{k-1}$, $\{h_j\}_{j=1}^{k-1}$, in sparse, distributed representation.
Ensure: The polynomial $S_k = \sum_{i=1}^{k-1} g_i \cdot h_j$, where $j = k - i$, $\deg(g_i) \leq i - r$, $\deg(h_j) \leq j - s$, and $\deg(S_k) \leq n - k$.

Algorithm 1. Local-Iterative

1: **for** $i = 1$ to $k - 1$ **do**
2:      Compute $p_i \leftarrow g_i \cdot h_j$.
3: **end for**
4: Compute $S_k = \sum_{i=1}^{k-1} p_i$.

As in [14,15,16], reasoning in the sparse distributed representation produces worst-case versus best case polynomial multiplication, depending on the structure of the output. In the worst case, a given multiplication $g_i \cdot h_j$ is sparse as it yields a product with $\Theta(\#g_i \cdot \#h_j)$ non-zero terms, an incidence of a memory bound computation. At best, the multiplication is dense as it yields a product with $\Theta(\#g_i + \#h_j)$ terms. When the product has significantly fewer terms due to cancelation of terms, the operation is said to suffer from expression swell.

When the input polynomial f is sparse, the polynomial multiplications in Hensel lifting are also highly likely to be sparse, and consequently memory bound. Fast polynomial arithmetic here no longer pays off, as actual performance becomes heavily dependent on locality [9]. A MAX-heap implementation performs the least number of monomial comparisons required for sparse multiplications in the sparse distributed representation [12,14,15,16]. Whilst a Max-heap is more tied to the physical implementation, the abstract structure intended is effectively a priority queue, all of whose available implementations achieve $O(\log N)$ work to perform *Insert* or *Extract-max*. In the rest of this paper, let $C_{M,B}(N)$ denote the number of I/O's required to perform an *Insert* or *Extract-max* onto a priority queue. We then have $C_{M,B}(N) = O(\log N)$ when the queue is implemented as a binary heap. This improves to $O\left(\frac{1}{B} \log N\right)$ using, say, the cache oblivious bucket heap [8], and optimally to $O\left(\frac{1}{B} \log_{M/B}(N)\right)$, using the cache oblivious funnel heap [7] or Arge heap [4]. Since the I/O complexity depends on both M and B, we will thereafter simply write $C(N)$, and observe that $C(ab) = C(a) + C(b)$ for any physical implementation of the priority queue.

Step 4 above is an instance of merging, which can be handled using (a) iterated merging or (b) multi-way merging. We elaborate on each of these options and formalise our results in Prop. 1.

Proposition 1. *Assume the sparse distributed representation for polynomials. In the worst case analysis when each polynomial multiplication $g_i h_j$ is sparse,*

Alg. 1 using a MAX priority queue for each iteration in Step 2 and additive merging for Step 4 requires the following costs:

- a. *iterated merging:*
    - Work: $O\left(k\bar{g}\bar{h}(k + \log \bar{g})\right)$
    - Space: $O(k\bar{g}\bar{h})$
    - I/O: $O\left(k\bar{g}\bar{h}[C(\bar{g}) + \frac{k}{B}]\right)$
- b. *multi-way merging:*
    - i. $k \in \Theta(M/B)$:
        * Work: $O\left(k\bar{g}\bar{h} \log(k\bar{g})\right)$
        * Space: $O(k\bar{g}\bar{h})$
        * I/O: $O\left(k\bar{g}\bar{h}[C(\bar{g}) + \frac{1}{B}]\right)$
    - ii. $k \gg M/B$:
        * Work: $O\left(k\bar{g}\bar{h}[\log \bar{g} + \log k \log_{M/B} k]\right)$
        * Space: $O(k\bar{g}\bar{h})$
        * I/O: $O\left(k\bar{g}\bar{h}[C(\bar{g}) + \frac{1}{B} \log_{M/B} k]\right)$

Proof. Given a polynomial pair (g_i, h_j), assume w.l.o.g. that $\#g_i < \#h_j$. In all of the following, let $\bar{g} = \max\{\#g_i\}$. Also, Let $\bar{p} = \max\{\#p_i\}_{i=1}^{k-1}$. Then p denotes the largest number of nonzero terms appearing in any one polynomial product. **Steps 1-3:** A sparse polynomial multiplication $g_i \cdot h_j$ using the methods of [12,14,15,16] will require a queue with size $O(\min(\#g_i, \#h_j)) = O(\bar{g})$. To conclude a single multiplication, the queue will process $O(\#g_i\#h_j)$ elements. This brings the work and I/O costs to $O(\#g_i\#h_j \log(\min(\#g_i, \#h_j))) = O(\bar{g}\bar{h} \log(\bar{g}))$ monomial comparisons and $O(\#g_i\#h_j C(\min(\#g_i, \#h_j))) = O(\bar{g}\bar{h} C\bar{g})$ I/O's. Accumulating the costs for all iterations of Step 2, we have that Steps 1-3 require

$$O\left(k\bar{g}\bar{h} \log(\bar{g})\right) \quad \text{work}, O\left(\bar{g}\right) \quad \text{space, and } O\left(k\bar{g}\bar{h} C(\bar{g})\right) \quad \text{I/O's} \quad (3)$$

Step 4: In this step all products $\{p_i = g_i h_j\}_{i=1}^{k-1}$, $j = k - i$, are merged into the final summand S_k. In the worst case analysis, each pair of products p_i and p_j, when merged, yields a summand with $\#p_i + \#p_j$ non-zero terms rather than $\max(\#p_i, \#p_j)$. Accumulating, we may derive $\#S_k = \sum_i \#p_i = O(k\bar{p})$.

(a) Using iterated merging:

The intermediary sum is constantly experiencing growth. Each polynomial product is read consecutively into memory. The working space corresponds to the amount of space required to store all polynomial products. Accumulating the expression swell we get that the total costs required to conclude Step 4 are:

$$O(k^2\bar{p}) \quad \text{work}, O(k\bar{p}) \quad \text{space, and } O\left(k^2\frac{\bar{p}}{B}\right) \quad \text{I/O's} \quad (4)$$

Here, the work and I/O costs for merging are respectively factors of $O(k/\log k)$ and $O(k/\log_{M/B} k)$ more than optimal (e.g. see [3]). One addresses this poor performance using multi-way merging strategies.

(b) Using multi-way merging:

Binary merging à la Mergesort is work efficient provided the number of streams to be merged is a power of 2, and all streams contain roughly the same number of elements to be merged. In our application, the number of streams is about $k \in \{1, \ldots, n/2\}$ and when the intemediary polynomials are sparse the resulting products $g_j h_j$ are far from having roughly the same number of non-zero elements. The conditions to achieve efficient binary merging thus do not hold.

When $k = \Theta(M/B)$, one can achieve work and I/O efficient multi-way merging by running a sequence of *Insert* and *Extract-max* operations on a complete (tournament) binary tree that is kept in-core (See [11], Ch. 14). Given $k - 1$ polynomial products to be merged, the tournament tree consists of $k - 1$ leaves, and thus the size of the tournament tree is $\Theta(k)$. Each of the input streams p_i is read consecutively in blocks B_i of size B. All blocks $\{B_i\}_{i=1}^{k-1}$ will reside in-core together with the binary tree. This approach produces the merged output using optimal work, as each term entering the tournament tree is compared against another $\Theta(\log k)$ number of times, and there are $O(k\bar{p})$ such terms. It also achieves optimal I/O since each term from p_i is read from out-of-core to in-core memory only once, bringing the total cost to read a single polynomial product into the tournament tree to $O(\bar{p}/B)$. The space to store the tournament tree is $\Theta(k)$, dominated by $O(k\bar{p})$, the space required to store all the polynomial products in Step 4. Summarising, when $k \in \Theta(M/B)$, Step 4 requires:

$$O(k\bar{p}\log k) \quad \text{work}, O(k\bar{p}) \quad \text{space, and } O\left(k\frac{\bar{p}}{B}\right) \quad \text{I/O's} \qquad (5)$$

The condition $k \in \Theta(M/B)$, however, is not a realistic assumption. For a sufficiently large input degree n, there will exist $k_0 \in \mathbb{N}$ where $k \gg M/B$ for all $k \geq k_0$. In this case, one can no longer fit all blocks $\{B_i\}_{i=1}^{k-1}$ in core with the tournament tree, causing the I/O performance to degenerate. To reclaim the benefits of this merging structure, one has to re-arrange all the $\{p_i\}$'s into groups consisting of $\Theta(M/B)$ streams each. All streams in any one group can now be merged efficiently using a tournament tree. One recurses if the number of groups is not small enough, and the base case is reached when there are $\Theta(M/B)$ streams to be merged. The number of recursive steps required is $\Theta(\log_{M/B} k)$. The total costs required to conclude Step 4 when $k \gg M/B$ are now:

$$O(k\bar{p} \log k \log_{M/B} k) \quad \text{work}, O(k\bar{p}) \quad \text{space, and } O\left(k\frac{\bar{p}}{B} \log_{M/B} k\right) \quad \text{I/O's}$$
$$(6)$$

Both work and I/O are a factor of $\Theta(\log_{M/B} k)$ more than when $k \in \Theta(M/B)$. Although a cache oblivious merger with optimal work and I/O exists (see the k-merger of [13], for example), these costs are defined in the amortised sense, and optimal performance is attained only when the streams to be merged have an equal number of nonzero terms. This clearly is not the case for our application.

We summarise all of the cases and indicate the equations that are needed to accumulate their respective costs as claimed in the proposition above. Additionally, we also use that $\bar{p} = \Theta(\bar{g}\bar{h})$, since in the worst case analysis each polynomial multiplication $g_i h_j$ is sparse.

- a. iterated merging (Equations 3 and 4)
- b. multi-way merging:
    - i. $k \in \Theta(M/B)$ (Equations 3 and 5)
    - ii. $k \gg M/B$ (Equations 3 and 6)

Qualitative Assessment: From the analytical perspective in Prop. 1 we deduce the following qualitative assessment of the bottlenecks associated with each lifting step:

1. In addition to each Hensel lifting step being susceptible to the sparsity of each given multiplication, merging all the polynomial products to compute S_k is also memory bound. Particularly, an intermediary result $S_t = \sum_{i=1}^{t} p_i$ during iterated merging may be denser than all its summands. Consequently, the working space is now $O(k\bar{g}\bar{h})$, which in turn causes the work and I/O performance of iterated merging to deteriorate asymptotically.

2. Whilst attempting to manage expression swell, multi-way merging is not guaranteed to attain the lower bounds unless $k \in \Theta(M/B)$. Given that k is the iteration index in the Hensel lifting process, we have $k \in O(n)$, where n is the degree of the input polynomial. Efficiency of multi-way merging thus hinges on extremely unrealistic assumptions on k.

3. To evade the unrealistic assumptions on k, regrouping polynomial products recursively to benefit from multi-way merging is a cache aware process, that needs to be tuned for each level of the memory hierarchy, and during each lifting step. The fact that there could be $O(n)$ lifting steps makes this process extremely prohibitive from a design and implementation perspective.

4. Also, the work of recursive multi-way merging when $k \gg M/B$ is $O(\log_{M/B} k)$ factor more than optimal.

5. Using a cache oblivious merger to handle the polynomial additions will not attain optimal I/O for merging, as the amortised cost of invoking a k-merger cannot be achieved unless all polynomial products to be merged have the same number of non-zero terms.

4   Overlapping Computations Using a Priority Queue

Consider again the sequence $\{(g_i, h_j)\}_{i=1}^{k-1}$, $j = k - i$, in the ring $\mathbb{F}[x]$. The polynomials g_i and h_j are in sparse distributed representation and we accordingly express each as a sum of non-zero terms appearing in decreasingly sorted order. Write $g_i = \sum_{u=1}^{\#g_i} g_u^{(i)}$, $h_j = \sum_{w=1}^{\#h_j} h_w^{(j)}$ such that $g_u^{(i)} = a_u^{(i)} X_u^{(i)}$, $h_w^{(j)} = b_w^{(j)} Y_w^{(j)}$. As before, assume w.l.o.g., that for each pair (g_i, h_j), we have $\#g_i < \#h_j$.

To overlap all computations required to produce $S_k = \sum_{i=1}^{k-1} g_i h_j$, expand S_k as $\sum_{i=1}^{k-1} \left(\sum_{u=1}^{\#g_i} g_u^{(i)} \right) h_j$. With this expansion we can now perform a merge on the outer and inner sums simultaneously. For this, we instantiate Q, a MAX priority queue whose elements are structures of the form $(XY, \mathbf{g}, \mathbf{h})$. Here, XY denotes a monomial product using two terms from some pair (g_i, h_j), and $\mathbf{g}$ and $\mathbf{h}$ denote pointers to those two terms in g_i and h_j respectively. The priority key of each element is determined by the rank of the corresponding monomial under the assumed monomial ordering. For example, the maximal element in the queue corresponds to the monomial with highest rank.

We initialise Q using monomials from all the polynomial pairs $\{(g_i, h_j)\}_{i=1}^{k-1}$ of highest rank. For a given pair in the list, we perform the product of all monomials of g_i, given by $\{g_u\}_{u=1}^{\#g_i}$, by the first monomial of h_j, given by $h_1^{(j)}$. We then proceed iteratively as follows. We extract the maximal element residing in the queue corresponding to a monomial product $X_u^{(i)} Y_w^{(j)}$ for some i and j. We perform the coefficient arithmetic needed to produce a term corresponding to this monomial, and accumulate the result into S_k. We repeat until no more monomial products of this particular rank reside in the queue. For each monomial product $X_u^{(i)} Y_w^{(j)}$ extracted in this iteration, we insert its successor element $X_u^{(i)} Y_{w+1}^{(j)}$, if $Y_{w+1}^{(j)}$ exists. We repeat until no more monomials can be inserted. The process is a generalisation of the single polynomial multiplication algorithm of [12,14] to the case of Hensel lifting. We summarise it below:

Algorithm 2. Global Priority Queue

1: **for** $i = 1, \ldots, k - 1$, and $j = k - i$ **do**
2:     **for** $u = 1, \ldots, \#g_i$ **do**
3:         Build a Max-priority queue Q using the elements $\{\{(X_u^{(i)} Y_1^{(j)}, \mathbf{g}^{(i)}, \mathbf{h}^{(j)})\}\}$
4:     **end for**
5: **end for**
6: Set $t \leftarrow 1$.
7: **repeat**
8:     Let the maximal element be denoted by $(XY, \mathbf{g}, \mathbf{h})$. Let α denote the rank of XY under the assumed monomial ordering.
9:     Set $t \leftarrow t + 1$, $a_t \leftarrow 0$, and $S_k \leftarrow a_t R_t$.
10:     Call *Extract-max* on Q.
11:     Set $R_t \leftarrow XY$, and perform the coefficient arithmetic required to produce a_t by reading the coefficients of terms pointed to by $\mathbf{g}$ and $\mathbf{h}$.
12:     Update $S_k \leftarrow S_k + a_t R_t$.
13:     **while** the maximal element in Q has rank equal to α **do**
14:         Repeat Steps 10–12 above
15:     **end while**
16:     For each queue element $(X_u^{(i)} Y_w^{(j)}, \mathbf{g}^{(i)}, \mathbf{h}^{(j)})$ extracted in Step 10 above, if $w < \#h_j$, insert into Q the successor element corresponding to the monomial $X_u^{(i)} Y_{w+1}^{(j)}$, and update the pointer $\mathbf{h}^{(j)}$ accordingly.
17: **until** no monomials can be inserted into Q.
18: Return S_k.

Alg. 2 is correct as a result of the following loop invariant:

Proposition 2. *Let α_r denote the rank of the r'th monomial appearing in the distributed representation of S_k sorted in decreasing monomial order. At the end of the r'th iteration, S_k is the sum of all terms having rank between α_1 and α_r.*

Proof. We proceed by induction on r. In the first iteration, Q is initialised using the monomials $\{X_1^{(i)}Y_1^{(j)}\}_{i=1}^{k-1}$, $j = k - i$, taken from all the products $\{g_i \cdot h_j\}_{i=1}^{k-1}$. Following the sparse distributed representation, each monomial $X_1^{(i)}Y_1^{(j)}$ has maximal rank in the representation of the corresponding polynomial product $(g_i h_j)$ it was derived from. As a result, the monomials returned via the sequence of *Extract-max* issued during the first iteration have rank higher than that of any other monomial still residing in or not yet inserted into the queue. This rank constitutes α_1. Now suppose that the loop invariant holds for the first r iterations. At the beginning of the $r + 1$'st iteration, each successor monomial recently inserted into the queue has rank lower than α_r (the rank of the monomials extracted in the previous iteration). The same holds for monomials already residing in the queue. As a result, the monomials returned via the sequence of *Extract-max* issued during the $r + 1$'st iteration have rank lower than α_r but higher than that of any other monomial still residing in or not yet inserted into the queue. This rank constitutes α_{r+1}.

Proposition 3. *Assume the sparse distributed representation for polynomials. In the worst case analysis when each polynomial multiplication $g_i h_j$ is sparse, Alg. 2 requires*

$$O(k\bar{g}\bar{h}\log(k\bar{g})) \textit{work, } O(k\bar{g}) \textit{space, and } O\left(k\bar{g}\bar{h}[C(k) + C(\bar{g})]\right) \textit{I/O's.} \quad (7)$$

Proof. The structure Q will have at most $\sum_{i=1}^{k-1} \#g_i \subseteq O(k\bar{g})$ elements. The queue will process $\sum_{i=1}^{k-1} \#g_i \#h_j \subseteq O(k\bar{g}\bar{h})$ elements using a sequence of *Insert* and *Extract-max* operations. The work to process all those elements is $O((k\bar{g}\bar{h})\log(k\bar{g}))$ monomial comparisons. We now assess the I/O complexity. Initially, we initialise Q using all monomials from $\{g_i\}_{i=1}^{k-1}$, which are read into the queue only once and also consecutively. This requires $\Theta\left(k\frac{\#g_i}{B}\right)$ I/O's. Thereafter, for each iteration of the loop where a batch of successor elements are inserted, accessing monomials of all the $\{h_j\}$'s may each incur a random access. Since the total number of insertions to Q is $O(k\bar{g}\bar{h})$, it follows that the I/O required to read the input monomials is also $O(k\bar{g}\bar{h})$, and the I/O required to process each of these elements into and out of Q is $O(k\bar{g}\bar{h}\,C(k\bar{g}))$, where the latter cost dominates. This concludes our proof.

Corollary 1. *Assume the sparse distributed representation for polynomials. In the worst case analysis when each polynomial multiplication $g_i h_j$ is sparse:*

- *(i) Alg. 2 achieves orders of magnitude reduction in space, work, and I/O, over Alg. 1 using iterated merging.*
- *(ii) When $k \gg M/B$, Alg. 2 achieves order of magnitude reduction in both space and work over Alg. 1 using multi-way merging. Additionally, if the queues in both algorithms are implemented using a cache oblivious structure with optimal I/O, Alg. 2 achieves the same I/O complexity whilst having the advantage of eliminating all cache aware aspects of Hensel lifting and making it entirely cache oblivious.*

- *(iii) When $k \in \Theta(M/B)$, Alg. 2 achieves order of magnitude reduction in space over Alg. 1 using multi-way merging, whilst maintaining the same work complexity. Additionally, if the queues in both algorithms are implemented using a cache oblivious data structure with optimal I/O, Alg. 2 achieves the same I/O complexity whilst having the advantage of eliminating all cache aware aspects of Hensel lifting and making it entirely cache oblivious.*

Proof. *(i)* Compare work and space showing in rows 1 and 2 of the table below. For the I/O complexity, since $C(k) \in O(\log k)$, we get that $C(k) = O\left(\frac{k}{B}\right)$ for sufficiently large values of k that satisfy $\frac{k}{B} = \Omega(\log k)$. For existing values of B on modern machines this condition is extremely realistic, especially that $k \in O(n)$.

(ii) Compare the work and space costs showing in rows 1 and 3 of the table below. For the I/O complexity, assume that the priority queues in both algorithms are implemented using a cache oblivious data structure with optimal I/O. Then $C(k) = O(\frac{1}{B}\log_{M/B} k)$, which renders the I/O complexity of both algorithms to be of equal order.

(iii) Compare work and space showing in rows 1 and 4 of the table below. For the I/O complexity, as in *(ii)* above we get that $C(k) = O(\frac{1}{B}\log_{M/B} k)$. But $k \in \Theta(M/B)$, and so $C(k) = O(\frac{1}{B})$, which renders the I/O complexity of both algorithms to be of equal order.

Table 1. Overlapping (Alg. 2) versus Serialised (Alg. 1) Approach

	Work	Space	I/O
Alg. 2	$O(k\bar{g}\bar{h}\,\log(k\bar{g}))$	$O(k\bar{g})$	$O\left(k\bar{g}\bar{h}[C(k) + C(\bar{g})]\right)$
Alg. 1 iterated merging	$O\left(k\bar{g}\bar{h}(k + \log\bar{g})\right)$	$O(k\bar{g}\bar{h})$	$O\left(k\bar{g}\bar{h}[C(\bar{g}) + \frac{k}{B}]\right)$
Alg. 1 multi-way merging $k \gg M/B$	$O\left(k\bar{g}\bar{h}[\log\bar{g} + \log k \log_{M/B} k]\right)$	$O(k\bar{g}\bar{h})$	$O\left(k\bar{g}\bar{h}[C(\bar{g}) + \frac{1}{B}\log_{M/B} k]\right)$
Alg. 1 multi-way merging $k \in \Theta(M/B)$	$O(k\bar{g}\bar{h}\,\log(k\bar{g}))$	$O(k\bar{g}\bar{h})$	$O\left(k\bar{g}\bar{h}[C(\bar{g}) + \frac{1}{B}]\right)$

Qualitative Assessment

1. The serialised approach was seen whereas the overlapping approach now becomes entirely cache-oblivious when a cache-oblivious priority queue is used.
2. The overlapping approach is no longer susceptible to expression swell arising due to growth in the intermediary polynomial products: particularly, the working space grows only with the index of the Hensel lifting step and the number of non-zero terms belonging to the shorter of each pair (g_i, h_j).
3. As a result, we have now eliminated superfluous work and I/O, which manifests itself in the asymptotic reductions observed in Corollary 1. We do so independently of the Hensel lifting iteration index k, and of fluctuations in the sparsity of the intermediary polynomial products. Since $k \in O(n)$, this feature of the overlapping approach becomes significant at large scale.

4. The only instance in which the overlapping approach achieves the same work and I/O complexity as the serialised approach is when $k \in \Theta(M/B)$. But even then, the overlapping approach now (i) achieves order of magnitude less working space and (ii) is cache-oblivious as opposed to cache-aware, which offers portability across different machines.

5   Experimental Results

We focus on the adaptation of the polytope method in [2] to have it use:

1. Alg. 2 with the global priority queue implemented as a MAX binary heap and polynomials encoded in the sparse distributed representation.
2. Alg. 1 employing local priority queues for the polynomial multiplications, where the queues are also implemented as MAX binary heaps. This is followed by iterated merging over the polynomial products.

These implementations are in C++ and compiled using g++ version 4.4.6 20120305 with optimization level -O3. We compare both these implementations against Magma's built-in function for factoring bivariate polynomials, which relies on the standard algorithms in [5,17]. Our input suite consists of random bivariate polynomials over $\mathbb{F}_3$ that all turn out to factor into two irreducibles. In several instances, we specifically generate random polynomials whose Newton polygon consist of the triangle $(0, n)$, $(n, 0)$, and $(0, 0)$, as in rows 1 – 7 below. Here, lifting is initiated from the horizontal edge. In rows 8-10, the Newton polygon of the input polynomial is allowed to have an arbitrary number of edges, and lifting is attempted across dominating sets of edges (see [1]).

We ran the experiments on an Intel(R) Xeon(R) CPU E5645 with 43GB of RAM, 12MB in L3 cache, and 256KB in L2 cache. We present sample results in the table below for extremely sparse bivariate polynomials. The parameter n denotes the total degree of the input polynomial, and t denotes the total number of its non-zero terms. The parameter F corresponds to the total number of boundary factorisations attempted before the two irreducible factors are produced. The timings below correspond to wall clock time in seconds. The gap in performance between the overlapping approach and the serialised one becomes severe starting $n = 10,000$. Looking at the 7'th row below we see that the serialised version took over 92 hours. Thus, starting from the 8'th row onwards we dispense with the serialised approach.

Our interpretations of the results are as follows. The version of the polytope method which uses the overlapping approach is almost always significantly faster than when using the serialised version with iterated merging. This confirms our results from Prop. 7. More significantly, despite that the Newton polygons in rows 1-6 have only three edges, this is not seen to be reflecting positively on the performance of the polytope method when using iterated merging. This confirms our earlier remark that the polytope method, even when the polygon has a few edges, would remain susceptible to fluctuation in the sparsity of the intermediary polynomials, unless one undertakes suitable approaches to address expression swell.

The version of the polytope method which uses the overlapping approach is also significantly faster than the built-in Magma function for bivariate polynomial factorisation. Starting from rows 7 onwards, Magma is no longer able to factor the input polynomials and runs out of memory. The fact that we are outperforming Magma in all of these instances can be attributed to the following facts. The polytope method is suited for sparse polynomials and with the overlapping approach can now handle them better, by achieving an order of magnitude less working space than all generic forms of Hensel lifting implementations, including Magma built-in functions.

Finally, and aside from the comparison against Alg. 1 or Magma, we emphasise the aspect of our implementation which allows its run-time to decrease significantly starting from row 7 onwards. In those rows, the degrees are substantially increasing but the polynomials are also becoming substantially sparser, which confirms that the polytope method will now adapt very favourably to sparse polynomials.

	n	t	F	Alg. 2-Binary Heap	Alg. 1-iterated merging	Magma
1.	100	$\approx 10^3$	1	0.002	0.02	0.03
2.	1000	$\approx 3 \times 10^5$	1	2.8''	0.06''	0.04''
3.	2000	$\approx 10^6$	1	4.2''	226.6''	127.2''
4.	4000	$\approx 2 \times 10^2$	8	32''	2,700''	1,085''
5.	5000	$\approx 2 \times 10^2$	5	64''	2,160''	739''
6.	6000	$\approx 8 \times 10^2$	3	810''	15,457''	2,728''
7.	10000	$\approx 3 \times 10^5$	7	17 hrs 19' 15''	92 hours	Segmentation Fault
8.	10000	$\approx 6 \times 10$	1	4 hrs 27' 42''	...	Segmentation Fault
9.	15000	$\approx 2 \times 10$	1	8 hrs 6' 44''	...	Segmentation Fault
10.	20,000	$\approx 2 \times 10$	1	31,569	...	Segmentation Fault

6 Conclusion

We presented an engineering of the polytope factoring method which made it possible to efficiently process sparse bivariate polynomials. For significant improvements to be attained by algorithms that are dedicated for sparse input, one must revise the classical analysis using insight from the sparse distributed representation. After one has gathered enough understanding of the performance under sparsity considerations, one must invest in advanced data structures and to optimise on the flow of computation for the benefit of data locality. We believe this direction of work has a far reaching practical impact as demonstrated in our empirical results.

Acknowledgments. We thank the Lebanese National Council for Scientific Research and the University Research Board – American University of Beirut for supporting this work.

References

1. Abu Salem, F.K., Gao, S., Lauder, A.G.B.: Factoring polynomials via polytopes. In: Proc. of ISSAC, pp. 4–11 (2004)
2. Abu Salem, F.K.: An efficient sparse adaptation of the polytope method over $\mathbb{F}_p$ and a record-high binary bivariate factorisation. J. Symb. Comp. 43(5), 311–341 (2008)
3. Aggarwal, A., Vitter, J.: The input/output complexity of sorting and related problems. Communications of the ACM 31(9), 1116–1127 (1988)
4. Arge, L., Bender, M.A., Demaine, E.D., Holland-Minkley, B., Munro, J.I.: Cache-oblivious priority queue and graph algorithm applications. In: Proc. of STOC, pp. 268–276 (2002)
5. Bernardin, L., Monagan, M.B.: Efficient multivariate factorization over finite fields. In: Mora, T., Mattson, H. (eds.) AAECC. LNCS, vol. 1255, pp. 15–28. Springer, Heidelberg (1997)
6. Bostan, A., Lecerf, G., Salvy, B., Schost, E., Wiebelt, B.: Complexity issues in bivariate polynomial factorization. In: Proc. of ISSAC, pp. 42–49 (2004)
7. Brodal, G.S., Fagerberg, R.: Funnel heap - a cache oblivious priority queue. In: Bose, P., Morin, P. (eds.) ISAAC 2002. LNCS, vol. 2518, pp. 219–228. Springer, Heidelberg (2002)
8. Brodal, G.S., Fagerberg, R., Meyer, U., Zeh, N.: Cache-oblivious data structures and algorithms for undirected breadth-first search and shortest paths. In: Hagerup, T., Katajainen, J. (eds.) SWAT 2004. LNCS, vol. 3111, pp. 480–492. Springer, Heidelberg (2004)
9. Fateman, R.: Comparing the speed of programs for sparse polynomial multiplication. ACM SIGSAM Bulletin 37(1), 4–15 (2003)
10. Gao, S., Lauder, A.G.B.: Hensel lifting and bivariate polynomial factorisation over finite fields. Math. Comp. 71, 1663–1676 (2002)
11. Goodrich, M., Tamassia, R.: Algorithm Design. John Wiley and Sons (2002)
12. Johnson, S.C.: Sparse polynomial arithmetic. ACM SIGSAM Bulletin 8, 63–71 (1974)
13. Frigo, H.P.M., Leiserson, C.E., Ramachandran, S.: Cache-oblivious algorithms. In: Proc. of FOCS, pp. 285–297 (1999)
14. Monagan, M., Pearce, R.: Polynomial division using dynamic arrays, heaps, and packed exponent vectors. In: Ganzha, V.G., Mayr, E.W., Vorozhtsov, E.V. (eds.) CASC 2007. LNCS, vol. 4770, pp. 295–315. Springer, Heidelberg (2007)
15. Monagan, M., Pearce, R.: Parallel sparse polynomial multiplication using heaps. In: Proc. ISSAC, pp. 263–269 (2009)
16. Monagan, M., Pearce, R.: Sparse polynomial pseudo division using a heap. J. Symb. Comp. 46(7), 807–822 (2011)
17. Von Hoeij, M.: Factoring polynomials and the knapsack problem. J. Number Theory 95(2), 167–189 (2002)
18. Von zur Gathen, J., Gerhard, J.: Modern Computer Algebra. Cambridge University Press (1999)
19. Wan, D.Q.: Factoring multivariate polynomials over large finite fields. Math. Comp. 54, 755–770 (1990)
20. Wang, P., Rothschild, L.: Factoring multivariate polynomials over the integers. Math. Comp. 29, 935–950 (1975)
21. Zassenhaus, H.: On Hensel factorization I. J. Number Theory 1, 291–311 (1969)

Solving Parametric Sparse Linear Systems by Local Blocking*

Tateaki Sasaki[1], Daiju Inaba[2], and Fujio Kako[3]

[1] University of Tsukuba,
Tsukuba-city, Ibaraki 305-8571, Japan
sasaki@math.tsukuba.ac.jp
[2] Japanese Association of Mathematics Certification,
Ueno 5-1-1, Tokyo 110-0005, Japan
d.inaba@su-gaku.net
[3] Dept. Info. Comp. Sci., Nara Women's University,
Nara-city, Nara 630-8506, Japan
kako@ics.nara-wu.ac.jp

Abstract. In solving parametric sparse linear systems, we want 1) to know relations on parametric coefficients which change the system largely, 2) to express the parametric solution in a concise form suitable for theoretical and numerical analysis, and 3) to find simplified systems which show characteristic features of the system. The block triangularization is a standard technique in solving the sparse linear systems. In this paper, we attack the above problems by introducing a concept of local blocks. The conventional block corresponds to a strongly connected maximal subgraph of the associated directed graph for the coefficient matrix, and our local blocks correspond to strongly connected non-maximal subgraphs. By determining local blocks in a nested way and solving subsystems from low to higher ones, we replace sub-expressions by solver parameters systematically, obtaining the solution in a concise form. Furthermore, we show an idea to form simple systems which show characteristic features of the whole system.

Keywords: parametric sparse linear system, application-oriented method, block triangularization, local blocks, strongly connected subgraph.

1 Introduction

The parametric sparse linear systems (PSLSs) are very important in many application fields. For example, when performing simulation of a machine or a plant, a system of parametric linear equations is given as a theoretical model of the machine or the plant, where most equations contain only several terms [1]. In [6], the leading author and Yamaguchi investigated parametric linear

---

* Work supported in part by Japan Society for the Promotion of Science under Grants 23500003.

V.P. Gerdt et al. (Eds.): CASC Workshop 2014, LNCS 8660, pp. 403–418, 2014.

systems from the viewpoint of reducing floating-point errors but they did not consider the sparseness. We consider the sparseness seriously in this paper. Let the given system be $A\boldsymbol{x} = \boldsymbol{c}$, where $A \in K[\boldsymbol{p}]^{l' \times l}$, $\boldsymbol{c} \in K[\boldsymbol{p}]^{l'}$, with K a number field of characteristic 0 and $\boldsymbol{p}$ a set of parameters; $\boldsymbol{p} = \{a, b, \ldots, g\}$ in this paper.

A standard method for solving sparse linear systems is the *block triangularization* (BTon) of the matrix A [4]. This method is popularly used in numerical computation, and it is applicable to parametric linear systems, too, because the BTon is nothing but to reorder the rows and columns. In the PSLS solving, however, we have other problems which we do not encounter in numeric systems. In the parametric linear systems, not only solvability but also solving way and the solution may change if parameters satisfy some relations, and we want to know such relations, as in the comprehensive Gröbner basis computation[8]. Much more important desires in application fields are, a) the solver expresses the solution in a concise form because the solution expressed in expanded rational function is often very large, and b) solving method is such that it extracts characteristic features of the system. Such characteristic features will be useful in determining optimal parameter values. In order to satisfy these desires, we must develop a new application-oriented method.

A fast algorithm of the BTon is based on the graph theory; the graph theoretical treatment of BTon is not only useful but also quite simple. An essential point is that each square block corresponds to a "strongly connected maximal subgraph" of an "associated directed graph" G_A of matrix A. In this paper, we propose a concept of "local block". Let B be a square block obtained by the BTon of A and let G_B be its associated directed graph. Each local block of B corresponds to a strongly connected non-maximal subgraph of G_B. Let $L_1, \ldots, L_k$ be local blocks of B; some of them may be contained in others. Each local block L_i corresponds to a PSLS w.r.t. local variables $\boldsymbol{x}_i$, where its solution is expressed in terms of other variables $\boldsymbol{x}'_i$. We will show that we can solve the system $B\boldsymbol{x} = \boldsymbol{c}$ by solving local systems successively.

In Sect. 2, we give toy examples to explain our method and show a simplified characteristic system (to be abbreviated to *characteristic system* below). In Sect. 3, we overview the BTon and the graph theory briefly. In Sect. 4, we introduce concepts of tightly-coupled graph and local block. Furthermore, we present an algorithm for computing strongly connected subgraphs corresponding to local blocks; this is the heart of this paper. In Sect. 5, we analyze the local blocking theoretically and prove that the PSLS can be solved by utilizing local blocks. We also explain how we form characteristic systems and how we replace subexpressions by solver parameters. In Sect. 6, we explain our method by applying it to a simple parametric sparse linear system.

2   Toy Examples and Degenerating Factor

We explain our method by toy examples. The first one is as follows; below by Eqi we denote the i-th equation.

$$\text{Example-1}: \quad \begin{pmatrix} a & b & 0 & 0 & 0 \\ -b & 2a & 0 & 0 & 1 \\ 0 & 0 & c & d & 0 \\ 0 & 0 & d & 2c & 2 \\ 0 & f & 0 & g & 3 \end{pmatrix} \begin{pmatrix} x_1 \\ x_2 \\ x_3 \\ x_4 \\ x_5 \end{pmatrix} = \begin{pmatrix} 3 \\ 2 \\ 1 \\ -1 \\ -2 \end{pmatrix}. \tag{2.1}$$

Let D_1 and D_2 be the following determinants.

$$D_1 = \begin{vmatrix} a & b \\ -b & 2a \end{vmatrix}, \qquad D_2 = \begin{vmatrix} c & d \\ d & 2c \end{vmatrix}. \tag{2.2}$$

We first solve Eq1 and Eq2 of (2.1) w.r.t. x_1 and x_2, obtaining

$$x_1 = (6a - 2b + bx_5)/D_1, \qquad x_2 = (2a + 3b - ax_5)/D_1. \tag{2.3}$$

We also solve Eq3 and Eq4 of (2.1) w.r.t. x_3 and x_4, obtaining

$$x_3 = (2c + d + 2dx_5)/D_2, \qquad x_4 = -(c + d + 2cx_5)/D_2. \tag{2.4}$$

Substituting these "local solutions" into Eq5, we obtain

$$x_5 = \frac{-2D_1D_2 + g(c+d)D_1 - f(2a+3b)D_2}{3D_1D_2 - 2cgD_1 - afD_2}. \tag{2.5}$$

Once x_5 is determined, $x_1, \ldots, x_4$ are determined easily from (2.3) and (2.4).
    Let A_1 be the coefficient matrix in (2.1), then we have

$$\det(A_1) = 3D_1D_2 - 2cgD_1 - afD_2. \tag{2.6}$$

The local solution in (2.3) (or (2.4)) shows that something will happen if $D_1 = 0$ (or $D_2 = 0$). In this case, the local solution in (2.3) is inadequate, so we consider (2.1) again. The system {Eq1, Eq2} with $a \neq 0$ is equivalent to

$$\{ax_1 + bx_2 = 3, \ D_1x_2 = 3b + 2a - ax_5\} \quad \text{(so long as } a \neq 0\text{)}.$$

Hence, if $D_1 = 0$ then $x_5 = (2a+3b)/a$. Substituting this into (2.4), we obtain x_3 and x_4, but x_1 and x_2 are not changed except that the denominator becomes aD_2. Thus, if $D_1 = 0$ then the corresponding local system and some solution are "degenerated".

Definition 1 (degenerating factor). *The local system such as {Eq1,Eq2} in Example-1 can be expressed as $S_i : L_i\boldsymbol{x}_i = L_i'\boldsymbol{x}_i' + \boldsymbol{c}_i$, where L_i is an $n_i \times n_i$ matrix and $\boldsymbol{x}_i$ and $\boldsymbol{x}_i'$ have no common variable. Considering S_i to be a linear system w.r.t. $\boldsymbol{x}_i$, we call $\det(L_i)$ a degenerating factor of S_i.*

By changing Example-1 a little bit, the system becomes very instructive.

$$\text{Example-2}: \quad \begin{pmatrix} a & b & 0 & e & 0 \\ -b & 2a & 0 & 0 & 0 \\ 0 & 0 & c & d & 0 \\ 0 & 0 & d & 2c & 2 \\ 0 & f & 0 & 0 & 3 \end{pmatrix} \begin{pmatrix} x_1 \\ x_2 \\ x_3 \\ x_4 \\ x_5 \end{pmatrix} = \begin{pmatrix} 3 \\ 2 \\ 1 \\ -1 \\ -2 \end{pmatrix}. \qquad (2.7)$$

As in Example-1, we replace $\{Eq1, Eq2\}$ and $\{Eq3, Eq4\}$ by

$$\{D_1 x_2 + b e x_4 = 2a + 3b, \quad -b x_1 + 2a x_2 = 2\}, \qquad (2.8)$$
$$\{c x_3 + d x_4 = 1, \quad D_2 x_4 + 2c x_5 = -c - d\}, \qquad (2.9)$$

respectively. Consider the following simplified system:

$$\{D_1 x_2 + b e x_4 = 2a + 3b, \quad D_2 x_4 + 2c x_5 = -c - d, \quad Eq5\}. \qquad (2.10)$$

After solving this system, we can determine x_1 and x_3. We can regard (2.10) as a simplified system which manifests a characteristic feature of the whole system.

3   An Overview of Block Triangularization

By the BTon, the matrix A is split into three types of blocks (we explain the case of upper-triangular decomposition): the top-left block is of *horizontal type* (H-type) because the number of columns is larger than that of rows, the central k blocks $(k \geq 1)$ are square matrices (S-type), and the bottom right block is of *vertical type* (V-type) because the number of columns is smaller than that of rows. The underdeterminedness and the overdeterminedness of the linear system are represented by H-type part and V-type part, respectively. In the linear system solving, the equations in each block are solved independently from those in other blocks, from the lowest block to upper. Let the i-th block contains variables $x_{l_i}, \ldots, x_{l_{i+1}-1}$. Once equations in the i-th block are solved, we substitute the solutions of these variables into upper equations.

The fast algorithm of the BTon is based on the graph theory. A graph G is composed of *vertices* $\{v_1, v_2, \ldots, v_n\}$ and *edges* connecting two vertices. If an edge connecting v_i and v_j is undirected then we express it as (v_i, v_j). If an edge is directed from v_i to v_j then we express it either as $(v_i \rightarrow v_j)$ or as $(v_j \leftarrow v_i)$. A *path* is a sequence of edges connected sequentially; by $(v_1, v_2, \ldots, v_k)$ we denote a path which passes vertices $v_1, v_2, \ldots, v_k$ in this order. In PSLS solving, we divide the vertices into *E-vertices* $\{E_1, E_2, \ldots, E_{l'}\}$ and *X-vertices* $\{x_1, x_2, \ldots, x_l\}$, where E_i denotes the i-th equation and x_j denotes the j-th variable. Given an equation E_i: $a_{i,j_1} x_{j_1} + \cdots + a_{i,j_s} x_{j_s} = c_i$, we connect the E-vertex E_i with X-vertices $x_{j_1}, \ldots, x_{j_s}$ by s edges, obtaining a *bipartite graph* G for matrix A.

An important concept on bipartite graph is "maximum matching". In the context of BTon, the maximum matching is to maximize the number of nonzero diagonal elements of the given matrix A by changing the order of equations (or, by changing the variable order). After determining a maximum matching, the algorithm of BTon performs the so-called "coarse decomposition" [3,5]. By this, A is decomposed into three matrices A_H, A_S and A_V of H-type, S-type and V-type, respectively. This decomposition and further fine decomposition of A_H and A_V are called Dulmage-Mendelsohn decomposition [3]. From now on, we assume that A is an $l \times l$ S-type matrix having l nonzero diagonal elements.

The final step of BTon is rather technical. First, the undirected bipartite graph is converted to a *directed graph* by changing each edge to be directed from the X-vertex to the E-vertex and shrinking each pair of matched vertices into a single vertex; we call the resulting directed graph *associated directed graph* of A. Then, the directed graph is decomposed into "strongly connected components". A directed graph G is called *strongly connected* if every vertex of G is reachable from any other. A subgraph G of a directed graph is called a *strongly connected component* (SCC) if G is strongly connected and is *maximal* in that no edge or vertex outside of G can be included in G without breaking its property of being strongly connected. By assumption on the matching of E-vertices and X-vertices, we express the i-th shrunken vertex by a rectangular box containing i in it.

Figure 1 illustrates the correspondence between the matrix B and the associated directed graph G_B; each edge in associated directed graph corresponds to a non-diagonal element.

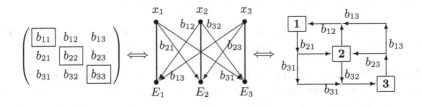

Fig. 1. Correspondence between matrix and shrunken graph

We note that the solvability of the system depends on the values of parameters. On the other hand, the BTon is independent of the parameter values. So, we introduce the following concept.

Definition 2 (formally solvable). *Let $S : B\boldsymbol{x} = \boldsymbol{c}$ be a linear system w.r.t. $\boldsymbol{x}$, where $B \in K[\boldsymbol{p}]^{m \times m}$ and $\boldsymbol{c} \in K[\boldsymbol{p}]^m$. We say S is formally solvable w.r.t. $\boldsymbol{x}$ if it is solvable for the generic values of the parameters.*

Theorem 1. *Let B be an $m \times m$ matrix over $K[\boldsymbol{p}]$, and let G be an associated directed graph of B. If G is strongly connected then the linear system $S : B\boldsymbol{x} = \boldsymbol{c}$, with $\boldsymbol{c} \in K[\boldsymbol{p}]^m$, is formally solvable w.r.t. $\boldsymbol{x}$.*

Proof The theorem is obvious for $m = 1$, so we assume that $m \geq 2$. Let $P = (v_{i_1}, \ldots, v_{i_n}, v_{i_1})$ be a cyclic path of G. Walking on P is translated into visiting terms of the linear system as $b_{i_1 i_1} x_{i_1} \rightarrow b_{i_2 i_1} x_{i_1} \rightarrow \cdots \rightarrow b_{i_n i_n} x_{i_n} \rightarrow b_{i_1 i_n} x_{i_n} \rightarrow b_{i_1 i_1} x_{i_1}$. Thus, we see that each variable x_i ($i \in \{i_1, \ldots, i_n\}$) appears at least in two equations of S and that each equation of S contains at least two nonzero terms. Hence, we can perform the elimination of all the variables appearing in S formally. □

4   Local Blocking of a Block of BT Matrix

By the BTon, solving the given system $Ax = c$ is reduced to solving each subsystem $B_i x_i = c_i$ successively, where B_i is an $m_i \times m_i$ block. In this section, we consider the subsystem solving by expressing the subsystem as $Bx = c$. Note that the associated directed graph G_B for B is strongly connected; although there may be many edges connecting equations in B_i with those in $B_{i-1}, \ldots, B_1$ in the bipartite graph for A, such edges are removed in G_B.

4.1   Definitions

Definition 3 (SCsubG, shrinking to big-vertex). *Let G be a strongly connected subgraph (SCsubG) of G_B, hence G does not contain any edge which is connected with a vertex outside of G. Shrinking of G is to replace G by a vertex V (we call it a big-vertex). G may contain big-vertices. In order to avoid confusion, we say the vertex given initially* input-vertex.

If V contains a vertex v explicitly then we say that v is contained in V. If V is contained in big-vertex $\bar{V}$ then we say that v is recursively contained in $\bar{V}$.

Definition 4 (tightly-coupled). *We say that two vertices are* tightly- coupled *if they are connected by two edges of opposite directions. We call a subgraph G a tightly-coupled subgraph (TCsubG) if, in G, each vertex is coupled tightly with some other one and is maximal in that no vertex of G is coupled tightly with any vertex which is not contained in G.*

Definition 5 (local block). *Let L be a submatrix of B, with $L \neq B$, and G_B and G_L be associated directed graphs of B and L, respectively. L is called a local block of B if i) G_L is a TCsubG of G_B, or ii) G_L is an SCsubG of G_B, where G_L may contain TCsubGs or smaller SCsubGs as big-vertices. We call forming L local blocking of B.*

Remark 1. *The minimality is not imposed on the local block, hence the local blocks are not uniquely defined. Even the tightly-coupled graph is not uniquely determined from the equations, as Fig. 2 shows; employing the maximum matching $\{(x_1 : E_2), (x_2 : E_3), (x_3 : E_4), (x_4 : E_1)\}$, we obtain a tightly-coupled subgraph. This is due to that the maximum matching is not unique.*

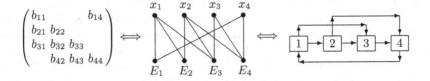

Fig. 2. Illustration of non-uniqueness of tightly-coupled graph

Let G_L be an SCsubG of G_B, containing input-vertices $v_{i_1}, \ldots, v_{i_n}$ recursively, so G_L corresponds to n equations $\mathrm{Eq}_{i_1}, \ldots, \mathrm{Eq}_{i_n}$. If G_L contains a big-vertex V then $G_L - V$ will be often not strongly connected, hence the system corresponding to $G_L - V$ alone is often not formally solvable. Even in such cases, we have the following corollary which is based on the fact that the solvability is independent of the order of eliminations. This is the reason why we search for SCsubGs in the followings.

Corollary 1 (corollary of Theorem 1). *Let G_L be the associated directed graph of the local block L. Even if G_L contains big-vertices, the local system corresponding to G_L is formally solvable. Computationally, it is better to solve the local systems corresponding to big-vertices of G_L earlier.*

4.2   Terminology for Graph Handling

The graph G_A with l vertices $\{v_1, v_2, \ldots, v_l\}$ is often input by an *adjacency list*. Let **Adj** be an array of size l. For any v_i ($i \in \{1, \ldots, l\}$), if G_A has s_i outgoing edges $(v_i \to v_{j_1}), \ldots, (v_i \to v_{j_{s_i}})$, then $\mathrm{Adj}[v_i] = (v_{j_1}, \ldots, v_{j_{s_i}})$.

The depth-first search (DFS) is a method of visiting all the vertices systematically. First, choose the initial vertex to visit, let it be v_1 (one may choose any vertex). Suppose $\mathrm{Adj}[v_1] = (v_{i_1}, \ldots, v_{i_s})$ then DFS visits v_{i_1} by walking the edge $(v_1 \to v_{i_1})$, and erases v_{i_1} from $\mathrm{Adj}[v_1]$. Suppose $\mathrm{Adj}[v_{i_1}] = (v_{j_1}, \ldots, v_{j_t})$ then DFS visits v_{j_1} by walking the edge $(v_{i_1} \to v_{j_1})$, and erases v_{j_1} from $\mathrm{Adj}[v_{j_1}]$, and so on. If all the vertices $v_{j_1}, \ldots, v_{j_t}$ are visited, DFS returns to v_1 and visits the vertex v_{i_2}; see Fig. 3 below.

Let the DFS walk an edge $(v \to w)$ of a directed graph. If DFS has not visited w yet then $(v \to w)$ is called a *forward edge*. If w has been visited by DFS then $(v \to w)$ is called a *back edge*. When the DFS walked the back edge, it returns to the vertex v immediately after erasing w from $\mathrm{Adj}[v]$. The SCsubG of a directed graph is characterized by that, for any two different vertices u and v, there is a cyclic path passing u and v, and if the edges are classified into forward and back edges by the DFS then the cycle contains at least one back edge. So, the back edge is very important in investigating graphs.

The *DFS-subtree* (often called a *branch*) of a directed graph, starting from a vertex b, is a subgraph the vertices of which are visited by the DFS until it comes back to b without walking any possible back edge $(v \to b)$.

The *dfsnum* is a sequential natural number starting from 1, given for each vertex of a directed graph: if a vertex v is visited i-th by the DFS, then the dfsnum of v is i. In constructing the famous SCC decomposition algorithm, Tarjan introduced a concept of *lowlink* which is given to each vertex v of a directed graph [7,2]. The lowlink(v) is the smallest of dfsnums of vertices reachable from v by passing any number of forward edges and lastly one or zero back edge; if there is no such back edge then lowlink(v) := dfsnum(v_i). That lowlink(v) = dfsnum(u) means that there is a path from v to u.

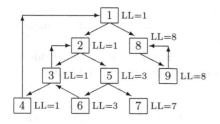

The DFS-tree for the adjacency list:
$$\text{Adj} = [(2,8), (3,5), (2,4), (1), (6,7),$$
$$(6,7), (3), (\,), (9), (8)],$$
where the vertex number is the same as dfsnum in this figure.

Fig. 3. Illustration of DFS, dfsnums and lowlinks

4.3   Computing SCsubGs Satisfying Some Restrictions

Our problem is, given an SCC G_B consisting of m vertices, how to compute SCsubGs contained in the SCC efficiently. We will find SCsubGs by finding cyclic paths in the DFS-tree. We have first considered to utilize Tarjan's lowlinks. Let $P = (u, \ldots, v, w)$ be a path walked by DFS, with $(v \rightarrow w)$ a back edge. Using lowlinks and adjacency list, we can find cyclic paths $(u, \ldots, v, w, \ldots, u)$ by a simple procedure, although some cyclic path may be long because lowlink(v) is often much smaller than dfsnum(v). However, in this method, we will obtain so many SCsubGs most of them will be unused. Therefore, we considered the second method to be explained below.

When we find a new SCsubG, we shrink it into a big-vertex immediately. By this, we can reduce the number of cyclic pahts to be checked greatly.

First of all, we restrict the required SCsubGs rather arbitrarily, as follows: the reason is to obtain only SCsubGs which are useful for PSLS solving.

Restriction-1. Each TCsubG must be the lowest-level unseparable SCsubG.

Restriction-2. Because of Corollary 1, each SCsubG must contain at least one input-vertex. Any input-vertex can be contained only in one big-vertex.

Restriction-3. It is desirable that the big-vertex consists of as neighboring input-vertices as possible. (This restriction is vague, so we will specify it concretely later.)

We explain these restrictions by Fig. 3, showing big-vertices by brackets. When the DFS visits the vertex 1 the second time, we obtain two new SC-subGs, cyclic paths $(1, 2, 3, 4, 1)$ and $(1, [2, 3, 2], 4, 1)$, but the former is abandoned because of Restriction-1. When the DFS visits the vertex 3 the second time, we obtain two new SCsubGs, cyclic paths $([2, 3, 2], 5, 6, [2, 3, 2])$ and $([1, [2, 3, 2], 4, 1], 5, 6, [1, [2, 3, 2], 4, 1])$ but Restriction-3 favors the former. Without these restrictions, one will obtain many useless SCsubGs.

Theorem 2. *Let T be a DFS-tree of a strongly connected graph, and S be a subtree of T. When the DFS has just finished visiting S, let T' be the maximal subtree of T being unvisited by the DFS. Then, $T - T'$ is strongly connected.*

Proof. As an induction hypothesis, we assume that the lemma is valid before the DFS visits S, and we consider visiting of S. If the lemma is valid then every vertex of S has an edge going to a vertex in $T - T'$. Suppose the lemma is invalid, hence S contains a vertex v which has no edge going to a vertex in $T - T'$. By definition of DFS, this fact is not changed if the DFS visits T', hence v is not strongly connected in T. This contradicts that T is strongly connected.   □

We prepare a stack **StkV** to stack (big-)vertices. When the DFS visits a vertex v with StkV $= (u, \dots)$, we stack v into StkV hence StkV becomes $(v, u, \dots)$.

First, we consider forming TCsubGs. Each time the DFS visits an unvisited vertex v, we check whether v is tightly-coupled with an unvisited vertex, and if so then we form the TCsubG by a procedure; see **4.4**.

Next, we consider non tightly-coupled SCsubGs. We consider that the DFS has walked to vertex b which is the head of subtree S, and it begins to walk S. Following Theorem 2, we set **Assumption-A**) all the vertices in $T - T' - S$ have been contained in big-vertices (this is assured below in subcase C33). Each time we meet a back-edge $(v \to w)$, we backtrack StkV and search for a possible cyclic path $P = (w \to \cdots \to v \to w)$ or a path $Q = (u \to \cdots \to v \to w)$ which may become a part of a possible cyclic path when w is contained in a big-vertex. Note that w is known but u is unknown.

We have three cases: C1) w is not contained in any big-vertex, C2) u and w are contained in a big-vertex V, C3) u and w are contained in big-vertices V_u and V_w, respectively, where $V_u \neq V_w$; needless to say, we detect cases C2 and C3 by the backtracking of StkV. In case C1, by Assumption-A, w must be in the path which the DFS is now walking. So, if the path P has an input-vertex then form a new SCsubG by P, else continue the DFS. In case C2, if Q contains an input-vertex then we form a new SCsubG composed of V and Q, else Q must contain a big-vertex so we enlarge V by including Q. When the DFS has visited all of S, we will do as in subcase C33 below.

In case C3, we have three subcases; see Fig. 4 below. C31) V_u and V_w are contained in a big-vertex V_W (V_W may be V_u or V_w), C32) V_u and V_w are contained in V_U and V_W, respectively, where V_U and V_W are different but have a common part (V_U may be V_u and/or V_W may be V_w), C33) V_u and V_w have no common part.

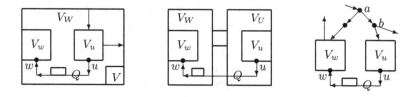

Fig. 4. Illustration of subcases C31(left), C32(middle) and C33(right)

Our main problem is how to satisfy Restriction-3. We specify Restriction-3 concretely as follows (one may relax the restriction to check more possibilities).

Restriction-3'. As V_U and V_W we consider only SCsubGs which contain V_u and V_w, respectively, and no recursively contain.

With this specification, we are rather easy to form SCsubGs satisfying Restriction-3' for subcases C31 and C32.

In subcase C31, if V_w (resp. V_u) contains V_u (resp. V_w) then enlarge V_w (resp. V_u) by including Q into V_w (resp. V_u), else if V_W contains both V_u and V_w and no more big-vertex then enlarge V_W by connecting V_u and V_w by Q, else there must be a path Q' which starts from V_w and goes to V_u hence form a new SCsubG composed of V_u, V_w and Q'; the new SCsubG is contained in V_W. In subcase C32, if $V_w \neq V_W$ and $V_u = V_U$ (V_W and V_u share a big-vertex) then enlarge V_W by including V_u and Q, else if $V_w = V_W$ and $V_u \neq V_U$ then enlarge V_U by including V_w and Q, else if Q has an input-vertex then form a new SCsubG composed of V_U, V_W and Q; the new SCsubG contains V_W and V_U.

In subcase C33, by Assumption-A we see that V_w and V_u are contained in subtrees $T-T'-S$ and S, respectively. Checking the inclusion relation will be pretty complicated. However, Theorem 2 makes the situation extremely simple.

Because of Restriction-3', StkV will often contain a sequence of input-vertices and/or big-vertices which are connected with $T-S-T'$ but not shrunk into big-vertices; let S' be the sequence. Theorem 2 tells that $T-T'$ is strongly connected, so we treat $T-T'$ as a *comprehensive SCsubG* without violating Restriction-3'. In the implementation, we do not generate comprehensive SCsubG but process S' as follows.

How to treat S'. If S' contains no input-vertex then do nothing, else, by virtue of Corollary 1, regard S' as an SCsubG and solve the local system corresponding to S' just after solving all the local systems constructed before S'. We call a graph for S' *complementary subgraph (ComplG)*.

4.4   Implementation

First, we note that the number of SCsubGs including TSsubGs is not greater than m, because each of them contains at least one input-vertex.

In the following procedures, by nil and 'id we denote the empty list and the identifier of the name id, respectively. The DFS staying at vertex v visits the next vertex by procedure visit(v,Adj) as follows: if Adj[v] $\neq$ nil then it returns the leftmost vertex of Adj[v] and deletes the vertex from Adj[v], else return nil. We prepare an array **New** of size m. Initially, New[v] = 'new for every v, and when the DFS visits vertex u then we set New[u] := 'old.

We prepared two procedures for TCsubG. chkTCpair(v,Adj) checks whether a vertex v is tightly-coupled with some vertex, say w, and if so then returns w by deleting v (resp., w) from Adj[w] (resp., Adj[v]). getTCvtxs(v,w,...) finds a set of vertices contained in the TCsubG starting from the tightly-coupled pair (v, w) and returns a set of vertices contained. Note that v and/or w may be coupled tightly with two or more vertices.

Procedure getTCvtxs(v, w, Adj,New,stk) ==
%% called just after chkTCpair(v,Adj)
%%   where, New[v] = 'old, stk = (v).
begin local u;
        New[w] := 'old; **stack** w into stk;
        while (u := chkTCpair(w,Adj)) $\neq$ nil **do**
            stk := getTCvtxs(w,u,Adj,New,stk);
    loop: **if** Adj[v] = nil **then return** stk;
        while (w := chkTCpair(v,Adj)) $\neq$ nil **do**
            stk := getTCvtxs(v,w,Adj,New,stk);
        go to loop;   **end.**

Below, by Bvtx[k] (Tvtx[k] for the TCsubG) we denote the big-vertex formed k-th. In computing SCsubGs as described in **4.3**, we need to know in which SCsubG a given vertex is contained and in which SCsubGs a given SCsubG is contained. In order to answer these questions quickly, we prepare arrays **SCG**, **inBig** and **BinBig**, all of size m. The SCG is for saving the SCsubG (= a sequence of vertices and big-vertices): SCG[i] := [SCsubG formed i-th]. The inBig tells us in which big-vertex the input-vertex v is contained: inBig[v] := k if v is contained in Bvtx[k]. The BinBig tells us in which big-vertex Bvtx[k] is contained: BinBig[k] := a list of indices of big-vertices which contain Bvtx[k]; we save only such big-vertices which are two-level upper. These arrays as well as current maximum index of Bvtx/Tvtx are stored in **SCGs**.

The main procedure for finding SCsubGs is findSCsubG:

> **Procedure** findSCsubG(v, Adj,New,StkV,SCGs) ==
>
> %% a procedure which calls this initially is necessary.
> **begin local** w, u, vtxs;
> $w :=$ chkTCpair(v, Adj);
>   **if** $w =$ nil **then goto** SCG;
> TCG: vtxs := formTCsubG(v, w, Adj, New,SCGs,StkV);
>   **for each** u **in** vtxs **do**
>   StkV := findSCsubG(u, Adj,New,StkV,SCGs);
>   SCG: **if** ($w :=$ visit(v, Adj)) $=$ nil **then goto** rtn;
>   **if** New(w) $=$ 'old **then goto** bakT;
>   New(w) := 'old;
>   StkV := findSCsubG(w, Adj,New,StkV,SCGs);
>   **goto** SCG;
>   bakT: btrackStkV(v, w, StkV,SCGs);
>   **goto** SCG;
>   rtn: StkV := formComplG(v,StkV,SCGs);
>   **return** StkV;   **end**.

Here, formTCsubG forms a TCsubG, saves information into SCGs, stacks the corresponding big-vertex into StkV, and returns a list of vertices outgoing from the TCsubG; btrackStkV backtracks StkV and performs complicated jobs described in **4.3**; formComplG forms a complementary SCsubG, if any, by backtracking StkV to vertex v or wrapped vertex (v Bvtx[*]); for "wrapped vertex", see below.

In the implementation, we encounter a problem. Suppose vertices v_{k_1} and v_{k_2} are outgoing from Bvtx[k]. If we stack these (big-)vertices into StkV as $(\cdots, v_{k_2}, \cdots, v_{k_1}, \text{Bvtx}[k], \cdots)$, it is not easy to recognize that v_{k_2} is connected with Bvtx[k]. Even more difficult case is that Bvtx[k] is contained in Bvtx[k'] then the StkV will be $(\cdots, v_{k_2}, \cdots, v_{k_1}, \text{Bvtx}[k'], \cdots)$. Our idea to solve this problem is to wrap v_{k_i} and v_{k_j} as (v_{k_i} Bvtx[k]) and (v_{k_j} Bvtx[k]), respectively, and stack them instead of v_{k_i} and v_{k_j}.

5   Solving PSLSs by Local Blocking

In this section, we consider solving $B\boldsymbol{x} = \boldsymbol{c}$ by local blocking, where B is an $m \times m$ matrix (b_{ij}), with diagonal elements $b_{11}, \ldots, b_{mm}$ corresponding to the vertices $v_1, \ldots, v_m$ of G_B. Let L_i be the i-th local block of B, composed of i_1-th, $\ldots, i_n$-th rows of B. The corresponding local system can be expressed as

$$L_i\boldsymbol{x}_i = L_i'\boldsymbol{x}_i' + \boldsymbol{c}_i, \quad \boldsymbol{x}_i = {}^t(x_{i_1}, \ldots, x_{i_n}), \quad \boldsymbol{x}_i' = {}^t(c_{i_1'}, \ldots, c_{i_{m-n}'}),$$
$$\text{where } L_i \in K[\boldsymbol{p}]^{n \times n}, \ L_i' \in K[\boldsymbol{p}]^{n \times (m-n)}, \quad \forall i_j' \notin \{i_1, \ldots, i_n\}. \tag{5.1}$$

Definition 6 (Blocked Variable, Unblocked Variable). *Let* L_i, $\boldsymbol{x}_i$, $\boldsymbol{x}_i'$, L_i' *and* $\boldsymbol{c}_i$ *be defined as in (5.1). We call each variable in* $\boldsymbol{x}_i$ *and* $\boldsymbol{x}_i'$ *blocked variable and unblocked variable, respectively, of* L_i.

Let G_L be an SCsubG, corresponding to the local block L_i. The local system in (5.1) is formally solvable w.r.t. blocked variables of L_i and the solution is expressed in terms of unblocked variables. However, G_L may contain big-vertices. In such cases, we solve the local systems corresponding to the big-vertices first, and solve the remaining system last, as we mentioned in Corollary 1.

Theorem 3. *The system $Bx = c$ is formally solvable even if $\det(L_i) = 0$, provided $\det(B) \neq 0$.*

Proof First, we see that $L_i' \neq 0$, because if $L_i' = 0$ then G_L becomes an SCC of G_B, contradicting that G_B is an SCC. Second, the row rank of augmented matrix $(L_i \mid L_i')$ is n, because if the rank is less than n then we have $\det(B) = 0$. Thus, we obtain the theorem. □

5.1 Forming Characteristic Systems

The DFS method is very simple and useful, but it has a serious drawback: the result depends strongly on the initial setting of adjacency list. On the other hand, the characteristic system should be settled by considering global structure of the PSLS. Furthermore, user's professional knowledge is indispensable to determine the characteristic system. In order to cope with these demands, we introduce a concept of "global vertex".

Definition 7 (global vertex). *The user can specify some vertices global. Let v_g be a global vertex which corresponds to variable x_g. Each equation containing a diagonal term $b_{gg}x_g$ is treated to be in a characteristic system.*

Actually, we determine a characteristic system as follows.

First, choose a global vertex as the vertex which the DFS visits first.
Then, modify the adjacency list so that a cyclic path obtained first by the DFS contains the global vertices naturally.

5.2 Actual Method for PSLS Solving

Summarizing, we solve PSLS as follows.

1. Before PSLS solving, determine a characteristic system with the help of users, but solve the system last.
2. First, solve all the tightly-coupled systems, because the variables corresponding to input-vertices of a TCsubG seem to be correlated strongly.
3. Then, solve other local systems in the order that systems in subtree visited earlier are solved earlier. In each SCsubG, solve the local systems corresponding to lower-level big-blocks earlier.
4. In solving local systems successively, replace sub-expressions appearing in the solutions systematically by the rules given below.

5. Output the determinant D of each local systems as a degenerating factor, but do not solve the local system for $D = 0$. Such a computation is quite heavy because the solving way branches quite often. One had better compute degenerated solutions only when such a solution is actually required.

The rules for sub-expression replacement are as follows.

Rule-1. In each local system, replace the determinant of coefficient matrix by solver parameter $D[i]$, where the index i show that $D[i]$ is generated i-th.

Rule-2. Let the local system to be solved be

$$
\begin{cases}
b_{j_1,i_1} x_{i_1} + \cdots + b_{j_1,i_n} x_{i_n} = c_{j_1} + b_{j_1,i'_1} x_{i'_1} + \cdots + b_{j_1,i'_{n'}} x'_{n'}, \\
\vdots \qquad \cdots \qquad \vdots \quad : \quad \vdots \qquad \cdots \qquad \vdots \\
b_{j_n,i_1} x_{i_1} + \cdots + b_{j_n,i_n} x_{i_n} = c_{j_n} + b_{j_n,i'_1} x_{i'_1} + \cdots + b_{j_n,i'_{n'}} x'_{n'},
\end{cases}
$$

Then, we express the solution of x_{i_k} $(1 \le k \le n)$ as

$$
x_{i_k} = \frac{C[k] + \tilde{C}[k, i'_{i_1}] x_{i'_1} + \cdots + \tilde{C}[k, i'_{i_n}] x_{i'_{n'}}}{D},
$$

where C and $\tilde{C}$ are solver parameters; if a sub-expression is a monomial then we do not perform the replacement.

Rule-3. After substituting local solutions for unblocked variables of another local system, we express the system by replacing sub-expressions: the resulting system is such that coefficients $b_{j,i}$, $b_{j,i'}$, c_j $(j_1 \le j \le j_n;\ i_1 \le i \le i_n; i'_1 \le i' \le i'_{n'})$ are replaced by $B_{j,i}$, $B_{j,i'}$, C_j, respectively (for simplicity, we employed the same suffixes as above).

6 Application to a Simple PSLS

We explain our method by applying it to a simple PSLS $Bx = c$, where B and c are as follows. The associated directed graph G_B for B is a little modification of the graph in Fig. 3 in **4.2**: we added two back-edges $(7 \to 4)$ and $(9 \to 7)$.

$$
B = \begin{array}{c}
\text{Eq1} \\ \text{Eq2} \\ \text{Eq3} \\ \text{Eq4} \\ \text{Eq5} \\ \text{Eq6} \\ \text{Eq7} \\ \text{Eq8} \\ \text{Eq9}
\end{array}
\left(\begin{array}{ccccccccc}
b_{11} & & & b_{14} & & & & & \\
b_{21} & b_{22} & b_{23} & & & & & & \\
& b_{32} & b_{33} & & & b_{36} & & & \\
& & b_{43} & b_{44} & & & b_{47} & & \\
& b_{52} & & & b_{55} & & & & \\
& & & & b_{65} & b_{66} & & & \\
& & & & b_{75} & & b_{77} & & b_{79} \\
b_{81} & & & & & & & b_{88} & b_{89} \\
& & & & & & & b_{98} & b_{99}
\end{array}\right),
\quad c = \begin{pmatrix} c_1 \\ c_2 \\ c_3 \\ c_4 \\ c_5 \\ c_6 \\ c_7 \\ c_8 \\ c_9 \end{pmatrix}.
$$

We first show how SCsubGs are found by the adjacency list which is obtained by modifying Adj given in Fig. 3 as $\text{Adj}[7] := (4)$ and $\text{Adj}[9] := (7, 8)$. We show how StkV changes as the DFS walks each edge.

1. DFS($1\rightarrow2$): StkV = (2 1).
2. DFS($2\rightarrow3$): Tvtx[1] := (2 3), StkV = (Tvtx[1] 1).
3. DFS($3\rightarrow4$): StkV := (4 (3 Tvtx[1]) Tvtx[1] 1).
4. back($4\rightarrow1$): Bvtx[2] := (1 Tvtx[1] 4), StkV = (Bvtx[2]).
5. DFS($2\rightarrow5$): StkV = (5 (2 Tvtx[1]) Bvtx[2]).
6. DFS($5\rightarrow6$): StkV = (6 5 (2 Tvtx[1]) Bvtx[2]).
7. back($6\rightarrow3$): Bvtx[3] := (5 6 Tvtx[1]), StkV = (Bvtx[3] Bvtx[2]).
8. DFS($5\rightarrow7$): StkV = (7 (5 Bvtx[3]) Bvtx[3] Bvtx[2]).
9. back($7\rightarrow4$): Bvtx[4] := (Bvtx[3] 7 Bvtx[3]), StkV = (Bvtx[4]).
10. DFS($1\rightarrow8$): StkV = (8 (1 Bvtx[2]) Bvtx[4]),
11. DFS($8\rightarrow9$): Tvtx[5] := (8 9), StkV = (Tvtx[5] (1 Bvtx[2]) Bvtx[4]).
12. back($9\rightarrow7$): StkV = (Tvtx[5] Bvtx[4]).

Next, we consider a characteristic system S_{char}. Specifying vertices $1, 9, 7, 4$ to be global, we obtain $S_{\text{char}} := \{\text{Eq1, Eq4, Eq7}\}$, as shown in Fig. 5.

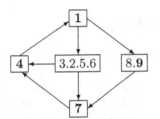

By modifying the Adj specified above as Adj[1] := $(8, 2)$, the DFS walks a cyclic path $(1, 9, 7, 4, 1)$ first, and the path gives the S_{char}.

Fig. 5. Characteristic system obtained by global vertices 1, 9, 7, 4.

Finally, we show how sub-expressions are replaced by solver parameters. We note that Cramer's formula gives us $x_i = N_i/D$, $1 \le i \le 9$, where $D, N_1, N_2, N_3, N_4, N_5, N_6, N_7, N_8, N_9$ contain 15, 44, 28, 37, 47, 29, 34, 51, 34, 31 terms, respectively. Let S_{23}, S_{89} and S_{2356} be local systems corresponding to Tvtx[1], Tvtx[5], Bvtx[3], respectively. We first solve S_{23} and S_{89}, by putting

$$D_{23} := b_{22}b_{33} - b_{23}b_{32}, \quad D_{89} := b_{88}b_{99} - b_{89}b_{98}.$$

$$S_{23} : \begin{cases} b_{22}x_2 + b_{23}x_3 = c_2 - b_{21}x_1, \\ b_{32}x_2 + b_{33}x_3 = c_3 - b_{36}x_6. \end{cases} \Rightarrow \begin{cases} x_2 = (C_2 - b_{33}b_{21}x_1 + b_{36}b_{23}x_6)/D_{23}, \\ x_3 = (C_3 + b_{32}b_{21}x_1 - b_{36}b_{22}x_6)/D_{23}, \end{cases}$$

$$S_{89} : \begin{cases} b_{88}x_8 + b_{89}x_9 = c_8 - b_{81}x_1, \\ b_{98}x_8 + b_{99}x_9 = c_9, \end{cases} \Rightarrow \begin{cases} x_8 = (C_8 - b_{99}b_{81}x_1)/D_{89}, \\ x_9 = (C_9 + b_{98}b_{81}x_1)/D_{89}, \end{cases}$$

where $C_2 := b_{33}c_2 - b_{23}c_3$, $C_3 := -b_{32}c_2 + b_{22}c_3$, $C_8 := b_{99}c_8 - b_{89}c_9$, and $C_9 := -b_{98}c_8 + b_{88}c_9$. Here, we obtain degenerating factors D_{23} and D_{89}.

We next solve S_{2356} which is composed of Eq5 and Eq6:

$$S_{2356} : \begin{cases} D_{23}b_{55}x_5 + b_{52}b_{36}b_{23}x_6 = -C_2 b_{52} + D_{23}c_5 + b_{52}b_{33}b_{21}x_1, \\ b_{65}x_5 + \quad b_{66}x_6 = c_6, \end{cases}$$

$$\Rightarrow \begin{cases} x_5 = (C_5 + b_{66}b_{52}b_{33}b_{21}x_1)/D_{56}, \\ x_6 = (C_6 - b_{65}b_{52}b_{33}b_{21}x_1)/D_{56}, \end{cases}$$

where $C_5 := -C_2 b_{66} b_{52} + D_{23} b_{66} c_5 - b_{52} b_{36} b_{23} c_6$, $C_6 := C_2 b_{65} b_{52} - D_{23} b_{65} c_5 + D_{23} b_{55} c_6$, and $D_{56} := D_{23} b_{66} b_{55} - b_{65} b_{52} b_{36} b_{23}$. Thus, once x_1 is determined, one can determine x_8, x_9, x_5, x_6, x_2, x_3 readily. The degenerating factor obtained here is D_{56}. By substituting local solutions of x_3, x_5, x_9 into Eq4 and Eq7, S_{char} becomes

$$\begin{cases} b_{11} x_1 + b_{14} x_4 = c_1, \\ B_{41} x_1 + D_{56} D_{23} b_{44} x_4 + D_{56} D_{23} b_{47} x_7 = C_4, \\ B_{71} x_1 + D_{89} D_{56} b_{77} x_7 = C_7, \end{cases}$$

$$x_1 = \begin{vmatrix} c_1 & b_{14} & 0 \\ C_4 & D_{56} D_{23} b_{44} & D_{56} D_{23} b_{47} \\ C_7 & 0 & D_{89} D_{56} b_{77} \end{vmatrix} / D_{147}, \quad \text{etc.},$$

$$D_{147} = \begin{vmatrix} b_{11} & b_{14} & 0 \\ B_{41} & D_{56} D_{23} b_{44} & D_{56} D_{23} b_{47} \\ B_{71} & 0 & D_{89} D_{56} b_{77} \end{vmatrix}.$$

7 Concluding Remarks and Acknowledgment

We have tested our algorithm only by small systems so far, and we are wondering about the effectiveness of the current version to big systems. In handling big systems, capabilities of handling modular systems will be necessary. Furthermore, we must treat linear equations containing derivatives w.r.t. the time. Recently, we have found a simple method of forming SCsubGs, in which the SCC algorithm is used recursively, and which seems to be applicable to big systems.

We thank Mr. T. Yamaguchi of Maplesoft for guiding us to industrial computations.

References

1. Cellier, F.E., Kofman, E.: Differential Algebraic Equations. In: Continuous System Simulation, ch. 7. Springer (2006)
2. Duff, I.S., Reid, J.K.: An implementation of Tarjan's algorithm for the block triangularization of a matrix. ACM Trans. Math. Soft. 4, 137–147 (1978)
3. Dulmage, A.C., Mendelsohn, N.S.: Coverings of bipartite graph. Canad. J. Math. 10, 517–534 (1958); A structure theorem of bipartite graphs of finite exterior dimension. Trans. Roy. Soc. Canad. Sec. III 53, 1–13 (1959)
4. Murota, K.: Matrices and Matroids for Systems Analysis. Springer, Berlin (2000)
5. Pothen, A., Fan, C.-J.: Computing the block triangular form of a sparse matrix. ACM Trans. Math. Soft. 16, 303–324 (1990)
6. Sasaki, T., Yamaguchi, T.: On Algebraic Preprocessing of Floating-point DAEs for Numerical Model Simulation. In: Proceedings of SYNASC 2013 (Symbolic Numeric Algorithms on Scientific Computing), SYNASC 2012, West University of Timisoara, Romania, pp. 81–88. IEEE (2013)
7. Tarjan, R.E.: Depth-first search and linear graph algorithms. SIAM J. Computing 1, 146–160 (1972)
8. Weispfenning, V.: Comprehensive Gröbner bases. J. Symb. Comp. 14, 1–29 (1992)

Analytical Calculations in Maple to Implement the Method of Adiabatic Modes for Modelling Smoothly Irregular Integrated Optical Waveguide Structures

Leonid A. Sevastyanov[1,2], Anton L. Sevastyanov[2],
and Anastasiya A. Tyutyunnik[2]

[1] Joint Institute for Nuclear Research,143500, Dubna MR, Russia
[2] Peoples Friendship University of Russia, 117198 Moscow, Russia
{leonid.sevast,alsevastyanov,nastya.tyutyunnik}@gmail.com

Abstract. This paper presents analytical calculations in CAS Maple. The calculations are used for the method of adiabatic waveguide modes, applied in mathematical modeling of smoothly irregular integrated-optical waveguides. Such structures drew researchers' attention at the end of the 20th century, when several models were proposed to describe the coherent laser radiation in such structures. But these models could not adequately characterize the phenomena of depolarization and hybridization of guided modes.

The proposed model of adiabatic waveguide modes, on the contrary, describes these experimentally observed phenomena, but on the other hand, we have more complicated analytical expressions. So we have developed a special program in Maple to computerize analytical calculations. The program is presented in this work.

Keywords: integrated optics, waveguide modes, adiabatic waveguide modes, thin-film waveguide Luneburg lens, numerical modeling, analytical calculations.

PACS: 02.30.Mv, 02.60.Cb, 02.60.Lj, 03.50.De, 03.65.-w, 42.25.-p, 42.25.Gy, 42.50.-p, 42.82.-m, 02.30.Hq , 02.30.Jr, 02.60.-x, 02.60.Cb, 02.70.-c

1 Introduction

The new scientific and technical direction emerged and developed successfully in the 60es of the 20th century (the department of Radiophysics, PFUR, was one of the pioneer groups [2]). The first and simplest models of monochromatic polarized light propagation in fiber and planar optical waveguides were proposed and investigated, some custom-made and serial samples of the first integrated-optical products mentioned in [6] were made.

Some custom-made samples of irregular integrated-optical waveguides have revealed the inadequacy of existing models. Consequently, integrated optical

V.P. Gerdt et al. (Eds.): CASC Workshop 2014, LNCS 8660, pp. 419–431, 2014.

waveguide components and devices designed on the base of these models, were very imperfect. Despite this, some of the samples were run in series of both elements and products including thin-film waveguide lenses and spectrum analyzers on board the U.S. aircraft [6].

In order to improve mass production technologies of integrated optical products, some attempts were made to improve mathematical models of irregular waveguides: — a method of expansion in the small parameter (smooth irregularity), a coupled-wave method, a cross-sections, incomplete Galerkin method. However, they did not describe the phenomena of hybridization and the depolarization of the propagating light observed in the experimental studies of fabricated samples of integrated optical waveguides.

One of the founders of integrated optics as scientific and technical direction, Professor LN Deryugin (the head of the Department of Radiophysics, the PFUR) set the task to develop an adequate model of the propagation of laser radiation in smoothly irregular integrated optical waveguides. A corresponding model was created by a scientific team of the JINR, GPI and PFUR researchers headed by L.A. Sevastiyanov, A.A. Egorov, and A.L. Sevastiyanov, the description and examples of using this model are published in [1,5,4,3,7].

Within the model the key position is given to the following:

a description of the eigenvalues and eigenmodes of an integrated optical waveguide;

a description of propagating in an integrated optical waveguide radiation excited by an external source.

The first task is successfully solved by the team, the second one has arisen on the agenda.

Here we should emphasize a significant "complexity" of the analytical expressions for the base units of the method of adiabatic waveguide modes compared to the corresponding values in the previous methods. In those methods, the analogues of the second problem could be solved by analytical calculations performed manually. In the case of the method of adiabatic waveguide modes (MAWM) needed to solve the second task, analytic calculations are more complex and cumbersome. It is not possible to conduct them manually; one of ways out is to perform these calculations in one of the CASs. We have chosen Maple.

In this paper, we present our first results in the way of analytical calculations to solve the second task. Namely, we describe the analytical calculation of the quantities involved in the formation of MAWM, more over the analytical calculations needed to solve the first problem. In spite of the fact we performed these calculations manually to simulate a number of samples of integrated optical waveguide elements, for modeling and designing future production models it would be necessary to perform the calculations for a large number of waveguide devices with different design solutions. These analytical calculations require automation. This work focuses on the automation of analytical calculations needed to solve the first task.

2   The Concept of the Method of Adiabatic Waveguide Modes

The electromagnetic field propagating in a smoothly irregular multilayer integrated optical waveguide (see Fig. 1) of laser radiation is described by Maxwell's equations, boundary equations at interfaces, and constitutive equations. In our case, the scalar Maxwell's equations follow from the vector ones:

$$rot\tilde{\mathbf{H}} = \frac{1}{c}\frac{\partial \mathbf{D}}{\partial t}, rot\tilde{\mathbf{E}} = -\frac{1}{c}\frac{\partial \mathbf{B}}{\partial t}, \tag{1}$$

and the boundary conditions for the normal components follow from the boundary conditions for the tangential components [8]. Constitutive equations in this case are assumed linear and isotropic: $\mathbf{D} = \varepsilon \tilde{\mathbf{E}}$, $\mathbf{B} = \mu \tilde{\mathbf{H}}$ where ε - permittivity of the medium; μ - permeability of the medium. Tangential boundary conditions at the interface of two media inside the waveguide can be written as

$$\tilde{\mathbf{H}}^{\tau}\Big|_{1} = \tilde{\mathbf{H}}^{\tau}\Big|_{2}, \quad \tilde{\mathbf{E}}^{\tau}\Big|_{1} = \tilde{\mathbf{E}}^{\tau}\Big|_{2}. \tag{2}$$

At an infinite distance from the waveguide the tangential components of the electromagnetic field satisfy the asymptotic conditions:

$$\left\|\tilde{\mathbf{E}}^{\tau}\right\|\Big|_{x\to\pm\infty} < +\infty, \left\|\tilde{\mathbf{H}}^{\tau}\right\|\Big|_{x\to\pm\infty} < +\infty. \tag{3}$$

In the most typical cases, the electrodynamical task is formulated as follows: a certain eigenmode with unit amplitude falls on the irregular region connecting

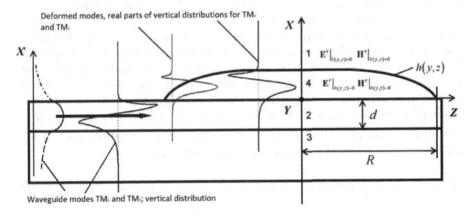

Fig. 1. Cross section of the integrated optical structure formed by regular three-layer waveguide (left panel) and a smoothly irregular four-layer waveguide (on the right side of the figure). Three-layer waveguide is formed by 1-3 media and four-layer by 1-4 media. Also shown are the profiles of 2-directed lower TM modes as well as the real parts of two lower weakly deformed waveguide quasi-TM modes.

two different regular waveguide regions. We need to calculate the amplitude of all the modes that diverge at both sides of the irregular region. This is a direct task. Set of all complex amplitudes is called a scattering matrix of the irregular region. Of particular interest is the inverse problem: the choice of the waveguide parameters, providing, for example, the lowest conversion losses at the given frequency band. The method of adiabatic waveguide modes is based on the fact that waveguide modes (classic and generalized) oscillate rapidly in the direction Ox and change slowly in the directions Oy and Oz. Therefore, the waveguide modes of smoothly irregular waveguides are constructed at first by averaging over the variable x. The resulting averaged solutions $\mathbf{E}\,(y, z)$ and $\mathbf{H}\,(y, z)$ are proportional to functions $\exp\left\{-i\varphi\,(y, z)\right\}\big/\sqrt{\beta\,(y, z)}$, where $\beta\,(y, z) = \sqrt{\beta_y^2\,(y, z) + \beta_z^2\,(y, z)}$, $\beta_y\,(y, z) = \partial\varphi/\partial y$, $\beta_z\,(y, z) = \partial\varphi/\partial z$. "The asymptotic expansion" of the averaged solutions on the fast variable (analogue of the asymptotic expansion in [8]) is sought in the form:

$$\tilde{\mathbf{E}}\,(x, y, z, t) = \mathbf{E}\,(x; y, z)\exp\left\{i\omega t \mp i\varphi\,(y, z)\right\}\big/\sqrt{\beta\,(y, z)}, \qquad (4)$$

$$\tilde{\mathbf{H}}\,(x, y, z, t) = \mathbf{H}\,(x; y, z)\exp\left\{i\omega t \mp i\varphi\,(y, z)\right\}\big/\sqrt{\beta\,(y, z)} \qquad (5)$$

Constructed by the method of "partial separation of variables" the expression for the adiabatic waveguide modes allows to solve on its basis both direct and inverse problems of numerical modeling of smoothly irregular integrated-optical waveguides.

Expressions (4) - (5) are substituted into Maxwell's equations (1) and after that we obtain with additional derivations the equations for the components $E_z\,(x)$ and $H_z\,(x)$ (parametrically dependent on variables y, z), and computational formulas for the components $E_x\,(x)$, $H_y\,(x)$ and $H_x\,(x)$, $E_y\,(x)$ (also parametrically dependent on the variables y, z). The expressions obtained for vertical field distributions of adiabatic modes show (and performed numerical studies [1,5,4,3,7] confirm) that in the course of propagation through the waveguide irregular region these modes are subject to depolarization, so that all six of their components are non-zero, thus, they become hybrid modes. Remind that the expansion of fields by regular waveguide modes in the method of comparison waveguides contains contributions from the TE and TM modes, its separate summations (both discrete and continuous). At the same time, the expansion of fields $\mathbf{F}^\beta = \left(\mathbf{E}^\beta, \mathbf{H}^\beta\right)^T$ in the method of adiabatic waveguide modes already contain "linked" terms of hybrid modes with coefficients $C_\beta^\pm\,(y, z)$:

$$\mathbf{F}^\beta(x, y, z, t) = \int dp(\beta)\mathbf{C}_\beta^\pm(y, z)\mathbf{F}^\beta(x; y, z)\exp\left(i\omega t \mp i\phi^\beta\,(y, z)\right)/\sqrt{\beta(y, z)} \quad (6)$$

Here the integral is taken over the spectral set $\{\beta\}$ for non-self-adjoint operator [8], which has a more complex structure in comparison with a real spectrum of a regular planar waveguide. In expansion (6), a finite set of complex points $\beta = \beta_n = \mathrm{Re}\beta_n + i\mathrm{Im}\beta_n, n = 1, ...N$ with point measures $\delta\,(\beta - \beta_n)\,d\beta$, corresponding to the set of root eigenvectors is present.

3 Basic Equations for an Analysis of Adiabatic Waveguide Modes

Substituting (4) and (5) into Maxwell equations (1) leads after a series of manipulations to the following relations for the dependence of the fast variable $\mathbf{E}\left(x; \beta\left(y, z\right)\right)$, $\mathbf{H}\left(x; \beta\left(y, z\right)\right)$. For the longitudinal components $E_z\left(x; \beta\left(y, z\right)\right)$, $H_z\left(x; \beta\left(y, z\right)\right)$, we get the system of ordinary differential equations of the second order:

$$
\frac{d^2 E_z^\beta}{dx^2} + \chi_\beta^2 E_z^\beta = -p_y^\beta \left(\chi_z^\beta\right)^2 \frac{\partial}{\partial y}\left(\left(\chi_z^\beta\right)^{-2}\right) E_z^\beta
$$
$$
- \frac{1}{i\varepsilon}\left[\left(\chi_z^\beta\right)^2 p_z^\beta \frac{\partial}{\partial y}\left(\left(\chi_z^\beta\right)^{-2}\right)\right]\frac{dH_z^\beta}{dx}, \tag{7}
$$

$$
\frac{d^2 H_z^\beta}{dx^2} + \chi_\beta^2 H_z^\beta = -p_y^\beta \left(\chi_z^\beta\right)^2 \frac{\partial}{\partial y}\left(\left(\chi_z^\beta\right)^{-2}\right) H_z^\beta
$$
$$
+ \frac{1}{i\mu}\left[\left(\chi_z^\beta\right)^2 p_z^\beta \frac{\partial}{\partial y}\left(\left(\chi_z^\beta\right)^{-2}\right)\right]\frac{dE_z^\beta}{dx}. \tag{8}
$$

For the transverse and vertical components of the field $E_x\left(x; \beta\left(y, z\right)\right)$, $E_y\left(x; \beta\left(y, z\right)\right)$, $H_x\left(x; \beta\left(y, z\right)\right)$, $H_y\left(x; \beta\left(y, z\right)\right)$, we obtain the expressions through the longitudinal components $E_z\left(x; \beta\left(y, z\right)\right)$, $H_z\left(x; \beta\left(y, z\right)\right)$ and their derivatives:

$$
\left(\chi_z^\beta\right)^2 H_y^\beta = \left(p_y^\beta p_z^\beta + \frac{\partial p_z^\beta}{\partial y}\right) H_z^\beta - i\varepsilon\frac{dE_z^\beta}{dx}, \left(\chi_z^\beta\right)^2 H_x^\beta = p_z^\beta \frac{dH_z^\beta}{dx} + i\varepsilon p_y^\beta E_z^\beta, \tag{9}
$$
$$
\left(\chi_z^\beta\right)^2 E_y^\beta = i\mu\frac{dH_z^\beta}{dx} + \left(p_y^\beta p_z^\beta + \frac{\partial p_z^\beta}{\partial y}\right) E_z^\beta, \left(\chi_z^\beta\right)^2 E_x^\beta = p_z^\beta \frac{dE_z^\beta}{dx} - i\mu p_y^\beta H_z^\beta. \tag{10}
$$

Here $\beta\left(y, z\right)$ is to be determined for the spectral parameter $\beta \in \{\beta\}$, which does no longer refer to a single polarization TE or TM, but relates to a hybrid of mixed polarization. In (7) - (10) we use the notation:

$$
\left(\chi_z^\beta\right)^2 = \varepsilon\mu + p_z^\beta p_z^\beta + \partial p_z^\beta/\partial z, \chi_\beta^2 = \left(\chi_z^\beta\right)^2 + p_y^\beta p_y^\beta + \partial p_y^\beta/\partial y,
$$

$$
p_y^\beta = -i\beta_y - (2\beta)^{-1}\partial\beta/\partial y, p_z^\beta = -i\beta_z - (2\beta)^{-1}\partial\beta/\partial z.
$$

After substituting the expressions (4) and (5) for the mode fields $\tilde{\mathbf{E}}^\beta\left(x, y, z\right)$, $\tilde{\mathbf{H}}^\beta\left(x, y, z\right)$ in the boundary conditions (2) - (3) reduced boundary conditions for vector functions of one (fast) variable on the layer interfaces are obtained

$$
\mathbf{E}^\tau|_1 = \mathbf{E}^\tau|_2, \mathbf{H}^\tau|_1 = \mathbf{H}^\tau|_2, \tag{11}
$$

and conditions at infinity are obtained:

$$
\|\mathbf{E}^\tau\|\|_{x\to\pm\infty} < +\infty, \|\mathbf{H}^\tau\|\|_{x\to\pm\infty} < +\infty. \tag{12}
$$

Thus, the task for partial differential equations is reduced to a simpler task aimed at solving an ODE system for coupled oscillators with the exact boundary conditions reduced from the exact original boundary conditions for Maxwell's equations. In general, the electromagnetic field is given by the decomposition (6) as a linear combination of guided and radiation modes, the substitution of an electromagnetic field into boundary conditions (2) – (3) leads to the general case of boundary conditions (11)–(12) for the desired vertical distributions of the mode fields $\mathbf{E}\left(x; \beta\left(y, z\right)\right)$, $\mathbf{H}\left(x; \beta\left(y, z\right)\right)$ with unknown coefficients $C_\beta^\pm\left(y, z\right)$.

For vertical distributions of the electromagnetic field of adiabatic modes, the task is aimed at finding (for each fixed value of the horizontal coordinates y, z) solutions (depending on the argument x) of the system of differential equations (7) – (10) satisfying the boundary conditions (11) – (12).

Formulas (7) – (12) were used to solve the vector problem of the electrodynamic guided modes propagation in a three-dimensional integrated optical multilayer waveguide in [1,5,4,3,7]. In all these publications, we solve the task of the propagation and transformation of one adiabatic waveguide mode.

4   Implementation of the Method of Adiabatic Waveguide Modes to Describe the Eigenvalue and Eigenmode

As a result of the implementation of our method for the solution of the electrodynamic problem it splits into three autonomous subproblems:

- The task of solving a nonlinear equation relating the phase $\varphi\left(y, z\right)$ and its derivatives $\partial\varphi/\partial y = \beta_y$ and $\partial\varphi/\partial z = \beta_z$ to the waveguide layer thickness profile $h\left(y, z\right)$ and its derivatives $\partial h/\partial y$ and $\partial h/\partial z$ generalizing the dispersion relation for a regular waveguide.
- The task of solving the second-order ordinary differential equations for the amplitudes $E_z\left(x\right)$ and $H_z\left(x\right)$ parametrically dependent on horizontal variables $\left(y, z\right)$, the right-hand sides of which define the interaction between them.
- The task of numerical integration of the factor $\exp\left\{-i \int\limits_{y_0, z_0}^{y, z}\left(\beta_y dy + \beta_z dz\right)\right\}$ entering the expressions for the electromagnetic fields along the two-dimensional rays given by equations $\frac{d}{ds}\left(\beta\frac{dy}{ds}\right) = \frac{\partial\beta}{\partial y}$, $\frac{d}{ds}\left(\beta\frac{dz}{ds}\right) = \frac{\partial\beta}{\partial z}$, $ds^2 = dy^2 + dz^2$.

The equations for the amplitudes $E_z\left(x\right)$ and $H_z\left(x\right)$ are the equations for coupled oscillators the right side frequencies of which are close to the frequencies of the oscillators. The direct calculation of the frequency shifts in papers [1,5] showed that the zero-approximations of frequency are real and equal for TE_n- and TM_n- modes with the same numbers $n = 1, 2, ...N$, but additives of the first order are various and imaginary.

A new algorithmic method proposed for modeling quasi-guided modes of smoothly irregular integrated optical waveguides allowed in [4,3] to synthesize thin-film generalized Luneburg waveguide lens (TGWL) and compare the results with those of Southwell [9], the most consistent in the synthesis of a Luneberg TGWL with limited aperture. The comparison showed that Southwell's results correspond to the model of comparison waveguides, which is less accurate than zero approximation of the adiabatic modes model.

Theoretical and numerical investigations carried out in our papers led to the understanding of the correct and closed formulation of the direct and inverse problems of waveguide propagation of polarized light in integrated optical waveguides with desired functional properties, such as an ideal amplitude-phase Fourier transformation, the transformation of one type of waveguide modes into another type of waveguide modes, irregular integrated-optical sensors on the base of leaky modes and others.

Most integrated optical devices satisfy the following condition: $\delta = \max \frac{|\nabla \beta|}{\beta}$. This condition allows us to use a series in the small parameter.

5   Analytical Calculations for the First Problem

We obtain the following ordinary differential equations for transverse components of electromagnetic field from the zero-order approximation of the asymptotic series in the small parameter:

$$\frac{d^2 \tilde{E}_y^{(0)}}{dx^2} + \left(\varepsilon \mu - \beta^2\right) \tilde{E}_y^{(0)} = 0, \quad \frac{d^2 \tilde{H}_y^{(0)}}{dx^2} + \left(\varepsilon \mu - \beta^2\right) \tilde{H}_y^{(0)} = 0 \qquad (13)$$

We obtain the following relations for the other components of the field:

$$\tilde{H}_z^0 = \frac{i}{\left(\varepsilon \mu - \beta_y^{\,2}\right)} \left(\varepsilon \frac{d\tilde{E}_y^0}{dx} + i\beta_y \beta_z \tilde{H}_y^0\right) \qquad (14)$$

$$\tilde{E}_x^0 = \frac{-i}{\left(\varepsilon \mu - \beta_y^{\,2}\right)} \left(\beta_y \frac{d\tilde{E}_y^0}{dx} + i\mu \beta_z \tilde{H}_y^0\right) \qquad (15)$$

$$\tilde{H}_x^0 = \frac{-i}{\left(\varepsilon \mu - \beta_y^{\,2}\right)} \left(\beta_y \frac{d\tilde{H}_y^0}{dx} - i\varepsilon \beta_z \tilde{E}_y^0\right) \qquad (16)$$

$$\tilde{E}_z^0 = \frac{-i}{\left(\varepsilon \mu - \beta_y^{\,2}\right)} \left(\mu \frac{d\tilde{H}_y^0}{dx} - i\beta_y \beta_z \tilde{E}_y^0\right) \qquad (17)$$

Let us consider the solution of the first problem for the integrated optical object: thin-film generalized waveguide lens Luneburg (TGWL) (see Fig. 2.). If $\beta > n_j$ general solution (13) for each dielectric layer can be represented as

$$\tilde{E}_y^j = A_j^+ e^{\gamma_j x} + A_j^- e^{-\gamma_j x} \qquad (18)$$

$$\tilde{H}_y^j = B_j^+ e^{\gamma_j x} + B_j^- e^{-\gamma_j x}. \qquad (19)$$

If $\left(n_j^2 - \beta^2\right) > 0$, that is $\beta < n_j$, the solution (13) can be represented as:

$$\tilde{E}_y^j = A_j^+ e^{i\chi_j x} + A_j^- e^{-i\chi_j x} \qquad (20)$$

$$\tilde{H}_y^j = B_j^+ e^{i\chi_j x} + B_j^- e^{-i\chi_j x}. \qquad (21)$$

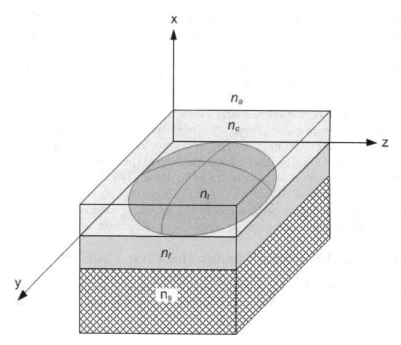

Fig. 2. A 4-layer smoothly irregular waveguide

Having obtained the solution for each dielectric layer we can now get the boundary conditions in an analytical form.

The boundary conditions are represented as the equality of the tangential components of the electromagnetic field at the interface of dielectric media. In the case of plane regular boundaries in the coordinate system connected with the main waveguide layer the tangential components are E_y, E_z , H_y , H_z . If the interface of dielectric media is not regular, the expressions for the tangential components are related to the geometry of the boundary through the derivatives $\partial h/\partial y$, $\partial h/\partial z$. We consider the case with an irregular boundary of the additional waveguide layer. In this case the boundary conditions are written using the expressions for the tangential component of the electric and magnetic fields.

$$E_y^\tau = \frac{\frac{\partial h}{\partial y} E_x + \left[1 + \left(\frac{\partial h}{\partial z}\right)^2\right] E_y - \frac{\partial h}{\partial y}\frac{\partial h}{\partial z} E_z}{1 + \left(\frac{\partial h}{\partial y}\right)^2 + \left(\frac{\partial h}{\partial z}\right)^2},$$

$$E_z^\tau = \frac{\frac{\partial h}{\partial z} E_x - \frac{\partial h}{\partial y}\frac{\partial h}{\partial z} E_y + \left[1 + \left(\frac{\partial h}{\partial y}\right)^2\right] E_z}{1 + \left(\frac{\partial h}{\partial y}\right)^2 + \left(\frac{\partial h}{\partial z}\right)^2},$$

$$H_y^\tau = \frac{\frac{\partial h}{\partial y} H_x + \left[1 + \left(\frac{\partial h}{\partial z}\right)^2\right] H_y - \frac{\partial h}{\partial y}\frac{\partial h}{\partial z} H_z}{1 + \left(\frac{\partial h}{\partial y}\right)^2 + \left(\frac{\partial h}{\partial z}\right)^2},$$

$$H_z^\tau = \frac{\frac{\partial h}{\partial z} H_x - \frac{\partial h}{\partial y}\frac{\partial h}{\partial z} H_y + \left[1 + \left(\frac{\partial h}{\partial y}\right)^2\right] H_z}{1 + \left(\frac{\partial h}{\partial y}\right)^2 + \left(\frac{\partial h}{\partial z}\right)^2}.$$

We have decided to use CAS Maple to obtain analytical expression for the given problem. At first we obtain the tangential components of the electric and magnetic fields in an explicit form as functions of the variable x. For example, for the substrate layer these expressions are represented as follows

```
> #Hz_s := x → B_s · exp(gama_s · (x − a_1));
> #Ez_s := x → A_s · exp(gama_s · (x − a_1));
>
> Ez_s := x → s1·A_s·exp(I·khi_s·(x − a_1)) + (s1 − 1)·A_s·exp(−I·khi_s·(x − a_1)) :
> Hz_s := x → s1·B_s·exp(I·khi_s·(x − a_1)) + (s1 − 1)·B_s·exp(−I·khi_s·(x − a_1)) :
> Ey_s := x → (I·k_0/khi²_sz (mu·diff(Hz_s(x), x) + I·k_0 β_y β_z Ez_s(x))) :
> Hy_s := x → (−I·k_0/khi²_sz (eps_s·diff(Ez_s(x), x) − I·k_0 β_y β_z·Hz_s(x))) :
```

We use the following expressions to obtain analytical form for the tangential components of the electric and magnetic fields on an irregular interface dielectric media:

```
> EyT_lf := x → dhdy·Ex_lf(x) + (1 + (dhdz)²)Ey_lf(x) − dhdy·dhdz·Ez_lf(x) :
> EzT_lf := x → dhdz·Ex_lf(x) + −dhdy·dhdzEy_lf(x) + (1 + (dhdz)²)Ez_lf(x) :
> HyT_lf := x → dhdy·Hx_lf(x) + (1 + (dhdz)²)Hy_lf(x) − dhdy·dhdz·Hz_lf(x) :
> HzT_lf := x → dhdz·Hx_lf(x) + −dhdy·dhdzHy_lf(x) + (1 + (dhdz)²)Hz_lf(x) :
```

The equations corresponding to the equality of the tangential components at the interface of the dielectric medium are obtained using the following commands:

```
> eq9 := simplify(eval((EzT_lf(x) − EzT_l(x))·khi²_lfz·khi²_lz = 0, x = a_3)) :
> eq10 := simplify(eval((EyT_lf(x) − EyT_l(x))·khi²_lfz·khi²_lz = 0, x = a_3)) :
> eq11 := simplify(eval((HzT_lf(x) − HzT_l(x))·khi²_lfz·khi²_lz = 0, x = a_3)) :
> eq12 := simplify(eval((HyT_lf(x) − HyT_l(x))·khi²_lfz·khi²_lz = 0, x = a_3)) :
```

As we obtained the equations, which correspond to the boundary conditions, we can form the matrix from the equations:

> eqs := [eq1, eq2, eq3, eq4, eq5, eq6, eq7, eq8, eq9, eq10, eq11, eq12, eq13, eq14, eq15, eq16] :

> var := $\left[A_s, A_{f1}, A_{f2}, A_{lf1}, A_{lf2}, A_{l1}, A_{l2}, A_c, B_s, B_{f1}, B_{f2}, B_{lf1}, B_{lf2}, B_{l1}, B_{l2}, B_c \right]$:

> M, A := GenerateMatrix(eqs, var) :

The indeterminate vector consists from amplitudes of electric and magnetic fields.

We have a 16×16 matrix for a four-dimensional integrated-optical structure. Increasing the number of dielectric layers, we increase the dimension of the system of linear algebraic equations and corresponding expressions become more cumbersome. We consider it is necessary to carry out such calculations using computer algebra systems. So it is necessary to computerize the obtaining of analytical expressions.

We cite the fragment of the matrix as an example below: the first eight rows and eight columns:

$$
\begin{vmatrix}
2s1-1 & & -e^{-Ikhi_{f}d_1} & -e^{Ikhi_{f}d_1} & 0 & 0 & 0 & 0 & 0 \\
-I\left(-2s1k_0khi_{f2}^2\beta_y\beta_z + k_0khi_{f2}^2\beta_y\beta_z\right) & -Ikhi_{zz}^2e^{-Ikhi_{f}d_1}k_0\beta_y\beta_z & -Ikhi_{zz}^2e^{Ikhi_{f}d_1}k_0\beta_y\beta_z & 0 & 0 & 0 & 0 & 0 \\
0 & 0 & 0 & 0 & 0 & 0 & 0 & 0 \\
-1eps_zkhi_{f2}^2khi_z & Ikhi_{zz}^2e^{-Ikhi_{f}d_1}eps_fkhi_f & -Ikhi_{zz}^2e^{Ikhi_{f}d_1}eps_fkhi_f & 0 & 0 & 0 & 0 & 0 \\
0 & 1 & 1 & -e^{-Ikhi_{y}h} & -e^{Ikhi_{y}h} & 0 & 0 & 0 \\
0 & Ikhi_{f2}^2k_0\beta_y\beta_z & Ikhi_{f2}^2k_0\beta_y\beta_z & -Ikhi_{f2}^2e^{-Ikhi_{y}h}k_0\beta_y\beta_z & -Ikhi_{f2}^2e^{Ikhi_{y}h}k_0\beta_y\beta_z & 0 & 0 & 0 \\
0 & 0 & 0 & 0 & 0 & 0 & 0 & 0 \\
0 & -Ikhi_{f2}^2eps_fkhi_f & Ikhi_{f2}^2eps_fkhi_f & Ikhi_{f2}^2e^{-Ikhi_{y}h}eps_{lf}khi_{lf} & -Ikhi_{f2}^2e^{Ikhi_{y}h}eps_{lf}khi_{lf} & 0 & 0 & 0
\end{vmatrix}
$$

6   Calculation of the Adiabatic Waveguide Modes in the Delphi Package

The aim of our study is designing irregular integrated optical structures with the assigned properties. As an example we consider a fairly complex object — thin-film generalized waveguide lens of Luneburg. In the framework of the model under consideration the greatest difficulty is presented by obtaining an analytical expression for the boundary conditions. These conditions are written down for the electric and magnetic fields at the boundaries of the interface of the homogeneous dielectric layers of the object under consideration. Their number and structure are directly related to the geometric structure of the object being designed.

With the help of the Maple package we managed to obtain the symbolic expressions to record the exact boundary conditions for the task under consideration. Our challenge is to find the solutions of equations (13) - (17) for the components of the electric and magnetic field along the Ox axis. Considering the type of solutions, there are two main parameters that determine the solution. These are the phase retardation coefficient β and the amplitude coefficients themselves $(\mathbf{A}, \mathbf{B})^T$ for the components of the electric and magnetic fields. The task of obtaining the solution is divided into two stages: obtaining β and obtaining amplitude coefficients. The system of linear algebraic equations obtained on the basis of boundary conditions is homogeneous. The condition for solvability of such system is the condition of its determinant equaling zero. This condition is expressed as follows $F_{Disp}\left(\beta, \beta_y, \beta_z; h, \partial h/\partial y, \partial h/\partial z; n_s, n_f, n_l, n_c, d\right) = 0$. Minimizing the corresponding functional allows us to obtain $\beta(y, z)$ distribution.

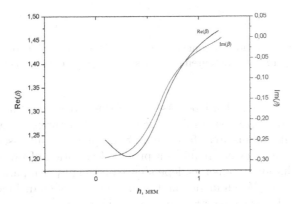

Fig. 3. Dispersion dependencies for the real and imaginary parts of β

Having obtained β, we can calculate all the elements of the resulting matrix. To calculate the amplitude coefficients it is necessary to solve the linear algebraic equations obtained earlier. In the framework of the Delphi package, we implemented stable algorithm for solving homogeneous linear algebraic equations. This algorithm is based on the minimization of Tikhonov's functional:

$$\left\|\hat{M}\left(\beta_m\right)(\mathbf{A}, \mathbf{B})^T\right\|^2 + \alpha\left\|(\mathbf{A}, \mathbf{B})^T - (\mathbf{A}_0, \mathbf{B}_0)^T\right\|^2 \longrightarrow min,$$

here $(\mathbf{A}_0, \mathbf{B}_0)^T$ means the amplitude coefficients of the electromagnetic field at the preceding point of the trajectory of the adiabatic waveguide mode propagation in the simulated object. Having phase retardation coefficient and amplitude coefficients at our disposal, we can calculate the electrical and magnetic field at any point of the object under consideration with a high degree of accuracy.

7   Conclusion

This paper touches upon the issue of scientific and technical directions of integrated optics. The topical areas urgency is confirmed by numerous works of both

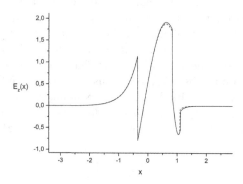

Fig. 4. Dependence of field component E_z on x at the nearby points on the trajectory. Fields are given in relative measurement units, x are given in micrometers.

theoretical and practical character. As part of this direction we have considered integrated-optical layered waveguides (waveguide structures). These structures are widely used in various areas from aircraft equipment to ultra-sensitive detectors. Designing such devices requires a precise description of propagating electromagnetic radiation parameters dependence on the physical parameters of the waveguide structure. When modelling the propagation of guided modes through integrated-optical waveguide structures two different tasks can be solved. The first one is related to the eigenvalues and eigenvectors problems. In the second task, the solution is sought as a linear combination of eigenvectors. In both cases, the direct tasks are solved.

In this work, we have considered smoothly irregular integrated-optical waveguides having a layered structure with irregular boundaries between isotropic media layers. Such a feature leads to a number of physical phenomena, such as the phenomena of depolarization and hybridization of waveguide modes while propagating through irregular regions. The account of these phenomena in modelling integrated optical waveguide structures is crucial for the design of devices based on them. This work is devoted to the application of analytical calculations in the implementation of the method of adiabatic waveguide modes in modelling smoothly irregular integrated-optical waveguides.

Moreover, analytical expressions for boundary conditions should be obtained on every boundary for each certain integral-optical device. It allows to reach a description (of sufficient quality) of such effects. A special algorithm for the automation of this process has been developed in an environment of computer algebra Maple. As the result of this algorithm, the analytical expressions were obtained and then converted into the Delphi code for further numeral calculations.

References

1. Ayryan, E.A., Egorov, A.A., Sevastyanov, L.A., Lovetskiy, K.P., Sevastyanov, A.L.: Mathematical modeling of irregular integrated optical waveguides. In: Adam, G., Buša, J., Hnatič, M. (eds.) MMCP 2011. LNCS, vol. 7125, pp. 136–147. Springer, Heidelberg (2012)
2. Deryugin, L.N., Marchuk, A.N., Sotin, V.E.: Properties of planar asymmetric dielectric waveguides on a substrate of dielectric. Izv. Vuzov, Radioelectronics 10(2), 134–142 (1967) (in Russian)
3. Egorov, A.A., Lovetskiy, K.P., Sevastianov, A.L., Sevastianov, L.A.: Simulation of guided modes (eigenmodes) and synthesis of a thin-film generalised waveguide Luneburg lens in the zero-order vector approximation. Quantum Electronics 40(9), 830–836 (2010)
4. Egorov, A.A., Sevastyanov, A.L., Ayrjan, E.A., Lovetskiy, K.P., Sevastianov, L.A.: Zero approximation of vector model for smoothly-irregular optical waveguide. Matem. Mod. 22(8), 42–54 (2010)
5. Egorov, A.A., Sevast'yanov, L.A.: Structure of modes of a smoothly irregular integrated-optical four-layer three-dimensional waveguide. Quantum Electronics 39(6), 566–574 (2009)
6. Hansperger, R.G.: Integrated Optics: Theory and Technology. Springer (1991)
7. Sevastianov, L.A., Egorov, A.A.: The theoretical analysis of waveguide propagation of electromagnetic waves in dielectric smoothly-irregular integrated structures. Optics and Spectroscopy 105(4), 576–584 (2008)
8. Sevastianov, L.A., Egorov, A.A., Sevastyanov, A.L.: Method of adiabatic modes in studying problems of smoothly irregular open waveguide structures. Physics of Atomic Nuclei. 76(2), 224–239 (2013)
9. Southwell, W.H.: Inhomogeneous optical waveguide lens analysis. JOSA 67(8), 1004–1009 (1977)

CAS Application to the Construction of the Collocations and Least Residuals Method for the Solution of the Burgers and Korteweg–de Vries–Burgers Equations*

Vasily P. Shapeev and Evgenii V. Vorozhtsov

Khristianovich Institute of Theoretical and Applied Mechanics,
Russian Academy of Sciences, Novosibirsk 630090, Russia
{shapeev,vorozh}@itam.nsc.ru

Abstract. In the present work, the computer algebra system (CAS) is applied for constructing a new version of the analytic-numerical method of collocations and least residuals (CLR) for solving the Burgers equation and the Korteweg–de Vries–Burgers equation. The CAS is employed at all stages from writing, deriving, and verifying the formulas of the method to their translation into arithmetic operators of the Fortran language. The verification of derived formulas of the method has been done on the test problem solutions. Comparisons of the results of numerical computations by the new CLR method with the exact solutions of test problems show a high accuracy of the developed method.

Keywords: Computer algebra system, Korteweg–de Vries–Burgers equation, derivation of the formulas of the analytic-numerical algorithm, interface between CAS and Fortran, computer code verification.

1 Introduction

As was shown previously by different researchers, in particular, in the works [14, 19, 20, 24, 38–43], the CASs provide the efficient means for the development and investigation of new numerical algorithms.

The method of collocations and least squares (CLS) was originally developed for the numerical solution of stationary problems of fluid dynamics. In the work [39], this method was applied for the numerical solution of the Stokes equations of viscous incompressible fluid. The CLS method was extended in [38] for the case of the numerical solution of two-dimensional Navier–Stokes equations of viscous incompressible fluid, and a square computational grid was used in the plane of two spatial coordinates. The CLS method was generalized in [19, 20] for the case of rectangular grid cells in the two-dimensional case. The versions of the CLS method with the accuracy orders from 2 to 8 were developed in [20],

---

* The work was partially supported by the Russian Foundation for Basic research (grant No. 13-01-00277).

V.P. Gerdt et al. (Eds.): CASC Workshop 2014, LNCS 8660, pp. 432–446, 2014.

and the capabilities of the method were checked on the solution of a well-known benchmark problem of the viscous incompressible fluid flow in a two-dimensional lid-driven cavity for the Reynolds number Re = 1000. The accuracy of the solution obtained by the CLS method of the sixth accuracy order was shown to be at the level of the best known results. The CLS method was extended in [41] for the case of three spatial variables, and the solution increments were found in the process of iterations. A modification of the CLR method was presented in [42], which was termed the "method of collocations and least rsiduals" (CLR).

The CLR method is a projection grid method. It searches for the solution in each cell of the difference grid in the form of a linear combination of basis elements of some functional space. The space of polynomials is mainly used as this space due to certain convenience. The CLR method differs from other numerical methods in that it reduces the numerical solution of the problem to the solution of an overdetermined system of linear algebraic equations (SLAE). The solution of the latter is found from the requirement of the minimization of the functional of the residual of problem equations on its numerical solution. Such a combination of the method of collocations with a "strong" requirement for the discrete problem solution leads to an improvement of its properties (smoothness, accuracy) as compared to the solutions obtained by a simple method of collocations. The CLR method in fact possesses also other improved properties in comparison with the method of collocations. In particular, the minimization of the functional of the numerical solution residual contributes to a suppression (damping) of various disturbances arising in the process of problem solution and accelerates its convergence at the iterative technique of its construction.

A version of the algorithm for convergence acceleration based on the well-known Krylov's subspaces was used in [43] at the solution of problems by the CLR method. This has enabled a reduction of the iterations number at problems solution by the factors from 11 to 17 in comparison with the case when the Krylov's subspaces were not applied. In the works [42, 43], the numerical solutions of the problem of a steady flow in a cubic lid-driven cavity were compared with the known most accurate results of solving this benchmark problem. It has turned out that the numerical solution of this problem by the CLR method agrees very well with the known high-accuracy solutions.

Along with the stationary problems there are in fluid dynamics many unsteady applied problems that is the problems in which the solutions depend not only on spatial variables but also on time. In this connection, it is of practical interest to generalize the CLR method for the case of solving the unsteady Navier–Stokes equations of the viscous incompressible fluid.

The Burgers equation [7] involves a nonlinear convective term and a viscosity term. In this sense, it has the features common with the unsteady Navier–Stokes equations. In this connection, the new numerical methods of solving the unsteady Navier–Stokes equations are often tested on a simpler mathematical model — the Burgers equation — in order to better understand such properties as the accuracy and stability. A typical difficulty at the numerical solution of the Burgers equation is the consideration of the case of a very small viscosity.

Various numerical methods were developed until now for the numerical solution of the Burgers' equation. These are, in particular, automatic differentiation method [5], Galerkin finite element method [12], cubic B-splines collocation method [2, 11, 33], spectral collocation method [22, 23], sinc differential quadrature method [25], polynomial based differential quadrature method [27], quartic B-splines differential quadrature method [26], quartic B-splines collocation method [37], quartic B-splines finite element method [1, 34, 36], fourth-order finite difference method [16], factorized diagonal Padé approximation [4], non-polynomial spline approach [35], explicit and exact-explicit finite difference methods [29], least-squares B-spline finite element method [30], reproducing kernel function method [44], fourth-order compact finite difference scheme [32], cubic B-spline quasi-interpolation method [21], wavelet-Taylor Galerkin method [28].

The Korteweg–de Vries–Burgers (KdVB) equation differs from the Burgers equation by the presence of a term involving the third derivative. The KdVB equation is applied for the mathematical description of wave processes in bubbly fluids [24, 31], for modelling weak plasma shocks propagation perpendicularly to a magnetic field [15], and for describing shallow water waves on viscous fluid [18].

We now enumerate several methods, which are applied for the numerical solution of the KdVB equation: finite difference methods [6, 24], B-spline finite element method [3], quintic spline method [13], spectral collocation method [10], finite Fourier transform technique [8].

In the present work, a symbolic-numeric CLR method is described for solving the Burgers equation and the KdVB equation. The symbolic stage has been implemented with the aid of CAS *Mathematica*. The accuracy of the new method has been investigated by the examples of computations of several test initial- and boundary-value problems for the above equations.

2   Description of the CLR Method

2.1   Statements of Problems

Consider the Burgers equation [7]

$$\frac{\partial U}{\partial t} + U\frac{\partial U}{\partial x} = \nu\frac{\partial^2 U}{\partial x^2} \tag{1}$$

and the Korteweg–de Vries–Burgers equation [24]

$$\frac{\partial U}{\partial t} + U\frac{\partial U}{\partial x} - \frac{1}{\mathrm{Re}}\frac{\partial^2 U}{\partial x^2} + \frac{1}{D_\sigma^2}\frac{\partial^3 U}{\partial x^3} = 0, \tag{2}$$

where x is the spatial coordinate, t is the time, ν is the viscosity coefficient, $\nu = \mathrm{const} > 0$, Re is the Reynolds number, D_σ is the dispersion. Equations (1) and (2) are solved in the interval $0 \leq x \leq X$ $(X > 0)$ under the following initial and boundary conditions:

$$U(x, t_0) = U_0(x), \quad 0 \leq x \leq X, \tag{3}$$
$$U(0, t) = g_1(t), \quad U(X, t) = g_2(t), \quad t \geq t_0, \tag{4}$$

where $U_0(x), g_1(t), g_2(t)$ are the given functions, t_0 is a given initial moment of time (for example, $t_0 = 0$). In the case of the solution of equation (2), the condition $\partial U(X, t)/\partial x = 0$ was posed in addition to conditions (3) and (4).

2.2   Local Coordinates and Basis Functions

Let us formulate a "discrete" problem approximating the original differential initial- and boundary-value problem. In the CLR method, a computational grid is generated in the interval $[0, X]$. In the present work, a uniform computational grid is used, in which the length of each cell along the x-axis amounts to $2h$. For writing the formulas of the CLR method it is convenient to introduce in each cell the local coordinate y by the formula

$$y = (x - x_{ci})/h, \quad i = 1, \ldots, I, \tag{5}$$

where x_{ci} is the value of the x-coordinate at the geometric center of the ith cell, I is the number of grid cells in the interval $[0, X]$, $I \geq 1$. By virtue of definition (5), the local y-coordinate varies in the interval $y \in [-1, 1]$. Introduce the notation $u(y, t) = U(hy + x_{ci}, t)$. After this substitution of the variable, equations (1) and (2) take the following form:

$$\frac{\partial u}{\partial t} + \frac{1}{h} u \frac{\partial u}{\partial y} = \frac{\nu}{h^2} \frac{\partial^2 u}{\partial y^2}, \tag{6}$$

$$\frac{\partial u}{\partial t} + \frac{1}{h} u \frac{\partial u}{\partial y} - \frac{1}{\mathrm{Re} h^2} \frac{\partial^2 u}{\partial y^2} + \frac{1}{D_\sigma^2 h^3} \frac{\partial^3 u}{\partial y^3} = 0. \tag{7}$$

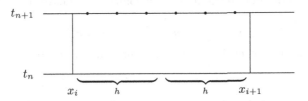

Fig. 1. Positions of collocation points

The solution of each of equations (6) and (7) is advanced in time with a variable step τ_n, where n is the time layer number, $n = 0, 1, \ldots$. Let us assume that the solution of equation (6) or (7) is known at the moment of time t_n. Let us construct the formulas of the method for computing the solution at the moment of time $t_{n+1} = t_n + \tau_n$. By analogy with [41, 42] we introduce N_c collocation points in each cell of the grid on the x-axis, where N_c is specified by the user, $N_c \geq 1$. Figure 1 shows the positions of collocation points for the case when $N_c = 6$. Let Δy be the distance between the collocation points, and let $y_1, \ldots, y_{N_c}$ be the local coordinates of collocation points in the ith cell. Then $y_1 = -1 + \frac{\Delta y}{2}, y_{j+1} = y_j + \Delta y, j = 1, \ldots, N_c - 1$. At such a technique for specifying the y_j coordinates the collocation point y_{N_c-1} in the $(i-1)$th cell lies from the collocation point y_1 in the ith cell at the distance Δy.

A specific form of each collocation equation depends on the technique of the approximation of differential equation (6) or (7) in time as well as on the technique of the linearization of the nonlinear term of the equation to be solved. At first consider the approximation of the Burgers equation (1) at the jth collocation point of the ith cell by the following implicit method:

$$u_{i,j}^{n+1,s+1} - u_{i,j}^n + \tau_n \left[\alpha \left(\frac{1}{h} u_{i,j}^{n+1,s} \frac{\partial u_{i,j}^{n+1,s+1}}{\partial y} - \frac{\nu}{h^2} \frac{\partial^2 u_{i,j}^{n+1,s+1}}{\partial y^2} \right) + (1 - \alpha) \times \right.$$

$$\left. \left(\frac{1}{h} u_{i,j}^n \frac{\partial u_{i,j}^n}{\partial y} - \frac{\nu}{h^2} \frac{\partial^2 u_{i,j}^n}{\partial y^2} \right) \right] = \tau_n \left[\alpha f_{i,j}(y_j, t_{n+1}) + (1 - \alpha) f_{i,j}(y_j, t_n) \right], \qquad (8)$$

where α is the weight parameter. At $\alpha = 1$, equation (8) gives the implicit Euler method, and at $\alpha = \frac{1}{2}$, (8) coincides with the Crank–Nicholson method [9]; $i = 1, \ldots, I$; $j = 1, \ldots, N_c$; $n = 0, 1, \ldots$, s is the iteration number for the iterations in nonlinearity, $s = 0, 1, \ldots$; $u_{i,j}^{n+1,0} = u_{i,j}^n$, $f_{i,j}(y, t)$ is a given right-hand side, it is generally different from zero if the convective term is linearized after Newton.

Represent the approximate solution in the ith cell on the x-axis in the form of a linear combination of basis functions φ_l

$$u_i^{n+1,s+1} = \sum_{l=1}^{4} b_{i,l}^{n+1,s+1} \varphi_l(y) \qquad (9)$$

with indeterminate coefficients, which will be found from the discrete problem solution and will determine the numerical solution. We have used the following basis functions:

$$\varphi_1 = 1, \quad \varphi_2 = y, \quad \varphi_3 = y^2, \quad \varphi_4 = y^3. \qquad (10)$$

That is the solution in each cell is represented in the form of a third-degree polynomial in y.

2.3   Derivation of the Overdetermined System of Collocation Equations and Matching Conditions

Substituting the y_j coordinates of collocation points in equation (8), we obtain N_c collocation equations in each cell:

$$\sum_{m=1}^{4} a_{i,j,m} b_{i,m}^{n+1,s+1} = f_{i,j}^{n+1,s}, \quad i = 1, \ldots, I, \quad j = 1, \ldots, N_c. \qquad (11)$$

They are linear algebraic equations for determining the coefficients in the solution representation (9).

Along with collocation equations (11) we use at the boundaries of each cell the conditions of matching the solution therein with the solution in two neighboring cells. We have applied the matching conditions of two kinds. The first kind is the matching of solutions:

$$u^+ = u^-. \tag{12}$$

On the left-hand side of this equality, one takes the solution in a cell under consideration, and on the right-hand side, one takes the solution in the neighboring cell. For example, if we consider the right boundary x_{i+1} of the ith cell, then the value of the local coordinate $y = 1$ at this boundary; and for the $(i + 1)$th cell, the point $x = x_{i+1}$ is the left boundary, therefore, we will have $y = -1$ at the same point in representation (9).

The second kind of the matching condition is the matching of the first solution derivative:

$$\frac{du^+}{dy} = \frac{du^-}{dy}. \tag{13}$$

One can restrict oneself only to the matching conditions (12) for solutions and avoid the inclusion of the matching condition (13) in the algebraic system for determining the coefficients $b_{i,l}^{n+1,s+1}$. The computational experiments have, however, shown that the inclusion of conditions (13) increases substantially the accuracy of numerical solutions obtained by the CLR method; furthermore, the convergence of iterations in nonlinearity accelerates considerably.

If the cell boundary coincides with one of the computational region boundaries $x = 0$ or $x = X$, then the boundary conditions are used instead of the matching conditions at this boundary. For example, we assume $u^{n+1,s+1} = g_1(t_{n+1})$ at point $x = 0$ according to (4).

Uniting the collocation equations (11) and the matching conditions (12), (13) into a single algebraic system, we obtain a linear algebraic system, the number of the equations of which in each cell is equal to $N_c + 4$. Because $N_c \geq 1$, it is clear that the system is overdetermined, that is the number of unknowns $b_{i,1}^{n+1,s+1}, \ldots, b_{i,4}^{n+1,s+1}$ is less than the number of equations.

The process of the computation at the time level t_{n+1} proceeds in the direction of the increasing numbers i. Therefore, one employs in the right-hand sides of matching conditions at the right boundary of each cell the solution obtained at the sth iteration in nonlinearity. The overdetermined system (11),(12),(13) was solved in each cell with respect to four unknowns $b_{i,l}^{n+1,s+1}$, $l = 1, \ldots, 4$ by an orthogonal method of QR expansion, where the Givens rotation matrix or the Householder matrix of reflections was taken as the matrix Q. The advantage of this method over the least-squares method was discussed previously in [42] and in more detail in [43]. The essence of this advantage lies in the fact that the application of the QR expansion for solving the SLAE does not deteriorate the condition number of the original system in contrast to the case of the application of the least-squares method.

2.4   Application of CAS for Generating the Fortran Subroutines of the Numerical Solution of the Problem

The use of CASs, which implement without difficulties the transformations of expressions in symbolic form, enables one to avoid many errors, which usually arise at an attempt to derive the big formulas "manually". In the given work, a program was written in the language of system *Mathematica*, with the aid of which all basic computational formulas of the CLR method versions were obtained and then tested. In the following, we briefly describe the main steps of the work of this program and present its corresponding fragments.

Step 1. The specification of expressions for basis functions φ_l according to (10). Here is the fragment of the *Mathematica* program, in which the basis functions are specified:

```
fi[[1]] = 1; fi[[2]] = y; fi[[3]] = y^2; fi[[4]] = y^3;
```

Step 2. Solution expansion in the given basis φ_l:

```
u = Sum[a[[m]]*fi[[m]], {m, 4}];
```

Here a[[m]] is the coefficient of the expansion in basis, it is assumed known at the foregoing sth iteration. The solution at the next iteration, which is to be found, is specified similarly:

```
u1 = Sum[b[[i]]*fi[[i]], {i, 4}];
```

Here b[[i]] are the expansion coefficients, which are to be found by the CLR method.

Step 3. Symbolic computation of the left- and right-hand sides of the collocation equation (8):

```
lapu1 = D[u1, {y, 2}]; convuu1 = u*D[u1, y];
equ1 = u1 - ujn + dt*(convuu1/h - anu*lapu1/h^2)- f);
equ1 = Expand[equ1];
```

Here anu $\overset{def}{=} \nu$, dt $\overset{def}{=} \tau_n$, ujn $\overset{def}{=} u_j^n$.

Step 4. Computation of the collocation equation coefficients and their storing in the external file colloc.txt.

```
SetDirectory["D:\\papers\\CASC"];
eNS1 = Table[0, {4}];
Do[e11 = Coefficient[equ1, b[[m]]]; eNS1[[m]] = e11, {m, 4}];
 "!! Collocation equation at the given y-point" >> colloc.txt;
 Do[eq = "        AR(j,"; eq = eq <> ToString[m] <> ") = ";
 e11 = FullSimplify[eNS1[[m]]]; e1f = FortranForm[e11];
 eq<>ToString[e1f] >>> colloc.txt, {m, 4}];
```

Here AR(j,m) is the element of the matrix of the overdetermined system, which stands at the intersection of the jth row and the mth column, $j = 1, \ldots, N_c$, $m = 1, \ldots, 4$.

Step 5. The verification of the correctness of computing the entries AR(j,m), m = 1,2,3,4. To this end, a sum of the products of the above entries and the corresponding expansion coefficients b1,...,b4 is composed. This sum is then compared with the original equation equ1 obtained at Step 3.

Step 6. Symbolic computations of matching conditions (12) on the faces y= -1, y = 1 of a computational grid cell and their storing in the Fortran form in th external file `match.txt`. Let us illustrate this step by the example of considering the case when the face y= 1 lies at the right boundary of the computational region, and we use in the right-hand side of the matching condition the solution value in accordance with the boundary condition:

```
matNS1 = Table[0, {4}];
Do[e11 = Coefficient[u1, b[[m]]]; matNS1[[m]] = Simplify[e11], {m, 4}];
Do[eq = "        AR(j,"; eq = eq <> ToString[m] <> ") = ";
e11 = matNS1[[m]]; e1f = FortranForm[e11];
eq <> ToString[e1f] >>> match.txt, {m, 4}];
"        BR(i) = g2(t)" >>> match.txt;
```

Here g2 is the notation for the function g_2 in the boundary condition; BR(i) is the right-hand side of the matching equation. The matching conditions (13) are calculated similarly in symbolic form and are written in the Fortran form into the same external file `match.txt`.

The result of the work of the Fortran code is a piecewise-polynomial solution in the form of a polynomial in the variable y with numerical coefficients, which is individual for each cell. The polynomial can be differentiated and integrated exactly without using the approximate numerical procedures, which introduce the extra errors in the results of these operations.

The CAS is used again for plotting the numerical solution graphs and their analysis. Such an interface between the CAS and the language of the numerical solution of problems enables the mathematician to avoid many errors at all stages of the work, reduces the necessary stress of his efforts, which is related to the required increased attention, and the amount of the routine work, and speeds it up on the whole.

3 Results of Numerical Computations

We have considered the following three smooth test solutions of equation (1):

$$U(x,t) = t + x^3; \tag{14}$$
$$U(x,t) = t^2 + x^4; \tag{15}$$
$$U(x,t) = \exp(t + x). \tag{16}$$

Since these solutions do not satisfy equation (1), we have solved instead of (1) the following equation with a nonzero right-hand side $f(x,t)$:

$$\frac{\partial U}{\partial t} + U\frac{\partial U}{\partial x} - \nu\frac{\partial^2 U}{\partial x^2} = f(x,t). \tag{17}$$

The right-hand sides $f(x,t)$ corresponding to functions (14),(15), and (16) are as follows:

$$f(x,t) = 1 + 3x^2(t + x^3) - 6\nu x;$$
$$f(x,t) = 2t + 4x^3(t^2 + x^4) - 12\nu x^2;$$
$$f(x,t) = e^{t+x}(1 - \nu + e^{t+x}).$$

The initial conditions were specified at $t = 0$. To this end, the value $t = 0$ was assumed in (14)–(16). For example, the initial function $U_0(x)$ corresponding to solution (14) has the form $U_0(x) = x^3$. The boundary conditions were specified similarly also from the exact solution so that $U(0, t) = t$, $U(X, t) = t + X^3$ in the case of the test solution (14).

The time step τ_n was specified at the numerical solution of the Burgers equation with regard for the following limitations caused by the requirement of the numerical solution stability.

1°. The limitation due to the convective term approximation [14]:

$$\max_j |u_j^n| \cdot \frac{\tau_n}{2h} \leq \kappa_1. \tag{18}$$

Here κ_1 is the Courant number, a typical stability condition for explicit difference schemes has the form $0 < \kappa_1 \leq 1$.

2°. A typical form of the stability condition of explicit difference schemes for the diffusion equation $u_t = \nu u_{xx}$ has the following form [14]:

$$\frac{\nu \tau_n}{4h^2} \leq \kappa_2, \tag{19}$$

where $0 < \kappa_2 < 1$. One can unite formulas (18) and (19) into a single one as follows:

$$\tau_n = \frac{2h\kappa_1}{\max_j |u_j^n|} + \frac{4h^2\kappa_2}{\nu}. \tag{20}$$

If one sets in (20) $\kappa_1 \neq 0$, $\kappa_2 = 0$, then one obtains that $\tau_n = O(h)$. If $\kappa_1 = 0$, $\kappa_2 \neq 0$, then $\tau_n = O(h^2)$. Besides (20), we used in our computations also the formula

$$\tau_n = 4h^3 \kappa_3 / \nu, \tag{21}$$

where $\kappa_3 > 0$. That is in this case $\tau_n = O(h^3)$.

The condition $\delta b^{n+1} < \varepsilon$ was used as a criterion for termination of iterations in nonlinearity, where

$$\delta b^{n+1} = \max_i \left(\max_{1 \leq l \leq 4} \left| b_{i,l}^{n+1,s+1} - b_{i,l}^{n+1,s} \right| \right), \tag{22}$$

ε is a small positive user-specified number. In all computations presented below, the value $\varepsilon = 10^{-14}$ was specified.

The error of the method on a specific uniform grid was computed with the use of a grid analog of the L_2 space norm:

$$\delta u^n = \left\{ \frac{1}{X} \sum_{i=1}^{I} [u_i^n - u_{\text{ex}}(x_i, t_n)]^2 \cdot 2h \right\}^{\frac{1}{2}}, \tag{23}$$

where $u_{\text{ex}}(x, t)$ is the exact solution.

Table 1. The errors $\delta u_1, \delta u_2, \delta u_3$ and the convergence orders p_1, p_2, p_3 on a sequence of grids, $\nu = 0.1$, $X = 2.0$, $\kappa_1 = 1.0$, $\kappa_2 = 0$ in (20), $N_c = 6$

I	δu_1	δu_2	δu_3	p_1	p_2	p_3
20	0.356E–13	0.247E–2	0.166E–2			
40	0.514E–13	0.118E–2	0.823E–3		1.07	1.01
80	0.168E–12	0.575E–3	0.411E–3		1.04	1.00
160	0.145E–11	0.284E–3	0.206E–3		1.02	1.00

The convergence order p of the CLR method was computed as in [41, 42] by the formula

$$p = \frac{\log[\delta u(h_{m-1})] - \log(\delta u(h_m))}{\log(h_{m-1}) - \log(h_m)},$$

where h_m, $m = 2, 3, \ldots$, are some values of step h such that $h_{m-1} \neq h_m$.

Table 2. The errors $\delta u_1, \delta u_2, \delta u_3$ and the convergence orders p_1, p_2, p_3 on a sequence of grids, $\nu = 0.1$, $X = 2.0$, $\kappa_1 = 0$, $\kappa_2 = 0.16$ in (20), $N_c = 6$

I	δu_1	δu_2	δu_3	p_1	p_2	p_3
20	0.296E–13	0.591E–2	0.420E–2			
40	0.601E–13	0.149E–2	0.118E–2		1.99	1.83
80	0.934E–13	0.375E–3	0.305E–3		1.99	1.95
160	0.447E–12	0.938E–4	0.769E–4		2.00	1.99

Table 3. The errors $\delta u_1, \delta u_2, \delta u_3$ and the convergence orders p_1, p_2, p_3 on a sequence of grids, $\nu = 0.1$, $X = 2.0$, $\kappa_3 = 2$ in (21), $N_c = 5$

I	δu_1	δu_2	δu_3	p_1	p_2	p_3
20	0.329E–13	0.368E–2	0.276E–2			
40	0.500E–13	0.465E–3	0.378E–3		2.98	2.87
80	0.216E–13	0.587E–4	0.477E–4		2.99	2.99
160	0.237E–11	0.745E–5	0.595E–5		2.98	3.00

Denote by $\delta u_1, \delta u_2$, and δu_3 the errors (23) obtained under the initial and boundary values corresponding to functions (14), (15), and (16). These errors were computed at the moment of time $t = 1.0$ and are presented in Tables 1, 2, and 3 for the case when $\alpha = 1$ in (8). The computations were done on three different grid sequences when the grid steps tend to zero according to the laws: $\tau = O(h)$ in the first sequence (see Table 1), the results for $\tau = O(h^2)$ and $\tau = O(h^3)$ in the second and third sequences are presented, respectively, in Tables 2 and 3. From an analysis of the convergence of the solution error we

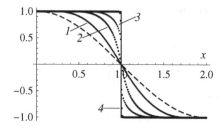

Fig. 2. Numerical solution of the Burgers equation by the CLR method: $(- - -)$ initial profile at $t = 0$; solid lines – numerical solutions at the moments of time $t = 0.25$ (curve 1), 0.5 (curve 2), 0.65 (curve 3), and 1.0 (curve 4)

draw the unambiguous conclusion that the error magnitude has the first order of smallness $O(\tau)$ in the time variable and the third order of smallness $O(h^3)$ in the spatial variable. Consequently, there are in the approximation error of the equation no terms of the first and second orders of smallness with respect to a small quantity h. The value κ_2 for computations with $\tau = O(h^2)$ was chosen in such a way that the size of the step τ_0 for the computation at the first time step on the grid of 20 cells be approximately equal to the step τ_0 obtained at the computations with $\tau = O(h)$ at $\kappa_1 = 1.0$.

Table 4. Crank–Nicholson scheme combined with the CLR method. The errors $\delta u_1, \delta u_2, \delta u_3$ and the convergence orders p_1, p_2, p_3 on a sequence of grids, $\nu = 0.1$, $X = 2.0$, $\kappa_1 = 1.0, \kappa_2 = 0$ in (20), $N_c = 6$.

I	δu_1	δu_2	δu_3	p_1	p_2	p_3
20	0.918E–14	0.112E–4	0.860E–5			
40	0.121E–13	0.572E–5	0.103E–5		0.97	3.06
80	0.400E–13	0.335E–5	0.618E–7		0.77	4.06

Table 4, which is similar to Table 1, presents some computational results for the case of the Crank–Nicholson scheme ($\alpha = 0.5$ in (8)). As was to be expected, the obtained solution errors are much less in the case of the CLR method combined with the Crank–Nicholson scheme (which has the second order of accuracy in time) than in the case of the implicit Euler method. This shows that the numerical method built here with the aid of a CAS is promising for its further extensions and applications.

As follows from Tables 1–4, the test solution (14) proved to be a proper function of the discrete problem. The machine accuracy of the solution was reached on it. It has served here for the verification of the formulas of the method.

The capabilities of the developed method were also checked on the solution of the well-known benchmark problem for equation (1) under the initial condition $u_0(x) = \cos(\pi x/2)$, $0 \leq x \leq 2$. In the case of $\nu = 0$ equation (1) is hyperbolic, and it has the characteristics at each (t, x) point of the solution domain — the straight lines with the slope $dx/dt = u$. In the exact solution, the constant value of u is transferred along each characteristic, which was on it at $t = 0$. Under the initial data indicated above, all characteristics in the left half of the computational region have the inclination angle with a positive tangent value, and

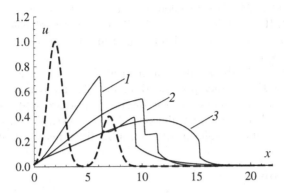

Fig. 3. Numerical solution of the KdVB equation by the CLR method, $\nu = 10^{-8}$ (Re= 10^8), $D_\sigma = 10$ in (2): (– – –) initial profile at $t = 0$; solid lines – numerical solutions at the moments of time $t = 10.98$ (curve 1), 21.17 (curve 2), 39.2 (curve 3)

all characteristics in the right half have an inclination with a negative tangent value because $u_0 > 0$ at $x < 1$ and $u_0 < 0$ at $x > 1$. In the left half at $t = 0$, a point with a smaller value of x has a higher initial value $u_0 > 0$ at it, and the inclination of the characteristic to the x-axis is smaller. As a result of such a solution behavior, the graph of quantity $u(x)$ at $t = $ const moves with increasing t as a wave towards the point $x = 1$, and in the right half, a wave with a negative value of u moves towards this point. In the limit as $t \to \infty$ the solution has the form of a step with a jump of the u value equal to 2. The idea of Hopf [17] was implemented in the present work with the use of the developed method that the generalized solutions of hyperbolic equations represent the limits of solutions with the same boundary conditions of the parabolic equations with the terms modelling the viscosity. In the given case, the discontinuous solution of equation (1) at $\nu = 0$ is a limit of the solution of the parabolic equation as $\nu \to 0$. According to this idea, numerical experiments were done for different values of ν as $\nu \to 0$. The numerical results obtained for the values $\nu \le 10^{-5}$ were visually indistinguishable on the graphs. The difference in tabular numerical solution values for the values $\nu \le 10^{-14}$ was below 10^{-12}. In addition, the numerical algorithm worked and produced the needed result also at $\nu = 0$ that is the method proposed here is applicable for the solution of both the parabolic Burgers equation and the hyperbolic Burgers equation without viscosity.

Figure 2 shows the graphs of the solution of equation (1) at different moments of time t and $\nu = 0$. The computation was done by the pseudo-unsteady method. Thus, the Hopf's idea was implemented in numerical experiment with a high accuracy. Note that in the constructed numerical solution, the step was "smeared" only over three cells, and the solution is monotone that is there are no oscillations. Most existing numerical methods do not satisfy simultaneously the contradictory requirements of the absence of oscillations and a small thickness of the shock smearing zone.

The developed method was also applied to the KdVB equation (2) to solve the well-known problem of the modelling of the interaction of two solitons having different amplitudes [24]. This problem was computed as experiment at different values of ν: 10^{-4}, 10^{-5}, 10^{-6}, 10^{-7}, and 10^{-8}. Note that in all cases, a sufficiently steep leading front of the soliton with a higher amplitude is observed, and there are no oscillations in its profile.

We present in Fig. 3 the results obtained at different moments of time only for the case of $\nu = 10^{-8}$, which presents difficulties for many numerical methods. One can see that in the location of the interaction of two solitons as well as in its neighborhood, there are no oscillations of the numerical solution, which are typical of the modelling of solitons dynamics by difference methods.

4   Conclusions

The CAS was applied in the present work for constructing a new version of the numerical solution of boundary-value problems for the Burgers and KdVB equations. The CAS has enabled us to derive the formulas of the method, to verify them and to write considerable parts of the program for solving the problem in the Fortran language during a relatively short time. The program parts generated by the CAS included the formulas for the entries of a SLAE matrix of the discrete problem approximating the differential problem and finding the solution in the form of a third-degree polynomial in spatial variable. In addition, the CAS was used here for a rapid plotting of many graphs of intermediate values of the problem solution to control the correctness of the Fortran computer code with the aid of tests. A convenient interface implemented for this purpose between the CAS and Fortran has enabled a rapid debugging of errors in the process of writing the computer code, facilitated greatly the work of the mathematician both at the stage of the derivation of the formulas of the method and at the stage of the algorithm improvement and the development of the computer code in the Fortran language.

References

1. Aksan, E.N.: Quadratic B-spline finite element method for numerical solution of the Burgers' equation. Appl. Math. Comput. 174, 884–896 (2006)
2. Ali, A.H.A., Gardner, G.A., Gardner, L.R.T.: A collocation solution for Burgers' equation using cubic B-spline finite elements. Comput. Methods Appl. Mech. Eng. 100, 325–337 (1992)
3. Ali, A.H.A., Gardner, L.R.T., Gardner, G.: Numerical studies of the Korteweg–de Vries–Burgers equation using B-spline finite elements. J. Math. Phys. Sci. 27, 37–53 (1993)
4. Altiparmak, K.: Numerical solution of Burgers' equation with factorized diagonal Padé approximation. Int. J. Numer. Methods Heat Fluid Flow 21(3), 310–319 (2011)
5. Asaithambi, A.: Numerical solution of the Burgers' equation by automatic differentiation. Appl. Math. Comput. 216, 2700–2708 (2010)
6. Berezin, Y.A.: Modeling of Nonlinear Wave Processes. Nauka, Novosibirsk (1982) (in Russian)

7. Burgers, J.M.: A mathematical model illustrating the theory of turbulence. In: Mises, R., von Kármán, T. (eds.) Advances in Applied Mechanics, pp. 171–199. Academic Press, New York (1948)

8. Canosa, J., Gazdag, J.: The Korteweg–de Vries–Burgers' equation. J. Comput. Phys. 23, 393–403 (1977)

9. Crank, J., Nicholson, P.: A practical method for numerical evaluation of solutions of partial differential equations of the heat-conduction type. Proc. Cambridge Philosophical Soc. 43(50), 50–67 (1947)

10. Darvishi, M.T., Khani, F., Kheybari, S.: A numerical solution of the KdV–Burgers' equation by spectral collocation method and Darvishi preconditionings. Int. J. Contemp. Math. Sciences 2(22), 1085–1095 (2007)

11. Dăg, I., Irk, D., Sahin, A.: B-Spline collocation methods for numerical solutions of the Burgers' equation. Math. Probl. Eng. 5, 521–538 (2005)

12. Dogan, A.A.: Galerkin, Finite element approach to Burgers' equation. Appl. Math. Comput. 157, 331–346 (2004)

13. El Sayed, T., El Danaf, A.: Numerical solution of the Korteweg–de Vries Burgers equation by using quintic spline method. Studia Univ. "Babeş-Bolyai", Mathematica 47(2), 41–54 (2002)

14. Ganzha, V.G., Vorozhtsov, E.V.: Numerical Solutions for Partial Differential Equations: Problem Solving Using Mathematica. CRC Press, Boca Raton (1996)

15. Grad, H., Hu, P.N.: Unified shock profile in plasma. Phys. Fluids 10, 2596–2602 (1967)

16. Hassanien, I.A., Salama, A.A., Hosham, H.A.: Fourth-order finite difference method for solving Burgers' equation. Appl. Math. Comput. 170, 781–800 (2005)

17. Hopf, E.: The partial differential equation $u_t + uu_x = \mu u_{xx}$. Comm. Pure Appl. Math. 3, 201–230 (1950)

18. Johnson, R.S.: Shallow water waves in a viscous fluid – the undular bore. Phys. Fluids 15, 1693–1699 (1970)

19. Isaev, V.I., Shapeev, V.P.: Development of the collocations and least squares method. Proc. Inst. Math. Mech. 261(suppl. 1), 87–106 (2008)

20. Isaev, V.I., Shapeev, V.P.: High-accuracy versions of the collocations and least squares method for the numerical solution of the Navier–Stokes equations. Computat. Math. and Math. Phys. 50, 1670–1681 (2010)

21. Jiang, Z., Wang, R.: An improved numerical solution of Burgers' equation by cubic B-spline Quasi-interpolation. J. Inform. Comput. Sci. 7(5), 1013–1021 (2010)

22. Khalifa, A.K., Noor, K.I., Aslam Noor, M.A.: Some numerical methods for solving Burgers equation. Int. J. Phys. Sci. 6(7), 1702–1710 (2011)

23. Khater, A.H., Temsah, R.S., Hassan, M.M.: A Chebyshev spectral collocation method for solving Burgers' type equations. J. Comput. Appl. Math. 222, 333–350 (2008)

24. Kiselev, S.P., Vorozhtsov, E.V., Fomin, V.M.: Foundations of Fluid Mechanics with Applications: Problem Solving Using Mathematica. Birkhäuser, Boston (1999)

25. Korkmaz, A.: Shock wave simulations using Sinc Differential Quadrature Method. Int. J. Comput. Aided Eng. Software 28(6), 654–674 (2011)

26. Korkmaz, A., Aksoy, A.M., Dăg, I.: Quartic B-spline Differential Quadrature Method. Int. J. Nonlinear Sci. 11(4), 403–411 (2011)

27. Korkmaz, A., Dăg, I.: Polynomial based differential quadrature method for numerical solution of nonlinear Burgers' equation. J. Franklin Inst. (2011), doi:10.1016/j.jfranklin.2011.09.008.

28. Kumar, B.V.R., Mehra, M.: Wavelet-Taylor Galerkin method for the Burgers equation. BIT Numer. Math. 45, 543–560 (2005)

29. Kutulay, S., Bahadir, A.R., Özdes, A.: Numerical solution of the one-dimensional Burgers' equation: explicit and exact-explicit finite difference methods. J. Comput. Appl. Math. 103, 251–261 (1999)
30. Kutulay, S., Esen, A., Dag, I.: Numerical solutions of the Burgers' equation by the least-squares quadratic B-spline finite element method. J. Comput. Appl. Math. 167, 21–33 (2004)
31. Kuznetsov, V.V., Nakoryakov, V.E., Pokusaev, B.G., Shreiber, I.R.: Propagation of perturbations in a gas-liquid mixture. J. Fluid Mech. 85(1), 85–96 (1978)
32. Liao, W.: An implicit fourth-order compact finite difference scheme for one-dimensional Burgers' equation. Appl. Math. Comput. 206, 755–764 (2008)
33. Mittal, R.C., Jain, R.K.: Numerical solutions of nonlinear Burgers' equation with modified cubic B-splines collocation method. Appl. Math. Comput. 218, 7839–7855 (2012)
34. Özis, T., Esen, A., Kutluay, S.: Numerical solution of Burgers' equation by quadratic B-spline finite elements. Appl. Math. Comput. 165, 237–249 (2005)
35. Ramadan, M.A., El-Danaf, T.S., Abd Alaal, F.E.I.: Application of the non-polynomial spline approach to the solution of the Burgers equation. Open Appl. Math. J. 1, 15–20 (2007)
36. Raslan, K.R.: A collocation solution for Burgers equation using quadratic B-spline finite elements. Int. J. Comput. Math. 80(7), 931–938 (2003)
37. Saka, B., Dag, I.: Quartic B-spline collocation method to the numerical solutions of the Burgers' equation. Chaos Solitons Fractals 32, 1125–1137 (2007)
38. Semin, L., Shapeev, V.: Constructing the numerical method for Navier–Stokes equations using computer algebra system. In: Ganzha, V.G., Mayr, E.W., Vorozhtsov, E.V. (eds.) CASC 2005. LNCS, vol. 3718, pp. 367–378. Springer, Heidelberg (2005)
39. Semin, L.G., Sleptsov, A.G., Shapeev, V.P.: Collocation and least-squares method for Stokes equations. Computat. Technologies 1(2), 90–98 (1996) (in Russian)
40. Shapeev, V.P., Isaev, V.I., Idimeshev, S.V.: The collocations and least squares method: application to numerical solution of the Navier-Stokes equations. In: CD-ROM Proceedings of the 6th ECCOMAS, Vienna Univ. of Tech. (September 2012) ISBN: 978-3-9502481-9-7
41. Shapeev, V.P., Vorozhtsov, E.V.: Symbolic-numeric implementation of the method of collocations and least squares for 3D Navier–Stokes equations. In: Gerdt, V.P., Koepf, W., Mayr, E.W., Vorozhtsov, E.V. (eds.) CASC 2012. LNCS, vol. 7442, pp. 321–333. Springer, Heidelberg (2012)
42. Shapeev, V.P., Vorozhtsov, E.V.: CAS application to the construction of the collocations and least residuals method for the solution of 3D Navier–Stokes equations. In: Gerdt, V.P., Koepf, W., Mayr, E.W., Vorozhtsov, E.V. (eds.) CASC 2013. LNCS, vol. 8136, pp. 381–392. Springer, Heidelberg (2013)
43. Shapeev, V.P., Vorozhtsov, E.V., Isaev, V.I., Idimeshev, S.V.: The method of collocations and least residuals for three-dimensional Navier-Stokes equations. Computational Methods and Programming 14, 306–322 (2013) (in Russian)
44. Xie, S.-S., Heo, S., Kim, S., Woo, G., Yi, S.: Numerical solution of one-dimensional Burgers' equation using reproducing kernel function. J. Comput. Appl. Math. 214, 417–434 (2008)

An Algorithm for Computing the Truncated Annihilating Ideals for an Algebraic Local Cohomology Class

Takafumi Shibuta[1,*] and Shinichi Tajima[2]

[1] Institute of Mathematics for Industry, Kyushu University
shibuta@imi.kyushu-u.ac.jp
[2] Graduate School of Pure and Applied Sciences, University of Tsukuba
tajima@math.tsukuba.ac.jp

Abstract. Let σ be an algebraic local cohomology class, and k a natural number. The purpose of this paper is to present an algorithm for computing the right D-ideal $\mathcal{A}nn^{(k)}(\sigma)$ generated by linear differential operators annihilating σ and of order less than or equal to k. This algorithm is based on Matlis duality theorem, and is applicable to the case where σ has parameters in its coefficients. Our main interest is where algebraic local cohomology classes σ is a generator of the dual space of the Milnor algebra of a hypersurface isolated singularity.

1 Introduction

Let $X = \mathbb{C}^n$ be the complex n-space with the coordinate $x = (x_1, \ldots, x_n)$. We use multi-index notation $x^\alpha = x_1^{\alpha_1} \cdots x_n^{\alpha_n}$, $|\alpha| = \alpha_1 + \cdots + \alpha_n$. We denote by $\mathcal{O}_X$ the sheaf of holomorphic functions on X, and by $\mathcal{D}_X$ the sheaf of linear partial differential operators whose coefficients are holomorphic. Let $f : (\mathbb{C}^n, O) \to (\mathbb{C}, 0)$ be a germ of holomorphic function at the origin $O \in \mathbb{C}^n$ defining an isolated singularity at the origin. We denote by $\mathcal{J}_f$ the Jacobian ideal $\langle \frac{\partial f}{\partial x_1}, \ldots, \frac{\partial f}{\partial x_n} \rangle \subset \mathcal{O}_{X,O}$ of f. An important topological invariant of the germ of $\{f(x) = 0\}$ at O is the Milnor number μ_f which is the $\mathbb{C}$-dimension of the Milnor algebra $\mathcal{O}_{X,O}/\mathcal{J}_f$. In [9,10], the the second author and Nakamura defined a new analytic invariant $\mu_f^{[k]}$ of f for $k \in \mathbb{Z}_{\geq 0}$ in terms of linear partial differential operators. where $\mu_f^{[0]}$ coincides with the Milnor number μ_f. We will recall the definition of $\mu_f^{[k]}$. Let $W_f = \{\psi \in \mathcal{H}^n_{[O]}(\Omega_X^n) \mid \psi g = 0, \forall g \in \mathcal{J}_f\}$. Then W_f is generated by a single element, say ω_f. As the local cohomology module $\mathcal{H}^n_{[O]}(\Omega_X^n)$ is a right $\mathcal{D}_{X,O}$-module, we can define the truncated annihilating ideals for ω_f

$$\mathcal{A}nn^{(k)}_{\mathcal{D}_{X,O}}(\omega_f) := \{P \in \mathcal{D}_{X,O} \mid \omega_f P = 0, \mathrm{ord}(P) \leq k\}$$

* The first author was supported by Grant-in-Aid for Young Scientists (B) 25800029

V.P. Gerdt et al. (Eds.): CASC Workshop 2014, LNCS 8660, pp. 447–459, 2014.

for $k \in \mathbb{Z}_{\geq 0}$ with are right $\mathcal{D}_{X,O}$-ideals. The invariant $\mu_f^{[k]}$ is defined to be

$$\mu_f^{[k]} = \dim_{\mathbb{C}} \operatorname{Hom}_{\mathcal{D}_{X,O}} \left(\frac{\mathcal{D}_{X,O}}{Ann_{\mathcal{D}_{X,O}}^{(k)}(\omega_f)}, \mathcal{H}_{[O]}^n(\Omega_X^n) \right)$$

which is the $\mathbb{C}$-dimension of the solution space of the holonomic system $Ann_{\mathcal{D}_{X,O}}^{(k)}(\omega_f)$ attached to $\mathcal{H}_{[O]}^n(\Omega_X^n)$. Since the choice of a generator ω_f is unique up to multiplication by an invertible element of $\mathcal{O}_{X,O}$, the isomorphism class of the right $\mathcal{D}_{X,O}$-module $\frac{\mathcal{D}_{X,O}}{Ann_{\mathcal{D}_{X,O}}^{(k)}(\omega_f)}$ is independent from choice of a generator ω_f, and thus so is $\mu_f^{[k]}$. Furthermore, it is known that $\mu_f^{[k]}$ is actually an analytic invariant for f [10].

In [8], it is proved that the $\mu_f^{[1]} = 1$ if and only if f is quasi-homogeneous, and in [7], it is also observed that if f is in an exceptional family of unimodal singularities,

$$\mu_f^{[1]} = \dim_{\mathbb{C}} \mathcal{O}_{X,O}/\mathcal{J}_f - \dim_{\mathbb{C}} \mathcal{O}_{X,O}/(f, \mathcal{J}_f) + 1$$

and $\mu_f^{[2]} = 1$.

We consider the following problem. Let $U \subset \mathbb{C}^r$ an Zariski open set, and

$$F : (\mathbb{C}^n \times \mathbb{C}^r, O \times U) \to (\mathbb{C}, 0)$$
$$(z, a) \mapsto f_a(x) = F(x, a)$$

a germ of holomorphic function such that for any $a \in U$, $f_a : (\mathbb{C}^n, O) \to (\mathbb{C}, 0)$ defines isolated singularity at the origin. Then, compute the stratification of the parameter space U according to $\mu_{f_a}^{[k]}$. Let $\gamma : [0,1] \to U$ be a C^1 curve on U, and consider the deformation $f_{\gamma(t)}$, $0 \leq t \leq 1$. If the Milnor's number of $f_{\gamma(t)}$ at the origin does not change in this family, this deformation called μ-constant deformation. Lê–Ramanujan [4] proved that μ-constant deformation is topologically trivial in case of $n \neq 3$. We are interested in how $\mu_{f_{\gamma(t)}}^{[k]}$ changes along μ-constant deformation. Our problem is the computation of the stratification of the parameter space U by the value of $\mu_{f_a}^{[k]}$.

For this purpose, we will present an algorithm for computing $Ann_{\mathcal{D}_{X,O}}^{(k)}(\omega_f)$ which is applicable to the parametric case. One may compute $Ann_{\mathcal{D}_{X,O}}^{(k)}(\omega_{f_a})$ by computing a Göbner basis of $Ann_{\mathcal{D}_{X,O}}(\omega_{f_a})$ with respect to a certain term order with parametric method as in [12]. There is an implementation of comprehensive Gröbner bases for Weyl algebra by Nabeshima [6]. However, Calculating a Gröbner basis is a very time-consuming process (see [3]), and in many cases the computation is not feasible with the current implementation. Our algorithm construct a system of generators of $Ann_{\mathcal{D}_{X,O}}^{(k)}(\omega_f)$ for $k = 0, 1, 2, \ldots$ iteratively. Even in case where it is hard to compute them for large k, it is worth computing them only for small k.

2   Preliminaries

We will briefly recall some basic definitions and facts that we will used throughout this paper. See [1] and [2] for details.

2.1   Local Cohomology Modules

Let $\mathcal{D}_X$ be the sheaf of linear differential operators on X. We denote by $\mathcal{H}^n_{[O]}(\Omega^n_X)$ the algebraic local cohomology group supported at the origin O of the sheaf Ω^n_X of holomorphic n-forms on X. We denote by $\widehat{\mathcal{O}}_{X,O}$ the $\mathfrak{m}_{X,O}$-adic completion of $\mathcal{O}_{X,O}$ where $\mathfrak{m}_{X,O}$ is the unique maximal ideal of $\mathcal{O}_{X,O}$. We note that $\widehat{\mathcal{O}}_{X,O}$ is isomorphic to the formal power series ring $\mathbb{C}[\![x_1, \ldots, x_n]\!]$. The local cohomology module $\mathcal{H}^n_{[O]}(\Omega^n_X)$ has a relative Čech cohomology representation

$$\mathcal{H}^n_{[O]}(\Omega^n_X) \cong \mathbb{C}[x_1^{-1}, \ldots, x_n^{-1}] \frac{dx}{x_1 \cdots x_n},$$

where $dx = dx_1 \wedge \cdots \wedge dx_n$, and has an $\widehat{\mathcal{O}}_{X,O}$-module structure defined by

$$x^\alpha \cdot \frac{dx}{x^{\beta+1}} = \begin{cases} \frac{dx}{x^{\beta-\alpha+1}} & \text{if } \beta - \alpha \in \mathbb{Z}^n_{\geq 0}, \\ 0 & \text{otherwise.} \end{cases}$$

Furthermore, $\mathcal{H}^n_{[O]}(\Omega^n_X)$ admits a right $\mathcal{D}_{X,O}$-module structure given by

$$(gdx)P = (P^*g)dx$$

for $gdx \in \mathcal{H}^n_{[O]}(\Omega^n_X)$ and $P \in \mathcal{D}_{X,O}$ where P^* is the formal adjoint operator of P. As $\frac{dx}{x_1 \cdots x_n} \frac{1}{\alpha!}(\frac{\partial}{\partial x})^\alpha = \frac{dx}{x^\alpha}$, $\mathcal{H}^n_{[O]}(\Omega_X)$ is generated by $\frac{dx}{x_1 \cdots x_n}$ as a right $\mathcal{D}_{X,O}$-module. Furthermore, $\mathcal{H}^n_{[O]}(\Omega_X)$ is a simple right $\mathcal{D}_{X,O}$-module since for any $0 \neq \xi \in \mathcal{H}^n_{[O]}(\Omega_X)$, there exists $h \in \mathcal{O}_{X,O}$ such that $h\xi = \frac{dx}{x_1 \cdots x_n}$. We set

$$W_f = \mathrm{Hom}_{\mathcal{O}_{X,O}}(\mathcal{O}_{X,O}/\mathcal{J}_f, \mathcal{H}^n_{[O]}(\Omega^n_X)) = \{\psi \in \mathcal{H}^n_{[O]}(\Omega^n_X) \mid \psi g = 0, \forall g \in \mathcal{J}_f\}.$$

As f has isolated singularity at the origin, $\mathcal{O}_{X,O}/\mathcal{J}_f$ is a complete intersection Artinian local ring, and thus W_f is of finite length and generated by a single element as an $\mathcal{O}_{X,O}$-module ([1] Proposition 3.2.12). We fix a generator ω_f of W_f. We can consider annihilators of ω_f in $\mathcal{D}_{X,O}$

$$\mathcal{A}nn_{\mathcal{D}_{X,O}}(\omega_f) = \{P \in \mathcal{D}_{X,O} \mid \omega_f P = 0\}.$$

This ideal is a right ideal of $\mathcal{D}_{X,O}$. Since $\mathcal{D}_{X,O}/\mathcal{A}nn_{\mathcal{D}_{X,O}}(\omega_f) \cong \mathcal{D}_{X,O}\omega_f = \mathcal{H}^n_{[O]}(\Omega_X)$ is a simple $\mathcal{D}_{X,O}$-module, $\mathcal{A}nn_{\mathcal{D}_{X,O}}(\omega_f)$ is a holonomic ideal.

2.2   Matlis Duality

Our algorithm is based of Matlis duality theorem. Here, we give a brief review of Matlis duality. Let R be a Noetherian complete local ring, and $E = E_R$ the injective hull of the residue field of R. We denote by $\ell_R(M) = \ell(M)$ the length of an R-module M. We note that in case of $R = \widehat{\mathcal{O}}_{X,O}$, $\ell(M) = \dim_{\mathbb{C}} M$ and $E_{\mathcal{O}_{X,O}}$ is isomorphic to $\mathcal{H}^n_{[O]}(\Omega^n_X)$ ([1] Proposition 3.5.4).

Definition 1. *For an R-module M, we write $M^\vee := \mathrm{Hom}_R(M, E)$. The functor $(-)^\vee = \mathrm{Hom}_R(-, E)$ is called the* Matlis duality functor.

For an R-module homomorphism $\varphi : M \to N$, $\varphi^\vee$ denotes the natural homomorphism $N^\vee \to M^\vee$. Since E is injective, the Matlis dual functor is an exact contravariant functor. Thus the following holds.

Lemma 1. $(M/\mathrm{Ker})^\vee \cong \mathrm{Image}\, \varphi^\vee$.

Now, we recall the Matlis duality theorem.

Theorem 1 (Matlis [5]). *Let M be a Noetherian R-module, and N an Artinian R-module. Then the following hold.*

1. $R^\vee \cong E$, and $E^\vee \cong R$.
2. $M^\vee$ is Artinian, and $N^\vee$ is Noetherian.
3. There are natural isomorphism $M^{\vee\vee} \cong M$ and $N^{\vee\vee} \cong N$.

If M is of finite length, then $\ell(M) = \ell(M^\vee)$. We note that an R-module M is Noetherian if and only if M is finitely generated. Any Noetherian R-module can be expressed by as a quotient module of a free R-module, and any Artinian R-module can be expressed as a submodule of a direct sum of copies of E. Let $F = R^{\oplus r}$ be a free R-module of rank r. For an R-submodule $M \subset F$, we may regard $(F/M)^\vee$ as a submodule of $E^{\oplus r}$ since $(F/M)^\vee \subset F^\vee = E^{\oplus r}$. On the other hand, for an R-submodule $V \subset E^{\oplus r}$, we may regard $V^\vee$ as a quotient module F/M of F for some $M \subset F$. Let $\langle \cdot, \cdot \rangle : F \times F^\vee \to E$ be the natural pairing map and $M \subset F$ an R-submodule. By Theorem 1, for $m \in F$, m is contained in M if and only if $\langle m, \eta \rangle = 0$ for any $\eta \in (F/M)^\vee$. In case where $R = \widehat{\mathcal{O}}_{X,O}$, this pairing is

$$\widehat{\mathcal{O}}_{X,O}^{\oplus r} \times \mathcal{H}_{[O]}^n(\Omega_X^n)^{\oplus r} \to \mathcal{H}_{[O]}^n(\Omega_X^n)$$

$$((g_1, \ldots, g_r)^T, (\eta_1, \ldots, \eta_r)) \mapsto \sum_{i=1}^r g_i \eta_i.$$

2.3 Standard Basis

Here, we recall the theory of standard basis of modules. A total order $\prec$ on the set of monomial $\{x^\alpha \mid \alpha \in \mathbb{Z}_{\geq 0}^n\}$ of $x = (x_1, \ldots, x_n)$ if a *local order* is the following conditions hold:

- For any $\alpha \in \mathbb{Z}_{\geq 0}^n$, $x^\alpha \prec 1$.
- For any $\alpha, \beta, \gamma \in \mathbb{Z}_{\geq 0}^n$, $x^\alpha \prec x^\beta$ implies $x^{\alpha+\gamma} \prec x^{\beta+\gamma}$.

For any set of monomials Λ, and a local order $\prec$, there exists the maximal element Λ with respect to $\prec$. Thus, for $g(x) = \sum_{\alpha \in \mathbb{Z}_{\geq 0}^n} c_\alpha x^\alpha \in \mathcal{O}_{X,O}$ $(c_\alpha \in \mathbb{C})$, we can define the leading term $\mathrm{LT}_\prec(g) := \max_\prec \{x^\alpha \mid c_\alpha \neq 0\}$ of g with respect to $\prec$.

Let $F = \mathcal{O}_{X,O}^{\oplus r}$ be a free $\mathcal{O}_{X,O}$-module of rank r with basis $\mathbf{e}_1, \ldots, \mathbf{e}_r$, and fix a local order $\prec$. A total order $\prec_F$ on the set of monomials $\{x^\alpha \mathbf{e}_i \mid \alpha \in \mathbb{Z}_{\geq 0}^n, 1 \leq i \leq r\}$ of F is a local order on F if the following conditions hold:

- For any $\alpha, \beta \in \mathbb{Z}_{\geq 0}^n$, and $1 \leq i \leq r$, $x^\alpha \prec x^\beta$ implies $x^\alpha \mathbf{e}_i \prec_F x^\beta \mathbf{e}_i$
- For any $\alpha, \beta, \gamma \in \mathbb{Z}_{\geq 0}^n$, and $1 \leq i, j \leq r$, $x^\alpha \mathbf{e}_i \prec_F x^\beta \mathbf{e}_j$ implies $x^{\alpha+\gamma} \mathbf{e}_i \prec_F x^{\beta+\gamma} \mathbf{e}_j$.

Any set of monomials of F admits a maximal element with respect to $\prec_F$, and thus we can define the leading monomial $\mathrm{LT}_{\prec_F}(m)$ of $m \in F$ similarly. For an $\mathcal{O}_{X,O}$-submodule $M \subset F$, $\mathbb{C}[x]$-submodule of the free $\mathbb{C}[x]$-module

$$\mathrm{LT}_{\prec_F}(M) := \langle \mathrm{LT}_{\prec_F}(m) \mid m \in M \rangle_{\mathbb{C}[x]} \subset \mathbb{C}[x]^{\oplus r}$$

is called the initial module of M with respect to $\prec_F$. A monomials not contained in $\mathrm{LT}_{\prec_F}(M)$ is called a *standard monomial* of M with respect to $\prec_F$. A subset $\{m_1, \ldots, m_s\}$ of M is called a *standard basis* of M with respect to $\prec_F$ if $\mathrm{LT}_{\prec_F}(M)$ is generated by $\mathrm{LT}_{\prec_F}(m_1), \ldots, \mathrm{LT}_{\prec_F}(m_s)$.

3   The Invariant $\mu_f^{[k]}$

Let $\{\mathcal{F}^k \mathcal{D}_{X,O}\}_{k \in \mathbb{Z}_{\geq 0}}$ be the order filtration on $\mathcal{D}_{X,O}$, that is,

$$\mathcal{F}^k \mathcal{D}_{X,O} = \left\{ \sum g_\alpha \left(\frac{\partial}{\partial x} \right)^\alpha \mid g_\alpha \in \mathcal{O}_{X,O}, |\alpha| \leq k \right\}.$$

Let $Ann_{\mathcal{D}_{X,O}}^{(k)}(\omega_f)$ be the right $\mathcal{D}_{X,O}$-ideal generated by $\mathcal{F}^k \mathcal{D}_{X,O} \cap Ann_{\mathcal{D}_{X,O}}(\omega_f)$ for $k \in \mathbb{Z}_{\geq 0}$. We call $Ann_{\mathcal{D}_{X,O}}^{(k)}(\omega_f)$, $k = 0, 1, 2, \ldots$ the truncated annihilating ideals of ω_f. For any k, it holds that $Ann_{\mathcal{D}_{X,O}}^{(k)}(\omega_f) \subset Ann_{\mathcal{D}_{X,O}}^{(k+1)}(\omega_f)$, and $Ann_{\mathcal{D}_{X,O}}^{(k)}(\omega_f) = Ann_{\mathcal{D}_{X,O}}(\omega_f)$ for sufficiently large k since $\mathcal{D}_{X,O}$ is Noetherian. As $Ann_{\mathcal{D}_{X,O}}^{(0)}(\omega_f)$ is generated by zero-dimensional ideal $\mathcal{J}_f$, $Ann_{\mathcal{D}_{X,O}}^{(k)}(\omega_f)$ is holonomic for all k.

Definition 2.

$$\mu_f^{[k]} := \dim_{\mathbb{C}} \mathrm{Hom}_{\mathcal{D}_{X,O}} \left(\frac{\mathcal{D}_{X,O}}{Ann_{\mathcal{D}_{X,O}}^{(k)}(\omega_f)}, \mathcal{H}_{[O]}^n(\Omega_X^n) \right)$$

In [10] it is proved that this definition is independent from the choice of a generator ω_f and $\mu_f^{[k]}$ is an analytic invariant for f. By definition, we have $\mu_f^{[0]} \geq \mu_f^{[1]} \geq \mu_f^{[2]} \geq \cdots$. As $Ann_{\mathcal{D}_{X,O}}^{(0)}(\omega_f) = \mathcal{J}_f \mathcal{D}_{X,O}$, $\mu_f^{[0]}$ coincides with the Milnor number μ_f of f, and $\mu_f^{[k]} = 1$ for k satisfying $Ann_{\mathcal{D}_{X,O}}^{(k)}(\omega_f) = Ann_{\mathcal{D}_{X,O}}(\omega_f)$. In [8], it is proved that the $\mu_f^{[1]} = 1$ if and only if f is quasi-homogeneous, and in [7], it is also observed that if f is in an exceptional family of unimodal singularities,

$$\mu_f^{[1]} = \dim_{\mathbb{C}} \mathcal{O}_{X,o}/\mathcal{J}_f - \dim_{\mathbb{C}} \mathcal{O}_{X,o}/(f, \mathcal{J}_f) + 1$$

and $\mu_f^{[2]} = 1$. An efficient method for computing a $\mathbb{C}$-basis of W_f and the standard basis of $\mathcal{J}_f$ using Grothendieck duality theorem is given in an efficient algorithm in [11]. Generalizing this method, we will present an algorithm for computing $Ann_{\mathcal{D}_{X,o}}^{(k)}(\omega_f)$. For $k \in \mathbb{Z}_{\geq 0}$, let $F_k = \mathcal{O}_{X,o}^{\oplus \binom{n+k}{k}}$ be a free $\mathcal{O}_{X,o}$-module which is isomorphic to $\mathcal{F}^k \mathcal{D}_{X,o}$ as a right $\mathcal{O}_{X,o}$-module, and we define

$$\mathcal{N}^{(k)}(\omega_f) := \left\{ (g_\alpha(x))_{\alpha \in \mathbb{Z}_{\geq 0}^n, |\alpha| \leq k} \;\middle|\; \sum_{\alpha \in \mathbb{Z}_{\geq 0}^n, |\alpha| \leq k} \left(\frac{\partial}{\partial x} \right)^\alpha g_\alpha(x) \in Ann_{\mathcal{D}_{X,o}}^{(k)}(\omega_f) \right\}$$

an $\mathcal{O}_{X,o}$-submodule of F_k. We will present an algorithm for computing a standard basis of $\mathcal{N}^{(k)}(\omega_f)$ by using Matlis duality theorem.

4   Matlis Dual and Standard Basis

Let $M \subset F$ be an $\mathcal{O}_{X,o}$-submodule of a free $\mathcal{O}_{X,o}$-module $F = \mathcal{O}_{X,o}^{\oplus r}$ such that $\ell(F/M) < \infty$, and fix a local order $\prec_F$ on F. Then, we can regard F/M as an $\hat{\mathcal{O}}_{X,o}$-module. In this case, we can compute a standard basis of M using a basis of $(F/M)^\vee$ as a $\mathbb{C}$-vector space.

In this paper, we call $\{ \frac{dx}{x^{\alpha+1}} \mathbf{e}_i^\vee \mid \alpha \in \mathbb{Z}_{\geq 0}^n, 1 \leq i \leq r \}$ the set of monomials of $F^\vee = E^{\oplus r} = \mathcal{H}_{[O]}^n(\Omega_X^n)^{\oplus r}$. We define a total oder $\prec_F^\vee$ on the set of monomials of $\mathcal{H}_{[O]}^n(\Omega_X^n)^{\oplus r}$ corresponding to $\prec_F$ as follows:

$$\frac{dx}{x^{\alpha+1}} \mathbf{e}_i^\vee \prec_F^\vee \frac{dx}{x^{\beta+1}} \mathbf{e}_j^\vee \overset{def}{\Longleftrightarrow} x^\alpha \mathbf{e}_i \succ_F x^\beta \mathbf{e}_j.$$

We call $\prec_F^\vee$ the term order on $\mathcal{H}_{[O]}^n(\Omega_X^n)^{\oplus r}$ corresponding to $\prec_F$. For

$$\eta = \sum_{\alpha \in \mathbb{Z}_{\geq 0}^n, 1 \leq i \leq r} c_{\alpha,i} \frac{dx}{x^{\alpha+1}} \mathbf{e}_i^\vee, \quad c_{\alpha,i} \in \mathbb{C},$$

we set

$$\mathrm{LT}_{\prec_F^\vee}(\eta) = \max_{\prec_F^\vee} \left\{ \frac{dx}{x^{\alpha+1}} \mathbf{e}_i^\vee \mid c_{\alpha,i} \neq 0 \right\}$$

By definition, for $m \in F$ and $\eta \in F^\vee$, $\mathrm{LT}_{\prec_F}(\langle m, \eta \rangle) = \langle \mathrm{LT}_{\prec_F}(m), \mathrm{LT}_{\prec_F^\vee}(\eta) \rangle$ if $\langle \mathrm{LT}_{\prec_F}(m), \mathrm{LT}_{\prec_F^\vee}(\eta) \rangle \neq 0$.

We say that a subset $\{v_1, \ldots, v_n\} \subset \mathcal{H}_{[O]}^n(\Omega_X^n)^{\oplus r}$ is *reduced* with respect to $\prec_F^\vee$ if for any i, $\mathrm{LT}_{\prec_F^\vee}(v_i)$ does not appear in v_j, $j \neq i$, with non-zero coefficient. Let $\eta_1, \ldots, \eta_\ell$ be the reduced $\mathbb{C}$-basis of $(F/M)^\vee$ with respect to $\prec_F^\vee$ where $\ell = \ell(F/M)$, and write

$$\eta_j = \frac{dx}{x^{\beta_j+1}} \mathbf{e}_{i_j}^\vee + \sum_{\alpha \in \mathbb{Z}_{\geq 0}, 1 \leq i \leq r} c_{\alpha,i}^{(j)} \frac{dx}{x^{\alpha+1}} \mathbf{e}_i^\vee,$$

with $\mathrm{LT}_{\prec^{\vee}_F}(\eta_j) = \frac{dx}{x^{\beta_j+1}}\mathbf{e}^{\vee}_{i_j}$. Since $\eta_1, \ldots, \eta_\ell$ is reduced with respect to $\prec^{\vee}_F$, the term $\mathrm{LT}_{\prec^{\vee}_F}(\eta_j)$ does not appear in $\eta_{j'}$ for $j' \neq j$.

Proposition 1. *Let the situation as above. Then the following hold.*

(1) $\{x^{\beta_j}\mathbf{e}_{i_j} \mid 1 \leq j \leq \ell\}$ *is the set of standard monomial of M with respect to* $\prec_F$.
(2) If $x^{\alpha}\mathbf{e}_i \in \mathrm{LT}_{\prec_F}(M)$, then $x^{\alpha}\mathbf{e}_i - \sum_{j=1}^{\ell} c^{(j)}_{\alpha,i} x^{\beta_j}\mathbf{e}_{i_j} \in M$.

Proof. (1) Since $\langle x^{\beta_j}\mathbf{e}_{i_j}, \mathrm{LT}_{\prec^{\vee}_F}(\eta_j)\rangle = \frac{1}{x_1 \cdots x_n} \neq 0$, it follows that for any element $m \in F$ with $\mathrm{LT}_{\prec_F}(m) = x^{\beta_j}\mathbf{e}_{i_j}$, it holds that $\langle m, \eta_j\rangle \neq 0$. As $\eta_j \in (F/M)^{\vee}$, this shows that there exists no element $m \in M$ such that $\mathrm{LT}_{\prec_F}(m) = x^{\beta_j}\mathbf{e}_{i_j}$. Thus $x^{\beta_j}\mathbf{e}_{i_j}$ is a standard monomial. Since the set of standard monomials of M forms a $\mathbb{C}$-basis of F/M, the number of standard monomials is $\ell = \dim_{\mathbb{C}} F/M$. Therefore, we conclude the assertion.

(2) Since the set of standard monomials of M forms a $\mathbb{C}$-basis of F/M, there exists $c^{(j)} \in \mathbb{C}$, $1 \leq j \leq \ell$, such that $x^{\alpha}\mathbf{e}_i - \sum_{j=1}^{\ell} c^{(j)} x^{\beta_j}\mathbf{e}_{i_j} \in M$. The coefficient of $\frac{1}{x_1 \cdots x_n}$ of $\langle x^{\alpha}\mathbf{e}_i - \sum_{j=1}^{\ell} c^{(j)} x^{\beta_j}\mathbf{e}_{i_j}, \eta_j\rangle$ is $c^{(j)}_{\alpha,i} - c^{(j)}$, which should be zero.   □

It is easy to compute $\mathrm{LT}_{\prec_F}(M)$ from the set of standard monomials. Thus Proposition 4 give a method to compute a standard basis of M with respect to $\prec$ from a $\mathbb{C}$-basis of $(F/M)^{\vee}$.

5   Algorithm for Computing $\mathcal{N}^{(k)}(\omega_f)$

We will present a method for computing a system of generators of $\mathcal{N}^{(k)}(\omega_f)$. Recall that for $k \in \mathbb{Z}_{\geq 0}$, $F_k = \mathcal{O}^{\oplus\binom{n+k}{k}}_{X,O}$ is a free $\mathcal{O}_{X,O}$-module, and

$$\mathcal{N}^{(k)}(\omega_f) = \left\{(g_\alpha(x))_{\alpha \in \mathbb{Z}^n_{\geq 0}, |\alpha| \leq k} \;\middle|\; \sum_{\alpha \in \mathbb{Z}^n_{\geq 0}, |\alpha| \leq k} \left(\frac{\partial}{\partial x}\right)^{\alpha} g_\alpha(x) \in Ann^{(k)}_{\mathcal{D}_{X,O}}(\omega_f)\right\}$$

a submodule F_k. By definition, $(F_k/\mathcal{N}^{(k)}(\omega_f))^{\vee}$ is generated by a single element $(\omega_f \cdot (\frac{\partial}{\partial x})^{\alpha})_{\alpha \in \mathbb{Z}^n_{\geq 0}, |\alpha| \leq k}$, and thus $\ell(F_k/\mathcal{N}^{(k)}(\omega_f)) < \infty$.

From now on, we fix a local order $\prec$ on $\mathcal{O}_{X,O}$. We also fix a generator ω_f of W_f corresponding to $\prec$ as follow.

Definition 3. *Let ω_f be the element in the reduced $\mathbb{C}$-basis of W_f whose leading term is maximal with respect to $\prec^{\vee}$.*

The element ω_f is actually a generator of W_f since $\mathfrak{m}_{X,O} \cdot \omega_f \notin W_f$ by the maximality of the leading term of ω_f,

We denote by $[P,Q]$ the commutator $PQ - QP$ of $P, Q \in \mathcal{D}_{X,O}$. The following result provides a constructive method for computing the standard bases of $\mathcal{N}^{(k)}(\omega_f)$.

Theorem 2 ([9] Proposition 10). *Let $P \in \mathcal{F}^k \mathcal{D}_{X,O}$ be a linear partial differential operator of order k. Then the following are equivalent.*

(i) *There exists $h \in \mathcal{O}_{X,O}$ such that $P + h \in Ann^{(k)}_{\mathcal{D}_{X,O}}(\omega_f)$.*

(ii) *For any $g \in \mathcal{J}_f$, $[P,g] \in Ann^{(k-1)}_{\mathcal{D}_{X,O}}(\omega_f)$.*

(iii) *For any $1 \le i \le n$, $[P, \frac{\partial f}{\partial x_i}] \in Ann^{(k-1)}_{\mathcal{D}_{X,O}}(\omega_f)$.*

For $k \ge 1$, let $F'_k = \mathcal{O}^{\oplus \binom{n+k}{k}-1}_{X,O}$, and $pr : F_k \to F'_k$ the projection omitting the component corresponding to order zero part. Then

$$pr(\mathcal{N}^{(k)}(\omega_f)) = \left\{ (f_\alpha(X))_{|\alpha| \le k, \alpha \ne 0} \mid \exists f_0 \in \mathcal{O}_{X,O}, \sum (f_\alpha(X))_{|\alpha| \le k} \in \mathcal{N}^{(k)}(\omega_f) \right\}$$

We fix a local order $\prec_{F'_k}$ on F'_k.

5.1   In the Case of $k = 1$

In the case of $k = 0$, $\mathcal{N}^{(0)}(\omega_f) = \mathcal{J}_f$, and an algorithm for computing the reduced $\mathbb{C}$-basis of $(\mathcal{O}_{X,O}/\mathcal{J}_f)^\vee = W_f$ and the standard basis of $\mathcal{J}_f$ is given in [9,11]. We will consider the case $k = 1$ when a $\mathbb{C}$-basis of $(\mathcal{O}_{X,O}/\mathcal{J}_f)^\vee = W_f$ is given. We note that

$$pr(\mathcal{N}^{(1)}(\omega_f)) =$$

$$\{(a_1, \ldots, a_n)^T \in \mathcal{O}^{\oplus n}_{X,O} \mid \exists h \in \mathcal{O}_{X,O}, \frac{\partial}{\partial x_1}a_1 + \cdots + \frac{\partial}{\partial x_1}a_n + h \in Ann^{(1)}_{\mathcal{D}_{X,O}}(\omega_f)\}$$

By Theorem 2, $(a_1, \ldots, a_n)^T \in pr(\mathcal{N}^{(1)}(\omega_f))$ if and only if

$$\begin{pmatrix} \frac{\partial^2 f}{\partial x_1^2} & \frac{\partial^2 f}{\partial x_1 \partial x_2} & \cdots & \frac{\partial^2 f}{\partial x_1 \partial x_n} \\ \frac{\partial^2 f}{\partial x_2 \partial x_1} & \frac{\partial^2 f}{\partial x_2^2} & \cdots & \frac{\partial^2 f}{\partial x_2 \partial x_n} \\ \vdots & \vdots & \ddots & \vdots \\ \frac{\partial^2 f}{\partial x_n \partial x_1} & \frac{\partial^2 f}{\partial x_n \partial x_2} & \cdots & \frac{\partial^2 f}{\partial x_n^2} \end{pmatrix} \begin{pmatrix} a_1 \\ a_2 \\ \vdots \\ a_n \end{pmatrix} \in (\mathcal{N}^{(0)}(\omega_f))^{\oplus n} = \mathcal{J}^{\oplus n}_f. \qquad (1)$$

The matrix appearing in the left-hand side is called Hessian matrix H_f of f. Therefore,

$$pr(\mathcal{N}^{(1)}(\omega_f)) = \mathrm{Ker}(\mathcal{O}^{\oplus n}_{X,O} \xrightarrow{H_f} (\mathcal{O}_{X,O}/\mathcal{J}_f)^{\oplus n}).$$

Since $(\mathcal{O}_{X,O}/\mathcal{J}_f)^\vee = W_f$,

$$(\mathcal{O}^{\oplus n}_{X,O}/pr(\mathcal{N}^{(1)}(\omega_f)))^\vee = \{(\eta_1, \ldots, \eta_n) \cdot H_f \mid (\eta_1, \ldots, \eta_n) \in W^{\oplus n}_f\}$$

by Lemma 1. Since we a $\mathbb{C}$-basis of W_f is given, one can compute a $\mathbb{C}$-basis of $(\mathcal{O}^{\oplus n}_{X,O}/pr(\mathcal{N}^{(1)}(\omega_f)))^\vee$. Thus the standard basis of $pr(\mathcal{N}^{(1)}(\omega_f))$ is computable by using Proposition 4. For each element $(a_1, \cdots, a_n)^T$ of the standard basis of $pr(\mathcal{N}^{(1)}(\omega_f))$, we can find $h \in \mathcal{O}_{X,O}$ such that $\omega_f(\frac{\partial}{\partial x_1}a_1 + \cdots + \frac{\partial}{\partial x_n}a_n) = h\omega_f$

by Theorem 2. Then $\mathcal{N}^{(1)}(\omega_f)$ is generated by elements of form $(a_1,\ldots,a_n,h)^T$ and $(0,\ldots,0,\frac{\partial f}{\partial x_i})^T$, $1 \le i \le n$.

We can also compute the $\mathbb{C}$-basis of $(\mathcal{O}_{X,O}^{n+1}/\mathcal{N}^{(1)}(\omega_f))^\vee$ as follows. Let η_1, $\ldots, \eta_t$ and $h_1\omega_f,\ldots,h_\mu\omega_f$ be $\mathbb{C}$-bases of $(\mathcal{O}_{X,O}^{\oplus n}/pr(\mathcal{N}^{(1)}(\omega_f)))^\vee$ and W_f respectively. Then

$$\{(\eta_i,0) \mid 1 \le i \le t\} \cup \left\{((\omega_f\left(\frac{\partial}{\partial x}\right)^\alpha h_i)_{|\alpha|\le k}) \mid 1 \le i \le \mu\right\}$$

is a $\mathbb{C}$-basis of $(\mathcal{O}_{X,O}^{n+1}/\mathcal{N}^{(1)}(\omega_f))^\vee$.

5.2   In the Case of $k \ge 2$

For the case of $k \ge 2$, by describing the condition of Theorem 2 (3) in terms of matrix similarly to (1), we can compute a system of generators of $\mathcal{N}^{(k)}(\omega_f)$. For $1 \le k$, $1 \le i \le n$, and $g = (g_\alpha)_{|\alpha|\le k} \in F_k$, the condition

$$\left[\sum_{|\alpha|\le k}\left(\frac{\partial}{\partial x}\right)^\alpha g_\alpha, \frac{\partial f}{\partial x_i}\right] \in Ann_{\mathcal{D}_{X,O}}^{(k-1)}(\omega_f)$$

can be express as

$$M_i^{(k)} \cdot g \in \mathcal{N}^{(k-1)}(\omega_f)$$

for some $\binom{n+k-1}{k-1} \times \left(\binom{n+k}{k} - 1\right)$ matrix $M_i^{(k)}$. For example, if $k = 1$,

$$M_i^{(1)} = \left(\frac{\partial^2 f}{\partial x_1 \partial x_i}, \frac{\partial^2 f}{\partial x_2 \partial x_i}, \cdots, \frac{\partial^2 f}{\partial x_n \partial x_i}\right),$$

and if $n = 2$ and $k = 2$, since

$$\left[\frac{\partial^2}{\partial x_1^2}a + \frac{\partial^2}{\partial x_1\partial x_2}b + \frac{\partial^2}{\partial x_2^2}c + \frac{\partial}{\partial x_1}d + \frac{\partial}{\partial x_2}e, \frac{\partial f}{\partial x_1}\right] =$$

$$\frac{\partial}{\partial x_1}\left(2\frac{\partial^2 f}{\partial x_1^2}a + \frac{\partial^2 f}{\partial x_1\partial x_2}b\right) + \frac{\partial}{\partial x_2}\left(\frac{\partial^2 f}{\partial x_1^2}b + 2\frac{\partial^2 f}{\partial x_1\partial x_2}c\right)$$

$$-\frac{\partial^3 f}{\partial x_1^3}a - \frac{\partial^3 f}{\partial x_1^2\partial x_2}b - \frac{\partial^3 f}{\partial x_1\partial x_2^2}c + \frac{\partial^2 f}{\partial x_1^2}d + \frac{\partial^2 f}{\partial x_1\partial x_2}e,$$

we have

$$M_1^{(2)} = \begin{pmatrix} 2\frac{\partial^2 f}{\partial x_1^2} & \frac{\partial^2 f}{\partial x_1\partial x_2} & 0 & 0 & 0 \\ 0 & \frac{\partial^2 f}{\partial x_1^2} & 2\frac{\partial^2 f}{\partial x_1\partial x_2} & 0 & 0 \\ -\frac{\partial^3 f}{\partial x_1^3} & -\frac{\partial^3 f}{\partial x_1^2\partial x_2} & -\frac{\partial^3 f}{\partial x_1\partial x_2^2} & \frac{\partial^2 f}{\partial x_1^2} & \frac{\partial^2 f}{\partial x_1\partial x_2} \end{pmatrix},$$

and by computing $[\frac{\partial^2}{\partial x_1^2}a + \frac{\partial^2}{\partial x_1\partial x_2}b + \frac{\partial^2}{\partial x_2^2}c + \frac{\partial}{\partial x_1}d + \frac{\partial}{\partial x_2}e, \frac{\partial f}{\partial x_2}]$, we have

$$M_2^{(2)} = \begin{pmatrix} 2\frac{\partial^2 f}{\partial x_1\partial x_2} & \frac{\partial^2 f}{\partial x_2^2} & 0 & 0 & 0 \\ 0 & \frac{\partial^2 f}{\partial x_1\partial x_2} & 2\frac{\partial^2 f}{\partial x_2^2} & 0 & 0 \\ -\frac{\partial^3 f}{\partial x_1^2\partial x_2} & -\frac{\partial^3 f}{\partial x_1\partial x_2^2} & -\frac{\partial^3 f}{\partial x_2^3} & \frac{\partial^2 f}{\partial x_1\partial x_2} & \frac{\partial^2 f}{\partial x_2^2} \end{pmatrix}.$$

Since

$$pr(\mathcal{N}^{(k)}(\omega_f)) = \{g = (g_\alpha)_{|\alpha| \le k} \mid M_i^{(k)} \cdot g \in \mathcal{N}^{(k-1)}(\omega_f)\}$$

$$= \bigcap_{i=1}^{n} \mathrm{Ker}(F'_k \xrightarrow{M_i(k)} F_{k-1}/\mathcal{N}^{(k-1)}(\omega_f))$$

Thus we conclude that

$$(F'_k/pr(\mathcal{N}^{(k)}(\omega_f)))^\vee = \sum_{i=1}^{n}\{\eta \cdot M_i^{(k)} \mid \eta \in (F_{k-1}/\mathcal{N}^{(k-1)}(\omega_f))^\vee\}.$$

5.3   Algorithm

Combining all, we obtain the following algorithm.

Algorithm 1. Input: $k \ge 1$, a $\mathbb{C}$-basis $\eta_1, \ldots, \eta_t$ of $(F_{k-1}/\mathcal{N}^{(k-1)}(\omega_f))^\vee$, and a $\mathbb{C}$-basis $h_1\omega_f, \ldots, h_\mu\omega_f$ of W_f.
Output: A system of generators G of $\mathcal{N}^{(k)}(\omega_f)$, and a $\mathbb{C}$-basis B of $(F_k/\mathcal{N}^{(k)}(\omega_f))^\vee$.

1: Compute matrices $M_i^{(k)}$, $1 \le i \le n$.
2: Let $V := \bigcup_{i=1}^{n}\{\eta_j \cdot M_i^{(k)} \mid 1 \le i \le n, 1 \le j \le t\}$ which is a $\mathbb{C}$-basis of $(F'_k/pr(\mathcal{N}^{(k)}(\omega_f)))^\vee$.
3: Compute the reduced $\mathbb{C}$-basis $\xi_1, \ldots, \xi_s$ of $(F'_k/pr(\mathcal{N}^{(k)}(\omega_f)))^\vee$ from V with respect to $\prec_{F'_k}^\vee$ by using Gaussian elimination method.
4: By using Proposition 4, compute the standard basis $S = \{g^{(1)}, \ldots, g^{(p)}\}$ of $pr(\mathcal{N}^{(k)}(\omega_f))$ with respect to $\prec_{F'_k}$.
5: For each $g^{(j)} = (g_\alpha^{(j)})_{|\alpha| \le k, \alpha \ne 0}$, compute $g_0^{(j)} \in \mathcal{O}_{X,O}$ such that $\omega_f \sum_\alpha (\frac{\partial}{\partial x})^\alpha g_\alpha = g_0^{(j)} \omega_f$.
6: Return $G := \{(g_\alpha^{(j)})_{|\alpha| \le k} \mid 1 \le j \le p\} \cup \{(0, \ldots, 0, \frac{\partial f}{\partial x_i})^T \mid 1 \le i \le n\}$, and $B = \{(\xi_1, 0), \ldots, (\xi_s, 0)\} \cup \{(\omega_f(\frac{\partial}{\partial x})^\alpha h_j)_{|\alpha| \le k} \mid 1 \le j \le \mu\}$.

The method presented in this paper is applicable when f has parameters in its coefficients.

6   The Size of the Matrix Appearing in Algorithm 1

The main part of our algorithm is the Gaussian elimination step. The complexity of Gaussian elimination method depends on the size of the input matrix. In this section, we give an estimation on the size of the matrix induced by V in Algorithm 1, which we denote by M_V.

Let $\Delta_k := \dim_\mathbb{C}(F_k/\mathcal{N}^{(k)}(\omega_f))$ and let N_k be the number of terms appearing in elements of the module $\sum_{|\alpha| \le k} \mathcal{O}_{X,O} \cdot \omega_f(\frac{\partial}{\partial x})^\alpha$. Then size of M_V is at most $n\Delta_{k-1} \times N_{k-1}(\binom{n+k}{n} - 1)$.

First, we estimate Δ_k which coincides with the $\mathbb{C}$-dimension of its Matlis dual. As $(F_k/\mathcal{N}^{(k)}(\omega_f))^\vee$ is generated by a single element $(\omega_f(\frac{\partial}{\partial x})^\alpha)_{\alpha \in \mathbb{Z}_{\geq 0}^n, |\alpha| \leq k}$ as $\mathcal{O}_{X,O}$ module, we have

$$\Delta_k = \dim_{\mathbb{C}} \mathcal{O}_{X,O} / Ann_{\mathcal{O}_{X,O}}\left(\omega_f\left(\frac{\partial}{\partial x}\right)^\alpha\right)_{|\alpha| \leq k}$$

$$= \dim_{\mathbb{C}} \mathcal{O}_{X,O} / \bigcap_{|\alpha| \leq k} Ann_{\mathcal{O}_{X,O}} \omega_f\left(\frac{\partial}{\partial x}\right)^\alpha.$$

Write

$$\omega_f = \sum_{\alpha \in \mathbb{Z}_{\geq 0}} \frac{c_\alpha dx}{x^{\alpha+1}},$$

where $c_\alpha \in \mathbb{C}$, and we set

$$\delta = \max\{|\alpha| \mid c_\alpha \neq 0\}.$$

Then it is easy to show that $\mathfrak{m}_{X,O}^\delta \cdot \omega_f = 0$ and $\mathfrak{m}_{X,O}^{\delta+|\alpha|} \cdot \omega_f(\frac{\partial}{\partial x})^\alpha = 0$ for $\alpha \in \mathbb{Z}_{\geq 0}^n$. Thus $\Delta_k \leq \dim_{\mathbb{C}} \mathcal{O}_{X,O} / \mathfrak{m}_{X,O}^{\delta+k} = \binom{n+\delta+k-1}{n}$.

Next, we will estimate N_k. For any $\frac{dx}{x^{\beta+1}}$ appearing in an element of the module $\sum_{|\alpha| \leq k} \mathcal{O}_{X,O} \cdot \omega_f(\frac{\partial}{\partial x})^\alpha$, since $\mathfrak{m}_{X,O}^{\delta+k}$ annihilates $\frac{dx}{x^{\beta+1}}$, it holds that $|\beta| \leq \delta+k$. Thus $N_k \leq \binom{n+\delta+k-1}{n}$.

Therefore, the size of M_V is at most $n\binom{n+\delta+k-2}{n} \times \binom{n+\delta+k-2}{n}(\binom{n+k}{n} - 1)$. If the number of variables n is fixed, this is of polynomial order in δ and k.

7   Example

Let

$$f_t(x,y) = x^4 + y^7 + tx^2y^5 + xy^6$$

with t a parameter. The Milnor number of f_t is 18 for all $t \in \mathbb{C}$. Let $\prec$ be the anti-graded lexicographic order on $\mathbb{C}\{x,y\}$, that is,

$$x^a y^b \prec x^c y^d \text{ if } a+b > c+d, \text{ or if } a+b = c+d \text{ and } b < d.$$

We take ω_{f_t} as in Definition 3. Then $\mathcal{N}^{(1)}(\omega_{f_t})$ is generated by the union of the following set and $\{(0, 0, \frac{\partial f_t}{\partial x})^T, (0, 0, \frac{\partial f_t}{\partial y})^T\}$.

1. If $t = 0$,

$\{(0, -2058yx - 2401y^2, (972y^3 + 10290)x + 1134y^4 + 14406y)^T,$

$(0, 686yx^2, (-432y^3 - 4116)x^2 + 126y^4x + 735y^5)^T,$

$(0, 98x^3, 18y^3x^2 - 21y^4x)^T,$

$(-84035yx, -6174x^2 + 28812yx, -1944y^2x^2 + (-7938y^3 - 100842)x - 6615y^4 + 252105y)^T,$

$(-24010x^2, 2058x^2 - 9604yx, 648y^2x^2 + (2646y^3 + 129654)x + 2205y^4)^T,$

$(-1715y^5, -686x^2 - 4802yx, 864y^2x^2 + (1638y^3 + 28812)x + 735y^4)^T\}.$

2. If $t \neq 0, \frac{3}{7}, \frac{36}{35}, \frac{36}{119}$,

$\{(0, (-24010t^2 + 6174t)x^2 + (-115248t + 37044)yx + (-100842t + 43218)y^2, (-8575t^4 - 131565t^3 + 82404t^2 -$
$12636t)y^2 x^2 + ((18865t^3 - 204183t^2 + 117936t - 17496)y^3 + 605052t - 185220)x + (-12005t^3 - 219177t^2 +$
$134946t - 20412)y^4 + (605052t - 259308)y)^T,$

$(0, -686yx^2, ((1225t^2 - 1953t + 432)y^3 + 4116)x^2 + (-196t - 126)y^4 x + (1715t - 735)y^5)^T,$

$(0, 98x^3, (-77t + 18)y^3 x^2 + (49t - 21)y^4 x)^T,$

$((-7058940t + 3025260)yx, (1680700t^2 - 1166886t + 222264)x^2 + (2924418t - 1037232)yx, (600250t^4 + 25725t^3 +$
$586089t^2 - 428652t + 69984)y^2 x^2 + ((-1320550t^3 + 4164363t^2 - 2087694t + 285768)y^3 - 11495988t + 3630312)x +$
$(840350t^3 + 2485035t^2 - 1620675t + 238140)y^4 + (21176820t - 9075780)y)^T,$

$(48020x^2, -4116x^2 + 19208yx, (11025t^2 + 1134t - 1296)y^2 x^2 + ((18228t - 5292)y^3 - 259308)x + (15435t -$
$4410)y^4)^T,$

$((12005t - 5145)y^5, -2058x^2 - 14406yx, (-3675t^2 - 6993t + 2592)y^2 x^2 + ((8575t^2 - 23226t + 4914)y^3 +$
$86436)x + (-15435t + 2205)y^4)^T\}.$

3. If $t = \frac{36}{35}$,

$\{(0, 185220x^2 + 792330yx + 588245y^2, 763992y^2 x^2 + (891324y^3 - 4249770)x + 1230390y^4 - 3529470y)^T,$

$(0, -3430yx^2, (-1404y^3 + 20580)x^2 - 1638y^4 x + 5145y^5)^T,$

$(0, 490x^3, -306y^3 x^2 + 147y^4 x)^T,$

$(2941225yx, -555660x^2 - 1368570yx, -659016y^2 x^2 + (-768852y^3 + 5690370)x - 1468530y^4 - 8823675y)^T,$

$(-120050x^2, 10290x^2 - 48020yx, -28836y^2 x^2 + (-33642y^3 + 648270)x - 28665y^4)^T,$

$(-12005y^5, 3430x^2 + 24010yx, 14148y^2 x^2 + (16506y^3 - 144060)x + 22785y^4)^T\}.$

4. If $t = \frac{36}{119}$,

$\{(0, 10705716x^2 - 70776678yx - 412863955y^2, -145800y^2 x^2 + (-578340y^3 + 70776678)x - 674730y^4 +$
$2477183730y)^T,$

$(0, -198254yx^2, (-13500y^3 + 1189524)x^2 - 53550y^4 x - 62475y^5)^T,$

$(0, 1666x^3, -90y^3 x^2 - 105y^4 x)^T,$

$(-412863955yx, -10705716x^2 + 70776678yx, 145800y^2 x^2 + (578340y^3 - 70776678)x + 674730y^4 + 1238591865y)^T,$

$(-6938890x^2, 594762x^2 - 2775556yx, -8100y^2 x^2 + (-32130y^3 + 37470006)x - 37485y^4)^T,$

$(-145775y^5, -198254x^2 - 1387778yx, 13500y^2 x^2 + (-127890y^3 + 8326668)x - 237405y^4)^T\}.$

5. If $t = \frac{3}{7}$,

$\{(0, -686x^2 - 4802yx, -360y^2 x^2 + (-1155y^3 + 28812)x - 1470y^4)^T,$

$(0, -33614y^3, (405y^3 + 4116)x^2 + (1575y^4 - 28812y)x + 2205y^5 + 201684y^2)^T,$

$(0, 98x^3, -15y^3 x^2)^T,$

$(-184473632x, 19765032x - 46118408y, (60750y^4 - 926100y)x^2 + (70875y^5 - 2593080y^2)x - 3025260y^3 +$
$830131344)^T,$

$(672280y^4, 941192x + 6588344y, (-4050y^4 + 61740y)x^2 + (-4725y^5 + 576240y^2)x + 1008420y^3 - 39530064)^T\}.$

Conclusion. In this paper, we give an algorithm for computing truncated annihilating ideals for an algebraic local cohomology class. It is easy to generalize our algorithm to the case where the function f contains parameters in coefficient. Our algorithm presents a new computational method for studying μ-constant deformations.

References

1. Bruns, W., Herzog, J.: Cohen-Macaulay rings. University Press, Cambridge (1993)
2. Greuel, G.-M., Pfister, G.: A Singular Introduction to Commutative Algebra. Springer, Berlin (2002); With contributions by Olaf Bachmann, Christoph Lossen and Hans Schonemann, With 1 CD-ROM (Windows, Macintosh, and UNIX)

3. Grigorév, D.Y., Chistov, A.L.: Complexity of the standard basis of a D-module. Algebra i Analiz 20(5), 41–82 (2008) (Russian. Russian summary)
4. Lê, D.T., Ramanujan, C.P.: The invariance of Milnor's number implies the invariance of the topological type. Amer. J. Math. 98, 67–78 (1976)
5. Matlis, E.: Injective modules over Noetherian rings. Pacific J. Math. 8, 511–528 (1958)
6. Nabeshima, K.: PGB: A Package for Computing Parametric Gröbner Bases and Related Objects. In: Conference posters of ISSAC 2007, pp. 104–105 (2007)
7. Nakamura, Y., Tajima, S.: Unimodal singularities and differential operators, Séminaires et Congrès 10, Sociétés Mathématiques de France, pp. 191–208 (2005)
8. Tajima, S., Nakamura, Y.: Algebraic local cohomology classes attached to quasi-homogeneous isolated hypersurface singularities. Publ. Res. Inst. Math. Sci. 41, 1–10 (2005)
9. Tajima, S., Nakamura, Y.: Annihilating ideals for an algebraic local cohomology class. Journal of Symbolic Computation 44, 435–448 (2009)
10. Tajima, S., Nakamura, Y.: Algebraic local cohomology classes attached to unimodal singularities. Publ. Res. Inst. Math. Sci. 48(2), 21–43 (2012)
11. Tajima, S., Nakamura, Y., Nabeshima, K.: Standard bases and algebraic local cohomology for zero dimensional ideals. Advanced Studies in Pure Mathematics 56, 341–361 (2009)
12. Weispfenning, V.: Comprehensive Gröbner bases. J. Symbolic Computation 14, 1–29 (1991)

Applications of the Newton Index
to the Construction of Irreducible Polynomials

Doru Ştefănescu

University of Bucharest, Romania
stef@rms.unibuc.ro

Abstract. We use properties of the Newton index associated to a polynomial with coefficients in a discrete valuation domain for generating classes of irreducible polynomials. We obtain factorization properties similar to the case of bivariate polynomials and we give new applications to the construction of families of irreducible polynomials over various discrete valuation domains. The examples are obtained using the package gp-pari.

1 Introduction

The construction of classes of irreducible polynomials is based on some few irreducibility criteria or is the result of factorization algorithms. One of the devices used for obtaining irreducibility criteria is to associate properly to a polynomial a Newton polygon and to deduce from the properties of the polygon useful information concerning the irreducibility. This was done by G. Dumas [10] in his extension of the irreducibility criteria of T. Schönemann [16] and G. Eisenstein [11]. In fact Dumas considered the product of two univariate polynomials F_1 and F_2 with integer coefficients and studied the relations among the slopes of the Newton polygons of the polynomials F_1, F_2 and their product $F = F_1 F_2$.

The Newton polygon method was subsequently used by various authors for the study of the irreducibily of the polynomials. Recently such results were obtained by A. Bishnoi–S. K. Khanduja–K. Sudesh [3], C. N. Bonciocat [7], C. N. Bonciocat–Y. Bugeaud–M. Cipu–M. Mignotte [8], D. Ştefănescu [17], [18] and S. H. Weintraub [20].

The Newton polygon was initially defined for bivariate polynomials. Another approach is to associate a Newton polygon to a univariate polynomial with the coefficients in a discrete valuation domain. However, the irreducibility criterion of G. Dumas [10] makes use of Newton polygons of univariate polynomials over the integers and of the valuation defined by powers of a prime p. This result was improved by O. Ore [13]. This idea was used by many authors, recently the irreducibility over valued fields was considered by A. Bishnoi–S. K. Khanduja–K. Sudesh [3], A. I. Bonciocat–C. N. Bonciocat [4], [5], C. N. Bonciocat [6], [9], and A. Zaharescu [19]. On the other hand, the Newton polygon was used by L. Panaitopol–D. Ştefănescu [14] for obtaining irreducibility criteria for bivariate

V.P. Gerdt et al. (Eds.): CASC Workshop 2014, LNCS 8660, pp. 460–471, 2014.

polynomials. The Newton polyhedra were considered by A. Lipkovski [12] for the study of absolute irreducibility of multivariate polynomials.

In this paper we consider properties of the Newton index for obtaining information on the factorization of a univariate polynomial with the coefficients in a discrete valuation field. A related method was first used by the author in [17], in the case of bivariate polynomials. However, the results cannot be applied directly to polynomials with coefficients in a valuation domain, so we restate Theorem 1 from [17] in this context, as Theorem 1. The Theorem 2 gives more information on the factorization of a general univariate polynomial over a discrete valuation domain. These results will be used for generating families of irreducible polynomials. In particular, we construct new classes of univariate irreducible polynomials over the integers and over fields of formal power series. Other applications are given to bivariate irreducible polynomials over algebraically closed fields of characteristic zero.

2   On the Newton Index

We consider a univariate polynomial $F(X) = \sum_{i=0}^{d} a_i X^{d-i}$ with coefficients in a discrete valuation domain (A, v). We remind that the Newton polygon $N(F)$ of the polynomial $F(X) = \sum_{i=0}^{d} a_i X^{d-i}$ is the lower convex hull of the set $\{(d - i, v(a_i)) \, ; \, a_i \neq 0\}$. The slopes of the Newton polygon are the slopes of some line segment. We note that the slope of the line joining the points $(d, v(a_0))$ and $(d - i, v(a_i))$ is $\dfrac{v(a_0) - v(a_i)}{i}$. The *Newton index* $e(F)$ of the polynomial F is the largest slope $e(F)$ of these lines. More precisely,

$$e(F) = \max_{1 \leq i \leq d} \frac{v(a_0) - v(a_i)}{i} \, .$$

G. Dumas [10] studied the relationship between the Newton indices of two polynomials and the index of their product. He considered the case of univariate integer polynomials with the valuation defined by powers of a prime p. If F_1 and F_2 are such polynomials, he established that the Newton polygon of the product $F_1 F_2$ can be obtained by translating the edges of the polygons $N(F_1)$ and $N(F_2)$ in such a way that they compose a convex polygonal path with the slopes of the edges ordered increasingly. The proof of Dumas is based only on properties of the Newton polygons and it remains true for the case of arbitrary discrete valuations. From the result of Dumas we obtain:

Proposition 1. *If $F_1, F_2 \in A[X] \setminus A$ then*

$$e(F_1 F_2) = \max\left(e(F_1), e(F_2)\right) .$$

In the case of bivariate polynomials Proposition 1 gives a relation between the degree indices of two polynomials and the degree index of their product. We remind that, in [17], to a bivariate polynomial $F(X, Y) = \sum_{i=0}^{d} P_i(X) Y^{d-i}$ we associated the degree-index

$$P_Y(F) = \max_{1 \leq i \leq d} \frac{\deg(P_i) - \deg(P_0)}{i} \, .$$

It was used for obtaining irreducibility criteria for bivariate generalized difference polynomials and their extensions by L. Panaitopol–D. Ştefănescu [14]. Among other generalizations of irreducibility tests on generalized difference polynomials we mention those of G. Angermüller [1], S. Bhatia–S. K. Khanduja [2], and A. Bishnoi–S. K. Khanduja–K. Sudesh [3], D. Ştefănescu [17] and [18].

The oldest polynomial irreducibility criterion that applies to a general family of polynomials was obtained by T. Schönemann [16] in 1846. A particular case is Eisenstein's criterion [11] published in 1850. G. Dumas [10] noted that Eisenstein's criterion is related to properties of the Newton polygon and obtained a generalization of the Schönemann–Eisenstein criterion. We remind its valuation approach:

Lemma 1 (G. Dumas, 1906). *Let* $F(X) = \sum_{i=0}^{d} a_i X^{d-i} \in A[X]$ *be a polynomial over a discrete valuation domain A, with the valuation field (K, v). If the following conditions*

 i) $v(a_0) = 0$,

 ii) $\frac{v(a_d)}{d} < \frac{v(a_i)}{i}$ *for all* $i \in \{1, 2, \ldots, d-1\}$,

 iii) $\gcd(v(a_d), d) = 1$,

are satisfied, the polynomial $F(X)$ is irreducible in $K[X]$.

Remark 1. The condition ii) in Lemma 1 means that the Newton index of the polynomial F is $e(F) = -v(a_d)/d$.

Remark 2. We consider now a generalized difference polynomial $F(X, Y) \in k[X, Y]$, where k is a field,

$$F(X, Y) = cY^d + \sum_{i=1}^{d} P_i(X)Y^{d-i},$$

with $c \in k \setminus \{0\}$, $\in \mathbb{N}^*$, $P_i(X) \in k[X]$ and

$$\frac{\deg(P_i)}{i} < \frac{\deg(P_d)}{d} \quad \text{for all} \quad i, \ 1 \le i \le d-1. \tag{1}$$

Putting, for a polynomial $P \in k[X]$, $v(P) = -\deg(P)$, we observe that $k[X, Y]$ can be organized as a discrete valuation domain. The relation (1) becomes exactly the condition ii) from Theorem 1. Because $v(c) = 0$ the Theorem of Dumas 1 states that the generalized difference polynomial $F(X, Y)$ is irreducible if $(\deg(P_d), d) = 1$. This proves a result established, using a different method, by G. Angermüller in [1].

We will look at factorization properties of univariate polynomials over a discrete valuation domain for which the hypotheses in Theorem 1 are not satisfied.

3   Factorization Conditions

Let (A, v) be a discrete valuation domain and $F(X) = \sum_{i=0}^{d} a_i X^{d-i} \in A[X]$. We will consider the case in which the Newton index could be attained for an index $s \ne d$ and for which $v(a_0)$ could be nonzero.

Theorem 1. *Let (A, v) be a discrete valuation domain, and let*

$$F(X) = a_0 X^d + a_1 X^{d-1} + \cdots + a_{d-1} X + a_d \in A[X],$$

with $a_0 a_d \neq 0$ and $d \geq 2$. We assume that there exists an index $s \in \{1, 2, \ldots, d\}$ such that

(a) $\dfrac{v(a_0) - v(a_s)}{s} > \dfrac{v(a_0) - v(a_i)}{i}$ *for $i \in \{1, 2, \ldots, d\}, i \neq s$,*

(b) $\dfrac{v(a_0) - v(a_s)}{s} - \dfrac{v(a_0) - v(a_d)}{d} = \dfrac{1}{ds}$,

(c) $\gcd(v(a_0) - v(a_s), s) = 1$.

Then the polynomial F is either irreducible in $A[X]$, or has a factor whose degree is a multiple of s.

Proof. The proof follows the same lines as that of Theorem 5 in [18], using valuations instead of degrees. We suppose that there exists a nontrivial factorization $F = F_1 F_2$ in $A[X]$. We have $d = \deg(F)$ and we put

$$d_1 = \deg(F_1), \quad d_2 = \deg(F_2).$$

We suppose that

$$F_1(X) = \sum_{i=0}^{d_1} a_{1i} X^{d_1 - i}, \quad F_2(X) = \sum_{i=0}^{d_2} a_{2i} X^{d_2 - i}.$$

We observe that $a_d = a_{1d_1} a_{2d_2}$ and, $a_0 = a_{10} a_{20}$.
Then we put

$$c = v(a_0) - v(a_s), \quad m = v(a_0) - v(a_d).$$

$$m_1 = v(a_{10}) - v(a_{1d_1}), \quad m_2 = v(a_{20}) - v(a_{2d_2}).$$

We observe that

$$d = d_1 + d_2, \quad m = m_1 + m_2.$$

¿From the condition (b) we obtain

$$cd - sm = 1. \tag{2}$$

By Proposition 1 we have $e(F) = \max\{e(F_1), e(F_2)\}$ and, by the hypothesis (a), it follows that

$$\frac{c}{s} = \frac{v(a_0) - v(a_s)}{s} = e(F) \geq e(F_1) \geq \frac{v(a_{10}) - v(a_{1d_1})}{d_1} = \frac{m_1}{d_1},$$

which gives

$$\frac{c}{s} - \frac{m_1}{d_1} \geq 0,$$

so

$$cd_1 - sm_1 \geq 0 \,.$$

Because $e(F) \geq e(F_2)$ we also have

$$cd_2 - sm_2 \geq 0 \,.$$

But we have

$$1 = cd - sm = (cd_1 - sm_1) + (cd_2 - sm_2) \,,$$

so one of the positive integers $cd_1 - sm_1$ and $cd_2 - sm_2$ must be 0.

Suppose, for example, that we have $cd - sm_1 = 0$. So $cd = sm_1$. But, by the condition (c), the integers c and s are coprime. Therefore, s must divide d. If $cd - sm_2 = 0$ we obtain that s must divide m_2. So, if the polynomial F is reducible, the degree of one of its divisors must be a multiple of s.   □

Corollary 1. *In the conditions of Theorem 1, if $d \geq 3$ and $s > d/2$, then the polynomial F is either irreducible, or has a divisor of degree s.*

Proof. By Theorem 1 the polynomial F is irreducible or it has a factor of degree a multiple of s. If F would have a factor of degree ks, with $k \geq 2$, then we would obtain

$$d > ks > k\frac{d}{2} \geq d \,,$$

a contradiction. Therefore, $k = 1$ or F is irreducible.   □

If the difference between the numbers in the left-hand side in condition (b) in Theorem 1 is larger than $\frac{1}{ds}$ we can also say something about the possible divisors of F. More precisely, we have the following result:

Theorem 2. *Let (A, v) be a discrete valuation domain, and let*

$$F(X) = a_0X^d + a_1X^{d-1} + \cdots + a_{d-1}X + a_d \in A[X] \,,$$

with $a_0a_d \neq 0$ and $d \geq 2$. We assume that there exists an index $s \in \{1, 2, \ldots, d\}$ such that

(a) $\dfrac{v(a_0) - v(a_s)}{s} > \dfrac{v(a_0) - v(a_i)}{i}$ *for $i \in \{1, 2, ..., d\}, i \neq s$;*

(b) $\dfrac{v(a_0) - v(a_s)}{s} - \dfrac{v(a_0) - v(a_d)}{d} = \dfrac{u}{ds}$, *with $u \geq 2$;*

(c) $\gcd(v(a_0) - v(a_s), s) = 1$.

Then one of the following conditions is satisfied:

   i. The polynomial F is irreducible in $A[X]$.

   ii. The polynomial F has a divisor whose degree is a multiple of s.

   iii. The polynomial F admits a factorization $F = F_1F_2$ and s divides $\beta d_1 - \alpha d_2$, for some $\alpha, \beta \in \{1, 2, \ldots, u-1\}$, where $d_1 = \deg(F_1)$, $d_2 = \deg(F_2)$.

Proof. We use the same notation as in the proof of Theorem 1. We obtain the relation

$$cd - sm = u. \tag{3}$$

We have $cd_1 - sm_1 \geq 0$, $cd_2 - sm_2 \geq 0$ and

$$(cd_1 - sm_1) + (cd_2 - sm_2) = u. \tag{4}$$

We look to the possible values of $cd_1 - sm_1$.

If $cd_1 - sm_1 = 0$ as in Theorem 1 we deduce that the degree of a divisor of the polynomial F must be divisible by s.

If $cd_1 - sm_1 = 1$ we have $cd_2 - sm_2 = u - 1$ and we obtain

$$c(d_2 - (u - 1)d_1)) = (m_2 - (u - 1)m_1),$$

therefore s divides $d_2 - (u - 1)d_1$.

In general, we suppose that

$$\begin{aligned} cd_1 - sm_1 &= \alpha, \\ cd_2 - sm_2 &= \beta, \end{aligned} \tag{5}$$

with $\alpha + \beta = u$.

From the relations (5) we obtain

$$c(\beta d_1 - \alpha d_2) = s(\beta m_1 - \alpha m_2).$$

But s and c are coprime, so s should divide $\beta d_1 - \alpha d_2$. Therefore, the case iii is satisfied. $\square$

4   Applications

We consider univariate polynomials over particular discrete valuation domains (the p–adic numbers, the integers, the formal power series) and bivariate polynomials with coefficients in an algebraically closed field of characteristic zero.

Theorems and 1 and 2 are suitable for constructing families of irreducible polynomials over $A[X]$, where $A = (A, v)$ is a discrete valuation domain. Given a nonconstant polynomial $F(X) = a_0 X^d + a_1 X^{d-1} + \cdots + a_{d-1} X + a_d \in A[X]$, with $a_0 a_d \neq 0$, $d \geq 2$ the method is summarized in the following steps:

- Compute the valuations $v(a_0)$, $v(a_1)$, $\ldots$, $v(a_d)$.
- Compute the Newton index $e(F) = \max_{1 \leq i \leq d} \{(v(a_0) - v(a_i))/i\}$ and the index s for which $e(F) = (v(a_0) - v(a_s))/s$.
- Compute $\gcd(v(a_0) - v(a_s))$.
- If $\gcd(v(a_0) - v(a_s)) \neq 1$, the irreducibility of the polynomial cannot be tested by this method.
- If $s = d$ we conclude that F is irreducible by the argument in the Theorem of Dumas.

– If $s \neq d$ we compute u such that

$$e(F) - \frac{v(a_0) - v(a_d)}{d} = \frac{u}{sd}.$$

– If $u \notin \{1, 2, \ldots, d-1\}$, the irreducibility cannot be tested by this method.
– If $u = 1$ we apply Theorem 1.
– If $u \in \{2, \ldots, d-1\}$ we apply Theorem 2.

Using the package `gp-pari` we computed the Newton indices and we found couples of numbers (s, u) that satisfy the hypotheses in Theorems 1 or 2.

4.1 Univariate Polynomials over p–adic Numbers

Let $r \in \mathbb{Z}_p$ be a p–adic number, $r = p^n \sum_{i=0}^{\infty} a_i p^i$, $a_i \in \{0, 1, \ldots, p-1\}$, $a_0 \neq 0$ ..
We define a discrete valuation by $v(r) = n$.

Example 1. Let $F(X) = X^d + aX^2 + bX + c \in \mathbb{Z}_p[X]$. Suppose that $d \geq 2$ and $v(a) = d$, $v(b) = d - 2$, $v(c) = d - 1$. We have

$$\frac{v(1) - v(a)}{d - 2} = \frac{-d}{d - 2} < 0,$$

$$\frac{v(1) - v(b)}{d - 1} = \frac{1}{d - 1} - 1,$$

$$\frac{v(1) - v(c)}{d} = \frac{1}{d} - 1.$$

It follows that $e(F) = \frac{v(1) - v(b)}{d-1}$ and

$$e(F) - \frac{v(1) - v(c)}{d} = \frac{1}{d(d-1)}.$$

We have $s = d - 1$ and, by Theorem 1, we conclude that F is either irreducible, or has a factor of degree $d - 1$, and hence also a linear factor. Therefore, if F has no p–adic roots it is irreducible over $\mathbb{Z}_p[X]$.

4.2 Univariate Polynomials over Formal Power Series

Let k be an algebraically closed field of characteristic zero. If $f(X) = \sum_{i=0}^{\infty} a_i X^i$ is a formal power series from $k[[X]]$ we put $v(f) = \mathrm{ord}(f) := \min_i \{i; a_i \neq 0\}$.

Example 2. Let $F(Y) = XY^d + f(X)Y^{d-1} + g(X)Y^2 + Y + h(x) \in k[[X]][Y]$, with $d \geq 3$,

$$f(X) = X^d + X^{d+1} + \ldots + X^{d+n} + \ldots),$$
$$g(X) = X^{d-2} + X^{d-1} + X^d,$$
$$h(X) = X^{d+1}(1 - X + X^2 - X^3 + \cdots).$$

We have $\frac{v(X)-v(f)}{1} = 1-d$, $\frac{v(X)-v(g)}{d-2} = \frac{3-d}{d-2}$, $\frac{v(X)-v(1)}{d-1} = \frac{1}{d-1}$, $\frac{v(1)-v(h)}{d} = -1$.
Therefore, $e(F) = \frac{v(1)-v(1)}{d-1} = \frac{1}{d-1}$ and we have $s = d-1$. We have by Theorem 1
that F is either irreducible in $K[X,Y]$, or has a factor whose degree is a multiple
of $d-1$. Hence F is either irreducible, or has a linear factor.

4.3   Univariate Polynomials over the Integers

We suppose that $F(X) \in \mathbb{Z}[X] \setminus \mathbb{Z}$ and we consider the valuation given by the
power with respect to a prime ≥ 2.

Example 3. Let $F(X) = (p^2 + p + 1)X^d + X^3 + p^{d-2}(p+1)X + p^d$, with $d \geq 4$
and p a prime. We have

$$v(a_0) = 0, \quad v(a_{d-3}) = 0, \quad v(a_{d-1}) = d-2, \quad v(a_d) = d.$$

$$e(F) = \max \left\{ \frac{-d+2}{d-1}, -1 \right\} = \frac{-d+2}{d-1} = \frac{v(a_0) - v(a_{d-1})}{d-1},$$

so we can apply Theorem 1. We have $s = d-1$ and $\gcd(v(a_0) - v(a_{d-1}), s) = \gcd(d-2, d-1) = 1$.

Therefore, the polynomial F is irreducible or has a divisor of degree $s = d-1$.
In this case, it should have also a linear divisor, so an integer root. Such roots
should be of the form $-p^t$, with $t \in \{0, 1, \ldots, p^d\}$, and this can be checked for
particular values of d and t.

4.4   Bivariate Polynomials

Let k be an algebraically closed field of characteristic zero and suppose that F is
a bivariate polynomial from $k[X,Y]$. We suppose that it has the representation

$$F(X,Y) = P_0(X)Y^d + P_1(X)Y^{d-1} + \ldots + P_{d-1}(X)Y + P_d(X),$$

where $P_i \in k[X]$, $P_0 \neq 0$.
   For $P \in k[X]$ we put $v(P) = -\deg(P)$, and this defines a discrete valuation
on $A := k[X]$. Because

$$v(P_0) - v(P_i) = \deg(P_i) - \deg(P_0)$$

the Newton index of the polynomial $F(X,Y) \in A[Y]$ becomes

$$e(F) = \max_{1 \leq i \leq d} \left\{ \frac{\deg(P_i) - \deg(P_0)}{i} \right\},$$

which is exactly the degree index considered bu L. Panaitopol–D. Ştefănescu
in [14]. The results within Section 3 have, therefore, polynomial approaches. For
example, by Theorem 1 we obtain:

Corollary 2 (D. Ştefănescu [18]). *Let k be an algebraically closed field of characteristic zero and let*

$$F(X,Y) = P_0(X)Y^d + P_1(X)Y^{d-1} + \ldots + P_{d-1}(X)Y + P_d(X), \quad P_0 P_d \neq 0.$$

If there exists an index $s \in \{1, 2, \ldots, d\}$ such that the following conditions are satisfied

(a) $\dfrac{\deg(P_i) - \deg(P_0)}{i} < \dfrac{\deg(P_s) - \deg(P_0)}{s}$ *for* $i \in \{1, 2, ..., d\}, i \neq s$;

(b) $\dfrac{\deg(P_s) - \deg(P_0)}{s} - \dfrac{\deg(P_d) - \deg(P_0)}{d} = \dfrac{1}{ds}$.

(c) $\gcd(\deg(P_s) - \deg(P_0), s) = 1$

the polynomial F is either irreducible in $A[X]$, or has a factor whose degree is a multiple of s.

Example 4. Let $F(X,Y) = X^m Y^d + XY^{d-1} + XY^{d-2} + Y^2 + p(X)Y + q(X)$ with $\deg(p) = \deg(q) = m+1$, $m \geq 1$, $d \geq 5$ and $q(0) \neq 0$. We have

$$\frac{\deg(P_1) - \deg(P_0)}{1} = \frac{1-m}{1},$$

$$\frac{\deg(P_2) - \deg(P_0)}{1} = \frac{1-m}{2},$$

$$\frac{\deg(P_{d-2}) - \deg(P_0)}{d-2} = \frac{-2}{d-3},$$

$$\frac{\deg(P_{d-1}) - \deg(P_0)}{d-1} = \frac{1}{d-1},$$

$$\frac{\deg(P_d) - \deg(P_0)}{d} = \frac{1}{d}.$$

We then apply Theorem 1 and obtain that the polynomial F is either irreducible or it has a divisor of degree $d-1$ with respect to Y. Therefore, F is irreducible or has a linear divisor with respect to Y.

Example 5. Let $F(X,Y) = (X^3+1)Y^d + X^2Y^{d-1} + (X^{d-2}+X+1)Y^3 - XY + X^{d+1} + 1$. We have

$$\frac{\deg(P_1) - \deg(P_0)}{1} = \frac{0-3}{1} = -3,$$

$$\frac{\deg(P_{d-3}) - \deg(P_0)}{d-3} = \frac{d-2}{d-3} > 1,$$

$$\frac{\deg(P_{d-1}) - \deg(P_0)}{d-1} = \frac{1-3}{d-1} = -\frac{2}{d-1},$$

$$\frac{\deg(P_d) - \deg(P_0)}{d} = \frac{4-3}{d} = \frac{1}{d}.$$

It follows that the Newton index is $e(F) = \frac{d-2}{d-3}$. We have $s = d-3$, $(d-2, d-3) = 1$ and

$$\frac{\deg(P_{d-3}) - \deg(P_0)}{d-3} - \frac{\deg(P_d) - \deg(P_0)}{d} = \frac{d-2}{d-3} - \frac{d+1}{d} = \frac{3}{d(d-3)}.$$

So we can apply Theorem 2. We have the following possibilities.

i. The polynomial F is irreducible in $k[X, Y]$.

ii. The polynomial F has a divisor whose degree with respect to Y is a multiple of $d - 3$. Therefore, there exists a divisor of degree 3 with respect to Y.

iii. There exists a nontrivial factorization $F = F_1 F_2$ such that $d - 3$ divides $\beta\, d_1 - \alpha\, d_2$, where $d_1 = \deg(F_1)$, $d_2 = \deg(F_2)$ and $\alpha, \beta \in \{1, 2, 3\}$. If we look at the proofs of Theorems 1 and 2 we can compute α and β.

In fact, from the relations

$$cd_1 - sm_1 = 1,$$
$$cd_2 - sm_2 = 2$$

we obtain $c(d_2 - 2d_1) = s(m_2 - 2m_1)$, so s must divide $d_2 - 2d_1$.

For our example we deduce that $d_2 - 2d_1$ must be divisible by 3. For particular values of d this condition is not satisfied. For example, for $d = 5$, we have $(d_2, d_1) \in \{(1, 4), (2, 3), (3, 2), (4, 1)\}$, so the cases to be considered are

$$1 - 2 \cdot 4 = -7,$$
$$2 - 2 \cdot 3 = -4,$$
$$3 - 2 \cdot 2 = -1,$$
$$4 - 2 \cdot 1 = 2,$$

and none of them is a multiple of 3.

Example 6. Let $F(X, Y) = p(X)Y^d + Y^{d-1} + q(X)Y^2 + r(X)$, with $\deg(p) = m \geq 1$, $\deg(q) = d + m - 1$, $\deg(r) = d + m + 1$, $d \geq 5$. We have

$$\frac{\deg(P_1) - \deg(P_0)}{1} = \frac{0 - m}{1} = -m,$$

$$\frac{\deg(P_{d-2}) - \deg(P_0)}{d-2} = \frac{d+m-1-m}{d-2} = \frac{d-1}{d-2},$$

$$\frac{\deg(P_d) - \deg(P_0)}{d} = \frac{d+m-2-m}{d} = \frac{d+1}{d}.$$

We obtain $e(F) = \frac{d-1}{d-2}$ and $d - 1$ and $d - 2$ are coprime. On the other hand,

$$e(F) - \frac{\deg(r)}{d} = \frac{d-1}{d-2} - \frac{d+1}{d} = \frac{2}{d(d-2)}$$

and we can apply Theorem 2. There are three possible cases:

i. The polynomial F is irreducible in $k[X, Y]$.

ii. The polynomial F has a divisor whose degree with respect to Y is a multiple of $d - 2$. So this divisor is od degree $d - 2$ with respect to Y. Therefore F could have a quadratic divisor with respect to Y.

iii. There exists a factorization $F = F_1 F_2$ and the difference of their degrees is a multiple of $d-2$. If we suppose $1 \leq d_1 \leq d_2 \leq d-1$ we obtain $0 \leq d_2-d_1 \leq d-2$. It follows that we have

$$d_1 = d_2 \quad \text{or} \quad d_2 - d_1 = d - 2.$$

The last condition is satisfied only if $d_1 = 1$ and $d_2 = d - 1$.

We conclude that the polynomial F is irreducible if it does not have quadratic divisors with respect to Y and satisfies one of the two conditions:

a. Its degree d is odd.

b. It does not have linear divisors with respect to Y.

5   Conclusion

In this paper we proposed a method for the construction of univariate irreducible polynomials over discrete valuation domains. We proved that our approach extends basic results on the irreducibility of univariate polynomials over the integers and on bivariate polynomials over an algebraically closed field. The method has applications also to polynomials in other discrete valuation domains. It requires the computation of families of numbers that satisfy some conditions. The use of computer packages allows us to obtain new classes of irreducible polynomials.

Future work will be done for applying these techniques for the construction of multivariate irreducible polynomials.

Acknowledgement. The author is grateful to the anonymous referees for valuable comments and suggestions.

References

1. Angermüller, G.: A generalization of Ehrenfeucht's irreducibility criterion. J. Number Theory 36, 80–84 (1990)
2. Bhatia, S., Khanduja, S.K.: Difference polynomials and their generalizations. Mathematika 48, 293–299 (2001)
3. Bishnoi, A., Khanduja, S.K., Sudesh, K.: Some extensions and applications of the Eisenstein irreducibility criterion. Developments in Mathematics 18, 189–197 (2010)
4. Bonciocat, A.I., Bonciocat, N.C.: Some classes of irreducible polynomials. Acta Arith. 123, 349–360 (2006)

5. Bonciocat, N.C.: A Capelli type theorem for multiplicative convolutions of polynomials. Math. Nachr. 281, 1240–1253 (2008)
6. Bonciocat, N.C.: On an irreducibility criterion of Perron for multivariate polynomials. Bull. Math. Soc. Sci. Math. Roumanie 53(101), 213–217 (2010)
7. Bonciocat, N.C.: Schönemann-Eisenstein-Dumas-type irreducibility conditions that use arbitrarily many prime numbers. arXiv:1304.0874v1
8. Bonciocat, N.C., Bugeaud, Y., Cipu, M., Mignotte, M.: Irreducibility criteria for sums of two relatively prime polynomials. Int. J. Number Theory 9, 1529–1539 (2013)
9. Bonciocat, N.C., Zaharescu, A.: Irreducible multivariate polynomials obtained from polynomials in fewer variables. J. Pure Appl. Algebra 212, 2338–2343 (2008)
10. Dumas, G.: Sur quelques cas d'irréducibilité des polynômes à coefficients rationnels. Journal de Math. Pures et Appl. 12, 191–258 (1906)
11. Eisenstein, G.: Über die Irreductibilität und einige andere Eigenschaften der Gleichung, von welcher die Theilung der ganzen Lemniscate abhängt. J. Reine Angew. Math. 39, 160–182 (1850)
12. Lipkovski, A.: Newton Polyhedra and Irreducibility. Math. Z. 199, 119–128 (1988)
13. Ore, O.: Zur Theorie der Eisensteinschen Gleichungen. Math. Z. 20, 267–279 (1924)
14. Panaitopol, L.D., Ştefănescu, D.: On the generalized difference polynomials. Pacific J. Math. 143, 341–348 (1990)
15. Rubel, L.A., Schinzel, A., Tverberg, H.: On difference polynomials and hereditary irreducible polynomials. J. Number Theory 12, 230–235 (1980)
16. Schönemann, T.: Von denjenigen Moduln, welche Potenzen von Primzahlen sind. J. Reine Angew. Math. 32, 93–105 (1846)
17. Ştefănescu, D.: Construction of classes of irreducible bivariate polynomials. In: Gerdt, V.P., Koepf, W., Mayr, E.W., Vorozhtsov, E.V. (eds.) CASC 2013. LNCS, vol. 8136, pp. 393–400. Springer, Heidelberg (2013)
18. Ştefănescu, D.: On the irreducibility of bivariate polynomials. Bull. Math. Soc. Sci. Math. Roumanie 56(104), 377–384 (2013)
19. Zaharescu, A.: Residual transcendental extentions of valuations, irreducible polynomials and trace series over p–adic fields. Bull. Math. Soc. Sci. Math. Roumanie 56(104), 125–131 (2013)
20. Weintraub, S.H.: A mild generalization of Eisenstein's criterion. Proc. Amer. Math. Soc. 141, 1159–1160 (2013)

Symbolic-Numeric Algorithm for Solving the Problem of Quantum Tunneling of a Diatomic Molecule through Repulsive Barriers

Sergue Vinitsky[1], Alexander Gusev[1], Ochbadrakh Chuluunbaatar[1,2],
Luong Le Hai[1,3], Andrzej Góźdź[4], Vladimir L. Derbov[5],
and Pavel Krassovitskiy[6]

[1] Joint Institute for Nuclear Research, Dubna, Moscow Region, Russia
`vinitsky@theor.jinr.ru`
[2] National University of Mongolia, UlaanBaatar, Mongolia
[3] Belgorod State University, Belgorod, Russia
[4] Institute of Physics, Maria Curie-Skłodowska University, Lublin, Poland
[5] Saratov State University, Saratov, Russia
[6] Institute of Nuclear Physics, Almaty, Kazakhstan

Abstract. Symbolic-numeric algorithm for solving the boundary-value problems that describe the model of quantum tunneling of a diatomic molecule through repulsive barriers is described. Two boundary-value problems (BVPs) in Cartesian and polar coordinates are formulated and reduced to 1D BVPs for different systems of coupled second-order differential equations (SCSODEs) that contain potential matrix elements with different asymptotic behavior. A symbolic algorithm implemented in CAS Maple to calculate the required asymptotic behavior of adiabatic basis, the potential matrix elements, and the fundamental solutions of the SCSODEs is elaborated. Comparative analysis of the potential matrix elements calculated in the Cartesian and polar coordinates is presented. Benchmark calculations of quantum tunneling of a diatomic molecule with the nuclei coupled by Morse potential through Gaussian barriers below dissociation threshold are carried out in Cartesian and polar coordinates using the finite element method, and the results are discussed.

Keywords: Symbolic-numeric algorithm, quantum tunneling problem, diatomic molecule, repulsive barriers, boundary-value problem, adiabatic representation, asymptotic solutions, finite element method.

1 Introduction

The study of tunneling of coupled particles through repulsive barriers [11] has revealed the effect of resonance quantum transparency: when the cluster size is comparable with the spatial width of the barrier, there are mechanisms that lead to greater transparency of the barrier. These mechanisms are related to the formation of the barrier resonances, provided that the potential energy of the composite system has local minima giving rise to metastable states of the moving cluster [10]. Currently this effect and its possible applications are a

V.P. Gerdt et al. (Eds.): CASC Workshop 2014, LNCS 8660, pp. 472–490, 2014.

subject of extensive study in relation with different quantum-physical problems, e.g., quantum diffusion of molecules [12], exciton resonance passage through a quantum heterostructure barrier [8], resonant formation of molecules from individual atoms [13], controlling the direction of diffusion in solids [1], and tunnelling of ions and clusters through repulsive barriers [7,6]. For the analysis of these effects, it is useful to develop model approaches based on approximations providing a realistic description of interactions between the atoms in the molecule as well as with the barriers, and to elaborate symbolic-numeric algorithms and software.

In this paper, we formulate and study the model of a diatomic molecule with the nuclei coupled via the effective Morse potential that penetrates through a Gaussian repulsive barrier, using Galerkin and Kantorovich expansion of the desired solution in Cartesian and polar coordinates, respectively. We formulate two boundary-value problems (BVP) and use different sets of basis functions to reduce the original problem to 1D BVPs for different systems of coupled second-order differential equations (SCSODEs) that contain potential matrix elements with different asymptotic behavior. In the first case, the potential matrix elements decrease exponentially, and in the second case, they decrease as inverse powers of the independent variable. In the second case, we must calculate the asymptotic behavior of the potential matrix elements to solve the boundary value problem. For this goal, we develop symbolic algorithms implemented in CAS Maple to calculate the required asymptotic behavior of the potential matrix elements as well as the fundamental solutions of SCSODEs. We present a comparative analysis of the potential matrix elements calculated in the Cartesian and polar coordinates, which are used to solve the quantum tunneling problem below the dissociation threshold. The necessity for two statements of the problem follows from the important practical applications of further self-consistent study of the system above the dissociation threshold, which is convenient in polar coordinates. The effect of quantum transparency, i.e., the resonance behavior of the transmission coefficient versus the energy of the molecule is analyzed.

The paper is organized as follows. In Sections 2 and 3, we formulate and solve the BVPs in Cartesian and polar coordinates. In Section 4, the leading terms of the asymptotic expressions of effective potentials and fundamental solutions are calculated using the elaborated algorithms in CAS Maple. In Section 5, we analyze the solution of the quantum tunneling problem below the dissociation threshold. In Conclusion, the prospects of future studies are discussed.

2   Model I. Quantum Tunneling in Cartesian Coordinates

We consider a 2D model of two identical particles with the mass m coupled by the pair potential $\tilde{V}(x_2 - x_1)$ and interacting with the external barrier potentials $\tilde{V}^b(x_1)$ and $\tilde{V}^b(x_2)$. Using the change of variables $x = x_2 - x_1$, $y = x_2 + x_1$,

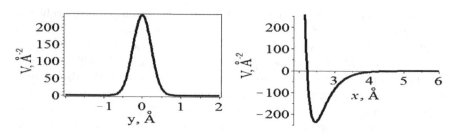

Fig. 1.   Gaussian-type barrier $V^b(x_i)$ = $\hat{D}\exp\left(-\frac{x_i^2}{2\sigma}\right)$,   at $\hat{D}$ = 236.510003758401Å^{-2} = $(m/\hbar^2)\tilde{V}_0$ = $(m/\hbar^2)D$, $\tilde{V}_0$ = D =   1280K, σ = $5.23 \cdot 10^{-2}$Å^2,   and the two-particle interaction potential, $V^M(x)$ = $\hat{D}\{\exp[-2(x-\hat{x}_{eq})\hat{\rho}] - 2\exp[-(x-\hat{x}_{eq})\hat{\rho}]\}$, $\hat{x}_{eq} = 2.47$Å, $\hat{\rho} = 2.96812423381643$Å^{-1}

$y \in (-\infty, \infty)$, $x \in (-\infty, \infty)$, we arrive at the Schrödinger equation for the wave function $\Psi(x, y)$ in the s-wave approximation

$$\left(-\frac{\hbar^2}{m}\frac{1}{f_1(y)}\frac{\partial}{\partial y}f_2(y)\frac{\partial}{\partial y} - \frac{\hbar^2}{m}\frac{1}{f_3(x)}\frac{\partial}{\partial x}f_4(x)\frac{\partial}{\partial x} + \tilde{V}(x,y) - \tilde{E}\right)\Psi(y,x) = 0. \quad (1)$$

where $\hbar$ is the Planck constant, $\tilde{E}$ is the total energy of the system, and the potential function $V(x, y)$ is defined by the formula

$$\tilde{V}(x,y) = \tilde{V}^M(x) + \tilde{V}^b(x_1) + \tilde{V}^b(x_2). \quad (2)$$

The equation describing the molecular subsystem has the form

$$\left(-\frac{\hbar^2}{m}\frac{1}{f_3(x)}\frac{\partial}{\partial x}f_4(x)\frac{\partial}{\partial x} + \tilde{V}^M(x) - \tilde{\varepsilon}\right)\phi(x) = 0. \quad (3)$$

The molecular subsystem is assumed to possess the continuous energy spectrum with the eigenvalues $\tilde{\varepsilon} \geq 0$ and eigenfunctions $\phi_{\tilde{\varepsilon}}(x)$ and the discrete energy spectrum, consisting of the finite number n of bound states with the eigenfunctions $\phi_j(x)$ and the eigenvalues $\tilde{\varepsilon}_j = -|\tilde{\varepsilon}_j|$, $j = 1, n$.

The asymptotic boundary conditions imposed on the solution for the 2D model in the s-wave approximation $\Psi(y, x) = \{\Psi_j(y, x)\}_{j=1}^{N_o}$ in the asymptotic region $\Omega_j^{as} = \{(x, y)||x|/|y| \ll 1\}$ with the direction $v = \rightarrow$ can be written in the obvious form

$$\Psi_j(y \to -\infty, x) \to \phi_j(x)\frac{\exp(\imath p_j y)}{\sqrt{p_j f_2(y)}} + \sum_{l=1}^{N_o}\phi_l(x)\frac{\exp(-\imath p_l y)}{\sqrt{p_l f_2(y)}}R_{lj},$$

$$\Psi_j(y \to +\infty, x) \to \sum_{l=1}^{N_o}\phi_l(x)\frac{\exp(\imath p_l y)}{\sqrt{p_l f_2(y)}}T_{lj}, \quad (4)$$

$$\Psi_j(y, x \to \pm\infty) \to 0,$$

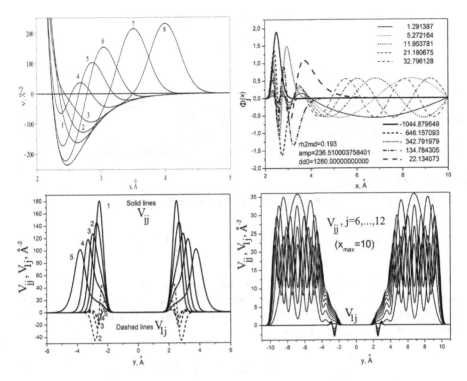

Fig. 2. Sections of the total potential energy $V(y; x) = V^M(y; x) + V^b(y; x)$ at $y = 2.2, 2.3, 2.4, 2.6, 2.8, 3, 3.5, 4$ (curves are noted by 1,...,8). The wave functions $\phi_j(r)$ of the bound states $j = 1, 5$ (solid lines) and pseudostates $j = 6, ..., 12$ (dashed lines) (corresponding energy eigenvalues given in K). The matrix elements $V_{jj}(y)$ (solid lines) and $V_{j1}(y)$ (dashed lines) (in Å^{-2})

where $f_1(y) = f_2(y) = 1$, $R_{lj}(\tilde{E})$ and $T_{lj}(\tilde{E})$ are the reflection and transmission amplitudes, $N_o \leq n$ is the number of open channels, p_i is the wave number, $p_i = \sqrt{(m/\hbar^2)(\tilde{E} - \tilde{\varepsilon}_i)} > 0$, below dissociation threshold $\tilde{E} < 0$, $\phi_j(x)$ and $\varepsilon_j < 0$ at $j = 1, n$ are the eigenfunctions and eigenvalues of the BVP for Eq. (3). The solution of Eq. (1) is sought for in the form of Galerkin expansion

$$\Psi_{i_o}(y, x) = \sum_{j=1}^{j_{max}} \phi_j(x) \chi_{ji_o}(y). \tag{5}$$

Here $\chi_{ji_o}(y)$ are unknown functions and the orthonormalized basis functions $\phi_j(x)$ in the interval $0 \leq x \leq x_{max}$ are defined as eigenfunctions of the BVP for the equation

$$\left(-\frac{1}{f_3(x)} \frac{\partial}{\partial x} f_4(x) \frac{\partial}{\partial x} + V^M(x) - \varepsilon_j \right) \phi_j(x) = 0, \tag{6}$$

with the boundary and orthonormalization conditions

$$\phi_j(0) = \phi_j(x_{\max}) = 0, \quad \int_0^{x_{\max}} f_3(x) dr \phi_i(x) \phi_j(x) = \delta_{ij}, \tag{7}$$

where $f_3(x) = f_4(x) = 1$, $V(x) = (m/\hbar^2)\tilde{V}(x)$, $\varepsilon_j = (m/\hbar^2)\tilde{\varepsilon}_j$. The desired set of numerical solutions of this BVP is calculated with the given accuracy by means of the program ODPEVP [4]. Hence, we calculate the set of n bound states having the eigenfunctions $\phi_j(x)$ and the eigenvalues ε_j, $j = 1, n$ and the desired set of pseudostates with the eigenfunctions $\phi_j(x)$ and the eigenvalues $\varepsilon_j \geq 0$, $j = n+1, j_{\max}$. The latter approximate the set of continuum eigensolutions $\varepsilon \geq 0$ of the BVP for Eq. (3).

The set of closed-channel Galerkin equations has the form

$$\left[-\frac{1}{f_1(y)} \frac{\partial}{\partial y} f_2(y) \frac{\partial}{\partial y} + \varepsilon_i - E \right] \chi_{ii_o}(y) + \sum_{j=1}^{j_{\max}} V_{ij}^b(y) \chi_{ji_o}(y) = 0. \tag{8}$$

Thus, the scattering problem (1)–(3) with the asymptotic boundary conditions (4) is reduced to the boundary-value problem for the set of close-coupling equations in the Galerkin form (8) for $f_1(y) = f_2(y) = 1$ with the boundary conditions at $y = y_{\min}$ and $y = y_{\max}$ [6]:

$$\frac{d\mathbf{F}(y)}{dy} \bigg|_{y=y_t} = \mathcal{R}(y_t)\mathbf{F}(y_t), \quad t = \min, \max, \tag{9}$$

where $\mathcal{R}(y_{\min})$ and $\mathcal{R}(y_{\max})$ are $j_{\max} \times j_{\max}$ symmetric matrix function of E, $\mathbf{F}(y) = \{\chi_{i_o}(y)\}_{i_o=1}^{N_o} = \{\{\chi_{ji_o}(y)\}_{j=1}^{j_{\max}}\}_{i_o=1}^{N_o}$ is the required $j_{\max} \times N_o$ matrix solution at the number of open channels $N_o = \max\limits_{E \geq \varepsilon_j} j \leq j_{\max}$. These matrices and the sought-for $N_o \times N_o$ matrices of the reflection and transmission amplitudes $\mathbf{R}$ and $\mathbf{T}$ are calculated using the third version of the program KANTBP [3].

In Eq. (8), the effective potentials $V_{ij}(y)$ are expressed by the integrals

$$V_{ij}^b(y) = \int_0^{x_{\max}} f_1(x) dx \phi_i(x) (V^b(\frac{x+y}{2}) + V^b(\frac{x-y}{2})) \phi_j(x). \tag{10}$$

For example, let us take the parameters of the molecule Be$_2$, namely, the reduced mass $\mu = m/2 = 4.506$Da, the average distance between the nuclei 2.47Å, the frequency of molecular vibrations expressed in temperature units $\hbar\omega = 398.72$K, the ground state of molecule $^1\Sigma_u^+$, the wave number of the order of 277.124cm^{-1} for the observable excited-to-ground state transitions (we use the relation 1K $=$ 0.69503476 cm^{-1} from [5]). These values were used to determine the parameters of the Morse potential $\tilde{V}^M(x)$ and $V^M(x) = (m/\hbar^2)\tilde{V}^M(x)$ of Eqs. (3) and (6)

$$\tilde{V}^M(x) = D\{\exp[-2(x - \hat{x}_{eq})\hat{\rho}] - 2\exp[-(x - \hat{x}_{eq})\hat{\rho}]\}, \tag{11}$$

where D is the depth of the interaction potential well and $\hat{\rho}$ describes the potential well width. The values of D and $\hat{\rho}$ are determined from the discrete spectrum of the BVP (6)–(7) which is approximated by the known discrete spectrum of Eq. (3)

$$\tilde{\varepsilon}_j = -D\left[1 - \varsigma(j - 1/2)\right]^2, \quad j = 1, ..., n = \left[\varsigma^{-1} + \frac{1}{2}\right]. \tag{12}$$

The discrete spectrum eigenfunctions $\phi_j(x)$ of the BVP (6)–(7) are approximated by the solutions $\tilde{\phi}_j(\zeta)$ of equation (3) in the new variable ζ:

$$\frac{d^2\tilde{\phi}_j(\zeta)}{d\zeta^2} + \frac{1}{\zeta}\frac{d\tilde{\phi}_j(\zeta)}{d\zeta} + \left(-\frac{1}{4} + \frac{j + s_j - 1/2}{\zeta} - \frac{s_j^2}{\zeta^2}\right)\tilde{\phi}_j(\zeta) = 0,$$

where $s_j = \sqrt{-\tilde{\varepsilon}_j}/\hat{\rho} = \sqrt{\hat{D}}/\hat{\rho} - j + 1/2$ and $\zeta = 2\sqrt{\hat{D}}\exp[-(x - \hat{x}_{eq})\hat{\rho}]/\hat{\rho}$, at $\zeta \in (0, +\infty)$ corresponding to the extended interval $x \in (-\infty, +\infty)$ and have the form

$$\tilde{\phi}_j(\zeta) = N_j \exp(-\frac{\zeta}{2})\zeta^{s_j} {}_1F_1(1 - j, 2s_j + 1, \zeta), \quad N_j^2 = \frac{\hat{\rho}\Gamma(2s_j + j)}{(j-1)!\Gamma(2s_j)\Gamma(2s_j + 1)}. \tag{13}$$

Having the average size of the molecule and the separation between the energy levels taken into account, one can parameterize the molecular potential to fit the observable quantities, namely, $D = 1280K$, $\hat{x}_{eq} = 2.47\text{Å}$, $\hat{\rho} = 2.968\text{Å}^{-1}$ is determined from the condition $(\tilde{\varepsilon}_2 - \tilde{\varepsilon}_1)/(2\pi\hbar c) = 277.124$ cm^{-1}, $\varsigma = \frac{\hat{\rho}\hbar}{\sqrt{mD}} = 0.193$ is the dimensionless constant of the problem, and $\hat{D} = (\frac{\sqrt{mD}}{\hbar})^2 = (\hat{\rho}/0.193)^2 = (2.968\text{Å}^{-1}/0.193)^2 = 236.5\text{Å}^{-2}$. In accordance with (12), the ground state energy of the molecule Be$_2$ is equal to $-\tilde{\varepsilon}_1 = -1044.88$K.

The set of pseudostates with the eigenfunctions $\phi_j(x)$ and the eigenvalues $\varepsilon_j \geq 0, j = n+1, j_{\max}$, approximated by the set of continuous spectrum solutions $\tilde{\phi}_k(\zeta)$ with fixed $k = \sqrt{\varepsilon} > 0$ that satisfy Eq. (3) written in the new variable ζ, i.e., the equation

$$\frac{d^2\tilde{\phi}_k(\zeta)}{d\zeta^2} + \frac{1}{\zeta}\frac{d\tilde{\phi}_k(\zeta)}{d\zeta} + \left(-\frac{1}{4} + \frac{\sqrt{\hat{D}}/\hat{\rho}}{\zeta} + \frac{s_k^2}{\zeta^2}\right)\tilde{\phi}_k(\zeta) = 0.$$

At fixed $s_k = \frac{k}{\hat{\rho}}$, these solutions take the form

$$\tilde{\phi}_k(\zeta) = \frac{N_k \exp(-\zeta/2)}{2i}(\exp(iw)\zeta^{-ik/\hat{\rho}} {}_1F_1(-\frac{\sqrt{D}}{\hat{\rho}} + \frac{1}{2} - \frac{ik}{\hat{\rho}}, 1 - \frac{2ik}{\hat{\rho}}, \zeta)$$

$$- \exp(-iw)\zeta^{ik/\hat{\rho}} {}_1F_1(-\frac{\sqrt{D}}{\hat{\rho}} + \frac{1}{2} + \frac{ik}{\hat{\rho}}, 1 + \frac{2ik}{\hat{\rho}}, \zeta)), \tag{14}$$

$$w = \arg(\Gamma(1 + \frac{2ik}{\hat{\rho}})) + \arg(\Gamma(-\frac{\sqrt{D}}{\hat{\rho}} + \frac{1}{2} - \frac{ik}{\hat{\rho}}))).$$

Asymptotically $\tilde{\phi}_k^{as}(x \to \infty) = \sin(kx + \delta(k))$, $\delta(k) = -kx_{eq} - s_k \ln(2\sqrt{\hat{D}}/\hat{\rho}) + w$ corresponds to the scattering phase.

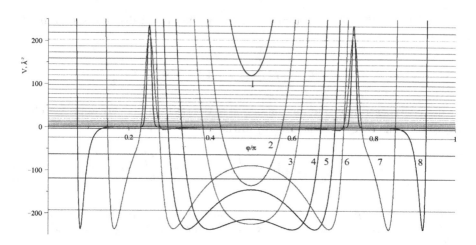

Fig. 3. Sections of the total potential energy $V(\rho;\varphi) = V^M(\rho;\varphi) + V^b(\rho;\varphi)$ in polar coordinates at $\rho = 2.2, 2.3, 2.4, 2.6, 2.8, 3, 5, 10$ (curves are noted by 1,...,8). Straight lines are energy levels at $\rho = 10$.

Since the bond in the molecule Be$_2$ is of the Van der Waals type, one can consider each constituent atom independently interacting with the external barrier potential. The latter should be chosen to have the height and the width typical of barriers in a real crystal lattice. Moreover, this potential should be a smooth function having the second derivative to apply high-accuracy numerical methods, like the Numerov method or the finite element method, for solving the BVP for the systems of second-order ordinary differential equations. We choose the repulsive barrier potential to be Gaussian:

$$\tilde{V}^b(x_i) = \tilde{V}_0 \exp\left(-\frac{x_i^2}{2\sigma}\right), \quad V^b(x_i) = \frac{m}{\hbar^2}\tilde{V}^b(x_i) = \hat{D}\exp\left(-\frac{x_i^2}{2\sigma}\right). \quad (15)$$

Here the parameters $\tilde{V}_0 = 1280K$, $\hat{D} = 236.510003758401\text{Å}^{-2} = (m/\hbar^2)\tilde{V}_0$, $\sigma = 5.23 \cdot 10^{-2}\text{Å}^2$ are determined by the model requirement that the width of the repulsive potential at the kinetic energy equal to that of the ground state is 1Å, so that the average distance 2.47Å between the atoms of Be is smaller than the distance 2.56Å between Cu atoms in the plane (111) of the crystal lattice cell. The potential barrier height $\tilde{V}_0$ of the order of 200 meV was estimated following the experimental observation of quantum diffusion of hydrogen atoms [9]. Fig. 1 illustrates the Gaussian and Morse potentials.

Figure 2 presents the sections of the total potential energy, the calculated eigenfunctions of the BVP (6) and the effective potentials $V_{ij}(y)$ of Eq. (10) calculated using these functions. Note that the wave functions $\phi_j(x)$ and the eigenvalues $\varepsilon_j(x)$ of the bound states $j = 1, 5$ (solid lines) approximate the known analytical ones of the BVP for Eq. (3) with the Morse potential (11) with four and seven significant digits, respectively. The states are localized in the well,

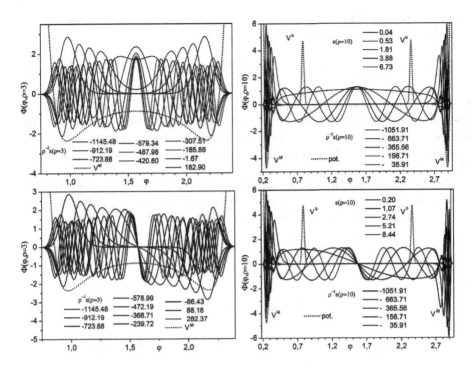

Fig. 4. Even and odd eigenfunctions of the parametric eigenvalue problem for the fast subsystem at $\rho = 3$ and $\rho = 10$ (corresponding energy eigenvalues given in K)

while the pseudostates $j = 6, ..., 12$ are approximated with the same accuracy and localized outside the well. The matrix elements between the bound states are localized in the vicinity of the barriers and the matrix elements between the pseudostates are localized beyond the barriers. The matrix elements between the bound states and pseudostates are small. The solution of the BVP (6), (7) was performed on the finite-element grids $\Omega_x = \{0(N_{elem} = 800)12\}$, with N_{elem} fourth-order Lagrange elements $p = 4$ between the nodes, using the program ODPEVP [4].

3   Model II. Quantum Tunneling in Polar Coordinates

Using the change of variables $x = \rho \sin\varphi$, $y = \rho \cos\varphi$, we can rewrite Eq. (1) in polar coordinates (ρ, φ) $\Omega_{\rho,\varphi} = (\rho \in (0, \infty), \varphi \in [0, \pi])$ in the dimensionless form

$$\left(-\frac{1}{\rho}\frac{d}{d\rho}\rho\frac{d}{d\rho} - \frac{1}{\rho^2}\frac{\partial^2}{\partial\varphi^2} + V(\rho, \varphi) - E \right) \Psi(\rho, \varphi) = 0, \tag{16}$$

where the potential function $V(\rho, \varphi) = V^M(\rho, \varphi) + V^b(\rho, \varphi)$ is defined by the formula in term of potentials (11) and (15)

$$V^M(\rho, \varphi) = V(\rho \sin \varphi), \quad V^b(\rho, \varphi) = V^b(\rho \frac{\sin(\varphi + \pi/4)}{\sqrt{2}}) + V^b(\rho \frac{\sin(\varphi - \pi/4)}{\sqrt{2}}). \quad (17)$$

Sections of the potential function $V(\rho, \varphi)$ at a set of slow variable values ρ are shown in Fig. 3. One can see that at large ρ, the width of the potential wells decreases as ρ increases. Therefore, at large ρ, the potential of two-center problem, symmetric with respect to $\varphi = \pi/2$, transforms into two one-center Morse potentials.

The asymptotic boundary conditions imposed on the solution for the 2D model in the s-wave approximation $\Psi(\rho, \varphi) = \{\Psi_j(\rho, \varphi)\}_{j=1}^{N_o}$ in the asymptotic region $\Omega_j^{as} = \{(\varphi, \rho) | \varphi/\rho \ll 1\}$ can be written in the obvious form

$$\Psi(\rho, \varphi, \varphi_0) = \sum_{i_o=1}^{N_o} \Psi_{ji_o}(\rho, \phi) \phi_{i_o}(-\varphi_0; \rho \to +\infty) \quad (18)$$

$$\Psi_{i_o}(\rho \to +\infty, \varphi) \to \sqrt{\frac{2}{\pi}} \sum_{j=1}^{N_o} \phi_j(\varphi; \rho) \left[\chi_{ji_o}^*(\rho) \delta_{ji_o} - \chi_{ji_o}(\rho) S_{ji_o}(E) \right], \quad (19)$$

$$\Psi_{i_o}(\rho, \phi \to 0) \to 0, \quad \Psi_{i_o}(\rho, \phi \to \pi) \to 0, \quad \chi_{ji_o}(\rho) = \frac{\exp(\imath(p_j\rho - \frac{\pi}{4}))}{2\sqrt{p_j\rho}},$$

where the angle φ_0 determines the direction of the incident wave propagation, in particular, $\varphi_0 = 0$ corresponds to $v = \to$ and $\varphi_0 = \pi$ corresponds to $v = \leftarrow$. $S_{ji_o}(E)$ are the elements of the $N_o \times N_o$ S-matrix, N_o is the number of open channels, p_i is the wave number, $p_i = \sqrt{(m/\hbar^2)(\tilde{E} - \tilde{\varepsilon}_i(\rho \to +\infty))} > 0$, below the dissociation threshold $\tilde{E} < 0$, $\phi_i(\varphi, \rho \to +\infty) = \sqrt{\rho}\phi_i(x)$, and $\varepsilon_i(\rho \to \infty)/\rho^2 = \varepsilon_i^{(0)} < 0$ are the eigenfunctions localized in the asymptotic region Ω_j^{as}, and the eigenvalues of the BVP for Eq. (21).

The solution of Eq. (16) is sought for in the form of Kantorovich expansion

$$\Psi_{i_o}(\rho, \varphi) = \sum_{j=1}^{j_{max}} \phi_j(\varphi; \rho) \chi_{ji_o}(\rho). \quad (20)$$

Here $\chi_{ji_o}(\rho)$ are unknown functions and the orthonormalized basis functions $\phi_j(\varphi; \rho)$ in the interval $\varphi \in [0, \pi]$ are defined as eigenfunctions of the BVP for the equation

$$\left(-\frac{\partial^2}{\partial\varphi^2} + \rho^2(V^M(\rho \sin \varphi) + V^b(\rho, \phi)) - \varepsilon_j(\rho) \right) \phi_j(\varphi; \rho) = 0, \quad (21)$$

with orthonormalization conditions

$$\int_0^\pi d\varphi \phi_i(\varphi; \rho) \phi_j(\varphi; \rho) = \delta_{ij}. \quad (22)$$

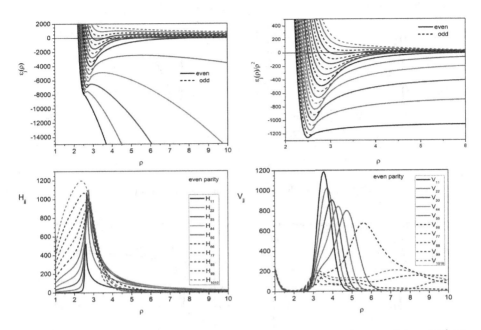

Fig. 5. Potential curves $\varepsilon_j(\rho)$ and even diagonal effective potentials $H_{jj}(\rho)$ and $V_{jj}^b(\rho)$ vs ρ (Å)

The solution of the BVPs (21), (22) was performed on the finite-element grids $\Omega_\varphi = \{\varphi_1(N_{elem} = 800)\pi/2\}$, if $\varphi_3 = (8 + \varphi x_{eq})/(\varphi\rho) > \pi/4$, $\Omega_\varphi = \{\varphi_1(N_{elem} = 300)\varphi_2(N_{elem} = 60)\varphi_4(N_{elem} = 40)\varphi_5(N_{elem} = 100)\pi/2\}$ with N_{elem} fourth-order Lagrange elements $p = 4$ between the nodes, using the program ODPEVP [4]. Here angles $\varphi_1 = (-3 + \varphi x_{eq})/(\varphi\rho)$ and $\varphi_2 = (4 + \varphi x_{eq})/(\varphi\rho)$ are marked left and right bounds of well (17) and angles $\varphi_4 = \pi/4 - 4\sqrt{\sigma}/\rho$ and $\varphi_5 = \pi/4 + 4\sqrt{\sigma}/\rho$ are marked left and right bounds of potential barrier (17).

First, let us put $V^b(\rho, \varphi) = 0$ in Eq. (21). In this case, we calculate the set of n bound states having the eigenfunctions $\phi_j(\varphi; \rho)$ and the eigenvalues $\varepsilon_j(\rho) < 0$ at $j, = 1, 2, ..., n$, and the desired set of pseudostates with the eigenfunctions $\phi_j(\varphi; \rho)$ and the eigenvalues $\varepsilon_j(\rho) \geq 0$ at $j = n+1, ..., j_{max}$. The latter approximate the set of continuum eigensolutions $\varepsilon(\rho) \geq 0$ of the BVP for Eq. (3). The eigenvalues have the following asymptotes: $\varepsilon_j(\rho \to \infty)/\rho^2 = \varepsilon_j$ at $j, = 1, 2, ..., n$ and $\varepsilon_j(\rho \to \infty)/\rho^2 = (j - n)^2/\rho^2 + O(1/\rho^3)$ at $j = n+1, ..., j_{max}$.

The eigenfunctions $\phi_j(\varphi; \rho)$, $j = 1, 20$ are shown in Fig. 4 at $\rho = 3$ and $\rho = 10$. Taking the above symmetry $V(\varphi, \rho) = V(\pi - \varphi, \rho)$ of the potential into account, the eigenfunctions are separated into two subsets, namely, the even $\phi_j^{\sigma=1}(\varphi; \rho)$ and odd $\phi_j^{\sigma=-1}(\varphi; \rho)$ ones. The linear combinations

$$\phi_j^{\to\leftarrow}(\varphi; \rho) = (\phi_j^{\sigma=1}(\varphi; \rho) \pm \phi_j^{\sigma=-1}(\varphi; \rho))/\sqrt{2}$$

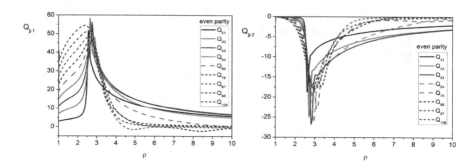

Fig. 6. Even effective potentials $Q_{ij}(\rho)$ vs ρ (Å)

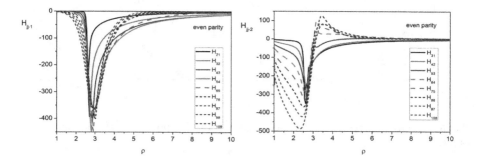

Fig. 7. Even effective potentials $H_{ij}(\rho)$ vs ρ (Å)

at large ρ have maxima in the vicinity of $\varphi = 0$ and $\varphi = \pi$, respectively, such that they correspond to the functions presented in Fig. 2. Taking this property into account, we arrive at the expressions [2]

$$\check{\mathbf{T}} = (-\check{\mathbf{S}}_{+1} + \check{\mathbf{S}}_{-1})/2, \quad \check{\mathbf{R}} = (-\check{\mathbf{S}}_{+1} - \check{\mathbf{S}}_{-1})/2, \tag{23}$$

which relate the even $\check{\mathbf{S}}_{+1}$ and odd $\check{\mathbf{S}}_{-1}$ elements of the matrix $\check{\mathbf{S}} = e^{\imath\pi/4}\mathbf{S}e^{\imath\pi/4}$ from Eq. (19) to the transmission $\check{\mathbf{T}}$ and reflection $\check{\mathbf{R}}$ amplitudes from Eq. (4).

The set of closed-channel Kantorovich self-adjoint equations has the form

$$\left[-\frac{1}{\rho}\frac{d}{d\rho}\rho\frac{d}{d\rho} + \frac{\varepsilon_i(\rho)}{\rho^2} - E\right]\chi_{ii_o}(\rho) + \sum_{j=1}^{j_{\max}} W_{ij}(\rho)\chi_{ji_o}(\rho) = 0. \tag{24}$$

where the potential matrix operator $W_{ij}(\rho)$ has the form

$$W_{ij}(\rho) = V_{ij}^b(\rho) + H_{ji}(\rho) + \frac{1}{\rho}\frac{d}{d\rho}\rho Q_{ji}(\rho) + Q_{ji}(\rho)\frac{d}{d\rho}. \tag{25}$$

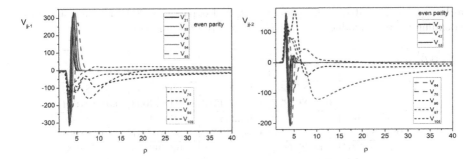

Fig. 8. Even effective potentials $V_{ij}(\rho)$ vs ρ (Å)

The potential curves $\varepsilon_j(\rho)$ (see Fig. 5) and the effective potentials $Q_{ij}(\rho) = -Q_{ji}(\rho)$, $H_{ij}(\rho) = H_{ji}(\rho)$ and $V_{ij}^b(\rho)$ (see Figs. 6–8) are determined by the integrals calculated using the program ODPEVP

$$Q_{ij}(\rho) = -\int_0^\pi d\varphi \phi_i(\varphi;\rho)\frac{d\phi_j(\varphi;\rho)}{d\rho}, H_{ij}(\rho) = \int_0^\pi d\varphi \frac{d\phi_i(\varphi;\rho)}{d\rho}\frac{d\phi_j(\varphi;\rho)}{d\rho}, \quad (26)$$

$$V_{ij}^b(\rho) = \int_0^\pi d\varphi \phi_i(\varphi;\rho)(V^b(\rho\frac{\sin(\varphi+\pi/4)}{\sqrt{2}}) + V^b(\rho\frac{\sin(\varphi-\pi/4)}{\sqrt{2}}))\phi_j(\varphi;\rho).$$

If we take the potential $V^b(\rho,\phi)$ in Eq. (21) into account by using the matrix elements $V_{ij}^b(\rho)$ from Eq.(26), then we put $V_{ij}^b(\rho) = 0$ in Eq.(25). Thus, the scattering problem for Eq. (16) with the asymptotic boundary conditions (19) is reduced to the boundary-value problem for the set of close-coupling equations in the Kantorovich form (18) with the boundary conditions at $\rho = \rho_{\min}$ and $\rho = \rho_{\max}$ [6]:

$$\frac{d\boldsymbol{F}(\rho)}{d\rho}\bigg|_{\rho=\rho_t} = (\mathcal{R}(\rho_t) + \mathbf{Q}(\rho_t))\boldsymbol{F}(\rho_t),\, t = \min, \max, \quad (27)$$

where $\mathcal{R}(\rho)$ is an unknown $j_{\max} \times j_{\max}$ symmetric matrix function, $\boldsymbol{F}(\rho) = \{\boldsymbol{\chi}_{i_o}(\rho)\}_{i_o=1}^{N_o} = \{\{\chi_{ji_o}(\rho)\}_{j=1}^{j_{\max}}\}_{i_o=1}^{N_o}$ is the required $j_{\max} \times N_o$ matrix solution, and N_o is the number of open channels, $N_o = \max_{E \geq \varepsilon_j} j \leq j_{\max}$, calculated using the program KANTBP 3.0 [3].

4 Asymptotic Form of Effective Potentials and Solutions

Algorithm 1. At large ρ, the width of the potential well is decreasing with increasing ρ (see Fig. 3). This allows linearization of the argument $\rho\sin\varphi - \hat{x}_{eq} \to \rho(\varphi - \arcsin(\hat{x}_{eq}/\rho))$ at $|x - \hat{x}_{eq}|/\rho \ll 1$ in the expression of the potential function $V^M(\rho\sin\varphi)$ and reformulation of Eq. (21) on the interval $\varphi = (0, \pi)$

$$\left(-\frac{\partial^2}{\partial\varphi^2} + \rho^2 V^M(\rho(\varphi - \arcsin(\hat{x}_{eq}/\rho))) - \varepsilon_j(\rho)\right)\phi_j(\varphi;\rho) = 0. \quad (28)$$

Table 1. The calculated coefficients $Q_{ij}^{(1)}$ $H_{ij}^{(2)}$ of expansions (31) (up rows) and corresponding numerical values Q_{ij} and H_{ij} at $\rho = 100$(down rows)

$Q_{ij}^{(1)}$ Q_{ij}	1	2	3	4	5
1	0	55.852657	−20.662584	9.913235	−4.888752
	0	0.55 863277	−0.20 664572	0.09 914008	−0.04 891971
2	−55.852657	0	66.253422	−30.004416	14.557626
	−0.55 863277	0	0.66 270932	−0.30 010937	0.14 568965
3	20.662584	−66.253422	0	62.290358	−28.724086
	0.20 664572	−0.66 270932	0	0.62 317875	−0.28 751980
4	−9.913235	30.004416	62.290358	0	43.265811
	−0.09 914008	0.30 010937	0.62 317875	0	0.43 320993
5	4.888752	−14.557626	28.724086	−43.265811	0
	0.04 891971	−0.14 568966	0.28 751983	−0.43 321006	0

$H_{ij}^{(2)}$ H_{ij}	1	2	3	4	5
1	692.635	−364.132	−462.085	397.196	−240.775
	0.0692 859	−0.0364 209	−0.0462 371	0.0397 441	−0.0241 084
2	−364.132	1718.621	−873.970	−219.292	253.669
	−0.0364 209	0.1719 273	−0.0874 195	−0.0219 721	0.0254 209
3	−462.085	−873.970	2210.843	−1250.672	244.905
	−0.0462 371	−0.0874 195	0.2211 927	−0.1251 191	0.0244 755
4	397.196	−219.292	−1250.672	2088.603	−1167.908
	0.0397 441	−0.0219 721	−0.1251 191	0.2090 243	−0.1169 414
5	−240.775	253.669	244.905	−1167.908	1209.648
	−0.0241 084	0.0254 209	0.0244 755	−0.1169 414	0.1212 568

This equation coincides with Eq. (6), (11), taking the notations

$$\hat{D} \to \hat{D}\rho^2, \hat{\rho} \to \hat{\rho}\rho, \hat{x}_{eq} \to \arcsin(\hat{x}_{eq}/\rho) \qquad (29)$$

into account.

As a result, we obtain the approximate eigenvalues $\varepsilon_j(\rho)$ that depend on ρ as a parameter, expressed as

$$\varepsilon_j(\rho) = \rho^2 \varepsilon_j^{(0)}, \quad \varepsilon_j^{(0)} = -\hat{D}\left[1 - \frac{\hat{\rho}(j - \frac{1}{2})}{\sqrt{\hat{D}}}\right]^2, \quad j = 1, ..., n = \left[\frac{\sqrt{\hat{D}}}{\hat{\rho}} + \frac{1}{2}\right]. \qquad (30)$$

These eigenvalues demonstrate correct asymptotic behavior $\tilde{\varepsilon}_j(\rho)/\rho^2 = \tilde{\varepsilon}_j$ describing the lower part of the discrete spectrum of problem (3). In the considered case, they correspond to the first five ($n = 5$) eigenvalues $\tilde{\varepsilon}_1, ..., \tilde{\varepsilon}_5$. The corresponding eigenfunctions $\phi_j(\varphi; \rho)$ at $j = 1, ..., n$, parametrically depending on the slow variable ρ via the new independent variable $\zeta = \zeta(\varphi; \rho) = 2\rho\sqrt{\hat{D}}\exp[-\hat{\rho}\rho(\varphi - \arcsin(\hat{x}_{eq}/\rho))]/\hat{\rho}$, $\zeta \in [0, +\infty)$ have the form

$$\tilde{\phi}_j(\zeta; \rho) = N_j(\rho)\exp(-\frac{\zeta}{2})\zeta^{s_j}{}_1F_1(1-j, 2s_j+1, \zeta),$$

$$N_j^2(\rho) = \frac{\rho\hat{\rho}\Gamma(2s_j+j)}{(j-1)!\Gamma(2s_j)\Gamma(2s_j+1)},$$

where $s_j = \sqrt{\hat{D}}/\hat{\rho} - j + 1/2$ is a positive parameter. In the considered case, the wave function outside the well at $|x - \hat{x}_{eq}|/\rho \gg 1$ is exponentially decreasing. This makes it possible to integrate the product of functions $\tilde{\psi}_j(\zeta(\varphi; \rho); \rho)$ and/or $\partial\tilde{\psi}_j(\zeta(\varphi; \rho); \rho)/\partial\rho|_{\phi=const}$ by ζ in the interval $\zeta \in (0, +\infty)$. The calculated eigenfunctions with $\rho = 10$ for $j = 1, ..., 5$ shown in Fig. 4 qualitatively agree with the bound states in Fig. 2. The matrix elements between the states of the lower part of the discrete spectrum $i, j = 1, ..., n = 5$ with the eigenvalues $\varepsilon_j(\rho)/\rho^2 = \varepsilon_j^{(0)}$ are expanded in inverse powers of ρ:

$$Q_{ij}(\rho) = \sum_{k=1}^{k_{max}} \frac{Q_{ij}^{(2k-1)}}{\rho^{2k-1}}, \; H_{ij}(\rho) = \sum_{k=1}^{k_{max}} \frac{H_{ij}^{(2k)}}{\rho^{2k}}, \; V_{ij}(\rho) = O(\exp(-\rho)), \quad (31)$$

and calculated up to the desired order k_{max} in CAS MAPLE. As an example, the calculated coefficients $Q_{ij}^{(1)}$ and $H_{ij}^{(2)}$ of expansions (31) are presented in Table 1. For comparison, the numerical values of matrix elements Q_{ij} and H_{ij} at $\rho = 100$ are also given in Table 1. One can see that with the first nonzero coefficients of these expansions, one gets the numerical approximation of the matrix elements with three significant digits.

For the states $i, j = n+1, ..., j_{max}$ with the eigenvalues $\varepsilon_j(\rho \to \infty) = (j-n)^2 + O(1/\rho) = \varepsilon^{(2)} + O(1/\rho) = k^2 + O(1/\rho)$ corresponding to pseudo states of the BVP (6), (7) we consider the approximation by the eigenfunctions of continuous spectrum (see Eq. (14) with the notations (29)) reduced to the finite interval $\varphi \in (0, \pi/2)$ by means of the procedure implemented in CAS MAPLE. The energy spectrum of even and odd states is evaluated basing on the conditions

$$\frac{d\tilde{\phi}_k(\varphi; \rho)}{d\varphi}\Big|_{\varphi=\pi/2} = 0 \text{ and } \tilde{\phi}_k(\pi/2; \rho) = 0$$

for even and odd states, respectively. The calculated eigenfunctions at $\rho = 10$ for $i = 6, ..., 10$ are in quantitative agreement with the numerical ones shown in Fig. 4 and in qualitative agreement with pseudo-states displayed in Fig. 2. Thus, the basis eigenfunctions of Galerkin expansion (5) correspond to the asymptotic ones for Kantorovich expansion (20) at large values of the parameter ρ.

The diagonal and nondiagonal barrier matrix elements $V_{ij}(\rho)$ shown in Figs. 5 and 8 should be compared with the corresponding ones displayed in Fig. 2. From this comparison, one can see that the matrix elements $V_{ij}(\rho)$ from (26) between discrete-spectrum states of BVP (21), (22) and the matrix elements

$V_{ij}(y)$ from (10) between a discrete spectrum state and a pseudo-state (6), (7) demonstrate qualitatively similar behavior in the coordinates y and ρ. Since $\rho = \sqrt{x^2 + y^2} > y$, the potentials $V_{ij}(\rho)$ are delocalized with respect to $V_{ij}(y)$. Due to slowly decreasing kinematic behavior of the potentials $Q_{ij}(\rho)$ and $H_{ij}(\rho)$ as ρ^{-1} and ρ^{-2}, respectively, compared to the exponentially decreasing $V_{ij}(y)$, one should take into account the leading terms of their asymptotic expressions in solving the BVP (24)-(26) generated by the Kantorovich expansion (18) in the calculation of scattering with five open channels.

Algorithm 2. Evaluation of the Asymptotic Solutions

Input. We calculate the asymptotic solution of the set of N ODEs at high values of the independent variable $\rho \gg 1$

$$\left[-\frac{1}{\rho}\frac{d}{d\rho}\rho\frac{d}{d\rho} + \frac{\varepsilon_i(\rho)}{\rho^2} + \mathcal{H}_{ii}(\rho) - 2E \right] \chi_{ii'}(\rho) \tag{32}$$

$$= \sum_{j=1, j\neq i}^{N} \left[-Q_{ij}(\rho)\frac{d}{d\rho} - \frac{1}{\rho}\frac{d}{d\rho}\rho Q_{ij}(\rho) - \mathcal{H}_{ij}(\rho) \right] \chi_{ji'}(\rho).$$

The coefficients of Eqs. (32), where $\mathcal{H}_{ij} = V_{ij}^b + H_{ij}$ are presented in the form of the inverse power series (31). In particular, $\varepsilon_i(\rho)/\rho^2 = \varepsilon_i^{(0)} + \varepsilon_i^{(2)}/\rho^2$.

Step 1. We construct the solution of Eqs. (32) in the form:

$$\chi_{ji'}(\rho) = \left(\phi_{ji'}(\rho) + \psi_{ji'}(\rho)\frac{d}{d\rho} \right) R_{i'}(\rho), \tag{33}$$

where $\phi_{ji'}(\rho)$ and $\psi_{ji'}(\rho)$ are unknown functions, $R_{i'}(\rho)$ is a known function. We choose $R_{i'}(\rho)$ as solutions of the auxiliary problem treated like an etalon equation:

$$\left[-\frac{1}{\rho}\frac{d}{d\rho}\rho\frac{d}{d\rho} + \frac{Z_{i'}^{(2)}}{\rho^2} - p_{i'}^2 \right] R_{i'}(\rho) = 0, \tag{34}$$

where $Z_{i'}^{(2)} = \varepsilon_{i'}^{(2)}$.

Step 2. At this step, we compute the coefficients $\phi_{i'}(\rho)$ and $\psi_{i'}(\rho)$ of the expansion (33) in the form of truncated expansion in inverse powers of ρ $(\phi_{ji'}^{(k'<0)} = \psi_{ji'}^{(k'<0)} = 0)$:

$$\phi_{ji'}(\rho) = \phi_{ji'}^{(0)} + \sum_{k'=1}^{k_{max}} \frac{\phi_{ji'}^{(k')}}{\rho^{k'}}, \quad \psi_{ji'}(\rho) = \psi_{ji'}^{(0)} + \sum_{k'=1}^{k_{max}} \frac{\psi_{ji'}^{(k')}}{\rho^{k'}}. \tag{35}$$

After the substitution of Eqs.(33)–(35) into Eq. (32) with the use of Eq.(34), we arrive at the set of recurrence relations at $k' \leq k_{max}$:

$$\left(\varepsilon_i^{(0)} - 2E + p_{i'}^2 \right) \phi_{ii'}^{(k')} - 2p_{i'}^2(k'-1)\psi_{ii'}^{(k'-1)} = -f_{ii'}^{(k')}, \tag{36}$$

$$\left(\varepsilon_i^{(0)} - 2E + p_{i'}^2 \right) \psi_{ii'}^{(k')} + 2(k'-1)\phi_{ii'}^{(k'-1)} = -g_{ii'}^{(k')},$$

where the right-hand sides $f_{ii'}^{(k)}$ and $g_{ii'}^{(k)}$ are defined by the relations

$$f_{ii'}^{(k')} = (-(k'-2)^2 - Z_{i'}^{(2)})\phi_{ii'}^{(k'-2)} + \sum_{k=2}^{k'} \mathcal{H}_{ii}^{(k)} \phi_{ii'}^{(k'-k)}$$

$$+Z_{i'}^{(2)}(2k'-4)\psi_{ii'}^{(k'-3)} + \sum_{k=1}^{k'} \sum_{j=1,j\neq i}^{N} \Big(2Q_{ij}^{(k)} Z_{i'}^{(2)} \psi_{ji'}^{(k'-k-2)}$$

$$-2p_{i'}^2 Q_{ij}^{(k)} \psi_{ji'}^{(k'-k)} + Q_{ij}^{(k)}(-2k'+k+3)\phi_{ji'}^{(k'-k-1)} + \mathcal{H}_{ij}^{(k)} \phi_{ji'}^{(k'-k)}\Big); \qquad (37)$$

$$g_{ii'}^{(k)} = (-(k'-1)^2 - Z_{i'}^{(2)})\psi_{ii'}^{(k'-2)} + \sum_{k=2}^{k'} \mathcal{H}_{ii}^{(k)} \psi_{ii'}^{(k'-k)}$$

$$+ \sum_{j=1,j\neq i}^{N} \sum_{k=1}^{k'} \Big(2Q_{ij}^{(k)} \phi_{ji'}^{(k'-k)} - Q_{ij}^{(k)}(2k'-1-k)\psi_{ji'}^{(k'-k-1)} + \mathcal{H}_{ij}^{(k)} \psi_{ji'}^{(k'-k)}\Big)$$

with the initial conditions $p_{i'}^2 = 2E - \varepsilon_{i'}^{(0)}$, $\phi_{ii'}^{(0)} = \delta_{ii'}$, $\psi_{ii'}^{(0)} = 0$.

Step 3. Here we calculate the coefficients $\phi_{ii'}^{(k')}$ and $\psi_{ii'}^{(k')}$ using the step-by-step procedure of solving Eqs. (36) for $2E \neq \epsilon_{i'}^{(0)}$, $i \neq i'$ and $k' = 2, \ldots, k_{\max}$:

$$\phi_{ii'}^{(k')} = \Big[\varepsilon_i^{(0)} - \varepsilon_{i'}^{(0)}\Big]^{-1} \Big[-f_{ii'}^{(k')} + 2p_{i'}^2(k'-1)\psi_{ii'}^{(k'-1)}\Big],$$

$$\psi_{ii'}^{(k')} = \Big[\varepsilon_i^{(0)} - \varepsilon_{i'}^{(0)}\Big]^{-1} \Big[-g_{ii'}^{(k')} - 2(k'-1)\phi_{ii'}^{(k'-1)}\Big],$$

$$\phi_{i'i'}^{(k'-1)} = -\left[2(k'-1)\right]^{-1} g_{i'i'}^{(k)}, \qquad (38)$$

$$\psi_{i'i'}^{(k'-1)} = \Big[2(k'-1)\Big(2E - \varepsilon_{i'}^{(0)}\Big)\Big]^{-1} f_{i'i'}^{(k)}.$$

The above described algorithm was implemented in MAPLE and FORTRAN to calculate the desired $\phi_{ii'}^{(k')}$ and $\psi_{ii'}^{(k')}$ in the **output** up to needed order of $k_{\max}$.

The choice of appropriate values $\rho_{\min}$ and $\rho_{\max}$ for the constructed expansions of the linearly independent solutions for $p_{i_o} > 0$ is controlled by the fulfilment of the Wronskian condition to the prescribed precision ε_{Wr}:

$$Wr(\mathbf{Q}(\rho); \boldsymbol{\chi}^*(\rho), \boldsymbol{\chi}(\rho)) = \frac{2\imath}{\pi}\mathbf{I}_{oo}, \qquad (39)$$

$$W(\mathbf{Q}, \boldsymbol{\chi}^*, \boldsymbol{\chi}) \equiv \rho\left(\boldsymbol{\chi}^{*T}\left(\frac{d\boldsymbol{\chi}}{d\rho} - \mathbf{Q}\boldsymbol{\chi}\right) - \boldsymbol{\chi}^T\left(\frac{d\boldsymbol{\chi}^*}{d\rho} - \mathbf{Q}\boldsymbol{\chi}^*\right)\right).$$

5   Analysis of Quantum Tunneling Problem

The solutions of the BVPs (8)–(15) and (24)–(27) were performed on the finite-element grids $\Omega_y = \{-12(N_{elem} = 120)12\}$ and $\Omega_\rho = \{0(N_{elem} = 1200)120\}$, respectively, with N_{elem} fourth-order Lagrange elements $p = 4$ between the

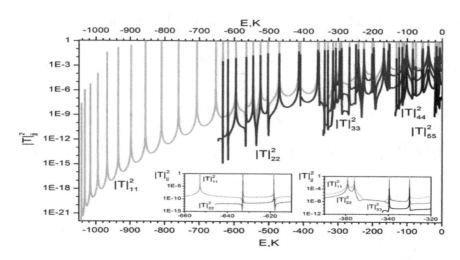

Fig. 9. The total probability of penetration from the first channels with the energies $E_1 = -1044.879649$, $E_2 = -646.1570935$, $E_3 = -342.7919791$, $E_4 = -134.7843058$, $E_5 = -22.13407384$ (in K) to all five open channels simulated by the Galerkin and Kantorovich expansions

nodes using the program KANTBP 3.0. The expansion of the desirable solution (5) over such orthogonal basis at ($j_{max} = 15$) with only ten closed channels taken into account allows the calculation of approximate solutions of the original 2D problem (1) at $E < 0$ with the required accuracy. Fig. 9 shows the resonance behavior of the total penetration probability with the transition from the first channels having the energies $E_1 = -1044.879649$, $E_2 = -646.1570935$, $E_3 = -342.7919791$, $E_4 = -134.7843058$, $E_5 = -22.13407384$ (in K) to all five open channels, simulated using the Galerkin expansion (5) as well as the Katorovich one (18). The total transmission probability is seen to demonstrate the resonance behavior, i.e., effect of quantum transparency. Some peaks are high and narrow, and the positions of peaks corresponding to transitions from different bound states are similar.

As the energy of the initial excited state increases, the transmission peaks demonstrate a shift towards higher energies, the set of peak positions keeping approximately the same as for the transitions from the ground state and the peaks just replacing each other. For example, the left epure shows that the positions of the 13th and 14th peaks for transitions from the first state coincide with the positions of the 1st and 2nd peaks for the transitions from the second state, while the right epure shows that the positions of the 25th and 26th peaks for transitions from the first state coincide with the positions of the 13th and 14th peaks for transitions from the second state and with the positions of the 1st and 2nd peaks for the transitions from the third state.

As one can see from Fig. 2, the diagonal matrix elements of the potential $V_{jj}^b(y)$ have the shapes of double barriers, and the nondiagonal matrix elements

$V_{ij}^b(y)$ are by more than four times smaller than $V_{jj}^b(\rho)$ and $V_{ij}^b(\rho)$ in Figs. 5 and 8. It means that the position of peaks corresponds to the real part of energy of the metastable states embedded in the continuum, which are mainly localized between double barriers.

6 Conclusions

We have demonstrated efficiency of symbolic-numeric algorithms for solving the boundary-value problems that describe the quantum tunneling of diatomic low-dimensional model systems, coupled via realistic molecular potentials, through repulsive barriers below a dissociation threshold. We presented a comparative analysis of the potential matrix elements and solutions with different asymptotic behavior calculated in the Cartesian and polar coordinates. The necessity for two statements of the problem follows from the important practical applications of further self-consistent study of the system above the dissociation threshold, which is convenient in polar coordinates. The effect of quantum transparency in resonance tunneling of diatomic molecules through repulsive potential barriers was revealed that produced by metastable states imbedded in continuum. The proposed models and elaborated symbolic-numerical algorithms, the quantum transparency effect itself, and the developed software can find further applications in barrier heavy-ion reactions and molecular quantum diffusion. The authors thank Prof. F.M. Penkov for collaboration. The work was supported partially by grants RFBR 14-01-00420 and 13-01-00668 and 0602/GF MES RK.

References

1. Bondar, D.I., Liu, W.-K., Ivanov, M.Y.: Enhancement and suppression of tunneling by controlling symmetries of a potential barrier. Phys. Rev. A 82, 052112-1–9 (2010)
2. Chuluunbaatar, O., Gusev, A.A., Derbov, V.L., Kaschiev, M.S., Melnikov, L.A., Serov, V.V., Vinitsky, S.I.: Calculation of a hydrogen atom photoionization in a strong magnetic field by using the angular oblate spheroidal functions. J. Phys. A 40, 11485–11524 (2007)
3. Chuluunbaatar, O., Gusev, A.A., Vinitsky, S.I., Abrashkevich, A.G.: KANTBP 2.0: New version of a program for computing energy levels, reaction matrix and radial wave functions in the coupled-channel hyperspherical adiabatic approach. Comput. Phys. Commun. 179, 685–693 (2008)
4. Chuluunbaatar, O., Gusev, A.A., Vinitsky, S.I., Abrashkevich, A.G.: ODPEVP: A program for computing eigenvalues and eigenfunctions and their first derivatives with respect to the parameter of the parametric self-adjoined Sturm-Liouville problem. Comput. Phys. Commun. 180, 1358–1375 (2009)
5. Fundamental Physical Constants, http://physics.nist.gov/constants
6. Gusev, A.A., Vinitsky, S.I., Chuluunbaatar, O., Gerdt, V.P., Rostovtsev, V.A.: Symbolic-numerical algorithms to solve the quantum tunneling problem for a coupled pair of ions. In: Gerdt, V.P., Koepf, W., Mayr, E.W., Vorozhtsov, E.V. (eds.) CASC 2011. LNCS, vol. 6885, pp. 175–191. Springer, Heidelberg (2011)

7. Vinitsky, S., Gusev, A., Chuluunbaatar, O., Rostovtsev, V., Le Hai, L., Derbov, V., Krassovitskiy, P.: Symbolic-numerical algorithm for generating cluster eigenfunctions: tunneling of clusters through repulsive barriers. In: Gerdt, V.P., Koepf, W., Mayr, E.W., Vorozhtsov, E.V. (eds.) CASC 2013. LNCS, vol. 8136, pp. 427–442. Springer, Heidelberg (2013)
8. Kavka, J.J., Shegelski, M.R.A., Hong, W.P.: Tunneling and reflection of an exciton incident upon a quantum heterostructure barrier. J. Phys.: Condens. Matter. 24, 365802-1–13 (2012)
9. Lauhon, L.J., Ho, W.: Direct observation of the quantum tunneling of single hydrogen atoms with a scanning tunneling microscope. Phys. Rev. Lett. 85, 4566–4569 (2000)
10. Pen'kov, F.M.: Metastable states of a coupled pair on a repulsive barrier. Phys. Rev. A 62, 044701-1–4 (2000)
11. Pen'kov, F.M.: Quantum transmittance of barriers for composite particles. JETP 91, 698–705 (2000)
12. Pijper, E., Fasolino, A.: Quantum surface diffusion of vibrationally excited molecular dimers. J. Chem. Phys. 126, 014708-1–10 (2007)
13. Shegelski, M.R.A., Hnybida, J., Vogt, R.: Formation of a molecule by atoms incident upon an external potential. Phys. Rev. A. 78, 062703-1–5 (2007)

Enumeration of Schur Rings Over Small Groups

Matan Ziv-Av

Ben-Gurion University of the Negev
matan@svgalib.org

Abstract. By optimizing the algorithms used in COCO and COCO-II, we enumerated all Schur rings over the groups of orders up to 63. A few statistical views of results with respect to Schur property, amount and type of generators and primitivity are presented.

Discussion of the details of the old algorithms and the improvements we implemented in order to achieve those results is included. We compare the results to similar computerized efforts (Hanaki and Miyamoto, Pech and Reichard, Heinze), as well as to theoretical classifications of Schur groups.

The computer based results may assist the theoretical efforts to classify all Schur groups, over abelian and non-abelian groups.

1   Introduction

Schur rings (S-rings for short) were introduced and investigated by I. Schur as a purely group theoretical concept. With the advances in algebraic graph theory, especially in the study of association schemes, which may be considered as a generalization of S-rings, those rings are now also used in combinatorial context.

In this paper we report on the results of a computerized enumeration of S-rings over all groups of orders up to 63.

We discuss the algorithmic improvements, as well as implementation details, that allowed this project to go further than similar projects.

The complete list of S-rings is available as a computer file. We present a few statistical views of this generated data, dividing the groups and the rings above them by such properties as Schurian, primitive, coherently generated.

In addition, we compare our results with the results of other computerized enumeration projects whose subject matter intersect this project.

We also view the enumeration results in light of recent (and a few older) theoretical results which are part of the effort to classify S-rings.

2   Preliminaries

Schur rings were introduced by I. Schur in 1933 ([13]), and were later developed by H. Wielandt ([15]).

Recall that the group ring $\mathbb{C}[H]$ consists of all formal linear combinations of elements of the group H with coefficients from the field $\mathbb{C}$.

V.P. Gerdt et al. (Eds.): CASC Workshop 2014, LNCS 8660, pp. 491–500, 2014.

A Schur ring over the group H is a subring $\mathcal{A}$ of the group ring $\mathbb{C}[H]$ such that there exists a partition P of H satisfying:

1. $\underline{P}$ is a basis of $\mathcal{A}$ (as a vector space over $\mathbb{C}$).
2. $\{e\} \in P$, where e is the identity element of H.
3. $X^{-1} \in P$ for all $X \in P$.

Here, for a subset X of H we define $X^{-1} = \{g^{-1}|g \in X\}$ and $\underline{X} = \sum_{x \in X} 1 \cdot x$, while for a set of subsets T we define $\underline{T} = \{\underline{X}|X \in T\}$.

Let (G, Ω) be a permutation group and H a regular subgroup of G. Then Ω may be identified with H. The stabilizer G_e of the identity element $e \in H$ defines an S-ring over H (see [15]). We denote this S-ring by $V(G, H)$.

An S-ring $\mathcal{A}$ is called *Schurian* if it is equal to $V(G, H)$ for a suitable overgroup (G, H) of a regular group (H, H). A group H is called a *Schur group* if all S-rings over H are Schurian. Schur [13] conjectured that all groups are Schur groups, or in other words, all S-rings are Schurian. The first examples of non-Schurian S-rings were presented by Wielandt together with the history of their discovery in [15].

Let H be a group and S a subset of H. The Cayley graph $Cay(H, S) = (H, R)$ is a graph with vertex set H and with arc set $R = \{\langle x, sx\rangle|x \in H, s \in S\}$. A Cayley graph $Cay(H, S)$ is undirected if $S = S^{-1}$ and is connected if $H = \langle S\rangle$.

A *color graph* is a pair $(\Omega, \mathcal{R})$, where $\mathcal{R} = \{R_i|i \in I\}$ is a partition of Ω^2.

Let $(X, \mathcal{R} = \{R_1, \ldots, R_r\})$ be a color graph such that:

CC1 $\forall i \in [1, r] \exists i' \in [1, r] R_i' = R_{i'}$, where $R_i' = \{(y, x)|(x, y) \in R_i\}$;

CC2 $\exists I' \subseteq [1, r] \bigcup_{i \in I'} R_i = \Delta$, where $\Delta = \{(x, x)|x \in X\}$;

CC3 $\forall i, j, k \in [1, r] \forall (x, y) \in R_k |\{z \in X|(x, z) \in R_i \wedge (z, y) \in R_j\}| = p_{ij}^k$,

then $\mathfrak{m} = (X, \mathcal{R})$ is called a *coherent configuration*. The relations in $\mathcal{R}$ are called *basic relations* of $\mathfrak{m}$. If $\mathcal{R} = \{R_0, \ldots, R_r\}$ are the basic relations of a coherent configuration $\mathfrak{m}$, then the graphs $\Gamma_i = (X, R_i)$ are called *basic graphs* of $\mathfrak{m}$, and their adjacency matrices $A_i = A(\Gamma_i)$ are called *basic matrices* of $\mathfrak{m}$.

If $\mathfrak{c}_1$ and $\mathfrak{c}_2$ are coherent configuration over the same set X, such that each basic relation of $\mathfrak{c}_1$ is a union of basic relations of $\mathfrak{c}_2$, then $\mathfrak{c}_1$ is a *merging* (or *fusion*) of $\mathfrak{c}_2$.

An *association scheme* is a coherent configuration with Δ as a basic relation.

Let $\mathcal{A}$ be an S-ring over group H, $\mathcal{A} = \{T_0, T_1, \ldots, T_s\}$, where $T_0 = \{e\}$, $T_1, \ldots, T_s$ are the basic sets of $\mathcal{A}$. It follows from the definitions that $\underline{T_i} \cdot \underline{T_j} = \sum_{k=0}^s p_{ij}^k \underline{T_k}$ for suitable non-negative integers p_{ij}^k, $0 \leq i, j, k \leq s$. The numbers p_{ij}^k are called *structure constants* of $\mathcal{A}$. We also associate with $\mathcal{A}$ the color graph $\mathfrak{m} = (H, R_i)$, where for $0 \leq i \leq s$, R_i is the arc set of the Cayley graph $Cay(H, T_i)$. This color graph is an association scheme. In particular, an association scheme that has this form is called a *translation association scheme*. We say that the rank of $\mathfrak{m}$ is equal to $s + 1$. An AS is called *symmetric* if each basic relation is symmetric. It is called *commutative* if $p_{ij}^k = p_{ji}^k$ for all $i, j, k \in [1, r]$. It is called *primitive* if all basic graphs (except for the one that only has loops) are connected.

The structure constants of the AS (also called *intersection numbers*) coincide with the structure constants of the corresponding S-ring.

Enumeration of the S-rings over a given group is a special case of the enumeration of mergings of an AS and even more generally of a coherent configuration.

The coherent closure of a graph $\Gamma = (V, E)$, see [7] for details, is the smallest rank coherent configuration $\mathfrak{m}$ that E is a union of basic relations of $\mathfrak{m}$. Similarly the coherent closure of a color graph $\mathfrak{c}$ is the smallest rank coherent configuration $\mathfrak{m}$ that each basic relation of $\mathfrak{c}$ is a union of relations of $\mathfrak{m}$.

For a given coherent configuration $\mathfrak{c}$ we can also consider the closure of a partition of the relations of $\mathfrak{c}$. This is the closure of the color graph generated from $\mathfrak{c}$ by uniting the basic relations in each cell of the partition. The coherent closure of such a partition is necessarily a merging of $\mathfrak{c}$.

The Weisfeiler-Leman algorithm [14] is an efficient (polynomial time) algorithm for the calculation of the coherent closure of a graph, or equivalently, of a partition of basic relations of a given coherent configuration. The algorithm works by repeatedly calculating products of elements of a putative coherent configuration and splitting the scheme further according to the result. When it is not split any more, the result is indeed a coherent configuration.

If an association scheme is the coherent closure of a graph (necessarily a union of basic graphs of the scheme), we say that it is *generated* by this graph, and the scheme is called *coherently cyclic*. If the arc set of the generating graph is a basic relation, then we say that the scheme is *strictly coherently cyclic*. If the basic graphs of a strictly coherently cyclic association scheme $\mathfrak{m}$ are the distance graphs of its generating graph Γ, then $\mathfrak{m}$ is called a *metric* association scheme, while Γ is called a *distance regular graph* [1].

3   Description of the Algorithms

A naive algorithm for enumerating the mergings of an association scheme consists of simple consideration of all partitions of its index set, and checking which of those are coherent mergings.

A major improvement for this algorithm was introduced in COCO [3]. Instead of considering all sets of basic relations, only good sets are considered. A good set is a set that is not split by a single step of the WL algorithm. In other words, a set of indexes $\{i_1, \ldots, i_k\}$ is good if the coefficient of each A_{i_j} in the product $(A_{i_1} + \cdots + A_{i_k})^2$ is the same. Then, instead of considering all partitions, only partitions into good sets need to be considered.

Another approach uses the WL algorithm for the enumeration. Starting with the trivial partition into two sets, we split each cell of a given partition in every possible way into two good sets, and calculate the coherent closure of this partition. If the closure is a new merging, we add it to the list, and run the same steps on this new merging.

COCO-II initiative (due to C. Pech and S. Reichard, see [7]) improves further on this by running the complete WL algorithm for each good set, making sure it is indeed a basic set of a merging.

Each of the two algorithms has two stages: the first is the enumeration of good sets, and the second is the search of partitions made from those good sets.

We optimized the first stage of the COCO-II algorithm by considering all candidate sets in an order such that the difference between two consecutive sets is in at most one element. This allows us to calculate the squares of the form $(A_{i_1} + \cdots + A_{i_k})^2$ more efficiently, by requiring only $O(k)$ multiplications, instead of $O(k^2)$.

We optimized the second stage by stopping the WL stabilization when we can be sure that no new scheme will result, or when we can be sure that if a new scheme results, this scheme will also result in another, faster way.

As implementation details, we note that COCO is written in C, while COCO-II is written in GAP ([4]. The advantage of GAP is its easy handling of groups, allowing for some reduction in the number of calculations needed by considering orbits of some types of automorphism groups. The disadvantage is that as an interpreter language it is relatively slow for some basic operations.

We implemented the first stage of the calculation (the search for good sets) in C, and the second stage in GAP. In addition, since the second stage includes repeated WL stabilizations of the same partitions, as well as multiplications of the same sets, we cache results of multiplications and stabilizations.

3.1   The Algorithm for the Symmetric Step of the First Stage

The basic candidates for symmetric good sets are either symmetric relations of the coherent configuration, or unions of an anti-symmetric relation with its transpose. Every set of basic candidates is a candidate for a good set, therefore for n basic candidates we have 2^n candidates.

If we pre-calculate a new tensor, we save a bit on calculations. We save further by pre-calculating all products of the form $(B_i + B_j)B_k$, thus saving a bit of time when calculating products of the form $(B_{i_1} + \cdots + B_{i_k})B_j$.

Calculating the next set in a sequence that ensures the symmetric difference is a singleton is equivalent to calculating a Hamiltonian cycle in the n-dimensional cube. Indeed, the n-dimensional cube is a graph with subsets of a

---

Data: Tensor of structure constants of a coherent configuration
Result: Symmetric good sets
calculate 3-dimensional symmetrized tensor
start with the empty set
while *not at last set* **do**

    calculate next set
    calculate square of next set
    if *square does not split set* **and** *set is not split by WL stabilization* **then**
       | Output set
    end

end

Algorithm 1. Enumeration of symmetric good sets

set of size n as vertices, and edge connecting two sets if their symmetric difference is of size one.

As was mentioned above, when calculating the square, we reduce the number of calculations needed by utilizing the square of the previous set, and the identity

$$\left(B_{i_1} + \cdots + B_{i_{k+1}}\right)^2 = \left(B_{i_1} + \cdots + B_{i_k}\right)^2 + \left(B_{i_1} + \cdots + B_{i_k}\right) B_{i_{k+1}} + \\ + B_{i_{k+1}} \left(B_{i_1} + \cdots + B_{i_k}\right) + B_{i_{k+1}}^2$$

The WL stabilization is not completed in case the tested set is split, since we only need to know if it is split.

3.2   The Algorithm for the Anti-symmetric Step of the First Stage

Here the basic candidates are the anti-symmetric relations. For every pair of an anti-symmetric relation and its transpose we have three options: taking none of them, taking one, or taking the other. Therefore for n pairs, we have 3^n candidates.

We implement a simple depth-first search in a ternary search tree, which again allows us to reduce the calculations needed for calculating the squares.

---

calculate 3-dimensional tensor
prepare list of pairs of anti-symmetric relations
DFS(empty set, 0, 0)

Function *DFS()* **is**
    Data: a set, the square of the set and a position in the list of
             anti-symmetric relations
    if *beyond end of list* **then**
    | return
    else
        add relation to set
        calculate square
        if *square does not split set* **and** *set is not split by WL stabilization* **then**
        | Output set
        end
        DFS(new set, new square, next position)
        add transpose relation to original set
        calculate square
        if *square does not split set* **and** *set is not split by WL stabilization* **then**
        | Output set
        end
        DFS(new set, new square, next position)
        DFS(original set, original square, next position)
    end
end

Algorithm 2. Enumeration of anti-symmetric good sets

3.3   Performance Comparison with COCO-II

COCO-II takes 942 seconds to calculate S-rings over all the 88 groups of orders between 5 and 31. Our program takes 158 seconds, running about six times faster.

For the four groups of order 44, COCO-II took 18:48, 24:34, 9:47 and 6:56 hours, while our program took 26:32, 48:57 258:12 and 18:21 minutes (the times are on different, but comparable systems), running about 10 times faster.

For the larger groups we ran the new program in parallel on a few systems, and we did not measure exact times.

COCO-II takes about one month to calculate all S-rings over the group A_5 of order 60, while our program takes about 20 hours (less than 1 hour on 30 CPU cores).

$AGL_1(8)$ of order 56 took about two months with COCO-II, and about 30 hours with our programs.

Based on these, we extrapolated COCO-II to take about four years for the non-abelian group of order 55. Our program took about 500 hour (about 5 hours on 100 CPU cores).

For those larger groups, the new program runs from 25 to 70 times faster.

4   Computer Results

The results consist of enumeration of all S-rings over the groups of order up to 63.

The total numbers of S-rings over groups of each order, up to isomorphism (as association schemes) are presented in Table 1. Considering isomorphism as association schemes means that two S-rings over different groups (necessarily of the same order) may be isomorphic. In the third column is the number of non-Schurian S-rings for each order. The number in the fourth column is the number of strictly coherently cyclic S-rings for each order.

The number of non-Schur groups of each order appears in Table 2. Only orders for which non-Schurian S-rings exist are listed. The groups are also divided into abelian and non-abelian groups.

Primitive S-rings are of a special interest. Of the discovered S-rings, there are 135 primitive S-rings of rank at least 3 and not prime order. Among them, 88 are rank 3 symmetric S-rings, that is, they correspond to primitive strongly regular graphs. Two are rank 3 anti-symmetric S-rings (corresponding to doubly regular tournaments), both of order 27.

The remaining 45 primitive S-rings over 15 different groups of orders 16, 21, 25, 27, 49, 55, 56 and 60 have ranks from 4 to 13.

There are 9 non-Schurian primitive S-rings. One of them is over the group $\mathbb{Z}_4 \times \mathbb{Z}_4$ and has rank 4. Another is over the group $\mathbb{Z}_2 \times D_8$ also having rank 4. The other 7 S-rings are over $\mathbb{Z}_7 \times \mathbb{Z}_7$, with ranks 4 and 5.

Table 1. Number of S-rings for each order (nS-non Schurian, scc-strictly coherently cyclic)

Ord	#	nS	scc
3	2	0	2
4	4	0	3
5	3	0	3
6	8	0	7
7	4	0	4
8	21	0	11
9	12	0	8
10	11	0	10
11	4	0	4
12	58	0	29
13	6	0	6
14	16	0	14
15	21	0	17
16	204	9	39
17	5	0	5

Ord	#	nS	scc
18	91	1	33
19	6	0	6
20	83	0	39
21	32	0	25
22	16	0	14
23	4	0	4
24	654	23	110
25	36	4	18
26	22	0	21
27	123	1	45
28	111	0	48
29	6	0	6
30	185	0	68
31	8	0	8
32	4212	553	159

Ord	#	nS	scc
33	27	0	21
34	17	0	16
35	41	0	29
36	1259	73	168
37	9	0	9
38	23	1	21
39	44	0	33
40	936	31	154
41	8	0	8
42	293	3	108
43	8	0	8
44	107	0	44
45	245	0	71
46	16	1	14
47	4	0	4

Ord	#	nS	scc
48	16426	3309	485
49	93	35	32
50	237	27	70
51	35	0	27
52	169	2	77
53	6	0	6
54	2020	276	186
55	48	0	35
56	1271	46	198
57	43	1	32
58	21	0	20
59	4	0	4
60	2780	47	341
61	12	0	12
62	32	1	30
63	385	10	119

Table 2. Number of non-Schur groups of each order

Ord	nA+nS	nA+S	A+nS	A+S	total
16	7	2	2	3	14
18	2	1	0	2	5
24	11	1	0	3	15
25	0	0	1	1	2
27	2	0	0	3	5
32	1	43	4	3	51
36	1	9	1	3	14
38	1	0	0	1	2
40	10	1	0	3	14
42	5	0	0	1	6

Ord	nA+nS	nA+S	A+nS	A+S	total
46	1	0	0	1	2
48	47	0	3	2	52
49	0	0	1	1	2
50	2	1	1	1	5
52	2	1	0	2	5
54	11	1	0	3	15
56	10	0	0	3	13
57	1	0	0	1	2
60	9	2	0	2	13
62	1	0	0	1	2
63	1	1	0	2	4

A group H is called a B-group if there exists no non-trivial primitive S-ring over H. Table 3 lists the non-B-groups of order up to 63 (ignoring groups of prime order). The number in the column headed by # is the number of the group in the small groups library of GAP.

Table 3. non-B-groups of order up to 63

Ord	#	Structure
9	2	E_9
16	2	C4 x C4
16	3	(C4 x C2) : C2
16	4	C4 : C4
16	6	C8 : C2
16	8	QD16
16	10	C4 x C2 x C2
16	11	C2 x D8
16	14	E_{16}
21	1	C7 : C3
25	2	E_{25}
27	3	(C3 x C3) : C3
27	4	C9 : C3
27	5	E_{27}

Ord	#	Structure
36	6	C3 x (C3 : C4)
36	9	(C3 x C3) : C4
36	10	S3 x S3
36	11	C3 x A4
36	12	C6 x S3
36	13	C2 x ((C3 x C3) : C2)
36	14	E_{36}
49	2	E_{49}
55	1	C11 : C5
56	11	$AGL_1(8)$
57	1	C19 : C3
60	5	A5
60	6	C3 x (C5 : C4)
60	7	C15 : C4
60	9	C5 x A4

5   Comparison with Other Calculations and with Theoretical Results

Hanaki and Miyamoto [5] enumerated all association schemes of order up to 34. This includes all S-rings. Pech and Reichard [11] enumerated all S-rings over groups of order up to 47. Our results coincide with both of those.

In [8] we presented the results of computer enumeration of S-rings over A_5 and $AGL_1(8)$ in more details.

Muzychuk [10] classified (on a theoretical level) all primitive S-rings over A_5, showing exactly two such S-rings exist. Our enumeration agrees with his results.

If an abelian group which is neither cyclic nor elementary abelian is Schur then it is in one of nine families [2]:

- $\mathbb{Z}_2 \times \mathbb{Z}_{2^k}$, $\mathbb{Z}_{2p} \times \mathbb{Z}_{2^k}$, $E_4 \times \mathbb{Z}_{p^k}$, $E_4 \times \mathbb{Z}_{pq}$, $E_{16} \times \mathbb{Z}_p$
- $\mathbb{Z}_3 \times \mathbb{Z}_{3^k}$, $\mathbb{Z}_6 \times \mathbb{Z}_{3^k}$, $E_9 \times \mathbb{Z}_q$, $E_9 \times \mathbb{Z}_{2q}$.

(p, q are distinct primes, $p \neq 2$, $k \geq 1$).

In addition, any group of the form $E_4 \times \mathbb{Z}_p$ (p an odd prime) is Schur.

Our enumeration shows that every group of order up to 63 which is in one of those nine families is a Schur group.

There are fewer theoretical results for non-abelian groups. One such result states that for a prime $p > 11$ such that $p \cong 3 \pmod 4$ there exists a non-Schurian S-ring of rank 4 over the dihedral group of order $2p$ [12].

Our enumeration shows that for all primes $p \leq 31$, the only non-Schurian S-rings over D_{2p} are the ones above.

For a group H, a subset $D \subseteq H$ is a (v, k, λ, μ)-*partial difference set* if $\underline{DD^{-1}} = k\underline{\{e\}} + \lambda \underline{D \setminus \{e\}} + \mu \underline{(H \setminus D) \setminus \{e\}}$.

The existence of such a nontrivial set over a group G is equivalent to existence of a primitive rank 3 symmetric S-ring over G.

In [6] there is a list of small groups over which a partial difference set exists. Our results agree with the results in [6].

6   Concluding Remarks

While computer assisted proofs are generally accepted in Mathematics, errors in calculations may happen, even when the algorithm itself is correct. For that reason it is always better to have two different implementations of the algorithm, or two different algorithms, agreeing on the results (Lam principle [9]). In that light, since our implementation is completely different from COCO-II, then the enumeration of S-rings over groups of order up to 47 can be considered safe, but for groups of orders 48 to 63, we still have only the results of a single program.

For the groups of order 64 (especially for E_{64}) an innovative approach is necessary, as the current algorithms cannot finish the calculations in a reasonable time.

While we presented some statistical information for the discovered S-rings, more can be computed, such as the number of coherently cyclic S-rings and the number of metric S-rings.

In Table 1 we counted isomorphism classes of association schemes. We can also consider the isomorphism classes in more details, counting the number of association schemes in each class. Other interesting information is the number of Cayley isomorphism classes inside each isomorphism class of association schemes. Here, a Cayley isomorphism is an isomorphism that takes into account the underlying group

The reader is invited to download the list of all S-rings (in GAP format) at [16].

Acknowledgments. I would like to thank M. Klin, for initiation of this enumeration project, and for the help he provided with the preparation of this text.

I also acknowledge C. Pech and S. Reichard for their work on COCO-II, and for letting me use of preliminary versions of this package.

I thank the anonymous reviewers for their comments which helped me to improve this text.

References

1. Brouwer, A.E., Cohen, A.M., Neumaier, A.: Distance Regular Graphs. Springer, Berlin (1989)
2. Evdokimov, S., Kovács, I., Ponomarenko, I.: On schurity of finite abelian groups. arXiv:1309.0989

3. Faradžev, I.A., Klin, M.H.: Computer package for computations with coherent configurations. In: Proc. ISSAC 1991, pp. 219–223. ACM Press, Bonn (1991)
4. http://www.gap-system.org
5. Classification of association schemes with small vertices, http://math.shinshu-u.ac.jp/~hanaki/as/
6. Heinze, A.: Applications of Schur Rings in Algebraic Combinatorics: Graphs, Partial Difference Sets and Cyclotomic Schemes. Ph.D thesis. Department of Mathematics, Carl von Ossietzky University of Oldenburg, Germany (2001)
7. Klin, M., Muzychuk, M., Ziv-Av, M.: Higmanian rank-5 association schemes on 40 points. Michigan Math. J. 58(1), 255–284 (2009)
8. Klin, M., Ziv-Av, M.: Enumeration of Schur Rings over the Group A_5. In: Gerdt, V.P., Koepf, W., Mayr, E.W., Vorozhtsov, E.V. (eds.) CASC 2013. LNCS, vol. 8136, pp. 219–230. Springer, Heidelberg (2013)
9. Lam, C.W.H.: The Search for a Finite Projective Plane of Order 10. American Mathematical Monthly 98(4), 305–318 (1991)
10. Muzychuk, M.E.: Structure of primitive S-rings over group A_5. In: VIII All-Union Symposium on Group Theory, Kiev, pp. 83–84 (1982)
11. Pech, C., Reichard, S.: m Enumerating Set Orbits. In: Klin, M., et al. (eds.) Algorithmic Algebraic Combinatorics and Gröbner Bases, pp. 31–65. Springer, Heidelberg (2009)
12. Ponomarenko, I., Vasil'ev, A.: On non-abelian Schur groups, http://math.nsc.ru/~vasand/Papers_eng/NaScRing.pdf
13. Schur, I.: Zur Theorie der einfach transitiven Permutationsgruppen. Sitzungsber. Preuss. Akad. Wiss., Phys.-Math. Kl., 598–623 (1933)
14. Weisfeiler, B.J., Leman, A.A.: A reduction of a graph to a canonical form and an algebra arising during this reduction. Nauchno - Technicheskaja Informatsia 9(Seria 2), 12–16 (1968) (Russian)
15. Wielandt, H.: Finite Permutation Groups. Acad. Press, New York (1964)
16. http://my.svgalib.org/s-rings/wschur.tar.gz

Author Index